A TEXTBOOK OF ENGINEERING MATHEMATICS

A TEXTBOOK OF

ENGINEERING MATHEMATICS

For

B. Tech. I Year (*I and II Semesters*)

(Strictly According to New Revised Syllabus)

Cochin University of Science and Technology, Kerala

By

N.P. Bali
Formerly, Principal
S.B. College, Gurgaon
Haryana

Dr. Remadevi.S
Head Department of Mathematics
Model Engineering College, Cochin
Kerala

UNIVERSITY SCIENCE PRESS

(An Imprint of Laxmi Publications Pvt. Ltd.)

BANGALORE • CHENNAI • COCHIN • GUWAHATI • HYDERABAD
JALANDHAR • KOLKATA • LUCKNOW • MUMBAI • RANCHI
NEW DELHI • BOSTON, USA

Published by :

UNIVERSITY SCIENCE PRESS

(An Imprint of Laxmi Publications Pvt. Ltd.)

113, Golden House, Daryaganj,
New Delhi-110002

Phone : 011-43 53 25 00
Fax : 011-43 53 25 28

www.laxmipublications.com
info@laxmipublications.com

First Edition : 2011; Second Edition : 2013

OFFICES

✆ **Bangalore**	080-26 75 69 30	✆ **Jalandhar**	0181-222 12 72
✆ **Chennai**	044-24 34 47 26	✆ **Kolkata**	033-22 27 43 84
✆ **Cochin**	0484-237 70 04, 405 13 03	✆ **Lucknow**	0522-220 95 78
✆ **Guwahati**	0361-254 36 69, 251 38 81	✆ **Mumbai**	022-24 91 54 15, 24 92 78 69
✆ **Hyderabad**	040-24 65 23 33	✆ **Ranchi**	0651-221 47 64

UEM-9570-350-ENGG MATH I & II (KE)-REM **C—**

Typeset at : Goswami Associates, Delhi. *Printed at* : Repro Knowledgecast Limited, Thane

CONTENTS

SYLLABUS

For B. Tech. Semester I & II
Cochin University of Science and Technology, Kerala

CE/CS/EB/EC/EE/EI/IT/ME/SE 1101 ENGINEERING MATHEMATICS–I

Module I: Ordinary Differential Equations

First order differential equations-exact differential equations, Bernoulli's equation—Methods of solution and Simple applications.

Linear differential equations of higher orders with constant co-efficient-Methods of solution of these equations. Cauchy's Linear differential equations. Simultaneous linear differential equations—Simple applications of linear differential equations in engineering problems— Electrical Circuits, Mechanical Systems.

Module II: Infinite Series

Integral test, comparison test, ratio test, Cauchy's root test, Raabe's test, series of positive and negative terms, concept of absolute convergence, alternating series, Leibnitz test (No proofs for any of the above tests).

Power Series

Taylor's and Maclaurin's series of functions, Leibnitz formula for the nth derivative of the product of two functions (No proof), use of Leibnitz formula for the determination of co-efficients of the power series.

Module III: Partial Differentiation

Partial differentiation—Concept of partial derivative—Chain rule—Total derivative—Euler's theorem for homogeneous functions, Differentials and their applications in errors and approximations, Jacobians—Maxima minima of functions of two variables (Proof of the result not required)—Simple applications.

Co-ordinate Systems

Rectangular co-ordinates—Polar co-ordinates—In plane and in Space—Cylindrical polar co-ordinates—Spherical polar co-ordinates.

Module IV: Integral Calculus

Application of definite integrals: Area, Volume, Arc length, Surface area.

Multiple integrals: Evaluation of double integrals—Change of order of integration. Evaluation of triple integrals—Change of Variables in integrals.

Applications of multiple integrals Plane Area, Surface area and Volumes of solids.

PREFACE TO THE SECOND EDITION

With the grace of the Almighty and the over whelming response given by the teachers and the students, the book is gaining increasing popularity. We express our sincerest thanks and sense of gratitude for their appreciation of the book.

The new edition has been revised with lot of care, dedication and patience. The book has been divided into four modules containing 13 chapters and covers the entire portion in the B.Tech. First Year Cochin University Engineering Mathematics–I revised new syllabus. Many new types of questions have been added, in examples as well as in exercises, to enhance the utility of the book. The book presents the subject matter in a very systematic, simple and lucid style, so that students can solve the questions on their own. To make the students faimiliar with the university pattern, all the questions set in university papers have been included.

The most outstanding and distinguishing feature of the book is the large number of typical solved examples followed by well graded problems.

All suggestions for further improvement of the book will be highly appreciated and gratefully acknowledged.

—Authors

PREFACE TO THE FIRST EDITION

This book is a part of the original book **"A Textbook of Engineering Mathematics"** (*with 28 chapters and covering the syllabi of engineering courses of all semesters of all the Indian Universities*) running its seventh edition and very well received by the students and teachers of all Indian Universities. The rapid sale of the seventh edition bears testimony to the overwhelming response. We thank them all for the appreciation. Some topics included in this book were taken from the book **"Differential Calculus"** and **"Integral Calculus"** by N.P. Bali.

The present form of the book is divided into four modules containing 14 chapters and covers the entire portion in the B.Tech. First Year Cochin University Engineering Mathematics-I syllabus. Problems taken from Cochin University Question Papers are included with solutions in each unit.

There is no dearth of books on Engineering Mathematics but the students find it difficult to solve most of the problems in the exercise in the absence of adequate number of solved examples. An outstanding and distinguishing feature of the book is the large number of typical solved examples followed by well-graded problems.

We have endeavoured to present the fundamental concepts in a comprehensive and lucid manner. We are indebted to all authors, Indian and Foreign, whose works we have frequently consulted.

All efforts have been made to keep the book free from errors. Answers to all exercises have been re-checked. All suggestions for improvement will be highly appreciated and gratefully acknowledged.

—Authors

MODULE–I

DIFFERENTIAL EQUATIONS AND APPLICATIONS

1

Differential Equations of First Order

1.1. DEFINITIONS

(*i*) A **differential equation** is an equation involving differentials or differential coefficients.

Thus,
$$\frac{dy}{dx} = x^2 - 1 \qquad ...(1)$$

$$\frac{d^2y}{dx^2} + 2\left(\frac{dy}{dx}\right)^2 + y = 0 \qquad ...(2)$$

$$(x^2 + y^2 - 3y)\, dx = (x^2 + 3x + y)\, dy \qquad ...(3)$$

$$y = x\frac{dy}{dx} + \frac{c}{\frac{dy}{dx}} \qquad ...(4)$$

$$\frac{d^3y}{dx^3} + 2\frac{d^2y}{dx^2} \cdot \frac{dy}{dx} + x^2\left(\frac{dy}{dx}\right)^3 = 0 \qquad ...(5)$$

$$\left[1 + \left(\frac{dy}{dx}\right)^2\right]^{3/2} = k\frac{d^2y}{dx^2} \qquad ...(6)$$

$$x\frac{\partial u}{\partial x} + y\frac{\partial u}{\partial y} = nu \qquad ...(7)$$

$$\frac{\partial^2 z}{\partial x^2} + \frac{\partial^2 z}{\partial y^2} = x + y \qquad ...(8)$$

are all differential equations.

(*ii*) Differential equations which involve only one independent variable and the differential co-efficients with respect to it are called **ordinary differential equations.**

Thus equations (1) to (6) are all ordinary differential equations.

(*iii*) Differential equations which involve two or more independent variables and partial derivatives with respect to them are called **partial differential equations.**

Thus equations (7) and (8) are partial differential equations.

(*iv*) The **order** of a differential equation is the order of the highest ordered derivative occurring in the differential equation.

Thus equations (1), (3) and (4) are of first order ; equations (2) and (6) are of the second order while equation (5) is of the third order.

(*v*) The **degree** of a differential equation is the degree of the highest ordered derivative which occurs in it provided the equation has been made free of the radical signs and fractional powers as far as the derivatives are concerned.

Thus, equations (1), (2), (3) and (5) are of the first degree.

Equation (4) is $$y\frac{dy}{dx} = x\left(\frac{dy}{dx}\right)^2 + c$$

It is of the second degree.

Equation (6) is $$\left[1+\left(\frac{dy}{dx}\right)^2\right]^3 = k^2\left(\frac{d^2y}{dx^2}\right)^2$$

It is of the second degree.

(*vi*) **Solution of a Differential Equation.** A solution (or integral) of a differential equation is a relation, free from derivatives, between the variables which satisfies the given equation.

Thus if $y = f(x)$ be the solution, then by replacing y and its derivatives with respect to x, the given differential equation will reduce to an identity.

For example, $$y = c_1 \cos x + c_2 \sin x$$

is the solution of the differential equation $\frac{d^2y}{dx^2} + y = 0$

Since $$\frac{dy}{dx} = -c_1 \sin x + c_2 \cos x$$

$$\frac{d^2y}{dx^2} = -c_1 \cos x - c_2 \sin x = -y.$$

The **general** (or **complete**) **solution** of a differential equation is that in which the number of independent arbitrary constants is equal to the order of the differential equation.

Thus, $y = c_1 \cos x + c_2 \sin x$ (involving two arbitrary constants c_1, c_2) is the general solution of the differential equation $\frac{d^2y}{dx^2} + y = 0$ of second order.

A **particular solution** of a differential equation is that which is obtained from its general solution by giving particular values to the arbitrary constants.

For example, $y = c_1e^x + c_2e^{-x}$ is the general solution of the differential equation $\frac{d^2y}{dx^2} - y = 0$, whereas $y = e^x - e^{-x}$ or $y = e^x$ are its particular solutions.

The solution of a differential equation of nth order is its particular solution if it contains less than n arbitrary constants.

1.2. GEOMETRICAL MEANING OF A DIFFERENTIAL EQUATION OF THE FIRST ORDER AND FIRST DEGREE

Let $$f\left(x, y, \frac{dy}{dx}\right) = 0 \qquad ...(1)$$

be a differential equation of the first order and first degree.

We know that the direction of a curve at a particular point is determined by drawing a tangent line at that point, *i.e.*, its slope is given by $\frac{dy}{dx}$ at that particular point.

Let A_0 (x_0, y_0) be any point in the plane. Let $m_0 = \frac{dy_0}{dx_0}$ be the slope of the curve at A_0 derived from (1). Take a neighbouring point A_1 (x_1, y_1) such that the slope of A_0A_1 is m_0. Let $m_1 = \frac{dy_1}{dx_1}$ be the slope of the curve at A_1 derived from (1). Take a neighbouring point A_2 (x_2, y_2) such that the slope of A_1A_2 is m_1. Continuing like this, we get a succession of points. If the points are taken sufficiently close to each other, they approximate a smooth curve C : $y = \phi(x)$ which is a solution of (1) corresponding to the initial point A_0 (x_0, y_0). Any point on C and the slope of the tangent at that point satisfy (1). If the moving point starts at any other point, not on C and moves as before, it will describe another curve. The equation of each such curve is a *particular solution* of the differential equation (1). The equation of the system of all such curves is the general solution of (1).

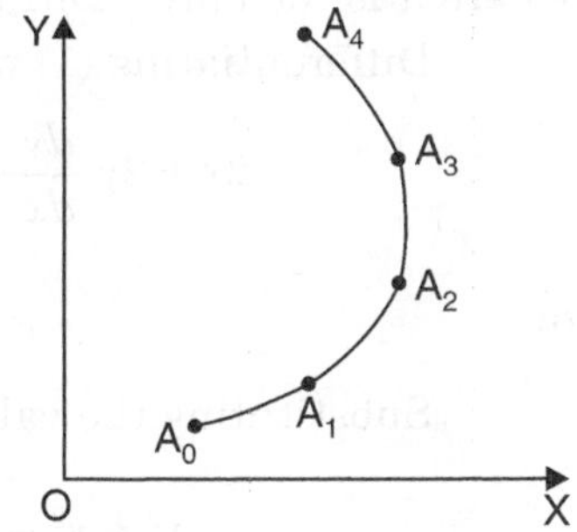

1.3. FORMATION OF A DIFFERENTIAL EQUATION

Differential equations are formed by elimination of arbitrary constants. To eliminate two arbitrary constants, we require two more equations besides the given relation, leading us to second order derivatives and hence a differential equation of the second order. Elimination of n arbitrary constants leads us to n^{th} order derivatives and hence a differential equation of the nth order.

ILLUSTRATIVE EXAMPLES

Example 1. *Eliminate the constants from the equation* $y = e^x$ *(A cos x + B sin x)* ...(1) *and obtain the differential equation.*

Sol. There are two arbitrary constants A and B in equation (1).

Differentiating (1) w.r.t. x, we have

$$\frac{dy}{dx} = e^x\,(\text{A} \cos x + \text{B} \sin x) + e^x\,(-\text{A} \sin x + \text{B} \cos x)$$

$$= y + e^x\,(-\text{A} \sin x + \text{B} \cos x) \qquad ...(2)$$

Differentiating again w.r.t. x, we have

$$\frac{d^2y}{dx^2} = \frac{dy}{dx} + e^x(-A\sin x + B\cos x) + e^x(-A\cos x - B\sin x)$$

$$= \frac{dy}{dx} + \left(\frac{dy}{dx} - y\right) - y \qquad \text{[Using (1) and (2)]}$$

or $$\frac{d^2y}{dx^2} - 2\frac{dy}{dx} + 2y = 0$$

which is the required differential equation.

Example 2. *Find the differential equation of all circles touching the axis of y at the origin and centres on the axis of x.*

Sol. The equation of such a circle is

$$(x-h)^2 + y^2 = h^2$$

or $$x^2 + y^2 - 2hx = 0 \qquad \ldots(1)$$

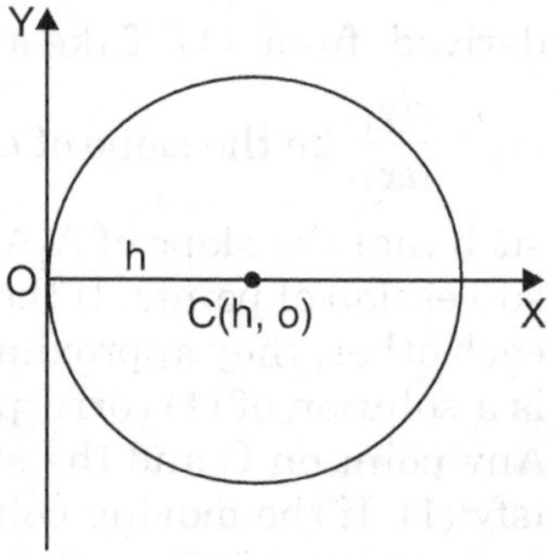

where h is the only arbitrary constant.

Differentiating (1) w.r.t. x, we have

$$2x + 2y\frac{dy}{dx} - 2h = 0$$

or $$h = x + y\frac{dy}{dx}$$

Substituting the value of h in (1), we have

$$x^2 + y^2 - 2x\left(x + y\frac{dy}{dx}\right) = 0$$

or $$2xy\frac{dy}{dx} + x^2 - y^2 = 0$$

which is the required differential equation.

Example 3. *Form the differential equation of all circles of radius a.*

Sol. The equation of any circle of radius a is $(x-h)^2 + (y-k)^2 = a^2$ ...(1)

where (h, k) the coordinates of the centre are arbitrary.

Differentiating (1) w.r.t. x, we have $2(x-h) + 2(y-k)\frac{dy}{dx} = 0$

or $$(x-h) + (y-k)\frac{dy}{dx} = 0 \qquad \ldots(2)$$

Differentiating again, we have $$1 + (y-k)\frac{d^2y}{dx^2} + \left(\frac{dy}{dx}\right)^2 = 0 \qquad \ldots(3)$$

From (3), $$y - k = -\frac{1 + \left(\frac{dy}{dx}\right)^2}{\frac{d^2y}{dx^2}}$$

and from (2), $$x - h = -(y-k)\frac{dy}{dx} = \frac{\frac{dy}{dx}\left[1+\left(\frac{dy}{dx}\right)^2\right]}{\frac{d^2y}{dx^2}}$$

Substituting the values of $(x - h)$ and $(y - k)$ in (1), we get

$$\frac{\left(\frac{dy}{dx}\right)^2\left[1+\left(\frac{dy}{dx}\right)^2\right]^2}{\left(\frac{d^2y}{dx^2}\right)^2} + \frac{\left[1+\left(\frac{dy}{dx}\right)^2\right]^2}{\left(\frac{d^2y}{dx^2}\right)^2} = a^2$$

or $$\left[1+\left(\frac{dy}{dx}\right)^2\right]^2\left[\left(\frac{dy}{dx}\right)^2+1\right] = a^2\left(\frac{d^2y}{dx^2}\right)^2$$

or $$\left[1+\left(\frac{dy}{dx}\right)^2\right]^3 = a^2\left(\frac{d^2y}{dx^2}\right)^2$$

which is the required differential equation.

TEST YOUR KNOWLEDGE

Eliminate the arbitrary constants and obtain the differential equations :

1. $y = cx + c^2$
2. $y = A + Bx + Cx^2$
3. $y = A\cos 2t + B\sin 2t$
4. $y = Ae^{3x} + Be^{2x}$
5. $y = Ae^x + Be^{-x} + C$
6. $y = ax^3 + bx^2$
7. $xy = Ae^x + Be^{-x} + x^2$
8. $x = A\cos(nt + \alpha)$
9. $y = ae^{2x} + be^{-3x} + ce^x$
10. $Ax^2 + By^2 = 1$
11. $y^2 - 2ay + x^2 = a^2$
12. $e^{2y} + 2ax\,e^y + a^2 = 0$.

Find the differential equations of :

13. All straight lines in a plane.
14. All circles of radius r whose centres lie on the x-axis.
15. All parabolas with x-axis as the axis and $(a, 0)$ as focus.
16. All conics whose axes coincide with the axes of co-ordinates.
17. All circles in a plane.
18. All circles in the first quadrant which touch the co-ordinate axes.
19. All circles passing through the origin and having centres on the x-axis.
20. All parabolas with latus rectum '$4a$' and axis parallel to the x-axis.

Answers

1. $x\frac{dy}{dx} + \left(\frac{dy}{dx}\right)^2 = y$
2. $\frac{d^3y}{dx^3} = 0$
3. $\frac{d^2y}{dt^2} + 4y = 0$

4. $\frac{d^2y}{dx^2} - 5\frac{dy}{dx} + 6y = 0$ 5. $\frac{d^3y}{dx^3} - \frac{dy}{dx} = 0$ 6. $x^2\frac{d^2y}{dx^2} - 4x\frac{dy}{dx} + 6y = 0$

7. $x\frac{d^2y}{dx^2} + 2\frac{dy}{dx} + x^2 - xy - 2 = 0$ 8. $\frac{d^2x}{dt^2} + n^2x = 0$ 9. $\frac{d^3y}{dx^3} - 7\frac{dy}{dx} + 6y = 0$

10. $xy\frac{d^2y}{dx^2} + x\left(\frac{dy}{dx}\right)^2 - y\frac{dy}{dx} = 0$ 11. $(x^2 - 2y^2)\left(\frac{dy}{dx}\right)^2 - 4xy\frac{dy}{dx} - x^2 = 0$

12. $(1 - x^2)\left(\frac{dy}{dx}\right)^2 + 1 = 0$ 13. $\frac{d^2y}{dx^2} = 0$ 14. $y^2\left[1 + \left(\frac{dy}{dx}\right)^2\right] = r^2$

15. $y\frac{dy}{dx} = 2a$ 16. $xy\frac{d^2y}{dx^2} + x\left(\frac{dy}{dx}\right)^2 = y\frac{dy}{dx}$

17. $\left[1 + \left(\frac{dy}{dx}\right)^2\right]\frac{d^3y}{dx^3} - 3\frac{dy}{dx}\left(\frac{d^2y}{dx^2}\right)^2 = 0$ 18. $(x - y)^2\left[1 + \left(\frac{dy}{dx}\right)^2\right] = \left(x + y\frac{dy}{dx}\right)^2$

19. $x^2 - y^2 + 2xy\frac{dy}{dx} = 0$ 20. $2a\frac{d^2y}{dx^2} + \left(\frac{dy}{dx}\right)^3 = 0.$

1.4. SOLUTION OF DIFFERENTIAL EQUATIONS OF THE FIRST ORDER AND FIRST DEGREE

All differential equations of the first order and first degree cannot be solved. Only those among them which belong to (or can be reduced to) one of the following categories can be solved by the standard methods.

(*i*) Equations in which variables are separable.

(*ii*) Homogeneous equations. (*iii*) Linear equations.

(*iv*) Exact equations.

1.5. LINEAR DIFFERENTIAL EQUATIONS

A differential equation is said to be linear if the dependent variable and its derivative occur only in the first degree and are not multiplied together.

The general form of a linear differential equation of the first order is $\frac{dy}{dx} + Py = Q$ ...(1)

where P and Q are functions of x only or may be constants.

Equation (1) is also known as *Leibnitz's linear equation.*

To solve it, we multiply both sides by $e^{\int P\,dx}$, by getting

$$\frac{dy}{dx}e^{\int P\,dx} + y\left(e^{\int P\,dx}\,P\right) = Q e^{\int P\,dx}$$

or

$$\frac{d}{dx}\left(ye^{\int P\,dx}\right) = Q e^{\int P\,dx}$$

Integrating both sides, we have $\quad ye^{\int P\,dx} = \int Qe^{\int P\,dx}\,dx + c$

which is the required solution.

Note 1. In the general form of a linear differential equation, the co-efficient of $\frac{dy}{dx}$ is unity.

The equation $R\frac{dy}{dx} + Sy = T$, where R, S and T are functions of x only or constants, must be divided by R to bring it to the general linear form.

Note 2. The factor $e^{\int P\,dx}$ on multiplying by which the LHS of (1) becomes the differential co-efficient of a single function is called the integrating factor (briefly written as I.F.) of (1).

Thus $\quad$ I.F. $= e^{\int P\,dx}$ and the solution is y (I.F.) $= \int Q(\text{I.F.})\,dx + c$.

Note 3. Sometimes a differential equation takes linear form if we regard x as dependent variable and y as independent variable. The equation can then be put as $\frac{dx}{dy} + Px = Q$, where P, Q are functions of y only or constants.

The integrating factor in this case is $e^{\int P\,dy}$ and the solution is x (I.F.) $= \int Q(\text{I.F.})\,dy + c$.

Note 4. $e^{\log f(x)} = f(x)$.

ILLUSTRATIVE EXAMPLES

Example 1. *Solve* $x(1 - x^2)\frac{dy}{dx} + (2x^2 - 1)\,y = x^3$.

Sol. Dividing by $x(1-x^2)$ to make the co-efficient of $\frac{dy}{dx}$ unity, the given equation becomes

$$\frac{dy}{dx} + \frac{2x^2 - 1}{x(1-x^2)}y = \frac{x^2}{1-x^2}$$

Comparing it with $\frac{dy}{dx} + Py = Q$, we have $P = \frac{2x^2 - 1}{x(1-x^2)}$, $Q = \frac{x^2}{1-x^2}$

Now
$$P = \frac{2x^2 - 1}{x(1-x)(1+x)} = -\frac{1}{x} + \frac{1}{2(1-x)} + \frac{1}{2(1+x)}$$

$$\int P\,dx = -\log x - \frac{1}{2}\log(1-x) - \frac{1}{2}\log(1+x) = -\log[x(1-x)^{1/2}(1+x)^{1/2}]$$

$$= -\log[x(1-x^2)^{1/2}] = \log\frac{1}{x\sqrt{1-x^2}}$$

$$\text{I.F.} = e^{\int P dx} = e^{\log\frac{1}{x\sqrt{1-x^2}}} = \frac{1}{x\sqrt{1-x^2}}$$

Thus the solution is $y\text{ (I.F.)} = \int Q\text{ (I.F.) } dx + c$

or $$y \cdot \frac{1}{x\sqrt{1-x^2}} = \int \frac{x^2}{1-x^2} \times \frac{1}{x\sqrt{1-x^2}}\, dx + c$$

$$= -\frac{1}{2}\int (1-x^2)^{-3/2}\,(-2x)\, dx + c = (1-x^2)^{-1/2} + c$$

or $$y = x + cx\sqrt{1-x^2}.$$

Example 2. *Solve* $\left(\frac{e^{-2\sqrt{x}}}{\sqrt{x}} - \frac{y}{\sqrt{x}}\right)\frac{dx}{dy} = 1.$

Sol. The given equation can be written as $\frac{dy}{dx} + \frac{y}{\sqrt{x}} = \frac{e^{-2\sqrt{2}}}{\sqrt{x}}$

Comparing it with $\frac{dy}{dx} + Py = Q$, we have $P = \frac{1}{\sqrt{x}}$, $Q = \frac{e^{-2\sqrt{x}}}{\sqrt{x}}$

$$\text{I.F.} = e^{\int P\,dx} = e^{\int x^{1/2} dx} = e^{2\sqrt{x}}$$

$\therefore$ The solution is $y\text{ (I.F.)} = \int Q\text{ (I.F.)}\, dx + c$ or $ye^{2\sqrt{x}} = \int \frac{e^{-2\sqrt{x}}}{\sqrt{x}} \cdot e^{2\sqrt{x}}\, dx + c$

or $$ye^{2\sqrt{x}} = \int x^{-1/2} dx + c = 2\sqrt{x} + c.$$

Example 3. *Solve* $(1 + y^2)dx = (\tan^{-1} y - x)dy.$

Sol. The given equation can be written as $(1 + y^2)\frac{dx}{dy} + x = \tan^{-1} y$

Dividing by $(1 + y^2)$, we get $\frac{dx}{dy} + \frac{x}{1+y^2} = \frac{\tan^{-1} y}{1+y^2}$

which is of the form $$\frac{dx}{dy} + Px = Q$$

Here $$P = \frac{1}{1+y^2},\quad Q = \frac{\tan^{-1} y}{1+y^2}$$

$$\text{I.F.} = e^{\int P\,dy} = e^{\int \frac{1}{1+y^2} dy} = e^{\tan^{-1} y}$$

$\therefore$ The solution is $x\text{ (I.F.)} = \int Q\text{ (I.F.)}\, dy + c$

or $$xe^{\tan^{-1} y} = \int \frac{\tan^{-1} y}{1+y^2} e^{\tan^{-1} y} dy + c = \int te^t dt + c, \text{ where } t = \tan^{-1} y$$

$$= te^t - \int 1 \cdot e^t\, dt + c = te^t - e^t + c = (\tan^{-1} y - 1)\, e^{\tan^{-1} y} + c$$

or $$x = \tan^{-1} y - 1 + ce^{-\tan^{-1} y}.$$

TEST YOUR KNOWLEDGE

Solve the following differential equations:

1. $\frac{dy}{dx} + \frac{y}{x} = x^3 - 3$
2. $x \log x \frac{dy}{dx} + y = 2 \log x$
3. $(x + 1) \frac{dy}{dx} - y = e^x (x + 1)^2$
4. $(x^2 + 1) \frac{dy}{dx} + 2xy = x^2$
5. $\cos^2 x \frac{dy}{dx} + y = \tan x$
6. $(1 + x^3) \frac{dy}{dx} + 6x^2y = 1 + x^2$
7. $\frac{dy}{dx} + y \cot x = \cos x$
8. $(1 + x^2) \frac{dy}{dx} + y = e^{\tan^{-1} x}$
9. $\frac{dy}{dx} + y \cot x = 4x \operatorname{cosec} x$, if $y = 0$ when $x = \frac{\pi}{2}$.
10. $\frac{dy}{dx} + y \cot x = 5e^{\cos x}$, if $y = -4$ when $x = \frac{\pi}{2}$
11. $\frac{dy}{dx} - y \tan x = 3e^{-\sin x}$, if $y = 4$ when $x = 0$
12. $(1 - x^2) \frac{dy}{dx} + 2xy = x \sqrt{1-x^2}$
13. $x \frac{dy}{dx} + y = e^x - xy$
14. $(1 + y^2) + (x - e^{-\tan^{-1} y}) \frac{dy}{dx} = 0$
15. $e^{-y} \sec^2 y \, dy = dx + x \, dy$
16. $(x + 2y^3) \frac{dy}{dx} = y$
17. $ye^y \, dx = (y^3 + 2xe^y) \, dy$
18. $\frac{dy}{dx} = \frac{y}{2y \log y + y - x}$
19. $x^2 \frac{dy}{dx} = 3x^2 - 2xy + 1$
20. $\sqrt{1-y^2} \, dx = (\sin^{-1} y - x) \, dy$

Answers

1. $10xy = 2x^5 - 15x^2 + c$
2. $y \log x = (\log x)^2 + c$
3. $y = (x + 1)(e^x + c)$
4. $y(x^2 + 1) = \frac{x^3}{3} + c$
5. $y = \tan x - 1 + ce^{-\tan x}$
6. $y(1 + x^3)^2 = x + \frac{x^3}{3} + \frac{x^4}{4} + \frac{x^6}{6} + c$
7. $y \sin x = \frac{1}{2} \sin^2 x + c$
8. $2ye^{\tan^{-1} x} = e^{2\tan^{-1} x} + c$
9. $y \sin x = 2x^2 - \frac{\pi^2}{2}$
10. $y \sin x = 1 - 5e^{\cos x}$
11. $y \cos x = 7 - 3e^{-\sin x}$
12. $y = \sqrt{1-x^2} + c(1 - x^2)$
13. $xy = \frac{1}{2} e^x + ce^{-x}$
14. $x = e^{-\tan^{-1} y}(\tan^{-1} y + c)$
15. $xe^y = \tan y + c$
16. $x = y^3 + cy$
17. $x = y^2 (c - e^{-y})$
18. $x = \frac{c}{y} + y \log y$.
19. $y = x + x^{-1} + cx^{-2}$
20. $x = \sin^{-1} y - 1 + ce^{-\sin^{-1} y}$.

1.6. EQUATIONS REDUCIBLE TO THE LINEAR FORM (BERNOULLI'S EQUATION)

(*i*) An equation of the form $\frac{dy}{dx} + \text{P}y = \text{Q}y^n$ ...(1)

where P and Q are functions of x only or constants is known as *Bernoulli's* equation. Though not linear, it can be made linear.

Dividing both sides of (1) by y^n, we have $y^{-n}\frac{dy}{dx} + \text{P}y^{1-n} = \text{Q}$ ...(2)

Put $y^{1-n} = z$ so that $(1-n)y^{-n}\frac{dy}{dx} = \frac{dz}{dx}$

or $$y^{-n}\frac{dy}{dx} = \frac{1}{1-n}\frac{dz}{dx}$$

Equation (2) becomes $\frac{1}{1-n}\frac{dz}{dx} + \text{P}z = \text{Q}$ or $\frac{dz}{dx} + (1-n)\text{P}z = (1-n)\text{Q}$

which is a linear differential equation with z as the dependent variable.

(*ii*) General equation reducible to linear form is $f'(y)\frac{dy}{dx} + \text{P}f(y) = \text{Q}$ ...(1)

where P and Q are functions of x only or constants.

Put $f(y) = z$ so that $f'(y)\frac{dy}{dx} = \frac{dz}{dx}$

Equation (1) becomes $\frac{dz}{dx} + \text{P}z = \text{Q}$, which is linear.

ILLUSTRATIVE EXAMPLES

Example 1. *Solve* $\frac{dy}{dx} + \frac{x}{1-x^2}y = x\sqrt{y}$.

Sol. Dividing by $\sqrt{y}$, the given equation becomes $y^{-1/2}\frac{dy}{dx} + \frac{x}{1-x^2}y^{1/2} = x$ ...(1)

Putting $y^{1/2} = z$ so that $\frac{1}{2}y^{-1/2}\frac{dy}{dx} = \frac{dz}{dx}$ or $y^{-1/2}\frac{dy}{dx} = 2\frac{dz}{dx}$

Equation (1) becomes $2\frac{dz}{dx} + \frac{x}{1-x^2}z = x$

or $$\frac{dz}{dx} + \frac{x}{2(1-x^2)}z = \frac{x}{2} \text{ which is linear in } z.$$

$$\text{I.F.} = e^{\int \frac{x}{2(1-x^2)}dx} = e^{-\frac{1}{4}\int \frac{-2x}{1-x^2}dx} = e^{-\frac{1}{4}\log(1-x^2)} = (1-x^2)^{-\frac{1}{4}}$$

$\therefore$ The solution is $z\,(\text{I.F.}) = \int \frac{x}{2}(\text{I.F.})\,dx + c$ or $z(1-x^2)^{-1/4} = \int \frac{x}{2}(1-x^2)^{-1/4}\,dx + c$

$$= -\frac{1}{4}\int (1-x^2)^{-\frac{1}{4}}(-2x)\,dx + c = -\frac{1}{4}\cdot\frac{(1-x^2)^{3/4}}{3/4} + c$$

or $$z = -\frac{1}{3}(1-x^2) + c(1-x^2)^{1/4}$$

or $$\sqrt{y} = -\frac{1}{3}(1-x^2) + c(1-x^2)^{1/4}. \qquad [\because z = \sqrt{y}]$$

Example 2. *Solve* $xy(1 + xy^2)\frac{dy}{dx} = 1.$

Sol. The given equation can be written as $\frac{dx}{dy} - yx = y^3x^2$

Dividing by x^2, we have $x^{-2}\frac{dx}{dy} - yx^{-1} = y^3$ ...(1)

Put $x^{-1} = z$ so that $-x^{-2}\frac{dx}{dy} = \frac{dz}{dy}$ or $x^{-2}\frac{dx}{dy} = -\frac{dz}{dy}$

Equation (1) becomes $-\frac{dz}{dy} - yz = y^3$

or $$\frac{dz}{dy} + yz = -y^3 \text{ which is linear in } z.$$

$$\text{I.F.} = e^{\int y\,dx} = e^{\frac{1}{2}y^2}$$

$\therefore$ The solution is $z\,(\text{I.F.}) = \int -y^3\,(\text{I.F.})\,dy + c$ or $z\,.\,e^{\frac{1}{2}y^2} = \int -y^3 e^{\frac{1}{2}y^2}\,dy + c$

$$= -\int y^2 e^{\frac{1}{2}y^2} \bullet y\,dy + c = -\int 2t\,e^t dt + c \qquad \text{where } t = \frac{1}{2}y^2$$

$$= -2\left[t\,e^t - \int 1 - e^t\,dt\right] + c = -2\,(te^t - e^t) + c = -2e^{\frac{1}{2}y^2}\left(\frac{1}{2}y^2 - 1\right) + c$$

or $$z = -2\left(\frac{1}{2}y^2 - 1\right) + c\,e^{-\frac{1}{2}y^2}$$

or $$\frac{1}{x} = 2 - y^2 + ce^{-\frac{1}{2}y^2}. \qquad \left(\because z = \frac{1}{x}\right)$$

Example 3. *Solve* $\frac{dy}{dx} + x\sin 2y = x^3\cos^2 y.$ [CUSAT 2010]

Sol. Dividing by $\cos^2 y$, we have $\sec^2 y\,\frac{dy}{dx} + x\,\frac{2\sin y\cos y}{\cos^2 y} = x^3$

or $$\sec^2 y\,\frac{dy}{dx} + 2x\tan y = x^3 \qquad ...(1)$$

Put $\tan y = z$ so that $\sec^2 y \dfrac{dy}{dx} = \dfrac{dz}{dx}$

Equation (1) becomes $\dfrac{dz}{dx} + 2xz = x^3$ which is linear in z.

$$\text{I.F.} = e^{\int 2x\,dx} = e^{x^2}$$

∴ The solution is

$$z \cdot e^{x^2} = \int x^3 \cdot e^{x^2} dx + c = \int x^2 e^{x^2} \cdot x\,dx + c$$

$$= \frac{1}{2}\int t e^t\,dt + c \text{ (where } t = x^2)$$

$$= \frac{1}{2}(t-1)e^t + c = \frac{1}{2}(x^2-1)e^{x^2} + c$$

or $$z = \frac{1}{2}(x^2-1) + ce^{-x^2}$$

or $$\tan y = \frac{1}{2}(x^2-1) + ce^{-x^2}. \qquad (\because z = \tan y)$$

Example 4. *Solve* $\dfrac{dy}{dx} + y \tan x = y^3 \sec x.$ [CUSAT 2009]

Sol. Dividing by y^3, the equation becomes $\dfrac{1}{y^3}\dfrac{dy}{dx} + \dfrac{1}{y^2}\tan x = \sec x$ ...(1)

Putting $\dfrac{1}{y^2} = z$ so that $\dfrac{-2}{y^3}\dfrac{dy}{dx} = \dfrac{dz}{dx}$ or $\dfrac{1}{y^3}\dfrac{dy}{dx} = -\dfrac{1}{2}\dfrac{dz}{dx}$

Equation (1) becomes

$$-\frac{1}{2}\frac{dz}{dx} + \tan x . z = \sec x$$

or $$\frac{dz}{dx} - 2\tan x . z = -2\sec x, \text{ which is linear in } z$$

$$\text{I.F.} = e^{-\int 2\tan x\,dx} = e^{-2\log\sec x} = \frac{1}{\sec^2 x} = \cos^2 x$$

∴ The solution is

$$z(\text{I.F.}) = \int(-2\sec x)\ \text{I.F.}\ dx + c$$

$$= \int -2\sec x\cos^2 x\,dx + c$$

$$= \int -2\cos x\,dx + c = -2\sin x + c$$

∴ $$\frac{1}{y^2}\cos^2 x = -2\sin x + c$$

or $$\cos^2 x = y^2(c - 2\sin x).$$

TEST YOUR KNOWLEDGE

Solve the following differential equations:

1. $x\dfrac{dy}{dx} + y = x^3y^6$

2. $2\dfrac{dy}{dx} = \dfrac{y}{x} + \dfrac{y^2}{x^2}$

3. $\dfrac{dy}{dx} - x^2y = y^2e^{-\frac{1}{3}x^3}$

4. $(x^3y^2 + xy)\, dx = dy$

5. (*a*) $e^y\left(\dfrac{dy}{dx} + 1\right) = e^x$ (*b*) $(x + 1)\dfrac{dy}{dx} + 1 = 2e^{-y}$

6. $\dfrac{dy}{dx} - \dfrac{\tan y}{1+x} = (1 + x)\, e^x \sec y$

7. $(xy^2 - e^{\frac{1}{x^3}})\, dx - x^2y\, dy = 0$

8. $\dfrac{dy}{dx} + y \tan x = y^3 \cos x$

9. $\dfrac{dy}{dx} + \dfrac{y\log y}{x} = \dfrac{y(\log y)^2}{x^2}$

10. $y - \cos x\dfrac{dy}{dx} = y^2(1 - \sin x)\cos x$, given that $y = 2$ when $x = 0$

11. $y\,(2xy + e^x)\, dx = e^x\, dy$

12. $\tan y\dfrac{dy}{dx} + \tan x = \cos y \cos^2 x.$

Answers

1. $\dfrac{1}{y^5} = \dfrac{5}{2}x^3 + cx^5$

2. $\dfrac{x}{y} = 1 + c\sqrt{x}$

3. $y(c - x) = e^{\frac{1}{3}x^3}$

4. $\dfrac{1}{y} = -x^2 + 2 + ce^{-\frac{1}{2}x^2}$

5. (*a*) $e^{x+y} = \dfrac{1}{2}e^{2x} + c$ (*b*) $(x + 1)\, e^y = 2x + c$

6. $\sin y = (1 + x)(e^x + c)$

7. $3y^2 = x^2\,(2e^{\frac{1}{x^3}} + c)$

8. $\cos^2 x = y^2\left(c - 2\sin x + \dfrac{2}{3}\sin^3 x\right)$

9. $\dfrac{1}{x\log y} = \dfrac{1}{2x^2} + c$

10. $2(\tan x + \sec x) = y(2\sin x + 1)$

11. $e^x = y\,(c - x^2)$

12. $\sec y = (c + \sin x)\cos x.$

1.7. EXACT DIFFERENTIAL EQUATIONS

A differential equation obtained from its primitive directly by differentiation, without any operation of multiplication, elimination or reduction etc. is said to be an exact differential equation.

Thus a differential equation of the form $M\,(x, y)dx + N\,(x, y)dy = 0$ is an exact differential equation if it can be obtained directly by differentiating the equation $u\,(x, y) = c$, which is its primitive.

i.e., if $$du = Mdx + Ndy$$

1.8. THEOREM

The necessary and sufficient condition for the differential equation $Mdx + Ndy = 0$ to be exact if

$$\frac{\partial M}{\partial y} = \frac{\partial N}{\partial x}.$$

The condition is necessary.

The equation $Mdx + Ndy = 0$ will be exact, if $du \equiv Mdx + Ndy$

But $$du = \frac{\partial u}{\partial x}dx + \frac{\partial u}{\partial y}dy$$

$\therefore$ $$Mdx + Ndy = \frac{\partial u}{\partial x}dx + \frac{\partial u}{\partial y}dy$$

Equating co-efficients of dx and dy, we get

$$M = \frac{\partial u}{\partial x} \quad \text{and} \quad N = \frac{\partial u}{\partial y}$$

$\therefore$ $$\frac{\partial M}{\partial y} = \frac{\partial^2 u}{\partial y \partial x} \quad \text{and} \quad \frac{\partial N}{\partial x} = \frac{\partial^2 u}{\partial x \partial y}$$

But $$\frac{\partial^2 u}{\partial y \partial x} = \frac{\partial^2 u}{\partial x \partial y}$$

$\therefore$ $$\frac{\partial M}{\partial y} = \frac{\partial N}{\partial x}$$

which is the necessary condition of exactness.

The condition is sufficient.

Let $$u = \int_{y \text{ constant}} M\,dx$$

$\therefore$ $$\frac{\partial u}{\partial x} = M \quad \text{and} \quad \frac{\partial^2 u}{\partial y \partial x} = \frac{\partial M}{\partial y}$$

But $$\frac{\partial^2 u}{\partial y \partial x} = \frac{\partial^2 u}{\partial x \partial y} \quad \text{and} \quad \frac{\partial M}{\partial y} = \frac{\partial N}{\partial x}$$

$\therefore$ $$\frac{\partial N}{\partial x} = \frac{\partial^2 u}{\partial x \partial y} = \frac{\partial}{\partial x}\left(\frac{\partial u}{\partial y}\right)$$

Integrating both sides w.r.t. x, treating y as constant, we have $N = \frac{\partial u}{\partial y} + f(y)$

$\therefore$ $$Mdx + Ndy = \frac{\partial u}{\partial x}dx + \left\{\frac{\partial u}{\partial y} + f(y)\right\}dy \qquad \left[\because M = \frac{\partial u}{\partial x}, N = \frac{\partial u}{\partial y} + f(y)\right]$$

$$= \left(\frac{\partial u}{\partial x}dx + \frac{\partial u}{\partial y}dy\right) + f(y)dy = du + f(y)dy = d\,[u + \int f(y)dy]$$

which shows that Mdx + Ndy is an exact differential and hence Mdx + Ndy = 0 is an exact differential equation.

Note. Since Mdx + Ndy = $d[u + \int f(y)\,dy]$

$\therefore$ Mdx + Ndy = 0 $\Rightarrow$ $d\,[u + \int f(y)dy] = 0$

Integrating $u + \int f(y)dy = c$

But $u = \int_{y \text{ constant}} \text{M}\,dx$ and $f(y)$ = terms of N not containing x

Hence the solution of Mdx + Ndy = 0 is

$$\int_{y \text{ constant}} \mathbf{M}dx + \int (\text{terms of N not containing } x)\, dy = c.$$

ILLUSTRATIVE EXAMPLES

Example 1. *Solve* $(5x^4 + 3x^2y^2 - 2xy^3)dx + (2x^3y - 3x^2y^2 - 5y^4)dy = 0.$

Sol. Here $\text{M} = 5x^4 + 3x^2y^2 - 2xy^3$ and $\text{N} = 2x^3 - 3x^2y^2 - 5y^4$

$$\therefore \quad \frac{\partial \text{M}}{\partial y} = 6x^2y - 6xy^2 = \frac{\partial \text{N}}{\partial x}$$

Thus the given equation is exact and its solution is

$$\int_{y \text{ constant}} \text{M}dx + \int (\text{terms of N not containing } x)\, dy = c$$

i.e.,
$$\int_{y \text{ constant}} (5x^4 + 3x^2y^2 - 2xy^3)\,dx + \int -5y^4 dy = c$$

or
$$x^5 + x^3y^2 - x^2y^3 - y^5 = c.$$

Example 2. *Solve* $[\cos x \tan y + \cos (x + y)]dx + [\sin x \sec^2 y + \cos (x + y)]dy = 0.$

Sol. Here $\text{M} = \cos x \tan y + \cos (x + y)$

and $\text{N} = \sin x \sec^2 y + \cos (x + y)$

$$\therefore \quad \frac{\partial \text{M}}{\partial y} = \cos x \sec^2 y - \sin(x + y) = \frac{\partial \text{N}}{\partial x}$$

Thus the given equation is exact and its solution is

$$\int_{y \text{ constant}} \text{M}dx + \int (\text{terms of N not containing } x)\, dy = c$$

i.e.,
$$\int_{y \text{ constant}} [\cos x \tan y + \cos (x + y)]\, dx = c$$

or
$$\sin x \tan y + \sin(x + y) = c.$$

TEST YOUR KNOWLEDGE

Solve the following differential equations:

1. $(x^2 - 4xy - 2y^2)dx + (y^2 - 4xy - 2x^2)dy = 0$
2. $(3x^2 + 6xy^2)dx + (6x^2y + 4y^3)dy = 0$
3. $(x^2 + y^2 - a^2)xdx + (x^2 - y^2 - b^2)ydy = 0$
4. $(1 + e^{x/y})dx + \left(1 - \frac{x}{y}\right) e^{x/y}\, dy = 0$
5. $(y \cos x + 1)dx + \sin x\, dy = 0$
6. $(\sec x \tan x \tan y - e^x)\, dx + \sec x \sec^2 y\, dy = 0$
7. $ye^{xy}\, dx + (xe^{xy} + 2y)dy = 0$
8. $(y^2 e^{xy^2} + 4x^3)dx + (2xye^{xy^2} - 3y^2)dy = 0$
9. $(2xy \cos x^2 - 2xy + 1)\, dx + (\sin x^2 - x^2)\, dy = 0$
10. $(\sin x \cos y + e^{2x})dx + (\cos x \sin y + \tan y)dy = 0$
11. $\left[y\left(1 + \frac{1}{x}\right) + \cos y\right] dx + (x + \log x - x \sin y)dy = 0$
12. $\frac{dy}{dx} + \frac{y \cos x + \sin y + y}{\sin x + x \cos y + x} = 0$
13. $x\, dy + y\, dx + \frac{xdy - ydx}{x^2 + y^2} = 0.$

$\left[\textbf{Hint.} \text{ The given equation is } d(xy) + d\left(\tan^{-1} \frac{y}{x}\right) = 0\right]$

Answers

1. $x^3 - 6x^2y - 6xy^2 + y^3 = c$
2. $x^3 + 3x^2y^2 + y^4 = c$
3. $x^4 + 2x^2y^2 - 2a^2x^2 - y^4 - 2b^2y^2 = c$
4. $x + ye^{x/y} = c$
5. $y \sin x + x = c$
6. $\sec x \tan y - e^x = c$
7. $e^{xy} + y^2 = c$
8. $e^{xy^2} + x^4 - y^3 = c$
9. $y \sin x^2 - x^2y + x = c$
10. $-\cos x \cos y + \frac{1}{2}e^{2x} + \log \sec y = c$
11. $y(x + \log x) + x \cos y = c$
12. $y \sin x + (\sin y + y)\, x = c$
13. $xy + \tan^{-1} \frac{y}{x} = c.$

1.9. EQUATIONS REDUCIBLE TO EXACT EQUATIONS

Differential equations which are not exact can sometimes be made exact after multiplying by a suitable factor (a function of x and/or y) called the integrating factor.

For example, consider the equation $y\, dx - x\, dy = 0$ ...(1)

Here $\quad M = y$ and $N = -x$

$$\frac{\partial M}{\partial y} \neq \frac{\partial N}{\partial x}, \text{ therefore the equation is not exact.}$$

(*i*) Multiplying the equation by $\frac{1}{y^2}$, it becomes $\frac{ydx - xdy}{y^2} = 0$ or $d\left(\frac{x}{y}\right) = 0$

which is exact.

(*ii*) Multiplying the equation by $\frac{1}{x^2}$, it becomes $\frac{ydx - xdy}{x^2} = 0$ or $d\left(\frac{y}{x}\right) = 0$

which is exact.

(*iii*) Multiplying the equation by $\frac{1}{xy}$, it becomes $\frac{dx}{x} - \frac{dy}{y} = 0$ or $d(\log x - \log y) = 0$

which is exact.

$$\therefore \qquad \frac{1}{y^2}, \frac{1}{x^2} \text{ and } \frac{1}{xy} \text{ are integrating factors of (1).}$$

If a differential equation has one integrating factor, it has an infinite number of integrating factors.

(*a*) **I.F. found by inspection.** In a number of problems, a little analysis helps to find the integrating factor. The following differentials are useful in selecting a suitable integrating factor.

(*i*) $ydx + xdy = d(xy)$ (*ii*) $\frac{xdy - ydx}{x^2} = d\left(\frac{y}{x}\right)$

(*iii*) $\frac{ydx - xdy}{y^2} = d\left(\frac{x}{y}\right)$ (*iv*) $\frac{xdy - ydx}{x^2 + y^2} = d\left(\tan^{-1}\frac{y}{x}\right)$

(*v*) $\frac{xdy - ydx}{xy} = d\left[\log\left(\frac{y}{x}\right)\right]$ (*vi*) $\frac{ydx + xdy}{xy} = d\,[\log(xy)]$

(*vii*) $\frac{xdy + ydy}{x^2 + y^2} = d\left[\frac{1}{2}\log\left(x^2 + y^2\right)\right]$ (*viii*) $\frac{xdy - ydx}{x^2 - y^2} = d\left(\frac{1}{2}\log\frac{x+y}{x-y}\right)$.

ILLUSTRATIVE EXAMPLES

Example 1. *Solve* $ydx - xdy + 3x^2y^2e^{x^3}\,dx = 0$.

Sol. Since $3x^2e^{x^3} = d(e^{x^3})$, the term $3x^2y^2e^{x^3}\,dx$ should not involve y^2.

This suggests that $\frac{1}{y^2}$ may be an I.F.

Multiplying throughout by $\frac{1}{y^2}$, we have $\frac{ydx - xdy}{y^2} + 3x^2e^{x^3}\,dx = 0$

$$d\left(\frac{x}{y}\right) + d(e^{x^3}) = 0, \text{ which is exact.}$$

Integrating, we get $\frac{x}{y} + e^{x^3} = c$

which is the required solution.

Example 2. *Solve* $xdy - ydx = x\sqrt{x^2 - y^2}\, dx$.

Sol. The given equation is $xdy - ydx = x^2\sqrt{1-\left(\frac{y}{x}\right)^2}\, dx$ or $\dfrac{\frac{xdy - ydx}{x^2}}{\sqrt{1-\left(\frac{y}{x}\right)^2}} = dx$

or $$d\left(\sin^{-1}\frac{y}{x}\right) = dx, \text{ which is exact.}$$

Integrating, we get $$\sin^{-1}\frac{y}{x} = x + c \quad \text{or} \quad y = x \sin (x + c)$$

which is the required solution.

(*b*) **I.F. for a homogeneous equation.** If Mdx + Ndy = 0 is a homogeneous equation in x and y, then $\dfrac{1}{Mx + Ny}$ is an I.F. provided $Mx + Ny \neq 0$.

Note. If Mx + Ny consists of only one term, use the above method of I.F. otherwise, proceed by putting $y = vx$.

Example. *Solve* $(x^2y - 2xy^2)\, dx - (x^3 - 3x^2y)\, dy = 0$.

Sol. The given equation is homogeneous in x and y with

$$M = x^2y - 2xy^2 \text{ and } N = -x^3 + 3x^2y$$

Now $$Mx + Ny = x^3y - 2x^2y^2 - x^2y + 3x^2y^2 = x^2y^2 \neq 0$$

$\therefore$ $$\text{I.F.} = \frac{1}{Mx + Ny} = \frac{1}{x^2y^2}$$

Multiplying throughout by $\dfrac{1}{x^2y^2}$, the given equation becomes

$$\left(\frac{1}{y} - \frac{2}{x}\right) dx - \left(\frac{x}{y^2} - \frac{3}{y}\right) dy = 0$$

which is exact. The solution is $$\int_{y \text{ constant}} \left(\frac{1}{y} - \frac{2}{x}\right) dx + \int \frac{3}{y} dy = c$$

or $$\frac{x}{y} - 2 \log x + 3 \log y = c.$$

(*c*) **I.F. for an equation of the form $f_1(xy)\, ydx + f_2(xy)\, xdy = 0$.**

If Mdx + Ndy = 0 is of the form $f_1(xy)\, ydx + f_2(xy)\, xdy = 0$, then $\dfrac{1}{Mx - Ny}$ is an I.F. provided

$$Mx - Ny \neq 0.$$

Example. *Solve* $y(xy + 2x^2y^2)\, dx + x(xy - x^2y^2)\, dy = 0$.

Sol. The given equation is of the form $f_1(xy)\, y\, dx + f_2(xy)\, x\, dy = 0$.

Here $$M = xy^2 + 2x^2y^3 \text{ and } N = x^2y - x^3y^2$$

Now $$Mx - Ny = x^2y^2 + 2x^3y^3 - x^2y^2 + x^3y^3 = 3x^3y^3 \neq 0$$

$$\therefore \qquad \text{I.F.} = \frac{1}{\text{M}x - \text{N}y} = \frac{1}{3x^3y^3}$$

Multiplying throughout by $\frac{1}{3x^3y^3}$, the given equation becomes

$$\left(\frac{1}{3x^2y} + \frac{2}{3x}\right) dx + \left(\frac{1}{3xy^2} - \frac{1}{3y}\right) dy = 0$$

which is exact. The solution is $\displaystyle\int_{y \text{ constant}} \left(\frac{1}{3x^2y} + \frac{2}{3x}\right) dx + \int -\frac{1}{3y} dy = c$

or $$-\frac{1}{3xy} + \frac{2}{3} \log x - \frac{1}{3} \log y = c$$

or $$-\frac{1}{xy} + 2 \log x - \log y = \text{C, where C} = 3c.$$

(*d*) **For the equation Mdx + Ndy = 0.**

(*i*) If $\dfrac{\dfrac{\partial \text{M}}{\partial y} - \dfrac{\partial \text{N}}{\partial x}}{\text{N}}$ is a function of x only, say $f(x)$, then $e^{\int f(x)\,dx}$ is an I.F.

(*ii*) If $\dfrac{\dfrac{\partial \text{N}}{\partial x} - \dfrac{\partial \text{M}}{\partial y}}{\text{M}}$ is a function of y only, we say $g(y)$, then $e^{\int g(y)\,dy}$ is an I.F.

ILLUSTRATIVE EXAMPLES

Example 1. *Solve* $(xy^2 - e^{\frac{1}{x^3}})dx - x^2y\, dy = 0.$

Sol. Here $\qquad \text{M} = xy^2 - e^{\frac{1}{x^3}}$ and $\text{N} = -x^2y$

$$\frac{\dfrac{\partial \text{M}}{\partial y} - \dfrac{\partial \text{N}}{\partial x}}{\text{N}} = \frac{2xy - (-2xy)}{-x^2y} = -\frac{4}{x}$$

which is a function of x only.

$$\therefore \qquad \text{I.F.} = e^{\int -\frac{4}{x} dx} = e^{-4 \log x} = \frac{1}{x^4}$$

Multiplying throughout by $\frac{1}{x^4}$, we have $\left(\dfrac{y^2}{x^3} - \dfrac{1}{x^4} e^{\frac{1}{x^3}}\right) dx - \dfrac{y}{x^2} dy = 0$

which is exact. The solution is $\displaystyle\int_{y \text{ constant}} \left(\frac{y^2}{x^3} - \frac{1}{x^4} e^{\frac{1}{x^3}}\right) dx = c$

or $$-\frac{y^2}{2x^2}+\frac{1}{3}\int -\frac{3}{x^4}e^{\frac{1}{x^3}}dx = c \quad \text{or} \quad -\frac{y^2}{2x^2}+\frac{1}{3}\int e^t dt = c\text{, where } t=\frac{1}{x^3}$$

or $$-\frac{y^2}{2x^2}+\frac{1}{3}e^t = c \quad \text{or} \quad -\frac{3y^2}{x^2}+2e^{\frac{1}{x^3}} = \text{C, where C} = 6c.$$

Example 2. *Solve* $(xy^3 + y)\, dx + 2(x^2y^2 + x + y^4)\, dy = 0.$

Sol. Here $\quad \text{M} = xy^3 + y$ and $\text{N} = 2x^2y^2 + 2x + 2y^4$

$$\frac{\frac{\partial \text{N}}{\partial x}-\frac{\partial \text{M}}{\partial y}}{\text{M}} = \frac{4xy^2+2-3xy^2-1}{xy^3+y} = \frac{xy^2+1}{y(xy^2+1)} = \frac{1}{y}$$

which is a function of y only.

$\therefore$ $$\text{I.F.} = e^{\int \frac{1}{y}dy} = e^{\log y} = y$$

Multiplying throughout by y, we have $(xy^4 + y^2)\, dx + 2\,(x^2y^3 + xy + y^5)\, dy = 0$

which is exact. The solution is $$\int_{y\ \text{constant}} (xy^4+y^2)\,dx + \int 2y^5 dy = c$$

or $$\frac{x^2y^4}{2} + xy^2 + \frac{y^6}{3} = c.$$

Example 3. *Solve* $(x^2 - 3xy + 2y^2)\, dx + x(3x - 2y)\, dy = 0.$ [CUSAT 2008]

Sol. $\text{M} = x^2 - 3xy + 2y^2$ and $\text{N} = x(3x - 2y)$

The given equation is homogeneous in x and y

Now, $\quad \text{M}x + \text{N}y = x^3 - 3x^2y + 2y^2x + 3x^2y - 2xy^2$
$\quad = x^3 \neq 0$

$\therefore$ $$\text{I.F.} = \frac{1}{\text{M}x+\text{N}y} = \frac{1}{x^3}$$

Multiplying throughout by $\frac{1}{x^3}$, the equation becomes

$$\left(\frac{1}{x}-\frac{3y}{x^2}+\frac{2y^2}{x^3}\right)dx + \left(\frac{3}{x}-\frac{2y}{x^2}\right)dy = 0$$

which is exact.

The solution is

$$\int_{y\ \text{constant}} \left(\frac{1}{x}-\frac{3y}{x^2}+\frac{2y^2}{x^3}\right)dx = c$$

$$\log x + \frac{3y}{x} - \frac{y^2}{x^2} = c.$$

(*e*) **I.F. for the equation of the form x^ay^b (my dx + nx dy) + x^cy^d (py dx + qx dy) = 0** where a, b, c, d, m, n, p, q are all constant is x^hy^k, where h, k are so chosen that after multiplication by x^hy^k the equation becomes exact.

Example. *Solve* $(2x^2y^2 + y)dx + (3x - x^3y)dy = 0$.

Sol. The equation can be written as $2(x^2y^2dx - x^3ydy) + (ydx + 3xdy) = 0$

or $$x^2y\,(2ydx - xdy) + x^0y^0\,(ydx + 3xdy) = 0$$

which is of the form mentioned above. Therefore, it has an I.F. of the form x^hy^k.

Multiplying the given equation by x^hy^k, we have

$$(2x^{h+2}y^{k+2} + x^h\, y^{k+1})dx + (3x^{h+1}y^k - x^{h+3}y^{k+1})dy = 0$$

For this equation to be exact, we must have

$$\frac{\partial M}{\partial y} = \frac{\partial N}{\partial x} \quad i.e.,\ 2(k + 2)x^{h+2}y^{k+1} + (k + 1)x^hy^k$$

$$= 3(h + 1)x^hy^k - (h + 3)x^{h+2}y^{k+1}$$

which holds when $2(k + 2) = -(h + 3)$ and $k + 1 = 3(h + 1)$

i.e., when $h + 2k + 7 = 0$ and $3h - k + 2 = 0$

Solving these equation, we have $h = -\dfrac{11}{7}$, $k = -\dfrac{19}{7}$

$$\therefore \qquad \text{I.F.} = x^{-\frac{11}{7}}\, y^{-\frac{19}{7}}$$

Multiplying the given equation by $x^{-\frac{11}{7}}\, y^{-\frac{19}{7}}$, we have

$$\left(2x^{\frac{3}{7}}\, y^{-\frac{5}{7}} + x^{-\frac{11}{7}}\, y^{-\frac{12}{7}}\right)dx + \left(3x^{-\frac{4}{7}}\, y^{-\frac{19}{7}} - x^{\frac{10}{7}}\, y^{-\frac{12}{7}}\right)dy = 0$$

which is exact. The solution is $$\int_{y\text{ constant}} \left(2x^{\frac{3}{7}}\, y^{-\frac{5}{7}} + x^{-\frac{11}{7}}\, y^{-\frac{12}{7}}\right)dx = c$$

or $$\frac{7}{5}x^{\frac{10}{7}}\, y^{-\frac{5}{7}} - \frac{7}{4}x^{-\frac{4}{7}}\, y^{-\frac{12}{7}} = c \quad \text{or} \quad 4x^{\frac{10}{7}}\, y^{-\frac{5}{7}} - 5x^{-\frac{4}{7}}\, y^{-\frac{12}{7}} = \text{C} \quad \text{where C} = \frac{20}{7}c.$$

Note. The values of h and k can also be determined from the relations

$$\frac{a + h + 1}{m} = \frac{b + k + 1}{n} \quad \text{and} \quad \frac{c + h + 1}{p} = \frac{d + k + 1}{q}.$$

TEST YOUR KNOWLEDGE

Solve the following differential equations:

1. $x\,dy - y\,dx = (x^2 + y^2)\,dx$
2. $y\,(2xy + e^x)\,dx - e^x\,dy = 0$
3. $(1 + xy)\,ydx + (1 - xy)\,xdy = 0$
4. $xdy - ydx = xy^2\,dx$
5. $(xye^{x/y} + y^2)\,dx - x^2e^{x/y}\,dy = 0$
6. $x^2ydx - (x^3 + y^3)\,dy = 0$
7. $(3xy^2 - y^3)\,dx - (2x^2y - xy^2)\,dy = 0$
8. $(x^2y^2 + xy + 1)\,ydx + (x^2y^2 - xy + 1)\,xdy = 0$

9. $y(2xy+1)dx + x(1+2xy-x^3y^3)dy = 0$

10. $(x^2+y^2+2x)\,dx + 2ydy = 0$

11. $(x^2+y^2+1)dx - 2xydy = 0$

12. $\left(y + \frac{y^3}{3} + \frac{x^2}{2}\right)dx + \frac{1}{4}(x+xy^2)dy = 0$

13. $(y^4+2y)dx + (xy^3+2y^4-4x)dy = 0$

14. $(3xy-2ay^2)dx + (x^2-2axy)dy = 0$

15. $ydx - xdy + \log x\,dx = 0$

16. $(xy^2+2x^2y^3)dx + (x^2y-x^3y^2)dy = 0$

17. $(2x^2y-3y^4)dx + (3x^3+2xy^3)dy = 0.$

Answers

1. $y = x\tan(x+c)$

2. $x^2 + \frac{e^x}{y} = c$

3. $-\frac{1}{xy} + \log\frac{x}{y} = c$

4. $\frac{x^2}{2} + \frac{x}{y} = c$

5. $e^{x/y} + \log x = c$

6. $\log y - \frac{x^3}{3y^3} = c$

7. $3\log x - 2\log y + \frac{y}{x} = c$

8. $xy + \log\frac{x}{y} - \frac{1}{xy} = c$

9. $\frac{1}{x^2y^2} + \frac{1}{3x^3y^3} + \log y = c$

10. $e^x(x^2+y^2) = c$

11. $x - \frac{y^2}{x} - \frac{1}{x} = c$

12. $3x^4y + x^4y^3 + x^6 = c$

13. $\left(y + \frac{2}{y^2}\right)x + y^2 = c$

14. $x^2(ay^2 - xy) = c$

15. $cx + y + \log x + 1 = 0$

16. $-\frac{1}{xy} + 2\log x - \log y = c$

17. $5x^{-\frac{36}{13}}y^{\frac{24}{13}} - 12x^{-\frac{10}{13}}y^{-\frac{15}{13}} = c.$

2

Applicactions of Differential Equations of First Order

In this chapter we shall discuss physical problems which lead to the differential equations of first order and first degree.

2.1. PHYSICAL APPLICATIONS

Let a body of mass m start moving from a fixed point O along a straight line OX under the action of a force F. Let P be the position of the body at any instant t, where OP = x, then

(*i*) the velocity (v) of the body = $\dfrac{dx}{dt}$

(*ii*) the acceleration (a) of the body = $\dfrac{dv}{dt}$ or $\dfrac{d^2x}{dt^2}$ or $v\,\dfrac{dv}{dx}$

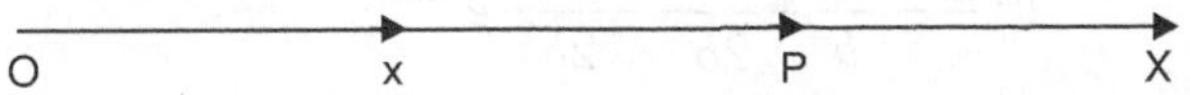

Also, by Newton's Second Law of motion, $F = ma = m\,\dfrac{dv}{dt}$ or $m\,\dfrac{d^2x}{dt^2}$ or $mv\,\dfrac{dv}{dx}$

where F is the effective force.

ILLUSTRATIVE EXAMPLES

(Resisted Motion)

Example 1. *A moving body is opposed by a force per unit mass of value cx and resistance per unit mass of value* bv^2*, where x and v are the displacement and velocity of the particle at that instant. Show that the velocity of the particle, if it starts from rest, is given by*

$$v^2 = \frac{c}{2b^2}\,(1 - e^{-2bx}) - \frac{cx}{b}.$$

Sol. By Newton's second law of motion, the equation of motion of the body is

$$v\,\frac{dv}{dx} = -\,cx - bv^2$$

or $$v\,\frac{dv}{dx} + bv^2 = -\,cx \qquad ...(1)$$

Putting $v^2 = z$ and $2v \frac{dv}{dx} = \frac{dz}{dx}$, eq. (1) becomes $\frac{1}{2}\frac{dz}{dx} + bz = -cx$

or $$\frac{dz}{dx} + 2bz = -2cx \qquad ...(2)$$

which is Leibnitz's linear equation

$$\text{I.F.} = e^{\int 2b\,dx} = e^{2bx}$$

$\therefore$ The solution of (2) is

$$z \,.\, e^{2bx} = \int -2cx \,.\, e^{2bx}\,dx + c_1 = -2c \int xe^{2bx}\,dx + c_1 \qquad \text{[Integrating by parts]}$$

$$= -2c\left[x \,.\, \frac{e^{2bx}}{2b} - \int 1 \,.\, \frac{e^{2bx}}{2b}\,dx\right] + c_1 = -\frac{cx}{b}e^{2bx} + \frac{c}{2b^2}e^{2bx} + c_1$$

$$\Rightarrow \quad v^2 \,.\, e^{2bx} = -\frac{cx}{b}e^{2bx} + \frac{c}{2b^2}e^{2bx} + c_1$$

$$\Rightarrow \quad v^2 = -\frac{cx}{b} + \frac{c}{2b^2} + c_1 e^{-2bx}. \qquad ...(3)$$

Initially, when $x = 0, v = 0$

$$\therefore \quad \frac{c}{2b^2} + c_1 = 0 \quad \text{or} \quad c_1 = -\frac{c}{2b^2}$$

Substituting the value of c_1 in (3), we have

$$v^2 = -\frac{cx}{b} + \frac{c}{2b^2} - \frac{c}{2b^2}e^{-2bx}$$

or $$v^2 = \frac{c}{2b^2}(1 - e^{-2bx}) - \frac{cx}{b}.$$

Example 2. *A particle of mass m is projected vertically upward under gravity, the resistance of the air being mk times the velocity. Show that the greatest height attained by the particle is* $\frac{V^2}{g}$ $[\lambda - \log(1 + \lambda)]$, *where V is the greatest velocity which the above mass will attain when it falls freely and* λV *is the initial velocity.*

Sol. Let v be the velocity of the particle at time t. The forces acting on the particle are

(*i*) its weight mg acting vertically downwards.

(*ii*) the resistance mkv of the air acting vertically downwards.

Accelerating force on the particle $= -mg - mkv$

$\therefore$ By Newton's second law, the equation of motion of the particle is

$$mv\frac{dv}{dx} = -mg - mkv$$

or $$v\frac{dv}{dx} = -g - kv \qquad ...(1)$$

When the particle falls freely (under gravity), equation (1) becomes (changing g to $-g$).

$$v\frac{dv}{dx} = g - kv \qquad ...(2)$$

When the particle attains the greatest velocity V, its acceleration is zero.

$\therefore$ From (2), $\quad 0 = g - kV$ or $k = \dfrac{g}{V}$

Putting this value of k in (1), we have $v \dfrac{dv}{dx} = -g - \dfrac{g}{V} v = -\dfrac{g}{V}(V + v)$

or $$\frac{v}{V + v} dv = -\frac{g}{V} dx$$

Integrating, $\displaystyle\int \frac{v}{V+v} dv = -\frac{g}{V}\int dx + c$ or $\displaystyle\int\left(1 - \frac{V}{V+v}\right) dv = -\frac{g}{V} x + c$

or $$v - V \log (V + v) = -\frac{g}{V} x + c \qquad ...(3)$$

Initially, when $x = 0$, $v = \lambda V$

$\therefore$ From (3), we have $\quad \lambda V - V \log (V + \lambda V) = c$ or $c = V [\lambda - \log V (1 + \lambda)]$

Substituting the value of c in (3), $v - V \log (V + v) = -\dfrac{g}{V} x + V [\lambda - \log V (1 + \lambda)] \quad ...(4)$

Let h be the greatest height attained by the particle, then $x = h$ when $v = 0$

$\therefore$ From (4), we have $-V \log V = -\dfrac{g}{V} h + V [\lambda - \log V (1 + \lambda)]$

or $$\frac{g}{V} h = V\lambda - V [\log V (1 + \lambda) - \log V] = V\lambda - V \log \frac{V(1+\lambda)}{V}$$

or $$h = \frac{V^2}{g} [\lambda - \log (1 + \lambda)].$$

Example 3. *A body of mass m, falling from rest, is subject to the force of gravity and an air resistance proportional to the square of the velocity (i.e., kv^2). If it falls through a distance x and possesses a velocity v at that instant, prove that* $\dfrac{2kx}{m} = \log \dfrac{a^2}{a^2 - v^2}$*, where $mg = ka^2$.*

Sol. The forces acting on the body are

(*i*) its weight mg acting vertically downwards.

(*ii*) the resistance kv^2 of the air acting vertically upwards.

Accelerating force on the body $= mg - kv^2 = ka^2 - kv^2 \qquad [\because mg = ka^2]$

$= k (a^2 - v^2)$

By Newton's second law, the equation of motion of the body is

$$mv \frac{dv}{dx} = k (a^2 - v^2) \quad \text{or} \quad \frac{v}{a^2 - v^2} dv = \frac{k}{m} dx$$

Integrating, $\displaystyle\int \frac{v}{a^2 - v^2} dv = \frac{k}{m}\int dx + c$ or $\displaystyle -\frac{1}{2}\int \frac{-2v}{a^2 - v^2} dv = \frac{k}{m} x + c$

or $$-\frac{1}{2} \log (a^2 - v^2) = \frac{k}{m} x + c \qquad ...(1)$$

Initially, when $x = 0$, $v = 0 \quad \therefore \quad c = -\dfrac{1}{2} \log a^2$

$\therefore$ From (1), we have $-\dfrac{1}{2}\log(a^2 - v^2) = \dfrac{k}{m}x - \dfrac{1}{2}\log a^2$

or $$\frac{2kx}{m} = \log a^2 - \log(a^2 - v^2) \quad \text{or} \quad \frac{2kx}{m} = \log\frac{a^2}{a^2 - v^2}.$$

Example 4. *A paratrooper and his parachute weigh 50 kg. At the instant parachute opens, he is travelling vertically downward at the speed of 20 m/s. If the air resistance varies directly as the instantaneous velocity and it is 20 Newtons when the velocity is 10 m/s, find the limiting velocity, the position and the velocity of the paratrooper at any time t.*

Sol. Let v m/s be the velocity of the paratrooper t seconds after the parachute opens. The forces acting on the paratrooper are :

(*i*) the weight 50 kg acting vertically downwards.

(*ii*) the air resistance kv acting vertically upwards.

Accelerating force = $(50 - kv)$ N

By Newton's second law, the accelerating force is

$$m\frac{dv}{dt} = \frac{W}{g}\frac{dv}{dt} = \frac{50}{g}\frac{dv}{dt}$$

$\therefore$ The equation of motion is $\dfrac{50}{g}\dfrac{dv}{dt} = 50 - kv$ ...(1)

When $v = 10$ m/s, the air resistance $kv = 20$ N

$\therefore$ $k \cdot 10 = 20 \Rightarrow k = 2$

Substituting the value of k in (1), we have $\dfrac{50}{g}\dfrac{dv}{dt} = 50 - 2v$

or $$\frac{dv}{25 - v} = \frac{g}{25}dt$$

Integrating, $-\log(25 - v) = \dfrac{gt}{25} + c$ ...(2)

When $t = 0$, $v = 20$ $\therefore$ $c = -\log 5$

Substituting the value of c in (2), we have $-\log(25 - v) = \dfrac{gt}{25} - \log 5$

or $$\log\frac{25 - v}{5} = -\frac{gt}{25} \quad \text{or} \quad \frac{25 - v}{5} = e^{-gt/25}$$

or $$v = 5(5 - e^{-gt/25}) \qquad ...(3)$$

which gives the velocity of the paratrooper at any time t.

The limiting velocity is the velocity when $t \to \infty$.

$\therefore$ From (3), the limiting velocity = 25 m/s.

Now, from (3), we have $\dfrac{dx}{dt} = 5(5 - e^{-gt/25})$ or $dx = 5(5 - e^{-gt/25})\,dt$

Integrating, $$x = 5\left(5t + \frac{25}{g}e^{-gt/25}\right) + c' \qquad ...(4)$$

Initially, when $t = 0, x = 0$

$\therefore$ $$c' = -\frac{125}{g}$$

Substituting the value of c' in (4), the position of the paratrooper at any time t is given by

$$x = 25t - \frac{125}{g}(1 - e^{-gt/25}).$$

Example 5. (Velocity of escape from the earth). *Determine the least velocity with which a particle must be projected vertically upwards so that it does not return to the earth. Assume that it is acted upon by the gravitational attraction of the earth only.*

Sol. Let r be the variable distance of the particle from the earth's centre.

By Newton's law of gravitation, the acceleration a of the particle is proportional to $\frac{1}{r^2}$.

$$\therefore \qquad a = v\frac{dv}{dr} = -\frac{\mu}{r^2} \qquad \ldots(1)$$

where v is the velocity of the particle when its distance from the earth's centre is r.

On the surface of the earth, $r = \mathrm{R}$, the radius of earth and $a = -g$

$$\therefore \qquad -g = -\frac{\mu}{\mathrm{R}^2} \quad \text{or} \quad \mu = g\mathrm{R}^2$$

$$\therefore \text{ From (1),} \qquad v\frac{dv}{dr} = -\frac{g\mathrm{R}^2}{r^2} \quad \text{or} \quad v\,dv = \frac{g\mathrm{R}^2}{r^2}\,dr$$

$$\text{Integrating} \qquad \int v\,dv = -g\mathrm{R}^2\int\frac{dr}{r^2} + c \quad \text{or} \quad \frac{v^2}{2} = \frac{g\mathrm{R}^2}{r} + c$$

$$\text{or} \qquad v^2 = \frac{2g\mathrm{R}^2}{r} + 2c \qquad \ldots(2)$$

Let v_0 be the velocity of projection from the surface of the earth.

Then $\qquad v = v_0 \quad \text{when} \quad r = \mathrm{R}$

$\therefore$ From (2), $\qquad v_0^2 = 2g\mathrm{R} + 2c \qquad \ldots(3)$

Subtracting (3) from (2) [to eliminate c]

$$v^2 - v_0^2 = \frac{2g\mathrm{R}^2}{r} - 2g\mathrm{R} \quad \text{or} \quad v^2 = \frac{2g\mathrm{R}^2}{r} + (v_0^2 - 2g\mathrm{R})$$

The particle will never return to earth if its velocity v during ascent remains positive. For, if v vanishes, the particle comes to rest and then descends, so that v becomes negative.

Now, as the particle rises upwards, $\frac{2g\mathrm{R}^2}{r}$ goes on decreasing. The velocity will remain positive

if $\qquad v_0^2 - 2g\mathrm{R} \geq 0.$

i.e., if $\qquad v_0 \geq \sqrt{2g\mathrm{R}}$

$\therefore$ The least velocity of projection $v_0 = \sqrt{2g\mathrm{R}}$.

A particle projected with this velocity will never return to the earth, *i.e.,* will escape from the earth. This velocity is called the *velocity of escape from the earth.*

Example 6. (Motion of a boat across a stream). *A boat is rowed with a velocity u across a stream of width a. If the velocity of the current is directly proportional to the product of the distances from the two banks, find the equation of the path of the boat and the distance down stream to the point where it lands.*

Sol. Let the point from where the boat starts be taken as the origin O and the axes as shown in the figure. At any time t, let the boat be at P(x, y), then its distances from the two banks are y and $(a - y)$.

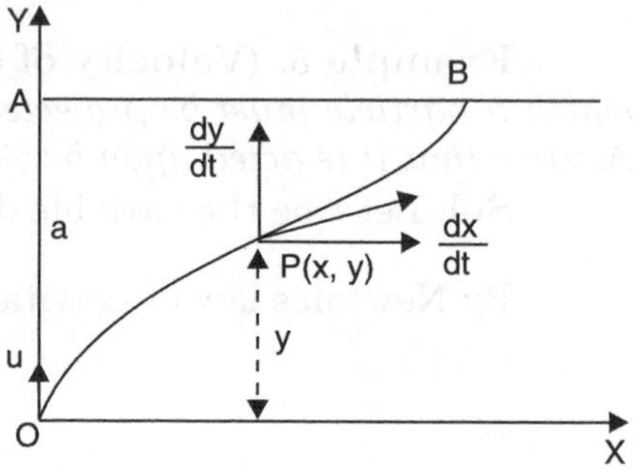

$$\frac{dx}{dt} = \text{velocity of the current} = ky\,(a - y)$$

$$\frac{dy}{dt} = \text{velocity of the boat} = u$$

$$\therefore \quad \frac{dy}{dx} = \frac{dy}{dt} \div \frac{dx}{dt} = \frac{u}{ky(a-y)} \qquad \ldots(1)$$

This is the direction of the resultant velocity of the boat at P and hence the direction of the tangent to the path of the boat at P.

From (1), $\quad y(a - y)\, dy = \dfrac{u}{k}\, dx$

Integrating, $\quad \dfrac{ay^2}{2} - \dfrac{y^2}{3} = \dfrac{u}{k}\, x + c$

When $x = 0$, $y = 0$, so that $c = 0$

$\therefore$ The equation of the path of the boat is $x = \dfrac{k}{6u}\, y^2\,(3a - 2y)$ $\qquad \ldots(2)$

The distance down-stream to the point where the boat lands, is obtained by putting $y = a$ in (2).

Thus $\qquad x = \dfrac{ka^3}{6u}$.

Example 7. (Atmospheric pressure). *Find the atmospheric pressure p kg/m² at a height z metres above the sea-level.*

Sol. Consider a vertical column of air of unit cross-section. Let an element of this column be bounded by two horizontal planes at height z and $z + \delta z$ above the sea-level. Let p and $p + \delta p$ be the pressures at heights z and $z + \delta z$ respectively. Let ρ be the average density of the element.

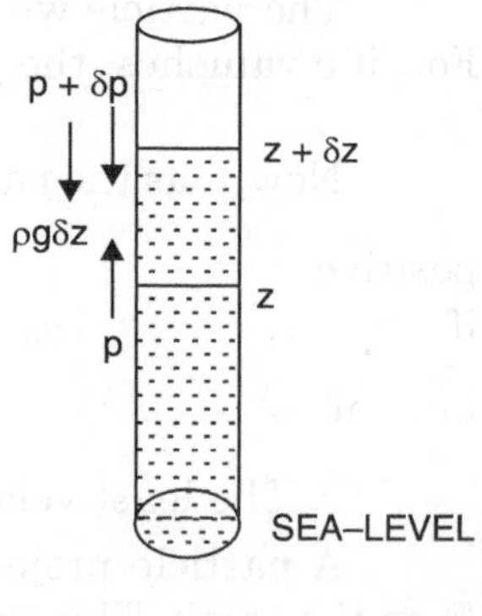

The thin column δz of air is in equilibrium under the action of forces

(*i*) p kg upwards

(*ii*) $(p + \delta)$ kg downwards

(*iii*) the weight $\rho g\, \delta z$ kg downwards

$$\therefore \quad p = p + \delta p + \rho g \delta z$$

$$\Rightarrow \quad \delta p + \rho g \delta z = 0 \quad \text{or} \quad \frac{\delta p}{\delta z} = -\rho g$$

Taking the limit as $\delta z \to 0$, we have $\frac{dp}{dz} = -\rho g$ ...(1)

which is the differential equation giving atmospheric pressure at any height z.

Now, we discuss equation (1) under two assumptions :

(*i*) when the temperature is constant

(*ii*) when the temperature varies.

(*i*) *When the temperature is constant.*

By Boyle's Law, $p = k\rho$ or $\rho = \frac{p}{k}$.

Substituting this value of ρ in (1), we have $\frac{dp}{dz} = -\frac{pg}{k}$ or $\frac{dp}{p} = -\frac{g}{k}dz$

Integrating, $$\int \frac{dp}{p} = -\frac{g}{k}\int dz + c \quad \text{or} \quad \log p = -\frac{g}{k}z + c \qquad ...(2)$$

At the sea-level, $z = 0$, $p = p_0$ (say), then $c = \log p_0$

$\therefore$ From (2), $$\log p = -\frac{g}{k}z + \log p_0 \quad \text{or} \quad \log \frac{p}{p_p} = -\frac{g}{k}z$$

$$\therefore \qquad p = p_0 e^{-\frac{gz}{k}}$$

(*ii*) *When the temperature varies*

Let $$p = k\rho^n, n \neq 1 \quad \text{or} \quad \rho = \left(\frac{p}{k}\right)^{\frac{1}{n}}$$

Substituting this value of ρ in (1), we have $\frac{dp}{dz} = -\left(\frac{p}{k}\right)^{\frac{1}{n}}$

or $$p^{-\frac{1}{n}}\,dp = -\,gk^{-\frac{1}{n}}\,dz$$

Integrating, $$\int p^{-\frac{1}{n}}\,dp = -\,gk^{-\frac{1}{n}}\int dz + c$$

or $$\frac{n}{n-1}p^{1-\frac{1}{n}} = -\,gk^{-\frac{1}{n}}\,z + c \qquad ...(3)$$

At the sea-level, $z = 0$, $p = p_0$ (say)

then $$c = \frac{n}{n-1}p_0^{1-\frac{1}{n}}$$

$\therefore$ From (3), $$\frac{n}{n-1}p^{1-\frac{1}{n}} = -\,gk^{-\frac{1}{n}}\,z + \frac{n}{n-1}p_0^{1-\frac{1}{n}}$$

or $$\frac{n}{n-1}\left(p_0^{1-\frac{1}{n}} - p^{1-\frac{1}{n}}\right) = gk^{-\frac{1}{n}}z.$$

Example 8. (Rotating cylinder containing liquid). *A cylindrical tank of radius r is filled with water. When the tank is rotated rapidly about its axis with angular velocity ω, centrifugal force tends to drive the water outward from the centre of the tank. Under steady conditions of uniform rotation, find the equation of the curve in which the free surface of the water is intersected by a plane through the axis of the cylinder.*

Sol. Consider a particle of mass m at the point P(x, y) on the plane curve cut out of the surface by an axial section. The forces acting on the particle are

(*i*) the weight mg acting vertically downwards.

(*ii*) the centrifugal force $m\omega^2 x$ acting horizontally outwards.

Since the motion is steady, P cannot move either up or down the surface of the water. Thus P moves just on the surface of the water and, therefore, there is no force along the tangent to the curve. The resultant R of mg and $m\omega^2 x$ is along the outward normal to the curve.

From the figure, R cos ψ = mg and R sin ψ = $m\omega^2 x$

$$\therefore \quad \frac{dy}{dx} = \tan\psi = \frac{m\omega^2 x}{mg} = \frac{\omega^2 x}{g} \quad ...(1)$$

which is the differential equation of the surface of the rotating liquid.

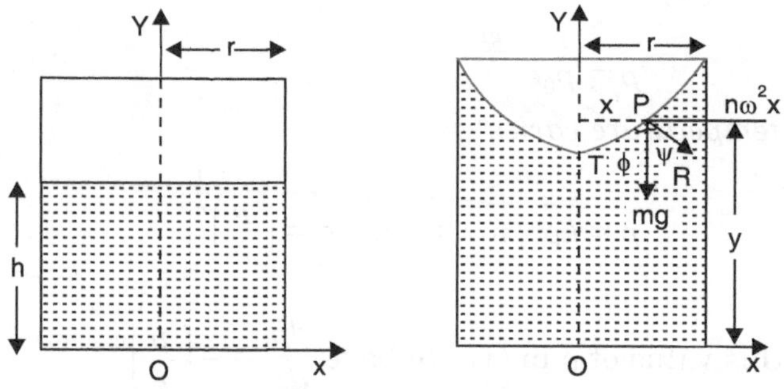

From (1), $$dy = \frac{\omega^2 x}{g}\, dx$$

Integrating, $$\int dy = \frac{\omega^2}{g}\int x dx + c \quad \text{or} \quad y = \frac{\omega^2 x^2}{2g} + c \quad ...(2)$$

To determine c, we use the fact that the volume of the liquid remains the same before and after rotation.

Thus the shaded areas in the two figures remain the same.

Now, the shaded area in non-rotational case = $2rh$.

Also, the shaded area in rotational case = $2\int_0^r y dx$

$$= 2\int_0^r \left(\frac{\omega^2 x^2}{2g} + c\right) dx = 2\left(\frac{\omega^2 r^3}{2g} + cr\right)$$

$$\therefore \quad 2\left(\frac{\omega^2 r^3}{6g} + cr\right) = 2rh \quad \Rightarrow \quad c = h - \frac{\omega^2 r^2}{6g}$$

From (2), we have $y = \frac{\omega^2 x^2}{2g} + h - \frac{\omega^2 r^2}{6g}$

or $$y = \frac{\omega^2}{2g}\left(x^2 - \frac{r^2}{3}\right) + h$$

which is the required equation of the curve.

TEST YOUR KNOWLEDGE

1. A particle is projected with velocity V along a smooth horizontal plane in a medium whose resistance per unit mass is μ times the cube of the velocity. Show that the distance it has described in time t is

$$\frac{1}{\mu V}[\sqrt{1 + 2\mu V^2 t} - 1].$$

2. A particle of unit mass moves in a straight line under retardation which is k times its velocity. Initially, the particle is at a distance a from a given point O in the line and is moving towards O with velocity u (> ak). Prove that it will reach O in time $\frac{1}{k}\log\frac{u}{u - ak}$.

3. A boat, with its load, weighs 981 kg. The force exerted upon the boat in the direction of motion, by the oars, is equivalent to a constant force of 15 Newtons and the resistance, in Newtons, to the motion is equal to twice the speed of the boat. If the boat starts from rest, determine its (*i*) limiting speed and (*ii*) speed at the end of 10 seconds.

   [**Hint.** The limiting or terminal speed is the speed when acceleration is zero].

4. A bullet enters a board of 0.1 m thickness with a velocity of 200 m/s, pierces it and leaves the board with a velocity of 80 m/s. Assuming that the resistance offered, by the board to the bullet is proportional to square of its velocity, find the time taken by the bullet to pierce the board.

5. A particle of unit mass is projected vertically upward with velocity V and the resistance of the air produces a retardation kv^2, where v is the velocity. Show that the velocity V′, with which the particle will return to the point of projection, is given by $\frac{1}{V'^2} = \frac{1}{V^2} + \frac{k}{g}$.

6. Assuming that the resistance to the movement of a ship through water is of the form $a^2 + b^2v^2$, where v is the velocity and a, b are constants, write down the differential equation for retardation of the ship moving with engines stopped. Prove that the time in which its speed falls off to one-half its original value u is given by

$$\frac{W}{abg}\tan^{-1}\frac{abu}{2a^2 + b^2u^2}$$ where W is the weight of the ship.

7. A particle of mass m moves under gravity in a medium whose resistance is k times its velocity, where k is a constant. If the particle is projected vertically upwards with a velocity v, show that the time to reach the highest point is $\frac{m}{k}\log\left(1 + \frac{kv}{mg}\right)$.

8. A body falling from rest is subjected to the force of gravity and an air resistance of $\frac{n^2}{g}$ times the square of velocity. Show that the distance travelled by the body in t seconds is $\frac{g}{n^2}\log\cosh nt$.

9. A body of mass m falls from rest under gravity in a field whose resistance is mk times the velocity of the body. Find the terminal velocity of the body and also the time taken to acquire one half of its limiting speed.

10. Two particles fall freely, one in a medium whose resistance is equal to k times the velocity and the other in a medium whose resistance is equal to k times the square of the velocity. If V_1 and V_2 are their maximum velocities respectively, show that $V_1 = V_2^2$.

11. Find the atmospheric pressure at a height h metres above sea-level. It is known that at sea-level this pressure is 1 kg/cm^2, while at a height of 500 metres above the sea-level, it is 0.923 kg/cm^2, temperature being assumed constant.

12. Upto a certain height in the atmosphere, it is found that the pressure p and the density ρ are connected by the relation $p = k\rho^n$ $(n > 1)$. If this relation continues to hold upto any height, show that the density would vanish at a finite height.

Answers

3. (*i*) 7.5 m/s, (*ii*) 1.36 m/s **9.** $\frac{g}{k}, \frac{1}{k}\log_e 2$; 0.00082 seconds

11. $p = e^{-0.00016h}$.

2.2. APPLICATION TO ELECTRIC CIRCUITS

If Q be the electrical charge on a condenser of capacity C and i be the current, then

(*a*) $i = \frac{dQ}{dt}$ or $Q = \int i\,dt$

(*b*) the potential drop across the resistance R is Ri.

(*c*) the potential drop across the inductance L is $L\frac{di}{dt}$.

(*d*) the potential drop across the capacitance C is $\frac{Q}{C}$.

Also, by Kirchhoff's Law, the total potential drop (voltage drop) in the circuit is equal to the applied voltage (E.M.F.).

ILLUSTRATIVE EXAMPLES

Example 1. *A constant electromotive force E volts is applied to a circuit containing a constant resistance R ohms in series and a constant inductance L henries. If the initial current is zero, show that the current builds up to half its theoretical maximum in (L log 2)/R seconds.*

Sol. Let i be the current in the circuit at any time t.

By Kirchhoff's law, we have $$L\frac{di}{dt} + Ri = E \quad \text{or} \quad \frac{di}{dt} + \frac{R}{L}i = \frac{E}{L} \qquad ...(1)$$

which is Leibnitz's linear equation, $\text{I.F.} = e^{\int \frac{R}{L}dt} = e^{\frac{Rt}{L}}$

$\therefore$ The solution of equation (1) is $i\,(\text{I.F.}) = \int \frac{E}{L}(\text{I.F.})\,dt + c$

or $$i.e^{\frac{Rt}{L}} = \int \frac{E}{L}.e^{\frac{Rt}{L}}\,dt + c = \frac{E}{L}.\frac{L}{R}e^{\frac{Rt}{L}} + c$$

or $$i = \frac{E}{R} + ce^{-\frac{Rt}{L}} \qquad ...(2)$$

Initially, when $t = 0, i = 0$ so that $c = -\frac{E}{R}$

Thus (2) becomes, $i = \frac{E}{R}\left(1 - e^{-\frac{Rt}{L}}\right)$ ...(3)

This equation gives the current in the circuit at any time t.

Clearly, i increases with t and attains the maximum value $\frac{E}{R}$.

Let the current in the circuit be half its theoretical maximum after a time T seconds. Then

$$\frac{1}{2}.\frac{E}{R} = \frac{E}{R}\left(1 - e^{-\frac{RT}{L}}\right) \quad \text{or} \quad e^{-\frac{RT}{L}} = \frac{1}{2}$$

or $$-\frac{RT}{L} = \log\frac{1}{2} = -\log 2$$

$\therefore$ $T = (L \log 2)/R.$

Example 2. *The equations of electromotive force in terms of current i for an electrical circuit having resistance R and a condenser of capacity C, in series, is* $E = Ri + \int \frac{i}{C}\,dt$. *Find the current i at any time t, when* $E = E_0 \sin \omega t$.

Sol. The given equation can be written as $R_i + \int \frac{i}{C}\,dt = E_0 \sin \omega t.$

Differentiating both sides w.r.t. t, we have $R\frac{di}{dt} + \frac{i}{C} = \omega E_0 \cos \omega t$

or $$\frac{di}{dt} + \frac{i}{RC} = \frac{\omega E_0}{R}\cos \omega t \qquad ...(1)$$

which is Leibnitz's linear equation.

$$\text{I.F.} = e^{\int \frac{1}{RC}dt} = e^{\frac{t}{RC}}$$

$\therefore$ The solution of equation (1) is

$$i.e^{\frac{t}{RC}} = \int \frac{\omega E_0}{R}\cos \omega t.\,e^{\frac{t}{RC}}\,dt = \frac{\omega E_0}{R}\int e^{\frac{t}{RC}}\cos \omega t\,dt$$

$$= \frac{\omega E_0}{R} \cdot \frac{e^{\frac{t}{RC}}}{\sqrt{\left(\frac{1}{RC}\right)^2 + \omega^2}} \cos\left(\omega t - \tan^{-1}\frac{\omega}{\frac{1}{RC}}\right) + k$$

$$\left[\because \int e^{ax} \cos bx\, dx = \frac{e^{ax}}{\sqrt{a^2+b^2}} \cos\left(bx - \tan^{-1}\frac{b}{a}\right)\right]$$

$$= \frac{\omega CE_0}{\sqrt{1+R^2C^2\omega^2}} e^{\frac{t}{RC}} \cos(\omega t - \phi) + k \text{ where } \tan\phi = RC\omega$$

or $$i = \frac{\omega CE_0}{\sqrt{1+R^2C^2\omega^2}} \cos(\omega t - \phi) + ke^{-\frac{t}{RC}}$$

which gives the current at any time t.

Example 3. *An RL circuit has an emf of 5V, a resistance of 50 Ω an inductance of 1H and no initial current. Find the current in the circuit at any time t.* [CUSAT 2008]

Sol. Let i be the current in the circuit at any time t.

By Kirchhoff's law, we have

$$L\frac{di}{dt} + Ri = E \quad \text{or} \quad \frac{di}{dt} + \frac{R}{L}i \pm \frac{E}{L}$$

Substituting R, E and L, we get

$$\frac{di}{dt} + 50i = 5 \qquad ...(1)$$

which is linear differential equation,

$$\text{I.F.} = e^{50t}$$

∴ The solution of equation (1) is

$$i\,(\text{I.F.}) = \int 5\,(\text{I.F.})\, dt + c$$

$$ie^{50t} = \int 5e^{50t}\, dt + c$$

⇒ $$ie^{50t} = \frac{1}{10} e^{50t} + c$$

Dividing by e^{50t},

$$i = \frac{1}{10} + ce^{-50t} \qquad ...(2)$$

at $t = 0$, $i = 0$

∴ Substitute in equation (2), we get

$$c = -\frac{1}{10} \quad \Rightarrow \quad i = \frac{1}{10}(1 - e^{-50t}).$$

TEST YOUR KNOWLEDGE

1. When a resistance R ohms is connected in series with an inductance L henries, an e.m.f. of E volts, the current i amperes at time t is given by $L\frac{di}{dt} + Ri = E$. If $E = 10 \sin t$ volts and $i = 0$, when $t = 0$, find i as a function of t.
2. A voltage Ee^{-at} is applied at $t = 0$ to a circuit containing inductance L and resistance R. Show that the current at any time t is $\frac{E}{R - aL}\left(e^{-at} - e^{-\frac{Rt}{L}}\right)$.
3. When a switch is closed in a circuit containing a battery E, a resistance R and an inductance L, the current i builds up at rate given by $L\frac{di}{dt} + Ri = E$. Find i as a function of t. How long will it be, before the current has reached one-half its maximum value if E = 6 volts, R = 100 ohms and L = 0.1 henry ?
4. Show that the differential equation for the current i in an electrical circuit containing an inductance L and a resistance R in series and acted on by an electromotive force $E \sin \omega t$ satisfies the equation. $L\frac{di}{dt} + Ri = E \sin \omega t$. Find the value of the current at any time t, if intially there is no current in the circuit.
5. Solve the equation $L\frac{di}{dt} + Ri = 200 \cos 300\, t$, when R = 100, L= 0.05 and find i given that $i = 0$ when $t = 0$. What value does i approach after a long time ?
6. A resistance R in series with inductance L is shunted by an equal resistance R with capacity C. An alternating e.m.f. $\sin pt$ produces currents i_1 and i_2 in the two branches. If initially there is no current, determine i_1 and i_2 from the equations :

$$L\frac{di_1}{dt} + Ri_1 = E \sin pt \quad \text{and} \quad \frac{i_2}{C} + R\frac{di_2}{dt} = p\, E \cos pt$$

Verify that if $R^2C = L$, the total current $i_1 + i_2$ will be $\frac{E}{R} \sin pt$.

Answers

1. $i = \frac{10}{L^2 + R^2}(R \sin t - L \cos t + Le^{-Rt/L})$

3. 0.0006931 sec.

4. $i = \frac{E}{\sqrt{R^2 + \omega^2 L^2}}[\sin(\omega t - \phi) + \sin \phi \,.\, e^{-Rt/L}]$ where $\phi = \tan^{-1}\frac{\omega L}{R}$

5. $i = \frac{40}{409}(20 \cos 300\, t + 3 \sin 300\, t) - \frac{800}{409} e^{-200t}; \frac{40}{\sqrt{409}}$

2.3. CONDUCTION OF HEAT

The fundamental principle involved in the problems of heat conduction is that the quantity of heat Q flowing per second across a slab of area A and thickness δx, whose faces are at temperature T and T + δT, is given by $Q = -kA\dfrac{dT}{dx}$ where k is the co-efficient of thermal conductivity and depends upon the material of the body.

Example. *A long hollow pipe has an inner diameter of 10 cm and outer diameter of 20 cm. The inner surface is kept at 200°C and the outer surface at 50°C. The thermal conductivity is 0.12. How much heat is lost per minute from a portion of the pipe 20 meters long ? Find the temperature at a distance x = 7.5 cm from the centre of the pipe.*

Sol. Here the isothermal surfaces are cylinders, the axis of each one of them is the axis of the pipe. **Consider one such cylinder of radius x cm and length 1 cm.** The surface area of this cylinder is $A = 2\pi x$ sq. cm. Let Q cal/sec be quantity of heat flowing across this surface, then

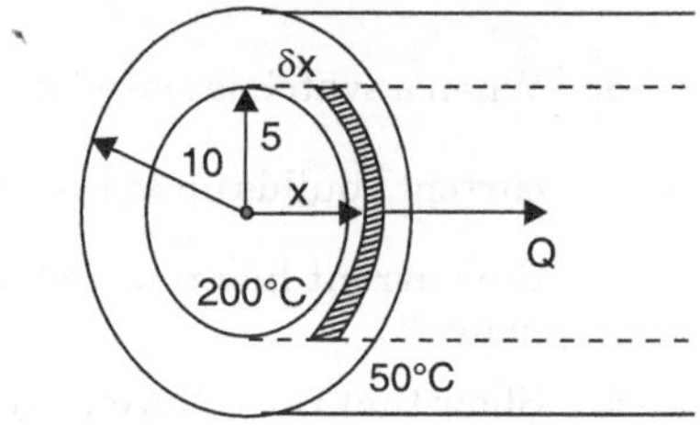

$$Q = -kA\frac{dT}{dx} = -k\,.\,2\pi x\frac{dT}{dx}$$

or $$dT = -\frac{Q}{2\pi k}\cdot\frac{dx}{x}$$

Integrating, we have $$T = -\frac{Q}{2\pi k}\log_e x + c \quad \text{...(1)}$$

Since T = 200, when $x = 5$

$$\therefore \quad 200 = -\frac{Q}{2\pi k}\log_e 5 + c \quad \text{...(2)}$$

Also T = 50, when $x = 10$

$$\therefore \quad 50 = \frac{Q}{2\pi k}\log_e 10 + c \quad \text{...(3)}$$

Subtracting (3) from (2), we have

$$150 = \frac{Q}{2\pi k}(\log_e 10 - \log_e 5)$$

or $$150 = \frac{Q}{2\pi k}\log_e 2 \quad \text{...(4)}$$

$$\therefore \quad Q = \frac{2\pi k \times 150}{\log_e 2} = \frac{300\pi \times 0.12}{\log_e 2} = 163 \text{ cal/sec.}$$

Hence the heat lost per minute through 20 metre length of the pipe

$$= 60 \times 2000Q = 120000 \times 163 = 1956000 \text{ cal.}$$

Now, let T = t, when $x = 7.5$

From (1), $$t = -\frac{Q}{2\pi k}\log_e 7.5 + c \quad \text{...(5)}$$

Subtracting (2) from (5), $$t - 200 = -\frac{Q}{2\pi k}(\log_e 7.5 - \log_e 5)$$

or $$t - 200 = -\frac{Q}{2\pi k}\log_e 1.5 \quad \text{...(6)}$$

Dividing (6) by (4), we have $\frac{t-200}{150} = -\frac{\log_e 1.5}{\log_e 2}$

or $t = 200 - 150 \times 0.58 = 113$

∴ When $x = 7.5$ cm, T = 113° C.

TEST YOUR KNOWLEDGE

1. Calculate the amount of heat passing through 1 sq. cm. of a refrigerator wall, if the thickness of the wall is 6 cm and the temperature inside the refrigerator is 0°C while outside it is 20°C. Assume $k = 0.0002$.
2. A pipe 10 cm in diameter contains steam at 100°C. It is covered with asbestos, 5 cm thick , for which $k = 0.0006$ and the outside surface is at 30°C. Find the amount of heat lost per hour from a metre long pipe.
3. A pipe 20 cm in diameter contains steam at 150°C and is protected with a covering 5 cm thick for which $k = 0.0025$. If the temperature of the outer surface of the covering is 40°C, find the temperature half-way through the covering under steady state conditions.
4. A steam pipe 20 cm in diameter contains steam at 150°C and is covered by a layer of insulation 5 cm thick. The outside temperature is kept at 60°C. By how much should the thickness of the covering be increased in order that the rate of heat loss should be decreased by 25% ?

Answers

1. 0.000667 cal/sec **2.** 140000 cal/hr **3.** 89.5°C
4. 2.16 cm

2.4. RATE OF GROWTH OR DECAY

If the rate of change of a quantity y is proportional to the quantity present at any instant, then we have the following differential equation : $\frac{dy}{dt} = ky$.

If k is positive, the problem is one of growth and if k is negative, the problem is one of decay.

Example. *Uranium disintegrates at a rate proportional to the amount present at any instant. If M_1 and M_2 grams of uranium are present at times T_1 and T_2 respectively, show that the half life of uranium is* $\frac{(T_2 - T_1)\log 2}{\log (M_1/M_2)}$.

Sol. Let M grams of uranium be present at any time t. Then the equation of distintegration of uranium is

$$\frac{dM}{dt} = -kM, \text{ where } k \text{ is a constant.}$$

or

$$\frac{dM}{M} = -k\,dt$$

Integrating, $\log M = -kt + c$ ...(1)

Initially, when $t = 0$, $M = M_0$ (say)

∴ From (1), $c = \log M_0$

Substituting the value of c in (1), we have $\log M = \log M_0 - kt$

or $$kt = \log M_0 - \log M \qquad ...(2)$$

Now, when $t = T_1$, $M = M_1$ and when $t = T_2$, $M = M_2$

$\therefore$ From (2), we have

$$kT_1 = \log M_0 - \log M_1 \qquad ...(3)$$

and $$kT_2 = \log M_0 - \log M_2 \qquad ...(4)$$

Subtracting (3) from (4), we get

$$k(T_2 - T_1) = \log M_1 - \log M_2 \quad \text{or} \quad k = \frac{\log \dfrac{M_1}{M_2}}{T_2 - T_1}$$

Let T be the half life of uranium *i.e.*, when $t = T$, $M = \frac{1}{2} M_0$.

$\therefore$ From (2), we get $kT = \log M_0 - \log \dfrac{M_0}{2} = \log 2$

$$\therefore \quad T = \frac{\log 2}{k} = \frac{(T_2 - T_1) \log 2}{\log \left(\dfrac{M_1}{M_2} \right)}.$$

TEST YOUR KNOWLEDGE

1. The number N of bacteria in a culture grew at a rate proportional to N. The value of N was initially 100 and increased to 332 in one hour. What was the value of N after $1\frac{1}{2}$ hours ?
2. In a culture of yeast, at each instant, the time rate change of active ferment is proportional to the amount present. If the active ferment doubles in two hours, how much can be expected at the end of 8 hours at the same rate of growth. Find also, how much time will elapse, before the active ferment grows to eight times its initial value.
3. Radium decomposes at a rate proportional to the amount present. If p per cent of the original amount disappears in l years, how much will remain at the end of $2l$ years ?
4. If 30% of a radioactive substance disappeared in 10 days, how long will it take for 90% of it to disappear.
5. Under certain conditions cane-sugar in water is converted into dextrose at a rate which is proportional to the amount unconverted at any time. If of 75 gm at time $t = 0$, 8 gm are converted during the first 30 minutes, find the amount converted in $1\frac{1}{2}$ hours.

Answers

1. 604.9
2. 16 times the original, 6 hours
3. $\left(10 - \dfrac{p}{10}\right)^2$
4. 64.5 days
5. 21.5 gm

Problem on Newton's Law of Cooling

Example. *According to Newton's law of cooling, the rate at which a substance cools in moving air is proportional to the difference between the temperature of the substance and that of the air. If the temperature of the air is 30°C and the substance cools from 100°C to 70°C in 15 minutes, find when the temperature will be 40°C.*

Sol. Let the unit of time be a minute and T the temperature of the substance of any instant t. Then by Newton's law of cooling, we have

$$\frac{dT}{dt} = -k\,(T-30) \quad \text{or} \quad \frac{dT}{T-30} = -k\,dt$$

Integrating, $\log(T-30) = -kt + c$ ...(1)

Initially, when $t = 0,\ T = 100$

$\therefore$ From (1), $c = \log 70$

Substituting the value of c in (1), we have $\log(T-30) = kt + \log 70$

or $kt = \log 70 - \log(T-30)$ ...(2)

Also, when $t = 15,\ T = 70$

$\therefore$ $15k = \log 70 - \log 40$ ...(3)

Dividing (2) by (3), we have $\dfrac{t}{15} = \dfrac{\log 70 - \log(T-30)}{\log 70 - \log 40}$ ...(4)

Now, when T = 40, we have from (4),

$$\frac{t}{15} = \frac{\log 70 - \log 10}{\log 70 - \log 40} = \frac{\log_e 7}{\log_e 7/4} = \frac{\log_{10} 7}{\log_{10}(7/4)} = 3.48$$

$\Rightarrow$ $t = 15 \times 3.48 = 52.20$

Hence the temperature will be 40°C after 52.2 minutes.

TEST YOUR KNOWLEDGE

1. Newton's law of cooling states that the temperature of an object changes at a rate proportional to the difference of a temperature between the object and the surroundings. Supposing water at a temperature 100°C cools to 80°C in 10 minutes, in a room maintained at a temperature of 30°C, find when the temperature of water will become 40°C.
2. Water at temperature 100°C cools in 10 minutes to 80°C in a room of temperature 25°C. Find (*i*) the temperature of water after 20 minutes, (*ii*) the time when the temperature is 40°C.

Answers

1. 57.9 minutes **2.** (*i*) 65.5°C (*ii*) 51.9 minutes

Problem on Chemical Reactions and Solutions

Example 1. *A tank contains 5000 litres of fresh water. Salt water which contains 100 gm of salt per litre flows into it at the rate of 10 litres per minute and the mixture kept uniform by stirring, runs out at the same rate. When will the tank contain 200000 gm of salt ?*

How long will it take for the quantity of salt in the tank of increase from 150000 gm to 250000 gm ?

Sol. Let Q gm be the quantity of salt present in the tank at time t, then $\frac{dQ}{dt}$ is the rate at which the salt content is changing and $\frac{dQ}{dt}$ = rate of salt entering the tank—rate of salt leaving the tank.

Now, the rate at which the salt increases due to the inflow = 100 × 10 = 1000 gm/min.

Let C gm be the concentration of salt at time t.

The rate at which the salt content decreases due to the outflow = C × 10 = 10 gm/min.

Since the rate of inflow is the same as the rate of outflow, there is no change in the volume of water at any instant.

$$\Rightarrow \qquad C = \frac{Q}{5000}$$

The rate of decrease of salt content $= 10 \times \frac{Q}{5000} = \frac{Q}{500}$ gm/min.

$$\therefore \qquad \frac{dQ}{dt} = 1000 - \frac{Q}{500} \quad \text{or} \quad \frac{dQ}{dt} = \frac{500000 - Q}{500}$$

$$\text{or} \qquad \frac{dt}{500000 - Q} = \frac{dt}{500}$$

Integrating, $\quad -\log(500000 - Q) = \frac{t}{500} + c \qquad \ldots(1)$

Initially, when $\quad t = 0, Q = 0$

$$\therefore \qquad c = -\log 500000$$

From (1), we have $\quad \frac{t}{500} = \log 500000 - \log(500000 - Q)$

$$\text{or} \qquad t = 500 \log \frac{500000}{500000 - Q} \qquad \ldots(2)$$

Let $\quad t = T$, when Q = 200000

From (2), $\quad T = 500 \log \frac{500000}{300000} = 500 \log_e \left(\frac{5}{3}\right)$

$$= 500 \times 2.303 \log_{10} \left(\frac{5}{3}\right) = 500 \times 2.303 \times 0.2219$$

$$= 255.5 \text{ minutes} = 4 \text{ hours } 15.52 \text{ minutes.}$$

Let $\quad t = T_1$, when Q = 150000 and $t = T_2$, when Q = 250000

From (2), we have $\quad T_1 = 500 \log \frac{500000}{350000} = 500 \log \frac{10}{7}$

$$T_2 = 500 \log \frac{500000}{250000} = 500 \log 2$$

$\therefore$ Required time $\quad = T_2 - T_1 = 500 \left(\log 2 - \log \frac{10}{7}\right) = 500 \log_e \left(\frac{7}{5}\right)$

$$= 500 \times 2.303 \log_{10} 1.4 = 500 \times 2.303 \times 0.1461$$
$$= 168.23 \text{ minutes} = 2 \text{ hours } 48.23 \text{ minutes.}$$

Example 2. *A tank contains 100 litres of fresh water. Two litres of brine, each containing 1 gm of dissolved salt, run into the tank per minute, and the mixture kept uniform by stirring runs out at the rate of 1 litres per minute. Find the amount of salt present when the tank contains 150 litres of brine.*

Sol. Let Q gm be the quantity of salt present in the brine at time t, then $\frac{dQ}{dt}$ is the rate at which the salt content is changing.

The rate at which the salt content increases due to the inflow = 2 × 1 = 2 gm/min.

Let C gm be the concentration of brine at time t.

The rate at which the salt content decreases due to the outflow = C × 1 = C gm/min

$$\therefore \quad \frac{dQ}{dt} = 2 - C \qquad \text{...(1)}$$

Now, the initial volume of liquid is 100 litres. In one minute, 2 litres of brine enter the tank and 1 litre of brine leaves the tank so that the volume of liquid in the tank increases at the rate of (2 – 1) = 1 litre/min.

∴ The volume of liquid at time t is $(100 + t)$ litres containing Q gm of salt.

$$\therefore \quad C = \frac{Q}{100 + t}$$

From (1), we have $\frac{dQ}{dt} = 2 - \frac{Q}{100 + t}$ or $\frac{dQ}{dt} + \frac{Q}{100 + t} = 2$ ...(2)

which is a linear equation in Q and t.

$$\text{I.F.} = e^{\int \frac{dt}{100 + t}} = e^{\log (100 + t)} = 100 + t$$

∴ The solution of (2) is $Q(100 + t) = \int 2(100 + t)\, dt + c$

or
$$(100 + t)\, Q = 2\left(100t + \frac{t^2}{2}\right) + c \qquad \text{...(3)}$$

Initially, when $t = 0$, $Q = 0$ so that $c = 0$

From (3),
$$Q = \frac{2\left(100t + \frac{t^2}{2}\right)}{100 + t} \qquad \text{...(4)}$$

Now, if V is the volume of liquid at time t, then $V = 100 + t$

∴ When $V = 150$ litres, $t = 150 - 100 = 50$ minutes

and salt content
$$Q = \frac{2\left[100 \times 50 + \frac{(50)^2}{2}\right]}{100 + 50} = 83.3 \text{ gm.}$$

TEST YOUR KNOWLEDGE

1. In a certain chemical reaction, the rate of conversion of a substance at time t is proportional to the quantity of substance still untransformed at that instant. The amount of substance remaining untransformed at the end of one hour and at the end of four hours are 60 gm and 21 gm respectively. How many grams of substance were present initially ?
2. A tank contains 1000 litres of fresh water. Salt water which contains 150 gm of salt per litre runs into it at the rate of 5 litres per minute and well stirred mixture runs out of it at the same rate. When will the tank contain 5000 gm of salt ?
3. A tank is initially filled with 100 litres of salt solution containing 1 gm of salt per litre. Fresh brine containing 2 gm of salt per litre runs into the tank at the rate of 5 litres per minute and the mixture, assumed to be kept uniform by stirring, runs out at the same rate. Find the amount of salt in the tank at any time, and determine how long it will take for this amount to reach 150 gm.
4. A tank contains 100 litres of an aqueous solution containing 10 kg of salt. Water is entering the tank at the rate of 3 litres per minute and the well stirred mixture runs out at 2 litres per minute. How much salt will the tank contain at the end of one hour ? After what time will the amount of salt in the tank be 625 gm ?

Answers

1.	85 gm	**2.** 6.77 minutes
3.	$100\,(2 - e^{-t/20})$; 13.9 minutes	**4.** 3.9 kg ; 5 hours.

3

Linear Differential Equations

3.1. DEFINITIONS

A **linear differential equation** is the differential equation in which the dependent variable and its derivatives occur only in the first degree and are not multiplied together.

Thus, the general linear differential equation of the nth order is of the form $\frac{d^n y}{dx^n} + P_1 \frac{d^{n-1}y}{dx^{n-1}} + P_2 \frac{d^{n-2}y}{dx^{n-2}} + \ldots\ldots + P_{n-1}\frac{dy}{dx} + P_n y = X$, where $P_1, P_2, \ldots\ldots, P_{n-1}, P_n$ and X are functions of x only.

A linear differential equation with constant co-efficients is of the form

$$\frac{d^n y}{dx^n} + a_1 \frac{d^{n-1}y}{dx^{n-1}} + a_2 \frac{d^{n-2}y}{dx^{n-2}} + \ldots\ldots + a_{n-1}\frac{dy}{dx} + a_n y = X \qquad \ldots(1)$$

where $a_1, a_2, \ldots\ldots, a_{n-1}, a_n$ are constants and X is either a constant or a function of x only.

3.2. THE OPERATOR D

The part $\frac{d}{dx}$ of the symbol $\frac{dy}{dx}$ may be regarded as an operator such that when it operates on y, the result is the derivative of y.

Similarly, $\frac{d^2}{dx^2}, \frac{d^3}{dx^3}, \ldots\ldots, \frac{d^n}{dx^n}$ may be regarded as operators.

For brevity, we write $\frac{d}{dx} \equiv D, \frac{d^2}{dx^2} \equiv D^2, \ldots\ldots, \frac{d^n}{dx^n} \equiv D^n$

Thus, the symbol D is a **differential operator** or simply an **operator.**

Written in symbolic form, equation (1) becomes

$$(D^n + a_1 D^{n-1} + a_2 D^{n-2} + \ldots\ldots + a_{n-1}D + a_n)y = X$$

or $$f(D)y = X$$

where $$f(D) = D^n + a_1 D^{n-1} + a_2 D^{n-2} + \ldots\ldots + a_{n-1}D + a_n$$

i.e., $f(D)$ is a polynomial in D.

The operator D can be treated as an algebraic quantity.

Thus $$D(u + v) = Du + Dv, \quad D(\lambda u) = \lambda\, Du$$
$$D^p D^q\, u = D^{p+q}\, u, \quad D^p D^q u = D^q D^p u$$

The polynomial $f(D)$ can be factorised by ordinary rules of algebra and the factors may be written in any order.

3.3. THEOREMS

Theorem 1. *If $y = y_1, y = y_2, \ldots, y = y_n$ are n linearly independent solutions of the differential equation*

$$(D^n + a_1 D^{n-1} + a_2 D^{n-2} + \ldots + a_n)y = 0 \qquad \ldots(i)$$

then $u = c_1 y_1 + c_2 y_2 + \ldots + c_n y_n$ is also its solution, where $c_1, c_2, \ldots, c_n$ are arbitrary constants.

Proof. Since $y = y_1, y = y_2, \ldots, y = y_n$ are solution of equation (*i*).

$$\therefore \left.\begin{aligned} &D^n y_1 + a_1 D^{n-1} y_1 + a_2 D^{n-2} y_1 + \ldots + a_n y_1 = 0 \\ &D^n y_2 + a_1 D^{n-1} y_2 + a_2 D^{n-2} y_2 + \ldots + a_n y_2 = 0 \\ &\ldots \quad \ldots \quad \ldots \quad \ldots \quad \ldots \\ &\ldots \quad \ldots \quad \ldots \quad \ldots \quad \ldots \\ &D^n y_n + a_1 D^{n-1} y_n + a_2 D^{n-2} y_n + \ldots + a_n y_n = 0 \end{aligned}\right\} \qquad \ldots(ii)$$

Now $D^n u + a_1 D^{n-1} u + a_2 D^{n-2} u + \ldots + a_n u$

$$\begin{aligned} &= D^n (c_1 y_1 + c_2 y_2 + \ldots + c_n y_n) \\ &\quad + a_1 D^{n-1}(c_1 y_1 + c_2 y_2 + \ldots + c_n y_n) + a_2 D^{n-2}(c_1 y_1 + c_2 y_2 + \ldots + c_n y_n) \\ &\quad + \ldots \quad \ldots \quad \ldots \quad \ldots + a_n(c_1 y_1 + c_2 y_2 + \ldots + c_n y_n) \\ &= c_1(D^n y_1 + a_1 D^{n-1} y_1 + a_2 D^{n-2} y_1 + \ldots + a_n y_1) \\ &\quad + c_2(D^n y_2 + a_1 D^{n-1} y_2 + a_2 D^{n-2} y_n + \ldots + a_n y_2) \\ &\quad + \ldots \quad \ldots + c_n(D^n y_n + a_1 D^{n-1} y_n + a_2 D^{n-2} y_n + \ldots + a_n y_n) \\ &= c_1(0) + c_2(0) + \ldots + c_n(0) \qquad \text{[From } (ii)] \\ &= 0 \end{aligned}$$

which shows that $u = c_1 y_1 + c_2 y_2 + \ldots\ldots + c_n y_n$ is also the solution of equation (*i*).

Since this solution contains n arbitrary constants, it is the general or complete solution of equation (*i*).

Theorem 2. *If $y = u$ is the complete solution of the equation $f(D)y = 0$ and $y = v$ is a particular solution (containing no arbitrary constants) of the equation $f(D)y = X$, then the complete solution of the equation $f(D)y = X$ is $y = u + v$.*

Proof. Since $y = u$ is the complete solution of the equation $f(D)y = 0$ ...(*i*)

$$\therefore \qquad f(D)u = 0 \qquad \ldots(ii)$$

Also $y = v$ is a particular solution of the equation $f(D)y = X$ ...(*iii*)

$$\therefore \qquad f(D)v = X \qquad \ldots(iv)$$

Adding (*ii*) and (*iv*), we have $f(D)(u + v) = X$

Thus $y = u + v$ satisfies the equation (*iii*), hence it is the **complete solution (C.S.)** because it contains n arbitrary constants.

The part $y = u$ is called the **complementary function (C.F.)** and the part $y = v$ is called the **particular integral (P.I.)** of the equation (*iii*).

$\therefore$ The complete solution of equation (*iii*), is **y = C.F. + P.I.**

Thus in order to solve the equation (iii), we first find the C.F. i.e., the C.S. of equation (i) and then the P.I. i.e., a particular solution of equation (iii).

3.4. AUXILIARY EQUATION (A.E.)

Consider the differential equation $(D^n + a_1D^{n-1} + a_2D^{n-2} + \ldots\ldots + a_n)y = 0$ ...(*i*)

Let $y = e^{mx}$ be a solution of (*i*), then $Dy = me^{mx}$, $D^2y = m^2e^{mx}$,......, $D^{n-2}y = m^{n-2}e^{mx}$

$$D^{n-1}y = m^{n-1}e^{mx},\ D^ny = m^ne^{mx}$$

Substituting the values of y, Dy, D^2y, ..., D^ny in (*i*), we get

$$(m^n + a_1m^{n-1} + a_2m^{n-2} + \ldots\ldots + a_n)\ e^{mx} = 0$$

or $$m^n + a_1m^{n-1} + a_2m^{n-2} + \ldots\ldots + a_n = 0, \text{ since } e^{mx} \neq 0 \qquad \ldots(ii)$$

Thus $y = e^{mx}$ will be a solution of equation (*i*) if m satisfies equation (*ii*).

Equation (*ii*) is called the auxiliary equation for the differential equation (*i*).

Replacing m by D in (*ii*), we get $D^n + a_1D^{n-1} + a_2D^{n-2} + \ldots\ldots + a_n = 0$ ...(*iii*)

Equation (*ii*) gives the same values of m *as* equation (*iii*) gives of D. In practice, we take equation (*iii*) as the auxiliary equation which is obtained by equating to zero the symbolic coefficient of y in equation (*i*).

Definition. The equation obtained by equating to zero the symbolic co-efficient of y is called the **auxiliary equation**, briefly written as A.E.

3.5. RULES FOR FINDING THE COMPLEMENTARY FUNCTION

Consider the equation $(D^n + a_1D^{n-1} + a_2D^{n-2} + \ldots\ldots + a_n)y = 0$ ...(*i*)

where all the a_i's are constant.

Its auxiliary equation is $D^n + a_1D^{n-1} + a_2D^{n-2} + \ldots\ldots + a_n = 0$ ...(*ii*)

Let $D = m_1, m_2, m_3, \ldots\ldots, m_n$ be the roots of the A.E. The solution of equation (*i*) depends upon the nature of roots of the A.E. The following cases arise :

Case I. *If all the roots of the A.E. are real and distinct,* then equation (*ii*) is equivalent to

$$(D - m_1)(D - m_2) \ldots\ldots (D - m_n) = 0 \qquad \ldots(iii)$$

Equation (*iii*) will be satisfied by the solutions of the equations

$$(D - m_1)y = 0,\ (D - m_2)y = 0, \ldots\ldots, (D - m_n)y = 0$$

Now, consider the equation $(D - m_1)y = 0$, *i.e.*, $\dfrac{dy}{dx} - m_1y = 0$

It is a linear equation and I.F. $= e^{\int -m_1 dx} = e^{-m_1x}$

$\therefore$ its solution is $y \,.\, e^{-m_1x} = \int 0 \,.\, e^{-m_1x}\, dx + c_1$ or $y = c_1e^{m_1x}$

Similarly, the solution of $(D - m_2)y = 0$ is $y = c_2e^{m_2x}$

....

the solution of $(D - m_n)y = 0$ is $y = c_ne^{m_nx}$

Hence the complete solution of equation (*i*) is

$$y = c_1e^{m_1x} + c_2e^{m_2x} + \ldots\ldots + c_ne^{m_nx} \qquad \ldots(iv)$$

Case II. *If two roots of the A.E. are equal,* let $m_1 = m_2$.

The solution obtained in equation (*iv*) becomes

$$y = (c_1 + c_2)\, e^{m_1 x} + c_3 e^{m_3 x} + \ldots\ldots + c_n e^{m_n x}$$

$$= c e^{m_1 x} + c_3 e^{m_3 x} + \ldots\ldots + c_n e^{m_n x}$$

It contains $(n - 1)$ arbitrary constants and is, therefore, not the complete solution of equation (*i*).

The part of the complete solution corresponding to the repeated root is the complete solution of

$$(D - m_1)(D - m_1)y = 0$$

Putting $(D - m_1)y = v$, it becomes $(D - m_1)\, v = 0$ *i.e.,* $\dfrac{dv}{dx} - m_1 v = 0$

As in case I, its solution is $v = c_1 e^{m_1 x}$

$\therefore$ $$(D - m_1)y = c_1 e^{m_1 x} \quad \text{or} \quad \frac{dy}{dx} - m_1 y = c_1 e^{m_1 x}$$

which is a linear equation and I.F. $= e^{-m_1 x}$

$\therefore$ its solution is $y\,.e^{-m_1 x} = \int c_1 e^{m_1 x} . e^{-m_1 x}\, dx + c_2 = c_1 x + c_2$

or $$y = (c_1 x + c_2)\, e^{m_1 x}$$

Thus, the complete solution of equation (*i*) is

$$y = (c_1 x + c_2)\, e^{m_1 x} + c_3 e^{m_3 x} + \ldots\ldots + c_n e^{m_n x}$$

If, however, three roots of the A.E. are equal, say $m_1 = m_2 = m_3$, then proceeding as above, the solution becomes

$$y = (c_1 x^2 + c_2 x + c_3)\, e^{m_1 x} + c_4 e^{m_4 x} + \ldots\ldots + c_n e^{m_n x}$$

Case III. *If two roots of the A.E. are imaginary,* let

$$m_1 = \alpha + i\beta \quad \text{and} \quad m_2 = \alpha - i\beta$$

The solution obtained in equation (*iv*) becomes

$$y = c_1 e^{(\alpha + i\beta)x} + c_2 e^{(\alpha - i\beta)x} + c_3 e^{m_3 x} + \ldots\ldots + c_n e^{m_n x}$$

$$= e^{\alpha x}(c_1 e^{i\beta x} + c_2 e^{-i\beta x}) + c_3 e^{m_3 x} + \ldots\ldots + c_n e^{m_n x}$$

$$= e^{\alpha x}[c_1(\cos \beta x + i \sin \beta x) + c_2 (\cos \beta x - i \sin \beta x)]$$

$$+ c_3 e^{m_3 x} + \ldots\ldots + c_n e^{m_n x} \quad [\because \text{ by Euler's Theorem, } e^{i\theta} = \cos\theta + i \sin\theta]$$

$$= e^{\alpha x}[(c_1 + c_2) \cos \beta x + i\,(c_1 - c_2) \sin \beta x] + c_3 e^{m_3 x} + \ldots\ldots + c_n e^{m_n x}$$

$$= e^{\alpha x}(C_1 \cos \beta x + C_2 \sin \beta x) + c_3 e^{m_3 x} + \ldots\ldots + c_n e^{m_n x}$$

[taking $c_1 + c_2 = C_1$, $i(c_1 - c_2) = C_2$]

Case IV. *If two pairs of imaginary roots be equal,* let

$$m_1 = m_2 = \alpha + i\beta \quad \text{and} \quad m_3 = m_4 = \alpha - i\beta$$

Then by case II, the complete solution is

$$y = e^{\alpha x}\,[(c_1x + c_2)\cos\beta x + (c_3x + c_4)\sin\beta x] + c_5e^{m_5x} + \ldots\ldots + c_ne^{m_nx}.$$

ILLUSTRATIVE EXAMPLES

Example 1. *Solve* $\dfrac{d^3y}{dx^3} - 7\dfrac{dy}{dx} - 6y = 0.$

Sol. Given equation in symbolic form is $(D^3 - 7D - 6)y = 0$

Its A.E. is $D^3 - 7D - 6 = 0$ or $(D + 1)(D + 2)(D - 3) = 0$

whence $D = -1, -2, 3$

Hence the C.S. is $y = c_1e^{-x} + c_2e^{-2x} + c_3e^{3x}.$

Example 2. *Solve* $(D^3 - 4D^2 + 4D)y = 0.$

Sol. The A.E. is $D^3 - 4D^2 + 4D = 0$ or $D(D^2 - 4D + 4) = 0$

or $D(D - 2)^2 = 0$

whence $D = 0, 2, 2$

Hence, the C.S is $y = c_1e^{0x} + (c_2x + c_3)e^{2x}$ or $y = c_1 + (c_2x + c_3)e^{2x}.$

Example 3. *Solve* $\dfrac{d^4y}{dx^4} + 13\dfrac{d^2y}{dx^2} + 36y = 0.$

Sol. Given equation in symbolic form is $(D^4 + 13D^2 + 36)y = 0$

Its A.E. is $D^4 + 13D^2 + 36 = 0$

or $(D^2 + 4)(D^2 + 9) = 0$ or $D = \pm 2i, \pm 3i$

Hence the C.S. is $y = e^{0x}(c_1\cos 2x + c_2\sin 2x) + e^{0x}(c_3\cos 3x + c_4\sin 3x)$

or $y = c_1\cos 2x + c_2\sin 2x + c_3\cos 3x + c_4\sin 3x.$

Example 4. *Solve* $\dfrac{d^4x}{dt^4} + 4x = 0.$ [CUSAT 2009]

Sol. Given equation in symbolic form is $(D^4 + 4)x = 0$, where $D = \dfrac{d}{dt}$

Its A.E. is $D^4 + 4 = 0$ or $(D^4 + 4D^2 + 4) - 4D^2 = 0$

or $(D^2 + 2)^2 - (2D)^2 = 0$ or $(D^2 + 2D + 2)(D^2 - 2D + 2) = 0$

whence $D = \dfrac{-2 \pm \sqrt{-4}}{2}$ and $\dfrac{2 \pm \sqrt{-4}}{2}$ *i.e.,* $D = -1 \pm i$ and $1 \pm i$

Hence the C.S. is $x = e^{-t}(c_1\cos t + c_2\sin t) + e^{t}(c_3\cos t + c_4\sin t).$

TEST YOUR KNOWLEDGE

Solve the following differential equations :

1. $\frac{d^2y}{dx^2} - 3\frac{dy}{dx} - 4y = 0.$

2. $\frac{d^2y}{dx^2} + (a+b)\frac{dy}{dx} + aby = 0.$

3. $\frac{d^2y}{dx^2} - 4\frac{dy}{dx} + y = 0.$

4. $\frac{d^2x}{dt^2} + 6\frac{dx}{dt} + 9x = 0.$

5. $\frac{d^3y}{dx^3} - 3\frac{d^2y}{dx^2} + 3\frac{dy}{dx} - y = 0.$

6. $\frac{d^3y}{dx^3} + 6\frac{d^2y}{dx^2} + 11\frac{dy}{dx} + 6y = 0.$

7. $\frac{d^4y}{dx^4} - 5\frac{d^2y}{dx^2} + 4y = 0.$

8. $\frac{d^4y}{dx^4} + 8\frac{d^2y}{dx^2} + 16y = 0.$

9. $(D^2 + 1)^3 (D^2 + D + 1)^2 y = 0.$

10. $\frac{d^2x}{dt^2} - 3\frac{dx}{dt} + 2x = 0$, given that when $t = 0$, $x = 0$ and $\frac{dx}{dt} = 0.$

11. $\frac{d^2y}{dx^2} + 4\frac{dy}{dx} + 29y = 0$, given that when $x = 0$, $y = 0$ and $\frac{dy}{dx} = 15.$

12. If $\frac{d^4x}{dt^4} = m^4x$, show that $x = c_1 \cos mt + c_2 \sin mt + c_3 \cosh mt + c_4 \sinh mt.$

Answers

1. $y = c_1 e^{4x} + c_2e^{-x}$

2. $y = c_1 e^{-ax} + c_2e^{-bx}$

3. $y = c_1e^{(2+\sqrt{3})x} + c_2e^{(2-\sqrt{3})x}$

4. $x = (c_1 + c_2t)\, e^{-3t}$

5. $y = (c_1 + c_2x + c_3x^2)\, e^x$

6. $y = c_1e^{-x} + c_2e^{-2x} + c_3e^{-3x}$

7. $y = c_1e^x + c_2e^{-x} + c_3e^{2x} + c_4e^{-2x}$

8. $y = (c_1 + c_2x) \cos 2x + (c_3 + c_4x) \sin 2x$

9. $y = (c_1 + c_2x + c_3x^2) \cos x + (c_4 + c_5x + c_6x^2) \sin x + e^{-\frac{1}{2}x}\left[(c_7 + c_8x) \cos \frac{\sqrt{3}}{2}x + (c_9 + c_{10}x) \sin \frac{\sqrt{3}}{2}x\right]$

10. $x = 0$

11. $y = 3e^{-2x} \sin 5x.$

3.6. THE INVERSE OPERATOR $\frac{1}{f(D)}$

Definition. $\frac{1}{f(D)}$ X is that function of x, *free from arbitrary constants,* which when operated upon by $f(D)$ gives X.

Thus $$f(D)\left\{\frac{1}{f(D)} X\right\} = X$$

$\therefore$ $f(D)$ and $\frac{1}{f(D)}$ are inverse operators.

Theorem 1. $\frac{1}{f(D)}$ *X is the particular integral of f(D)y = X.*

Proof. The given equation is $f(D)y = X$ ...(1)

Putting $y = \frac{1}{f(D)} X$ in (1), we have $f(D)\left\{\frac{1}{f(D)} X\right\} = X$ or $X = X$, which is true.

$$\therefore \quad y = \frac{1}{f(D)} X \text{ is a solution of (1).}$$

Since it contains no arbitrary constants, it is the particular integral of $f(D)\,y = X$.

Theorem 2. *Prove that,* $\frac{1}{D} X = \int X\, dx.$

Proof. Let $\frac{1}{D} X = y$

Operating both sides by D, we have $D\left(\frac{1}{D} X\right) = Dy$ or $X = \frac{dy}{dx}$

Integrating both sides w.r.t. x,

$$y = \int X\, dx,$$

no arbitrary constant being added since $y = \frac{1}{D} X$ contains no arbitrary constant.

$$\therefore \quad \frac{1}{D} X = \int X\, dx.$$

Theorem 3. *Prove that,* $\frac{1}{D-a} X = e^{ax} \int X e^{-ax}\, dx.$

Proof. Let $\frac{1}{D-a} X = y$

Operating on both sides by $(D - a)$, $(D - a)\left(\frac{1}{D-a} X\right) = (D - a)y$

or $$X = \frac{dy}{dx} - ay \quad i.e., \quad \frac{dy}{dx} - ay = X$$

which is a linear equation and I.F. $= e^{\int -a dx} = e^{-ax}$

$\therefore$ Its solution is $ye^{-ax} = \int X e^{-ax}\, dx$, no constant being added or $y = e^{ax} \int X e^{-ax}\, dx$

Hence $$\frac{1}{D-a} X = e^{ax} \int e^{-ax} X\, dx.$$

3.7. RULES FOR FINDING THE PARTICULAR INTEGRAL

Consider the differential equation, $(D^n + a_1 D^{n-1} + a_2 D^{n-2} + \ldots\ldots + a_{n-1} D + a_n)y = X$

It can be written as $f(D)y = X$

$$\therefore \quad \text{P.I.} = \frac{1}{f(D)} X$$

Case I. When $X = e^{ax}$

Since

$$D\, e^{ax} = a\, e^{ax}$$

$$D^2\, e^{ax} = a^2\, e^{ax}$$

....

....

$$D^{n-1}\, e^{ax} = a^{n-1}\, e^{ax}$$

$$D^n e^{ax} = a^n\, e^{ax}$$

$$\therefore \quad (D^n + a_1 D^{n-1} + a_2 D^{n-2} + \ldots\ldots + a_{n-1} D + a_n)\, e^{ax} = (a^n + a_1 a^{n-1} + a_2 a^{n-2} + \ldots\ldots + a_{n-1} a + a_n)\, e^{ax}$$

or

$$f(D)\, e^{ax} = f(a)\, e^{ax}$$

Operating on both sides by $\dfrac{1}{f(D)}$.

$$\frac{1}{f(D)}(f(D)e^{ax}) = \frac{1}{f(D)}(f(a)e^{ax}) \quad \text{or} \quad e^{ax} = f(a)\, \frac{1}{f(D)}\, e^{ax}$$

Dividing both sides by $f(a)$, $\dfrac{1}{f(a)}\, e^{ax} = \dfrac{1}{f(D)}\, e^{ax}$, provided $f(a) \neq 0$

Hence $$\mathbf{\frac{1}{f(D)}\, e^{ax} = \frac{1}{f(a)}\, e^{ax},} \qquad \textbf{provided } \mathbf{f(a) \neq 0.}$$

Case of failure. If $f(a) = 0$, the above method fails.

Since $f(a) = 0$, $D = a$ is a root of A.E. $f(D) = 0$

$\therefore$ $D - a$ is a factor of $f(D)$.

Let $$f(D) = (D - a)\, \phi(D), \text{ where } \phi(a) \neq 0 \qquad \ldots(i)$$

Then $$\frac{1}{f(D)} e^{ax} = \frac{1}{(D-a)\,\phi(D)} e^{ax} = \frac{1}{D-a} \cdot \frac{1}{\phi(D)} e^{ax} = \frac{1}{D-a} \cdot \frac{1}{\phi(a)} e^{ax}$$

$$= \frac{1}{\phi(a)} \cdot \frac{1}{D-a} e^{ax} = \frac{1}{\phi(a)} e^{ax} \int e^{ax} \,.\, e^{-ax} dx \qquad \text{[by Theorem 3]}$$

$$= \frac{1}{\phi(a)} e^{ax} \int 1\, dx = x \,.\, \frac{1}{\phi(a)} e^{ax} \qquad \ldots(ii)$$

Differentiating both sides of (*i*) w.r.t. D, we have $f'(D) = (D - a)\, \phi'(D) + \phi(D)$

$\Rightarrow$ $$f'(a) = \phi(a)$$

$\therefore$ From (*ii*), we have $$\mathbf{\frac{1}{f(D)} e^{ax} = x \,.\, \frac{1}{f'(a)} e^{ax},} \qquad \textbf{provided } \mathbf{f'(a) \neq 0.}$$

If $$f'(a) = 0, \text{ then } \frac{1}{f(D)} e^{ax} = x^2 \,.\, \frac{1}{f''(a)} e^{ax}, \qquad \text{provided } f''(a) \neq 0$$

and so on.

ILLUSTRATIVE EXAMPLES

Example 1. *Find the P.I. of (4D² + 4D – 3) y = e^{2x}.*

Sol. $$\text{P.I.} = \frac{1}{4D^2+4D-3}e^{2x} = \frac{1}{4(2)^2+4(2)-3}e^{2x} \qquad \text{(replacing D by 2)}$$

$$= \frac{1}{21}e^{2x}.$$

Example 2. *Find the P.I. of (D² + 3D + 2)y = 5.*

Sol. $$\text{P.I.} = \frac{1}{D^2+3D+2}(5e^{0x}) \qquad [\because\ e^{0x} = 1]$$

$$= 5 \cdot \frac{1}{0+0+2}e^{0x} \qquad \text{(replacing D by 0)}$$

$$= \frac{5}{2}.$$

Example 3. *Find the P.I. of (D³ – 3D² + 4)y = e^{2x}.*

Sol. $$\text{P.I.} = \frac{1}{D^3-3D^2+4}e^{2x}.$$

Hence the denominator vanishes when D is replaced by 2. It is a case of failure. We multiply the numerator by x and differentiate the denominator w.r.t. D.

$$\therefore \quad \text{P.I.} = x \cdot \frac{1}{3D^2-6D}e^{2x}$$

It is again a case of failure. We multiply the numerator by x and differentiate the denominator w.r.t. D.

$$\therefore \quad \text{P.I.} = x^2 \cdot \frac{1}{6D-6}e^{2x} = x^2 \cdot \frac{1}{6(2)-6}e^{2x} = \frac{x^2}{6}e^{2x}.$$

Case II. When X = sin (ax + b) or cos (ax + b)

$$D \sin(ax+b) = a\cos(ax+b)$$

$$\mathbf{D^2 \sin(ax+b) = (-a^2)\sin(ax+b)}$$

$$D^3 \sin(ax+b) = -a^3\cos(ax+b)$$

$$D^4 \sin(ax+b) = a^4\sin(ax+b)$$

or $$\mathbf{(D^2)^2 \sin(ax+b) = (-a^2)^2\sin(ax+b)}$$

....

....

In general, $(D^2)^n \sin(ax+b) = (-a^2)^n \sin(ax+b)$

$\therefore$ $f(D^2)\sin(ax+b) = f(-a^2)\sin(ax+b)$

Operating on both sides by $\frac{1}{f(D^2)}$,

$$\frac{1}{f(D^2)}(f(D^2)\sin(ax+b)) = \frac{1}{f(D^2)}(f(-a^2)\sin(ax+b))$$

or $$\sin(ax+b) = f(-a^2)\frac{1}{f(D^2)}\sin(ax+b).$$

Dividing both sides by $f(-a^2)$,

$$\frac{1}{f(-a^2)}\sin(ax+b) = \frac{1}{f(D^2)}\sin(ax+b), \qquad \text{provided } f(-a^2) \neq 0.$$

Hence $$\mathbf{\frac{1}{f(D^2)}\sin(ax+b) = \frac{1}{f(-a^2)}\sin(ax+b), \qquad \text{provided } f(-a^2) \neq 0}$$

Similarly, $$\mathbf{\frac{1}{f(D^2)}\cos(ax+b) = \frac{1}{f(-a^2)}\cos(ax+b), \qquad \text{provided } f(-a^2) \neq 0}$$

Case of failure. If $f(-a^2) = 0$, the above method fails.

Since $\cos(ax+b) + i\sin(ax+b) = e^{i(ax+b)}$ | Euler's Theorem

$$\therefore \quad \frac{1}{f(D^2)}[\cos(ax+b) + i\sin(ax+b)] = \frac{1}{f(D^2)}e^{i(ax+b)}$$

[If we replace D by ia, $f(D^2) = f(-a^2) = 0$, so that it is a case of failure]

$$= x \cdot \frac{1}{f'(D^2)}e^{i(ax+b)} = x \cdot \frac{1}{f'(D^2)}[\cos(ax+b) + i\sin(ax+b)]$$

Equating real parts

$$\mathbf{\frac{1}{f(D^2)}\cos(ax+b) = x \cdot \frac{1}{f'(D^2)}\cos(ax+b), \qquad \text{provided } f'(-a^2) \neq 0}$$

Equating imaginary parts

$$\mathbf{\frac{1}{f(D^2)}\sin(ax+b) = x \cdot \frac{1}{f'(D^2)}\sin(ax+b), \qquad \text{provided } f'(-a^2) \neq 0}$$

If $f'(-a^2) = 0$, then

$$\frac{1}{f(D^2)}\sin(ax+b) = x^2 \cdot \frac{1}{f''(D^2)}\sin(ax+b), \qquad \text{provided } f''(-a^2) \neq 0$$

$$\frac{1}{f(D^2)}\cos(ax+b) = x^2 \cdot \frac{1}{f''(D^2)}\cos(ax+b), \qquad \text{provided } f''(-a^2) \neq 0$$

and so on.

ILLUSTRATIVE EXAMPLES

Example 1. *Find the P.I. of $(D^3 + 1)y = \sin(2x + 3)$.*

Sol. $$\text{P.I.} = \frac{1}{D^3+1}\sin(2x+3) = \frac{1}{D(-2^2)+1}\sin(2x+3)$$

[Putting $D^2 = -2^2$]

$$= \frac{1}{1-4D}\sin(2x+3)$$

Multiplying and dividing by (1 + 4D), we get

$$\text{P.I.} = \frac{1+4D}{(1-4D)(1+4D)} \sin(2x+3) = \frac{1+4D}{1-16D^2} \sin(2x+3)$$

$$= \frac{1+4D}{1-16(-2^2)} \sin(2x+3) \qquad [\text{Putting } D^2 = -2^2]$$

$$= \frac{1}{65} [\sin(2x+3) + 4D \sin(2x+3)]$$

$$= \frac{1}{65} [\sin(2x+3) + 8\cos(2x+3)] \qquad \left[\because D = \frac{d}{dx}\right]$$

Example 2. *Find the P.I. of $(D^2 + 4)y = \cos 2x$.*

Sol. $$\text{P.I.} = \frac{1}{D^2+4} \cos 2x$$

Here the denominator vanishes when D^2 is replaced by $-2^2 = -4$. It is a case of failure. We multiply the numerator by x and differentiate the denominator w.r.t. D.

$$\therefore \qquad \text{P.I.} = x \cdot \frac{1}{2D} \cos 2x = \frac{x}{2}\int \cos 2x \, dx \qquad \left[\because \frac{1}{D} f(x) = \int f(x) dx\right]$$

$$= \frac{x}{4} \sin 2x.$$

Case III. When X = x^m, *m* being a positive integer.

Here $$\text{P.I.} = \frac{1}{f(D)} x^m$$

Take out the lowest degree term from $f(D)$ to make the first term unity (so that Binomial Theorem for a negative index is applicable).

The remaining factor will be of the form $1 + \phi(D)$ or $1 - \phi(D)$

Take this factor in the numerator. It takes the form $[1 + \phi(D)]^{-1}$ or $[1 - \phi(D)]^{-1}$

Expand it in ascending powers of D as far as the term containing D^m, since $D^{m+1}(x^m) = 0$, $D^{m+2}(x^m) = 0$ and so on.

Operate on x^m term by term.

Example. *Find the P.I. of $(D^2 + 5D + 4)y = x^2 + 7x + 9$.*

Sol. $$\text{P.I.} = \frac{1}{D^2+5D+4}(x^2+7x+9) = \frac{1}{4\left(1+\frac{5D}{4}+\frac{D^2}{4}\right)}(x^2+7x+9)$$

$$= \frac{1}{4}\left[1+\left(\frac{5D}{4}+\frac{D^2}{4}\right)\right]^{-1}(x^2+7x+9)$$

$$= \frac{1}{4}\left[1-\left(\frac{5D}{4}+\frac{D^2}{4}\right)+\left(\frac{5D}{4}+\frac{D^2}{4}\right)^2 - \ldots\ldots\right](x^2+7x+9)$$

$$= \frac{1}{4}\left(1-\frac{5D}{4}-\frac{D^2}{4}+\frac{25D^2}{16} \ldots\ldots\right)(x^2+7x+9)$$

$$= \frac{1}{4}\left(1 - \frac{5D}{4} + \frac{21D^2}{16} \ldots\ldots\right)(x^2 + 7x + 9)$$

$$= \frac{1}{4}\left[(x^2 + 7x + 9) - \frac{5}{4}D(x^2 + 7x + 9) + \frac{21}{16}D^2(x^2 + 7x + 9)\right]$$

$$= \frac{1}{4}\left[(x^2 + 7x + 9) - \frac{5}{4}(2x + 7) + \frac{21}{16}(2)\right] = \frac{1}{4}\left(x^2 + \frac{9}{2}x + \frac{23}{8}\right).$$

Case IV. When $X = e^{ax} V$, where V is a function of x.

Let u be a function of x, then by successive differentiation, we have

$$D(e^{ax} u) = e^{ax} Du + a e^{ax} u = e^{ax}(D + a)u$$

$$D^2(e^{ax} u) = D[e^{ax}(D + a)u] = e^{ax}(D^2 + aD)u + ae^{ax}(D + a)u$$

$$= e^{ax}(D^2 + 2aD + a^2)u = e^{ax}(D + a)^2 u$$

Similarly, $D^3(e^{ax} u) = e^{ax}(D + a)^3 u$

In general, $D^n(e^{ax} u) = e^{ax}(D + a)^n u$

$\therefore$ $f(D)(e^{ax} u) = e^{ax} f(D + a) u$

Operating on both sides by $\frac{1}{f(D)}$,

$$\frac{1}{f(D)}[f(D)(e^{ax} u)] = \frac{1}{f(D)}[e^{ax} f(D + a)u]$$

$$\Rightarrow \qquad e^{ax} u = \frac{1}{f(D)}[e^{ax} f(D + a)u] \qquad \ldots(i)$$

Now let $f(D + a)u = V$, *i.e.*, $u = \frac{1}{f(D + a)} V$

$\therefore$ From (*i*), we have $e^{ax} \frac{1}{f(D + a)} V = \frac{1}{f(D)}(e^{ax} V)$

or

$$\mathbf{\frac{1}{f(D)}(e^{ax} V) = e^{ax} \frac{1}{f(D + a)} V.}$$

Thus e^{ax} which is on the right of $\frac{1}{f(D)}$ may be taken out to the left provided D is replaced by $D + a$.

Example. *Find the P.I. of $(D^2 - 4D + 3)y = e^x \cos 2x$.*

Sol. $$\text{P.I.} = \frac{1}{D^2 - 4D + 3} e^x \cos 2x = e^x \frac{1}{(D + 1)^2 - 4(D + 1) + 3} \cos 2x$$

$$= e^x \frac{1}{D^2 - 2D} \cos 2x = e^x \frac{1}{-2^2 - 2D} \cos 2x \qquad [\text{Putting } D^2 = -2^2]$$

$$= -\frac{1}{2} e^x \frac{1}{2 + D} \cos 2x = -\frac{1}{2} e^x \frac{2 - D}{(2 + D)(2 - D)} \cos 2x$$

$$= -\frac{1}{2} e^x \frac{2 - D}{4 - D^2} \cos 2x = -\frac{1}{2} e^x \frac{2 - D}{4 - (-2^2)} \cos 2x$$

$$= -\frac{1}{16} e^x (2 \cos 2x - D \cos 2x) = -\frac{1}{16} e^x (2 \cos 2x + 2 \sin 2x)$$

$$= -\frac{1}{8} e^x (\cos 2x + \sin 2x).$$

Case V. When X is any other function of x.

Resolve $f(D)$ into linear factors.

Let $f(D) = (D - m_1)(D - m_2) \ldots\ldots (D - m_n)$ X

$$\text{Then P.I.} = \frac{1}{f(D)} X = \frac{1}{(D - m_1)(D - m_2) \ldots\ldots (D - m_n)} X$$

$$= \left(\frac{A_1}{D - m_1} + \frac{A_2}{D - m_2} + \ldots\ldots + \frac{A_n}{D - m_n} \right) X \qquad \text{(Partial Fractions)}$$

$$= A_1 \frac{1}{D - m_1} X + A_2 \frac{1}{D - m_2} X + \ldots\ldots + A_n \frac{1}{D - m_n} X$$

$$= A_1 e^{m_1 x} \int X e^{-m_1 x} dx + A_2 e^{m_2 x} \int X e^{-m_2 x} dx + \ldots\ldots + A_n e^{m_n x} \int X e^{-m_n x} dx$$

$$\left[\because \frac{1}{D - m} X = e^{mx} \int X e^{-mx} dx \right].$$

ILLUSTRATIVE EXAMPLES

Example 1. *Solve* $(D^3 - 6D^2 + 11D - 6)y = e^{-2x} + e^{-3x}$.

Sol. A.E. is $D^3 - 6D^2 + 11D - 6 = 0$ or $(D - 1)(D - 2)(D - 3) = 0$

where $D = 1, 2, 3$

$\therefore$ C.F. $= c_1 e^x + c_2 e^{2x} + c_3 e^{3x}$

$$\text{P.I.} = \frac{1}{D^3 - 6D^2 + 11D - 6} (e^{-2x} + e^{-3x})$$

$$= \frac{1}{D^3 - 6D^2 + 11D - 6} e^{-2x} + \frac{1}{D^3 - 6D^2 + 11D - 6} e^{-3x}$$

$$= \frac{1}{(-2)^3 - 6(-2)^2 + 11(-2) - 6} e^{-2x} + \frac{1}{(-3)^3 - 6(-3)^2 + 11(-3) - 6} e^{-3x}$$

$$= -\frac{1}{60} e^{-2x} - \frac{1}{120} e^{-3x} = -\frac{1}{120} (2e^{-2x} + e^{-3x})$$

Hence the C.S. is y = C.F. + P.I.

i.e., $$y = c_1 e^x + c_2 e^{2x} + c_3 e^{3x} - \frac{1}{120} (2e^{-2x} + e^{-3x}).$$

Example 2. *Solve* $(D - 2)^2 y = 8(e^{2x} + \sin 2x + x^2)$.

Sol. A.E. is $(D - 2)^2 = 0$ whence $D = 2, 2$

$\therefore$ C.F. $= (c_1 + c_2 x)e^{2x}$

$$\text{P.I.} = \frac{1}{(D-2)^2}\,[8(e^{2x} + \sin 2x + x^2)]$$

$$= 8\left[\frac{1}{(D-2)^2}\,e^{2x} + \frac{1}{(D-2)^2}\sin 2x + \frac{1}{(D-2)^2}\,x^2\right]$$

Now $\frac{1}{(D-2)^2}\,e^{2x} = x\,.\,\frac{1}{2(D-2)}\,e^{2x}$ | Case of failure

$= x^2\,.\,\frac{1}{2}\,e^{2x}$ | Case of failure

$$= \frac{x^2}{2}\,e^{2x}$$

$\frac{1}{(D-2)^2}\sin 2x = \frac{1}{D^2 - 4D + 4}\sin 2x = \frac{1}{-2^2 - 4D + 4}\sin 2x$ [Putting $D^2 = -2^2$]

$$= -\frac{1}{4D}\sin 2x = -\frac{1}{4}\int \sin 2x\,dx = -\frac{1}{4}\left(-\frac{\cos 2x}{2}\right) = \frac{1}{8}\cos 2x$$

$$\frac{1}{(D-2)^2}x^2 = \frac{1}{(2-D)^2}x^2 = \frac{1}{4\left(1-\frac{D}{2}\right)^2}x^2 = \frac{1}{4}\left(1-\frac{D}{2}\right)^{-2}x^2$$

$$= \frac{1}{4}\left[1 - 2\left(-\frac{D}{2}\right) + \frac{(-2)(-3)}{2}\left(\frac{D}{2}\right)^2 \ldots\ldots\right]x^2$$

$$= \frac{1}{4}\left[1 + D + \frac{3}{4}D^2 + \ldots\ldots\right]x^2$$

$$= \frac{1}{4}\left(x^2 + 2x + \frac{3}{2}\right)$$

$\therefore$ $\text{P.I.} = 8\left[\frac{x^2}{2}e^{2x} + \frac{1}{8}\cos 2x + \frac{1}{4}\left(x^2 + 2x + \frac{3}{2}\right)\right]$

$$= 4x^2 e^{2x} + \cos 2x + 2x^2 + 4x + 3$$

Hence the C.S. is $y = (c_1 + c_2x)\,e^{2x} + 4x^2 e^{2x} + \cos 2x + 2x^2 + 4x + 3.$

Example 3. *Solve* $(D + 2)(D - 1)^2\,y = e^{-2x} + 2\sinh x.$

Sol. A.E. is $(D + 2)(D - 1)^2 = 0$ so that $D = -2, 1, 1$

$\therefore$ $\text{C.F.} = c_1e^{-2x} + (c_2 + c_3x)\,e^x$

$$\text{P.I.} = \frac{1}{(D+2)(D-1)^2}\,(e^{-2x} + 2\sinh x)$$

$= \frac{1}{(D+2)(D-1)^2}\,(e^{-2x} + e^x - e^{-x})$ $\left[\because\ \sinh x = \frac{e^x - e^{-x}}{2}\right]$

Now $\frac{1}{(D+2)(D-1)^2} e^{-2x} = \frac{1}{D+2}\left[\frac{1}{(D-1)^2} e^{-2x}\right] = \frac{1}{D+2}\left[\frac{1}{(-2-1)^2} e^{-2x}\right]$

$$= \frac{1}{9} \cdot \frac{1}{D+2} e^{-2x} \qquad \text{| Case of failure}$$

$$= \frac{1}{9} x \cdot \frac{1}{1} e^{-2x} = \frac{x}{9} e^{-2x}$$

$$\frac{1}{(D+2)(D-1)^2} e^{x} = \frac{1}{(D-1)^2}\left[\frac{1}{D+2} e^{x}\right] = \frac{1}{(D-1)^2}\left[\frac{1}{1+2} e^{x}\right]$$

$$= \frac{1}{3} \cdot \frac{1}{(D-1)^2} e^{x} \qquad \text{| Case of failure}$$

$$= \frac{1}{3} \cdot x \frac{1}{2(D-1)} e^{x} \qquad \text{| Case of failure}$$

$$= \frac{1}{3} \cdot x^2 \cdot \frac{1}{2} e^{x} = \frac{1}{6} x^2 e^{x}$$

$$\frac{1}{(D+2)(D-1)^2} e^{-x} = \frac{1}{(-1+2)(-1-1)^2} e^{-x} = \frac{1}{4} e^{-x}$$

$$\therefore \qquad \text{P.I.} = \frac{x}{9} e^{-2x} + \frac{x^2}{6} e^{x} + \frac{1}{4} e^{-x}$$

Hence the C.S. is $\quad y = c_1 e^{-2x} + (c_2 + c_3 x)e^{x} + \frac{x}{9} e^{-2x} + \frac{x^2}{6} e^{x} + \frac{1}{4} e^{-x}.$

Example 4. *Solve* $\frac{d^2y}{dx^2} - 4y = x \sinh x.$

Sol. Given equation in symbolic form is $(D^2 - 4)y = x \sinh x$

A.E. is $\quad D^2 - 4 = 0 \quad$ so that $\quad D = \pm 2$

$\therefore \quad \text{C.F.} = c_1 e^{2x} + c_2 e^{-2x}$

$$\text{P.I.} = \frac{1}{D^2-4} x \sinh x = \frac{1}{D^2-4} x \left(\frac{e^x - e^{-x}}{2}\right)$$

$$= \frac{1}{2}\left[\frac{1}{D^2-4} e^{x} \cdot x - \frac{1}{D^2-4} e^{-x} \cdot x\right]$$

$$= \frac{1}{2}\left[e^{x} \frac{1}{(D+1)^2-4} x - e^{-x} \frac{1}{(D-1)^2-4} x\right]$$

$$= \frac{1}{2}\left[e^{x} \frac{1}{D^2+2D-3} x - e^{-x} \frac{1}{D^2-2D-3} x\right]$$

$$= \frac{1}{2}\left[e^x \frac{1}{-3\left(1 - \frac{2D}{3} - \frac{D^2}{3}\right)} x - e^{-x} \frac{1}{-3\left(1 + \frac{2D}{3} - \frac{D^2}{3}\right)} x \right]$$

$$= -\frac{1}{6}\left[e^x \left\{1 - \left(\frac{2D}{3} + \frac{D^2}{3}\right)\right\}^{-1} x - e^{-x}\left\{1 + \left(\frac{2D}{3} - \frac{D^2}{3}\right)\right\}^{-1} x \right]$$

$$= -\frac{1}{6}\left[e^x \left(1 + \frac{2D}{3} ......\right) x - e^{-x}\left(1 - \frac{2D}{3} ......\right) x \right] = -\frac{1}{6}\left[e^x \left(x + \frac{2}{3}\right) - e^{-x}\left(x - \frac{2}{3}\right)\right]$$

$$= -\frac{x}{3}\left(\frac{e^x - e^{-x}}{2}\right) - \frac{2}{9}\left(\frac{e^x + e^{-x}}{2}\right) = -\frac{x}{3}\sinh x - \frac{2}{9}\cosh x$$

Hence the C.S. is $y = c_1 e^{2x} + c_2 e^{-2x} - \frac{x}{3}\sinh x - \frac{2}{9}\cosh x$.

Example 5. *Solve* $\frac{d^4 y}{dx^4} - y = \cos x \cosh x$.

Sol. Given equation in symbolic form is $(D^4 - 1)y = \cos x \cosh x$

A.E. is $D^4 - 1 = 0$ or $(D^2 - 1)(D^2 + 1) = 0$ so that $D = \pm 1, \pm i$

$\therefore$ C.F. $= c_1 e^x + c_2 e^{-x} + e^{0x}(c_3 \cos x + c_4 \sin x)$

$= c_1 e^x + c_2 e^{-x} + c_3 \cos x + c_4 \sin x$

$$\text{P.I.} = \frac{1}{D^4 - 1}\cos x \cosh x = \frac{1}{D^4 - 1}\cos x \left(\frac{e^x + e^{-x}}{2}\right)$$

$$= \frac{1}{2}\left[\frac{1}{D^4 - 1} e^x \cos x + \frac{1}{D^4 - 1} e^{-x}\cos x\right]$$

$$= \frac{1}{2}\left[e^x \frac{1}{(D+1)^4 - 1}\cos x + e^{-x}\frac{1}{(D-1)^4 - 1}\cos x\right]$$

$$= \frac{1}{2}\left[e^x \frac{1}{D^4 + 4D^3 + 6D^2 + 4D}\cos x + e^{-x}\frac{1}{D^4 - 4D^3 + 6D^2 - 4D}\cos x\right]$$

$$= \frac{1}{2}\left[e^x \frac{1}{(-1^2)^2 + 4D(-1^2) + 6(-1^2) + 4D}\cos x \right.$$

$$\left. + e^{-x}\frac{1}{(-1^2)^2 - 4D(-1^2) + 6(-1^2) - 4D}\cos x\right]$$

$$= \frac{1}{2}\left[e^x \frac{1}{-5}\cos x + e^{-x}\frac{1}{-5}\cos x\right] = -\frac{1}{5}\left(\frac{e^x + e^{-x}}{2}\right)\cos x = -\frac{1}{5}\cosh x \cos x$$

Hence the C.S. is $y = c_1 e^x + c_2 e^{-x} + c_3 \cos x + c_4 \sin x - \frac{1}{5}\cos x \cosh x$.

Example 6. *Solve* $\frac{d^2y}{dx^2} - 2\frac{dy}{dx} + y = xe^x \sin x.$

Sol. Given equation in symbolic form is $(D^2 - 2D + 1)y = xe^x \sin x$

A.E. is $D^2 - 2D + 1 = 0$ or $(D - 1)^2 = 0$ so that $D = 1, 1$

$$\therefore \quad \text{C.F.} = (c_1 + c_2x)e^x$$

$$\text{P.I.} = \frac{1}{(D-1)^2} e^x . x \sin x = e^x . \frac{1}{(D+1-1)^2} x \sin x$$

$$= e^x \frac{1}{D^2} x \sin x = e^x \frac{1}{D}\int x \sin x \, dx$$

Integrating by parts,

$$\text{P.I.} = e^x \frac{1}{D}\left[x(-\cos x) - \int 1(-\cos x)\,dx\right] = e^x \frac{1}{D}(-x\cos x + \sin x)$$

$$= e^x \int(-x\cos x + \sin x)\,dx = e^x\left[-\left\{x \sin x - \int 1.\sin x\,dx\right\} - \cos x\right]$$

$$= e^x[-x\sin x - \cos x - \cos x] = -e^x(x \sin x + 2\cos x)$$

Hence the C.S. is $y = (c_1 + c_2x)e^x - e^x(x \sin x + 2 \cos x)$.

Example 7. *Solve* $\frac{d^2y}{dx^2} + y = \text{cosec } x.$

Sol. Given equation in symbolic form is $(D^2 + 1)y = \text{cosec } x$

A.E. is $\quad D^2 + 1 = 0 \Rightarrow D = \pm i$

$$\therefore \quad \text{C.F.} = c_1 \cos x + c_2 \sin x$$

$$\text{P.I.} = \frac{1}{D^2+1}\text{cosec } x = \frac{1}{(D+i)(D-i)}\text{cosec } x$$

$$= \frac{1}{2i}\left(\frac{1}{D-i} - \frac{1}{D+i}\right)\text{cosec } x \qquad \text{(Partial Fractions)}$$

$$= \frac{1}{2i}\left(\frac{1}{D-i}\text{cosec } x - \frac{1}{D+i}\text{cosec } x\right)$$

Now $\frac{1}{D-i}\text{cosec } x = e^{ix}\int \text{cosec } x \, e^{-ix}\,dx \qquad \left[\because \frac{1}{D-a}X = e^{ax}\int Xe^{-ax}\,dx\right]$

$$= e^{ix}\int \text{cosec } x(\cos x - i \sin x)\,dx = e^{ix}\int(\cot x - i)\,dx$$

$$= e^{ix}(\log \sin x - ix)$$

Changing i to $-i$, we have $\frac{1}{D+i}\text{cosec } x = e^{-ix}(\log \sin x + ix)$

$$\therefore \quad \text{P.I.} = \frac{1}{2i}[e^{ix}(\log \sin x - ix) - e^{-ix}(\log \sin x + ix)]$$

$$= \log \sin x \left(\frac{e^{ix} - e^{-ix}}{2i}\right) - x\left(\frac{e^{ix} + e^{-ix}}{2}\right)$$

$$= \log \sin x\ (\sin x) - x \cos x = \sin x \log \sin x - x \cos x$$

Hence the C.S. is $y = c_1 \cos x + c_2 \sin x + \sin x \log \sin x - x \cos x$.

Example 8. *Solve* $\frac{d^2y}{dx^2} + a^2y = \tan ax$.

Sol. Given equation in symbolic form is $(D^2 + a^2)y = \tan ax$

A.E. is $D^2 + a^2 = 0 \Rightarrow D = \pm ia$

$\therefore$ C.F. $= c_1 \cos ax + c_2 \sin ax$

$$\text{P.I.} = \frac{1}{D^2 + a^2} \tan ax = \frac{1}{(D + ia)(D - ia)} \tan ax$$

$$= \frac{1}{2ia}\left[\frac{1}{D - ia} - \frac{1}{D + ia}\right] \tan ax \qquad \text{(Partial Fractions)}$$

$$= \frac{1}{2ia}\left[\frac{1}{D - ia} \tan ax - \frac{1}{D + ia} \tan ax\right]$$

Now $\frac{1}{D - ia} \tan ax = e^{iax} \int \tan ax \,.\, e^{-iax}\, dx$

$$= e^{iax} \int \tan ax\ (\cos ax - i \sin ax)\ dx = e^{iax} \int \left(\sin ax - i\frac{\sin^2 ax}{\cos ax}\right) dx$$

$$= e^{iax} \int \left(\sin ax - i\frac{1 - \cos^2 ax}{\cos ax}\right) dx = e^{iax} \int [\sin ax - i(\sec ax - \cos ax)]\ dx$$

$$= e^{iax}\left[-\frac{\cos ax}{a} - \frac{i}{a} \log (\sec ax + \tan ax) + i\frac{\sin ax}{a}\right]$$

$$= -\frac{1}{a} e^{iax} [(\cos ax - i \sin ax) + i \log (\sec ax + \tan ax)]$$

$$= -\frac{1}{a} e^{iax} [e^{-iax} + i \log (\sec ax + \tan ax)] = -\frac{1}{a} [1 + ie^{iax} \log (\sec ax + \tan ax)]$$

Changing i to $-i$, we have $\frac{1}{D + ia} \tan ax = -\frac{1}{a} [1 - ie^{-iax} \log (\sec ax + \tan ax)]$

$$\therefore \quad \text{P.I.} = \frac{1}{2ia}\left[-\frac{1}{a}\left\{1 + ie^{iax} \log (\sec ax + \tan ax)\right\} + \frac{1}{a}\left\{1 - ie^{-iax} \log (\sec ax + \tan ax)\right\}\right]$$

$$= -\frac{1}{a^2} \log (\sec ax + \tan ax) \left(\frac{e^{iax} + e^{-iax}}{2}\right) = -\frac{1}{a^2} \log (\sec ax + \tan ax) \,.\, \cos ax$$

Hence the C.S. is $y = c_1 \cos ax + c_2 \sin ax - \frac{1}{a^2} \cos ax \log (\sec ax + \tan ax)$.

TEST YOUR KNOWLEDGE

Solve the following differential equations :

1. $\frac{d^3y}{dx^3} + y = 3 + 5e^x$

2. $\frac{d^2y}{dx^2} - 4y = (1 + e^x)^2$

3. $\frac{d^2y}{dx^2} + 4\frac{dy}{dx} + 5y = -2\cosh x$

4. $\frac{d^2y}{dx^2} - 2\frac{dy}{dx} + 5y = \sin 3x$

5. $\frac{d^3y}{dx^3} + \frac{d^2y}{dx^2} + \frac{dy}{dx} + y = \sin 2x$

6. $\frac{d^3y}{dx^3} + y = \sin 3x - \cos^2\frac{x}{2}$

7. $(D^2 - 4D + 3)y = \sin 3x \cos 2x$

8. $(D^2 - 3D + 2)y = 6e^{-3x} + \sin 2x$

9. $\frac{d^2y}{dx^2} + 4y = e^x + \sin 2x$

10. $\frac{d^3y}{dx^3} - 2\frac{d^2y}{dx^2} + 4\frac{dy}{dx} = e^{2x} + \sin 2x$

11. $\frac{d^2y}{dx^2} - 4y = x^2$

12. $\frac{d^3y}{dx^3} - \frac{d^2y}{dx^2} - 6\frac{dy}{dx} = 1 + x^2$

13. $\frac{d^2y}{dx^2} + \frac{dy}{dx} = x^2 + 2x + 4$

14. $\frac{d^2y}{dx^2} + y = e^{2x} + \cosh 2x + x^3$

15. $(D^2 - 3D + 2)y = 2e^x \cos\frac{x}{2}$

16. $\frac{d^2y}{dx^2} - 3\frac{dy}{dx} + 2y = xe^{3x} + \sin 2x$

17. $\frac{d^4y}{dx^4} - y = e^x \cos x$

18. $(D^2 - 2D)y = e^x \sin x$

19. $(D^2 + 4D + 8)y = 12e^{-2x}\sin x \sin 3x$

20. $\frac{d^2y}{dx^2} + 2y = x^2e^{3x} + e^x\cos 2x$

21. $(D^3 + 2D^2 + D)y = x^2e^{2x} + \sin^2 x$

22. $(D^2 - 4D + 4)y = 8x^2 e^{2x}\sin 2x$

23. $(D - 1)^2(D + 1)^2y = \sin^2\frac{x}{2} + e^x + x$

24. $\frac{d^2y}{dx^2} + 4y = x\sin x$

25. $(D^2 - 1)y = x\sin x + (1 + x^2)e^x$

26. $\frac{d^2y}{dx^2} + a^2y = \sec ax$

27. $\frac{d^2y}{dx^2} + 4y = 4\tan 2x$

28. $\frac{d^2y}{dx^2} + 3\frac{dy}{dx} + 2y = e^{e^x}$

29. $\frac{d^4y}{dx^4} - y = e^{-x}\cos x$ [*CUSAT 2008*]

30. $\frac{d^2y}{dx^2} + 3\frac{dy}{dx} + 2y = 2\sin^2 x + 2x^2$. [*CUSAT 2008*]

Answers

1. $y = c_1e^{-x} + e^{\frac{1}{2}x}\left(c_2\cos\frac{\sqrt{3}}{2}x + c_3\sin\frac{\sqrt{3}}{2}x\right) + 3 + \frac{5}{2}e^x$

2. $y = c_1e^{2x} + c_2e^{-2x} - \frac{1}{4} - \frac{2}{3}e^x + \frac{1}{4}xe^{2x}$

3. $y = e^{-2x}(c_1\cos x + c_2\sin x) - \frac{1}{10}e^x - \frac{1}{2}e^{-x}$

4. $y = e^x(c_1 \cos 2x + c_2 \sin 2x) + \frac{1}{26}(3 \cos 3x - 2 \sin 3x)$

5. $y = c_1 e^{-x} + c_2 \cos x + c_3 \sin x + \frac{1}{15}(2 \cos 2x - \sin 2x)$

6. $y = c_1 e^{-x} + e^{\frac{1}{2}x}\left(c_2 \cos \frac{\sqrt{3}}{2} x + c_3 \sin \frac{\sqrt{3}}{2} x\right) + \frac{1}{730}(\sin 3x + 27 \cos 3x) - \frac{1}{2} - \frac{1}{4}(\cos x - \sin x)$

7. $y = c_1 e^x + c_2 e^{3x} + \frac{1}{884}(10 \cos 5x - 11 \sin 5x) + \frac{1}{20}(\sin x + 2 \cos x)$

8. $y = c_1 e^x + c_2 e^{2x} + \frac{3}{10} e^{-3x} + \frac{1}{20}(3 \cos 2x - \sin 2x)$

9. $y = c_1 \cos 2x + c_2 \sin 2x + \frac{1}{5} e^x - \frac{x}{4} \cos 2x$

10. $y = c_1 + e^x (c_2 \cos \sqrt{3}\, x + c_3 \sin \sqrt{3}\, x) + \frac{1}{8}(e^{2x} + \sin 2x)$

11. $y = c_1 e^{2x} + c_2 e^{-2x} - \frac{1}{4}\left(x^2 + \frac{1}{2}\right)$

12. $y = c_1 + c_2 e^{3x} + c_3 e^{-2x} - \frac{1}{18}\left(x^3 - \frac{x^2}{2} + \frac{25}{6} x\right)$

13. $y = c_1 + c_2 e^{-x} + \frac{x^3}{3} + 4x$

14. $y = c_1 \cos x + c_2 \sin x + \frac{1}{5} e^{2x} + \frac{1}{5} \cosh 2x + x^3 - 6x$

15. $y = c_1 e^x + c_2 e^{2x} - \frac{8}{5} e^x \left(2 \sin \frac{x}{2} + \cos \frac{x}{2}\right)$

16. $y = c_1 e^x + c_2 e^{2x} + \frac{1}{4} e^{3x}(2x - 3) + \frac{1}{20}(3 \cos 2x - \sin 2x)$

17. $y = c_1 e^x + c_2 e^{-x} + c_3 \cos x + c_4 \sin x - \frac{1}{5} e^x \cos x$

18. $y = c_1 + c_2 e^{2x} - \frac{1}{2} e^x \sin x$

19. $y = e^{-2x}(c_1 \cos 2x + c_2 \sin 2x) + \frac{1}{2} e^{-2x}(3x \sin 2x + \cos 4x)$

20. $y = c_1 \cos \sqrt{2}\, x + c_2 \sin \sqrt{2}\, x + \frac{e^{3x}}{11}\left(x^2 - \frac{12}{11} x + \frac{50}{121}\right) + \frac{e^x}{17}(4 \sin 2x - \cos 2x)$

21. $y = c_1 + (c_2 + c_3 x)e^{-x} + \frac{x}{2} + \frac{e^{2x}}{18}\left(x^2 - \frac{7}{3} x + \frac{11}{6}\right) + \frac{1}{100}(3 \sin 2x + 4 \cos 2x)$

22. $y = (c_1 + c_2 x)e^{2x} - e^{2x}[4x \cos 2x + (2x^2 - 3) \sin 2x]$

23. $y = (c_1 + c_2 x)e^x + (c_3 + c_4 x)e^{-x} + \frac{1}{2} - \frac{1}{8} \cos x + \frac{x^2}{8} e^x + x$

24. $y = c_1 \cos 2x + c_2 \sin 2x + \frac{1}{9}(3x \sin x - 2 \cos x)$

25. $y = c_1 e^x + c_2 e^{-x} - \frac{1}{2}(x \sin x + \cos x) + \frac{1}{12} x e^x (2x^2 - 3x + 9)$

26. $y = c_1 \cos ax + c_2 \sin ax + \frac{1}{a}\left(x \sin ax + \cos ax \frac{\log \cos ax}{a}\right)$

27. $y = c_1 \cos 2x + c_2 \sin 2x - \cos 2x \log (\sec 2x + \tan 2x)$

28. $y = c_1e^{-x} + c_2e^{-2x} + e^{-2x} \,.\, e^{e^x}$ **29.** $c_1e^x + c_2e^{-x} + c_3 \cos x + c_4 \sin x - \frac{1}{5} e^{-x} \cos x$.

30. $c_1e^{-2x} + c_2e^{-x} + \frac{1}{2} + \frac{2 \cos x - 6 \sin 2x}{40} + x^2 - 3x + \frac{7}{2}$.

3.8. METHOD OF VARIATION OF PARAMETERS TO FIND P.I.

Consider the linear equation of **second order** with constant co-efficients

$$\frac{d^2y}{dx^2} + a_1 \frac{dy}{dx} + a_2y = X \qquad \text{...(1)}$$

Let its C.F. be $y = c_1y_1 + c_2y_2$ so that y_1 and y_2 satisfy the equation

$$\frac{d^2y}{dx^2} + a_1 \frac{dy}{dx} + a_2y = 0 \qquad \text{...(2)}$$

Now, let us assume that the P.I. of (1) is $y = uy_1 + vy_2$ ...(3)

where u and v are unknown functions of x.

Differentiating (3) w.r.t. x, we have $y' = uy_1' + vy_2' + u'y_1 + v'y_2 = uy_1' + vy_2'$ ...(4)

assuming that u, v satisfy the equation $u'y_1 + v'y_2 = 0$ ...(5)

Differentiating (4) w.r.t. x, we have $y'' = uy_1'' + u'y_1' + vy_2'' + v'y_2'$

Substituting the values of y, y' and y'' in (1), we get

$$(uy_1'' + u'y_1' + vy_2'' + v'y_2') + a_1(uy_1' + vy_2') + a_2(uy_1 + vy_1) = X$$

or $$u(y_1'' + a_1y_1' + a_2y_1) + v(y_2'' + a_1y_2' + a_2y_2) + u'y_1' + v'y_2' = X$$

or $$u'y_1' + v'y_2' = X \qquad \text{...(6)}$$

since y_1 and y_2 satisfy (2).

Solving (5) and (6), we get $u' = \begin{vmatrix} 0 & y_2 \\ X & y_2' \end{vmatrix} \div \begin{vmatrix} y_1 & y_2 \\ y_1' & y_2' \end{vmatrix} = -\frac{y_2X}{W}$

and $v' = \begin{vmatrix} y_1 & 0 \\ y_1' & X \end{vmatrix} \div \begin{vmatrix} y_1 & y_2 \\ y_1' & y_2' \end{vmatrix} = \frac{y_1X}{W}$

where $W = \begin{vmatrix} y_1 & y_2 \\ y_1' & y_2' \end{vmatrix}$ is called the Wronskian of y_1, y_2.

Integrating, $u = -\int \frac{y_2X}{W}\, dx, \quad v = \int \frac{y_1X}{W}\, dx$

Substituting in (3), the P.I. is known.

Note 1. As the solution is obtained by varying the arbitrary constants c_1, c_2 of the C.F., the method is known as *variation of parameters*.

Note 2. Method of variation of parameters is to be used if instructed to do so.

Example. *Apply the method of variation of parameters to solve* $\frac{d^2y}{dx^2} + 4y = 4 \sec^2 2x$.

Sol. Given equation in symbolic form is $(D^2 + 4)y = 4 \sec^2 2x$

Its A.E. is $D^2 + 4 = 0$ so that $D = \pm 2i$

$\therefore$ C.F. is $y = c_1 \cos 2x + c_2 \sin 2x$

Here $y_1 = \cos 2x$, $y_2 = \sin 2x$ and $X = 4 \sec^2 2x$

$$\therefore \qquad W = \begin{vmatrix} y_1 & y_2 \\ y_1' & y_2' \end{vmatrix} = \begin{vmatrix} \cos 2x & \sin 2x \\ -2\sin 2x & 2\cos 2x \end{vmatrix} = 2$$

$$\text{P.I.} = -y_1 \int \frac{y_2 X}{W}\, dx + y_2 \int \frac{y_1 X}{W}\, dx$$

$$= -\cos 2x \int \frac{\sin 2x \,.\, 4\sec^2 2x}{2}\, dx + \sin 2x \int \frac{\cos 2x \,.\, 4\sec^2 2x}{2}\, dx$$

$$= -2\cos 2x \int \sec 2x \tan 2x \, dx + 2\sin 2x \int \sec 2x \, dx$$

$$= -2\cos 2x \,.\, \frac{\sec 2x}{2} + 2\sin 2x \,.\, \frac{1}{2}\log(\sec 2x + \tan 2x)$$

$$= -1 + \sin 2x \log(\sec 2x + \tan 2x)$$

Hence the C.S. is $y = c_1 \cos 2x + c_2 \sin 2x - 1 + \sin 2x \log(\sec 2x + \tan 2x)$.

TEST YOUR KNOWLEDGE

Solve by the method of variation of parameters :

1. $\dfrac{d^2y}{dx^2} + y = \operatorname{cosec} x.$
2. $\dfrac{d^2y}{dx^2} + y = \sec x$
3. $\dfrac{d^2y}{dx^2} + 4y = \tan 2x$
4. $\dfrac{d^2y}{dx^2} + y = x \sin x$
5. $y'' - 2y' + 2y = e^x \tan x.$

Answers

1. $y = c_1 \cos x + c_2 \sin x - x \cos x + \sin x \log \sin x$
2. $y = c_1 \cos x + c_2 \sin x + \cos x \log(\cos x) + x \sin x$
3. $y = c_1 \cos 2x + c_2 \sin 2x - \dfrac{1}{4} \cos 2x \log(\sec 2x + \tan 2x)$
4. $y = c_1 \cos x + c_2 \sin x + \dfrac{x}{4} \sin x - \dfrac{x^2}{4} \cos x$
5. $y = e^x (c_1 \cos x + c_2 \sin x) - e^x \cos x \log(\sec x + \tan x).$

3.9. CAUCHY'S HOMOGENEOUS LINEAR EQUATION

An equation of the form

$$x^n \frac{d^n y}{dx^n} + a_1 x^{n-1} \frac{d^{n-1} y}{dx^{n-1}} + a_2 x^{n-2} \frac{d^{n-2} y}{dx^{n-2}} + \ldots\ldots + a_{n-1} x \frac{dy}{dx} + a_n y = X \qquad \ldots(i)$$

where a_i's are constants and X is a function of x, is called Cauchy's homogeneous linear equation.

Such equations can be reduced to linear differential equations with constant co-efficients by the substitution $x = e^z$ or $z = \log x$

so that $$\frac{dy}{dx} = \frac{dy}{dz} \cdot \frac{dz}{dx} = \frac{dy}{dz} \cdot \frac{1}{x} \quad \text{or} \quad x\frac{dy}{dx} = \frac{dy}{dz} = \text{D}y, \text{ where D} = \frac{d}{dz}$$

$$\frac{d^2y}{dx^2} = \frac{d}{dx}\left(\frac{1}{x} \cdot \frac{dy}{dz}\right) = -\frac{1}{x^2}\frac{dy}{dz} + \frac{1}{x} \cdot \frac{d^2y}{dz^2} \cdot \frac{dz}{dx}$$

$$= -\frac{1}{x^2}\frac{dy}{dz} + \frac{1}{x^2}\frac{d^2y}{dz^2} \qquad \left(\because \frac{dz}{dx} = \frac{1}{x}\right)$$

or $$x^2\frac{d^2y}{dx^2} = \frac{d^2y}{dz^2} - \frac{dy}{dz} = \text{D}^2y - \text{D}y = \text{D}(\text{D} - 1)y$$

Similarly, $x^3\dfrac{d^3y}{dx^3} = \text{D}(\text{D} - 1)(\text{D} - 2)y$ and so on.

Substituting these values in equation (*i*), we get a linear differential equation with constant co-efficients, which can be solved by the methods already discussed.

ILLUSTRATIVE EXAMPLES

Example 1. *Solve* $x^3\dfrac{d^3y}{dx^3} + 2x^2\dfrac{d^2y}{dx^2} + 2y = 10\left(x + \dfrac{1}{x}\right)$.

Sol. Given equation is a Cauchy's homogeneous linear equation.

Put $x = e^z$ *i.e.,* $z = \log x$

so that $$x\frac{dy}{dx} = \text{D}y,\ x^2\frac{d^2y}{dx^2} = \text{D}(\text{D} - 1)y$$

$$x^3\frac{d^3y}{dx^3} = \text{D}(\text{D} - 1)(\text{D} - 2)y, \text{ where D} \equiv \frac{d}{dz}$$

Substituting these values in the given equation, it reduces to

$$[\text{D}(\text{D} - 1)(\text{D} - 2) + 2\text{D}(\text{D} - 1) + 2]y = 10(e^z + e^{-z})$$

or $$(\text{D}^3 - \text{D}^2 + 2)y = 10(e^z + e^{-z})$$

which is a linear equation with constant co-efficients.

Its A.E. is $\text{D}^3 - \text{D}^2 + 2 = 0$ or $(\text{D} + 1)(\text{D}^2 - 2\text{D} + 2) = 0$

$$\therefore \quad \text{D} = -1, \frac{2 \pm \sqrt{4 - 8}}{2} = -1, 1 \pm i$$

$$\therefore \quad \text{C.F.} = c_1e^{-z} + e^z(c_2 \cos z + c_3 \sin z) = \frac{c_1}{x} + x[c_2 \cos(\log x) + c_3 \sin(\log x)]$$

$$\text{P.I.} = 10\frac{1}{\text{D}^3 - \text{D}^2 + 2}(e^z + e^{-z}) = 10\left(\frac{1}{\text{D}^3 - \text{D}^2 + 2}e^z + \frac{1}{\text{D}^3 - \text{D}^2 + 2}e^{-z}\right)$$

$$= 10\left(\frac{1}{1^3 - 1^2 + 2} e^z + z \cdot \frac{1}{3D^2 - 2D} e^{-z}\right) = 10\left(\frac{1}{2} e^z + z \cdot \frac{1}{3(-1)^2 - 2(-1)} e^{-z}\right)$$

$$= 5e^z + 2ze^{-z} = 5x + \frac{2}{x} \log x$$

Hence the C.S. is $y = \frac{c_1}{x} + x[c_2 \cos(\log x) + c_3 \sin(\log x)] + 5x + \frac{2}{x} \log x.$

Example 2. *Solve* $x^2 \frac{d^2y}{dx^2} - x\frac{dy}{dx} - 3y = x^2 \log x.$

Sol. Given equation is a Cauchy's homogeneous linear equation.

Put $x = e^z$ *i.e.,* $z = \log x$ so that $x\frac{dy}{dx} = Dy$, $x^2 \frac{d^2y}{dx^2} = D(D-1)y$, where $D \equiv \frac{d}{dz}$.

Substituting these values in the given equation, it reduces to

$$[D(D-1) - D - 3]y = ze^{2z} \quad \text{or} \quad (D^2 - 2D - 3)y = ze^{2z}$$

which is a linear equation with constant co-efficients.

Its A.E. is $D^2 - 2D - 3 = 0$ or $(D - 3)(D + 1) = 0$

$\therefore \quad D = 3, -1$

$$\text{C.F.} = c_1e^{3z} + c_2e^{-z} = c_1x^3 + \frac{c_2}{x}$$

$$\text{P.I.} = \frac{1}{D^2 - 2D - 3}(e^{2z} \cdot z) = e^{2z}\frac{1}{(D+2)^2 - 2(D+2) - 3} z = e^{2z}\frac{1}{D^2 + 2D - 3} z$$

$$= e^{2z}\frac{1}{-3\left(1 - \frac{2D}{3} - \frac{D^2}{3}\right)} z = -\frac{1}{3}e^{2z}\left[1 - \left(\frac{2D}{3} + \frac{D^2}{3}\right)\right]^{-1} z$$

$$= -\frac{1}{3}e^{2z}\left[1 + \left(\frac{2D}{3} + \frac{D^2}{3}\right) + \ldots\ldots\right] z = -\frac{1}{3}e^{2z}\left(z + \frac{2}{3}\right) = -\frac{x^2}{3}\left(\log x + \frac{2}{3}\right)$$

Hence the C.S. is $y = c_1x^3 + \frac{c_2}{x} - \frac{x^2}{3}\left(\log x + \frac{2}{3}\right).$

Example 3. *Solve* $x^2 \frac{d^2y}{dx^2} + x\frac{dy}{dx} + y = \log x \sin(\log x).$ (CUSAT 2009)

Sol. Given equation is a Cauchy's homogeneous linear equation.

Put $x = e^z$ *i.e.,* $z = \log x$ so that $x\frac{dy}{dx} = Dy$, $x^2 \frac{d^2y}{dx^2} = D(D-1)y$, where $D \equiv \frac{d}{dz}$.

Substituting these values in the given equation, it reduces to

$$[D(D-1) + D + 1]y = z \sin z$$

or

$$(D^2 + 1)y = z \sin z$$

Its A.E. is $D^2 + 1 = 0$ so that $D = \pm i$

$$\text{C.F.} = c_1 \cos z + c_2 \sin z = c_1 \cos(\log x) + c_2 \sin(\log x)$$

$$\text{P.I.} = \frac{1}{D^2+1} z \sin z = \text{Imaginary part of } \frac{1}{D^2+1} ze^{iz}$$

$$= \text{I.P. of } e^{iz} \frac{1}{(D+i)^2+1} z = \text{I.P. of } e^{iz} \frac{1}{D^2+2iD} z$$

$$= \text{I.P. of } e^{iz} \frac{1}{2iD\left(1+\frac{D}{2i}\right)} z = \text{I.P. of } e^{iz} \frac{1}{2iD\left(1-\frac{iD}{2}\right)} z$$

$$= \text{I.P. of } \frac{1}{2i} e^{iz} \frac{1}{D}\left(1-\frac{iD}{2}\right)^{-1} z = \text{I.P. of } \frac{1}{2i} e^{iz} \frac{1}{D}\left(1+\frac{iD}{2}+\ldots\ldots\right) z$$

$$= \text{I.P. of } \frac{1}{2i} e^{iz} \frac{1}{D}\left(z+\frac{i}{2}\right) = \text{I.P. of } \frac{1}{2i} e^{iz} \int\left(z+\frac{i}{2}\right) dz$$

$$= \text{I.P of} -\frac{1}{2} e^{iz}\left(\frac{z^2}{2}+\frac{i}{2}z\right) = \text{I.P. of } e^{iz}\left(-\frac{i}{4}z^2+\frac{z}{4}\right)$$

$$= \text{I.P. of } (\cos z + i \sin z)\left(-\frac{i}{4}z^2+\frac{z}{4}\right) = -\frac{z^2}{4}\cos z + \frac{z}{4}\sin z$$

$$= -\frac{1}{4}(\log x)^2 \cos(\log x) + \tfrac{1}{4}\log x \sin(\log x)$$

Hence the C.S. is

$$y = c_1 \cos(\log x) + c_2 \sin(\log x) - \tfrac{1}{4}(\log x)^2 \cos(\log x) + \tfrac{1}{4}\log x \sin(\log x).$$

Example 4. *Solve :* $x^2 \frac{d^2y}{dx^2} - x\frac{dy}{dx} + y = \log x.$ (CUSAT 2007)

Sol. Given equation is Cauchy's homogeneous linear equation.

Put $x = e^z$ *i.e.*, $z = \log x$ so that $x\frac{dy}{dx} = Dy$,

$$x^2\frac{d^2y}{dx^2} = D(D-1)y, \text{ where } D \equiv \frac{d}{dz}$$

Substituting these values in the given equation, it reduces to

$$[D(D-1) - D + 1]\, y = z$$

or $(D-1)^2 y = z$, which is a linear equation with constant co-efficients

Its A.E. is $(D-1)^2 = 0$ so that $D = 1, 1$

$$\text{C.F.} = (c_1 + c_2 z)\, e^z = (c_1 + c_2 \log x)\, x$$

$$\text{P.I.} = \frac{1}{(D-1)^2} z = (1-D)^{-2} z$$

$$= [1 + 2D + 3D^2 + \ldots..]\, z$$

$$= z + 2 = \log x + 2$$

Hence the C.S. is $y = (c_1 + c_2 \log x)\, x + \log x + 2.$

Example 5. *Solve* : $x^2 \dfrac{d^2y}{dx^2} - 2x \dfrac{dy}{dx} - 4y = x^4$. (CUSAT 2008)

Sol. Given equation is a Cauchy's homogeneous linear equation.

Put $x = e^z$ *i.e.*, $z = \log x$ so that $x \dfrac{dy}{dx} = \text{D}y$,

$$x^2 \frac{d^2y}{dx^2} = \text{D}(\text{D} - 1)y, \text{ where } \text{D} \equiv \frac{d}{dz}$$

Substituting these values in the given equation, it reduces to

$$[\text{D}(\text{D} - 1) - 2\text{D} - 4]\, y = e^{4z}$$

or $$(\text{D}^2 - 3\text{D} - 4)y = e^{4z},$$

which is a linear equation with constant co-efficients.

Its A.E. is $\text{D}^2 - 3\text{D} - 4 = 0$ or $(\text{D} - 4)(\text{D} + 1) = 0$

$\therefore$ $\text{D} = 4, -1$

$$\text{C.F.} = c_1 e^{4z} + c_2 e^{-z} = c_1 x^4 + \frac{c_2}{x}$$

$$\text{P.I.} = \frac{e^{4z}}{\text{D}^2 - 3\text{D} - 4}$$ the denominator vanishes

when D is replaced by 4.

$$\therefore \quad \text{P.I.} = z \frac{e^{4z}}{2\text{D} - 3} = z \frac{e^{4z}}{2(4) - 3} = \frac{ze^{4z}}{5} = \frac{x^4 \log x}{5}$$

Hence the C.S. is $y = c_1 x^4 + \dfrac{c_2}{x} + \dfrac{x^4 \log x}{5}$.

TEST YOUR KNOWLEDGE

Solve :

1. $x^2 \dfrac{d^2y}{dx^2} + 9x \dfrac{dy}{dx} + 25y = 50$

2. $x^2 \dfrac{d^2y}{dx^2} - 2y = x^2 + \dfrac{1}{x}$

3. $x^2 \dfrac{d^2y}{dx^2} + 2x \dfrac{dy}{dx} - 20y = (x + 1)^2$

4. $x^2 \dfrac{d^3y}{dx^3} - 4x \dfrac{d^2y}{dx^2} + 6 \dfrac{dy}{dx} = 4$
[**Hint.** Multiply throughout by x]

5. (*i*) $x^4 \dfrac{d^3y}{dx^3} + 2x^3 \dfrac{d^2y}{dx^2} - x^2 \dfrac{dy}{dx} + xy = 1$. (*ii*) $x^2 \dfrac{d^2y}{dx^2} - 4x \dfrac{dy}{dx} + 6y = x^2$.

6. The radial displacement u in a rotating disc at a distance r from the axis is given by $r^2 \dfrac{d^2u}{dr^2} + r \dfrac{du}{dr} - u + kr^3 = 0$, where k is a constant. Solve the equation under the conditions $u = 0$ when $r = 0$, $u = 0$ when $r = a$.

7. $x^2 \dfrac{d^2y}{dx^2} - x \dfrac{dy}{dx} + y = \log x$

8. $x^2 \dfrac{d^2y}{dx^2} - x \dfrac{dy}{dx} + 2y = x \log x$

9. $x^2 \frac{d^2y}{dx^2} + 2x \frac{dy}{dx} - 12y = x^3 \log x$

10. $x^2 \frac{d^2y}{dx^2} - 2x \frac{dy}{dx} - 4y = x^2 + 2 \log x$

11. $x^2 \frac{d^2y}{dx^2} - 3x \frac{dy}{dx} + 5y = \sin (\log x)$

12. $x^3 \frac{d^3y}{dx^3} + 3x^2 \frac{d^2y}{dx^2} + x \frac{dy}{dx} + 8y = 65 \cos (\log x)$

13. $x^2 \frac{d^2y}{dx^2} - 3x \frac{dy}{dx} + 5y = x^2 \sin (\log x)$

14. $x^2 \frac{d^2y}{dx^2} - 3x \frac{dy}{dx} + y = \frac{\log x \sin (\log x) + 1}{x}$

15. $\frac{d^2y}{dx^2} + \frac{1}{x} \cdot \frac{dy}{dx} = \frac{12 \log x}{x^2}$

Answers

1. $y = x^{-4} [c_1 \cos (3 \log x) + c_2 \sin (3 \log x)] + 2$

2. $y = c_1 x^2 + \frac{c_2}{x} + \frac{1}{3}\left(x^2 - \frac{1}{x}\right) \log x$

3. $y = c_1 x^4 + c_2 x^{-5} - \frac{x^2}{14} - \frac{x}{9} - \frac{1}{20}$

4. $y = c_1 + c_2 x^3 + c_3 x^4 + \frac{2}{3} x$

5. (*i*) $y = (c_1 + c_2 \log x)x + c_3 x^{-1} + \frac{1}{4x} \log x$ (*ii*) $y = c_1 x^2 + c_2 x^3 - x^2 \log x$

6. $u = \frac{kr}{8} (a^2 - r^2)$

7. $y = (c_1 + c_2 \log x)x + \log x + 2$

8. $y = x[c_1 \cos (\log x) + c_2 \sin (\log x)] + x \log x$

9. $y = c_1 x^3 + c_2 x^{-4} + \frac{x^3}{98} \log x (7 \log x - 2)$

10. $y = c_1 x^{-1} + c_2 x^4 - \frac{x^2}{6} - \frac{1}{2} \log x + \frac{3}{8}$

11. $y = x^2[c_1 \cos (\log x) + c_2 \sin (\log x)] + \frac{1}{8} [\sin (\log x) + \cos (\log x)]$

12. $y = c_1 x^{-2} + x[c_2 \cos (\sqrt{3} \log x) + c_3 \sin (\sqrt{3} \log x)] + 8 \cos (\log x) - \sin (\log x)$

13. $y = x^2[c_1 \cos (\log x) + c_2 \sin (\log x)] - \frac{1}{2} x^2 \log x \cos (\log x)$

14. $y = x^2 [c_1 \cosh (\sqrt{3} \log x) + c_2 \sinh (\sqrt{3} \log x)]$
$$+ \frac{1}{61x}\left[\log x \{5 \sin (\log x) + 6 \cos (\log x)\} + \frac{2}{61} \{27 \sin (\log x) + 191 \cos (\log x)\}\right] + \frac{1}{6x}$$

15. $y = c_1 \log x + c_2 + 2 (\log x)^3$

3.10. LEGENDRE'S LINEAR EQUATION

An equation of the form

$$(a + bx)^n \frac{d^ny}{dx^n} + a_1(a + bx)^{n-1} \frac{d^{n-1}y}{dx^{n-1}} + \ldots\ldots + a_{n-1} (a + bx) \frac{dy}{dx} + a_n y = X \quad \ldots(i)$$

where a_i's are constants and X is a function of x, is called Legendre's linear equation.

Such equations can be reduced to linear differential equations with constant co-efficients, by the substitution $a + bx = e^z$ *i.e.*, $z = \log (a + bx)$ so that $\frac{dy}{dx} = \frac{dy}{dz} \cdot \frac{dz}{dx} = \frac{b}{a + bx} \frac{dy}{dz}$

or $$(a + bx)\frac{dy}{dx} = b\frac{dy}{dz} = b\,\mathrm{D}y, \quad \text{where } \mathrm{D} = \frac{d}{dz}$$

$$\frac{d^2y}{dx^2} = \frac{d}{dx}\left(\frac{b}{a+bx}\frac{dy}{dz}\right) = -\frac{b^2}{(a+bx)^2}\frac{dy}{dz} + \frac{b}{a+bx}\cdot\frac{d^2y}{dz^2}\cdot\frac{dy}{dx}$$

$$= -\frac{b^2}{(a+bx)^2}\frac{dy}{dz} + \frac{b}{a+bx}\frac{d^2y}{dz^2}\cdot\frac{b}{a+bx} = \frac{b^2}{(a+bx)^2}\left(\frac{d^2y}{dz^2} - \frac{dy}{dz}\right)$$

or $$(a + bx)^2\frac{d^2y}{dx^2} = b^2(\mathrm{D}^2y - \mathrm{D}y) = b^2\,\mathrm{D}(\mathrm{D} - 1)y$$

Similarly, $(a + bx)^3\dfrac{d^3y}{dx^3} = b^3\,\mathrm{D}(\mathrm{D} - 1)(\mathrm{D} - 2)y.$

Substituting these values in equation (*i*), we get a linear differential equation with constant co-efficients, which can be solved by the methods already discussed.

Example. *Solve* $(3x + 2)^2\dfrac{d^2y}{dx^2} + 3(3x + 2)\dfrac{dy}{dx} - 36y = 3x^2 + 4x + 1.$

Sol. Given equation is a Legendre's linear equation.

Put $3x + 2 = e^z$ *i.e.*, $z = \log(3x + 2)$ so that $(3x + 2)\dfrac{dy}{dx} = 3\mathrm{D}y.$

$$(3x + 2)^2\frac{d^2y}{dx^2} = 3^2\,\mathrm{D}(\mathrm{D} - 1)y, \quad \text{where } \mathrm{D} \equiv \frac{d}{dz}.$$

Substituting these values in the given equation, it reduces to

$$[3^2\,\mathrm{D}(\mathrm{D} - 1) + 3.3\mathrm{D} - 36]y = 3\left(\frac{e^z - 2}{3}\right)^2 + 4\left(\frac{e^z - 2}{3}\right) + 1$$

or $$9(\mathrm{D}^2 - 4)y = \frac{1}{3}e^{2z} - \frac{1}{3} \quad \text{or} \quad (\mathrm{D}^2 - 4)y = \frac{1}{27}(e^{2z} - 1)$$

which is a linear equation with constant co-efficients.

Its A.E. is $\mathrm{D}^2 - 4 = 0 \quad \therefore \quad \mathrm{D} = \pm 2$

$$\text{C.F.} = c_1e^{2z} + c_2e^{-2z} = c_1(3x + 2)^2 + c_2(3x + 2)^{-2}$$

$$\text{P.I.} = \frac{1}{27}\cdot\frac{1}{\mathrm{D}^2 - 4}(e^{2z} - 1) = \frac{1}{27}\left[\frac{1}{\mathrm{D}^2 - 4}e^{2z} - \frac{1}{\mathrm{D}^2 - 4}e^{0z}\right]$$

$$= \frac{1}{27}\left[z\cdot\frac{1}{2\mathrm{D}}e^{2z} - \frac{1}{0 - 4}e^{0z}\right] = \frac{1}{27}\left[\frac{z}{2}\int e^{2z}\,dz + \frac{1}{4}\right]$$

$$= \frac{1}{27}\left[\frac{z}{4}e^{2z} + \frac{1}{4}\right] = \frac{1}{108}(ze^{2z} + 1) = \frac{1}{108}[(3x + 2)^2\log(3x + 2) + 1]$$

Hence the C.S. is $y = c_1(3x + 2)^2 + c_2(3x + 2)^{-2} + \dfrac{1}{108}[(3x + 2)^2\log(3x + 2) + 1].$

TEST YOUR KNOWLEDGE

Solve :

1. $(1+x)^2 \dfrac{d^2y}{dx^2} + (1+x)\dfrac{dy}{dx} + y = 4\cos\log(1+x)$
2. $(1+2x)^2 \dfrac{d^2y}{dx^2} - 6(1+2x)\dfrac{dy}{dx} + 16y = 8(1+2x)^2$
3. $(2x+3)^2 \dfrac{d^2y}{dx^2} - 2(2x+3)\dfrac{dy}{dx} - 12y = 6x.$

Answers

1. $y = c_1 \cos[\log(1+x)] + c_2 \sin[\log(1+x)] + 2\log(1+x) + \sin[\log(1+x)]$
2. $y = (1+2x)^2[c_1 + c_2 \log(1+2x) + \{\log(1+2x)\}^2]$
3. $y = c_1(2x+3)^{-1} + c^2(2x+3)^3 - \frac{3}{4}(2x+3) + 3.$

3.11. SIMULTANEOUS LINEAR EQUATIONS WITH CONSTANT CO-EFFICIENTS

Now we discuss differential equations in which there is one independent variable and two or more than two dependent variables. Such equations are called *simultaneous linear equations*. To solve such equations completely, we must have as many simultaneous equations as the number of dependent variables. Here, we shall consider simultaneous linear equations with constant co-efficients only.

Let x, y be the two dependent variables and t the independent variable. Consider the simultaneous equations

$$f_1(D)x + f_2(D)y = T_1 \quad \ldots(1) \quad \text{and} \quad \phi_1(D)x + \phi_2(D)y = T_2 \qquad \ldots(2)$$

where $D \equiv \dfrac{d}{dt}$ and T_1, T_2 are functions of t.

To eliminate y, operating on both sides of (1) by $\phi_2(D)$ and on both sides of (2) by $f_2(D)$ and subtracting, we get

$$[f_1(D)\,\phi_2(D) - \phi_1(D)\,f_2(D)]\,x = \phi_2(D)T_1 - f_2(D)T_2 \quad \text{or} \quad f(D)\,x = T$$

which is a linear equation in x and t and can be solved by the methods already discussed. Substituting the value of x in either (1) or (2), we get the value of y.

Note. We can also eliminate x to get a linear equation in y and t.

ILLUSTRATIVE EXAMPLES

Example 1. *Solve* $\dfrac{dx}{dt} + 4x + 3y = t$

$$\frac{dy}{dt} + 2x + 5y = e^t.$$ [CUSAT 2008, 2010]

Sol. Writing D for $\dfrac{d}{dt}$, the given equations become $(D+4)x + 3y = t$ ...(1)

and $2x + (D+5)y = e^t$ ...(2)

To eliminate y, operating on both sides of (1) by (D + 5) and on both sides of (2) by 3 and subtracting, we get

$$[(D + 4)(D + 5) - 6]x = (D + 5)t - 3e^t$$

or $$(D^2 + 9D + 14)x = 1 + 5t - 3e^t$$

Its A.E. is $$D^2 + 9D + 14 = 0$$

or $$(D + 2)(D + 7) = 0 \quad \therefore \quad D = -2, -7$$

$$\text{C.F.} = c_1e^{-2t} + c_2e^{-7t}$$

$$\text{P.I.} = \frac{1}{D^2 + 9D + 14}(1 + 5t - 3e^t)$$

$$= \frac{1}{D^2 + 9D + 14}e^{0t} + 5\frac{1}{D^2 + 9D + 14}t - 3\frac{1}{D^2 + 9D + 14}e^t$$

$$= \frac{1}{0^2 + 9(0) + 14}e^{0t} + 5 \cdot \frac{1}{14\left(1 + \frac{9D}{14} + \frac{D^2}{14}\right)}t - 3\frac{1}{1^2 + 9(1) + 14}e^t$$

$$= \frac{1}{14} + \frac{5}{14}\left[1 + \left(\frac{9D}{14} + \frac{D^2}{14}\right)\right]^{-1} t - \frac{1}{8}e^t = \frac{1}{14} + \frac{5}{14}\left[1 - \left(\frac{9D}{14} + \frac{D^2}{14}\right) + \ldots\ldots\right] t - \frac{1}{8}e^t$$

$$= \frac{1}{14} + \frac{5}{14}\left(t - \frac{9}{14}\right) - \frac{1}{8}e^t = \frac{1}{14} + \frac{5}{14}t - \frac{45}{196} - \frac{1}{8}e^t = \frac{5}{14}t - \frac{31}{196} - \frac{1}{8}e^t$$

$$\therefore \quad x = c_1e^{-2t} + c_2e^{-7t} + \frac{5}{14}t - \frac{31}{196} - \frac{1}{8}e^t$$

Now $$\frac{dx}{dt} = -2c_1e^{-2t} - 7c_2e^{-7t} + \frac{5}{14} - \frac{1}{8}e^t$$

Substituting the values of x and $\frac{dx}{dt}$ in (1), we have

$$3y = t - \frac{dx}{dt} - 4x$$

$$= t + 2c_1e^{-2t} + 7c_2e^{-7t} - \frac{5}{14} + \frac{1}{8}e^t - 4c_1e^{-2t} - 4c_2e^{-7t} - \frac{10}{7}t + \frac{31}{49} + \frac{1}{2}e^t$$

$$\therefore \quad y = \frac{1}{3}\left[-2c_1e^{-2t} + 3c_2e^{-7t} - \frac{3}{7}t + \frac{27}{98} + \frac{5}{8}e^t\right]$$

Hence $$x = c_1e^{-2t} + c_2e^{-7t} + \frac{5}{14}t - \frac{31}{196} - \frac{1}{8}e^t$$

$$y = -\frac{2}{3}c_1e^{-2t} + c_2e^{-7t} - \frac{1}{7}t + \frac{9}{98} + \frac{5}{24}e^t.$$

Example 2. *Solve* $\frac{d^2x}{dt^2} + 4x + 5y = t^2$

$$\frac{d^2y}{dt^2} + 5x + 4y = t + 1.$$

Sol. Writing D for $\frac{d}{dt}$, the given equations become $(D^2 + 4)x + 5y = t^2$ ...(1)

and $5x + (D^2 + 4)y = t + 1$ ...(2)

To eliminate y, operating on both sides of (1) by $(D^2 + 4)$ and on both sides of (2) by 5 and subtracting, we get

$$[(D^2 + 4)^2 - 25]x = (D^2 + 4)t^2 - 5(t + 1)$$

or $$(D^4 + 8D^2 - 9)x = 2 + 4t^2 - 5t - 5$$

or $$(D^4 + 8D^2 - 9)x = 4t^2 - 5t - 3$$

Its A.E. is $D^4 + 8D^2 - 9 = 0$

or $$(D^2 + 9)(D^2 - 1) = 0 \quad \therefore \quad D = \pm 1, \pm 3i$$

$$\text{C.F.} = c_1e^t + c_2e^{-t} + c_3 \cos 3t + c_4 \sin 3t$$

$$\text{P.I.} = \frac{1}{D^4 + 8D^2 - 9}(4t^2 - 5t - 3) = \frac{1}{-9\left(1 - \frac{8D^2}{9} - \frac{D^4}{9}\right)}(4t^2 - 5t - 3)$$

$$= -\frac{1}{9}\left[1 - \left(\frac{8D^2}{9} + \frac{D^4}{9}\right)\right]^{-1}(4t^2 - 5t - 3)$$

$$= -\frac{1}{9}\left[1 + \left(\frac{8D^2}{9} + \frac{D^4}{9}\right) + \ldots\ldots\right](4t^2 - 5t - 3)$$

$$= -\frac{1}{9}\left[4t^2 - 5t - 3 + \frac{8}{9}(8)\right] = -\frac{1}{9}\left(4t^2 - 5t + \frac{37}{9}\right)$$

$\therefore$ $$x = c_1e^t + c_2e^{-t} + c_3 \cos 3t + c_4 \sin 3t - \frac{4}{9}t^2 + \frac{5}{9}t - \frac{37}{81}$$

Now $$\frac{dx}{dt} = c_1e^t - c_2e^{-t} - 3c_3 \sin 3t + 3c_4 \cos 3t - \frac{8}{9}t + \frac{5}{9}$$

$$\frac{d^2x}{dt^2} = c_1e^t + c_2e^{-t} - 9c_3 \cos 3t - 9c_4 \sin 3t - \frac{8}{9}$$

Substituting the values of x and $\frac{d^2x}{dt^2}$ in (1), we have

$$5y = t^2 - 4x - \frac{d^2x}{dt^2}$$

$$= t^2 - 4c_1e^t - 4c_2e^{-t} - 4c_3 \cos 3t - 4c_4 \sin 3t$$

$$+ \frac{16}{9}t^2 - \frac{20}{9}t + \frac{148}{81} - c_1e^t - c_2e^{-t} + 9c_3 \cos 3t + 9c_4 \sin 3t + \frac{8}{9}$$

$$\therefore \qquad y = \frac{1}{5}\left[-5c_1e^t - 5c_2e^{-t} + 5c_3\cos 3t + 5c_4\sin 3t + \frac{25}{9}t^2 - \frac{20}{9}t + \frac{220}{81}\right]$$

Hence
$$x = c_1e^t + c_2e^{-t} + c_3\cos 3t + c_4\sin 3t - \frac{1}{9}\left(4t^2 - 5t + \frac{37}{9}\right)$$

$$y = -c_1e^t - c_2e^{-t} + c_3\cos 3t + c_4\sin 3t + \frac{1}{9}\left(5t^2 - 4t + \frac{44}{9}\right).$$

Example 3. *Solve* $\frac{dx}{dt} + 5x + y = e^t$ *and* $\frac{dy}{dt} - x + 3y = e^{2t}$. (CUSAT 2007)

Sol. Write D for $\frac{d}{dt}$, the given equations become

$$(D + 5)\,x + y = e^t \qquad \ldots(1)$$

and
$$-x + (D + 3)y = e^{2t} \qquad \ldots(2)$$

To eliminate y, operating on both sides of (1) by (D + 3) and substituting, we get

$$[(D + 5)(D + 3) + 1]\,x = (D + 3)e^t - e^{2t}$$

or
$$(D^2 + 8D + 16)\,x = 4e^t - e^{2t}$$

Its A.E. is $D^2 + 8D + 16 = 0 \quad \therefore \quad D = -4, -4.$

$$\text{C.F.} = (c_1 + c_2 t)\,e^{-4t}$$

$$\text{P.I.} = \frac{1}{D^2 + 8D + 16}\,4e^t - e^{2t}$$

$$= \frac{1}{(D+4)^2}(4e^t - e^{2t})$$

$$= \frac{4}{25}e^t - \frac{e^{2t}}{36}$$

$$\therefore \qquad x = (c_1 + c_2t)\,e^{-4t} + \frac{4}{25}e^t - \frac{e^{2t}}{36}$$

Now,
$$\frac{dx}{dt} = (c_1 + c_2t)(-4e^{-4t}) + c_2\,e^{-4t} + \frac{4}{25}e^t - \frac{2}{36}e^{2t}$$

Substituting the values of x and $\frac{dx}{dt}$ in (1), we have

$$y = e^t - \frac{dx}{dt} - 5x$$

$$= e^t + 4(c_1 + c_2t)e^{-4t} - c_2\,e^{-4t} - \frac{4}{25}e^t + \frac{2}{36}e^{2t} - 5(c_1 + c_2t)\,e^{-4t} - \frac{20}{25}e^t + \frac{5}{36}e^{2t}$$

$$= -(c_1 + c_2\,t)\,e^{-4t} - c_2\,e^{-4t} + \frac{1}{25}e^t + \frac{7}{36}e^{2t}$$

$$= -(c_2 + c_1 + c_2t)\,e^{-4t} + \frac{e^t}{25} + \frac{7}{36}e^{2t}.$$

Example 4. *Solve the simultaneous equations :*

$$t\frac{dx}{dt} + y = 0,\; t\frac{dy}{dt} + x = 0 \text{ given } x(1) = 1,\; y(-1) = 0.$$

Sol. The given equations are $t\dfrac{dx}{dt} + y = 0$ ...(1)

$$t\frac{dy}{dt} + x = 0 \qquad ...(2)$$

Differentiating (1) w.r.t. t, we have

$$t\frac{d^2x}{dt^2} + \frac{dx}{dt} + \frac{dy}{dt} = 0$$

Multiplying throughout by t

$$t^2\frac{d^2x}{dt^2} + t\frac{dx}{dt} + t\frac{dy}{dt} = 0$$

or $$t^2\frac{d^2x}{dt^2} + t\frac{dx}{dt} - x = 0 \qquad ...(3) \quad \text{[Using (2)]}$$

which is Cauchy's homogeneous linear equation.

Putting $t = e^u$ *i.e.*, $u = \log t$, so that $t\dfrac{d}{dt} = \dfrac{d}{du} = \text{D}$, equation (3) becomes

$$[\text{D}(\text{D} - 1) + \text{D} - 1]x = 0 \quad \text{or} \quad (\text{D}^2 - 1)\,x = 0$$

Its A.E. is $\text{D}^2 - 1 = 0$, whence $\text{D} = \pm 1$

$$\therefore \qquad x = c_1 e^u + c_2 e^{-u} = c_1 t + \frac{c_2}{t} \qquad ...(4)$$

From (1), $$y = -t\frac{dx}{dt} = -t\left(c_1 - \frac{c_2}{t^2}\right) = -c_1 t + \frac{c_2}{t} \qquad ...(5)$$

Since $x(1) = 1$, $\therefore$ from (4), we have $1 = c_1 + c_2$

Also $y(-1) = 0$, $\therefore$ from (5), we have $0 = c_1 - c_2$

Solving $$c_1 = c_2 = \frac{1}{2}$$

Hence $$x = \frac{1}{2}\left(t + \frac{1}{t}\right),\ y = \frac{1}{2}\left(-t + \frac{1}{t}\right).$$

Example 5. *Solve the following simultaneous equations :*

$$\frac{dx}{dt} = 2y, \quad \frac{dy}{dt} = 2z, \quad \frac{dz}{dt} = 2x.$$

Sol. The given equations are

$$\frac{dx}{dt} = 2y \quad ...(1) \qquad \frac{dy}{dt} = 2z \quad ...(2) \qquad \frac{dz}{dt} = 2x \quad ...(3)$$

Differentiating (1) w.r.t. t, $$\frac{d^2x}{dt^2} = 2\frac{dy}{dt} = 2(2z) \qquad \text{[Using (2)]}$$

Differentiating again w.r.t. t, $$\frac{d^3x}{dt^3} = 4\frac{dz}{dt} = 4(2x)$$

or $$(D^3 - 8)x = 0, \quad \text{where } D \equiv \frac{d}{dt}$$

Its A.E. is $$D^3 - 8 = 0 \quad \text{or} \quad (D - 2)(D^2 + 2D + 4) = 0$$

whence $D = 2, \dfrac{-2 \pm 2i\sqrt{3}}{2}$ or $D = 2, -1 \pm i\sqrt{3}$

$$\therefore \quad x = c_1 e^{2t} + c_2 e^{-t} \cos(\sqrt{3}\, t - c_3)$$

From (1), $$y = \frac{1}{2}\frac{dx}{dt}$$

$$= \frac{1}{2}[2c_1 e^{2t} - c_2 e^{-t} \cos(\sqrt{3}t - c_3) - c_2\sqrt{3}e^{-t} \sin(\sqrt{3}t - c_3)]$$

$$= c_1 e^{2t} + c_2 e^{-t}\left[\cos\frac{2\pi}{3}\cos(\sqrt{3}t - c_3) - \sin\frac{2\pi}{3}\sin(\sqrt{3}t - c_3)\right]$$

$$= c_1 e^{2t} + c_2 e^{-t} \cos\left(\sqrt{3}t - c_3 + \frac{2\pi}{3}\right)$$

From (2), $$z = \frac{1}{2}\frac{dy}{dt}$$

$$= \frac{1}{2}\left[2c_1 e^{2t} - c_2 e^{-t} \cos\left(\sqrt{3}t - c_3 + \frac{2\pi}{3}\right) - c_2\sqrt{3}\, e^{-t} \sin\left(\sqrt{3}t - c_3 + \frac{2\pi}{3}\right)\right]$$

$$= c_1 e^{2t} + c_2 e^{-t}\left[\cos\frac{2\pi}{3}\cos\left(\sqrt{3}t - c_3 + \frac{2\pi}{3}\right) - \sin\frac{2\pi}{3}\sin\left(\sqrt{3}t - c_3 + \frac{2\pi}{3}\right)\right]$$

$$= c_1 e^{2t} + c_2 e^{-t} \cos\left(\sqrt{3}t - c_3 + \frac{4\pi}{3}\right).$$

Note. $c_1 \cos \beta x + c_2 \sin \beta x$ can be replaced by $c_1 \cos(\beta x - c_2)$

or $c_1 \cos(\beta x + c_2)$ or $c_1 \sin(\beta x - c_2)$ or $c_1 \sin(\beta x + c_2)$.

TEST YOUR KNOWLEDGE

Solve the following simultaneous equations :

1. $\dfrac{dx}{dt} = 7x - y, \dfrac{dy}{dt} = 2x + 5y.$
2. $\dfrac{dx}{dt} + y = \sin t, \dfrac{dy}{dt} + x = \cot t$; given that $x = 2$ and $y = 0$ when $t = 0$.
3. $\dfrac{dx}{dt} + 5x - 2y = t, \dfrac{dy}{dt} + 2x + y = 0$; given that $x = y = 0$ when $t = 0$.
4. $(D + 1)x + (2D + 1)y = e^t, (D - 1)x + (D + 1)y = 1.$
5. $\dfrac{dx}{dt} + 2x + 3y = 0, 3x + \dfrac{dy}{dt} + 2y = 2e^{2t}.$
6. $(D - 1)x + Dy = 2t + 1, (2D + 1)x + 2Dy = t.$
7. $\dfrac{d^2x}{dt^2} - 3x - 4y = 0, \dfrac{d^2y}{dt^2} + x + y = 0.$
8. $\dfrac{d^2x}{dt^2} - \dfrac{dy}{dt} = 2x + 2t, \dfrac{dx}{dt} + 4\dfrac{dy}{dt} = 3y.$

9. $\frac{d^2y}{dt^2} + \frac{dy}{dt} - 2y = \sin t, \ \frac{dx}{dt} + x - 3y = 0.$

10. A mechanical system with two degrees of freedom satisfies the equations

$$2\frac{d^2x}{dt^2} + 3\frac{dy}{dt} = 4, 2\frac{d^2y}{dt^2} - 3\frac{dx}{dt} = 0.$$

Obtain expressions for x and y in terms of t, given $x, y, \frac{dx}{dt}, \frac{dy}{dt}$ all vanish at $t = 0$.

11. Solve the following equations : $\frac{d^2x}{dt^2} + y = \sin t, \ \frac{d^2y}{dt^2} + x = \cos t.$

Answers

1. $x = e^{6t}(c_1 \cos t + c_2 \sin t), y = e^{6t}[(c_1 - c_2)\cos t + (c_1 + c_2)\sin t]$

2. $x = e^t + e^{-t}, y = e^{-t} - e^t + \sin t$

3. $x = -\frac{1}{27}(1 + 6t)e^{-3t} + \frac{1}{27}(1 + 3t), y = -\frac{2}{27}(2 + 3t)e^{-3t} + \frac{2}{27}(2 - 3t)$

4. $x = \frac{1}{2}[e^t - 1 - ac_1e^{at} + bc_2e^{bt}], y = c_1e^{at} + c_2e^{bt} + \frac{1}{2}$, where $a = \frac{1}{2}(3 + \sqrt{17}), b = \frac{1}{2}(3 - \sqrt{17})$

5. $x = c_1e^t + c_2e^{-5t} - \frac{6}{7}e^{2t}, y = c_2e^{-5t} - c_1e^t + \frac{8}{7}e^{2t}$

6. $x = -t - \frac{2}{3}, y = \frac{1}{2}t^2 + \frac{4}{3}t + c$

7. $x = (c_1 + c_2t)e^{-t} + (c_3 + c_4t)e^t, y = -\frac{1}{2}[c_1 + c_2(1 + t)]e^{-t} + \frac{1}{2}[c_4(1 - t) - c_3]e^t$

8. $x = (c_1 + c_2t)e^t + c_3e^{-\frac{3}{2}t} - t, y = [c_2(3 - t) - c_1]e^t - \frac{1}{6}c_3e^{-\frac{3}{2}t}$

9. $x = \frac{3}{2}c_1e^t - 3c_2e^{-2t} + c_3e^{-t} + \frac{3}{10}e^t(\cos t - 2\sin t), y = c_1e^t + c_2e^{-2t} - \frac{1}{10}(\cos t + 3\sin t)$

10. $x = \frac{8}{9}\left(1 - \cos\frac{3t}{2}\right), y = \frac{4}{3}t - \frac{8}{9}\sin\frac{3t}{2}.$

11. $x = c_1e^t + c_2e^{-t} + c_3 \cos t + c_4 \sin t + \frac{t}{4}(\sin t - \cos t)$

$y = -c_1e^t - c_2e^{-t} + c_3 \cos t + c_4 \sin t + \frac{1}{4}(2 + t)(\sin t - \cos t).$

4

Applications of Linear Differential Equations

4.1. INTRODUCTION

In this chapter, we shall study the applications of linear differential equations to various physical problems. Such equations play a dominant role in unifying seemingly different theories of mechanical and electrical systems just by renaming the variables. This analogy has an important practical application. Since electrical circuits are easier to assemble, less expensive and accurate measurements can be made of electrical quantities, we construct equivalent electrical models of complicated mechanical systems to study their displacements. We shall also study the slightly less striking applications such as deflection of beams, whirling of shafts, electrical transmission lines etc.

4.2. SIMPLE HARMONIC MOTION (S.H.M.)

A particle is said to execute **simple harmonic motion** if it moves in a straight line such that its acceleration is always directed towards a fixed point in the line is proportional to the distance of the particle from the fixed point.

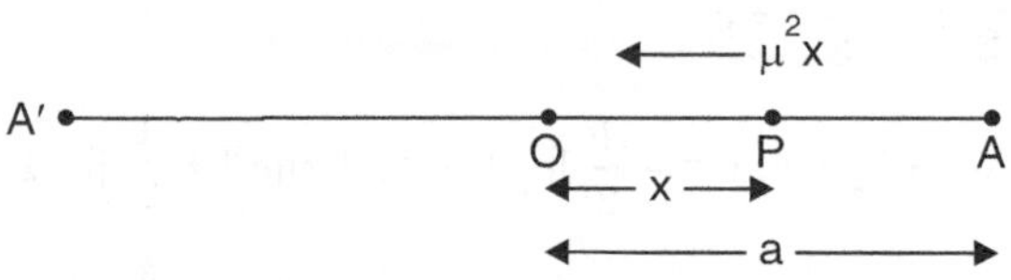

Let O be the fixed point in the line A′A. Let P be the position of the particle at any time t, where

$$OP = x.$$

Since the acceleration is always directed towards O, *i.e.*, the acceleration is in the direction opposite to that in which x increases, the equation of motion of the particle is $\dfrac{d^2x}{dt^2} = -\mu^2 x$

or $$(D^2 + \mu^2)x = 0, \text{ where } D \equiv \frac{d}{dt} \quad ...(1)$$

It is a linear differential equation with constant co-efficients.

Its A.E. is $D^2 + \mu^2 = 0$ so that $D = \pm i\mu$

$\therefore$ The solution of (1) is $x = c_1 \cos \mu t + c_2 \sin \mu t$ ...(2)

Velocity of particle at $P = \dfrac{dx}{dt} = -c_1\mu \sin \mu t + c_2\mu \cos \mu t$ ...(3)

If the particle starts from rest at A, where OA = a

then from (2), (at $t = 0, x = a$) $c_1 = a$

and from (3), $\left(\text{at } t = 0, \frac{dx}{dt} = 0\right) c_2 = 0$

$\therefore$ $$x = a \cos \mu t \quad \text{...(4)}$$

and $$\frac{dx}{dt} = -a\mu \sin \mu t \quad \text{...(5)}$$

$$= -a\mu\sqrt{1-\cos^2 \mu t} = -a\mu\sqrt{1-\frac{x^2}{a^2}} \qquad \left(\because \cos \mu t = \frac{x}{a}\right)$$

$$= -\mu\sqrt{a^2 - x^2}. \quad \text{...(6)}$$

Equation (4) gives the displacement of the particle from the fixed point O at any time t.

Equation (6) gives the velocity of the particle at any time t, when its displacement from the fixed point O is x.

Equation (6) also shows that the velocity is directed towards O and decreases as x increases.

Now, equations (4) and (5) remain unaltered when t is replaced by $t + \frac{2\pi}{\mu}$, *i.e.,* when t is increased by $\frac{2\pi}{\mu}$ showing thereby that the particle occupies the same position and has the same velocity after a time $\frac{2\pi}{\mu}$. The quantity $\frac{2\pi}{\mu}$, usually denoted by T, is called the **particle time** *i.e.,* the time of complete oscillation.

Nature of Motion. At A, $x = a$ and $v = 0$. Since acceleration is directed towards O, the particle moves towards O. The acceleration gradually decreases and vanishes at O, when the particle has acquired maximum velocity. Thus the particle moves further towards A′ under retardation and comes to rest at A′, where OA′ = OA. It moves back towards O under acceleration and acquires maximum velocity at O. Thus the particle moves further towards A under retardation and comes to rest at A. It retraces its path and goes on oscillating between A and A′. The point O is called the **centre of motion** or the **mean position.** The maximum distance a which the particle covers on either side of the mean position is called the **amplitude** of the motion.

The number of complete oscillations per second is called the **frequency** of motion. If n is the frequency,

then $$n = \frac{1}{T} = \frac{\mu}{2\pi}.$$

ILLUSTRATIVE EXAMPLES

Example 1. *A particle is executing simple harmonic motion with amplitude 20 cm and time 4 seconds. Find the time required by the particle in passing between points which are at distances 15 cm and 5 cm from the centre of force and are on the same side of it.*

Sol. Here $a = 20$ cm, T = 4 seconds

Since $$T = \frac{2\pi}{\mu} \quad \therefore \quad \mu = \frac{\pi}{2}$$

Let t_1 and t_2 seconds be the times when the particle is at distances 15 cm and 5 cm respectively from the centre of force.

Using $x = a \cos \mu t$, we have $15 = 20 \cos \frac{\pi}{2} t_1 \Rightarrow t_1 = \frac{2}{\pi} \cos^{-1} \frac{3}{4}$

and $5 = 20 \cos \frac{\pi}{2} t_2$ or $t_2 = \frac{2}{\pi} \cos^{-1} \frac{1}{4}$

Required time $= t_2 - t_1 = \frac{2}{\pi}\left(\cos^{-1} \frac{1}{4} - \cos^{-1} \frac{3}{4}\right) = 0.38$ sec.

Example 2. *A particle moving in a straight line with S.H.M. has velocities v_1 and v_2 when its distances from the centre are x_1 and x_2 respectively. Show that the period of motion is* $2\pi \sqrt{\frac{x_1^2 - x_2^2}{v_2^2 - v_1^2}}$.

Sol. The velocity v of the particle when it is at a distance x from the mean position is given by

$$v^2 = \mu^2(a^2 - x^2) \text{ where } a \text{ is the amplitude.}$$

$$\therefore \quad v_1^2 = \mu^2(a^2 - x_1^2) \quad \ldots(1) \quad \text{and} \quad v_1^2 = \mu^2(a^2 - x_2^2) \quad \ldots(2)$$

Subtracting (1) from (2), we get $v_2^2 - v_1^2 = \mu^2(x_1^2 - x_2^2)$ or $\mu^2 = \frac{v_2^2 - v_1^2}{x_1^2 - x_2^2}$

Periodic time $= \frac{2\pi}{\mu} = 2\pi \sqrt{\frac{x_1^2 - x_2^2}{v_2^2 - v_1^2}}$.

Example 3. *At the end of three successive seconds, the distances of a point moving with S.H.M. from its mean position are x_1, x_2, x_3 respectively. Show that the time of a complete oscillation is*

$$2\pi / \cos^{-1}\left(\frac{x_1 + x_3}{2x_2}\right).$$

Sol. Let the moving point be at distances x_1, x_2, x_3 from the mean position at the end of t, $t + 1$, $t + 2$ seconds respectively.

Using $x = a \cos \mu t$, we have $x_1 = a \cos \mu t \quad \ldots(1)$

$$x_2 = a \cos \mu(t + 1) \quad \ldots(2)$$

$$x_3 = a \cos \mu(t + 2) \quad \ldots(3)$$

Adding (1) and (3), we get

$$x_1 + x_3 = a\,[\cos \mu(t + 2) + \cos \mu t]$$

$$= a \,.\, 2 \cos \frac{\mu(t + 2) + \mu t}{2} \cos \frac{\mu(t + 2) - \mu t}{2}$$

$$= 2a \cos \mu(t + 1) \cos \mu = 2x_2 \cos \mu \quad \text{[Using (2)]}$$

$$\Rightarrow \qquad \mu = \cos^{-1}\left(\frac{x_1 + x_3}{2x_2}\right)$$

Hence the time of a complete oscillation $= \dfrac{2\pi}{\mu} = 2\pi / \cos^{-1}\left(\dfrac{x_1 + x_3}{2x_2}\right)$.

Example 4. *In the case of a stretched elastic horizontal string which has one end fixed and a particle of mass m attached to the other, find the equation of motion of the particle given that l is the natural length of the string and e is its elongation due to a weight mg. Also find the displacements of the particle when initially $s = s_0$, $v = 0$.*

Sol. Let OA = l be the elastic horizontal string with the end O fixed and a particle of mass m attached at A.

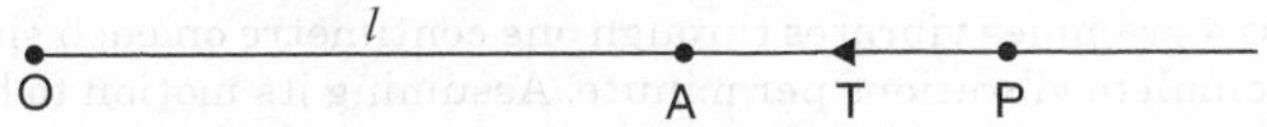

Let P be the position of the particle at any time t, where OP = s, so that the elongation AP = $s - l$.

Now, for the elongation e, tension $= mg$

$\therefore$ For the elongation $(s - l)$, tension $= \dfrac{mg\,(s - l)}{e}$

Since tension is the only horizontal force acting on the particle, its equation of motion is

$$m \frac{d^2 s}{dt^2} = -\text{T}$$

or
$$m \frac{d^2 s}{dt^2} = -\frac{mg\,(s - l)}{e} \quad \text{or} \quad \frac{d^2 s}{dt^2} = -\frac{g\,(s - l)}{e} \qquad \ldots(1)$$

From (1), we have $\dfrac{d^2 s}{dt^2} + \dfrac{g}{e} s = \dfrac{gl}{e}$ or $\left(\text{D}^2 + \dfrac{g}{e}\right) s = \dfrac{gl}{e}$ where $\text{D} \equiv \dfrac{d}{dt}$

Its A.E. is $\text{D}^2 + \dfrac{g}{e} = 0$, whence $\text{D} = \pm i\sqrt{\dfrac{g}{e}}$

$$\therefore \qquad \text{C.F.} = c_1 \cos\left(\sqrt{\frac{g}{e}}\right) t + c_2 \sin\left(\sqrt{\frac{g}{e}}\right) t$$

and
$$\text{P.I.} = \frac{1}{\text{D}^2 + \frac{g}{e}} \cdot \frac{gl}{e} = \frac{gl}{e} \cdot \frac{1}{\text{D}^2 + \frac{g}{e}} e^{0t} = \frac{gl}{e} \cdot \frac{1}{0 + \frac{g}{e}} e^{0t} = l$$

$\therefore$ The complete solution of (1) is $s = c_1 \cos\left(\sqrt{\dfrac{g}{e}}\right) t + c_2 \sin\left(\sqrt{\dfrac{g}{e}}\right) t + l$ $\qquad \ldots(2)$

When $t = 0$, $s = s_0$, so that from (2),

$$s_0 = c_1 + 0 + l \quad \text{or} \quad c_1 = s_0 - l$$

Also, from (2) $\quad \dfrac{ds}{dt} = -c_1\sqrt{\dfrac{g}{e}} \sin\left(\sqrt{\dfrac{g}{e}}\right) t + c_2\sqrt{\dfrac{g}{e}} \cos\left(\sqrt{\dfrac{g}{e}}\right) t$

When $t = 0, \frac{ds}{dt} = v = 0 \quad \therefore \quad c_2 = 0$

$\therefore$ From (2), we get $s = (s_0 - l) \cos\left(\sqrt{\frac{g}{e}}\right) t + l.$

TEST YOUR KNOWLEDGE

1. A particle is executing S.H.M. with amplitude 5 metres and time 4 seconds. Find the time required by the particle in passing between points which are at distances 4 and 2 metres from the centre of force and are on the same side of it.

2. A particle of mass 4 grammes vibrates through one centimetre on each side of the middle point of its making 330 complete vibrations per minute. Assuming its motion to be S.H.M. show that the maximum force upon the particle is $484\pi^2$ dynes.

3. A point executing S.H.M. passes through two points A and B, 2 metres apart, with the same velocity having occupied 4 seconds is passing from A to B. After another 4 seconds, it returns to B. Find the period and amplitude.

4. A particle of mass of 4 gm executing S.H.M. has velocities 8 cm/sec and 6 cm/sec respectively when it is at distances 3 cm and 4 cm from the centre of its path. Find its period and amplitude. Find also the force acting on the particle when it is a distance 1 cm from the centre.

5. At the end of three successive seconds, the distances of a point moving with S.H.M. from its mean position, measured in the same direction are 1, 5, 5. Show that the period of complete oscillation is $\frac{2\pi}{\theta}$, where $\cos\theta = \frac{3}{5}$.

6. A particle is performing S.H.M. of period T about a centre O and it passes through the position P(OP = b) with velocity v in the direction OP. Prove that the time which elapses before its return to P is $\frac{T}{\pi} \tan^{-1}\left(\frac{vT}{2\pi b}\right)$.

7. A particle moves with S.H.M. in a straight line. In the first second starting from rest, it travels a distance 'a' and in the next second it travels a distance 'b' in the same direction. Prove that the amplitude of motion is $\frac{2a^2}{3a - b}$.

8. An elastic string of natural length $2a$ and modulus λ is stretched between two points A and B distant $4a$ apart on a smooth horizontal table. A particle of mass m is attached to the middle of the string. Show that it can vibrate in line AB with period $2\pi/\omega$, where $\omega^2 = \frac{2\lambda}{am}$.

9. An elastic string of natural length $2l$ can just support a certain weight when it is stretched till its whole length is $3l$. One end of the string is now attached to a point on a smooth horizontal table and the same weight is attached to the other end. Prove that if the weight is pulled to any distance and then let go, the string will become slack again after a time $\frac{\pi}{2}\sqrt{\frac{l}{g}}$.

Answers

1. 0.33 second 3. 16 sec, $\sqrt{2}$ m 4. π sec, 5 cm, 16 dynes.

4.3. MECHANICAL AND ELECTRICAL OSCILLATORY CIRCUITS

(a) **Free Oscillations**

(i) *Mechanical System.* Consider a spring OA suspended vertically from a fixed support at O. Let a body of mass m be suspended from the end A, the mass of the body being so large in comparison with the mass of the spring, that the latter may be neglected. Let $e(=\text{AB})$ be the elongation produced by the mass m hanging in equilibrium, then B is called the position of **static equilibrium** and e is called the **static extension.**

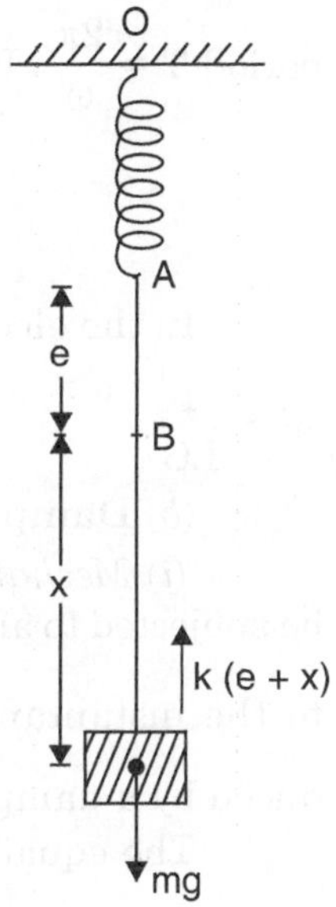

Let k be the stiffness of the spring, *i.e.*, the restoring force per unit stretch of the spring due to elasticity.

For equilibrium at B, $mg = \text{T} = ke$ ...(1)

Let the mass be displaced through a further distance x from the equilibrium position. The acceleration of the mass m at this position is $\dfrac{d^2x}{dt^2}$ and the forces acting upon it are its weight mg downwards and the restoring force $k(e + x)$ upwards.

$\therefore$ The equation of motion of mass m is

$$m\frac{d^2x}{dt^2} = mg - k(e + x) = -kx \qquad \text{[Using (1)]}$$

or

$$\frac{d^2x}{dt^2} + \frac{k}{m}x = 0$$

Writing $\omega^2 = \dfrac{k}{m}$, it becomes $\dfrac{d^2x}{dt^2} + \omega^2x = 0$ ...(2)

(ii) *Equivalent Electrical Circuit (L-C circuit).* Consider an electrical circuit containing an inductance L and capacitance C. Let q be the electrical charge on the condenser plate and i be the current in the circuit at any time t. The voltage drops across L and C being $\text{L}\dfrac{di}{dt}$ and $\dfrac{q}{\text{C}}$ respectively and since there is no applied E.M.F. in the circuit, we have by Kirchhoff's Law,

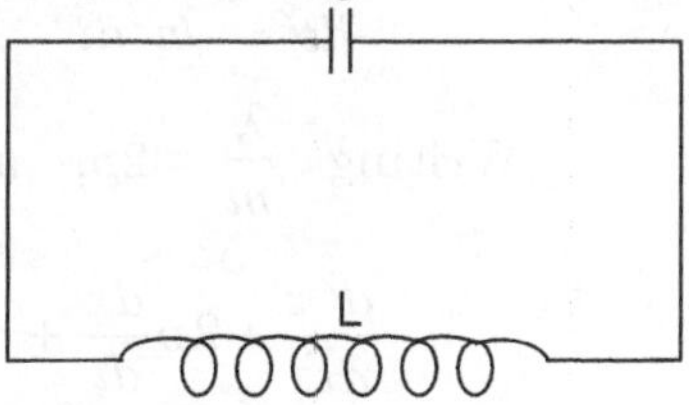

$$\text{L}\frac{di}{dt} + \frac{q}{\text{C}} = 0 \qquad ...(3)$$

Since $i = \dfrac{dq}{dt}$, equation (3) becomes $\text{L}\dfrac{d^2q}{dt^2} + \dfrac{q}{\text{C}} = 0$ or $\dfrac{d^2q}{dt^2} + \dfrac{q}{\text{LC}} = 0$

Writing $\omega^2 = \dfrac{1}{\text{LC}}$, it becomes $\dfrac{d^2q}{dt^2} + \omega^2q = 0$ ...(4)

which is of the same form as equation (2).

Solution and its Physical Interpretation

The solution of equation (2) is $x = c_1 \cos \omega t + c_2 \sin \omega t$ which can also be written as

$$x = A \cos (\omega t + B) \qquad \ldots(5)$$

where c_1, c_2 or A, B are arbitrary constants to be determined from initial conditions of the problem, electrical or mechanical.

Equation (5) represents oscillatory variations of the variable x with amplitude A and period $T = \frac{2\pi}{\omega}$. In mechanical system, it represents the S.H.M. of the mass m with period

$$2\pi \sqrt{\frac{m}{k}} = 2\pi \sqrt{\frac{e}{g}} \qquad \left(\text{since } \omega^2 = \frac{k}{m} \text{ and } mg = ke\right)$$

In the electrical circuit, it represents and oscillatory current with period $2\pi\sqrt{LC}$, since $\omega^2 = \frac{1}{LC}$.

(*b*) **Damped Free Oscillations**

(*i*) *Mechanical System.* If the motion of the mass m be subjected to an additional force of resistance, proportional to the instantaneous velocity of the mass, say $\lambda \frac{dx}{dt}$, produced by a damper, the oscillations are said to be *damped*.

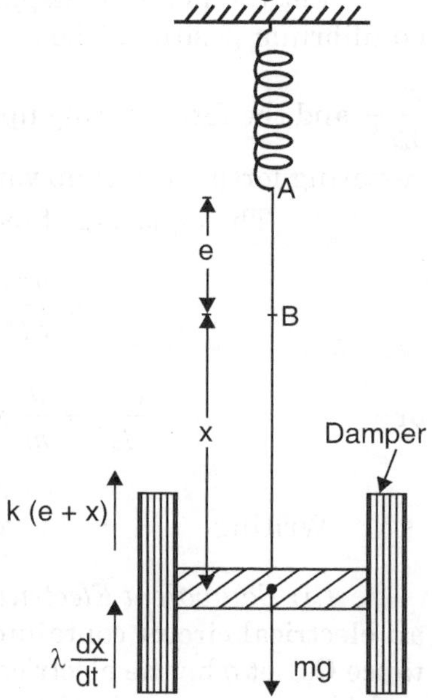

The equation of motion of the mass m is

$$m \frac{d^2x}{dt^2} = mg - k(e + x) - \lambda \frac{dx}{dt}$$

$$= -kx - \lambda \frac{dx}{dt} \qquad (\because \quad ke = mg)$$

or

$$\frac{d^2x}{dt^2} + \frac{\lambda}{m}\frac{dx}{dt} + \frac{k}{m}x = 0$$

Writing $\frac{\lambda}{m} = 2p$, and $\frac{k}{m} = \omega^2$, it becomes

$$\frac{d^2x}{dt^2} + 2p\frac{dx}{dt} + \omega^2 x = 0 \qquad \ldots(1)$$

(*ii*) *Equivalent Electrical Circuit (L-C-R circuit).* Consider the discharge of a condenser C through an inductance L and the resistance R. Let q be the charge and i the current in the circuit at any time t. The voltage drops across L, C and R are respectively.

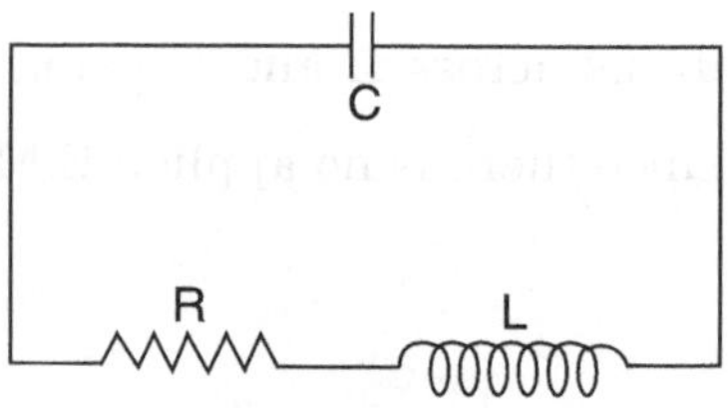

$$L \frac{d^2q}{dt^2}, \frac{q}{C} \quad \text{and} \quad R \frac{dq}{dt}$$

∴ By Kirchhoff's Law, we have

$$L \frac{d^2q}{dt^2} + R\frac{dq}{dt} + \frac{q}{C} = 0 \quad \text{or} \quad \frac{d^2q}{dt^2} + \frac{R}{L}\frac{dq}{dt} + \frac{1}{LC}q = 0$$

Writing $\frac{R}{L} = 2p$ and $\frac{1}{LC} = \omega^2$, it becomes $\frac{d^2q}{dt^2} + 2p\frac{dq}{dt} + \omega^2 q = 0$ ...(2)

which is of the same form as equation (1).

Solution and its Physical Interpretations

Equation (1) is a linear differential equation with constant co-efficients.

Its symbolic form is $(D^2 + 2pD + \omega^2)x = 0$ where $D \equiv \frac{d}{dt}$.

Its A.E. is $D^2 + 2pD + \omega^2 = 0$

whence $D = \frac{-2p \pm \sqrt{4p^2 - 4\omega^2}}{2} = -p \pm \sqrt{p^2 - \omega^2}$.

Thus the general solution of equation (1) depends upon the relative values of p and ω.

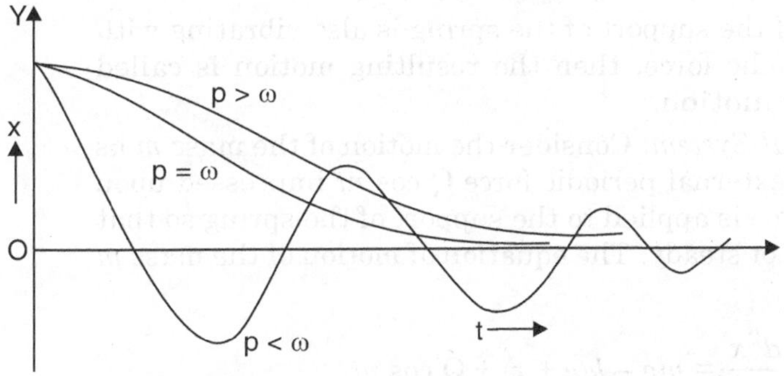

Case I. When $p > \omega$, the roots of the auxiliary equation are real and distinct.

The solution of equation (1) is $x = c_1 e^{\left(-p + \sqrt{p^2 - \omega^2}\right)t} + c_2 e^{\left(-p - \sqrt{p^2 - \omega^2}\right)t}$

or $x = e^{-pt}\left(c_1 e^{\sqrt{p^2 - \omega^2}\,t} + c_2 e^{-\sqrt{p^2 - \omega^2}\,t}\right)$

where c_1, c_2 are arbitrary constants to be determined from the initial conditions of the given problem.

The solution shows that x is always positive and hence the motion is non-oscillatory. Further, as $t \to \infty$, $x \to 0$. The restoring force in this case is so great that the non-oscillatory motion is called **over-damped** or **dead-beat motion.**

Case II. When $p = \omega$, the roots of the auxiliary equation are equal, each being equal to $-p$.

The solution of equation (1) is $x = (c_1 + c_2 t)\, e^{-pt}$

Here also x is always positive and the motion is non-oscillatory.

Further, as $t \to \infty$, $x \to 0$. The motion of the mass m in this case is called **critically damped motion,** because it separates the non-oscillatory motion of case I from the oscillatory motion of case III.

Case III. When $p < \omega$, the roots of the auxiliary equation are imaginary *i.e.,*

$$D = -p \pm i\sqrt{\omega^2 - p^2}$$

The solution of equation (1) is $x = e^{-pt}(c_1 \cos \sqrt{\omega^2 - p^2}\ t + c_2 \sin \sqrt{\omega^2 - p^2}\ t)$

or $$x = A\ e^{-pt} \cos (\sqrt{\omega^2 - p^2}\ t + B)$$

where c_1, c_2 and A, B are constants.

The variation of x is now oscillatory due to the presence of the trigonometric factor.

The periodic time of the oscillation is $T = \dfrac{2\pi}{\sqrt{\omega^2 - p^2}}$ which is greater than the periodic time in case of free oscillations (without damping) which is $\dfrac{2\pi}{\omega}$. Thus the effect of damping is to increase the period of oscillation and the motion ultimately dies away. Such a motion is called **damped oscillatory motion.**

(*c*) **Forced Oscillations** (*Without damping*)

If the point of the support of the spring is also vibrating with some external periodic force, then the resulting motion is called **forced oscillatory motion.**

(*i*) *Mechanical System.* Consider the motion of the mass m as in case (*a*) with an external periodic force Q cos nt impressed upon the system. This force is applied to the support of the spring so that now the support is not steady. The equation of motion of the mass m is

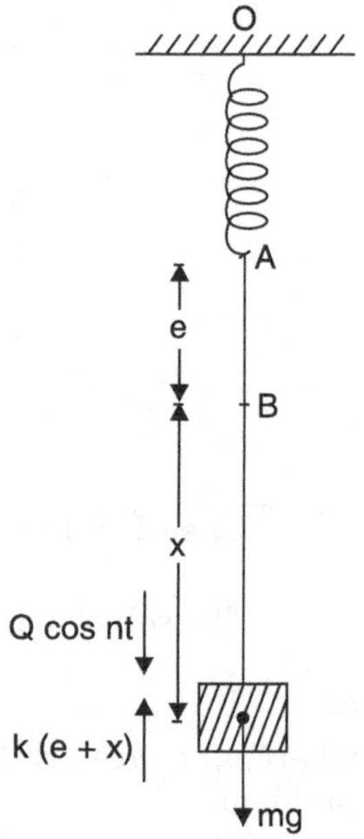

$$m \frac{d^2x}{dt^2} = mg - k(e + x) + Q \cos nt$$

Since $mg = ke$, it becomes $m \dfrac{d^2x}{dt^2} = -kx + Q \cos nt$

or $$\frac{d^2x}{dt^2} + \frac{k}{m} x = \frac{Q}{m} \cos nt$$

Writing $\omega^2 = \dfrac{k}{m}$ and $E = \dfrac{Q}{m}$, it becomes

$$\frac{d^2x}{dt^2} + \omega^2 x = E \cos nt \qquad \text{...(1)}$$

(*ii*) *Equivalent Electrical Circuit (L-C circuit with e.m.f. Q cos nt).* As in case (*a*), the equation for an L-C circuit now becomes

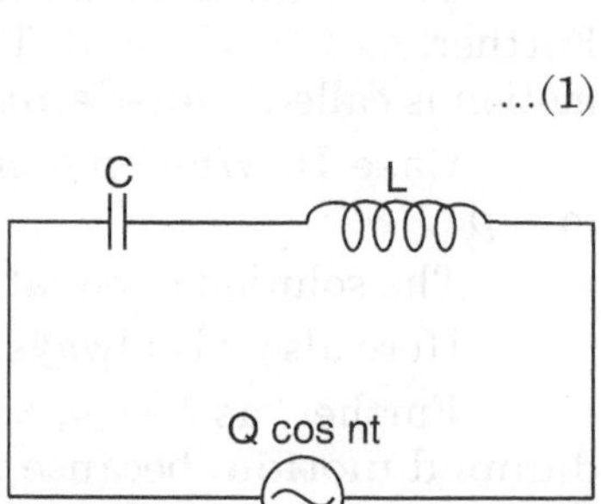

$$L \frac{di}{dt} + \frac{q}{C} = Q \cos nt$$

Since $i = \dfrac{dq}{dt}$, it becomes $L \dfrac{d^2q}{dt^2} + \dfrac{q}{C} = Q \cos nt$

or $$\frac{d^2q}{dt^2} + \frac{1}{LC} q = \frac{Q}{L} \cos nt$$

Writing $\omega^2 = \frac{1}{LC}$ and $E = \frac{Q}{L}$, it becomes

$$\frac{d^2q}{dt^2} + \omega^2 q = E \cos nt \qquad ...(2)$$

which is of the same form as equation (1).

Solution and its Physical Interpretation

Equation (1) is a linear differential equation with constant co-efficients.

Its symbolic form is $(D^2 + \omega^2)q = E \cos nt$, where $D \equiv \frac{d}{dt}$.

Its A.E. is $D^2 + \omega^2 = 0$ which gives $D = \pm i\omega$

Its $\quad$ C.F. $= c_1 \cos \omega t + c_2 \sin \omega t$

and $\quad$ P.I. $= E \frac{1}{D^2 + \omega^2} \cos nt$

Evaluation of P.I. depends upon the relative values of ω and n.

Thus, we have to consider two cases : $\omega \neq n$ and $\omega = n$.

Case I. When $\omega \neq n$

$$\text{P.I.} = \frac{E}{\omega^2 - n^2} \cos nt$$

$\therefore$ The complete solution of (1) is $x = c_1 \cos \omega t + c_2 \sin \omega t + \frac{E}{\omega^2 - n^2} \cos nt$

which may be written as $x = A \cos (\omega t + B) + \frac{E}{\omega^2 - n^2} \cos nt$

This shows that the variation of x consists of two oscillatory motions : the one due to the C.F. gives free oscillations of period $\frac{2\pi}{\omega}$ and the second due to the P.I. gives forced oscillations of period $\frac{2\pi}{n}$.

Also, if the frequency of free oscillations is very high, *i.e.*, ω is large, then the amplitude $\frac{E}{\omega^2 - n^2}$ of the forced oscillations is small.

Case II. When $\omega = n$

$$\text{P.I.} = Et \frac{1}{2D} \cos \omega t = \frac{Et}{2} \int \cos \omega t \, dt = \frac{Et}{2\omega} \sin \omega t$$

$\therefore$ The complete solution of (1) is $x = c_1 \cos \omega t + c_2 \sin \omega t + \frac{Et}{2\omega} \sin \omega t$

$$= c_1 \cos \omega t + \left(c_2 + \frac{Et}{2\omega} \right) \sin \omega t \qquad ...(3)$$

Let $\quad c_1 = r \sin \phi$ and $c_2 + \frac{Et}{2\omega} = r \cos \phi$

then $$r = \sqrt{c_1^2 + \left(c_2 + \frac{Et}{2\omega}\right)^2} \quad \text{and} \quad \phi = \tan^{-1}\left(\frac{c_1}{c_2 + \dfrac{Et}{2\omega}}\right)$$

Equation (3) reduces to $x = r \sin(\omega t + \phi)$

which shows that motion is oscillatory with period $\frac{2\pi}{\omega}$ and amplitude $= \sqrt{c_1^2 + \left(c_2 + \frac{Et}{2\omega}\right)^2}$.

Clearly, amplitude increases with time, so that after a long time the amplitude becomes abnormally large. This phenomenon in which the impressed frequency due to external periodic force becomes equal to the natural frequency of the system is known as resonance. A practical instance of the case of resonance is the soldiers breaking step while marching over a bridge, for the simple reason that, if by chance the steps be in rhythm with the natural frequency of the bridge, it may collapse due to resonance.

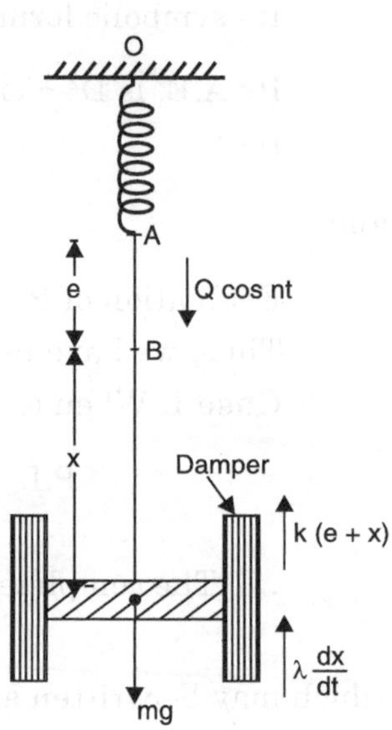

(*d*) **Forced Oscillations** (*with damping*)

(*i*) *Mechanical System*. In the above case (*c*), if in addition, there is a damping force, which is proportional to the instantaneous velocity of the mass, say $\lambda \frac{dx}{dt}$, then the equation of motion of mass m is

$$m\frac{d^2x}{dt^2} = mg - k(e + x) - \lambda\frac{dx}{dt} + Q\cos nt$$

Since $mg = ke$, it becomes

$$m\frac{d^2x}{dt^2} = -kx - \lambda\frac{dx}{dt} + Q\cos nt$$

or $$\frac{d^2x}{dt^2} + \frac{\lambda}{m}\frac{dx}{dt} + \frac{k}{m}x = \frac{Q}{m}\cos nt$$

Writing $\frac{\lambda}{m} = 2p$, $\frac{k}{m} = \omega^2$ and $\frac{Q}{m} = E$, it becomes

$$\frac{d^2x}{dt^2} + 2p\frac{dx}{dt} + \omega^2 x = E\cos nt \quad \text{...(1)}$$

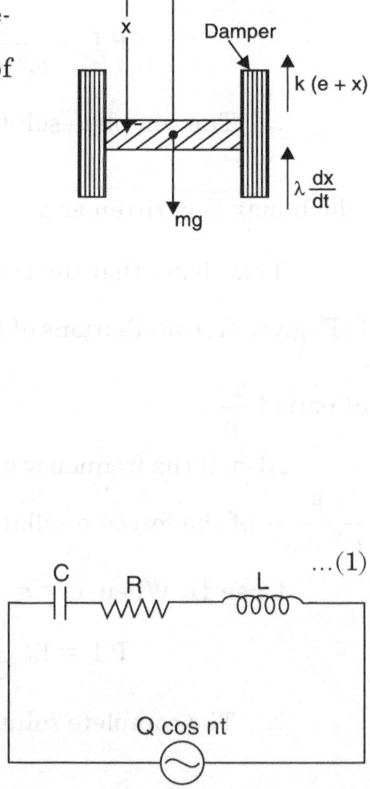

(*ii*) *Equivalent Electrical System (L-C-R circuit with e.m.f. Q cos nt).*

The differential equation of the L-C-R circuit which contains an alternating e.m.f. Q cos nt can be written as

$$L + \frac{d^2q}{dt^2} + R\frac{dq}{dt} + \frac{q}{C} = Q\cos nt$$

or $$\frac{d^2q}{dt^2} + \frac{R}{L}\frac{dq}{dt} + \frac{1}{LC}q = \frac{Q}{L}\cos nt$$

Writing $\frac{R}{L} = 2p$, $\frac{1}{LC} = \omega^2$ and $\frac{Q}{L} = E$, it becomes

$$\frac{d^2q}{dt^2} + 2p\frac{dq}{dt} + \omega^2 q = E\cos nt \qquad \ldots(2)$$

which is of the same form as equation (1).

Solution and its Physical Interpretation

Equation (1) is a linear differential equation with constant co-efficients.

Its symbolic form is $(D^2 + 2pD + \omega^2)x = E\cos nt$

Its A.E. is $\qquad D^2 + 2pD + \omega^2 = 0$

which gives
$$D = \frac{-2p \pm \sqrt{4p^2 - 4\omega^2}}{2} = -p \pm \sqrt{p^2 - \omega^2}$$

$$\therefore \quad \text{C.F.} = e^{-pt}\left(c_1 e^{\sqrt{p^2-\omega^2}\,t} + c_2 e^{-\sqrt{p^2-\omega^2}\,t}\right)$$

$$\text{P.I.} = E\,.\,\frac{1}{D^2 + 2pD + \omega^2}\cos nt = E\,.\,\frac{1}{-n^2 + 2pD + \omega^2}\cos nt$$

$$= E\,.\,\frac{1}{(\omega^2 - n^2) + 2pD}\cos nt = E\,.\,\frac{(\omega^2 - n^2) - 2pD}{(\omega^2 - n^2)^2 - 4p^2D^2}\cos nt$$

$$= E\,.\,\frac{(\omega^2 - n^2)\cos nt + 2pn\sin nt}{(\omega^2 - n^2)^2 + 4p^2n^2}$$

Now, let $\omega^2 - n^2 = r\cos\phi$ and $2pn = r\sin\phi$ so that $r = \sqrt{(\omega^2 - n^2)^2 + 4p^2n^2}$

and
$$\phi = \tan^{-1}\left(\frac{2pn}{\omega^2 - n^2}\right)$$

$$\therefore \quad \text{P.I.} = E\,.\,\frac{r\cos(nt - \phi)}{r^2} = \frac{E}{r}\cos(nt - \phi)$$

$\therefore$ The complete solution of equation (1) is

$$x = e^{-pt}\left(c_1 e^{\sqrt{p^2-\omega^2}\,t} + c_2 e^{-\sqrt{p^2-\omega^2}\,t}\right) + \frac{E\cos\left\{nt - \tan^{-1}\left(\frac{2pn}{\omega^2 - n^2}\right)\right\}}{\sqrt{\left(\omega^2 - n^2\right)^2 + 4p^2n^2}}$$

The C.F. represents free oscillations of the system which die out as $t \to \infty$ due to the presence of the factor e^{-pt}.

The P.I. represents the forced oscillations of the system having a constant amplitude

$$= \frac{E}{\sqrt{(\omega^2 - n^2)^2 + 4p^2n^2}}$$

and the period $= \frac{2\pi}{n}$ which is the same as that of the impressed force.

Thus as t increases, the free oscillations (given by the C.F.) die out while the forced oscillations (given by the P.I.) persist giving the steady state of motion.

ILLUSTRATIVE EXAMPLES

Example 1. *A body weighing 4.9 kg is hung from a spring. A pull of 10 kg will stretch the spring to 5 cm. The body is pulled down 6 cm below the static equilibrium position and then released. Find the displacement of the body from its equilibrium position at time t seconds, the maximum velocity and the period of oscillation.*

Sol. Let O be the fixed end and A the free end of the spring. Let k be the stiffness of the spring (or the spring constant).

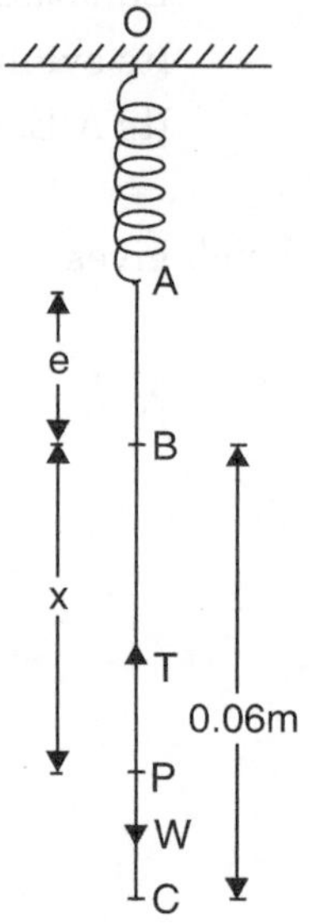

Since a pull of 10 kg wt. stretches the spring by 0.05 metres

$$\therefore \quad 10 = k \times 0.05 \quad \Rightarrow \quad k = 200 \text{ kg/m}$$

Let $e(= \text{AB})$ be the elongation produced by the body weighing 4.9 kg hanging in equilibrium, then $\quad 4.9 = ke = 200e$

$$\therefore \quad e = \frac{4.9}{200} = 0.0245 \text{ m}$$

Now the body is pulled down to C, where BC = 0.06 m. After t sec. of its release from C, let the body be at P, where BP = x metres. The forces acting on the body are

(*i*) its weight W = 4.9 kg downwards

(*ii*) the restoring force $\text{T} = k(e + x) = 200(0.0245 + x)$ kg upwards.

The equation of motion of the body is $m\frac{d^2x}{dt^2} = \text{W} - \text{T}$

i.e.,
$$\frac{4.9}{g}\frac{d^2x}{dt^2} = 4.9 - 200(0.0245 + x)$$

or
$$\frac{4.9}{9.8}\frac{d^2x}{dt^2} = -200 \qquad \text{(Taking } g = 9.8 \text{ m/sec}^2\text{)}$$

or
$$\frac{d^2x}{dt^2} + 400x = 0$$

Its A.E. is $\quad D^2 + 400 = 0, \quad$ where $D = \pm 20i$

$\therefore$ Its solution is $\quad x = c_1 \cos 20t + c_2 \sin 20t \qquad ...(1)$

Initially, when $\quad t = 0, x = 0.06$

$\therefore$ From (1), $\quad 0.06 = c_1$

Also,
$$\frac{dx}{dt} = -20c_1 \sin 20t + 20c_2 \cos 20t$$

Initially, when $\quad t = 0, \frac{dx}{dt} = 0$

$\therefore$ From (1), $\quad 0 = c_2$

Substituting the values of c_1 and c_2 in equation (1), we have $x = 0.06 \cos 20t$ which gives the displacement of the body from its equilibrium position at time t.

Maximum velocity $\quad = \omega$ (amplitude) $= 20 \times 0.06 = 1.2$ m/sec.

Period of oscillation $\quad = \frac{2\pi}{\omega} = \frac{2\pi}{\omega} = \frac{\pi}{10} = 0.314$ sec.

Example 2. *Show that the frequency of free vibrations in a closed electrical circuit with inductance L and capacity C in series is* $\dfrac{30}{\pi\sqrt{LC}}$ *per minute.*

Sol. Let i be the current and q the charge in the condenser plate at any time t. The voltage drops across L and C are $L\dfrac{di}{dt} = L\dfrac{d^2q}{dt^2}$ and $\dfrac{q}{C}$ respectively. Since there is no applied E.M.F. in the circuit, we have by Kirchhoff's Law.

$$L\frac{d^2q}{dt^2} + \frac{q}{C} = 0 \quad \text{or} \quad \frac{d^2q}{dt^2} = -\frac{1}{LC}q$$

Writing $\dfrac{1}{LC} = \omega^2$, it becomes $\dfrac{d^2q}{dt^2} = -\omega^2 q$

It represents oscillatory current with period $\dfrac{2\pi}{\omega} = 2\pi\sqrt{LC}$.

$$\therefore \quad \text{Frequency} = \frac{1}{T} \text{ per second} = \frac{60}{2\pi\sqrt{LC}} \text{ per minute} = \frac{30}{\pi\sqrt{LC}} \text{ per minute.}$$

Example 3. *A mass of 200 gm is tied at the end of a spring which extends to 4 cm under a force 196,000 dynes. The spring is pulled 5 cm and released. Find the displacement, t seconds after release, if there be a damping force of 2000 dynes per cm per second.*

What should be the damping force for the dead beat motion ?

Sol. Let O be the fixed end and A the free end of the spring.

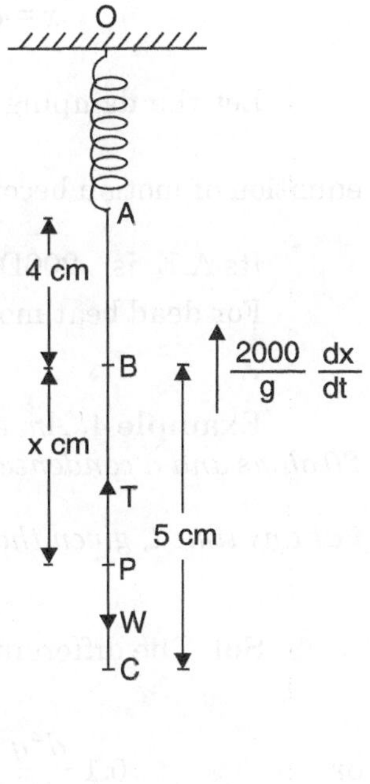

Since a force $\dfrac{196000}{980} = 200$ gm wt stretches the spring by 4 cm.

$$\therefore \quad 200 = k \times 4 \quad \text{or} \quad k = 50 \text{ gm/cm}$$

where k is the restoring force.

Let B be the equilibrium position when a mass of 200 gm is attached to A, then

$$200 = k \times AB \quad \text{or} \quad 50\,AB = 200$$

$$\therefore \quad AB = 4 \text{ cm.}$$

Now the 200 gm weight is pulled down to C, where BC = 5 cm. If t seconds after its release from C, the weight be at P, where BP = x cm, then the forces acting on it are :

(*i*) its weight W = 200 gm acting downwards.

(*ii*) the tension $T = k \times AP = 50(4 + x)$ gm upwards

(*iii*) the damping force $\dfrac{2000}{g}\dfrac{dx}{dt}$ gm upwards

$$\therefore \quad \text{The equation of motion is } \frac{200}{g}\frac{d^2x}{dt^2} = W - T - \frac{2000}{g}\frac{dx}{dt}$$

$$= 200 - 50(4 + x) - \frac{2000}{g}\frac{dx}{dt} = -50x - \frac{2000}{g}\frac{dx}{dt}$$

or $$\frac{d^2x}{dt^2} + 10\frac{dx}{dt} + \frac{g}{4}x = 0$$

or $$\frac{d^2x}{dt^2} + 10\frac{dx}{dt} + 245x = 0 \qquad (\because g = 980 \text{ cm/sec}^2) \qquad ...(1)$$

which is a linear differential equation with constant co-efficients.

Its A.E. is $D^2 + 10D + 245 = 0$ so that $D = \frac{-10 \pm \sqrt{100 - 980}}{2} = -5 \pm i\sqrt{220}$

$\therefore$ The solution of (1) is $x = e^{-5t}(c_1 \cos \sqrt{220}\, t + c_2 \sin \sqrt{220}\, t)$ ...(2)

Initially, when $t = 0, x = 5 \quad \therefore \quad c_1 = 5$

Also $$\frac{dx}{dt} = -5e^{-5t}(c_1 \cos \sqrt{200}\, t + c_2 \sin \sqrt{220}\, t)$$
$$+ e^{-5t}(-\sqrt{220}\, c_1 \sin \sqrt{220}\, t + \sqrt{220}\, c_2 \cos \sqrt{220}\, t)$$

Since $\frac{dx}{dt} = 0$ when $t = 0$, we have $0 = -5(c_1) + \sqrt{220}\, c_2$

or $$c_2 = \frac{5c_1}{\sqrt{220}} = \frac{25}{\sqrt{220}}$$

$\therefore$ From (2), the displacement, t seconds after release, is given by

$$x = e^{-5t}\left(5 \cos \sqrt{220}\, t + \frac{25}{\sqrt{220}} \sin \sqrt{220}\, t\right)$$

Let the damping force for the dead beat motion be p dynes per cm per second, then

equation of motion becomes $$\frac{200}{g}\frac{d^2x}{dt^2} = -50x - \frac{p}{g}\frac{dx}{dt} \quad \text{or} \quad 200\frac{d^2x}{dt^2} + p\frac{dx}{dt} + 49000x = 0$$

Its A.E. is $200D^2 + pD + 49000 = 0$

For dead beat motion, roots of A.E. must be equal

$$\therefore \qquad p^2 = 4 \times 200 \times 49000 \;\Rightarrow\; p = 2800\sqrt{5} = 6261.$$

Example 4. *An electric circuit consists of an inductance of 0.1 henry, a resistance of 20 ohms and a condenser of capacitance 25 micro-farads. Find the charge q and the current i at any time t, given that at t = 0, q = 0.05 coulomb,* $i = \frac{dq}{dt} = 0$ *when t = 0.*

Sol. The differential equation for the circuit can be written as $L\frac{d^2q}{dt^2} + R\frac{dq}{dt} + \frac{q}{C} = 0$

or $$0.1\frac{d^2q}{dt^2} + 20\frac{dq}{dt} + \frac{q}{25 \times 10^{-6}} = 0 \qquad [\because \ 1 \text{ microfarad} = 10^{-6} \text{ farads}]$$

or $$\frac{d^2q}{dt^2} + 200\frac{dq}{dt} + 400{,}000q = 0 \qquad ...(1)$$

Its A.E. is $D^2 + 200D + 400{,}000 = 0$ so that

$$D = \frac{-200 \pm \sqrt{40000 - 1600000}}{2} = -100 \pm 100\sqrt{39}\, i$$

$\therefore$ Its solution is $\quad q = e^{-100t}\,[c_1 \cos(100\sqrt{39}\, t) + c_2 \sin(100\sqrt{39}\, t)]$ ...(2)

Differentiating w.r.t. t, we have

$$\frac{dq}{dt} = -100e^{-100t}\,[c_1 \cos(100\sqrt{39}\, t) + c_2 \sin(100\sqrt{39}\, t)$$

$$+ e^{-100t}\,[-100\sqrt{39}\, c_1 \sin(100\sqrt{39}\, t) + 100\sqrt{39}\, c_2 \cos(100\sqrt{39}\, t)] \qquad ...(3)$$

Since $q = 0.05$ when $t = 0$, $\therefore$ from (2), $c_1 = 0.05$

Also $\dfrac{dq}{dt} = 0$ when $t = 0$, $\therefore$ from (3), $0 = -100c_1 + 100\sqrt{39}\, c_2$

or $\quad c_2 = \dfrac{c_1}{\sqrt{39}} = \dfrac{0.05}{\sqrt{39}} = 0.008$

Hence $\quad q = e^{-100t}\,[0.05 \cos(624.5\, t) + 0.008 \sin(624.5\, t)]$

and $\quad i = \dfrac{dq}{dt} = -100e^{-100t}\,[(c_1 - \sqrt{39}\, c_2)\cos 624.5\, t + (\sqrt{39}\, c_1 + c_2)\sin 624.5t]$

$= -0.32\, e^{-100t} \sin 624.5\, t.$ $\qquad [\because \sqrt{39}\, c_2 = c_1]$

Example 5. *A spring which stretches by an amount e under a force mk^2e is suspended from a support P and has a mass m at its lower end. At time t = 0, the mass is at rest in its equilibrium position at a point A below P. A vertical oscillation is now given to the support P such that at any time t (> 0) its displacement below its initial position is a sin nt. Show that the displacement x of the mass below A is given by*

$$\frac{d^2x}{dt^2} + k^2x = k^2 a \sin nt.$$

Hence show that if $n \neq k$, the displacement is given by $x = \dfrac{ka}{k^2 - n^2}(k \sin nt - n \sin kt)$.

What happens when n = k ?

Sol. If λ is the restoring force, then $mk^2e = \lambda e$ so that $\lambda = mk^2$

If e is the elongation produced by the mass m hanging in equilibrium, then $mg = \lambda e$.

Due to vertical oscillation of the support, if at any time $t(> 0)$ the mass is at B, where AB = x, then the stretch of the spring is $(e + x - a \sin nt)$

The equation of motion is

$$m\frac{d^2x}{dt^2} = mg - \lambda(e + x - a \sin nt)$$

$$= -\lambda x + \lambda a \sin nt \qquad (\because mg = \lambda e)$$

$$= -mk^2x + mk^2 a \sin nt \text{ or } \frac{d^2x}{dt^2} + k^2x = k^2 a \sin nt \qquad ...(1)$$

Its A.E. is $D^2 + k^2 = 0$ whence $D = \pm ik$

$$\text{C.F.} = c_1 \cos kt + c_2 \sin kt$$

$$\text{P.I.} = \frac{1}{D^2 + k^2} k^2 \; a \sin nt = k^2 a \frac{1}{D^2 + k^2} \sin nt.$$

Case I. When $n \neq k$

$$\text{P.I.} = k^2 a \,.\, \frac{1}{k^2 - n^2} \sin nt$$

$\therefore$ The complete solution of (1) is $x = c_1 \cos kt + c_2 \sin kt + \dfrac{k^2 a}{k^2 - n^2} \sin nt$ ...(2)

Initially, when $t = 0, x = 0$ $\therefore$ $c_1 = 0$

Differentiating (2) w.r.t. t, $\dfrac{dx}{dt} = -kc_1 \sin kt + kc_2 \cos kt + \dfrac{k^2 an}{k^2 - n^2} \cos nt$

Since $\dfrac{dx}{dt} = 0$ when $t = 0$, we have $kc_2 + \dfrac{k^2 an}{k^2 - n^2} = 0$ or $c_2 = -\dfrac{kan}{k^2 - n^2}$

Substituting the values of c_1 and c_2 in (2), the displacement of the mass at any time t is given by

$$x = -\frac{kan}{k^2 - n^2} \sin kt + \frac{k^2 a}{k^2 - n^2} \sin nt$$

or
$$x = \frac{ka}{k^2 - n^2} (k \sin nt - n \sin kt).$$

Case II. When $n = k$

$$\text{P.I.} = n^2 a \,.\, \frac{1}{D^2 + a^2} \sin nt = n^2 a \,.\, t \,.\, \frac{1}{2D} \sin nt$$

$$= \frac{n^2 at}{2} \left(-\frac{\cos nt}{n} \right) = -\frac{nat}{2} \cos nt$$

$\therefore$ The complete solution of (1) is $x = c_1 \cos nt + c_2 \sin nt - \dfrac{nat}{2} \cos nt$ (since $k = n$) ...(3)

Initially, when $t = 0, x = 0$ $\therefore$ $c_1 = 0$

Differentiating (3) w.r.t. t, $\dfrac{dx}{dt} = -nc_1 \sin nt + nc_2 \cos nt - \dfrac{na}{2} (\cos nt - nt \sin nt)$

Since $\dfrac{dx}{dt} = 0$ when $t = 0$, we have $nc_2 - \dfrac{na}{2} = 0$ $\therefore$ $c_2 = \dfrac{a}{2}$

Substituting the values of c_1 and c_2 in (3), the displacement of the mass at any time t is given by

$$x = \frac{a}{2} \sin nt - \frac{nat}{2} \cos nt \quad \text{or} \quad x = \frac{a}{2} (\sin nt - nt \cos nt) = \frac{ar}{2} \sin (nt - \phi)$$

where $r = \sqrt{1 + n^2 t^2}$, $\phi = \tan^{-1} nt$

The amplitude $\frac{ar}{2} = \frac{a}{2}\sqrt{1+n^2t^2}$ involves t and increases with t.

Thus the phenomenon of resonance occurs.

Example 6. *A mass M suspended from the end of a helical spring is subjected to a periodic force $f = F \sin \omega t$ in the direction of its length. The force f is measured positive vertically downwards and at zero time M is at rest. If the spring stiffness is S, prove that the displacement of M at time t from the commencement of motion is given by* $x = \frac{F}{M(p^2-\omega^2)}\left(\sin \omega t - \frac{\omega}{p}\sin pt\right)$, *where* $p^2 = \frac{S}{M}$ *and damping effects are neglected.*

Sol. Let x be the displacement from the equilibrium position at any time t, then the equation of motion of the mass M is

$$M\frac{d^2x}{dt^2} = -Sx + F\sin \omega t \quad \text{or} \quad \frac{d^2x}{dt^2} + \frac{S}{M}x = \frac{F}{M}\sin \omega t$$

or $$\frac{d^2x}{dt} + p^2x = \frac{F}{M}\sin \omega t \qquad \text{...(1)} \left(\because \frac{S}{M} = p^2\right)$$

Its A.E. is $D^2 + p^2 = 0$ whence $D = \pm ip$

Its C.F. $= c_1 \cos pt + c_2 \sin pt$

and $$\text{P.I.} = \frac{F}{M}\cdot\frac{1}{D^2+p^2}\sin \omega t = \frac{F}{M}\cdot\frac{1}{p^2-\omega^2}\sin \omega t$$

$\therefore$ The complete solution of equation (1) is

$$x = c_1 \cos pt + c_2 \sin pt + \frac{F}{M}\cdot\frac{1}{p^2-\omega^2}\sin \omega t \qquad \text{...(2)}$$

Initially, when $t = 0, x = 0 \quad \therefore \quad c_1 = 0$

Differentiating (2) w.r.t. t, $\frac{dx}{dt} = -pc_1 \sin pt + pc_2 \cos pt + \frac{F}{M}\cdot\frac{\omega}{p^2-\omega^2}\cos \omega t$

Since $\frac{dx}{dt} = 0$ when $t = 0$

$$\therefore \quad pc_2 + \frac{F}{M}\cdot\frac{\omega}{p^2-\omega^2} = 0 \quad \text{or} \quad c_2 = -\frac{\omega}{p}\cdot\frac{F}{M(p^2-\omega^2)}.$$

Substituting the values of c_1 and c_2 in (2), the displacement of the mass at any time t is given by

$$x = -\frac{\omega}{p}\cdot\frac{F}{M(p^2-\omega^2)}\sin pt + \frac{F}{M}\cdot\frac{1}{p^2-\omega^2}\sin \omega t$$

or $$x = \frac{F}{M(p^2-\omega^2)}\left(\sin \omega t - \frac{\omega}{p}\sin pt\right).$$

Example 7. *An uncharged condenser of capacity C is charged by applying an e.m.f. $E \sin \frac{t}{\sqrt{LC}}$, through leads of self-inductance L and negligible resistance. Prove that at time t, the charge on one of the plates is*

$$\frac{EC}{2}\left[\sin \frac{t}{\sqrt{LC}} - \frac{t}{\sqrt{LC}} \cos \frac{t}{\sqrt{LC}}\right].$$

Sol. Let q be the charge on the condenser at any time t. The differentiate equation for the circuit is

$$L\frac{d^2q}{dt^2} + \frac{q}{C} = E \sin \frac{t}{\sqrt{LC}} \qquad \ldots(1)$$

Its A.E. is $\quad LD^2 + \frac{1}{C} = 0 \quad$ or $\quad D^2 = -\frac{1}{LC}$ so that $D = \pm \frac{i}{\sqrt{LC}}$

$$\text{C.F.} = c_1 \cos \frac{t}{\sqrt{LC}} + c_2 \sin \frac{t}{\sqrt{LC}}$$

and

$$\text{P.I.} = \frac{1}{LD^2 + \frac{1}{C}} E \sin \frac{t}{\sqrt{LC}} \qquad \text{(Case of failure)}$$

$$= Et \,.\, \frac{1}{2LD} \sin \frac{t}{\sqrt{LC}} = \frac{Et}{2L}\left(-\sqrt{LC} \cos \frac{t}{\sqrt{LC}}\right)$$

$$= -\frac{Et}{2}\sqrt{\frac{C}{L}} \cos \frac{t}{\sqrt{LC}}$$

$\therefore$ The complete solution of (1) is

$$q = c_1 \cos \frac{t}{\sqrt{LC}} + c_2 \sin \frac{t}{\sqrt{LC}} - \frac{Et}{2}\sqrt{\frac{C}{L}} \cos \frac{t}{\sqrt{LC}} \qquad \ldots(2)$$

Initially, when $\quad t = 0, q = 0 \quad \therefore \quad c_1 = 0$

Differentiating (2) w.r.t. t

$$\frac{dq}{dt} = -\frac{c_1}{\sqrt{LC}} \sin \frac{t}{\sqrt{LC}} + \frac{c_2}{\sqrt{LC}} \cos \frac{t}{\sqrt{LC}} - \frac{E}{2}\sqrt{\frac{C}{L}}\left[\cos \frac{t}{\sqrt{LC}} - \frac{t}{\sqrt{LC}} \sin \frac{t}{\sqrt{LC}}\right].$$

Initially, $\frac{dq}{dt} = i = 0$ when $t = 0$

$$\therefore \quad \frac{c_2}{\sqrt{LC}} - \frac{E}{2}\sqrt{\frac{C}{L}} = 0 \quad \text{or} \quad c_2 = \frac{EC}{2}.$$

Substituting the values of c_1 and c_2 in equation (2), the charge q on the condenser plate, at any time t, is given by

$$q = \frac{EC}{2} \sin \frac{t}{\sqrt{LC}} - \frac{Et}{2}\sqrt{\frac{C}{L}} \cos \frac{t}{\sqrt{LC}}$$

or

$$q = \frac{EC}{2}\left[\sin \frac{t}{\sqrt{LC}} - \frac{t}{\sqrt{LC}} \cos \frac{t}{\sqrt{LC}}\right].$$

Example 8. *In an L-C-R circuit, the charge q on a plate of the condenser is given by* $L\frac{d^2q}{dt^2} + R\frac{dq}{dt} + \frac{q}{C} = E \sin \omega t$ *where* $i = \frac{dq}{dt}$. *The circuit is tuned to resonance so that*

$$\omega^2 = \frac{1}{LC}. \text{ If } R^2 < \frac{4L}{C} \text{ and } q = 0 = i$$

when t = 0, show that $q = \frac{E}{R\omega}\left[-\cos \omega t + e^{-\frac{Rt}{2L}}\left(\cos pt + \frac{R}{2Lp}\sin pt\right)\right]$

and $$i = \frac{E}{R}\left[-\sin \omega t - \frac{1}{p\sqrt{LC}}e^{-\frac{Rt}{2L}}\sin pt\right] \quad \text{where } p^2 = \frac{1}{LC} - \frac{R^2}{4L}.$$

Sol. The given differential equation is $L\frac{d^2q}{dt^2} + R\frac{dq}{dt} + \frac{q}{C} = E \sin \omega t$

or $$\left(LD^2 + RD + \frac{1}{C}\right)q = E \sin \omega t, \quad \text{where } D \equiv \frac{d}{dt} \qquad ...(1)$$

Its A.E. is $LD^2 + RD + \frac{1}{C} = 0$ so that

$$D = \frac{-R \pm \sqrt{R^2 - \frac{4L}{C}}}{2L} = \frac{-R \pm i\sqrt{\frac{4L}{C} - R^2}}{2L}, \text{ since } R^2 < \frac{4L}{C}$$

$$= -\frac{R}{2L} \pm i\sqrt{\frac{1}{LC} - \frac{R^2}{4L^2}} = -\frac{R}{2L} \pm ip, \quad \text{since } \frac{1}{LC} - \frac{R^2}{4L^2} = p^2$$

Its $$\text{C.F.} = e^{-\frac{Rt}{2L}}(c_1 \cos pt + c_2 \sin pt) \quad \text{and} \quad \text{P.I.} = \frac{1}{LD^2 + RD + \frac{1}{C}} E \sin \omega t$$

$$= E \cdot \frac{1}{-L\omega^2 + RD + \frac{1}{C}} \sin \omega t = \frac{E}{R} \cdot \frac{1}{D} \sin \omega t, \text{ since } \omega^2 = \frac{1}{LC}$$

$$= -\frac{E}{R\omega} \cos \omega t$$

$\therefore$ The complete solution of equation (1) is

$$q = e^{-\frac{Rt}{2L}}(c_1 \cos pt = c_2 \sin pt) - \frac{E}{R\omega} \cos \omega t \qquad ...(2)$$

Initially, when $t = 0, q = 0 \quad \therefore \quad c_1 = \frac{E}{R\omega}$

Differentiating (2) w.r.t. t

$$\frac{dq}{dt} = e^{-\frac{Rt}{2L}}(-pc_1 \sin pt + pc_2 \cos pt) - \frac{R}{2L}e^{-\frac{Rt}{2L}}(c_1 \cos pt + c_2 \sin pt) + \frac{E}{R}\sin \omega t$$

Initially, when $t = 0, \quad \frac{dq}{dt} = i = 0$

$$\therefore \qquad pc_2 - \frac{R}{2L} c_1 = 0 \quad \text{or} \quad c_2 = \frac{R}{2pL} \cdot \frac{E}{R\omega} = \frac{E}{2pL\omega}$$

Substituting the values of c_1 and c_2 in equation (2), we get

$$q = e^{-\frac{Rt}{2L}} \left(\frac{E}{R\omega} \cos pt + \frac{E}{2pL\omega} \sin pt \right) - \frac{E}{R\omega} \cos \omega t$$

or
$$q = \frac{E}{R\omega} \left[-\cos \omega t + e^{-\frac{Rt}{2L}} \left(\cos pt + \frac{R}{2Lp} \sin pt \right) \right] \qquad \ldots(3)$$

Differentiating (3) w.r.t. t

$$\frac{dq}{dt} = \frac{E}{R\omega} \left[\omega \sin \omega t - \frac{R}{2L} e^{-\frac{Rt}{2L}} \left(\cos pt + \frac{R}{2Lp} \sin pt \right) + e^{-\frac{Rt}{2L}} \left(-p \sin pt + \frac{R}{2L} \cos pt \right) \right]$$

or
$$i = \frac{E}{R\omega} \left[\omega \sin \omega t - e^{-\frac{Rt}{2L}} \left(\frac{R^2}{4L^2 p} + p \right) \sin pt \right]$$

$$= \frac{E}{R\omega} \left[\omega \sin \omega t - e^{-\frac{Rt}{2L}} \cdot \frac{R^2 + 4L^2 p^2}{4L^2 p} \sin pt \right]$$

$$= \frac{E}{R} \left[\sin \omega t - e^{-\frac{Rt}{2L}} \frac{1}{LC\, p\omega} \sin pt \right] \quad \text{since} \quad \frac{R^2 + 4L^2 p^2}{4L^2} = \frac{1}{LC}$$

$$= \frac{E}{R} \left[\sin \omega t - e^{-\frac{Rt}{2L}} \cdot \frac{\sqrt{LC}}{LC\, p} \sin pt \right] \quad \text{since} \quad \omega^2 = \frac{1}{LC}$$

or
$$i = \frac{E}{R} \left[\sin \omega t - \frac{1}{p\sqrt{LC}} e^{-\frac{Rt}{2L}} \sin pt \right].$$

TEST YOUR KNOWLEDGE

1. A body weighing 20 kg is hung from a spring. A pull of 40 kg wt. will stretch the spring to 10 cm. The body is pulled down to 20 cm below the static equilibrium position and then released. Find the displacement of the body from its equilibrium position at time t seconds, the maximum velocity and the period of oscillation.
2. One end of an elastic spring is fixed at A, to the other end is fastened a particle heavy enough to stretch the spring to double its normal length l. Show that if the particle is dropped from A, it will descend a distance $(2 + \sqrt{3})l$ before coming to rest.
3. An elastic string of natural length l has one extremity fixed at a point A and the other attached to a stone, the weight of which in equilibrium position would extend the string to a length l_1. Show that the stone dropped from A will come to instantaneous rest after a further depth $\sqrt{l_1^2 - l^2}$ below the equilibrium position.

4. The differential equation for a circuit in which self-inductance and capacitance neutralize each other is

$$L\frac{d^2i}{dt^2}+\frac{i}{C}=0$$

Find the current i as a function of t, given that I is the maximum current and $i = 0$ when $t = 0$.

5. A spring fixed at the upper end supports a weight of 980 gm at its lower end. The spring stretches $\frac{1}{2}$ cm under a load of 10 gm and the resistance (in gm. wt.) to the motion of the weight is numerically equal to $\frac{1}{10}$ of the speed of the weight in cm/sec. The weight is pulled down $\frac{1}{4}$ cm below its equilibrium position and then released. Find the expression for the distance of weight from its equilibrium position at time t during its first upward motion.

6. A spring is such that it would be stretched by 72 mm by mass of 15 kg. A mass of 30 kg is attached and brought to rest. The resistance of medium is numerically equal to 120 times the instantaneous velocity. Find the equation of motion of the weight, if it is pulled down 14 cm and given an upward velocity of 3 metres/sec.

7. The differential equation $\frac{d^2x}{dt^2}+2k\frac{dx}{dt}+n^2x=0$; $(k < n)$ represents the damped harmonic oscillations of a particle. Solve this equation and show that the ratio of the amplitude of any oscillation to that of the preceding one is constant, *i.e.*, its amplitudes form a G.P.

8. A constant e.m.f. E at $t = 0$ is applied to a circuit consisting of an inductance L, resistance R and capacitance C in series. The initial values of the current and the charge being zero, find the current at any time t, if $CR^2 < 4L$. Show that the amplitudes of the successive vibrations are in geometrical progression.

9. A spring is stretched 77 mm by a mass of 1.5 kg. This mass is attached to the free end of the spring and brought to rest. The mass is then pulled down 4 cm and released. Determine the displacement of the mass below the equilibrium position, at any time t, if a periodic force 7 sin 6t acts on the spring.

10. An e.m.f. E sin pt is applied at $t = 0$ to a circuit containing a capacitance C and inductance L. The current i satisfies the equation $L\frac{di}{dt}+\frac{1}{C}\int i\,dt = E\sin pt$. If $p^2=\frac{1}{LC}$ and initially the current i and the charge q are zero, show that the current at time t is $\frac{Et}{2L}\sin pt$, where $i=\frac{dq}{dt}$.

11. A body executes damped forced vibrations given by the equation $\frac{d^2x}{dt^2}+2k\frac{dx}{dt}+b^2x=e^{-kt}\sin\omega t$.

Solve the equation for both the cases when $\omega^2 \neq b^2-k^2$ and when $\omega^2=b^2-k^2$.

12. Solve the equation $\frac{d^2x}{dt^2}+2k\frac{dx}{dt}+\mu x=f\cos pt$ which represents damped forced oscillations $(\mu > k)$ and in the result point out the terms that represent free oscillations and forced oscillations respectively, and also the damping factor; and show that ultimately the free oscillations die away and forced oscillations alone persist.

13. Solve the equation $\frac{d^2x}{dt^2}+2h\frac{dx}{dt}+(h^2+p^2)\,x=ke^{-ht}\cos pt$ completely. Show that the particular integral represents an oscillation of variable amplitude which is maximum when $ht = 1$.

14. In an L-C-R circuit, the charge q on a plate of a condenser is given by $L\frac{d^2q}{dt^2} + R\frac{dq}{dt} + \frac{q}{C} = E\sin pt$.

The circuit is tuned to resonance so that $p^2 = \frac{1}{LC}$. If initially the current i and the charge q be zero, show that, for small values of $\frac{R}{L}$, the current in the circuit at time t is given by $\frac{Et}{2L}\sin pt$.

15. If an e.m.f. $E\sin\omega t$ is applied to a circuit containing a resistance R, an inductance L and a condenser of capacity C, the charge on the condenser at time t satisfies the equation $L\frac{d^2q}{dt^2} + R\frac{dq}{dt} + \frac{q}{C} = E\sin\omega t$. If $R = 2\sqrt{LC}$, solve the differential equation for q.

16. A circuit consists of an inductance L and a condenser of capacity C in series. An alternating e.m.f. $E\sin nt$ is applied to the circuit at time $t = 0$, the initial current and the charge on the condenser being zero. Prove that the current at time t is $i = \frac{nE}{L(n^2 - \omega^2)}(\cos\omega t - \cos nt)$, where $CL\omega^2 = 1$.

Prove also that if $n = \omega$, the current at time t is $\frac{Et\sin\omega t}{2L}$.

Answers

1. $x = 0.2\cos 14t$ metres, 2.8 m/sec, 0.45 sec.

4. $i = I\sin\frac{t}{\sqrt{LC}}$

5. $x = e^{-0.05t}[0.25\cos(4.5t) + 0.003\sin(4.5t)]$

6. $x = e^{-0.5t}(0.14\cos 8t - 0.37\sin 8t)$

8. $i = \frac{2E}{k}e^{-\frac{RT}{2L}}\sin\left(\frac{kt}{2L}\right)$, where $k = \sqrt{\frac{4L - CR^2}{C}}$

9. $x = 0.04\cos 8t - \frac{1}{8}\sin 8t + \frac{1}{6}\sin 6t$

11. $x = c_1e^{-kt}\cos(\sqrt{b^2 - k^2}\,t + c_2) + \frac{e^{-kt}}{b^2 - k^2 - \omega^2}\sin\omega t$, $x = c_1e^{-kt}\cos(\omega t + c^2) - \frac{e^{-kt}\,.\,t\cos\omega t}{2\omega}$

13. $x = c_1e^{-ht}\cos(pt + c_2) + \frac{k}{2p}te^{-ht}\sin pt$

15. $q = (c_1 + c_2t)\,e^{-\frac{Rt}{2L}} + EC\left[\frac{(1 - \omega^2 CL)\sin\omega t - 2\sqrt{LC}\cos\omega t}{(1 + \omega^2 CL)^2}\right]$.

4.4. SIMPLE PENDULUM

If a heavy particle is attached to one end of a light inextensible string, the other end of which is fixed, and oscillates under gravity in a vertical plane, then the system is called a **simple pendulum.**

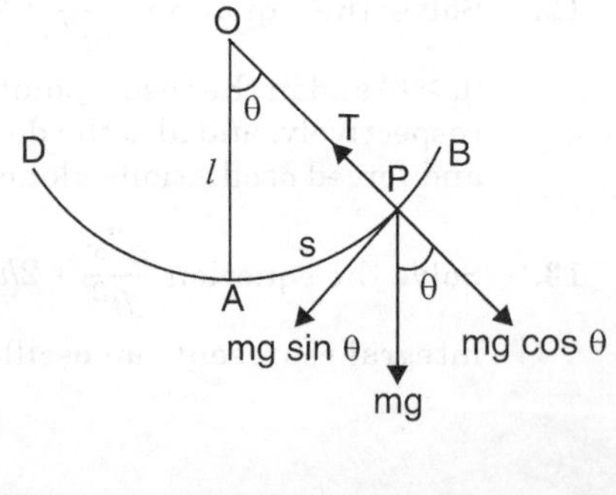

Let O be the fixed point, l the length of the string and m, the mass of the bob (heavy mass). Let P be the position of the bob at any time t. Let arc AP = s and $\angle AOP = \theta$, where OA is the vertical line through O.

The forces acting on the bob are

(*i*) its weight mg acting vertically downward

(*ii*) the tension T in the string.

The components of weight along and perpendicular to the path of motion are $mg \sin \theta$ and $mg \cos \theta$ respectively. The component $mg \cos \theta$ is balanced by the tension in the string.

$\therefore$ The equation of motion of the bob along the tangent is $m \dfrac{d^2 s}{dt^2} = -\, mg \sin \theta$

or $$\frac{d^2(l\theta)}{dt^2} = -\, g \sin \theta \qquad (\because \ s = l\theta)$$

or $$\frac{d^2\theta}{dt^2} = -\frac{g}{l} \sin \theta = -\frac{g}{l}\left(\theta - \frac{\theta^3}{3!} + \ldots\ldots\right) = -\frac{g}{l}\,\theta \text{ to a first approx.}$$

or $$\frac{d^2\theta}{dt^2} + \omega^2\theta = 0, \qquad \text{where } \omega^2 = \frac{g}{l}$$

Its A.E. is $\quad D^2 + \omega^2 = 0, \quad$ whence $D = \pm\, i\omega$.

$\therefore$ Its solution is $\quad \theta = c_1 \cos \omega t + c_2 \sin \omega t$

or $$\theta = c_1 \cos \sqrt{\frac{g}{l}}\, t + c_2 \sin \sqrt{\frac{g}{l}}\, t$$

The motion of the bob is simple harmonic and the time of an oscillation is $\dfrac{2\pi}{\omega} = 2\pi \sqrt{\dfrac{l}{g}}$.

The motion of the bob from one extreme position B to the other extreme position B′ is called *a beat or a swing.*

$\therefore$ The time of one beat $= \dfrac{1}{2}$ (time of one oscillation) $= \pi \sqrt{\dfrac{l}{g}}$

A simple pendulum which beats once every second is called a **second's pendulum.** Thus a second's pendulum beats 86400 times a day, the number of seconds in 24 hours.

Since the time for one beat in a second's pendulum is 1 second, taking $g = 981$ cm/sec^2, we have

$$1 = \pi \sqrt{\frac{l}{981}} \quad \text{or} \quad l = \frac{981}{\pi^2} = 99.4 \text{ cm}$$

Hence the length of a second's pendulum is 99.4 cm.

4.5. GAIN OR LOSS OF BEATS

Let a simple pendulum of length l make n beats in a time t, so that

$$\theta = n\pi \sqrt{\frac{l}{g}} \quad \text{or} \quad n = \frac{t}{\pi}\sqrt{\frac{g}{l}}$$

Taking logs, $$\log n = \log\left(\frac{t}{\pi}\right) + \frac{1}{2} \cdot (\log g - \log l)$$

Taking differentials on both sides, we get

$$\frac{dn}{n} = \frac{1}{2}\left(\frac{dg}{g} - \frac{dl}{l}\right) \quad \text{or} \quad dn = \frac{n}{2}\left(\frac{dg}{g} - \frac{dl}{l}\right)$$

which is the change in the number of beats.

If only g changes, l remaining constant, then $dn = \frac{n}{2}.\frac{dg}{g}$ $(\because \quad dl = 0)$

If only l changes, g remaining constant, then $dn = -\frac{n}{2}.\frac{dl}{l}$.

ILLUSTRATIVE EXAMPLES

Example 1. *Find how many seconds a clock would lose per day, if the length of its pendulum were increased in the ratio 900 : 901.*

Sol. Let l be the original length and $l + dl$, the increased length of the pendulum, then

$$\frac{l}{l+dl} = \frac{900}{901} \quad \text{or} \quad \frac{l+dl}{l} = \frac{901}{900}$$

$$\therefore \quad \frac{dl}{l} = \frac{901}{900} - 1 = \frac{1}{900}$$

Let n be the number of beats per day, then $n = 86400$.

If dn is the change in the number of beats, then $dn = -\frac{n}{2}.\frac{dl}{l} = -\frac{86400}{2} \times \frac{1}{900} = -48.$

Since dn is negative, the clock will lose 48 seconds per day.

Example 2. *A second's pendulum which gains 10 seconds per day at one place loses 10 seconds per day at another ; compare the accelerations due to gravity at the two places.*

Sol. Let g be the acceleration due to gravity when the pendulum beats seconds.

Let $g + g_1$ be the acceleration due to gravity at the place where it gains 10 seconds per day, then

$$dn = 10, \; n = 86400$$

Since $$dn = \frac{n}{2}.\frac{g_1}{g}$$ (here dg has been replaced by g_1)

$$\therefore \quad 10 = \frac{86400}{2}.\frac{g_1}{g} \quad \text{or} \quad \frac{g_1}{g} = \frac{1}{4320}$$

Adding 1 to both sides $$\frac{g + g_1}{g} = \frac{4321}{4320} \qquad \text{...(1)}$$

Let $g + g_2$ be the acceleration due to gravity at the place where it loses 10 seconds per day, then

$$dn = -10.$$

$$\therefore \quad -10 = \frac{86400}{2}.\frac{g_2}{g} \quad \text{or} \quad \frac{g_2}{g} = \frac{-1}{4320}$$

or $$\frac{g+g_2}{g} = \frac{4319}{4320} \qquad \text{...(2)}$$

Dividing (1) by (2), we have $\frac{g+g_1}{g+g_2} = \frac{4321}{4319}$ which is the required ratio.

Example 3. *A pendulum of length l hangs against a wall inclined at an angle θ to the horizontal. Show that the time of complete oscillation is* $2\pi\sqrt{\dfrac{l}{g \sin \theta}}$.

Sol. Let the position of the bob of mass m, at any time t, be P and O be the point of suspension so that OP = l and ∠AOP = α, where OA is the line of greatest slope through O.

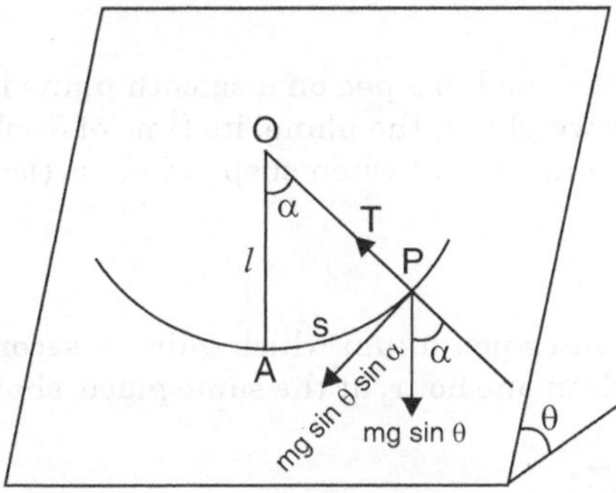

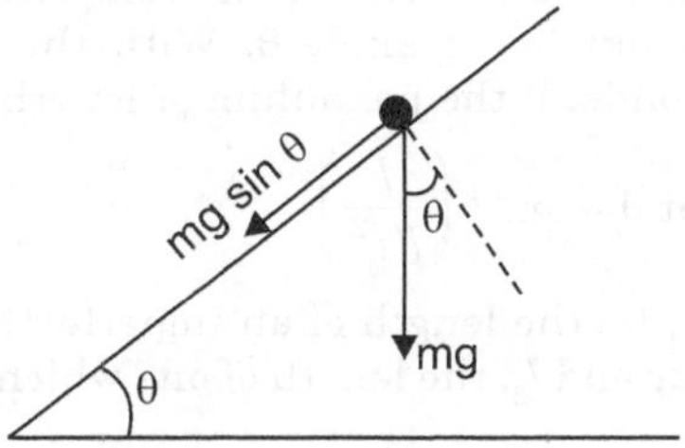

The component of weight of the bob along the plane is mg sin θ. The equation of motion of the bob along the tangent at P is $m\,\frac{d^2s}{dt^2} = -\,mg \sin\theta \sin\alpha$

or $$\frac{d^2(l\alpha)}{dt^2} = -\,g \sin\theta\left(\alpha - \frac{\alpha^3}{3!} + \ldots\right) \qquad [\because s = l\alpha]$$

or $$l\,\frac{d^2 a}{dt^2} = -\,g \sin\theta.\ \alpha \text{ to a first approx.}$$

or $$\frac{d^2\alpha}{dt^2} = -\,\frac{g\sin\theta}{l}.\alpha \quad\text{or}\quad \frac{d^2\alpha}{dt^2} = -\,\omega^2\alpha,\ \text{where } \omega^2 = \frac{g\sin\theta}{l}$$

∴ The motion of the bob is simple harmonic and the time of one oscillation

$$= \frac{2\pi}{\omega} = 2\pi\sqrt{\frac{l}{g\sin\theta}}.$$

TEST YOUR KNOWLEDGE

1. A pendulum oscillating seconds at one place is taken to another place where it loses 2 seconds per day. Compare the accelerations due to gravity at the two places.
2. A clock with a second's pendulum is gaining 2 minutes a day. Prove that the length of the pendulum must be increased by 0.28 cm to make it go correctly.
3. Calculate the number of beats lost per day by a second's pendulum when the length of the string is increased by $\frac{1}{10{,}000}$ of itself.

4. If a pendulum clock loses 9 minutes per week, find in mm, what change is required in the length of the pendulum in order that the clock may keep correct time ?

5. A clock is taken from one place on the earth's surface to another. If the value of g is thus increased by one per cent, find what increase must be made in the length of the pendulum in order that the clock may keep correct time.

6. A pendulum whose length is l makes m oscillations in 24 hours. When its length is slightly altered, it makes $m + n$ oscillations in 24 hours. Show that the diminution in length is $\frac{2nl}{m}$ nearly.

7. A second's pendulum was too long on a day by a quantity k, it was then over corrected next day so as to become too short by k. Prove that the number of minutes gained in these two days is $1080 \frac{k^2}{l^2}$, where l is the true length of the pendulum.

8. A pendulum of length l has one end of the string fastened to a peg on a smooth plane inclined to horizon to an angle θ. With the string and the weight on the plane, its time of oscillation is t seconds. If the pendulum of length l' oscillates in one second when suspended vertically, prove that $\theta = \sin^{-1}\left(\frac{l}{l' t^2}\right)$.

9. If l_1 be the length of an imperfectly adjusted second's pendulum which gains n seconds in one hour and l_2, the length of one which loses n seconds in one hour, at the same place, show that the true length of second's pendulum is $\frac{4l_1 l_2}{l_1 + l_2 + 2\sqrt{l_1 l_2}}$.

10. A simple pendulum has a period T. When the string is lengthened by a small fraction $1/n$ of its length, the period becomes T′. Show that approximately. $\frac{1}{n} = \frac{2(\text{T}' - \text{T})}{\text{T}}$.

11. A simple pendulum of length l is oscillating through a small angle θ in a medium in which the resistance is proportional to the velocity. Find the differential equation of its motion. Discuss the motion and find the period of oscillation.

12. The differential equation of a simple pendulum is $\frac{d^2x}{dt^2} + \omega_0^2 x = \text{F}_0 \sin nt$, where ω_0 and F_0 are constants. If initially $x = 0$, $\frac{dx}{dt} = 0$, determine the motion when $\omega_0 = n$.

Answers

1. 21600 : 21599 **2.** 1.7 mm **5.** 1%

11. $\frac{d^2\theta}{dt^2} + 2k\frac{d\theta}{dt} + \omega\theta = 0$, where $2k = \frac{\lambda}{m}$, $\omega = \frac{g}{l}$. Motion is oscillatory when $k < \omega$ and period $= \frac{2\pi}{\sqrt{\omega^2 - k^2}}$

12. $x = \frac{\text{F}_0}{2n^2}(\sin nt - nt \cos nt)$.

4.6. DEFLECTION OF BEAMS

Consider a uniform beam as made up of fibres running lengthwise. When the beam is deflected, the fibres in some region are elongated and those in some region are contracted. In between these two, there is a region of the beam where the fibres are neither compressed nor stretched. This region is called the *neutral surface* of the beam and the curve of any particular fibre in this surface is called the *elastic curve* or *deflection curve* of the beam. The line in which any plane section of the beam cuts the neutral surface is called the *neutral axis* of that section.

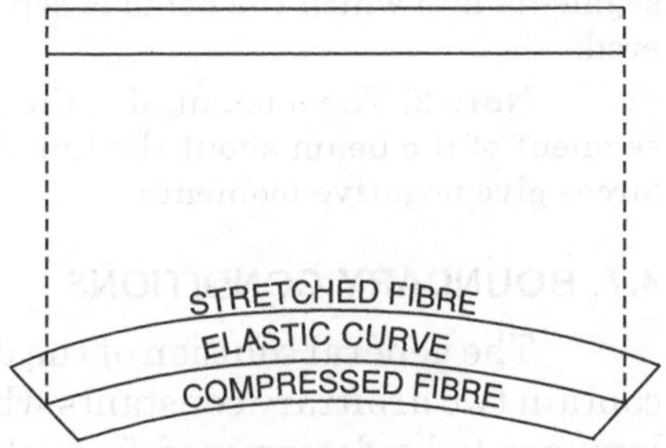

Consider a cross-section of the beam cutting the elastic curve in P and the neutral surface in the line AB, the neutral axis of this section.

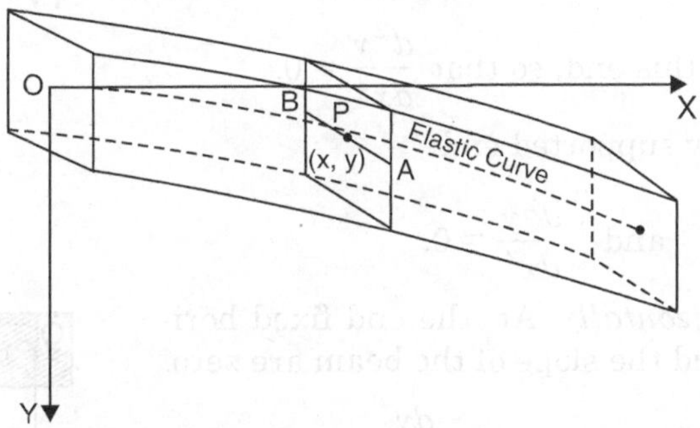

It is well-known from the mechanics of structure that the forces above and below the neutral surface acting in opposite directions have a tendency to restore the beam to its original position ; creating an interval bending moment M given by $\frac{EI}{R}$,

where E = modulus of elasticity of the material of the beam

I = moment of inertia of the cross-section of the beam about the neutral axis AB

R = radius of curvature of the elastic curve at P(x, y).

Now $$R = \frac{\left[1+\left(\frac{dy}{dx}\right)^2\right]^{3/2}}{\frac{d^2y}{dx^2}}$$

If the deflection (slope) of the beam is small, $\left(\frac{dy}{dx}\right)^2$ is very small and can be neglected, so that

$$R = \frac{1}{\frac{d^2y}{dx^2}}$$

Thus $$M = EI\,\frac{d^2y}{dx^2} \qquad ...(1)$$

which is the differential equation of the elastic curve.

Note 1. The moment M with respect to AB of all external forces acting on either of the two segments into which the beam is separated by the cross-section is independent of the segment considered.

Note 2. The amount M is the algebraic sum of the moments of the external forces acting on the segment of the beam about the line AB. The upward forces give positive moments and the downward forces give negative moments.

4.7. BOUNDARY CONDITIONS

The general solution of the differential equation (1) will contain two arbitrary constants which, in any particular problem, are to be determined from the boundary (or end) conditions given below :

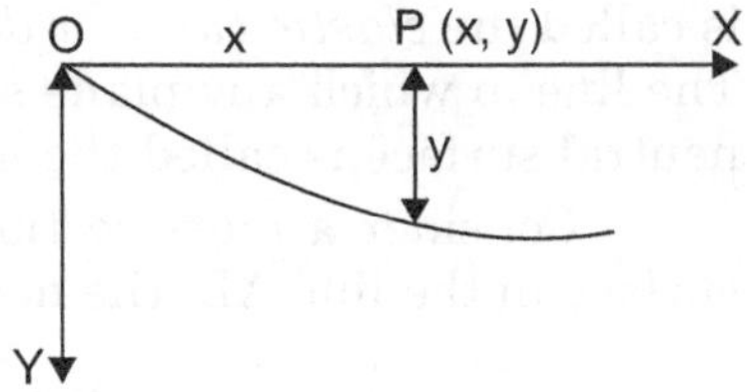

(i) *End freely supported.* At the freely supported end O, there is no deflection of the beam, so that $y = 0$. Also there is no bending moment at this end, so that $\frac{d^2y}{dx^2} = 0$.

Hence, at the freely supported end

$$y = 0 \quad \text{and} \quad \frac{d^2y}{dx^2} = 0.$$

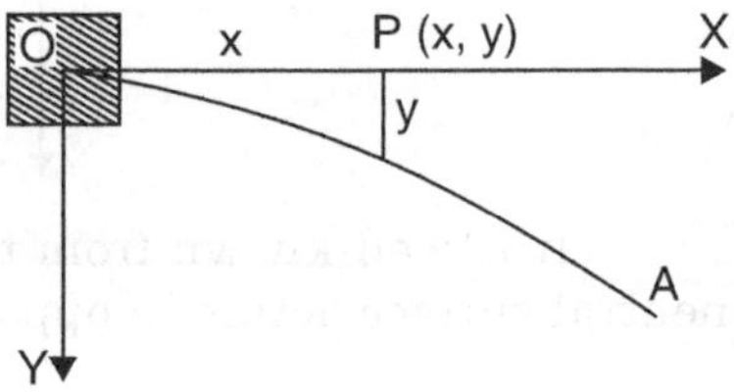

(ii) *End fixed horizontally.* At the end fixed horizontally, the deflection and the slope of the beam are zero.

$$\therefore \qquad y = 0 \quad \text{and} \quad \frac{dy}{dx} = 0.$$

(iii) *End perfectly free.* At the free end A in the above figure, there is no bending moment or shear force,

so that
$$\frac{d^2y}{dx^2} = 0 \quad \text{and} \quad \frac{d^3y}{dx^3} = 0.$$

ILLUSTRATIVE EXAMPLES

Example 1. *The deflection of a strut of length l with one end (x = 0) built in and the other supported and subjected to end thrust P, satisfies the equation* $\frac{d^2y}{dx^2} + a^2y = \frac{a^2R}{P}(l - x)$. *Prove that the deflection curve is* $y = \frac{R}{P}\left(\frac{\sin ax}{a} - l\cos ax + l - x\right)$ *where al = tan al.*

Sol. The given equation is $\frac{d^2y}{dx^2} + a^2y = \frac{a^2\text{R}}{\text{P}}(l - x)$

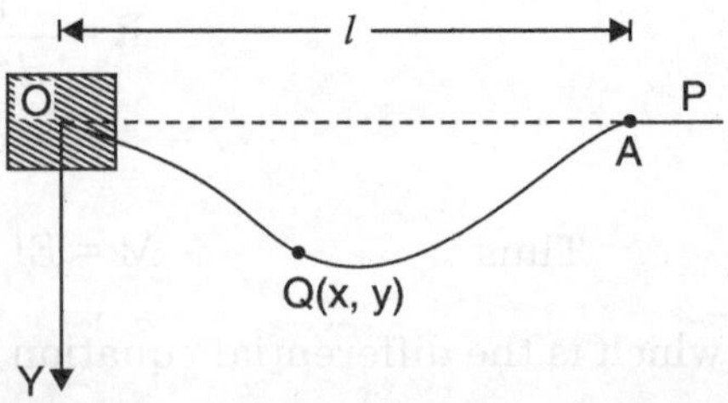

or
$$(\text{D}^2 + a^2)y = \frac{a^2\text{R}}{\text{P}}(l - x) \qquad \ldots(1)$$

Its A.E. is $\text{D}^2 + a^2 = 0$ so that $\text{D} = \pm ia$

$\therefore$ C.F. $= c_1 \cos ax + c_2 \sin ax$

and P.I. $= \dfrac{1}{D^2 + a^2} \dfrac{a^2 R}{P} (l - x)$

$$= \frac{a^2 R}{P} \cdot \frac{1}{a^2\left(1 + \frac{D^2}{a^2}\right)} (l - x) = \frac{R}{P}\left(1 + \frac{D^2}{a^2}\right)^{-1} (l - x)$$

$$= \frac{R}{P}\left(1 - \frac{D^2}{a^2} + \ldots\ldots\right)(l - x) = \frac{R}{P}(l - x)$$

$\therefore$ The complete solution of (1) is $y = c_1 \cos ax + c_2 \sin ax + \dfrac{R}{P}(l - x)$ ...(2)

Differentiating (2) w.r.t. x, $\dfrac{dy}{dx} = -ac_1 \sin ax + ac_2 \cos ax - \dfrac{R}{P}$ ...(3)

Since the end O is build in, $y = 0$ and $\dfrac{dy}{dx} = 0$ at $x = 0$

$\therefore$ From (2), $0 = c_1 + \dfrac{Rl}{P} \Rightarrow c_1 = -\dfrac{Rl}{P}$

From (3), $0 = ac_2 - \dfrac{R}{P} \Rightarrow c_2 = \dfrac{R}{aP}$

Substituting the values of c_1 and c_2 in (2), we have $y = \dfrac{R}{P}\left(\dfrac{\sin ax}{a} - l \cos ax + l - x\right)$ ...(4)

which is the equation of the deflection curve.

Also, at the end A, $y = 0$ when $x = l$.

$\therefore$ From (4), $0 = \dfrac{R}{P}\left(\dfrac{\sin al}{a} - l \cos al\right)$

or $\dfrac{\sin al}{a} = l \cos al \quad \therefore \quad al = \tan al.$

Example 2. *A light horizontal strut AB of length l is freely pinned at A and B and is under the action of equal and opposite compressive forces P at each of its ends and carries a load W at its centre.*

Prove that the deflection at the centre is $\dfrac{W}{2P}\left(\dfrac{1}{n}\tan\dfrac{nl}{2} - \dfrac{l}{2}\right)$, *where* $n^2 = \dfrac{P}{EI}$.

Sol. At each end there is a vertical reaction $\dfrac{W}{2}$. At any point Q(x, y) of the beam, the interval bending moment is EI $\dfrac{d^2y}{dx^2}$ and this must be equal to the moment of external forces taken to the left (or right) of the section QQ′.

$\therefore$ The differential equation for the elastic curve of the beam is $EI \frac{d^2y}{dx^2} = -\frac{W}{2}x - Py$

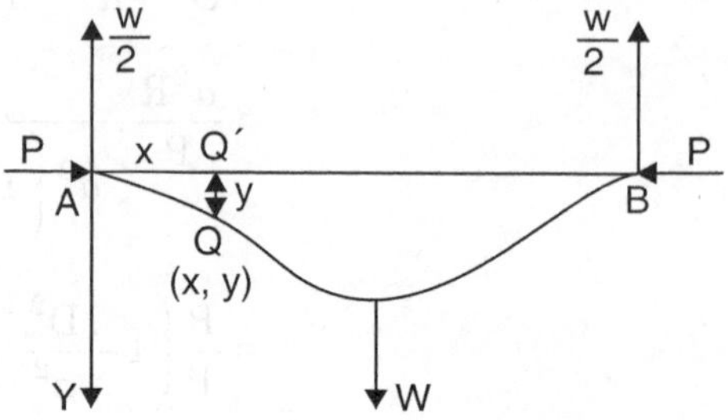

or $$\frac{d^2y}{dx^2} + \frac{P}{EI}y = -\frac{W}{2EI}x$$

or $$\frac{d^2y}{dx^2} + n^2y = -\frac{n^2W}{2P}x \quad ...(1) \quad \left(\because \frac{P}{EI} = n^2\right)$$

Its A.E. is $D^2 + n^2 = 0$ so that $D = \pm in$.

$\therefore$ C.F. $= c_1 \cos nx + c_2 \sin nx$

and $$\text{P.I.} = \frac{1}{D^2 + n^2}\left(-\frac{n^2W}{2P}x\right)$$

$$= -\frac{n^2W}{2P}.\frac{1}{n^2\left(1+\frac{D^2}{n^2}\right)}x = -\frac{W}{2P}\left(1+\frac{D^2}{n^2}\right)^{-1}x$$

$$= -\frac{W}{2P}\left(1-\frac{D^2}{n^2}+......\right)x = -\frac{Wx}{2P}$$

$\therefore$ The complete solution of (1) is $y = c_1 \cos nx + c_2 \sin nx - \frac{W}{2P}x$ ...(2)

Differentiating (2) w.r.t. x, $\frac{dy}{dx} = -nc_1 \sin nx + nc_2 \cos nx - \frac{W}{2P}$ ...(3)

At the end A, $x = 0, y = 0 \quad \therefore$ From (2), $c_1 = 0$

At $x = \frac{l}{2}, \frac{dy}{dx} = 0.$

$\therefore$ From (3), $0 = nc_2 \cos\frac{nl}{2} - \frac{W}{2P}$ $(\because c_1 = 0)$

or $$c_2 = \frac{W}{2Pn\cos\frac{nl}{2}}$$

Substituting the values of c_1 and c_2 in (2), we have

$$y = \frac{W}{2Pn\cos\frac{nl}{2}}\sin nx - \frac{W}{2P}x \quad \text{or} \quad y = \frac{W}{2P}\left(\frac{\sin nx}{n\cos\frac{nl}{2}} - x\right)$$

Deflection at the centre $= y\left(\text{at } x = \frac{l}{2}\right) = \frac{W}{2P}\left(\frac{\sin\frac{nl}{2}}{n\cos\frac{nl}{2}} - \frac{l}{2}\right) = \frac{W}{2P}\left(\frac{1}{n}\tan\frac{nl}{2} - \frac{l}{2}\right).$

TEST YOUR KNOWLEDGE

1. Find the equation of the elastic curve and its maximum deflection for the horizontal, simply supported, uniform beam of length $2l$ metres, having uniformly distributed load w kg/metre.
2. A horizontal strut of length l is clamped horizontally at one end and carries a vertical load W at the other end. If the horizontal thrust be P, prove that the deflection at the free end is $\frac{W}{nP}$ $(\tan nl - nl)$, where $n^2 = \frac{P}{EI}$.
3. A long uniform strut of length l is clamped at one end, the other end being free. If a thrust P be applied at the free end, at a distance 'a' from the neutral axis, prove that the deflection at the free end is a $(\sec nl - 1)$, where $n^2 = \frac{P}{EI}$.
4. A horizontal tie-rod is freely pinned at each end. It carries a uniform load w kg per unit length and has a horizontal pull P. Find the central deflection and the maximum bending moment, taking the origin at one of its ends.
5. A horizontal tie-rod of length $2l$ with concentrated load W at the centre and ends freely hinged, satisfies the differential equation EI $\frac{d^2y}{dx^2} = Py - \frac{W}{2}x$. With conditions $x = 0$, $y = 0$ and $x = l$, $\frac{dy}{dx} = 0$, prove that the deflection δ and the bending moment M at the centre $(x = l)$ are given by $\delta = \frac{W}{2P_n}(nl - \tanh nl)$ and $M = -\frac{W}{2n}\tanh nl$, where $n^2EI = P$.
6. A light horizontal strut of length l is clamped at one end and carries a vertical load W at the free end. If the horizontal thrust at the free is P, show that the strut satisfies the differential equation EI $\frac{d^2y}{dx^2} = (\delta - y)P + W(l - x)$, where y is the displacement of a point at a distance x from the fixed end and δ the deflection at the free end.

   Prove that the deflection at the free end is given by $\frac{W}{nP}(\tan nl - nl)$, where $n^2EI = P$.

Answers

1. $y = \frac{w}{24EI}(4lx^3 - x^4 - 8l^3x)$; $\frac{5wl^4}{24EI}$.
4. $\frac{w}{Pa^2}\left(\text{sech}\frac{al}{2} - 1\right) + \frac{wl^2}{8P}$; $\frac{w}{a^2}\left(\text{sech}\frac{al}{2} - 1\right)$.

4.8. APPLICATIONS OF SIMULTANEOUS LINEAR DIFFERENTIAL EQUATIONS

Now we discuss differential equations involving more than one dependent variables each of which is a function of 2 single independent variable.

ILLUSTRATIVE EXAMPLES

Example 1. *A particle is projected with velocity u making an angle α with the horizontal. Neglecting air resistance, show that the equation of its path is the parabola*

$$y = x \tan \alpha - \frac{gx^2}{2u^2 \cos^2 \alpha}.$$

Find the time of flight, the greatest height attained and the range on the horizontal plane.

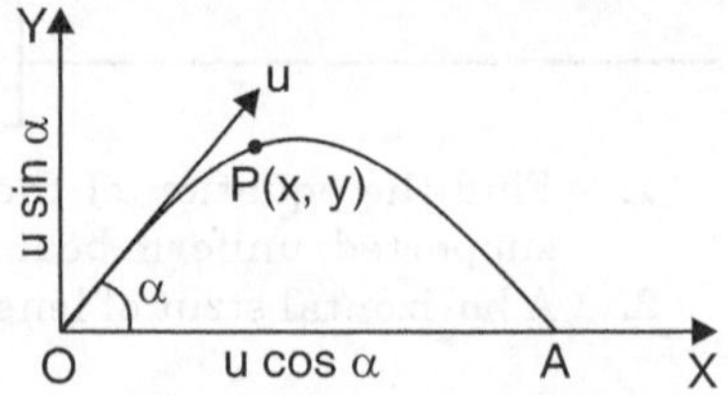

Sol. Let a particle of mass m be projected from a point O with velocity u in a direction making an angle α with the horizontal. Let the horizontal and the vertical lines through O in the plane of motion of the particle be taken as the axes of x and y respectively. Let P(x, y) be the position of the particle at time t.

The horizontal and vertical components of u are $u \cos \alpha$ and $u \sin \alpha$ respectively.

The only force acting on the particle is its weight mg acting vertically downwards.

∴ The equations of motion are

Parallel to x-axis $\quad m \dfrac{d^2x}{dt^2} = 0$ or $\dfrac{d^2x}{dt^2} = 0$ ...(1)

Parallel to y-axis $\quad m \dfrac{d^2y}{dt^2} = - mg$ or $\dfrac{d^2y}{dt^2} = - g$ ...(2)

Integrating (1) w.r.t. t, $\quad \dfrac{dx}{dt} = c_1$

Initially, when $\quad t = 0, \dfrac{dx}{dt} = u \cos \alpha \quad \therefore \quad c_1 = u \cos \alpha$

∴ $\quad \dfrac{dx}{dt} = u \cos \alpha$ ...(3)

Integrating w.r.t. t, $\quad x = (u \cos \alpha)t + c_2$

Initially, when $\quad t = 0, x = 0 \quad \therefore \quad c_2 = 0$

$$x = (u \cos \alpha)t \qquad ...(4)$$

Integrating (2) w.r.t. t, $\quad \dfrac{dx}{dt} = - gt + c_3$

Initially, when $\quad t = 0, \dfrac{dy}{dt} = u \sin \alpha$ so that $c_3 = u \sin \alpha$

∴ $\quad \dfrac{dy}{dt} = u \sin \alpha - gt$ ...(5)

Integrating w.r.t. t, $\quad y = (u \sin \alpha)t - \dfrac{1}{2} gt^2 + c_4$

Initially, when $\quad t = 0, y = 0 \quad \therefore \quad c_4 = 0$

∴ $\quad y = (u \sin \alpha)t - \dfrac{1}{2} gt^2$ ...(6)

Equations (4) and (6) give the position of the particle at any time t. The equation of the path described by the particle is obtained by eliminating the parameter t between equations (4) and (6).

From (4), $\quad t = \dfrac{x}{u \cos \alpha}$

Substituting this value of t in equation (6), $y = (u \sin \alpha) . \dfrac{x}{u \cos \alpha} - \dfrac{1}{2} g \left(\dfrac{x^2}{u^2 \cos^2 \alpha} \right)$

or $$y = x \tan \alpha - \frac{gx^2}{2u^2 \cos^2 \alpha} \qquad ...(7)$$

which is the equation of the path of the projectile. Clearly, the path is a parabola.

The *time of flight* of the projectile is the time taken by the particle to reach the horizontal plane through O.

At the point A, $y = 0$

$\therefore$ From (6), $$0 = (u \sin \alpha)\, t - \frac{1}{2} gt^2 \quad \text{or} \quad t = \frac{2u \sin \alpha}{g}$$

At the highest point, the vertical component of velocity vanishes, *i.e.*, $\frac{dy}{dt} = 0$

so that $$u \sin \alpha - gt = 0 \quad \text{or} \quad t = \frac{u \sin \alpha}{g}$$

$$\therefore \text{ The greatest height } = y \left(\text{when } \frac{dy}{dt} = 0 \right)$$

$$= (u \sin \alpha)t - \frac{1}{2} gt^2 \left(\text{when } t = \frac{u \sin \alpha}{g} \right)$$

$$= u \sin \alpha \,.\, \frac{u \sin \alpha}{g} - \frac{1}{2} g \,.\, \frac{u^2 \sin^2 \alpha}{g} = \frac{u^2 \sin^2 \alpha}{g} - \frac{u^2 \sin^2 \alpha}{2g} = \frac{u^2 \sin^2 \alpha}{2g}$$

The *range* R(= OA) is the horizontal distance covered by the particle in the time of flight.

$$\therefore \qquad R = u \cos \alpha \times \frac{2u \sin \alpha}{g} = \frac{u^2 \sin 2\alpha}{g}.$$

Example 2. *The equations of motion of an electron under certain conditions are*

$$m \frac{d^2 x}{dt^2} + eH \frac{dy}{dt} = eE \qquad ...(1)$$

and $$m \frac{d^2 y}{dt^2} - eH \frac{dx}{dt} = 0 \qquad ...(2)$$

With conditions $x = \frac{dx}{dt} = y = \frac{dy}{dt} = 0$, *when* $t = 0$, *find the path of the electron.*

Sol. Multiplying (2) by an arbitrary constant λ and adding to (1), we get

$$m \frac{d^2}{dt^2} (x + \lambda y) + eH \frac{d}{dt} (y - \lambda x) = eE$$

or $$m \frac{d^2}{dt^2} (x + \lambda y) - eH\lambda \frac{d}{dt} \left(x - \frac{1}{\lambda} y \right) = eE \qquad ...(3)$$

Now choose λ such that $$x - \frac{1}{\lambda} y = x + \lambda y$$

then $$\lambda = -\frac{1}{\lambda} \quad \text{or} \quad \lambda^2 = -1 \text{ so that } \lambda = \pm i$$

Putting $x + \lambda y = u$ in (3), we get

$$m\frac{d^2u}{dt^2} - eH\lambda\frac{du}{dt} = eE \quad \text{or} \quad \frac{d^2u}{dt^2} - \omega\lambda\frac{du}{dt} = \frac{eE}{m} \quad \text{...(4)}$$

where $$\omega = \frac{eH}{m}$$

Its A.E. is $D^2 - \omega\lambda D = 0$ whence $D = 0, \omega\lambda$

$\therefore$ $$\text{C.F.} = A + Be^{\omega\lambda t} \text{ and P.I.} = \frac{1}{D^2 - \omega\lambda D}\frac{eE}{m} = \frac{eE}{m}\cdot\frac{1}{D^2 - \omega\lambda D}e^{0t}$$

$$= \frac{\omega Et}{H}\cdot\frac{1}{2D - \omega\lambda}e^{0t} \qquad \left(\because \omega = \frac{eH}{m}\right)$$

$$= \frac{\omega Et}{H}\cdot\frac{1}{-\omega\lambda} = -\frac{Et}{H\lambda}$$

$\therefore$ The complete solution of (4) is

$$u = A + Be^{\omega\lambda t} - \frac{Et}{H\lambda} \quad \text{or} \quad x + \lambda y = A + Be^{\omega\lambda t} - \frac{Et}{H\lambda}$$

Putting $\lambda = i$ and $-i$, the solutions of the given equations are

$$x + iy = c_1 + c_2\, e^{i\omega t} - \frac{Et}{iH}$$

and $$x - iy = c_3 + c_4\, e^{-i\omega t} + \frac{Et}{iH}$$

i.e., $$x + iy = c_1 + c_2\, e^{i\omega t} + \frac{iEt}{H} \quad \text{...(5)}$$

and $$x - iy = c_3 + c_4\, e^{-i\omega t} - \frac{iEt}{H} \quad \text{...(6)}$$

Differentiating (5) and (6) w.r.t. t, we have $\frac{dx}{dt} + i\frac{dy}{dt} = i\omega c_2 e^{i\omega t} + \frac{iE}{H}$ ...(7)

and $$\frac{dx}{dt} - i\frac{dy}{dt} = -i\omega c_4 e^{-i\omega t} - \frac{iE}{H} \quad \text{...(8)}$$

Since $x = y = \frac{dx}{dt} = \frac{dy}{dt} = 0$, when $t = 0$, we have from equations (5), (6), (7) and (8).

$$0 = c_1 + c_2, \qquad 0 = c_3 + c_4$$

$$0 = i\omega c_2 + \frac{iE}{H}, \qquad 0 = -i\omega c_4 - \frac{iE}{H}$$

which give $$c_1 = -c_2 = \frac{E}{H\omega} \quad \text{and} \quad c_3 = -c_4 = \frac{E}{H\omega}$$

∴ Equations (5) and (6) become

$$x + iy = \frac{E}{H\omega} - \frac{E}{H\omega}e^{i\omega t} + \frac{iEt}{H} \quad ...(9)$$

$$x - iy = \frac{E}{H\omega} - \frac{E}{H\omega}e^{-i\omega t} - \frac{iEt}{H} \quad ...(10)$$

Adding (9) and (10), we have

$$2x = \frac{2E}{H\omega} - \frac{E}{H\omega}(e^{i\omega t} + e^{-i\omega t})$$

or
$$x = \frac{E}{H\omega} - \frac{E}{H\omega}\left(\frac{e^{i\omega t} + e^{-i\omega t}}{2}\right) \quad \text{or} \quad x = \frac{E}{H\omega}(1 - \cos \omega t)$$

Subtracting (10) from (9), we have

$$2iy = -\frac{E}{H\omega}(e^{i\omega t} - e^{-i\omega t}) + \frac{2iEt}{H}$$

or
$$y = -\frac{E}{H\omega}\left(\frac{e^{i\omega t} - e^{-i\omega t}}{2i}\right) + \frac{Et}{H} \quad \text{or} \quad y = \frac{E}{H\omega}(\omega t - \sin \omega t).$$

TEST YOUR KNOWLEDGE

1. An inclined plane makes angle α with the horizontal. A projectile is launched from the bottom of the inclined plane with speed V in a direction making an angle β with the horizontal. Set up the differential equations and find (*i*) the range on the incline, (*ii*) the maximum range up the incline.
2. A particle of unit mass is projected with velocity u at an inclination α above the horizon in a medium whose resistance is k times the velocity. Show that its direction will again make an angle α with the horizon after a time

$$\frac{1}{k}\log\left(1 + \frac{2ku}{g}\sin\alpha\right).$$

3. The motion of a particle is governed by the equations $\ddot{x} - n\dot{y} = 0$, $\ddot{y} + n\dot{x} = n^2 a$, when $x = y = \dot{x} = \dot{y} = 0$ at $t = 0$, find x and y in terms of t.
4. Two particles, each of mass m gm are suspended from two springs of same stiffness k. After the system comes to rest, the lower mass is pulled 1 cm downwards and released. Discuss the motion.
5. In a chemical transformation of certain substance, following equations occur $\frac{dx}{dt} + ax = 0$, $\frac{dz}{dt} = by$, $x + y + z = c$

   where a, b, c are constants. Obtain a differential equation for z. Hence prove that if $z = -\frac{dz}{dt} = 0$

   when $t = 0$, then $z = c + \frac{c}{a - b}(be^{-at} - ae^{-bt})$.
6. Two coils of a transformer are identical with resistance R, inductance L, mutual inductance M and a voltage E is impressed on the primary. Determine the currents in the coils at any instant, assuming that there is no current in either initially.

7. The currents i_1 and i_2 in two coupled circuits are given by $L\frac{di_1}{dt} + Ri_1 + R(i_1 - i_2) = E$

$$L\frac{di_2}{dt} + Ri_2 - R(i_1 - i_2) = 0$$

where L, R, E are constants. Find i_1 and i_2 in terms of t given that $i_1 = i_2 = 0$ at $t = 0$.

Answers

1. (*i*) $\frac{2u^2 \sin(\alpha - \beta)\cos\alpha}{g\cos^2\beta}$; (*ii*) $\frac{u^2}{g(1+\sin\beta)}$ **3.** $x = a(nt - \sin nt),\ y = a(1 - \cos nt)$

4. Combination of two harmonic motions

6. $i_1 = \frac{E}{2R}\left[2 - e^{-\frac{Rt}{L+M}} - e^{-\frac{Rt}{L-M}}\right]$; $i_2 = \frac{E}{2R}\left(e^{-\frac{Rt}{L-M}} - e^{-\frac{Rt}{L+M}}\right)$.

7. $i_1 = \frac{E}{R}\left(\frac{2}{3} - \frac{1}{2}e^{-\frac{Rt}{L}} - \frac{1}{6}e^{-\frac{3Rt}{L}}\right)$; $i_2 = \frac{E}{R}\left(\frac{1}{3} - \frac{1}{2}e^{-\frac{Rt}{L}} + \frac{1}{6}e^{-\frac{3Rt}{L}}\right)$.

MODULE–II

INFINITE SERIES AND POWER SERIES

5

Infinite Series

5.1. SEQUENCE

A sequence is a function whose domain is the set N of all natural numbers whereas the range may be any set S. In other words, a sequence in a set S is a rule which assigns to each natural number a unique element of S.

5.2. REAL SEQUENCE

A real sequence is a function whose domain is the set N of all natural numbers and range a subset of the set R of real numbers.

Symbolically $f : N \to R$ (or $x : N \to R$ or $a : N \to R$)

is a real sequence.

Note. If $x : N \to R$ be a sequence, the image of $n \in N$ instead of denoting it by $x(n)$, we shall generally denote it by x_n. Thus x_1, x_2, x_3 etc. are the real numbers associated to 1, 2, 3, etc. by this mapping. Also, the sequence $x : N \to R$ is denoted by $\{x_n\}$ or $< x_n >$.

$x_1, x_2, ...$ are called the first, second ...... terms of the sequence. The mth and n^{th} terms x_m and x_n for $m \neq n$ are treated as distinct event if $x_m = x_n$ *i.e.*, the terms occurring at different positions are treated as distinct terms even if they have the same value.

5.3. RANGE OF A SEQUENCE

The set of all **distinct** terms of a sequence is called its range.

Note. In a sequence $\{x_n\}$, since $n \in N$ and N is an infinite set, **the number of terms of a sequence is always infinite.** The range of a sequence may be a finite set. *e.g.*, if $x_n = (-1)^n$ then $\{x_n\} = \{-1, 1, -1, 1, ...\}$

The range of sequence $\{x_n\} = \{-1, 1\}$ which is a finite set.

5.4. CONSTANT SEQUENCE

A sequence $\{x_n\}$ defined by $x_n = c \in R \quad \forall \ n \in N$ is called a constant sequence.

e.g., $\{x_n\} = \{c, c, c, ...\}$ is a constant sequence with range $= \{c\}$.

5.5. BOUNDED AND UNBOUNDED SEQUENCES

Bounded above Sequence. A sequence $\{a_n\}$ is said to be bounded above if there exists a real number K such that $a_n \le K \quad \forall \ n \in N$.

Bounded below Sequence. A sequence $\{a_n\}$ is said to be bounded below if there exists a real number K such that $a_n \ge k \quad \forall \ n \in N$.

Bounded Sequence. A sequence $\{a_n\}$ is said to be bounded when it is bounded both above and below.

$\Rightarrow$ A sequence $\{a_n\}$ is bounded if $\exists$ two real numbers k and K ($k \leq$ K) such that
$$k \leq a_n \leq K \quad \forall \ n \in N.$$

Choosing M = maximum $\{|\ k\ |, |\ K\ |\}$, we can also define a sequence $\{a_n\}$ to be bounded if

$$|\ a_n\ | \leq M \quad \forall \quad n \in N.$$

Unbounded Sequence. If $\exists$ no real number M such that $|\ a_n\ | \leq M \ \forall \ n \in N$, then the sequence $\{a_n\}$ is said to be unbounded.

Examples (1). The sequence $\{a_n\}$ defined by $a_n = \dfrac{1}{n}$.

Here $$\{a_n\} = \left\{\frac{1}{1}, \frac{1}{2}, \frac{1}{3}, \ldots\ldots\right\}$$

$\because$ $$0 < a_n \leq 1 \quad \forall \quad n \in N$$

$\therefore$ $\{a_n\}$ is bounded.

(2) The sequence $\{a_n\}$ defined by $a_n = 2^{n-1}$

Here $$\{a_n\} = \{1, 2, 2^2, 2^3, \ldots\ldots\}.$$

Although $a_n \geq 1, \quad \forall \quad n \in N, \exists$ no real number K such that $a_n \leq K$.

$\therefore$ The sequence is unbounded above.

5.6. CONVERGENT, DIVERGENT AND OSCILLATING SEQUENCES

Convergent Sequence. A sequence $\{a_n\}$ is said to be convergent if $\underset{n \to \infty}{\text{Lt}} \ a_n$ is finite.

For example, consider the sequence $\dfrac{1}{2}, \dfrac{1}{2^2}, \dfrac{1}{2^3}, \ldots\ldots, \dfrac{1}{2^n} \ldots\ldots$

Here $a_n = \dfrac{1}{2^n}$, $\underset{n \to \infty}{\text{Lt}} \ a_n = \underset{n \to \infty}{\text{Lt}} \dfrac{1}{2^n} = 0$, which is finite.

$\Rightarrow$ The sequence $\{a_n\}$ is convergent.

Divergent Sequence. A sequence $\{a_n\}$ is said to be divergent if $\underset{n \to \infty}{\text{Lt}} \ a_n$ is not finite, *i.e.*, if

$$\underset{n \to \infty}{\text{Lt}} \ a_n = +\infty \text{ or } -\infty.$$

For example

(*i*) Consider the sequence $\{n^2\}$

Here $a_n = n^2$, $\underset{n \to \infty}{\text{Lt}} \ a_n = +\infty \ \Rightarrow$ The sequence $\{n^2\}$ is divergent.

(*ii*) Consider the sequence $\{-2^n\}$.

Here $$a_n = -2^n, \ \underset{n \to \infty}{\text{Lt}} \ a_n = \underset{n \to \infty}{\text{Lt}} \ (-2^n) = -\infty$$

$\Rightarrow$ The sequence $\{-2^n\}$ is divergent.

Oscillatory Sequence. If a sequence $\{a_n\}$ neither converges to a finite number nor diverges to $+\infty$ or $-\infty$, it is called an oscillatory sequence. Oscillatory sequences are of two types :

(*i*) A bounded sequence which does not converge is said to **oscillate finitely.**

For example, consider the sequence $\{(-1)^n\}$.

Here $a_n = (-1)^n$

It is a bounded sequence. $\underset{n\to\infty}{\text{Lt}}\ a_{2n} = \underset{n\to\infty}{\text{Lt}}\ (-1)^{2n} = 1$

$$\underset{n\to\infty}{\text{Lt}}\ a_{2n+1} = \underset{n\to\infty}{\text{Lt}}\ (-1)^{2n+1} = -1.$$

Thus $\underset{n\to\infty}{\text{Lt}}\ a_n$ does not exist $\Rightarrow$ the sequence does not converge.

Hence this sequence oscillates finitely.

(*ii*) An unbounded sequence which does not diverge is said to **oscillate infinitely.**

For example, consider the sequence $\{(-1)^n\, n\}$.

Here $a_n = (-1)^n\, n.$

It is an unbounded sequence.

$$\underset{n\to\infty}{\text{Lt}}\ a_{2n} = \underset{n\to\infty}{\text{Lt}}\ (-1)^{2n} \cdot 2n = \underset{n\to\infty}{\text{Lt}}\ 2n = +\infty$$

$$\underset{n\to\infty}{\text{Lt}}\ a_{2n+1} = \underset{n\to\infty}{\text{Lt}}\ (-1)^{2n+1}(2n+1) = \underset{n\to\infty}{\text{Lt}}\ -(2n+1) = -\infty.$$

Thus the sequence does not diverge.

Hence this sequence oscillates infinitely.

Note. When we say $\underset{n\to\infty}{\text{Lt}}\ a_n = l$, it means $\underset{n\to\infty}{\text{Lt}}\ a_{2n} = \underset{n\to\infty}{\text{Lt}}\ a_{2n+1} = l$

Similarly, $\underset{n\to\infty}{\text{Lt}}\ a_n = +\infty$ means $\underset{n\to\infty}{\text{Lt}}\ a_{2n} = \underset{n\to\infty}{\text{Lt}}\ a_{2n+1} = +\infty.$

5.7. MONOTONIC SEQUENCES

(*i*) A sequence $\{a_n\}$ is said to be **monotonically increasing** if $a_{n+1} \ge a_n \ \forall\ n \in \text{N}.$

i.e., if $a_1 \le a_2 \le a_3 \le \ldots\ldots \le a_n \le a_{n+1} \le \ldots\ldots$

(*ii*) A sequence $\{a_n\}$ is said to be **monotonically decreasing** if $a_{n+1} \le a_n \ \forall\ n \in \text{N}.$

i.e., if $a_1 \ge a_2 \ge a_3 \ge \ldots\ldots \ge a_n \ge a_{n+1} \ge \ldots\ldots$

(*iii*) A sequence $\{a_n\}$ is said to be **monotonic** if it is either monotonically increasing or monotonically decreasing.

(*iv*) A sequence $\{a_n\}$ is said to be **strictly monotonically increasing** if

$$a_{n+1} > a_n \ \forall\ n \in \text{N}.$$

(*v*) A sequence $\{a_n\}$ is said to be **strictly monotonically decreasing** if

$$a_{n+1} < a_n \ \forall\ n \in \text{N}.$$

(*vi*) A sequence $\{a_n\}$ is said to be **strictly monotonic** if it is either strictly monotonically increasing or strictly monotonically decreasing.

5.8. LIMIT OF A SEQUENCE

A sequence $\{a_n\}$ is said to approach the limit l (say) when $n \to \infty$, if for each $\varepsilon > 0$, $\exists$ a +ve integer m (depending upon ε) such that $|\, a_n - l\,| < \varepsilon \ \forall\ n \ge m.$

In symbols, we write $\underset{n\to\infty}{\text{Lt}}\ a_n = l.$

Note. $|\, a_n - l\,| < \varepsilon\ \forall\ n \ge m \Rightarrow l - \varepsilon < a_n < l + \varepsilon$ for $n = m, m+1, m+2, \ldots\ldots$

5.9. EVERY CONVERGENT SEQUENCE IS BOUNDED

Let the sequence $\{a_n\}$ be convergent. Let it tend to the limit l.

Then given $\varepsilon > 0$, $\exists$ a + ve integer m, such that

$$| a_n - l | < \varepsilon \quad \forall \quad n \geq m$$

$$\Rightarrow \qquad l - \varepsilon < a_n < l + \varepsilon \quad \forall \quad n \geq m.$$

Let k and K be the least and the greatest of $a_1, a_2, a_3, \ldots\ldots, a_{m-1}, l - \varepsilon, l + \varepsilon$

Then $$k \leq a_n \leq \text{K} \quad \forall \quad n \in \text{N},$$

$\Rightarrow$ the sequence $\{a_n\}$ is bounded.

The converse is not always true *i.e.*, a sequence may be bounded, yet it may not be convergent. *e.g.*, consider $a_n = (-1)^n$, then the sequence $\{a_n\}$ is bounded but not convergent since it does not have a unique limiting point.

5.10. CONVERGENCE OF MONOTONIC SEQUENCES

Theorem I. *The necessary and sufficient condition for the convergence of a monotonic sequence is that it is bounded.*

A monotonic increasing sequence which is bounded above converges.

A monotonic decreasing sequence which is bounded below converges.

Theorem II. *If a monotonic increasing sequence is not bounded above, it diverges to* $+\infty$.

Theorem III. *If a monotonic decreasing sequence is not bounded below, it diverges to* $-\infty$.

Theorem IV. *If* $\{a_n\}$ *and* $\{b_n\}$ *are two convergent sequences, then sequence* $\{a_n + b_n\}$ *is also convergent.*

Or

If $\text{Lt } a_n = A$ *and* $\text{Lt } b_n = B$, *then* $\text{Lt } (a_n + b_n) = A + B$.

Theorem V. *If* $\{a_n\}$ *and* $\{b_n\}$ *are two convergent sequences such that* $\text{Lt } a_n = A$ *and* $\text{Lt } b_n = B$, *then*

(*i*) sequence $\{a_n b_n\}$ is also convergent and converges to AB.

(*ii*) sequence $\left\{\dfrac{a_n}{b_n}\right\}$ is also convergent and converges to $\dfrac{\text{A}}{\text{B}}$, $(\text{B} \neq 0)$.

Theorem VI. *The sequence* $\{| a_n |\}$ *converges to zero if and only if the sequence* $\{a_n\}$ *converges to zero.*

Theorem VII. *If a sequence* $\{a_n\}$ *converges to a and* $a_n \geq 0 \ \forall \ n$, *then* $a \geq 0$.

Theorem VIII. *If* $a_n \to a$, $b_n \to b$ *and* $a_n \leq b_n \ \forall \ n$, *then* $a \leq b$.

Theorem IX. *If* $a_n \to l$, $b_n \to l$, *and* $a_n \leq c_n \leq b_n$, $\forall \ n$, *then* $c_n \to l$. (*Squeeze Principle*)

ILLUSTRATIVE EXAMPLES

Example 1. *Give an example of a monotonic increasing sequence which is (i) convergent, (ii) divergent.*

Sol. (*i*) Consider the sequence $\dfrac{1}{2}, \dfrac{2}{3}, \dfrac{3}{4}, \ldots\ldots, \dfrac{n}{n+1}, \ldots\ldots$

Since $\dfrac{1}{2} < \dfrac{2}{3} < \dfrac{3}{4} < \ldots\ldots$ the sequence is monotonic increasing.

$$a_n = \frac{n}{n+1}, \quad \underset{n\to\infty}{\text{Lt}}\ a_n = \underset{n\to\infty}{\text{Lt}}\ \frac{n}{n+1} = \underset{n\to\infty}{\text{Lt}}\ \frac{1}{1+\frac{1}{n}} = 1$$

which is finite.

$\therefore$ The sequence is convergent.

(*ii*) Consider the sequence 1, 2, 3, ..., n, ...

Since $1 < 2 < 3 <$ ......, the sequence is monotonic increasing,

$$a_n = n, \quad \underset{n\to\infty}{\text{Lt}}\ a_n = \underset{n\to\infty}{\text{Lt}}\ n = \infty$$

$\therefore$ The sequence diverges to $+\infty$.

Example 2. *Give an example of a monotonic decreasing sequence which is*

(*i*) *convergent*, (*ii*) *divergent.*

Sol. (*i*) Consider the sequence $1, \frac{1}{2}, \frac{1}{3}, ..., \frac{1}{n}$, ...

Since $1 > \frac{1}{2} > \frac{1}{3} >$ ......, the sequence is monotonic decreasing.

$$a_n = \frac{1}{n}, \quad \underset{n\to\infty}{\text{Lt}}\ a_n = \underset{n\to\infty}{\text{Lt}}\ \frac{1}{n} = 0.$$

$\therefore$ The sequence converges to 0.

(*ii*) Consider the sequence $-1, -2, -3, ..., -n$, ...

Since $-1 > -2 > -3 >$ ......, the sequence is monotonic decreasing.

$$a_n = -n, \quad \underset{n\to\infty}{\text{Lt}}\ a_n = \underset{n\to\infty}{\text{Lt}}\ (-n) = -\infty$$

$\therefore$ The sequence diverges to $-\infty$.

Example 3. *Discuss the convergence of the sequence* $\{a_n\}$ *where*

(*i*) $a_n = \frac{n+1}{n}$ (*ii*) $a_n = \frac{n}{n^2+1}$ (*iii*) $a_n = 1 + \frac{1}{3} + \frac{1}{3^2} + ... + \frac{1}{3^n}$.

Sol. (*i*) $$a_n = \frac{n+1}{n}$$

$$a_{n+1} - a_n = \frac{n+2}{n+1} - \frac{n+1}{n} = \frac{-1}{n(n+1)} < 0 \ \forall\ n$$

$$\Rightarrow \quad a_{n+1} < a_n \ \forall\ n$$

$\Rightarrow$ $\{a_n\}$ is a decreasing sequence.

Also $$a_n = \frac{n+1}{n} = 1 + \frac{1}{n} > 1 \ \forall\ n$$

$\Rightarrow$ $\{a_n\}$ is bounded below by 1,

$\because$ $\{a_n\}$ is decreasing and bounded below, it is convergent.

$$\underset{n\to\infty}{\text{Lt}}\ a_n = \underset{n\to\infty}{\text{Lt}}\left(1 + \frac{1}{n}\right) = 1.$$

(*ii*) $$a_n = \frac{n}{n^2+1}$$

$$a_{n+1} - a_n = \frac{n+1}{(n+1)^2+1} - \frac{n}{n^2+1} = \frac{(n+1)(n^2+1) - n(n^2+2n+2)}{(n^2+2n+2)(n^2+1)}$$

$$= \frac{-n^2 - n + 1}{(n^2 + 2n + 2)(n^2 + 1)} < 0 \quad \forall\, n \;\Rightarrow\; a_{n+1} < a_n \quad \forall\, n$$

$\Rightarrow$ $\{a_n\}$ is a decreasing sequence.

Also $\qquad a_n = \dfrac{n}{n^2+1} > 0 \quad \forall\, n \;\Rightarrow\; \{a_n\}$ is bounded below by 0.

$\because$ $\{a_n\}$ is decreasing and bounded below, it is convergent.

$$\underset{n\to\infty}{\text{Lt}}\; a_n = \underset{n\to\infty}{\text{Lt}} \frac{n}{n^2+1} = \underset{n\to\infty}{\text{Lt}} \frac{\frac{1}{n}}{1+\frac{1}{n^2}} = 0.$$

(iii) $\qquad a_n = 1 + \dfrac{1}{3} + \dfrac{1}{3^2} + \ldots\ldots + \dfrac{1}{3^n}$

= sum of $(n + 1)$ terms of a G.P. whose first term is 1 and common ratio is $\frac{1}{3}$

$$= \frac{1\left(1 - \frac{1}{3^{n+1}}\right)}{1 - \frac{1}{3}} \qquad \left|\; S_n = \frac{a(1-r^n)}{1-r}\right.$$

$$= \frac{3}{2}\left(1 - \frac{1}{3^{n+1}}\right)$$

Now $\qquad a_{n+1} = 1 + \dfrac{1}{3} + \dfrac{1}{3^2} + \ldots\ldots + \dfrac{1}{3^n} + \dfrac{1}{3^{n+1}}$

$\therefore \qquad a_{n+1} - a_n = \dfrac{1}{3^{n+1}} > 0 \quad \forall\, n \;\Rightarrow\; a_{n+1} > a_n \quad \forall\, n$

$\Rightarrow$ $\{a_n\}$ is an increasing sequence.

Also $\qquad a_n = \dfrac{3}{2}\left(1 - \dfrac{1}{3^{n+1}}\right) < \dfrac{3}{2} \quad \forall\, n \;\Rightarrow\; \{a_n\}$ is bounded above by $\dfrac{3}{2}$.

$\because$ $\{a_n\}$ is increasing and bounded above, it is convergent.

$$\underset{n\to\infty}{\text{Lt}}\; a_n = \underset{n\to\infty}{\text{Lt}} \frac{3}{2}\left(1 - \frac{1}{3^{n+1}}\right) = \frac{3}{2}.$$

5.11. INFINITE SERIES

If $\{u_n\}$ is a sequence of real numbers, then the expression $u_1 + u_2 + u_3 + \ldots\ldots + u_n \ldots\ldots$ [*i.e.*, the sum of the terms of the sequence, which are infinite in number] is called an infinite series.

The infinite series $u_1 + u_2 + \ldots\ldots + u_n + \ldots\ldots$ is usually denoted by $\sum_{n=1}^{\infty} u_n$ or more briefly, by Σu_n.

5.12. SERIES OF POSITIVE TERMS

If all the terms of the series $\Sigma u_n = u_1 + u_2 + \ldots\ldots + u_n + \ldots\ldots$ are positive *i.e.*, if $u_n > 0$, $\forall\, n$, then the series Σu_n is called a series of positive terms.

5.13. ALTERNATING SERIES

A series in which the terms are alternate positive and negative is called an alternating series. Thus, the series $\Sigma (-1)^{n-1} u_n = u_1 - u_2 + u_3 - u_4 + \ldots\ldots + (-1)^{n-1} u_n + \ldots\ldots$, where $u_n > 0 \;\forall\, n$ is an alternating series.

5.14. PARTIAL SUMS

If $\Sigma u_n = u_1 + u_2 + u_3 + \ldots\ldots + u_n + \ldots\ldots$ is an infinite series, where the terms may be + ve or – ve, then

$S_n = u_1 + u_2 + \ldots\ldots + u_n$ is called the nth partial sum of Σu_n. Thus, the nth partial sum of an infinite series is the sum of its first n terms.

$S_1, S_2, S_3, \ldots\ldots$ are the first, second, third, ... partial sums of the series.

Since $n \in N$, $\{S_n\}$ is a sequence called the **sequence of partial sums** of the infinite series Σu_n.

$\therefore$ To every infinite series Σu_n, there corresponds a sequence $\{S_n\}$ of its partial sums.

5.15. BEHAVIOUR OF AN INFINITE SERIES

An infinite series Σu_n converges, diverges or oscillates (finitely or infinitely) according as the sequence $\{S_n\}$ of its partial sums converges, diverges or oscillates (finitely or infinitely).

(*i*) The series Σu_n converges (or is said to be convergent) if the sequence $\{S_n\}$ of its partial sums converges.

Thus, Σu_n is convergent if $\underset{n\to\infty}{\text{Lt}}\, S_n = \text{finite}$.

(*ii*) The series Σu_n diverges (or is said to be divergent) if the sequence $\{S_n\}$ of its partial sums diverges.

Thus, Σu_n is divergent if $\underset{n\to\infty}{\text{Lt}}\, S_n = +\infty \text{ or } -\infty$.

(*iii*) The series Σu_n oscillates finitely if the sequence $\{S_n\}$ of its partial sums oscillates finitely.

Thus, Σu_n oscillates finitely if $\{S_n\}$ is bounded and neither converges nor diverges.

(*iv*) The series Σu_n oscillates infinitely if the sequence $\{S_n\}$ of its partial sums oscillates infinitely.

Thus, Σu_n oscillates infinitely if $\{S_n\}$ is unbounded and neither converges nor diverges.

ILLUSTRATIVE EXAMPLES

Example 1. *Discuss the convergence or otherwise of the series*

$$\frac{1}{1.2} + \frac{1}{2.3} + \frac{1}{3.4} + \ldots\ldots + \frac{1}{n(n+1)} + \ldots\ldots \text{ to } \infty.$$

Sol. Here $\quad u_n = \dfrac{1}{n(n+1)} = \dfrac{1}{n} - \dfrac{1}{n+1}$

Putting $\quad n = 1, 2, 3, \ldots\ldots, n$

$$u_1 = \frac{1}{1} - \frac{1}{2}$$

$$u_2 = \frac{1}{2} - \frac{1}{3}$$

$$u_3 = \frac{1}{3} - \frac{1}{4}$$

....

....

$$u_n = \frac{1}{n} - \frac{1}{n+1}$$

Adding $$S_n = 1 - \frac{1}{n+1}$$

$$\underset{n\to\infty}{\text{Lt}} S_n = 1 - 0 = 1$$

$\Rightarrow$ $\{S_n\}$ converges to 1 $\Rightarrow$ Σu_n converges to 1.

Note. For another method, see Comparison Test.

Example 2. *Show that the series* $1^2 + 2^2 + 3^2 + \ldots\ldots + n^2 + \ldots\ldots$ *diverges to* $+\infty$.

Sol. $$S_n = 1^2 + 2^2 + 3^2 + \ldots.. + n^2 = \frac{n(n+1)(2n+1)}{6}$$

$$\underset{n\to\infty}{\text{Lt}} S_n = +\infty$$

$\Rightarrow$ $\{S_n\}$ diverges to $+\infty$

$\Rightarrow$ The given series diverges to $+\infty$.

Article 1. *The geometric series* $1 + x + x^2 + x^3 + \ldots\ldots$ *to* ∞

(*i*) *converges if* $-1 < x < 1$ *i.e.,* $|x| < 1$ (*ii*) *diverges if* $x \geq 1$

(*iii*) *oscillates finitely if* $x = -1$ (*iv*) *oscillates infinitely if* $x < -1$.

Proof. (*i*) **When** $|x| < 1$

Since $|x| < 1$, $x^n \to 0$ as $n \to \infty$

$$S_n = 1 + x + x^2 + \ldots\ldots \text{ to } n \text{ terms} = \frac{1(1-x^n)}{1-x} = \frac{1}{1-x} - \frac{x^n}{1-x}$$

$$\underset{n\to\infty}{\text{Lt}} S_n = \frac{1}{1-x} \Rightarrow \text{ the sequence } \{S_n\} \text{ is convergent}$$

$\Rightarrow$ the given series is convergent.

(*ii*) **When** $x \geq 1$

Sub-case I. When $x = 1$

$$S_n = 1 + 1 + 1 + \ldots\ldots \text{ to } n \text{ terms} = n$$

$$\underset{n\to\infty}{\text{Lt}} S_n = \infty \Rightarrow \text{ the sequence } \{S_n\} \text{ diverges to } \infty.$$

$\Rightarrow$ the given series diverges to ∞.

Sub-case II. When $x > 1$, $x^n \to \infty$ as $n \to \infty$

$$S_n = 1 + x + x^2 + \ldots\ldots \text{ to } n \text{ terms} = \frac{1(x^n - 1)}{x-1}$$

$$\underset{n\to\infty}{\text{Lt}} S_n = \infty \Rightarrow \text{ the sequence } \{S_n\} \text{ diverges to } \infty$$

$\Rightarrow$ the given series diverges to ∞.

(*iii*) **When** $x = -1$

$$S_n = 1 - 1 + 1 - 1 + \ldots\ldots \text{ to } n \text{ terms}$$
$$= 1 \text{ or } 0 \text{ according as } n \text{ is odd or even.}$$

$\Rightarrow \quad \underset{n\to\infty}{\text{Lt}}\ S_n = 1 \text{ or } 0 \quad \Rightarrow \quad$ the sequence $\{S_n\}$ oscillates finitely.

$\Rightarrow$ the given series oscillates finitely.

(*iv*) **When $x < -1$**

$$x < -1 \Rightarrow -x > 1$$

Let $r = -x$, then $r > 1$

$\therefore \quad r^n \to \infty$ as $n \to \infty$

$$S_n = 1 + x + x^2 + x^3 + \ldots\ldots \text{ to } n \text{ terms} = \frac{1-x^n}{1-x} = \frac{1-(-r)^n}{1+r} \qquad [\because x = -r]$$

$$= \frac{1-r^n}{1+r} \quad \text{or} \quad \frac{1+r^n}{1+r} \text{ according as } n \text{ is even or odd}$$

$$\underset{n\to\infty}{\text{Lt}}\ S_n = \frac{1-\infty}{1+r} \quad \text{or} \quad \frac{1+\infty}{1+r} = -\infty \text{ or } +\infty$$

$\Rightarrow$ the sequence $\{S_n\}$ oscillates infinitely

$\Rightarrow$ the given series oscillates infinitely.

Article 2. *If a series Σu_n is convergent, then $\underset{n\to\infty}{Lt}\ u_n = 0$.*

Proof. Let S_n denote the *n*th partial sum of the series Σu_n.

Then Σu_n is convergent $\Rightarrow$ $\{S_n\}$ is convergent

$\Rightarrow \underset{n\to\infty}{\text{Lt}}\ S_n$ is finite and unique $= s$ (say). $\Rightarrow \underset{n\to\infty}{\text{Lt}}\ S_{n-1} = s$

Now $\quad S_n - S_{n-1} = u_n$

$$\therefore \quad \underset{n\to\infty}{\text{Lt}}\ u_n = \underset{n\to\infty}{\text{Lt}}\ (S_n - S_{n-1}) = \underset{n\to\infty}{\text{Lt}}\ S_n - \underset{n\to\infty}{\text{Lt}}\ S_{n-1} = s - s = 0.$$

Hence Σu_n is convergent $\Rightarrow \underset{n\to\infty}{\text{Lt}}\ u_n = 0$.

The converse of the above theorem is not always true, *i.e.*, the *n*th term may tend to zero as $n \to \infty$ even if the series is not convergent.

For example, the series $1 + \frac{1}{2} + \frac{1}{3} + \ldots\ldots + \frac{1}{n} + \ldots\ldots$ diverges, though

$$\underset{n\to\infty}{\text{Lt}}\ u_n = \underset{n\to\infty}{\text{Lt}}\ \frac{1}{n} = 0.$$

Note 1. Σu_n is convergent $\Rightarrow \underset{n\to\infty}{\text{Lt}}\ u_n = 0$.

Note 2. $\underset{n\to\infty}{\text{Lt}}\ u_n = 0 \Rightarrow \Sigma u_n$ may or may not be convergent.

Note 3. $\underset{n\to\infty}{\text{Lt}}\ u_n \neq 0 \Rightarrow \Sigma u_n$ is not convergent.

Article 3. *A positive term series either converges or diverges to $+\infty$.*

Proof. Let Σu_n be a positive term series and S_n be its *n*th partial sum.

Then $\quad S_{n+1} = u_1 + u_2 + \ldots\ldots + u_n + u_{n+1} = S_n + u_{n+1}$

$\Rightarrow \quad S_{n+1} - S_n = u_{n+1} > 0 \ \forall\ n \qquad [\because u_n > 0\ \forall\ n]$

$\Rightarrow \quad S_{n+1} > S_n \ \forall\ n$

$\Rightarrow$ $\{S_n\}$ is a monotonic increasing sequence.

Two cases arise. The sequence $\{S_n\}$ may be bounded or unbounded above.

Case I. *When $\{S_n\}$ is bounded above.*

Since $\{S_n\}$ is monotonic increasing and bounded above, it is convergent $\Rightarrow$ Σu_n is convergent.

Case II. *When $\{S_n\}$ is not bounded above.*

Since $\{S_n\}$ is monotonic increasing and not bounded above, it diverges to $+\infty$ $\Rightarrow$ Σu_n diverges to $+\infty$.

Hence a positive term series either converges or diverges to $+\infty$.

Cor. *If* $u_n > 0$ $\forall$ n *and* $\underset{n\to\infty}{Lt}\ u_n \neq 0$, *then the series* Σu_n *diverges to* $+\infty$.

Proof. $u_n > 0$ $\forall$ n $\Rightarrow$ Σu_n is a series of +ve terms.

$\Rightarrow$ Σu_n either converges or diverges to $+\infty$

Since $\qquad \underset{n\to\infty}{\text{Lt}}\ u_n \neq 0$ (given)

$\therefore$ Σu_n does not converge.

Hence Σu_n diverges to $+\infty$.

Article 4. *(a) The necessary and sufficient condition for the convergence of a positive term series Σu_n is that the sequence $\{S_n\}$ of its partial sums is bounded above.*

Proof. (*i*) Suppose the sequence $\{S_n\}$ is bounded above. Since the series Σu_n is of positive terms, the sequence $\{S_n\}$ is monotonically increasing. Since every monotonically increasing sequence which is bounded above, converges, therefore $\{S_n\}$ and hence Σu_n converges.

(*ii*) Conversely, suppose Σu_n converges. Then the sequence $\{S_n\}$ of its partial sums also converges. Since every convergent sequence is bounded, $\{S_n\}$ is bounded. In particular, $\{S_n\}$ is bounded above.

(b) Cauchy's General Principle of Convergence of Series

The necessary and sufficient condition for the infinite series $\sum_{n=1}^{\infty} u_n$ to converge is that given $\varepsilon > 0$, however small, there exists a positive integer m such that $| u_{m+1} + u_{m+2} + \ldots\ldots + u_n | < \varepsilon$ $\forall$ $n > m$.

Example. *Prove with the help of Cauchy's general principle of convergence that the series*

$$\sum_{n=1}^{\infty} \frac{1}{n} = 1 + \frac{1}{2} + \frac{1}{3} + \ldots\ldots + \frac{1}{n} + \ldots\ldots \textit{ does not converge.}$$

Sol. If possible, suppose the given series is convergent.

Take $\qquad \varepsilon = \dfrac{1}{2}$

By Cauchy's general principle of convergence, there exists a positive integer m such that

$$\left| \frac{1}{m+1} + \frac{1}{m+2} + \ldots\ldots + \frac{1}{n} \right| < \frac{1}{2} \quad \forall \quad n > m$$

or

$$\frac{1}{m+1} + \frac{1}{m+2} + \ldots\ldots + \frac{1}{n} < \frac{1}{2} \quad \forall \quad n > m \qquad \ldots(1)$$

By taking $n = 2m$, we observe that $\frac{1}{m+1}+\frac{1}{m+2}+......+\frac{1}{n}$

$$= \frac{1}{m+1}+\frac{1}{m+2}+......+\frac{1}{2m} > \frac{1}{2m}+\frac{1}{2m}+......+\frac{1}{2m} = \left(\frac{m}{2m}=\frac{1}{2}\right)$$

i.e.,
$$\frac{1}{m+1}+\frac{1}{m+2}+......+\frac{1}{n} > \frac{1}{2}$$

where $n = 2m > m$. This contradicts (1).

$\Rightarrow$ Our supposition is wrong.

$\Rightarrow$ The given series does not converge.

Article 5. *If m is a given positive integer, then the two series* $u_1 + u_2 + ... + u_{m+1} + u_{m+2} + ...$ *and* $u_{m+1} + u_{m+2} + ...$ *converge or diverge together.*

Proof. Let S_n and s_n denote the nth partial sums of the two series.

Then
$$S_n = u_1 + u_2 + ... + u_n$$
$$s_n = u_{m+1} + u_{m+2} + ... + u_{m+n}$$
$$= (u_1 + u_2 + ... + u_{m+n}) - (u_1 + u_2 + ... + u_m)$$
$$= S_{m+n} - S_m \Rightarrow s_n = S_{m+n} - S_m \qquad ...(1)$$

S_m being the sum of a finite number of terms of Σu_n is a fixed finite quantity.

(*i*) If $S_{m+n} \to$ a finite limit as $n \to \infty$, then from (1), so does s_n.

(*ii*) If $S_{m+n} \to +\infty$ as $n \to \infty$, so does s_n.

(*iii*) If $S_{m+n} \to -\infty$ as $n \to \infty$, so does s_n.

(*iv*) If S_{m+n} does not tend to any limit (finite or infinite), so does s_n.

$\Rightarrow$ The sequences $\{S_n\}$ and $\{s_n\}$ converge or diverge together.

$\Rightarrow$ The two given series converge or diverge together. Hence the result.

Note. The above theorem shows that the convergence, divergence or oscillation of a series is not affected by addition or omission of a finite number of its terms.

Article 6. *If* Σu_n *and* Σv_n *converge to u and v respectively, then* $\Sigma(u_n + v_n)$ *converges to* $(u + v)$.

Proof. Let
$$U_n = u_1 + u_2 + ... + u_n$$
$$V_n = v_1 + v_2 + ... + v_n$$
and
$$S_n = (u_1 + v_1) + (u_2 + v_2) + ... + (u_n + v_n)$$
Then
$$S_n = (u_1 + u_2 + ... + u_n) + (v_1 + v_2 + ... + v_n) = U_n + V_n.$$

Since Σu_n converges to u, $\underset{n\to\infty}{\text{Lt}}\ U_n = u$

Σv_n converges to v, $\underset{n\to\infty}{\text{Lt}}\ V_n = v$

$$\therefore \quad \underset{n\to\infty}{\text{Lt}}\ S_n = \underset{n\to\infty}{\text{Lt}}\ (U_n + V_n) = \underset{n\to\infty}{\text{Lt}}\ U_n + \underset{n\to\infty}{\text{Lt}}\ V_n = u + v.$$

$\Rightarrow$ $\Sigma(u_n + v_n)$ converges to $(u + v)$.

Article 7. Comparison Tests

Test I. *If* Σu_n *and* Σv_n *are series of positive terms and* Σv_n *is convergent and there is a positive constant k such that* $u_n \le kv_n$, $\forall\ n > m$, *then* Σu_n *is also convergent.*

Proof. Let $U_n = u_1 + u_2 + ... + u_n$ and $V_n = v_1 + v_2 + ... + v_n$

Now $\quad u_n \le kv_n\ \forall\ n > m$

$\Rightarrow \quad u_{m+1} \le kv_{m+1}$

$$u_{m+2} \le kv_{m+2}$$

....

Adding $u_{m+1} + u_{m+2} + ... + u_n \le k(v_{m+1} + v_{m+2} + ... + v_n)$

$\Rightarrow \quad U_n - U_m \le k(V_n - V_m) \quad \forall \quad n > m$

$\Rightarrow \quad U_n \le kV_n + (U_m - kV_m) \quad \forall \quad n > m$

$\Rightarrow \quad U_n \le kV_n + k_0 \quad \forall \quad n > m \quad ...(1)$

where $k_0 = U_m - kV_m$ is a fixed number. Since Σv_n is convergent, the sequence $\{V_n\}$ is convergent and hence bounded above.

$\therefore$ From (1), the sequence $\{U_n\}$ is bounded above.

$\because$ Σu_n is a series of + ve terms.

$\{U_n\}$ is monotonic increasing.

$\because$ $\{U_n\}$ is monotonic increasing sequence and is bounded above.

$\therefore$ It is convergent.

$\Rightarrow$ Σu_n is convergent.

Test II. *If Σu_n and Σv_n are two series of positive terms and Σv_n is divergent and there is a positive constant k such that $u_n > kv_n$, $\forall\ n > m$, then Σu_n is also divergent.*

Proof. Let $\quad U_n = u_1 + u_2 + ... + u_n$

and $\quad V_n = v_1 + v_2 + ... + v_n$

Now $\quad u_n > kv_n \quad \forall \quad n > m$

$\Rightarrow \quad u_{m+1} > kv_{m+1}$

$$u_{m+2} > kv_{m+2}$$

....

....

Adding $u_{m+1} + u_{m+2} + ... + u_n > k(v_{m+1} + v_{m+2} + ... + v_n)$

$\Rightarrow \quad U_n - U_m > k(V_n - V_m) \quad \forall \quad n > m$

$\Rightarrow \quad U_n > kV_n + (U_m - kV_m) \quad \forall \quad n > m$

$\Rightarrow \quad U_n > kV_n + k_0 \quad \forall \quad n > m \quad ...(1)$

where $k_0 = U_m - kV_m$ is a fixed number.

Since Σv_n is divergent, the sequence $\{V_n\}$ is divergent.

$\Rightarrow$ For each positive real number k_1, however large, there exists a +ve integer m' such that

$$V_n > k_1 \quad \forall \quad n > m'$$

Let $m^* = \max.\{m, m'\}$, then $\quad V_n > k_1 \quad \forall \quad n > m^*$

From (1), $\quad U_n > kk_1 + k_0 = K \quad \forall \quad n > m^*$

$\Rightarrow$ $\{U_n\}$ is divergent

$\Rightarrow$ Σu_n is divergent.

Test III. *If Σu_n and Σv_n are two positive term series and there exist two positive constants H and K (independent of n) and a positive integer m such that $H < \frac{u_n}{v_n} < K\ \forall\ n > m$ then the two series Σu_n and Σv_n converge or diverge together.*

Proof. Since Σv_n is a series of + ve terms, $v_n > 0, \quad \forall\ n$

$$\therefore \quad H < \frac{u_n}{v_n} < K \quad \forall\ n > m$$

$\Rightarrow \quad Hv_n < u_n < Kv_n \quad \forall\, n > m$ ...(1)

Case I. *When Σv_n is convergent*

From (1), $u_n < Kv_n \quad \forall\, n > m$ and Σv_n is convergent.

$\Rightarrow \Sigma u_n$ is convergent. [See Test I]

Case II. *When Σv_n is divergent*

From (1), $u_n > Hv_n \quad \forall\, n > m$ and Σv_n is divergent.

$\Rightarrow \Sigma u_n$ is divergent. [See Test II]

Case III. *When Σu_n is convergent*

From (1), $Hv_n < u_n \quad \forall\, n > m$

$$\Rightarrow \quad v_n < \frac{1}{H} u_n \quad \forall\, n > m \quad (\because H > 0)$$

Since Σu_n is convergent $\therefore$ Σv_n is convergent. [See Test I]

Case IV. *When Σu_n is divergent*

From (1), $Kv_n > u_n \quad \forall\, n > m$

$$\Rightarrow \quad v_n > \frac{1}{K} u_n. \quad \forall\, n > m \quad (\because K > 0)$$

Since Σu_n is divergent $\therefore$ Σv_n is divergent. [See Test II]

Particular Case of Test III (When m = 0)

If Σu_n and Σv_n are two positive term series and there exist two positive constants H and K (independent of n) such that $H < \dfrac{u_n}{v_n} < K \quad \forall\, n$,

then the two series Σu_n and Σv_n converge or diverge together.

Test IV. *Let Σu_n and Σv_n be two positive term series.*

(*i*) *If* $\underset{n \to \infty}{\text{Lt}} \dfrac{u_n}{v_n} = l$ *(finite and non-zero), then Σu_n and Σv_n both converge or diverge together.*

(*ii*) *If* $\underset{n \to \infty}{\text{Lt}} \dfrac{u_n}{v_n} = 0$ *and Σv_n converges, then Σu_n also converges.*

(*iii*) *If* $\underset{n \to \infty}{\text{Lt}} \dfrac{u_n}{v_n} = \infty$ *and Σv_n diverges, then Σu_n also diverges.*

Proof. (*i*) Since $u_n > 0$, $v_n > 0$ $\therefore$ $\dfrac{u_n}{v_n} > 0$

$$\therefore \quad \underset{n \to \infty}{\text{Lt}} \frac{u_n}{v_n} \geq 0$$

But $\underset{n \to \infty}{\text{Lt}} \dfrac{u_n}{v_n} = l \neq 0 \Rightarrow l > 0$

Now $\underset{n \to \infty}{\text{Lt}} \dfrac{u_n}{v_n} = l$

$\Rightarrow$ given $\varepsilon > 0$, $\exists$ a +ve integer m such that $\left| \dfrac{u_n}{v_n} - l \right| < \varepsilon \quad \forall\, n > m$

$\Rightarrow \quad l - \varepsilon < \dfrac{u_n}{v_n} < l + \varepsilon \quad \forall\ n > m$

$\Rightarrow \quad (l - \varepsilon)\ v_n < u_n < (l + \varepsilon)\ v_n \quad \forall\ n > m \quad (\because\ v_n > 0)$

Choose $\varepsilon > 0$ such that $l - \varepsilon > 0$.

Let $l - \varepsilon = H$, $l + \varepsilon = K$, where H, K are > 0

$\therefore \quad Hv_n < u_n < Kv_n \quad \forall\ n > m \quad \ldots(1)$

Case I. *When Σu_n is convergent*

From (1), $\quad Hv_n < u_n \quad \forall\ n > m$

$\Rightarrow \quad v_n < \dfrac{1}{H} u_n \quad \forall\ n > m \quad (\because\ H > 0)$

Since Σu_n is convergent, Σv_n is also convergent.

Case II. *When Σu_n is divergent.*

From (1), $\quad Kv_n > u_n \quad \forall\ n > m$

$\Rightarrow \quad v_n > \dfrac{1}{K} u_n \quad \forall\ n > m \quad (\because\ K > 0)$

Since Σu_n is divergent, Σv_n is also divergent.

Case III. *When Σv_n is convergent.*

From (1), $\quad u_n < Kv_n \quad \forall\ n > m$

Since Σv_n is convergent, Σu_n is also convergent.

Case IV. *When Σv_n is divergent.*

From (1), $\quad u_n > Hv_n \quad \forall\ n > m$

Since Σv_n is divergent, Σu_n is also divergent.

Hence Σu_n and Σv_n converge or diverge together.

(*ii*) Here $\quad \underset{n \to \infty}{\text{Lt}} \dfrac{u_n}{v_n} = 0$

$\therefore \quad$ Given $\varepsilon > 0$, $\exists$ a +ve integer m such that $\left| \dfrac{u_n}{v_n} - 0 \right| < \varepsilon \quad \forall\ n > m$

$\Rightarrow \quad -\varepsilon < \dfrac{u_n}{v_n} < \varepsilon \quad \forall\ n > m$

$\Rightarrow \quad u_n < \varepsilon\ v_n \quad \forall\ n > m \quad (\because\ v_n > 0)$

Since Σv_n is convergent, Σu_n is also convergent.

(*iii*) Here $\quad \underset{n \to \infty}{\text{Lt}} \dfrac{u_n}{v_n} = \infty$

$\therefore \quad$ Given $M > 0$, however large, $\exists$ a +ve integer m such that $\dfrac{u_n}{v_n} > M \quad \forall\ n > m$

$\Rightarrow \quad u_n > Mv_n \quad \forall\ n > m$

Since Σv_n is divergent, Σu_n is also divergent.

Test V. *Let Σu_n and Σv_n be two positive term series.*

(*i*) *If* $\dfrac{u_n}{u_{n+1}} > \dfrac{v_n}{v_{n+1}} \quad \forall\ n > m$ *and Σv_n is convergent, then Σu_n is also convergent.*

(ii) If $\dfrac{u_n}{u_{n+1}} < \dfrac{v_n}{v_{n+1}}$ $\forall\, n > m$ *and* Σv_n *is divergent, then* Σu_n *is also divergent.*

Proof. (*i*) $$\frac{u_n}{u_{n+1}} > \frac{v_n}{v_{n+1}} \qquad \forall\, n > m$$

$$\Rightarrow \quad \frac{u_{m+1}}{u_{m+2}} > \frac{v_{m+1}}{v_{m+2}}$$

$$\frac{u_{m+2}}{u_{m+3}} > \frac{v_{m+2}}{v_{m+3}}$$

$$\frac{u_{m+3}}{u_{m+4}} > \frac{v_{m+3}}{v_{m+4}}$$

....

....

$$\frac{u_{n-1}}{u_n} > \frac{v_{n-1}}{v_n}$$

Multiplying the corresponding sides of the above inequalities, we have

$$\frac{u_{m+1}}{u_n} > \frac{v_{m+1}}{v_n} \qquad \forall\, n > m$$

$$\Rightarrow \quad u_n < \left(\frac{u_{m+1}}{v_{m+1}}\right) v_n \qquad \forall\, n > m$$

$$\Rightarrow \quad u_n < kv_n \qquad \forall\, n > m,$$

where $k = \dfrac{u_{m+1}}{v_{m+1}}$ is a fixed +ve quantity.

Since Σv_n is convergent, so is Σu_n.

(*ii*) Using $$\frac{u_n}{u_{n+1}} < \frac{v_n}{v_{n+1}} \qquad \forall\, n > m$$

and proceeding as in part (*i*), we have $\dfrac{u_{m+1}}{u_n} < \dfrac{v_{m+1}}{v_n}$ $\forall\, n > m$

$$\Rightarrow \quad u_n > \left(\frac{u_{m+1}}{v_{m+1}}\right) v_n \qquad \forall\, n > m$$

$$\Rightarrow \quad u_n > kv_n \qquad \forall\, n > m,$$

where $k = \dfrac{u_{m+1}}{v_{m+1}}$ is a fixed +ve quantity.

Since Σv_n is divergent, so is Σu_n.

Article 8. *An Important Test for Comparison.* $\sum \dfrac{1}{n^p}$. *[Hyper harmonic series or p-series]*

The series $\sum \dfrac{1}{n^p} = \dfrac{1}{1^p} + \dfrac{1}{2^p} + \dfrac{1}{3^p} + \ldots\ldots + \dfrac{1}{n^p} + \ldots\ldots$ *to* ∞ *converges if* $p > 1$ *and diverges if* $p \le 1$.

Proof. Case I. When p > 1

$$\frac{1}{1^p} = 1$$

$$\frac{1}{2^p} + \frac{1}{3^p} < \frac{1}{2^p} + \frac{1}{2^p} = \frac{2}{2^p} = \frac{1}{2^{p-1}} \qquad \left[\because \frac{1}{3^p} < \frac{1}{2^p}\right]$$

$$\frac{1}{4^p} + \frac{1}{5^p} + \frac{1}{6^p} + \frac{1}{7^p} < \frac{1}{4^p} + \frac{1}{4^p} + \frac{1}{4^p} + \frac{1}{4^p}$$

$$= \frac{4}{4^p} = \frac{1}{4^{p-1}} = \frac{1}{(2^{p-1})^2} \qquad \left[\because \frac{1}{5^p} < \frac{1}{4^p}, \frac{1}{6^p} < \frac{1}{4^p} \text{ etc.}\right]$$

Similarly, the sum of next eight terms

$$= \frac{1}{8^p} + \frac{1}{9^p} + \ldots\ldots + \frac{1}{15^p} < \frac{1}{8^p} + \frac{1}{8^p} + \ldots\ldots + \frac{1}{8^p} = \frac{8}{8^p}$$

$$= \frac{1}{8^{p-1}} = \frac{1}{(2^{p-1})^3} \text{ and so on.}$$

Now $\sum \frac{1}{n^p} = \frac{1}{1^p} + \frac{1}{2^p} + \frac{1}{3^p} + \ldots\ldots + \frac{1}{n^p} + \ldots\ldots$

$$= \frac{1}{1^p} + \left(\frac{1}{2^p} + \frac{1}{3^p}\right) + \left(\frac{1}{4^p} + \frac{1}{5^p} + \frac{1}{6^p} + \frac{1}{7^p}\right) + \left(\frac{1}{8^p} + \frac{1}{9^p} + \ldots\ldots + \frac{1}{15^p}\right) + \ldots\ldots \quad \ldots(1)$$

$\therefore$ Each term of (1) after the first is less than the corresponding term in

$$1 + \frac{1}{2^{p-1}} + \frac{1}{(2^{p-1})^2} + \frac{1}{(2^{p-1})^3} + \ldots\ldots \quad \ldots(2)$$

But (2) is a G.P. whose common ratio $= \frac{1}{2^{p-1}} < 1$ $\qquad (\because \ p > 1)$

$\therefore$ (2) is convergent $\Rightarrow$ (1) is convergent.

Hence the given series is convergent.

Case II. When p = 1

$$\sum \frac{1}{n^p} = \sum \frac{1}{n} = 1 + \frac{1}{2} + \frac{1}{3} + \frac{1}{4} + \ldots\ldots$$

$$1 + \frac{1}{2} = 1 + \frac{1}{2}$$

$$\frac{1}{3} + \frac{1}{4} > \frac{1}{4} + \frac{1}{4} = \frac{2}{4} = \frac{1}{2}$$

$$\frac{1}{5} + \frac{1}{6} + \frac{1}{7} + \frac{1}{8} > \frac{1}{8} + \frac{1}{8} + \frac{1}{8} + \frac{1}{8} = \frac{4}{8} = \frac{1}{2} \text{ and so on.}$$

Now $\sum \frac{1}{n} = 1 + \frac{1}{2} + \frac{1}{3} + \frac{1}{4} + \ldots\ldots + \frac{1}{n} + \ldots\ldots$

$$= 1 + \frac{1}{2} + \left(\frac{1}{3} + \frac{1}{4}\right) + \left(\frac{1}{5} + \frac{1}{6} + \frac{1}{7} + \frac{1}{8}\right) + \ldots\ldots \quad \ldots(1)$$

Each term of (1) after the second is greater than the corresponding term in

$$1 + \frac{1}{2} + \frac{1}{2} + \frac{1}{2} + \ldots\ldots \quad \ldots(2)$$

But after the second term (2) is a G.P. whose common ratio = 1.

$\therefore$ (2) is divergent. $\Rightarrow$ (1) is divergent.

Hence the given series is divergent.

Case III. When p < 1

$$p < 1 \;\Rightarrow\; n^p < n \;\Rightarrow\; \frac{1}{n^p} > \frac{1}{n} \quad \forall\, n$$

But the series $\sum \frac{1}{n}$ is divergent (Case II).

Hence $\sum \frac{1}{n^p}$ is also divergent.

ILLUSTRATIVE EXAMPLES

Example 1. *Examine the convergence of the series:*

(i) $\frac{3}{5} + \frac{4}{5^2} + \frac{3}{5^3} + \frac{4}{5^4} + \ldots\ldots$ to ∞ *(ii)* $1 + \frac{1}{4^{2/3}} + \frac{1}{9^{2/3}} + \frac{1}{16^{2/3}} + \ldots\ldots$.

Sol. (*i*) $\frac{3}{5} + \frac{4}{5^2} + \frac{3}{5^3} + \frac{4}{5^4} + \ldots\ldots$ to ∞

$$= \left(\frac{3}{5} + \frac{3}{5^3} + \ldots\ldots \text{ to } \infty\right) + \left(\frac{4}{5^2} + \frac{4}{5^4} + \ldots\ldots \text{ to } \infty\right) = \Sigma u_n + \Sigma v_n \text{ (say)}$$

Now Σu_n is a G.P. with common ratio $= \frac{1}{5^2}$ which is numerically less than 1,

$\therefore$ Σu_n is convergent.

Σv_n is also a G.P. with common ratio $= \frac{1}{5^2}$ which is numerically less than 1.

$\therefore$ Σv_n is convergent.

$\therefore$ The given series *viz.* $\Sigma(u_n + v_n)$ is also convergent.

(*ii*) $1 + \frac{1}{4^{2/3}} + \frac{1}{9^{2/3}} + \frac{1}{16^{2/3}} + \ldots\ldots$ to $\infty = 1 + \frac{1}{(2^2)^{2/3}} + \frac{1}{(3^2)^{2/3}} + \frac{1}{(4^2)^{2/3}} + \ldots\ldots$ to ∞

$$= \frac{1}{1^{4/3}} + \frac{1}{2^{4/3}} + \frac{1}{3^{4/3}} + \frac{1}{4^{4/3}} + \ldots\ldots \text{ to } \infty = \sum \frac{1}{n^{4/3}} = \sum \frac{1}{n^p} \text{ with } p = \frac{4}{3} > 1$$

$\therefore$ By *p*-series test, the given series is convergent.

Example 2. *Test the convergence of the series* : $\frac{1}{1.2.3} + \frac{3}{2.3.4} + \frac{5}{3.4.5} + \ldots\ldots$

Sol. Here $$u_n = \frac{\text{T}_n \text{ of } 1, 3, 5, \ldots\ldots}{n(n+1)(n+2)} = \frac{2n-1}{n(n+1)(n+2)} = \frac{n\left(2 - \frac{1}{n}\right)}{n^3\left(1+\frac{1}{n}\right)\left(1+\frac{2}{n}\right)}$$

$$= \frac{2 - \frac{1}{n}}{n^2\left(1+\frac{1}{n}\right)\left(1+\frac{2}{n}\right)}$$

Let us compare Σu_n with Σv_n, where $v_n = \frac{1}{n^2}$

$$\frac{u_n}{v_n} = \frac{2 - \dfrac{1}{n}}{\left(1+\dfrac{1}{n}\right)\left(1+\dfrac{2}{n}\right)}$$

$$\underset{n\to\infty}{\text{Lt}} \frac{u_n}{v_n} = \underset{n\to\infty}{\text{Lt}} \frac{2 - \dfrac{1}{n}}{\left(1+\dfrac{1}{n}\right)\left(1+\dfrac{2}{n}\right)} = \frac{2}{(1)(1)} = 2 \text{ which is finite and} \neq 0.$$

$\therefore$ Σu_n and Σv_n converge or diverge together.

Since $\Sigma v_n = \Sigma \dfrac{1}{n^2}$ is of the form $\Sigma \dfrac{1}{n^p}$ with $p = 2 > 1$.

$\therefore$ Σv_n is convergent $\Rightarrow$ Σu_n is convergent.

Example 3. *Test the convergence of the following series:*

(*i*) $\dfrac{1}{\sqrt{1}+\sqrt{2}} + \dfrac{1}{\sqrt{2}+\sqrt{3}} + \dfrac{1}{\sqrt{3}+\sqrt{4}} + \ldots\ldots$ (*ii*) $\sqrt{\dfrac{1}{4}} + \sqrt{\dfrac{2}{6}} + \sqrt{\dfrac{3}{8}} + \ldots\ldots + \sqrt{\dfrac{n}{2(n+1)}} + \ldots\ldots$

Sol. (*i*) Here $u_n = \dfrac{1}{\sqrt{n}+\sqrt{n+1}} = \dfrac{1}{\sqrt{n}\left[1+\sqrt{1+\dfrac{1}{n}}\right]}$

Let us compare Σu_n with Σv_n, where $v_n = \dfrac{1}{\sqrt{n}}$

$$\frac{u_n}{v_n} = \frac{1}{1+\sqrt{1+\dfrac{1}{n}}}$$

$$\underset{n\to\infty}{\text{Lt}} \frac{u_n}{v_n} = \underset{n\to\infty}{\text{Lt}} \frac{1}{1+\sqrt{1+\dfrac{1}{n}}} = \frac{1}{1+1} = \frac{1}{2} \quad \text{which is finite and} \neq 0.$$

$\therefore$ Σu_n and Σv_n converge or diverge together.

Since $\Sigma v_n = \Sigma \dfrac{1}{n^{1/2}}$ is of the form $\Sigma \dfrac{1}{n^p}$ with $p = \dfrac{1}{2} < 1$

$\therefore$ Σv_n is divergent $\Rightarrow$ Σu_n is divergent.

(*ii*) Here $u_n = \sqrt{\dfrac{n}{2(n+1)}} = \sqrt{\dfrac{1}{2\left(1+\dfrac{1}{n}\right)}}$

$$\underset{n\to\infty}{\text{Lt}}\ u_n = \underset{n\to\infty}{\text{Lt}} \frac{1}{\sqrt{2\left(1+\dfrac{1}{n}\right)}} = \frac{1}{\sqrt{2}} \neq 0$$

$\Rightarrow$ Σu_n does not converge.

Since the given series is a series of +ve terms, it either converges or diverges. Since it does not converge, it must diverge.

Hence the given series is divergent.

Example 4. *Test the convergence of the series:* $1 + \frac{1}{2^2} + \frac{2^2}{3^3} + \frac{3^3}{4^4} + \frac{4^4}{5^5} + \ldots\ldots$

Sol. *Leaving aside the first term* ($\because$ *addition or deletion of a finite number of terms does not alter the nature of the series*), we have $u_n = \frac{n^n}{(n+1)^{n+1}} = \frac{n^n}{n^{n+1}\left(1+\frac{1}{n}\right)^{n+1}} = \frac{1}{n\left(1+\frac{1}{n}\right)^{n+1}}$

Take $$v_n = \frac{1}{n}.$$

$$\underset{n\to\infty}{\text{Lt}} \frac{u_n}{v_n} = \underset{n\to\infty}{\text{Lt}} \frac{1}{\left(1+\frac{1}{n}\right)^{n+1}} = \underset{n\to\infty}{\text{Lt}} \frac{1}{\left(1+\frac{1}{n}\right)^n} \cdot \frac{1}{\left(1+\frac{1}{n}\right)}$$

$$= \frac{1}{e} \cdot \frac{1}{1} \qquad \left[\because \underset{n\to\infty}{\text{Lt}} \left(1+\frac{1}{n}\right)^n = e\right]$$

$$= \frac{1}{e} \text{ which is finite and } \neq 0.$$

$\therefore$ Σu_n and Σv_n converge or diverge together.

Since $\Sigma v_n = \Sigma \frac{1}{n}$ is of the form $\Sigma \frac{1}{n^p}$ with $p = 1$

$\therefore$ Σv_n is divergent. $\Rightarrow$ Σu_n is divergent.

Example 5. *Examine the convergence of the series:* $\frac{\sqrt{2}-1}{3^3-1} + \frac{\sqrt{3}-1}{4^3-1} + \frac{\sqrt{4}-1}{5^3-1} + \ldots\ldots$

Sol. Here $$u_n = \frac{\sqrt{n+1}-1}{(n+2)^3-1} = \frac{\sqrt{n}\left(\sqrt{1+\frac{1}{n}} - \frac{1}{\sqrt{n}}\right)}{n^3\left[\left(1+\frac{2}{n}\right)^3 - \frac{1}{n^3}\right]} = \frac{\sqrt{1+\frac{1}{n}} - \frac{1}{\sqrt{n}}}{n^{5/2}\left[\left(1+\frac{2}{n}\right)^3 - \frac{1}{n^3}\right]}$$

Take $$v_n = \frac{1}{n^{5/2}}$$

$$\underset{n\to\infty}{\text{Lt}} \frac{u_n}{v_n} = \underset{n\to\infty}{\text{Lt}} \frac{\sqrt{1+\frac{1}{n}} - \frac{1}{\sqrt{n}}}{\left(1+\frac{2}{n}\right)^3 - \frac{1}{n^3}} = \frac{\sqrt{1+0}-0}{(1+0)^3-0} = 1 \quad \text{which is finite and } \neq 0.$$

$\therefore$ Σu_n and Σv_n converge or diverge together.

Since $\Sigma u_n = \Sigma \frac{1}{n^{5/2}}$ is of the form $\Sigma \frac{1}{n^p}$ with $p = \frac{5}{2} > 1$.

Σv_n is convergent. $\Rightarrow$ Σu_n is convergent.

Example 6. *Examine the convergence of the series:*

(i) $\sum \frac{\sqrt{n+1}-\sqrt{n}}{n^p}$ *(ii)* $\sum \left(\sqrt[3]{n^3+1} - n\right)$.

Sol. (*i*) Here $$u_n = \frac{\sqrt{n+1}-\sqrt{n}}{n^p} = \frac{\sqrt{n+1}-\sqrt{n}}{n^p} \times \frac{\sqrt{n+1}+\sqrt{n}}{\sqrt{n+1}+\sqrt{n}} \quad \text{(Rationalising)}$$

$$= \frac{(n+1)-n}{n^p \cdot \sqrt{n}\left(\sqrt{1+\frac{1}{n}}+1\right)} = \frac{1}{n^{p+1/2}\left(\sqrt{1+\frac{1}{n}}+1\right)}$$

Take $$v_n = \frac{1}{n^{p+1/2}}$$

$$\underset{n\to\infty}{\text{Lt}} \frac{u_n}{v_n} = \underset{n\to\infty}{\text{Lt}} \frac{1}{\sqrt{1+\frac{1}{n}}+1} = \frac{1}{1+1} = \frac{1}{2}$$ which is finite and $\neq 0$.

$\therefore$ Σu_n and Σv_n converge or diverge together.

Since $\sum v_n = \sum \frac{1}{n^{p+1/2}}$ is convergent if $p + \frac{1}{2} > 1$ and divergent if $p + \frac{1}{2} \leq 1$.

i.e., convergent if $p > \frac{1}{2}$ and divergent if $p \leq \frac{1}{2}$.

$\therefore$ Σu_n is convergent if $p > \frac{1}{2}$ and divergent if $p \leq \frac{1}{2}$.

(*ii*) Here $$u_n = (n^3+1)^{1/3} - n = \left[n^3\left(1+\frac{1}{n^3}\right)\right]^{1/3} - n$$

$$= n\left(1+\frac{1}{n^3}\right)^{1/3} - n = n\left[\left(1+\frac{1}{n^3}\right)^{1/3} - 1\right]$$

$$= n\left[1+\frac{1}{3}\cdot\frac{1}{n^3}+\frac{\frac{1}{3}\left(\frac{1}{3}-1\right)}{2!}\cdot\left(\frac{1}{n^3}\right)^2+\ldots\ldots-1\right]$$

$$= \frac{n}{n^3}\left[\frac{1}{3}-\frac{1}{9}\cdot\frac{1}{n^3}\ldots\ldots\right] = \frac{1}{n^2}\left[\frac{1}{3}-\frac{1}{9}\cdot\frac{1}{n^3}\ldots\ldots\right]$$

Take $$v_n = \frac{1}{n^2}.$$

$$\underset{n\to\infty}{\text{Lt}} \frac{u_n}{v_n} = \underset{n\to\infty}{\text{Lt}} \left(\frac{1}{3}-\frac{1}{9}\cdot\frac{1}{n^3}\ldots\ldots\right) = \frac{1}{3}$$

which is finite and $\neq 0$.

$\therefore$ Σu_n and Σv_n converge or diverge together.

Since $\sum v_n = \sum \frac{1}{n^2}$ is of the form $\sum \frac{1}{n^p}$ with $p = 2 > 1$

$\therefore$ Σv_n is convergent $\Rightarrow$ Σu_n is convergent.

Note. Rationalisation is effective only when square roots are involved whereas this method of Binomial Expansion is general.

TEST YOUR KNOWLEDGE

Test the convergence or divergence of the following series :

1. $\frac{1}{2}+\frac{1}{3^2}+\frac{1}{2^3}+\frac{1}{3^4}+ \ldots\ldots$ to ∞

2. $\frac{1}{1.2}+\frac{1}{2.3}+\frac{1}{3.4}+ \ldots\ldots$ to ∞

3. $\frac{1}{1.4}+\frac{1}{2.5}+\frac{1}{3.6}+ \ldots\ldots$ to ∞

4. $\frac{1}{1.2}+\frac{2}{3.4}+\frac{3}{5.6}+ \ldots\ldots$ to ∞

5. $\frac{1}{1.3}+\frac{2}{3.5}+\frac{3}{5.7}+ \ldots\ldots$ to ∞

6. $\frac{1}{\sqrt{1.2}}+\frac{1}{\sqrt{2.3}}+\frac{1}{\sqrt{3.4}}+ \ldots\ldots$ to ∞

7. $\frac{1}{1^p}+\frac{1}{3^p}+\frac{1}{5^p}+ \ldots\ldots$ to ∞

8. $\frac{2}{1^p}+\frac{3}{2^p}+\frac{4}{3^p}+ \ldots\ldots$ to ∞

9. $\sum_{n=1}^{\infty} \frac{n+1}{n(2n-1)}$

10. $\sum_{n=1}^{\infty} \frac{1}{n^p(n+1)^p}$

11. $\frac{2^p}{1^q}+\frac{3^p}{2^q}+\frac{4^p}{3^q}+ \ldots\ldots$ to ∞
(*p* and *q* are positive numbers)

12. $\sum_{n=1}^{\infty} \frac{\sqrt{n}}{n^2+1}$

13. $\sum \frac{2n^3+5}{4n^5+1}$

14. $\sum \frac{\sqrt{n^2-1}}{n^3+1}$

15. $\frac{1}{5}+\frac{\sqrt{2}}{7}+\frac{\sqrt{3}}{9}+\frac{\sqrt{4}}{11}+ \ldots\ldots$ to ∞

16. $\sum \left(\sqrt{n^2+1}-n\right)$

17. $\sum \left(\sqrt{n^3+1}-\sqrt{n^3}\right)$

18. $\sum \left(\sqrt{n^4+1}-\sqrt{n^4-1}\right)$ [*CUSAT 2010*]

19. $\frac{\sqrt{2}-\sqrt{1}}{1}+\frac{\sqrt{3}-\sqrt{2}}{2}+\frac{\sqrt{4}-\sqrt{3}}{3}+ \ldots\ldots$

20. $\sum (\sqrt[3]{n+1}-\sqrt[3]{n})$

21. $\sum \frac{1}{\sqrt{n}} \sin \frac{1}{n}$

22. $\sum \cot^{-1} n^2$

23. $\sum \left(\frac{1}{n}-\log \frac{n+1}{n}\right)$

24. $\sum_{n=2}^{\infty} \frac{1}{n \log n}$.

Answers

1. Convergent
2. Convergent
3. Convergent
4. Divergent
5. Divergent
6. Divergent
7. Convergent for $p > 1$, divergent for $p \le 1$
8. Convergent for $p > 2$, divergent for $p \le 2$
9. Divergent
10. Convergent for $p > \frac{1}{2}$, divergent for $p \le \frac{1}{2}$
11. Convergent for $q > p + 1$, divergent for $q \le p + 1$
12. Convergent
13. Convergent
14. Convergent
15. Divergent
16. Divergent
17. Convergent
18. Convergent
19. Convergent
20. Divergent
21. Convergent
22. Convergent
23. Convergent
24. Divergent.

Article 9. D' Alembert's Ratio Test

Statement. *If* Σu_n *is a positive term series, and* $\underset{n\to\infty}{Lt} \frac{u_{n+1}}{u_n} = l$, *then*

(*i*) Σu_n *is convergent if* $l < 1$. (*ii*) Σu_n *is divergent if* $l > 1$.

Note. If $l = 1$, the test fails, *i.e.*, no conclusion can be drawn about the convergence or divergence of the series. The series may converge, it may diverge.

Proof. Since $\underset{n\to\infty}{\text{Lt}} \frac{u_{n+1}}{u_n} = l$,

$\therefore$ given $\varepsilon > 0$, however small, there exists a positive integer m such that

$$\left| \frac{u_{n+1}}{u_n} - l \right| < \varepsilon \quad \forall\, n \geq m$$

$$\Rightarrow \qquad l - \varepsilon < \frac{u_{n+1}}{u_n} < 1 + \varepsilon \quad \forall\, n \geq m \qquad \ldots(i)$$

(*i*) **When $l < 1$.**

Choose $\varepsilon > 0$ such that $l < l + \varepsilon < 1$

Put $l + \varepsilon = r$, then $0 < r < 1$.

From (*i*), $\frac{u_{n+1}}{u_n} < r \;\; \forall\, n \geq m \;\Rightarrow\; u_{n+1} < ru_n \;\; \forall\, n \geq m \qquad (\because\; u_n > 0)$

Putting $n = m, m + 1, m + 2, \ldots\ldots$, we get

$$u_{m+1} < ru_m$$
$$u_{m+2} < ru_{m+1} < r^2 u_m$$
$$u_{m+3} < ru_{m+2} < r^3 u_m \quad \text{and so on.}$$

Adding $u_{m+1} + u_{m+2} + u_{m+3} + \ldots\ldots < u_m (r + r^2 + r^3 + \ldots\ldots)$

$\Rightarrow$ each term of the given series Σu_n after leaving the first m terms (*i.e.*, a finite number of terms) is less than the corresponding term of a geometric series which is convergent ($\because$ its common ratio $r < 1$). Hence the given series is also convergent.

(*ii*) **When $l > 1$.**

Choose $\varepsilon > 0$ such that $l - \varepsilon > 1$

Put $l - \varepsilon = R$, then $R > 1$.

From (*i*), $\frac{u_{n+1}}{u_n} > R \;\; \forall\, n \geq m \;\Rightarrow\; u_{n+1} > Ru_n \;\; \forall\; n \geq m \qquad (\because\; u_n > 0)$

Putting $n = m, m + 1, m + 2, \ldots\ldots$, we get

$$u_{m+1} > Ru_m$$
$$u_{m+2} > Ru_{m+1} > R^2 u_m$$
$$u_{m+3} > Ru_{m+2} > R^3 u_m \quad \text{and so on.}$$

Adding $u_{m+1} + u_{m+2} + u_{m+3} + \ldots\ldots > u_m (R + R^2 + R^3 + \ldots\ldots)$

$\Rightarrow$ each term of the given series Σu_n after leaving the first m terms (*i.e.*, a finite number of terms) is greater than the corresponding term of a geometric series which is divergent ($\because$ its common ratio $R > 1$). Hence the given series is also divergent.

Practical Form of D'Alembert's Ratio Test

In practice, Ratio Test is used in the following form :

If Σu_n is a positive term series, and $\underset{n\to\infty}{\text{Lt}} \dfrac{u_n}{u_{n+1}} = l$, then

(*i*) Σu_n is convergent if $l > 1$ (*ii*) Σu_n is divergent if $l < 1$.

ILLUSTRATIVE EXAMPLES

Example 1. *Discuss the convergence of the following series:*

(*i*) $1 + \dfrac{2^p}{2!} + \dfrac{3^p}{3!} + \dfrac{4^p}{4!} + \ldots\ldots, (p > 0)$ (*ii*) $\dfrac{1}{2} + \dfrac{1}{3} + \dfrac{1}{5} + \ldots\ldots + \dfrac{1}{2^{n-1}+1} + \ldots\ldots$

(*iii*) $\dfrac{1^2 . 2^2}{1!} + \dfrac{2^2 . 3^2}{2!} + \dfrac{3^2 . 4^2}{3!} + \dfrac{4^2 . 5^2}{4!} + \ldots\ldots$

Sol. (*i*) Here $u_n = \dfrac{n^p}{n!}$ $\left[\because 1 = \dfrac{1^p}{1!}\right]$

$\therefore$ $u_{n+1} = \dfrac{(n+1)^p}{(n+1)!}$

$$\frac{u_n}{u_{n+1}} = \frac{n^p}{n!} \frac{(n+1)!}{(n+1)^p} = \frac{n^p . (n+1)\, n!}{n!\,(n+1)^p} = \frac{n^p}{(n+1)^{p-1}}$$

$$= \frac{n^p}{n^{p-1}\left(1+\dfrac{1}{n}\right)^{p-1}} = \frac{n}{\left(1+\dfrac{1}{n}\right)^{p-1}}$$

$$\underset{n\to\infty}{\text{Lt}} \frac{u_n}{u_{n+1}} = \underset{n\to\infty}{\text{Lt}} \frac{n}{\left(1+\dfrac{1}{n}\right)^{p-1}} = \infty > 1$$

$\therefore$ By D'Alembert's Ratio Test, Σu_n is convergent.

(*ii*) Here $u_n = \dfrac{1}{2^{n-1}+1}$ $\therefore$ $u_{n+1} = \dfrac{1}{2^n+1}$

$$\frac{u_n}{u_{n+1}} = \frac{2^n+1}{2^{n-1}+1} = \frac{2^n\left(1+\dfrac{1}{2^n}\right)}{2^{n-1}\left(1+\dfrac{1}{2^{n-1}}\right)} = 2 . \frac{1+\dfrac{1}{2^n}}{1+\dfrac{1}{2^{n-1}}}$$

$$\underset{n\to\infty}{\text{Lt}} \frac{u_n}{u_{n+1}} = \underset{n\to\infty}{\text{Lt}}\; 2 . \frac{1+\dfrac{1}{2^n}}{1+\dfrac{1}{2^{n-1}}} = 2 > 1$$

$\therefore$ By D'Alembert's Ratio Test, Σu_n is convergent.

(*iii*) Here $u_n = \dfrac{n^2 (n+1)^2}{n!} \quad \therefore \quad u_{n+1} = \dfrac{(n+1)^2 (n+2)^2}{(n+1)!}$

$$\frac{u_n}{u_{n+1}} = \frac{n^2 (n+1)^2}{n!} \cdot \frac{(n+1)!}{(n+1)^2 (n+2)^2} = \frac{n^2 . (n+1)\, n!}{n!\,(n+2)^2}$$

$$= \frac{n^3\left(1+\frac{1}{n}\right)}{n^2\left(1+\frac{2}{n}\right)^2} = n \cdot \frac{1+\frac{1}{n}}{\left(1+\frac{2}{n}\right)^2}$$

$$\underset{n\to\infty}{\text{Lt}} \frac{u_n}{u_{n+1}} = \underset{n\to\infty}{\text{Lt}}\; n \cdot \frac{1+\frac{1}{n}}{\left(1+\frac{2}{n}\right)^2} = \infty > 1$$

$\therefore$ By D'Alembert's Ratio Test, Σu_n is convergent.

Example 2. *Test the convergence of the following series:*

(*i*) $\sum \dfrac{n^3 + a}{2^n + a}$ (*ii*) $\sum \dfrac{n!\, 2^n}{n^n}$.

Sol. (*i*) Here $u_n = \dfrac{n^3 + a}{2^n + a} \quad \therefore \quad u_{n+1} = \dfrac{(n+1)^3 + a}{2^{n+1} + a}$

$$\frac{u_n}{u_{n+1}} = \frac{n^3 + a}{(n+1)^3 + a} \cdot \frac{2^{n+1} + a}{2^n + a}$$

$$= \frac{n^3\left(1+\frac{a}{n^3}\right)}{n^3\left[\left(1+\frac{1}{n}\right)^3 + \frac{a}{n^3}\right]} \cdot \frac{2^{n+1}\left(1+\frac{a}{2^{n+1}}\right)}{2^n\left(1+\frac{a}{2^n}\right)} = \frac{1+\frac{a}{n^3}}{\left(1+\frac{1}{n}\right)^3 + \frac{a}{n^3}} \cdot \frac{2\left(1+\frac{a}{2^{n+1}}\right)}{1+\frac{a}{2^n}}$$

$$\underset{n\to\infty}{\text{Lt}} \frac{u_n}{u_{n+1}} = \frac{1+0}{1+0} \cdot 2 \cdot \frac{1+0}{1+0} = 2 > 1$$

$\therefore$ By D'Alembert's Ratio Test, Σu_n is convergent.

(*ii*) Here $u_n = \dfrac{n!\, 2^n}{n^n} \quad \therefore \quad u_{n+1} = \dfrac{(n+1)!\, 2^{n+1}}{(n+1)^{n+1}}$

$$\frac{u_n}{u_{n+1}} = \frac{n!\, 2^n}{(n+1)!\, 2^{n+1}} \cdot \frac{(n+1)^{n+1}}{n^n} = \frac{1}{2(n+1)} \cdot \frac{(n+1)^{n+1}}{n^n}$$

$$= \frac{1}{2} \cdot \frac{(n+1)^n}{n^n} = \frac{1}{2} \cdot \left(\frac{n+1}{n}\right)^n = \frac{1}{2}\left(1+\frac{1}{n}\right)^n$$

$$\underset{n\to\infty}{\text{Lt}} \frac{u_n}{u_{n+1}} = \underset{n\to\infty}{\text{Lt}} \frac{1}{2}\left(1+\frac{1}{n}\right)^n = \frac{e}{2}$$

Now $2 < e < 3 \;\Rightarrow\; 1 < \dfrac{e}{2} < \dfrac{3}{2}$

$$\underset{n\to\infty}{\text{Lt}} \frac{u_n}{u_{n+1}} = \frac{e}{2} > 1 \;\Rightarrow\; \Sigma u_n \text{ is convergent.}$$

Example 3. *Discuss the convergence of the series:* $\sum \frac{n!}{n^n}$.

Sol. Here $$u_n = \frac{n!}{n^n}$$

$$u_{n+1} = \frac{(n+1)!}{(n+1)^{n+1}} = \frac{(n+1)\,n!}{(n+1)^{n+1}} = \frac{n!}{(n+1)^n}$$

$$\frac{u_n}{u_{n+1}} = \frac{(n+1)^n}{n^n} = \left(\frac{n+1}{n}\right)^n = \left(1+\frac{1}{n}\right)^n$$

$$\underset{n\to\infty}{\text{Lt}} \frac{u_n}{u_{n+1}} = \underset{n\to\infty}{\text{Lt}} \left(1+\frac{1}{n}\right)^n = e > 1 \qquad (\because \quad 2 < e < 3)$$

$\therefore$ By D'Alembert's Ratio Test, Σu_n is convergent.

Example 4. *Discuss the convergence of the series:* $\sum \frac{\sqrt{n}}{\sqrt{n^2+1}} x^n$.

Sol. Here $$u_n = \sqrt{\frac{n}{n^2+1}}\, x^n$$

$\therefore$ $$u_{n+1} = \sqrt{\frac{n+1}{(n+1)^2+1}}\,.\, x^{n+1}$$

$$\frac{u_n}{u_{n+1}} = \sqrt{\frac{n}{n+1}\cdot\frac{n^2+2n+2}{n^2+1}}\cdot\frac{1}{x} = \sqrt{\frac{1}{1+\frac{1}{n}}\cdot\frac{1+\frac{2}{n}+\frac{2}{n^2}}{1+\frac{1}{n^2}}}\cdot\frac{1}{x}$$

$$\underset{n\to\infty}{\text{Lt}} \frac{u_n}{u_{n+1}} = \underset{n\to\infty}{\text{Lt}} \sqrt{\frac{1}{1+\frac{1}{n}}\cdot\frac{1+\frac{2}{n}+\frac{2}{n^2}}{1+\frac{1}{n^2}}}\cdot\frac{1}{x} = \frac{1}{x}$$

$\therefore$ By D'Alembert's Ratio Test, $\sum u_n$ converges if $\frac{1}{x} > 1$ *i.e.,* $x < 1$

and diverges if $\frac{1}{x} < 1$ *i.e.,* $x > 1$

When $x = 1$, the Ratio Test fails.

When $x = 1$, $u_n = \sqrt{\frac{n}{n^2+1}} = \sqrt{\frac{n}{n^2\left(1+\frac{1}{n^2}\right)}} = \frac{1}{\sqrt{n}}\cdot\frac{1}{\sqrt{1+\frac{1}{n^2}}}$

Take $v_n = \frac{1}{\sqrt{n}}$, $\underset{n\to\infty}{\text{Lt}} \frac{u_n}{v_n} = \underset{n\to\infty}{\text{Lt}} \frac{1}{\sqrt{1+\frac{1}{n^2}}} = 1$

which is finite and $\neq 0$.

$\therefore$ By Comparison Test, Σu_n and Σv_n converge or diverge together.

Since $\sum v_n = \sum \frac{1}{\sqrt{n}}$ is of the form $\sum \frac{1}{n^p}$ with $p = \frac{1}{2}$ (< 1),

Σv_n diverges $\Rightarrow$ Σu_n diverges.

Hence the given series Σu_n converges if $x < 1$ and diverges if $x \geq 1$.

Example 5. *Examine the convergence or divergence of the following series :*

$$\frac{1}{2\sqrt{1}}+\frac{x^2}{3\sqrt{2}}+\frac{x^4}{4\sqrt{3}}+\frac{x^6}{5\sqrt{4}}+\ldots\ldots$$

Sol. Here $\qquad u_n = \dfrac{x^{2n-2}}{(n+1)\sqrt{n}} \quad \therefore \quad u_{n+1} = \dfrac{x^{2n}}{(n+2)\sqrt{n+1}}$

$$\frac{u_n}{u_{n+1}} = \frac{(n+2)\sqrt{n+1}}{(n+1)\sqrt{n}} \cdot \frac{1}{x^2} = \frac{1+\frac{2}{n}}{1+\frac{1}{n}} \cdot \sqrt{1+\frac{1}{n}} \cdot \frac{1}{x^2}$$

$$\underset{n\to\infty}{\text{Lt}} \frac{u_n}{u_{n+1}} = \frac{1}{x^2}$$

By D'Alembert's Ratio Test, $\sum u_n$ converges if $\dfrac{1}{x^2} > 1$, *i.e.*, $x^2 < 1$

and diverges if $\dfrac{1}{x^2} < 1$ *i.e.*, $x^2 > 1$.

When $x^2 = 1$, $u_n = \dfrac{1}{(n+1)\sqrt{n}} = \dfrac{1}{n^{3/2}\left(1+\frac{1}{n}\right)}$

Take $v_n = \dfrac{1}{n^{3/2}}$ By Comparison Test Σu_n is convergent $\qquad (\because \ p > 1)$

Hence Σu_n is convergent if $x^2 \leq 1$ and divergent if $x^2 > 1$.

Example 6. *Examine the convergence or divergence of the following series:*

$$1+\frac{2}{5}x+\frac{6}{9}x^2+\frac{14}{17}x^3+\ldots\ldots+\frac{2^{n+1}-2}{2^{n+1}+1}x^n+\ldots\ldots \ (x > 0).$$

Sol. Here, leaving the first term, $u_n = \dfrac{2^{n+1}-2}{2^{n+1}+1}x^n$

$$\therefore \qquad u_{n+1} = \frac{2^{n+2}-2}{2^{n+2}+1}x^{n+1}$$

$$\frac{u_n}{u_{n+1}} = \frac{2^{n+1}-2}{2^{n+1}+1} \cdot \frac{2^{n+2}+1}{2^{n+2}-2} \cdot \frac{1}{x} = \frac{2^{n+1}\left(1-\frac{2}{2^{n+1}}\right)}{2^{n+1}\left(1+\frac{1}{2^{n+1}}\right)} \cdot \frac{2^{n+2}\left(1+\frac{1}{2^{n+2}}\right)}{2^{n+2}\left(1-\frac{2}{2^{n+2}}\right)} \cdot \frac{1}{x}$$

$$\underset{n\to\infty}{\text{Lt}} \frac{u_n}{u_{n+1}} = \underset{n\to\infty}{\text{Lt}} \frac{1-\frac{1}{2^n}}{1+\frac{1}{2^{n+1}}} \cdot \frac{1+\frac{1}{2^{n+2}}}{1-\frac{1}{2^{n+1}}} \cdot \frac{1}{x} = \frac{1}{x}$$

$\therefore$ By D'Alembert's Ratio Test, $\sum u_n$ converges if $\dfrac{1}{x} > 1$, *i.e.*, $x < 1$

and diverges if $\dfrac{1}{x} < 1$ *i.e.*, $x > 1$.

When $x = 1$, $u_n = \dfrac{2^{n+1} - 2}{2^{n+1} + 1} = \dfrac{2^{n+1}\left(1 - \dfrac{2}{2^{n+1}}\right)}{2^{n+1}\left(1 + \dfrac{1}{2^{n+2}}\right)} = \dfrac{1 - \dfrac{1}{2^n}}{1 + \dfrac{1}{2^{n+1}}}$

$\underset{n\to\infty}{\text{Lt}}\ u_n = 1 \neq 0 \Rightarrow \Sigma u_n$ does not converge. Being a series of + ve terms, it must diverge.

Hence Σu_n is convergent if $x < 1$ and divergent if $x \geq 1$.

Example 7. *Test for convergence the positive term series :*

$$1 + \frac{\alpha+1}{\beta+1} + \frac{(\alpha+1)(2\alpha+1)}{(\beta+1)(2\beta+1)} + \frac{(\alpha+1)(2\alpha+1)(3\alpha+1)}{(\beta+1)(2\beta+1)(3\beta+1)} + \ldots\ldots$$

Sol. Leaving the first term $u_n = \dfrac{(\alpha+1)(2\alpha+1)\ldots\ldots(n\alpha+1)}{(\beta+1)(2\beta+1)\ldots\ldots(n\beta+1)}$

$$\therefore \qquad u_{n+1} = \frac{(\alpha+1)(2\alpha+1)\ldots\ldots(n\alpha+1)[(n+1)\alpha+1]}{(\beta+1)(2\beta+1)\ldots\ldots(n\beta+1)[(n+1)\beta+1]}$$

$$\frac{u_n}{u_{n+1}} = \frac{(n+1)\beta+1}{(n+1)\alpha+1} = \frac{\left(1+\dfrac{1}{n}\right)\beta + \dfrac{1}{n}}{\left(1+\dfrac{1}{n}\right)\alpha + \dfrac{1}{n}}$$

$$\underset{n\to\infty}{\text{Lt}} \frac{u_n}{u_{n+1}} = \underset{n\to\infty}{\text{Lt}} \frac{\left(1+\dfrac{1}{n}\right)\beta + \dfrac{1}{n}}{\left(1+\dfrac{1}{n}\right)\alpha + \dfrac{1}{n}} = \frac{\beta}{\alpha}$$

$\therefore$ By D'Alembert's Ratio Test, Σu_n converges if $\dfrac{\beta}{\alpha} > 1$ *i.e.,* $\beta > \alpha > 0$

and diverges if $\dfrac{\beta}{\alpha} < 1$ *i.e.,* $\beta < \alpha$ or $\alpha > \beta > 0$

When $\alpha = \beta$, the Ratio Test fails.

When $\alpha = \beta$, $u_n = 1 \quad \therefore \quad \underset{n\to\infty}{\text{Lt}}\ u_n = 1 \neq 0$

$\Rightarrow \Sigma u_n$ does not converge. Being a series of +ve terms, it must diverge.

Hence the given series is convergent if $\beta > \alpha > 0$ and divergent if $\alpha \geq \beta > 0$.

TEST YOUR KNOWLEDGE

Discuss the convergence of the following series:

1. $1 + \dfrac{2^2}{2!} + \dfrac{3^2}{3!} + \dfrac{4^2}{4!} + \ldots\ldots$ to ∞

2. $1 + \dfrac{2!}{2^2} + \dfrac{3!}{3^3} + \dfrac{4!}{4^4} + \ldots\ldots$ to ∞

3. $\dfrac{1}{1+2} + \dfrac{2}{1+2^2} + \dfrac{3}{1+2^3} + \ldots\ldots$ to ∞

4. $\dfrac{2}{1} + \dfrac{2.5.8}{1.5.9} + \dfrac{2.5.8.11}{1.5.9.13} + \ldots\ldots$ to ∞

5. $\dfrac{2!}{3} + \dfrac{3!}{3^2} + \dfrac{4!}{3^3} + \ldots\ldots$ to ∞

6. $\left(\dfrac{1}{3}\right)^2 + \left(\dfrac{1.2}{3.5}\right)^2 + \left(\dfrac{1.2.3}{3.5.7}\right)^2 + \left(\dfrac{1.2.3.4}{3.5.7.9}\right)^2 + \ldots\ldots$ to ∞

7. $\sum \dfrac{2^{n-1}}{3^n + 1}$

8. $\sum \dfrac{1}{n!}$

9. $\sum \dfrac{n^2(n+1)^2}{n!}$

10. $\sum \dfrac{x^n}{3^n \cdot n^2}, x > 0$

11. $\sum \dfrac{x^n}{n}, x > 0$

12. $\sum \dfrac{n}{n^2+1} x^n, x > 0$

13. $\sum \sqrt{\dfrac{n+1}{n^3+1}} \cdot x^n, x > 0$

14. $x + 2x^2 + 3x^3 + 4x^4 + \ldots\ldots$ to ∞

15. $1 + \dfrac{x}{2} + \dfrac{x^2}{5} + \dfrac{x^3}{10} + \ldots\ldots \dfrac{x^n}{n^2+1} + \ldots\ldots$ to ∞

16. $\dfrac{x}{1.3} + \dfrac{x^2}{3.5} + \dfrac{x^3}{5.7} + \ldots\ldots$ to ∞

17. $x + \dfrac{3}{5}x^2 + \dfrac{8}{10}x^3 + \dfrac{15}{17}x^4 + \ldots\ldots + \dfrac{n^2-1}{n^2+1}x^n + \ldots\ldots$ to ∞

18. $\dfrac{x}{2\sqrt{3}} + \dfrac{x^2}{3\sqrt{4}} + \dfrac{x^3}{4\sqrt{5}} + \ldots\ldots$ to ∞

19. $\sum \dfrac{3^n - 2}{3^n + 1} \cdot x^{n-1}, x > 0$

20. $\sum\limits_{n=1}^{\infty} \dfrac{n^2}{3^n}$

21. $\sum\limits_{n=1}^{\infty} \dfrac{x^n}{(2n)!}$

Answers

1. Convergent
2. Convergent
3. Convergent
4. Convergent
5. Divergent
6. Convergent
7. Convergent
8. Convergent
9. Convergent
10. Convergent for $x \le 3$, divergent for $x > 3$.
11. Convergent for $x < 1$, divergent for $x \ge 1$
12. Convergent for $x < 1$, divergent for $x \ge 1$
13. Convergent for $x < 1$, divergent for $x \ge 1$
14. Convergent for $x < 1$, divergent for $x \ge 1$
15. Convergent for $x \le 1$, divergent for $x > 1$
16. Convergent for $x \le 1$, divergent for $x > 1$
17. Convergent for $x < 1$, divergent for $x \ge 1$
18. Convergent for $x \le 1$, divergent for $x > 1$
19. Convergent for $x < 1$, divergent for $x \ge 1$
20. Convergent
21. Convergent.

Article 10. Cauchy's Root Test

Statement. *If Σu_n is a positive term series and* $\underset{n \to \infty}{\text{Lt}} (u_n)^{1/n} = l$, *then*

(i) Σu_n is convergent if $l < 1$ *(ii) Σu_n is divergent if $l > 1$.*

Note. If $l = 1$, the test fails *i.e.*, no conclusion can be drawn about the convergence or divergence of the series. The series may converge, it may diverge.

Proof. Since $\underset{n \to \infty}{\text{Lt}} (u_n)^{1/n} = l$,

$\therefore$ given $\varepsilon > 0$, however small, there exists a +ve integer m such that

$$| (u_n)^{1/n} - l | < \varepsilon \qquad \forall \ n \ge m$$

$$\Rightarrow \qquad l - \varepsilon < (u_n)^{1/n} < l + \varepsilon \qquad \forall \ n \ge m$$

$$\Rightarrow \qquad (l - \varepsilon)^n < u_n < (l + \varepsilon)^n \qquad \forall \ n \ge m \qquad \ldots(1)$$

(i) **When $l < 1$**

Choose $\varepsilon > 0$ such that $l < l + \varepsilon < 1$

Put $l + \varepsilon = r$, then $0 < r < 1$

From (1), $u_n < r^n \quad \forall \ n \ge m$

Putting $n = m, m + 1, m + 2, \ldots\ldots$, we get $u_m < r^m$, $u_{m+1} < r^{m+1}$, $u_{m+2} < r^{m+2}$, ... and so on.

Adding $\qquad u_m + u_{m+1} + u_{m+2} + \ldots\ldots + < r^m + r^{m+1} + r^{m+2} + \ldots\ldots$

$\Rightarrow$ each term of the given series Σu_n after leaving the first $(m-1)$ terms, (*i.e.*, a finite number of terms) is less than the corresponding term of a geometric series which is convergent ($\because$ its common ratio $r < 1$). Hence the given series is also convergent.

(*ii*) **When $l > 1$**

Choose $\varepsilon > 0$ such that $l - \varepsilon > 1$

Put $l - \varepsilon = R$, then $R > 1$

From (1), $\qquad u_n > R^n \qquad \forall \quad n \geq m$

Putting $n = m, m+1, m+2, \ldots\ldots$, we get $u_m < R^m$, $u_{m+1} > R^{m+1}$, $u_{m+2} < R^{m+2}$, ... and so on.

Adding $\qquad u_m + u_{m+1} + u_{m+2} + \ldots\ldots > R^m + R^{m+1} + R^{m+2} + \ldots\ldots$

$\Rightarrow$ each term of the series Σu_n after leaving the first $(m-1)$ terms, (*i.e.*, a finite number of terms) is greater than the corresponding term of a geometric series which is divergent. ($\because$ its common ratio $R > 1$). Hence the given series is also divergent.

ILLUSTRATIVE EXAMPLES

Example 1. *Test the convergence of the following series:*

$$\sum \left(\frac{n}{n+1}\right)^{n^2} \quad or \quad \sum \left(1+\frac{1}{n}\right)^{-n^2}.$$

Sol. Here $\qquad u_n = \left(\frac{n}{n+1}\right)^{n^2}$

$$\therefore \qquad (u_n)^{1/n} = \left[\left(\frac{n}{n+1}\right)^{n^2}\right]^{1/n} = \left(\frac{n}{n+1}\right)^n = \left(\frac{n+1}{n}\right)^{-n} = \left[\left(1+\frac{1}{n}\right)^n\right]^{-1}$$

$$\underset{n\to\infty}{\text{Lt}} (u_n)^{1/n} = \underset{n\to\infty}{\text{Lt}} \left[\left(1+\frac{1}{n}\right)^n\right]^{-1} = e^{-1} = \frac{1}{e} < 1 \qquad (\because \quad e = 2.7)$$

$\therefore$ By Cauchy's Root Test, the given series Σu_n is convergent.

Example 2. *Examine the convergence of the series:*

$$\left(\frac{2^2}{1^2}-\frac{2}{1}\right)^{-1} + \left(\frac{3^3}{2^3}-\frac{3}{2}\right)^{-2} + \left(\frac{4^4}{3^4}-\frac{4}{3}\right)^{-3} + \ldots\ldots \qquad \text{[CASAT 2010]}$$

Sol. Here $\qquad u_n = \left[\left(\frac{n+1}{n}\right)^{n+1} - \frac{n+1}{n}\right]^{-n}$

$$(u_n)^{1/n} = \left[\left(\frac{n+1}{n}\right)^{n+1} - \frac{n+1}{n}\right]^{-1} = \left[\left(1+\frac{1}{n}\right)^{n+1} - \left(1+\frac{1}{n}\right)\right]^{-1}$$

$$= \left[\left(1+\frac{1}{n}\right)^n \cdot \left(1+\frac{1}{n}\right) - \left(1+\frac{1}{n}\right)\right]^{-1}$$

$$\underset{n\to\infty}{\text{Lt}}\ (u_n)^{1/n} = \underset{n\to\infty}{\text{Lt}}\left[\left(1+\frac{1}{n}\right)^n\left(1+\frac{1}{n}\right)-\left(1+\frac{1}{n}\right)\right]^{-1}$$

$$= (e.1-1)^{-1} = \frac{1}{e-1} < 1 \qquad (\because\ e = 2.7)$$

$\therefore$ By Cauchy's Root Test, Σu_n is convergent.

TEST YOUR KNOWLEDGE

Discuss the convergence of the following series:

1. $\sum \frac{1}{n^n}$
2. $\sum_{n=2}^{\infty} \frac{1}{(\log n)^n}$
3. $\sum \left(\frac{n+1}{3n}\right)^n$
4. $\sum \frac{(n-\log n)^n}{2^n \cdot n^n}$
5. $\sum \left(1+\frac{1}{\sqrt{n}}\right)^{-n^{3/2}}$
6. $\sum \left(\frac{nx}{n+1}\right)^n$
7. $\sum 5^{-n-(-1)^n}$
8. $\sum \frac{(1+nx)^n}{n^n}$
9. $\sum \frac{(n+1)^n x^n}{n^{n+1}}$
10. $\frac{1}{2}+\frac{2}{3}x+\left(\frac{3}{4}\right)^2 x^2+\left(\frac{4}{5}\right)^3 x^3+\ldots\ldots$ to ∞.

Answers

1. Convergent
2. Convergent
3. Convergent
4. Convergent
5. Convergent
6. Convergent for $x < 1$, divergent for $x \geq 1$.
7. Convergent
8. Convergent for $x < 1$, divergent for $x \geq 1$
9. Convergent for $x < 1$, divergent for $x \geq 1$
10. Convergent for $x < 1$, divergent for $x \geq 1$.

Article 11. Raabe's Test

Statement. *If Σu_n is a series of positive terms and $\underset{n\to\infty}{Lt}\ n\left(\frac{u_n}{u_{n+1}}-1\right) = l$, then the series is convergent if $l > 1$ and divergent if $l < 1$.*

Proof. Let us compare the given series Σu_n with an auxiliary series $\Sigma v_n = \Sigma \frac{1}{n^p}$ which we know converges if $p > 1$ and diverges if $p \leq 1$.

Now
$$\frac{v_n}{v_{n+1}} = \frac{\frac{1}{n^p}}{\frac{1}{(n+1)^p}} = \left(\frac{n+1}{n}\right)^p = \left(1+\frac{1}{n}\right)^p = 1+\frac{p}{n}+\frac{p(p-1)}{2!}\cdot\frac{1}{n^2}+\ldots\ldots$$

Case I. *Let $\Sigma v_n = \Sigma \frac{1}{n^p}$ be convergent, so that $p > 1$.*

Then Σu_n will also converge if $\frac{u_n}{u_{n+1}} > \frac{v_n}{v_{n+1}}$

or if
$$\frac{u_n}{u_{n+1}} > 1+\frac{p}{n}+\frac{p(p-1)}{2!}\cdot\frac{1}{n^2}+\ldots\ldots$$

or if $$n\left(\frac{u_n}{u_{n+1}}-1\right)>p+\frac{p(p-1)}{2!}\cdot\frac{1}{n}+\ldots\ldots$$

or if $$\underset{n\to\infty}{\text{Lt}}\ n\left(\frac{u_n}{u_{n+1}}-1\right)>p \quad \text{or} \quad \text{if } l>p$$

But p is itself greater than 1, $\therefore$ Σu_n is convergent if $l > 1$.

Case II. *Let Σv_n be divergent, so that $p \le 1$.*

Then Σu_n will also diverge if $\dfrac{u_n}{u_{n+1}}<\dfrac{v_n}{v_{n+1}}$

or if $$\frac{u_n}{u_{n+1}}<1+\frac{p}{n}+\frac{p(p-1)}{2!}\cdot\frac{1}{n^2}+\ldots\ldots$$

or if $$n\left(\frac{u_n}{u_{n+1}}-1\right)<p+\frac{p(p-1)}{2!}\cdot\frac{1}{n}+\ldots\ldots$$

or if $$\underset{n\to\infty}{\text{Lt}}\ n\left(\frac{u_n}{u_{n+1}}-1\right)<p \quad \text{or} \quad \text{if } l<p.$$

But p itself ≤ 1. Thus the given series Σu_n diverges if $l < 1$. This proves the result.

Note 1. If $\underset{n\to\infty}{\text{Lt}}\ n\left(\dfrac{u_n}{u_{n+1}}-1\right)=1$, then Raabe's test fails.

Note 2. Raabe's test is used when D'Alembert's Ratio test fails and when in the ratio test, $\dfrac{u_n}{u_{n+1}}$ does not involve the number e. When $\dfrac{u_n}{u_{n+1}}$ involves e, we apply logarithmic test after the ratio test and not Raabe's test.

Article 12. Logarithmic Test

Statement. *A positive term series Σu_n converges or diverges according as*

$$\underset{n\to\infty}{Lt}\ n \log\frac{u_n}{u_{n+1}} > 1 \text{ or } < 1.$$

Proof. Let us compare the given series Σu_n with an auxiliary series $\Sigma v_n = \Sigma\dfrac{1}{n^p}$ which we know converges if $p > 1$ and diverges if $p \le 1$.

Now $$\frac{v_n}{v_{n+1}}=\frac{(n+1)^p}{n^p}=\left(1+\frac{1}{n}\right)^p.$$

Case I. *Let Σv_n be convergent, so that $p > 1$.*

Then Σu_n will also be convergent if $\dfrac{u_n}{u_{n+1}}>\dfrac{v_n}{v_{n+1}}$

or if $$\frac{u_n}{u_{n+1}}>\left(1+\frac{1}{n}\right)^p$$

or if $$\log\frac{u_n}{u_{n+1}}>\log\left(1+\frac{1}{n}\right)^p=p\log\left(1+\frac{1}{n}\right)$$

or if $\log \dfrac{u_n}{u_{n+1}} > p\left[\dfrac{1}{n} - \dfrac{1}{2n^2} + \dfrac{1}{3n^3} - \ldots\ldots\right]$ $\left[\because \log(1+x) = x - \dfrac{x^2}{2} + \dfrac{x^3}{3} + \ldots\ldots\right]$

or if $n \log \dfrac{u_n}{u_{n+1}} > p\left[1 - \dfrac{1}{2n} + \dfrac{1}{3n^2} - \ldots\ldots\right]$ or if $\underset{n\to\infty}{\text{Lt}}\ n \log \dfrac{u_n}{u_{n+1}} > p$

or if $\underset{n\to\infty}{\text{Lt}}\ n \log \dfrac{u_n}{u_{n+1}} > 1$ $\mid \because\ p > 1$ (given)

Case II. *Let Σv_n be divergent, so that $p \le 1$.*

Then Σv_n also diverges if $\dfrac{u_n}{u_{n+1}} < \dfrac{v_n}{v_{n+1}}$ or if $\log \dfrac{u_n}{u_{n+1}} < \log \dfrac{v_n}{v_{n+1}}$

or if $\log \dfrac{u_n}{u_{n+1}} < \log\left(1+\dfrac{1}{n}\right)^p = p \log\left(1+\dfrac{1}{n}\right)$

or if $\log \dfrac{u_n}{u_{n+1}} < p\left(\dfrac{1}{n} - \dfrac{1}{2n^2} + \dfrac{1}{3n^3} - \ldots\ldots\right)$

or if $n \log \dfrac{u_n}{u_{n+1}} < p\left[1 - \dfrac{1}{2n} + \dfrac{1}{3n^2} - \ldots\ldots\right]$ or if $\underset{n\to\infty}{\text{Lt}}\ n \log \dfrac{u_n}{u_{n+1}} < p$

or if $\underset{n\to\infty}{\text{Lt}}\ n \log \dfrac{u_n}{u_{n+1}} < 1$ $\mid \because\ p \le 1$

Thus the series Σu_n converges or diverges according as $\underset{n\to\infty}{\text{Lt}}\ n \log \dfrac{u_n}{u_{n+1}} > 1$ or < 1.

Note 1. The test fails if $\underset{n\to\infty}{\text{Lt}}\ n \log \dfrac{u_n}{u_{n+1}} = 1$.

Note 2. The test is applied after the failure of Ratio test and generally when in Ratio test, $\dfrac{u_n}{u_{n+1}}$ involves '*e*'.

Article 13. Gauss Test

Statement. *If for the series Σu_n of positive terms, $\dfrac{u_n}{u_{n+1}}$ can be expanded in the form*

$$\frac{u_n}{u_{n+1}} = 1 + \frac{\lambda}{n} + O\left(\frac{1}{n^2}\right)$$

then Σu_n converges if $\lambda > 1$ and diverges if $\lambda \le 1$.

Note. The test never fails as we know that the series diverges for $\lambda = 1$. Moreover the test is applied after the failure of Ratio test and when it is possible to expand $\dfrac{u_n}{u_{n+1}}$ in powers of $\dfrac{1}{n}$ by Binomial Theorem or by any other method.

ILLUSTRATIVE EXAMPLES

Example 1. *Discuss the convergence of the series :* $\frac{1}{2}+\frac{1.3}{2.4}+\frac{1.3.5}{2.4.6}+......$

Sol. Here $$u_n=\frac{1.3.5......(2n-1)}{2.4.6.......2n}.$$

$$\therefore \quad u_{n+1}=\frac{1.3.5......(2n-1)(2n+1)}{2.4.6.......(2n)(2n+2)}$$

$$\therefore \quad \frac{u_n}{u_{n+1}}=\frac{1.3.5.......(2n-1)}{2.4.6......2n}\times\frac{2.4.6.......2n(2n+2)}{1.3.5......(2n-1)(2n+1)}$$

$$=\frac{2n+2}{2n+1}=\frac{1+\frac{1}{n}}{1+\frac{1}{2n}}\to 1 \text{ as } n\to\infty.$$

$\therefore$ D'Alembert's Ratio test fails.

$$n\left[\frac{u_n}{u_{n+1}}-1\right]=n\left[\frac{2n+2}{2n+1}-1\right]=\frac{n}{2n+1}=\frac{1}{2+\frac{1}{n}}$$

$$\therefore \quad \underset{n\to\infty}{\text{Lt}}\ n\left[\frac{u_n}{u_{n+1}}-1\right]=\tfrac{1}{2}<1.$$

$\therefore$ By Raabe's test, Σu_n diverges.

Example 2. *Discuss the convergence of the series* $\frac{1^2}{2^2}+\frac{1^2.2^2}{2^2.4^2}+\frac{1^2.3^2.5^2}{2^2.4^2.6^2}+......$

Sol. Here $$u_n=\frac{1^2.3^2.5^2......(2n-1)^2}{2^2.4^2.6^2......(2n)^2}$$

and $$u_{n+1}=\frac{1^2.3^2.5^2......(2n-1)^2(2n+1)^2}{2^2.4^2.6^2......(2n)^2(2n+2)^2}$$

$$\therefore \quad \frac{u_n}{u_{n+1}}=\frac{(2n+2)^2}{(2n+1)^2}=\frac{4n^2\left(1+\frac{1}{n}\right)^2}{4n^2\left(1+\frac{1}{2n}\right)^2}=\frac{\left(1+\frac{1}{n}\right)^2}{\left(1+\frac{1}{2n}\right)^2}$$

$$\therefore \quad \underset{n\to\infty}{\text{Lt}}\ \frac{u_n}{u_{n+1}}=1.$$

Hence the ratio test fails.

$$n\left(\frac{u_n}{u_{n+1}}-1\right)=n\left[\frac{(2n+2)^2}{(2n+1)^2}-1\right]$$

$$=n\left[\frac{4n^2+8n+4-(4n^2+4n+1)}{(2n+1)^2}\right]=n\frac{(4n+3)}{(2n+1)^2}=\frac{4n^2+3n}{(2n+1)^2}$$

$$= \frac{1+\frac{3}{4n}}{\left(1+\frac{1}{2n}\right)^2} \to 1 \text{ as } n \to \infty$$

$\therefore$ Raabe's test also fails.

When D'Alembert ratio test fails, we can directly apply Gauss test.

Now, $$\frac{u_n}{u_{n+1}} = \frac{(2n+2)^2}{(2n+1)^2} = \frac{\left(1+\frac{1}{n}\right)^2}{\left(1+\frac{1}{2n}\right)^2} = \left(1+\frac{1}{n}\right)^2\left(1+\frac{1}{2n}\right)^{-2}$$

$$= \left(1+\frac{2}{n}+\frac{1}{n^2}\right)\left(1-\frac{2}{2n}+\frac{3}{4n^2}-\ldots\ldots\right) = 1+\frac{1}{n}+\frac{1}{n^2}\left(1-2+\frac{3}{4}\right)+\ldots\ldots$$

$$= 1+\frac{1}{n}-\frac{1}{4n^2}+\ldots\ldots = 1+\frac{1}{n}+O\left(\frac{1}{n^2}\right)$$

Comparing it with $\frac{u_n}{u_{n+1}} = 1+\frac{\lambda}{n}+O\left(\frac{1}{n^2}\right)$

we have $\lambda = 1$. Thus by Gauss test, the series Σu_n diverges.

Example 3. *Discuss the convergence of the series:*

$$\frac{x}{1}+\frac{1}{2}\cdot\frac{x^3}{3}+\frac{1.3}{2.4}\cdot\frac{x^5}{5}+\frac{1.3.5}{2.4.6}\cdot\frac{x^7}{7}+\ldots\ldots (x>0).$$

Sol. Neglecting the first term, we have $u_n = \frac{1.3.5\ldots\ldots(2n-1)}{2.4.6\ldots\ldots(2n)}\cdot\frac{x^{2n+1}}{2n+1}$

and $$u_{n+1} = \frac{1.3.5.\ldots\ldots(2n-1)(2n+1)}{2.4.6\ldots\ldots(2n)(2n+2)}\cdot\frac{x^{2n+3}}{2n+3}$$

$\therefore$ $$\frac{u_n}{u_{n+1}} = \frac{2n+2}{2n+1}\cdot\frac{2n+3}{2n+1}\cdot\frac{1}{x^2} = \frac{2n\left(1+\frac{1}{n}\right).2n\left(1+\frac{3}{2n}\right)}{2n\left(1+\frac{1}{2n}\right).2n\left(1+\frac{1}{2n}\right)}\cdot\frac{1}{x^2}$$

$$= \frac{\left(1+\frac{1}{n}\right)\left(1+\frac{3}{2n}\right)}{\left(1+\frac{1}{2n}\right)^2}\cdot\frac{1}{x^2}$$

$$\underset{n\to\infty}{\text{Lt}}\frac{u_n}{u_{n+1}} = \underset{n\to\infty}{\text{Lt}}\frac{\left(1+\frac{1}{n}\right)\left(1+\frac{3}{2n}\right)}{\left(1+\frac{1}{2n}\right)^2}\cdot\frac{1}{x^2} = \frac{1}{x^2}$$

$\therefore$ By Ratio Test, Σu_n is convergent if $\frac{1}{x^2} > 1$ *i.e.,* $x^2 < 1$ and divergent if $\frac{1}{x^2} < 1$ *i.e.,* $x^2 > 1$.

If $x^2 = 1$, then Ratio Test fails.

When $x^2 = 1$, we have $\dfrac{u_n}{u_{n+1}} = \dfrac{(2n+2)(2n+3)}{(2n+1)^2} = \dfrac{4n^2+10n+6}{4n^2+4n+1}$

$$\underset{n\to\infty}{\text{Lt}}\ n\left(\frac{u_n}{u_{n+1}} - 1\right) = \underset{n\to\infty}{\text{Lt}}\ n\left(\frac{4n^2+10n+6}{4n^2+4n+1} - 1\right)$$

$$= \underset{n\to\infty}{\text{Lt}}\ \frac{6n^2+5n}{4n^2+4n+1} = \underset{n\to\infty}{\text{Lt}}\ \frac{6+\frac{5}{n}}{4+\frac{4}{n}+\frac{1}{n^2}} = \frac{6}{4} = \frac{3}{2} > 1.$$

$\therefore$ By Raabe's Test, the series converges.

Hence Σu_n is convergent if $x^2 \le 1$ and divergent if $x^2 > 1$.

Example 4. *Discuss the convergence of the series:* $1 + \dfrac{x}{2} + \dfrac{2!}{3^2}x^2 + \dfrac{3!}{4^2}x^3 + \dfrac{4!}{5^4}x^4 + \ldots\ldots$

Sol. Neglecting the first term, we have

$$u_n = \frac{n!}{(n+1)^n}x^n \quad \text{and} \quad u_{n+1} = \frac{(n+1)!}{(n+2)^{n+1}} \cdot x^{n+1}$$

$$\therefore \qquad \underset{n\to\infty}{\text{Lt}}\ \frac{u_n}{u_{n+1}} = \underset{n\to\infty}{\text{Lt}}\ \frac{n!}{(n+1)^n} \cdot \frac{(n+2)^{n+1}}{(n+1)!} \cdot \frac{1}{x}$$

$$= \underset{n\to\infty}{\text{Lt}}\ \frac{1}{n^n\left(1+\frac{1}{n}\right)^n} \cdot \frac{n^{n+1}\left(1+\frac{2}{n}\right)^{n+1}}{(n+1)} \cdot \frac{1}{x}$$

$$= \underset{n\to\infty}{\text{Lt}}\ \frac{\left(1+\frac{2}{n}\right)^n\left(1+\frac{2}{n}\right)}{\left(1+\frac{1}{n}\right)^n\left(1+\frac{1}{n}\right)} \cdot \frac{1}{x} = \frac{e^2}{e} \cdot \frac{1}{x} = \frac{e}{x}$$

$$\left[\because\ \underset{n\to\infty}{\text{Lt}}\left(1+\frac{a}{n}\right)^n = \underset{n\to\infty}{\text{Lt}}\left[\left(1+\frac{a}{n}\right)^{n/a}\right]^a = e^a\right]$$

$\therefore$ By D'Alembert's ratio test, the series converges if $\dfrac{e}{x} > 1$ or if $x < e$ and diverges if $\dfrac{e}{x} < 1$ or if $x > e$.

If $x = e$, the ratio test fails, $\therefore \quad \underset{n\to\infty}{\text{Lt}}\ \dfrac{u_n}{u_{n+1}} = 1$.

Now when $\qquad x = e$

$$\frac{u_n}{u_{n+1}} = \frac{\left(1+\frac{2}{n}\right)^{n+1}}{\left(1+\frac{1}{n}\right)^{n+1}} \cdot \frac{1}{e}\,.$$

Since the expression $\frac{u_n}{u_{n+1}}$ *involves the number e, so we do not apply Raabe's test but apply logarithmic test.*

$$\therefore \quad \log\frac{u_n}{u_{n+1}} = (n+1)\log\left(1+\frac{2}{n}\right) - (n+1)\log\left(1+\frac{1}{n}\right) - \log e$$

$$= (n+1)\left[\log\left(1+\frac{2}{n}\right) - \log\left(1+\frac{1}{n}\right)\right] - 1$$

$$= (n+1)\left[\left(\frac{2}{n}-\frac{1}{2}\cdot\frac{4}{n^2}+\frac{1}{3}\cdot\frac{8}{n^3}-\ldots\ldots\right)-\left(\frac{1}{n}-\frac{1}{2n^2}+\frac{1}{3n^3}+\ldots\ldots\right)\right]-1$$

$$= (n+1)\left[\frac{1}{n}-\frac{3}{2n^2}+\ldots\ldots\right]-1$$

$$= 1-\frac{3}{2n}+\frac{1}{n}-\frac{3}{2n^2}+\ldots\ldots-1 = -\frac{1}{2n}-\frac{3}{2n^2}+\ldots\ldots$$

$$\therefore \quad \underset{n\to\infty}{\text{Lt}}\, n\log\frac{u_n}{u_{n+1}} = \underset{n\to\infty}{\text{Lt}}\, n\left[-\frac{1}{2n}-\frac{3}{2n^2}+\ldots\ldots\right] = \underset{n\to\infty}{\text{Lt}}\left(-\frac{1}{2}-\frac{3}{2n}+\ldots\ldots\right) = -\tfrac{1}{2} < 1$$

$\therefore$ By logarithmic test, the series *diverges.*

Hence the given series Σu_n converges if $x < e$ and diverges if $x \geq e$.

Example 5. *Discuss the convergence of the series:*

$$1+\frac{\alpha\beta}{1.\gamma}x+\frac{\alpha(\alpha+1)\beta(\beta+1)}{1.2.\gamma(\gamma+1)}x^2+\frac{\alpha(\alpha+1)(\alpha+2)\beta(\beta+1)(\beta+2)}{1.2.3.\gamma(\gamma+1)(\gamma+2)}x^3+\ldots\ldots$$

Sol. Neglecting the first term,

$$u_n = \frac{\alpha(\alpha+1)\ldots\ldots(\alpha+n-1)\,\beta(\beta+1)\ldots\ldots(\beta+n-1)}{1.2.3\ldots\ldots n.\gamma(\gamma+1)\ldots\ldots(\gamma+n-1)}\cdot x^n$$

$$u_{n+1} = \frac{\alpha(\alpha+1)\ldots\ldots(\alpha+n-1)(\alpha+n)\,\beta(\beta+1)\ldots\ldots(\beta+n-1)(\beta+n)}{1.2.3\ldots\ldots n(n+1)\,\gamma(\gamma+1)(\gamma+2)\ldots\ldots(\gamma+n-1)(\gamma+n)}\cdot x^{n+1}$$

$$\therefore \quad \frac{u_n}{u_{n+1}} = \frac{(n+1)(\gamma+n)}{(\alpha+n)(\beta+n)}\cdot\frac{1}{x} = \frac{\left(1+\frac{1}{n}\right)\left(1+\frac{\gamma}{n}\right)}{\left(1+\frac{\alpha}{n}\right)\left(1+\frac{\beta}{n}\right)}\cdot\frac{1}{x}$$

and $$\underset{n\to\infty}{\text{Lt}}\frac{u_n}{u_{n+1}} = \frac{1}{x}.$$

Therefore, By D'Alembert's Ratio test, the series Σu_n converges if $\frac{1}{x} > 1$

i.e., if $x < 1$ and diverges if $\frac{1}{x} < 1$ or if $x > 1$.

If $x = 1$, $\underset{n\to\infty}{\text{Lt}}\frac{u_n}{u_{n+1}} = 1$ $\therefore$ Ratio test fails.

Putting $x = 1$ in $\frac{u_n}{u_{n+1}}$, we have $\frac{u_n}{u_{n+1}} = \frac{\left(1+\frac{1}{n}\right)\left(1+\frac{\gamma}{n}\right)}{\left(1+\frac{\alpha}{n}\right)\left(1+\frac{\beta}{n}\right)}$

$$= \left(1+\frac{1}{n}\right)\left(1+\frac{\gamma}{n}\right)\left(1+\frac{\alpha}{n}\right)^{-1}\left(1+\frac{\beta}{n}\right)^{-1}$$

$$= \left(1+\frac{1}{n}\right)\left(1+\frac{\gamma}{n}\right)\left(1-\frac{\alpha}{n}+\frac{\alpha^2}{n^2}+......\right)\left(1-\frac{\beta}{n}+\frac{\beta^2}{n^2}......\right)$$

[Expand by Binomial Theorem]

$$= \left(1+\frac{1}{n}+\frac{\gamma}{n}+\frac{\gamma}{n^2}\right)\left(1-\frac{\alpha}{n}-\frac{\beta}{n}+\frac{\alpha\beta}{n^2}+\frac{\alpha^2}{n^2}+\frac{\beta^2}{n^2}......\right)$$

$$= 1 + \frac{1}{n}(1+\gamma-\alpha-\beta) + O\left(\frac{1}{n^2}\right).$$

$\therefore$ By Gauss test, the series Σu_n converges if $1 + \gamma - \alpha - \beta > 1$ *i.e.,* if $\gamma > \alpha + \beta$ and diverges if $1 + \gamma - \alpha - \beta \le 1$ *i.e.,* if $\gamma \le \alpha + \beta$.

Thus the given series *converges if $x < 1$ and diverges if $x > 1$. If $x = 1$, then the series converges if $\gamma > \alpha + \beta$ and diverges if $\gamma \le \alpha + \beta$.*

TEST YOUR KNOWLEDGE

Discuss the convergence of the following series:

1. $1 + \frac{2^2}{3^2} + \frac{2^2 . 4^2}{3^2 . 5^2} + \frac{2^2 . 4^2 . 6^2}{3^2 . 5^2 . 7^2} + ...$ to ∞

2. $\frac{1^2}{4^2} + \frac{1^2 . 5^2}{4^2 . 8^2} + \frac{1^2 . 5^2 . 9^2}{4^2 . 8^2 . 12^2} + ...$ to ∞

3. $1 + \frac{3}{7}x + \frac{3 . 6}{7 . 10}x^2 + \frac{3 . 6 . 9}{7 . 10 . 13}x^3 + ...$ to ∞

4. $1 + \frac{2}{1}.\frac{1}{2} + \frac{2 . 4}{1 . 3}.\frac{1}{3} + \frac{2 . 4 . 6}{1 . 3 . 5}.\frac{1}{4} + ...$ to ∞

5. $\frac{1^2}{2^2} + \frac{1^2 . 3^2}{2^2 . 4^2}x + \frac{1^2 . 3^2 5^2}{2^2 . 4^2 . 6^2}x^2 + ...$ to ∞

6. $x + \frac{2^2 x^2}{2!} + \frac{3^3 x^3}{3!} + \frac{4^4 x^4}{4!} + \frac{5^5 x^5}{5!} + ...$ to ∞

7. $1 + \frac{2x}{2!} + \frac{3^2 x^2}{3!} + \frac{4^3 x^3}{4!} + ...$ to ∞

8. $\frac{a+x}{1!} + \frac{(a+2x)^2}{2!} + \frac{(a+3x)^3}{3!} + ...$ to ∞

9. $1^p + \left(\frac{1}{2}\right)^p + \left(\frac{1 . 3}{2 . 4}\right)^p + \left(\frac{1 . 3 . 5}{2 . 4 . 6}\right)^p + ...$ to ∞

10. $\frac{a}{b} + \frac{a(a+1)}{b(b+1)} + \frac{a(a+1)(a+2)}{b(b+1)(b+2)} + ...$ to ∞

11. $\frac{a}{b} + \frac{a(a+d)}{b(b+d)}x + \frac{a(a+d)(a+2d)}{b(b+d)(b+2d)}x^2 + ...$ to ∞ $(a > 0, b > 0, x > 0)$

12. $\sum \frac{n!}{x(x+1)(x+2)......(x+n-1)}$

13. $1 + \frac{1}{2}x + \frac{1 . 3}{2 . 4}x^2 + \frac{1 . 3 . 5}{2 . 4 . 6}x^3 + ...$ to ∞ $(x > 0)$

14. $\frac{x}{1 . 2} + \frac{x^2}{3 . 4} + \frac{x^3}{5 . 6} + \frac{x^4}{7 . 8} + ...$ to ∞ $(x > 0)$

15. $\frac{x}{1.2} + \frac{x^2}{2.3} + \frac{x^3}{3.4} + \frac{x^4}{4.5} + \ldots\ldots$ to ∞

16. $\frac{1}{2} + \frac{2}{3}x + \left(\frac{3}{4}\right)^2 x^2 + \left(\frac{4}{5}\right)^3 x^3 + \ldots\ldots$ to ∞ $(x > 0)$

17. $\sum \frac{1.3.5\ldots\ldots(2n-1)}{2.4.6\ldots\ldots(2n)} \cdot \frac{x^{2n}}{2n}$

18. $\sum \frac{(n\,!)^2}{(2n)\,!} x^{2n}$ $(x > 0)$.

19. $\sum \frac{4.7.10\ldots\ldots(3n+1)}{1.2.3\ldots\ldots n} x^n$

20. $1 + \frac{(1\,!)^2}{2\,!} x^2 + \frac{(2\,!)^2}{4\,!} x^4 + \frac{(3\,!)^2}{6\,!} x^6 + \ldots\ldots$ to ∞ $(x > 0)$

21. $x^2 (\log 2)^q + x^3 (\log 3)^q + x^4 (\log 4)^q + \ldots\ldots$ to ∞

Answers

1. Divergent
2. Convergent
3. Convergent for $x \le 1$, divergent for $x > 1$
4. Divergent
5. Convergent for $x < 1$, divergent for $x \ge 1$.
6. Convergent for $x < \frac{1}{e}$, divergent for $x \ge \frac{1}{e}$
7. Convergent for $x \le \frac{1}{e}$, divergent for $x > \frac{1}{e}$
8. Convergent for $x < \frac{1}{e}$, divergent for $x \ge \frac{1}{e}$
9. Convergent for $p > 2$, divergent for $p \le 2$
10. Convergent for $b > a + 1$, divergent for $b \le a + 1$
11. Convergent for $x < 1$ or $x = 1$ and $b > a + d$, divergent for $x > 1$ or $x = 1$ and $b \le a + d$
12. Convergent for $x > 2$, divergent for $x \le 2$
13. Convergent for $x < 1$, divergent for $x \ge 1$
14. Convergent for $x \le 1$, divergent for $x > 1$
15. Convergent for $x \le 1$, divergent for $x > 1$
16. Convergent for $x < 1$, divergent for $x \ge 1$
17. Convergent for $x^2 \le 1$, divergent for $x^2 > 1$
18. Convergent for $x^2 < 4$, divergent for $x^2 \ge 4$
19. Convergent for $x < \frac{1}{3}$, divergent for $x \ge \frac{1}{3}$
20. Convergent for $x^2 < 4$, divergent for $x^2 \ge 4$
21. Convergent for $x < 1$, divergent for $x \ge 1$.

Article 14. Cauchy's Integral Test

Statement. *If for $x \ge 1$, $f(x)$ is a non-negative, monotonic decreasing function of x such that $f(n) = u_n$ for all positive integral values of n, then the series Σu_n and the integral $\int_1^\infty f(x)\,dx$ converge or diverge together.*

Proof. Let r be a +ve integer. Choose x such that $r + 1 \ge x \ge r \ge 1$

Since $f(x)$ is a monotonic decreasing function of x.

$$\therefore \quad f(r+1) \le f(x) \le f(r) \Rightarrow u_{r+1} \le f(x) \le u_r \qquad [\because \ f(n) = u_n,\ n \in \mathrm{N}]$$

$$\Rightarrow \quad \int_r^{r+1} u_{r+1}\,dx \le \int_r^{r+1} f(x)\,dx \le \int_r^{r+1} u_r\,dx$$

$$\Rightarrow \quad u_{r+1} \int_r^{r+1} dx \le \int_r^{r+1} f(x)\,dx \le u_r \int_r^{r+1} dx$$

$$\Rightarrow \quad u_{r+1} \Big[\, x \,\Big]_r^{r+1} \le \int_r^{r+1} f(x)\,dx \le u_r \Big[\, x \,\Big]_r^{r+1}$$

$$\Rightarrow \quad u_{r+1} \le \int_r^{r+1} f(x)\,dx \le u_r \qquad \ldots(1)$$

Putting $r = 1, 2, 3, \ldots\ldots, n$ in succession in (1), we have

$$u_2 \le \int_1^2 f(x)\,dx \le u_1$$

$$u_3 \le \int_2^3 f(x)\,dx \le u_2$$

....

....

$$u_{n+1} \le \int_n^{n+1} f(x)\,dx \le u_n$$

Adding the above inequalities, we have $u_2 + u_3 + \ldots\ldots + u_{n+1} \le \int_1^2 f(x)\,dx + \int_2^3 f(x)\,dx + \ldots\ldots$

$$+ \int_n^{n+1} f(x)\,dx \le u_1 + u_2 + \ldots\ldots + u_n$$

$$\Rightarrow \quad S_{n+1} - u_1 \le \int_1^{n+1} f(x)\,dx \le S_n \quad \text{where} \quad S_n = \sum_1^n u_n = u_1 + u_2 + \ldots\ldots + u_n.$$

Proceeding to the limit as $n \to \infty$

$$\underset{n\to\infty}{\text{Lt}}\ S_{n+1} - u_1 \le \underset{n\to\infty}{\text{Lt}} \int_1^{n+1} f(x)\,dx \le \underset{n\to\infty}{\text{Lt}}\ S_n$$

$$\Rightarrow \quad \underset{n\to\infty}{\text{Lt}}\ S_{n+1} - u_1 \le \int_1^{\infty} f(x)\,dx \le \underset{n\to\infty}{\text{Lt}}\ S_n \qquad \ldots(2)$$

(*i*) If $\int_1^\infty f(x)\,dx$ converges, then $\int_1^\infty f(x)\,dx$ = a fixed finite number = I (say).

Then from (2), we have $\underset{n\to\infty}{\text{Lt}}\ S_{n+1} - u_1 \le I$

$$\Rightarrow \quad \underset{n\to\infty}{\text{Lt}}\ S_{n+1} \le I + u_1 = \text{a fixed finite number}$$

$\Rightarrow$ $\{S_n\}$ is a convergent sequence

$\Rightarrow$ the series Σu_n is convergent.

(*ii*) If $\int_1^\infty f(x)\,dx$ diverges, then $\int_1^\infty f(x)\,dx = +\infty$

From (2), $\quad \underset{n\to\infty}{\text{Lt}}\ S_n \ge \underset{n\to\infty}{\text{Lt}} \int_1^\infty f(x)\,dx = +\infty$

$\Rightarrow$ $\{S_n\}$ is a divergent sequence

$\Rightarrow$ the series Σu_n is divergent.

Hence Σu_n and $\int_1^\infty f(x)\,dx$ converge or diverge together.

Note. If $x \ge k$, then Σu_n and $\int_k^\infty f(x)\,dx$ converge or diverge together.

ILLUSTRATIVE EXAMPLES

Example 1. *Test for convergence the series:* $\sum \frac{1}{n^2+1}$.

Sol. Here $$u_n = \frac{1}{n^2+1} = f(n)$$

$\therefore$ $$f(x) = \frac{1}{x^2+1}$$

For $x \geq 1$, $f(x)$ is +ve and monotonic decreasing.

$\therefore$ Cauchy's Integral Test is applicable.

Now $$\int_1^\infty f(x)\,dx = \int_1^\infty \frac{dx}{x^2+1} = \left[\tan^{-1} x\right]_1^\infty = \frac{\pi}{2} - \frac{\pi}{4} = \frac{\pi}{4} = \text{finite}$$

$\Rightarrow$ $\int_1^\infty f(x)\,dx$ converges and hence by Integral Test, Σu_n also converges.

Example 2. *Show that the series* $\sum_1^\infty \frac{1}{n^p}$ *converges if* $p > 1$ *and diverges if* $0 < p \leq 1$.

Sol. Here $$u_n = \frac{1}{n^p} = f(n)$$

$\therefore$ $$f(x) = \frac{1}{x^p}$$

For $x \geq 1$, $f(x)$ is + ve and monotonic decreasing.

$\therefore$ Cauchy's Integral Test is applicable.

Case I. When $p \neq 1$

$$\int_1^\infty f(x)\,dx = \int_1^\infty \frac{1}{x^p}\,dx = \int_1^\infty x^{-p}\,dx = \left[\frac{x^{-p+1}}{-p+1}\right]_1^\infty$$

Sub-Case 1. When $p > 1$, $p - 1$ is +ve, so that $\int_1^\infty f(x)\,dx = -\frac{1}{p-1}\left[\frac{1}{x^{p-1}}\right]_1^\infty$

$$= -\frac{1}{p-1}[0-1] = \frac{1}{p-1} = \text{finite}$$

$\Rightarrow$ $\int_1^\infty f(x)\,dx$ converges $\Rightarrow$ Σu_n is convergent.

Sub-Case 2. When $0 < p < 1$, $1 - p$ is +ve, so that

$$\int_1^\infty f(x)\,dx = \frac{1}{1-p}\left[x^{1-p}\right]_1^\infty = \frac{1}{1-p}(\infty - 1) = \infty$$

$\Rightarrow \quad \int_1^\infty f(x)\,dx$ diverges $\Rightarrow \quad \Sigma u_n$ is divergent.

Case II. When $p = 1$, $f(x) = \dfrac{1}{x}$

$$\int_1^\infty f(x)dx = \int_1^\infty \frac{1}{x}\,dx = \Big[\log x\Big]_1^\infty = \infty - \log 1 = \infty - 0 = \infty$$

$\Rightarrow \quad \int_1^\infty f(x)\,dx$ diverges $\Rightarrow \quad \Sigma u_n$ is divergent.

Hence Σu_n converges if $p > 1$ and diverges if $p \le 1$.

Example 3. *Discuss the convergence of the series:* $\displaystyle\sum_{n=2}^{\infty} \frac{1}{n(\log n)^p}$, $(p > 0)$.

Sol. Here $\quad u_n = \dfrac{1}{n(\log n)^p} = f(x) \quad \therefore \quad f(x) = \dfrac{1}{x(\log x)^p}$

For $x \ge 2$, $p > 0$, $f(x)$ is +ve and monotonic decreasing.

$\therefore$ By Cauchy's Integral Test $\displaystyle\sum_{n=2}^{\infty} u_n$ and $\displaystyle\int_2^\infty f(x)\,dx$ converge or diverge together.

Case I. When $p \ne 1$

$$\int_2^\infty f(x)\,dx = \int_2^\infty (\log x)^{-p} \cdot \frac{1}{x}\,dx = \left[\frac{(\log x)^{-p+1}}{-p+1}\right]_2^\infty$$

Sub-Case 1. When $p > 1$, $p - 1$ is +ve, so that $\displaystyle\int_2^\infty f(x)\,dx = -\frac{1}{p-1}\left[\frac{1}{(\log x)^{p-1}}\right]_2^\infty$

$$= -\frac{1}{p-1}\left[0 - \frac{1}{(\log 2)^{p-1}}\right] = \frac{1}{(p-1)(\log 2)^{p-1}} = \text{finite}$$

$\Rightarrow \quad \displaystyle\int_2^\infty f(x)\,dx$ converges $\Rightarrow \quad \displaystyle\sum_{n=2}^{\infty} u_n$ converges.

Sub-Case 2. When $p < 1$, $1 - p$ is +ve, so that

$$\int_2^\infty f(x)\,dx = \frac{1}{1-p}\Big[(\log x)^{1-p}\Big]_2^\infty = \frac{1}{1-p}\,[\infty - (\log 2)^{1-p}] = \infty$$

$\Rightarrow \quad \displaystyle\int_2^\infty f(x)\,dx$ diverges $\Rightarrow \quad \displaystyle\sum_{n=2}^{\infty} u_n$ diverges.

Case II. When $p = 1$, $f(x) = \dfrac{1}{x\log x}$

$$\int_2^\infty f(x)\,dx = \int_2^\infty \frac{dx}{x\log x} = \int_2^\infty \frac{\frac{1}{x}}{\log x}\,dx = \Big[\log\log x\Big]_2^\infty = \infty - \log\log 2 = \infty$$

$\Rightarrow \quad \int_2^\infty f(x)\,dx$ diverges $\Rightarrow \quad \sum_{n=2}^{\infty} u_n$ diverges.

Hence Σu_n converges if $p > 1$ and diverges if $0 < p \le 1$.

TEST YOUR KNOWLEDGE

Using the integral test, discuss the convergence of the following series:

1. $\sum \frac{1}{2n+3}$
2. $\sum \frac{1}{n(n+1)}$
3. $\sum \frac{1}{\sqrt{n}}$
4. $\sum \frac{1}{(n+1)^2}$
5. $\sum \frac{2n^3}{n^4+3}$
6. $\sum \frac{n}{(n^2+1)^2}$
7. $\sum_{n=2}^{\infty} \frac{1}{n\sqrt{n^2-1}}$
8. $\sum n e^{-n^2}$
9. $\sum_{n=3}^{\infty} \frac{1}{n \log n (\log \log n)^p}$, $p > 0$.

Answers

1. Divergent 2. Convergent 3. Divergent 4. Convergent
5. Divergent 6. Convergent 7. Convergent 8. Convergent
9. Convergent for $p > 1$, divergent for $p \le 1$.

Article 15. Leibnitz's Test on Alternating Series

Statement. *The alternating series* $\Sigma(-1)^{n-1} u_n = u_1 - u_2 + u_3 - u_4 + \ldots\ldots$ $(u_n > 0 \;\forall\; n)$ *converges if*

(*i*) $u_n > u_{n+1} \;\forall\; n$ *and* (*ii*) $\underset{n\to\infty}{\text{Lt}}\; u_n = 0$.

Proof. Let S_n denote the nth partial sum of the series $\Sigma\,(-1)^{n-1}\,.\,u_n$.

$$\begin{aligned} S_{2n} &= u_1 - u_2 + u_3 - u_4 + u_5 - \ldots\ldots - u_{2n-2} + u_{2n-1} - u_{2n} \\ &= u_1 - (u_2 - u_3) - (u_4 - u_5) \ldots\ldots - (u_{2n-2} - u_{2n-1}) - u_{2n} \\ &= u_1 - [(u_2 - u_3) + (u_4 - u_5) + \ldots\ldots + (u_{2n-2} - u_{2n-1}) + u_{2n}] \\ &< u_1 \qquad [\because\; u_n > u_{n+1} \text{ and } u_n > 0 \text{ for all } n] \end{aligned}$$

$\Rightarrow$ The sequence $\{S_{2n}\}$ is bounded above.

Also $\quad S_{2n+2} = S_{2n} + u_{2n+1} - u_{2n+2}$

$\Rightarrow \quad S_{2n+2} - S_{2n} = u_{2n+1} - u_{2n+2} > 0$ for all $n \Rightarrow S_{2n+2} > S_{2n}$

$\Rightarrow$ The sequence $\{S_{2n}\}$ is monotonically increasing.

Since every monotonically increasing sequence which is bounded above converges, therefore, the sequence $\{S_{2n}\}$ converges. Let it converge to S, then $\underset{n\to\infty}{\text{Lt}}\; S_{2n} = S$

$$\begin{aligned} \text{Now} \quad \underset{n\to\infty}{\text{Lt}}\; S_{2n+1} &= \underset{n\to\infty}{\text{Lt}}\; (S_{2n} + u_{2n+1}) = \underset{n\to\infty}{\text{Lt}}\; S_{2n} + \underset{n\to\infty}{\text{Lt}}\; u_{2n+1} \\ &= S + 0 \qquad [\because\; \underset{n\to\infty}{\text{Lt}}\; u_n = 0] \\ &= S \end{aligned}$$

$\therefore$ The sequences $\{S_{2n}\}$ and $\{S_{2n+1}\}$ converge to the same real number S.

$\therefore$ Given $\varepsilon > 0$ there exists positive integers m_1 and m_2 such that

$$|\, S_{2n} - S \,| < \varepsilon \;\;\forall\;\; 2n > m_1 \quad \text{and} \quad |\, S_{2n+1} - S \,| < \varepsilon \;\;\forall\;\; 2n + 1 > m_2$$

Let $m = \max. \{m_1, m_2\}$, then

$$| S_{2n} - S | < \varepsilon \quad \forall \quad n > m \quad \text{and} \quad | S_{2n+1} - S | < \varepsilon \quad \forall \quad n > m$$

$\Rightarrow$ $$| S_n - S | < \varepsilon \quad \forall \quad n > m$$

$\therefore$ The sequence $\{S_n\}$ converges to S.

Hence the given series is convergent.

Note. The alternating series will not be convergent if any one of the two conditions is not satisfied.

ILLUSTRATIVE EXAMPLES

Example 1. *Examine the convergence of the series:* $1 - \frac{1}{2} + \frac{1}{3} - \frac{1}{4} + \ldots\ldots$

Sol. It is an alternating series.

(*i*) $u_n = \frac{1}{n}, \quad u_{n+1} = \frac{1}{n+1}$

$$\because \quad \frac{1}{n} > \frac{1}{n+1} \quad \forall\, n \quad \therefore \quad u_n > u_{n+1} \quad \forall\, n$$

(*ii*) $\underset{n \to \infty}{\text{Lt}}\ u_n = \underset{n \to \infty}{\text{Lt}}\ \frac{1}{n} = 0$

$\therefore$ Both the conditions of Leibnitz's Test are satisfied.

Hence the given series is convergent.

Example 2. *Examine the convergence of the series:*

(*a*) $2 - \frac{3}{2} + \frac{4}{3} - \frac{5}{4} + \ldots\ldots$

(*b*) $\frac{1}{2^3} - \frac{1}{3^3}(1+2) + \frac{1}{4^3}(1+2+3) - \frac{1}{5^3}(1+2+3+4) + \ldots\ldots$

Sol. (*a*) It is an alternating series

(*i*) $u_n = \frac{n+1}{n}, \quad u_{n+1} = \frac{n+2}{n+1}$

$$u_n - u_{n+1} = \frac{n+1}{n} - \frac{n+2}{n+1} = \frac{(n+1)^2 - n(n+2)}{n(n+1)} = \frac{1}{n(n+1)} > 0 \quad \forall\, n$$

$\Rightarrow$ $$u_n > u_{n+1} \quad \forall\, n$$

(*ii*) $\underset{n \to \infty}{\text{Lt}}\ u_n = \underset{n \to \infty}{\text{Lt}}\ \frac{n+1}{n} = \underset{n \to \infty}{\text{Lt}}\left(1 + \frac{1}{n}\right) = 1 \neq 0.$

Since the second condition of Leibnitz's Test is not satisfied, the series is not convergent.

(*b*) It is an alternating series.

(*i*) $u_n = \frac{1}{(n+1)^3}[1 + 2 + 3 + \ldots\ldots + n] = \frac{1}{(n+1)^3} \cdot \frac{n(n+1)}{2} = \frac{1}{2} \cdot \frac{n}{(n+1)^2}$

$$u_{n+1} = \frac{1}{2} \cdot \frac{n+1}{(n+2)^2}$$

$$u_n - u_{n+1} = \frac{1}{2}\left[\frac{n}{(n+1)^2} - \frac{n+1}{(n+2)^2}\right] = \frac{1}{2}\frac{n(n+2)^2-(n+1)^3}{(n+1)^2(n+2)^2}$$

$$= \frac{1}{2}\cdot\frac{n^2+n-1}{(n+1)^2(n+2)^2} > 0 \;\; \forall\, n$$

$$\Rightarrow \qquad u_n > u_{n+1} \;\; \forall\, n$$

(ii) $\underset{n\to\infty}{\text{Lt}}\; u_n = \underset{n\to\infty}{\text{Lt}}\; \dfrac{n}{2(n+1)^2} = \underset{n\to\infty}{\text{Lt}}\; \dfrac{\frac{1}{n}}{2\left(1+\frac{1}{n}\right)^2} = 0.$

Since both the conditions of Leibnitz's Test are satisfied, the given series is convergent.

Example 3. *Test the convergence of the following series:* $\displaystyle\sum_{n=1}^{\infty} \frac{(-1)^{n-1}.\, n}{2n-1}$.

Sol. The given series is $\displaystyle\sum_{n=1}^{\infty} \frac{(-1)^{n-1}.\, n}{2n-1} = \frac{1}{1} - \frac{2}{3} + \frac{3}{5} - \frac{4}{7} + \ldots\ldots$

It is an alternating series

(i) $\qquad u_n = \dfrac{n}{2n-1}, \quad u_{n+1} = \dfrac{n+1}{2n+1}$

$$u_n - u_{n+1} = \frac{1}{4n^2-1} > 0 \;\; \forall \;\; n \;\; \Rightarrow \;\; u_n > u_{n+1} \;\; \forall \;\; n$$

(ii) $\underset{n\to\infty}{\text{Lt}}\; u_n = \underset{n\to\infty}{\text{Lt}}\; \dfrac{n}{2n-1} = \underset{n\to\infty}{\text{Lt}}\; \dfrac{1}{2-\frac{1}{n}} = \dfrac{1}{2} \neq 0.$

Here the second condition of Leibnitz's Test is not satisfied. Hence the given series is not convergent.

TEST YOUR KNOWLEDGE

Examine the convergence of the following series :

1. $1 - \dfrac{1}{\sqrt{2}} + \dfrac{1}{\sqrt{3}} - \dfrac{1}{\sqrt{4}} + \ldots\ldots$ to ∞

2. $1 - \dfrac{1}{2^2} + \dfrac{1}{3^2} - \dfrac{1}{4^2} + \ldots\ldots$ to ∞

3. $1 - \dfrac{1}{2\sqrt{2}} + \dfrac{1}{3\sqrt{3}} - \dfrac{1}{4\sqrt{4}} + \ldots\ldots$ to ∞

4. $\dfrac{1}{a} - \dfrac{1}{a+b} + \dfrac{1}{a+2b} - \dfrac{1}{a+3b} + \ldots\ldots$ to ∞ $\quad (a > 0, b > 0)$

5. $\dfrac{1}{\log 2} - \dfrac{1}{\log 3} + \dfrac{1}{\log 4} - \dfrac{1}{\log 5} + \ldots\ldots$ to ∞

6. $\displaystyle\sum_{n=1}^{\infty} \frac{(-1)^{n-1}}{n!}$

7. $\displaystyle\sum_{n=1}^{\infty} \frac{(-1)^{n-1}}{3n-2}$

8. $\sum_{n=1}^{\infty} \frac{n(-1)^{n-1}}{5^n}$

9. $\frac{1}{\sqrt{2}+1} - \frac{1}{\sqrt{3}+1} + \frac{1}{\sqrt{4}+1} - \frac{1}{\sqrt{5}+1} + \ldots\ldots$ to ∞

10. $\sum_{n=1}^{\infty} (-1)^{n-1} \frac{n}{n^2+1}$

11. $\sum_{n=1}^{\infty} \frac{\cos n\pi}{n^2+1}$ **[Hint.** $\cos n\pi = (-1)^n$]

Answers

1. Convergent	**2.** Convergent	**3.** Convergent	**4.** Convergent
5. Convergent	**6.** Convergent	**7.** Convergent	**8.** Convergent
9. Convergent	**10.** Convergent	**11.** Convergent.	

5.16. ABSOLUTE CONVERGENCE OF A SERIES

Def. *If a convergent series whose terms are not all positive, remains convergent when all its terms are made positive, then it is called an* **absolutely convergent series,** *i.e.,*

The series Σu_n is said to be absolutely convergent if $\Sigma \mid u_n \mid$ is a convergent series.

A series is said to be conditionally convergent if it is convergent but does not converge absolutely.

Example. *Test whether the following series are absolutely convergent or conditionally convergent ?*

(*a*) $1 - \frac{1}{2^2} + \frac{1}{3^2} - \frac{1}{4^2} + \ldots\ldots$ (CUSAT 2009) (*b*) $\sum_{n=1}^{\infty} \frac{(-1)^{n-1}}{2n-1}$.

Sol. (*a*) The series is $1 - \frac{1}{2^2} + \frac{1}{3^2} - \frac{1}{4^2} + \ldots\ldots$

(*i*) In this alternating series, each term is less than the preceeding term numerically.

(*ii*) Moreover $u_n = \frac{1}{n^2}$ which $\to 0$ as $n \to \infty$.

Hence the series satisfies both the conditions of the test on alternating series and so the given series *converges.*

Again when all the term of the series are made positive, the series becomes

$$\sum \mid u_n \mid = 1 + \frac{1}{2^2} + \frac{1}{3^2} + \frac{1}{4^2} + \ldots\ldots = \Sigma \frac{1}{n^2}$$

which we know is a convergent series. ($\because$ here $p = 2 > 1$)

Thus the given series converges absolutely.

(*b*) $\sum_{n=1}^{\infty} \frac{(-1)^{n-1}}{2n-1} = \Sigma (-1)^{n-1} u_n$ (say)

Putting $n = 1, 2, 3, \ldots\ldots$, the series becomes $1 - \frac{1}{3} + \frac{1}{5} - \frac{1}{7} + \ldots\ldots$

The series is clearly an alternating series.

(*i*) The terms go on decreasing numerically and

(*ii*) $\underset{n \to \infty}{\text{Lt}}\ u_n = \underset{n \to \infty}{\text{Lt}}\ \frac{1}{2n-1} = 0$

$\therefore$ By Leibnitz's Test, the series *converges.*

But when all terms are made positive, the series becomes,

$$\sum |u_n| = 1 + \frac{1}{3} + \frac{1}{5} + \frac{1}{7} + \ldots\ldots$$

Here $\quad u_n = \dfrac{1}{2n-1}$. Take $v_n = \dfrac{1}{n}$.

$$\therefore \quad \underset{n\to\infty}{\text{Lt}} \frac{u_n}{v_n} = \underset{n\to\infty}{\text{Lt}} \frac{n}{2n-1} = \underset{n\to\infty}{\text{Lt}} \left[\frac{1}{2-\frac{1}{n}} \right] = \frac{1}{2} = \text{finite} \neq 0$$

Hence by Comparison Test series Σu_n and Σv_n behave alike.

But $\Sigma v_n = \Sigma \dfrac{1}{n}$ is a divergent series ($\because$ here $p = 1$), $\therefore$ Σu_n also diverges.

Hence the given series converges, and the series of absolute terms diverges, therefore the given series converges conditionally.

5.17. EVERY ABSOLUTELY CONVERGENT SERIES IS CONVERGENT

Let $\sum\limits_{n=1}^{\infty} u_n$ be an absolutely convergent series.

Then $\sum\limits_{n=1}^{\infty} |u_n|$ is convergent.

By Cauchy's general principle of convergence, given $\varepsilon > 0$, $\exists$ a positive integer m such that

$$|u_{m+1}| + |u_{m+2}| + \ldots\ldots + |u_n| < \varepsilon \quad \forall \quad n > m$$

or $$|u_{m+1}| + |u_{m+1}| + \ldots\ldots + |u_n| < \varepsilon \quad \forall \quad n > m \qquad \ldots(1)$$

Now, by triangle inequality, we have

$$|u_{m+1} + u_{m+2} + \ldots\ldots + u_n| \leq |u_{m+1}| + |u_{m+2}| + \ldots\ldots + |u_n|$$
$$< \varepsilon \quad \forall \quad n > m \qquad \text{[Using (1)]}$$

$\therefore$ By Cauchy's general principle of convergence, the series $\sum\limits_{n=1}^{\infty} u_n$ is convergent.

Hence $\sum |u_n|$ is convergent $\Rightarrow$ Σu_n is convergent.

Note 1. Absolute convergence $\Rightarrow$ Convergence, but convergence need not imply absolute convergence *i.e.*, the converse of above theorem need not be true.

For example, consider the series $\sum \dfrac{(-1)^{n-1}}{n} = 1 - \dfrac{1}{2} + \dfrac{1}{3} - \dfrac{1}{4} + \ldots\ldots$

It is convergent. [See Example 1 with Leibnitz's Test]

But the series $\sum \left| \dfrac{(-1)^{n-1}}{n} \right| = \sum \dfrac{1}{n}$ is divergent.

Note 2. The divergence of $\Sigma\ |u_n|$ does not imply the divergence of Σu_n.

For example, $\sum \left| \dfrac{(-1)^{n-1}}{n} \right| = \sum \dfrac{1}{n}$ is divergent whereas $\sum \dfrac{(-1)^{n-1}}{n}$ is convergent.

Example. *Prove that the series* $\frac{\sin x}{1^3} - \frac{\sin 2x}{2^3} + \frac{\sin 3x}{3^3} - \ldots\ldots$ *converges absolutely.*

(CUSAT 2009)

Sol. The given series is $\sum_{n=1}^{\infty} u_n = \sum_{n=1}^{\infty} \frac{(-1)^{n-1} \sin nx}{n^3}$

Since $$| u_n | = \frac{|\sin nx|}{n^3} \le \frac{1}{n^3} \ \forall\, n \quad \text{and} \quad \sum \frac{1}{n^3} \text{ converges.}$$

$\therefore$ By Comparison Test, the series $\sum | u_n |$ converges.

$\Rightarrow$ The given series converges absolutely.

TEST YOUR KNOWLEDGE

1. Prove that the series $\sum_{n=1}^{\infty} \frac{(-1)^{n-1} \cos^2 nx}{n\sqrt{n}}$ converges absolutely.
2. Test the following series for conditional convergence
$$1 - \frac{1}{2^{1/3}} + \frac{1}{3^{1/3}} + \frac{1}{4^{1/3}} + \ldots\ldots .$$
3. Find the nature of the following series :

   (*i*) $1 - \frac{1}{\sqrt{2}} + \frac{1}{\sqrt{3}} - \frac{1}{\sqrt{4}} \ldots\ldots .$ (*ii*) $\sum_{n=1}^{\infty} (-1)^{n+1} \frac{1}{n!}$.

   (*iii*) $\frac{1}{1+2} - \frac{2}{1+2^2} + \frac{3}{1+2^3} - \ldots\ldots .$

Answers

2. Conditionaly, convergent
3. (*i*) Conditionaly convergent (*ii*) Absolutely convergent (*iii*) Absolutely convergent.

6
Power Series

6.1. INTRODUCTION

A series of the form $a_0 + a_1 x + a_2 x^2 + ... + a_n x^n + ... = \sum_{n=0}^{\infty} a_n x^n$ is called a power series in x.

A series of the form $a_0 + a_1 (x - a) + a_2 (x - a)^2 + ... + a_n (x - a)^n + ... = \sum_{n=0}^{\infty} a_n (x - a)^n$ is called a power series in $(x - a)$, where a_0, a_1 ..... and a are +ve or –ve constants.

Examples.

(*i*) $1 + \frac{x}{1!} + \frac{x^2}{2!} + ...$ is an exponential series.

(*ii*) $1 - x + 2x^2 - 3x^3$ ..... .

(*iii*) $1 + 2\left(x - \frac{\pi}{4}\right) + 2\left(x - \frac{\pi}{4}\right)^2 + ...$.

The convergence of the power series depends upon the value of x.

6.2. INTERVAL OF CONVERGENCE

Interval of convergence of a power series is the interval of x say (a, b) $(a < x < b)$ such that the series converges for all values of x in this interval and diverges for all values of x outside this interval.

Example 1. *Find the interval of convergence of the series* $x - \frac{x^2}{\sqrt{2}} + \frac{x^3}{\sqrt{3}} - \frac{x^4}{\sqrt{4}} + ...$

(CUSAT 2008)

Sol. The given series is $\sum_{n=1}^{\infty} u_n = \sum_{n=1}^{\infty} \frac{(-1)^{n-1} x^n}{\sqrt{n}}$

Here $$| u_n | = \frac{| x^n |}{\sqrt{n}} = \frac{| x |^n}{\sqrt{n}} \quad \text{and} \quad | u_{n+1} | = \frac{| x |^{n+1}}{\sqrt{n+1}}$$

$$\therefore \quad \frac{| u_n |}{| u_{n+1} |} = \sqrt{\frac{n+1}{n}} \cdot \frac{1}{| x |} = \sqrt{1 + \frac{1}{n}} \cdot \frac{1}{| x |}$$

$$\lim_{n\to\infty} \frac{|u_n|}{|u_{n+1}|} = \frac{1}{|x|}$$

$\therefore$ By ratio test, the series $\sum |u_n|$ is convergent if $\frac{1}{|x|} > 1$ *i.e.*, if $|x| < 1$ *i.e.*, if $-1 < x < 1$ and divergent if $\frac{1}{|x|} < 1$ *i.e.*, if $|x| > 1$ *i.e.*, if $x > 1$ or $x < -1$.

Ratio test fails when $|x| = 1$, *i.e.*, when $x = 1$ or -1.

When $x = 1$, the series becomes $1 - \frac{1}{\sqrt{2}} + \frac{1}{\sqrt{3}} - \frac{1}{\sqrt{4}} + \ldots\ldots$

which is an alternating series and is convergent.

When $x = -1$, the series becomes $-\left(1 + \frac{1}{\sqrt{2}} + \frac{1}{\sqrt{3}} + \frac{1}{\sqrt{4}} + \ldots\ldots\right) = -\sum_{n=1}^{\infty} \frac{1}{\sqrt{n}}$

which is divergent by p-series test.

Hence the given series converges for $-1 < x \le 1$.

Example 2. *Find the interval of convergence of the series* $1 - \frac{x}{2} + \frac{x^2}{3} - \frac{x^3}{4} + \ldots\ldots$

(CUSAT 2007)

Sol. The given series is $\sum_{1}^{\infty} u_n = \sum_{1}^{\infty} \frac{(-1)^{n-1} x^{n-1}}{n}$

Here $$|u_n| = \frac{|x^{n-1}|}{|n|} = \frac{|x|^{n-1}}{n}$$

and $$|u_{n+1}| = \frac{|x|^n}{n+1}$$

$\therefore$ $$\frac{|u_n|}{|u_{n+1}|} = \frac{n+1}{n} \frac{1}{|x|} = \left(1 + \frac{1}{n}\right) \cdot \frac{1}{|x|}$$

$$\text{Lim}_{n\to\infty} \frac{|u_n|}{|u_{n+1}|} = \frac{1}{|x|}$$

$\therefore$ By Ratio test, the series $\sum |u_n|$ is convergent if $\frac{1}{|x|} > 1$ *i.e.*, if $|x| < 1$

i.e., if $-1 < x < 1$ and divergent if $\frac{1}{|x|} < 1$ *i.e.*, if $|x| > 1$ *i.e.*, if $x > 1$ or $x < -1$

Ratio test fails when $|x| = 1$, *i.e.*, when $x = 1$ or $x = -1$.

When $x = 1$, the series becomes

$1 - \frac{1}{2} + \frac{1}{3} - \frac{1}{4}$ ...... which is an alternating series and is convergent.

When $x = -1$,

The series becomes

$1+\frac{1}{2}+\frac{1}{3}+\frac{1}{4}$...... which is divergent by p-series test.

Hence the given series is convergent for $-1 < x \le 1$ and the interval of convergence is $(-1, 1]$.

Example 3. *Find the interval of convergence of Binomial series.*

Sol. The Binomial series is

$$1+rx+\frac{r(r-1)}{2!}x^2+\frac{r(r-1)(r-2)}{3!}x^3+\ldots\ldots$$

$\therefore$ The series is

$$\sum_1^{\infty} u_n = \sum_1^{\infty} \frac{r(r-1)\ldots\ldots(r-n+1)}{n!} x^n$$

Here
$$|u_n| = \frac{|r(r-1)\ldots\ldots(r-n+1)|}{|n!|}|x|^n$$

$$= \frac{|r(r-1)\ldots\ldots(r-n+1)|}{n!}|x|^n$$

$$|u_{n+1}| = \frac{|r(r-1)\ldots\ldots(r-n)|}{(n+1)!}|x|^{n+1}$$

$$\lim_{n\to\infty}\frac{|u_n|}{|u_{n+1}|} = \lim_{n\to\infty}\left[\frac{n+1}{|r-n|}\frac{1}{|x|}\right]$$

$$= \lim_{n\to\infty}\left(\frac{1+\frac{1}{n}}{\left|\frac{r}{n}-1\right|}\right)\frac{1}{|x|} = \frac{1}{|-1|}\frac{1}{|x|} = \frac{1}{|x|}$$

$\therefore$ By ratio test, the series $\sum |u_n|$ is convergent if $\frac{1}{|x|} > 1$ *i.e.*, if $|x| < 1$

i.e., if $-1 < x < 1$. Hence the interval of convergence is $(-1, 1)$.

TEST YOUR KNOWLEDGE

1. Find the interval to convergence for the following series

(*i*) $x-\frac{x^2}{2}+\frac{x^3}{3}-\frac{x^4}{4}+\ldots\ldots$ [logarithmic series]

(*ii*) $1+x+\frac{x^2}{2!}+\frac{x^3}{3!}+\ldots\ldots$ [exponential series]

(*iii*) $x-\frac{x^3}{3}+\frac{x^5}{5}-\frac{x^7}{7}+\ldots\ldots$

(*iv*) $x-\frac{x^2}{2^2}+\frac{x^3}{3^2}-\frac{x^4}{4^2}+\ldots\ldots$

Answers

1. (*i*) $-1 < x \leq 1$ (*ii*) all x (*iii*) $-1 < x \leq 1$ (*iv*) $-1 < x \leq 1$.

6.3. LEIBNITZ'S THEOREM

Leibnitz's Theorem helps us to find the *n*th derivative of the product of two functions. Using Leibnitz's Theorem we can find the Maclaurin's series expansion of functions.

Statement. *If $y = uv$, where u and v are functions of x, having derivatives of nth order, then*

$$D(uv) = y_n = {}^nC_0 u_n v + {}^nC_1 u_{n-1} v_1 + {}^nC_2 u_{n-2} v_2 + \ldots\ldots + {}^nC_r u_{n-r} v_r + \ldots\ldots + {}^nC_n u v_n$$

where suffixes of u and v denote the number of times they are differentiated and ${}^nC_r$ denotes the number of combinations of n different things taken r at a time. **[V.Imp.]**

Proof. We shall prove the theorem by mathematical induction.

Step I. Let $y = uv$

By actual differentiation, we have

$$y_1 = u_1 v + u v_1 = {}^1C_0 u_1 v + {}^1C_1 u v_1$$

Thus, the theorem is true for $n = 1$.

Step II. Let us as assume that the theorem is true for $n = m$ so that, we have

$$y_m = {}^mC_0 u_m v + {}^mC_1 u_{m-1} v_1 + {}^mC_2 u_{m-2} v_2 + \ldots\ldots + {}^mC_{r-1} u_{m-r+1} v_{r-1} + {}^mC_r u_{m-r} v_r + \ldots\ldots + {}^mC_m u v_m$$

Step III. Differentiating both sides, we have

$$\begin{aligned} y_{m+1} &= {}^mC_0(u_{m+1}v + u_m v_1) + {}^mC_1(u_m v_1 + u_{m-1} v_2) \\ &\quad + {}^mC_2 (u_{m-1} v_2 + u_{m-2} v_3) + \ldots\ldots + {}^mC_{r-1}(u_{m-r+2} + v_{r-1} + u_{m-r+1} v_r) \\ &\quad + {}^mC_r(u_{m-r+1} v_r + u_{m-r} v_{r+1}) + \ldots\ldots + {}^mC_m(u_1 v_m + u v_{m+1}) \\ &= {}^mC_0 u_{m+1} v + ({}^mC_0 + {}^mC_1)\, u_m v_1 + ({}^mC_1 + {}^mC_2)\, u_{m-1} v_2 + \ldots\ldots \\ &\quad + ({}^mC_{r-1} + {}^mC_r) u_{m-r+1} v_r + \ldots\ldots + {}^mC_m u v_{m+1} \end{aligned}$$

Now $\quad {}^mC_0 = 1 = {}^{m+1}C_0;\ {}^mC_m = 1 = {}^{m+1}C_{m+1}$

Also ${}^mC_{r-1} + {}^mC_r = {}^{m+1}C_r$

Putting $r = 1, 2, \ldots\ldots$

$${}^mC_0 + {}^mC_1 = {}^{m+1}C_1\ ;\ {}^mC_1 + {}^mC_2 = {}^{m+1}C_2\ ;\ \text{and so on}$$

$$\therefore \quad y_{m+1} = {}^{m+1}C_0 u_{m+1} v + {}^{m+1}C_1 u_m v_1 + {}^{m+1}C_2 u_{m-1} v_2 \ldots\ldots + {}^{m+1}C_r u_{m-r+1} v_r + \ldots\ldots + {}^{m+1}C_{m+1} u v_{m+1}$$

which is exactly of the same form as the given formula with $n = m + 1$.

$\therefore$ If the theorem is true for $n = m$, then it is also true for $n = m + 1$.

Hence, by the principle of mathematical induction, the theorem is true for every positive integer n.

Note. If x^m, where m is a + ve integer, is one of the factors, taking $\mathbf{v = x^m}$ simplifies the process of writing the *n*th derivative.

Example 1. *Find the nth derivative of*

(*a*) $x^3 e^{ax}$ (*b*) $x^2 \sin x$

(*c*) $x^3 \cos x$ (*d*) $x^3 \sin ax$

(*e*) $x^2 e^x \cos x$ (*f*) $x^2 x^{3x} \cos 4x$.

Sol. (*a*) Let

$u = e^{ax}$	$v = x^3$
$u_n = a^n e^{ax}$	$v_1 = 3x^2$
$u_{n-1} = a^{n-1} e^{ax}$	$v_2 = 6x$
$u_{n-2} = a^{n-2} e^{ax}$	$v_3 = 6$
$u_{n-3} = a^{n-3} e^{ax}$	$v_4 = 0$

Now by Leibnitz's Theorem, we have

$$\frac{d^n}{dx^n}(uv) = {}^n\mathrm{C}_0 u_n v + {}^n\mathrm{C}_1 u_{n-1} v_1 + {}^n\mathrm{C}_2 u_{n-2} v_2 + \ldots + {}^n\mathrm{C}_n u v_n$$

$$\therefore \quad \frac{d^n}{dx^n}(e^{ax} . x^3) = {}^n\mathrm{C}_0 a^n e^{ax} x^3 + {}^n\mathrm{C}_1\, a^{n-1} e^{ax} . 3x^2 + {}^n\mathrm{C}_2 a^{n-2} e^{ax} . 6x + {}^n\mathrm{C}_3 . a^{n-3} e^{ax} . 6$$

$$= e^{ax} . a^{n-3}\left[a^3x^3 + 3na^2x^2 + 6\frac{n(n-1)}{2!} . ax + 6 . \frac{n(n-1)(n-2)}{3!}\right]$$

$$= e^{ax} . a^{n-3}\,[a^3x^3 + 3na^2x^2 + 3n(n-1) . ax + n(n-1)(n-2)]$$

(*b*) Let

$u = \sin x$	$v = x^2$
$u_n = \sin\left(x + \frac{n\pi}{2}\right)$	$v_1 = 2x$
$u_{n-1} = \sin\left[x + (n-1)\frac{\pi}{2}\right]$	$v_2 = 2$
$u_{n-2} = \sin\left[x + (n-2)\frac{\pi}{2}\right]$	$v_3 = 0$

$\therefore$ By Leibnitz's Theorem,

$$\frac{d^n}{dx^n}(\sin x . x^2) = {}^n\mathrm{C}_0 \sin\left(x + \frac{n\pi}{2}\right)x^2$$

$$+ {}^n\mathrm{C}_1 . \sin\left(x + (n-1)\frac{\pi}{2}\right) . 2x + {}^n\mathrm{C}_2 \sin\left[x + (n-2)\frac{\pi}{2}\right]2$$

$$= x^2 \sin\left(x + \frac{n\pi}{2}\right) + 2nx \sin\left(x + (n-1)\frac{\pi}{2}\right) + 2 . \frac{n(n-1)}{2 . 1} \sin\left[x + (n-2)\frac{\pi}{2}\right]$$

$$= x^2 \sin\left(x + \frac{n\pi}{2}\right) + 2nx \sin\left(x + (n-1)\frac{\pi}{2}\right) + n(n-1) \sin\left(x + (n-2) . \frac{\pi}{2}\right)$$

(*c*) *Please try yourself.*

$$\left[\textbf{Ans. } x^3 \cos\left(x + \frac{n\pi}{2}\right) + 3nx^2 \cos\left(x + (n-1)\frac{\pi}{2}\right) + 3n(n-1)\, x \cos\left(x + (n-2)\frac{\pi}{2}\right)\right.$$

$$\left. + n(n-1)(n-2) \cos\left(x + (n-3)\frac{\pi}{2}\right)\right]$$

(*d*) Let

$$\begin{array}{l|l} u = \sin ax & v = x^3 \\ u_n = a^n \sin\left(ax + \dfrac{n\pi}{2}\right) & v_1 = 3x^2 \\ u_{n-1} = a^{n-1} \sin\left(ax + (n-1) \,.\, \dfrac{\pi}{2}\right) & v_2 = 6x \\ u_{n-2} = a^{n-2} \sin\left(ax + (n-2) \,.\, \dfrac{\pi}{2}\right) & v_3 = 6 \\ u_{n-3} = a^{n-3} \sin\left(ax + (n-3) \,.\, \dfrac{\pi}{2}\right) & v_4 = 0 \end{array}$$

$\therefore$ By Leibnitz' Theorem,

$$\frac{d^n}{dx^n}(\sin ax \,.\, x^3) = {}^n\text{C}_0 a^n \sin\left(ax + \frac{n\pi}{2}\right) . \, x^3$$

$$+ {}^n\text{C}_1 a^{n-1} \sin\left(ax + (n-1) \,.\, \frac{\pi}{2}\right) . \, 3x^2 + {}^n\text{C}_2 a^{n-3} \sin\left(ax + (n-2) \,.\, \frac{\pi}{2}\right) . \, 6x$$

$$+ {}^n\text{C}_3 a^{n-3} \sin\left(ax + (n-3) \,.\, \frac{\pi}{2}\right) . \, 6$$

$$= a^n x^3 \sin\left(ax + \frac{n\pi}{2}\right) + 3n . a^{n-1} \, x^2 \sin\left(ax + (n-1)\frac{\pi}{2}\right)$$

$$+ \frac{n\,(n-1)}{2.1} \,.\, a^{n-2} \,.\, 6x \sin\left(ax + (n-2) \,.\, \frac{\pi}{2}\right)$$

$$+ \frac{n\,(n-1)\,(n-2)}{3.2.1} \,.\, a^{n-3} \,.\, 6 \sin\left(ax + (n-3) \,.\, \frac{\pi}{2}\right)$$

$$= a^n x^3 \sin\left(ax + \frac{n\pi}{2}\right) + 3na^{n-1} \, x^2 \sin\left(ax + (n-1) \,.\, \frac{\pi}{2}\right)$$

$$+ 3n\,(n-1)\, a^{n-2} \,.\, x \sin\left(ax + (n-2) \,.\, \frac{\pi}{2}\right)$$

$$+ n\,(n-1)\,(n-2) \,.\, a^{n-3} \sin\left(ax + (n-3) \,.\, \frac{\pi}{2}\right).$$

(*e*) Let

$$\begin{array}{l|l} u = e^x \cos x & v = x^2 \\ u_n = r^n e^x \cos\,(x + n\theta) & v_1 = 2x \\ u_{n-1} = r^{n-1} e^x \cos\,[x + (n-1)\,\theta] & v_2 = 2 \\ u_{n-2} = r^{n-2}\, e^x \cos\,[x + (n-2)\theta] & v_3 = 0 \end{array}$$

where $r = \sqrt{1^2 + 1^2} = 2^{1/2}$

$$\theta = \tan^{-1}\frac{1}{1} = \frac{\pi}{4}$$

$\therefore$ By Leibnitz's Theorem,

$$\frac{d^n}{dx^n}(e^x \cos x \,.\, x^2) = \frac{d^n}{dx^n}(uv) = {}^n\mathrm{C}_0 u_n v + {}^n\mathrm{C}_1 u_{n-1} v_1 + {}^n\mathrm{C}_2 u_{n-2}\, v_2$$

$$= 2^{n/2}\, e^x \cos\left(x + n\frac{\pi}{4}\right). x^2 + n \,.\, 2^{\frac{n-1}{2}}\, e^x \cos\left(x + (n-1)\frac{\pi}{4}\right). 2x$$

$$+ \frac{n\,(n-1)}{2.1} . 2^{\frac{n-2}{2}}\, e^x \cos\left(x + (n-2)\frac{\pi}{4}\right). 2$$

$$= 2^{n/2}\, e^x \left[x^2 \cos\left(x + n\frac{\pi}{4}\right) + \sqrt{2}n\, x \cos\left\{x + (n-1)\frac{\pi}{4}\right\}\right.$$

$$\left. + \frac{n\,(n-1)}{2} \cos\left\{x + (n-2)\frac{\pi}{4}\right\}\right]$$

(*f*) Let

$$u = e^{3x} \cos 4x \qquad\qquad v = x^2$$
$$u_n = r^n\, e^{3x} \cos\,(4x + n\theta) \qquad\qquad v_1 = 2x$$
$$v_{n-1} = r^{n-1} e^{3x} \cos\,[4x + (n-1)\,\theta] \qquad\qquad v_2 = 2$$
$$u_{n-2} = r^{n-2}\, e^{3x} \cos\,[4x + (n-2)\,\theta] \qquad\qquad v_3 = 0$$

where $r = \sqrt{3^2 + 4^2} = 5,\ \theta = \tan^{-1}\frac{4}{3}$

$\therefore$ By Leibnitz's Theorem,

$$\frac{d^n}{dx^n}(e^{3x} \cos 4x \,.\, x^2) = \frac{d^n}{dx^n}(uv)$$

$$= {}^n\mathrm{C}_0 u_n v + {}^n\mathrm{C}_1 u_{n-1} v_1 + {}^n\mathrm{C}_2 u_{n-2} v_2$$

$$= r^n e^{3x} \cos\,(4x + n\theta) \,.\, x^2 + n r^{n-1}\, e^{3x} \cos\,[4x + (n-1)\,\theta] \,.\, 2x$$

$$+ \frac{n\,(n-1)}{2\,!} . r^{n-2}\, e^{3x} \cos\,[4x + (n-2)\,\theta] . 2$$

$$= r^{n-2}\, e^{3x}\, [r^2 x^2 \cos\,(4x + n\theta) + 2nr\, x \cos\,(4x + (n-1)\,\theta)$$
$$+ n\,(n-1) \cos\,(4x + (n-2)\,\theta)]$$

$$= 5^{n-2} \,.\, e^{3x} \left[25x^2 \cos\left(4x + n \tan^{-1}\frac{4}{3}\right)\right.$$

$$+ 10\, nx \cos\left(4x + (n-1) \tan^{-1}\frac{4}{3}\right)$$

$$\left. + n\,(n-1) \cos\left(4x + (n-2) \tan^{-1}\frac{4}{3}\right)\right]$$

Example 2. *If $y = x^2 \sin x$, prove that*

$$\frac{d^n y}{dx^n} = (x^2 - n^2 + n)\, sin\left(x + \frac{n\pi}{2}\right) - 2nx\, cos\left(x + \frac{n\pi}{2}\right).$$

Sol. Proceeding as in Example 1 (*b*),

$$\frac{d^n y}{dx^n} = x^2 \sin\left(x + \frac{n\pi}{2}\right) + 2nx \sin\left(x + (n-1)\frac{\pi}{2}\right) + n(n-1)\sin\left(x + (n-2)\frac{\pi}{2}\right)$$

$$= x^2 \sin\left(x + \frac{n\pi}{2}\right) + 2nx \sin\left[\left(x + \frac{n\pi}{2}\right) - \frac{\pi}{2}\right] + n(n-1)\sin\left[\left(x + \frac{n\pi}{2}\right) - \pi\right]$$

$$= x^2 \sin\left(x + \frac{n\pi}{2}\right) - 2nx \sin\left[\frac{\pi}{2} - \left(x + \frac{n\pi}{2}\right)\right] - n(n-1)\sin\left[\pi - \left(x + \frac{n\pi}{2}\right)\right]$$

$$= x^2 \sin\left(x + \frac{n\pi}{2}\right) - 2nx \cos\left(x + \frac{n\pi}{2}\right) - n(n-1)\sin\left(x + \frac{n\pi}{2}\right)$$

$$= (x^2 - n^2 + n)\sin\left(x + \frac{n\pi}{2}\right) - 2nx \cos\left(x + \frac{n\pi}{2}\right).$$

Example 3. *Find the nth derivative of*

(*a*) $x^2 \log x$ (*b*) $e^x \log x$

(*c*) $x \log x$ (*d*) $x^3 \log x$.

Sol. (*a*) Let $u = \log x$ $v = x^2$

$u_n = \dfrac{(-1)^{n-1}(n-1)!}{x^n}$	$v_1 = 2x$
$u_{n-1} = \dfrac{(-1)^{n-2}(n-2)!}{x^{n-1}}$	$v_2 = 2$
$u_{n-2} = \dfrac{(-1)^{n-3}(n-3)!}{x^{n-2}}$	$v_3 = 0$

∴ By Leibnitz's Theorem,

$$\frac{d^n}{dx^n}(\log x \,.\, x^2) = {}^nC_0 \,.\, \frac{(-1)^{n-1}(n-1)!}{x^n}\, x^2$$

$$+ {}^nC_1 \,.\, \frac{(-1)^{n-2}(n-2)!}{x^{n-1}} \,.\, 2x + {}^nC_2 \,.\, \frac{(-1)^{n-3}(n-3)!}{x^{n-2}} \,.\, 2$$

$$= \frac{(-1)^{n-1}(n-1)!}{x^{n-2}} + 2n\,\frac{(-1)^{n-2}(n-2)!}{x^{n-2}} + n(n-1)\,.\,\frac{(-1)^{n-3}(n-3)!}{x^{n-2}}$$

$$= \frac{(-1)^{n-1}\, n!}{nx^{n-2}} + 2\,.\,\frac{(-1)^{n-2}\,.\, n!}{(n-1)\, x^{n-2}} + \frac{(-1)^{n-3}\, n!}{(n-2)\, x^{n-2}}$$

$$= \frac{(-1)^{n-1}\, n!}{x^{n-2}}\left[\frac{1}{n} - \frac{2}{n-1} + \frac{1}{n-2}\right].$$

(*b*) Let

$u = e^x$	$v = \log x$
$u_n = e^x$	$v_1 = \frac{1}{x}$
$u_{n-1} = e^x$	$v_2 = -\frac{1}{x^2}$
$u_{n-2} = e^x$	$v_3 = \frac{2}{x^3}$
....	....
$u_2 = u_1 = e^x$	$v_n = \frac{(-1)^{n-1}(n-1)!}{x^n}$

$\therefore$ By Leibnitz's Theorem,

$$\frac{d^n}{dx^n}(e^x \log x) = {}^n\text{C}_0 e^x . \log x + {}^n\text{C}_1 e^x . \left(\frac{1}{x}\right) + {}^n\text{C}_2\, e^x . \left(-\frac{1}{x^2}\right)$$

$$+ ...... + {}^n\text{C}_n . e^x \frac{(-1)^{n-1}(n-1)!}{x^n}$$

$$= e^x \left[\log x + \frac{n}{x} - \frac{n(n-1)}{2} . \frac{1}{x^2} + ... + \frac{(-1)^{n-1}(n-1)!}{x^n}\right]$$

(*c*) *Please try yourself.* **Ans.** $\frac{(-1)^n (n-2)!}{x^{n-1}}$

(*d*) *Please try yourself.* **Ans.** $\frac{(-1)^{n-1} n!}{x^{n-3}}\left[\frac{1}{n} - \frac{3}{n-1} + \frac{3}{n-2} - \frac{1}{n-3}\right]$

Example 4. *If* $y = x^2 e^x$*, show that*

$$\frac{d^n y}{dx^n} = \frac{1}{2} n(n-1)\frac{d^2 y}{dx^2} - n(n-2)\frac{dy}{dx} + \frac{1}{2}(n-1)(n-2)y$$

Sol. $y = x^2 e^x$

Let

$u = e^x$	$v = x^2$
$u_n = e^x$	$v_1 = 2x$
$u_{n-1} = e^x$	$v_2 = 2$
$u_{n-2} = e^x$	$v_3 = 0$

$\therefore$ By Leibnitz's theorem,

$$\frac{d^n y}{dx^n} = e^x . x^2 + n.e^x.2x + \frac{n(n-1)}{2!} . e^x.2 = e^x . [x^2 + 2nx + n(n-1)]$$

Putting $n = 1, 2$

$$\frac{dy}{dx} = e^x [x^2 + 2x]; \quad \frac{d^2 y}{dx^2} = e^x [x^2 + 4x + 2]$$

$$\therefore \quad \text{R.H.S.} = \frac{1}{2} n(n-1)\frac{d^2y}{dx^2} - (n-2)\frac{dy}{dx} + \frac{1}{2}(n-1)(n-2)\,y$$

$$= \frac{1}{2} n(n-1)\, e^x (x^2 + 4x + 2) - n(n-2)\,.\, e^x (x^2 + 2x) + \frac{1}{2}(n-1)(n-2)\, x^2 e^x$$

$$= \frac{1}{2} e^x [n(n-1)(x^2 + 4x + 2) - 2n(n-2)(x^2 + 2x) + (n-1)(n-2)\, x^2]$$

$$= \frac{1}{2} e^x [x^2(n^2 - n - 2n^2 + 4n + n^2 - 3n + 2) + x(4n^2 - 4n - 4n^2 + 8n) + 2n(n-1)]$$

$$= \frac{1}{2} e^x [2x^2 + 4nx + 2n(n-1)]$$

$$= e^x [x^2 + 4nx + n(n-1)] = \frac{d^ny}{dx^n} = \text{L.H.S.}$$

Example 5. *If $y = x^2e^x$, show by Leibnitz's theorem that*

$$\frac{d^8y}{dx^8} = 28\frac{d^2y}{dx^2} - 48\frac{dy}{dx} + 21y.$$

Sol. Please try yourself.

Example 6. *If $y = x^n \log x$, prove that $y_{n+1} = \dfrac{n!}{x}$.*

Sol. $\quad y = x^n \log x$

$$\therefore \quad y_1 = x^n \,.\, \frac{1}{x} + nx^{n-1} \,.\, \log x \qquad \ldots(i)$$

or $\quad xy_1 = x^n + nx^n \log x$

or $\quad xy_1 = x^n + ny \qquad [\because \text{ of } (i)]$

Differentiating n times by Leibnitz's Theorem,

$y_{n+1}x + n \,.\, y_n \,.\, 1 = n! + ny_n$

$$\therefore \quad xy_{n+1} = n! \quad \text{or} \quad y_{n+1} = \frac{n!}{x}.$$

☞**Note.** To differentiate $(1 - x^2)y_2$, n times by Leibnitz's Theorem.

Let

$u = y_2$	$v = 1 - x^2$
$u_n = y_{n+2}$	$v_1 = -2x$
$u_{n-1} = y_{n+1}$	$v_2 = -2$
$u_{n-2} = y_n$	$v_3 = 0$

$$\therefore \quad \frac{d^n}{dx^n}[(1-x^2)y_2] = y_{n+2}\,(1-x^2) + n.\,y_{n+1}\,(-2x) + \frac{n(n-1)}{2!}\,y_n\,(-2)$$

$$= (1 - x^2)y_{n+2} - 2nxy_{n+1} - n(n-1)y_n.$$

Example 7. *Differentiate n times the equation*

$$(1 - x^2)y_2 - xy_1 + a^2y = 0.$$

Sol. $(1 - x^2)y_2 - xy_1 + a^2y = 0$

Differentiating every term n times by Leibnitz's Theorem

$$\left[y_{n+2}(1-x^2) + ny_{n+1}(-2x) + \frac{n(n-1)}{2!}\cdot y_n(-2)\right]$$

$$- [y_{n+1} \cdot x + ny_n \cdot 1] + a^2y_n = 0$$

or $$(1 - x^2)y_{n+2} + (-2nx - x)y_{n+1} + (-n^2 + n - n + a^2)y_n = 0$$

or $$(1 - x^2)y_{n+2} - (2n + 1)xy_{n+1} - (n^2 - a^2)y_n = 0.$$

Example 8. *If $y = a \cos (\log x) + b \sin (\log x)$, show that*

$$x^2y_{n+2} + (2n + 1)\, xy_{n+1} + (n^2 + 1)y_n = 0$$

Sol. $$y = a \cos (\log x) + b \sin (\log x)$$

$\Rightarrow$ $$y_1 = -a \sin (\log x) \cdot \frac{1}{x} + b \cos (\log x) \cdot \frac{1}{x}$$

or $$xy_1 = -a \sin (\log x) + b \cos (\log x)$$

Differentiating again,

$$xy_2 + y_1.1 = -a \cos (\log x) \cdot \frac{1}{x} - b \sin (\log x) \cdot \frac{1}{x}$$

or $$x^2y_2 + xy_1 = -[a \cos (\log x) + b \sin (\log x)]$$

or $$x^2y_2 + xy_1 = -y \qquad | \because \text{ of } (i)$$

or $$x^2y_2 + xy_1 + y = 0.$$

Differentiating every term n times by Leibnitz's Theorem,

$$\left[y_{n+2}\, x^2 + ny_{n+1}\, 2x + \frac{n(n-1)}{2!}\, y_n.2\right] + [y_{n+1}.x + ny_n \cdot 1] + y_n = 0$$

or $$x^2y_{n+2} + (2nx + x)y_{n+1} + (n^2 - n + n + 1)y_n = 0$$

or $$x^2y_{n+2} + (2n + 1)xy_{n+1} + (n^2 + 1)y_n = 0.$$

Example 9. *If $y = \sin^{-1} x$, prove that*

$$(1 - x^2)y_{n+2} - (2n + 1)xy_{n+1} - n^2y_n = 0. \qquad \text{[CUSAT 2009]}$$

Sol. $$y = \sin^{-1} x$$

$\Rightarrow$ $$y_1 = \frac{1}{\sqrt{1-x^2}}$$

Squaring and cross-multiplying, $(1 - x^2)y_1^2 = 1$

Differentiating again, $(1 - x^2) \cdot 2y_1y_2 + y_1^2(-2x) = 0$

Dividing both sides by $2y_1$, $(1 - x^2)y_2 - xy_1 = 0$

Differentiating every term n times by Leibnitz's Theorem,

$$\left[y_{n+2}(1-x^2) + ny_{n+1}(-2x) + \frac{n(n-1)}{2!}\, y_n(-2)\right] - [y_{n+1}x + ny_n \cdot 1] = 0$$

or $$(1 - x^2)y_{n+2} - (2n + 1)xy_{n+1} + (- n^2 + n - n)y_n = 0$$

or $$(1 - x^2)y_{n+2} - (2n + 1)xy_{n+1} - n^2y_n = 0.$$

Example 10. *If $y = (\sin^{-1} x)^2$, prove that*

$$(1 - x^2)y_{n+2} - (2n + 1)xy_{n+1} - n^2y_n = 0.$$

Sol. $$y_1 = (\sin^{-1} x)^2$$

$\Rightarrow$ $$y_1 = 2 \sin^{-1} x \cdot \frac{1}{\sqrt{1-x^2}}$$

Squaring and cross-multiplying,

$$(1 - x^2)y_1^2 = 4 (\sin^{-1} x)^2 = 4y \quad [\because \text{ of } (i)]$$

Differentiating again,

$$(1 - x^2) \cdot 2y_1y_2 + y_1^2 (- 2x) = 4y_1$$

Dividing both sides by $2y_1$, $(1 - x^2)y_2 - xy_1 = 2$

Differentiating every term n times by Leibnitz's Theorem,

$$\left[y_{n+2} (1-x^2) + ny_{n+1} (-2x) + \frac{n(n-1)}{2!} \cdot y_n(-2)\right] - (y_{n+1} \cdot x + ny_n \cdot 1) = 0$$

or $$(1 - x^2)y_{n+2} - (2n + 1)xy_{n+1} + (- n^2 + n - n)y_n = 0$$

or $$(1 - x^2)\, y_{n+2} - (2n + 1)xy_{n+1} - n^2y_n = 0.$$

Example 11. *If $y = e^{a \sin^{-1} x}$, show that*

$$(1 - x^2)y_{n+2} - (2n + 1)xy_{n+1} - (n^2 + a^2)y_n = 0. \quad \text{[CUSAT 2010]}$$

Sol. $$y = e^{a \sin^{-1} x} \quad ...(i)$$

$\Rightarrow$ $$y_1 = e^{a \sin^{-1} x} \cdot \frac{a}{\sqrt{1-x^2}} = \frac{ay}{\sqrt{1-x^2}} \quad [\because \text{ of } (i)]$$

Squaring and cross-multiplying, $(1 - x^2)y_1^2 = a^2y^2$

Differentiating again, $(1 - x^2)\, 2y_1y_2 + y_1^2 (- 2x) = a^2 \cdot 2yy_1$

Dividing both sides by $2y_1$, $(1 - x^2)y_2 - xy_1 = a^2y$

Differentiating every term n times by Leibnitz's Theorem,

$$\left[y_{n+2} (1-x^2) + ny_{n+1} (-2x) + \frac{n(n-1)}{2!}\, y_n (-2)\right] - [y_{n+1}x + ny_n \cdot 1] = a^2y_n$$

or $$(1 - x^2)y_{n+2} - (2n + 1)xy_{n+1} + (- n^2 + n - n - a^2)y_n = 0$$

or $$(1 - x^2)y_{n+2} - (2n + 1)xy_{n+1} - (n^2 + a^2)y_n = 0.$$

Example 12. *If $y = e^{m \cos^{-1} x}$, prove that*

$$(1 - x^2)y_{n+2} - (2n + 1)xy_{n+1} - (n^2 + m^2)y_n = 0.$$

Sol. Please try yourself.

Example 13. *If $y = \sin (m \sin^{-1} x)$, prove that*

$$(1 - x^2)y_{n+2} = (2n + 1)xy_{n+1} + (n^2 - m^2)y_n.$$

Also show that $\underset{x \to 0}{\text{Lt}} \dfrac{y_{n+2}}{y_n} = n^2 - m^2.$

Sol. $y = \sin (m \sin^{-1} x)$ ...(*i*)

$\Rightarrow$ $y_1 = \cos (m \sin^{-1} x) \dfrac{m}{\sqrt{1 - x^2}}$

Squaring and cross-multiplying,

$$(1 - x^2)y_1^2 = m^2 \cos^2 (m \sin^{-1} x) = m^2 [1 - \sin^2 (m \sin^{-1} x)]$$

$\therefore$ $(1 - x^2)y_1^2 = m^2 (1 - y^2)$ [$\because$ of (*i*)]

Differentiating again, $(1 - x)^2\, 2y_1y_2 + y_1^2 (- 2x) = m^2 (- 2yy_1)$

Dividing both sides by $2y_1$, $(1 - x^2)y_2 - xy_1 + m^2y = 0$

Differentiating every term n times by Leibnitz's Theorem,

$$\left[y_{n+2}(1 - x^2) + ny_{n+1}(-2x) + \frac{n(n-1)}{2!} y_n (-2)\right] - [y_{n+1} . x + ny_n . 1] + m^2y_n = 0$$

or $(1 - x^2)y_{n+2} - (2n + 1)xy_{n+1} + (- n^2 + n - n + m^2)y_n = 0$

or $(1 - x^2)y_{n+2} - 2(2n + 1)xy_{n+1} - (n^2 - m^2)y_n = 0$

or $(1 - x^2)y_{n+2} = (2n + 1)xy_{n+1} + (n^2 - m^2)y_n.$

Second Part. Proceeding to the limit as $x \to 0$, we have

$$\underset{x \to 0}{\text{Lt}} (1 - x^2)y_{n+2} = \underset{x \to 0}{\text{Lt}} (n^2 - m^2)y_n$$

or $\underset{x \to 0}{\text{Lt}}\ y_{n+2} = (n^2 - m^2) \underset{x \to 0}{\text{Lt}}\ y_n$

or $\underset{x \to 0}{\text{Lt}} \dfrac{y_{n+2}}{y_n} = n^2 - m^2.$

Example 14. *If $y = (x^2 - 1)^n$, prove that $(x^2 - 1)y_{n+2} + 2xy_{n+1} - n(n + 1)y_n = 0.$*

Hence if $P_n = \dfrac{d^n}{dx^n}(x^2 - 1)^n$, *show that* $\dfrac{d}{dx}\left\{(1 - x^2)\dfrac{dP_n}{dx}\right\} + n(n + 1)P_n = 0.$

Sol. $y = (x^2 - 1)^n$ ...(*i*)

$\Rightarrow$ $y_1 = n(x^2 - 1)^{n-1} . 2x = \dfrac{2nx (x^2 - 1)^n}{x^2 - 1} = \dfrac{2nxy}{x^2 - 1}$ [$\because$ of (*i*)]

$\Rightarrow$ $(x^2 - 1)y_1 = 2nxy$

Differentiating again,

$$(x^2 - 1) . y_2 + y_1 (2x) = 2n (xy_1 + y.1)$$

i.e., $(x^2 - 1)\, y_2 - 2(n - 1)xy_1 - 2ny = 0$

Differentiating every term n times by Leibnitz's Theorem,

$$\left[y_{n+2}(x^2-1)+n.y_{n+1}.2x+\frac{n(n-1)}{2!}.y_n 2\right]-2(n-1)\,[y_{n+1}.x+ny_n.1]-2ny_n=0$$

$$(x^2-1)y_{n+2}+(2nx-2nx+2x)y_{n+1}+(n^2-n-2n^2+2n-2n)y_n=0$$

or $$(x^2-1)y_{n+2}+2xy_{n+1}-n(n+1)y_n=0.$$

Second Part. $$P_n=\frac{d^n}{dx^n}(x^2-1)^n=\frac{d^n}{dx^n}(y)=y_n$$

$$\Rightarrow \quad \frac{dP_n}{dx}=\frac{d}{dx}(y_n)=y_{n+1}$$

$$\therefore \quad \frac{d}{dx}\left\{(1-x^2)\frac{dP_n}{dx}\right\}=\frac{d}{dx}\{(1-x^2)y_{n+1}\}$$

$$=(1-x^2)y_{n+2}-2xy_{n+1}$$

$$=-[(x^2-1)y_{n+2}+2xy_{n+1}]$$

$$=-n(n+1)y_n \qquad \text{(from first part)}$$

$$=-n(n+1)P_n \qquad [\because\ y_n=P_n]$$

Hence $\dfrac{d}{dx}\left\{(1-x^2)\dfrac{dP_n}{dx}\right\}+n(n+1)P_n=0.$

Example 15. *If* $y^{1/m}+y^{-1/m}=2x$, *prove that* $(x^2-1)y_{n+2}+(2n+1)xy_{n+1}+(n^2-m^2)y_n=0.$

Sol. $y^{1/m}+y^{-1/m}=2x$

Put $y^{1/m}=z$, then we have

$$z+\frac{1}{z}=2x \text{ or } z^2-2xz+1=0$$

$$\therefore \quad z=\frac{2x\pm\sqrt{4x^2-4}}{2}=x\pm\sqrt{x^2-1}$$

i.e., $$y^{1/m}=x\pm\sqrt{x^2-1} \text{ or } y=(x\pm\sqrt{x^2-1})^m$$

If $y=(x+\sqrt{x^2-1})^m$, then

$$y_1=m(x+\sqrt{x^2-1})^{m-1}\left(1+\frac{1}{2}\frac{1}{\sqrt{x^2-1}}.2x\right)$$

$$=m(x+\sqrt{x^2-1})^{m-1}\left(\frac{x+\sqrt{x^2-1}}{\sqrt{x^2-1}}\right)$$

$$= \frac{m(x+\sqrt{x^2-1})^m}{\sqrt{x^2-1}} = \frac{my}{\sqrt{x^2-1}}$$

If $$y = (x-\sqrt{x^2-1})^m$$

$$y_1 = m(x-\sqrt{x^2-1})^{m-1}\left(1-\frac{1}{2}\cdot\frac{1}{\sqrt{x^2-1}}\cdot 2x\right)$$

$$= m(x-\sqrt{x^2-1})^{m-1}\left(\frac{\sqrt{x^2-1}-x}{\sqrt{x^2-1}}\right)$$

$$= -\frac{m(x-\sqrt{x^2-1})^{m-1}[x-\sqrt{x^2-1}]}{\sqrt{x^2-1}}$$

$$= \frac{-m(x-\sqrt{x^2-1})^m}{\sqrt{x^2-1}} = -\frac{my}{\sqrt{x^2-1}}$$

Squaring, we have, in either case

$$y_1^2 = \frac{m^2y^2}{x^2-1} \quad \text{or} \quad (x^2-1)y_1^2 = m^2y^2$$

Differentiating again, $(x^2-1) \cdot 2y_1y_2 + y_1^2(2x) = m^2.2y.y_1$

Dividing both sides by $2y_1$, $(x^2-1)y_2 + xy_1 - m^2y = 0$

Differentiating every term n times by Leibnitz's Theorem,

$$\left[y_{n+2}(x^2-1) + n \cdot y_{n+1} \cdot 2x + \frac{n(n-1)}{2!} \cdot y_n \cdot 2\right] + [y_{n+1} \cdot x + ny_n \cdot 1] - m^2y_n = 0$$

or $$(x^2-1)y_{n+2} + (2n+1)xy_{n+1} + (n^2-n+n-m^2)y_n = 0$$

or $$(x^2-1)y_{n+2} + (2n+1)xy_{n+1} + (n^2-m^2)y_n = 0.$$

Example 16. *If* $y = [\log(x+\sqrt{1+x^2})]^2$, *prove that*

$$(1+x^2)y_{n+2} + (2n+1)xy_{n+1} + n^2y_n = 0.$$

Sol. $$y = [\log(x+\sqrt{1+x^2})]^2 \qquad ...(i)$$

$\Rightarrow$ $$y_1 = 2\log(x+\sqrt{1+x^2}) \cdot \frac{1}{x+\sqrt{1+x^2}} \cdot \left(1+\frac{1}{2}\cdot\frac{1}{\sqrt{1+x^2}}\cdot 2x\right)$$

$$= 2\log(x+\sqrt{1+x^2}) \cdot \frac{1}{x+\sqrt{1+x^2}} \cdot \frac{x+\sqrt{1+x^2}}{\sqrt{1+x^2}}$$

$$= \frac{2}{\sqrt{1+x^2}}\log[x+\sqrt{1+x^2}]$$

Squaring and cross-multiplying,

$$(1 + x^2)y_1^2 = 4 [\log (x + \sqrt{1+x^2})]^2 = 4y \qquad | \because \text{ of } (i)$$

Differentiating again, $(1 + x^2) . 2y_1y_2 + y_1^2 . 2x = 4y_1$

Dividing by $2y_1$, $$(1 + x^2)\, y_2 + xy_1 = 2$$

Differentiating every term n times by Leibnitz's Theorem,

$$\left[y_{n+2} . (1+x^2) + ny_{n+1} . 2x + \frac{n(n-1)}{2!} y_n . 2\right] + [y_{n+1} . x + ny_n . 1] = 0$$

or $$(1 + x^2)y_{n+2} + (2n + 1)xy_{n+1} + n^2y_n = 0.$$

Another Statement of Example 16.

If $y = (\sinh^{-1} x)^2$, prove that $(1 + x^2)y_{n+2} + (2n + 1)xy_{n+1} + n^2y_n = 0.$

Reason. $\sinh^{-1} x = \log (x + \sqrt{1+x^2})$.

Example 17. *If $\cos^{-1}\left(\frac{y}{b}\right) = \log\left(\frac{x}{n}\right)^n$, prove that*

$$x^2y_{n+2} + (2n + 1)\, xy_{n+1} + 2n^2y_n = 0.$$

Sol. $\cos^{-1}\left(\frac{y}{b}\right) = \log\left(\frac{x}{n}\right)^n = n \log \frac{x}{n}$

$$\Rightarrow \quad \frac{y}{b} = \cos\left(n \log \frac{x}{n}\right)$$

$$\Rightarrow \quad y = b \cos\left(n \log \frac{x}{n}\right) \qquad ...(i)$$

$$\therefore \quad y_1 = -\, b \sin\left(n \log \frac{x}{n}\right) . n . \frac{1}{x/n} . \frac{1}{n} = -\frac{bn}{x} \sin\left(n \log \frac{x}{n}\right)$$

or $$xy_1 = -\, bn \sin\left(n \log \frac{x}{n}\right)$$

Differentiating again,

$$xy_2 + y_1 . 1 = -\, bn \cos\left(n \log \frac{x}{n}\right) n . \frac{1}{x/n} . \frac{1}{n}$$

or $$xy_2 + y_1 = -\, bn . \frac{n}{x} \cos\left(n \log \frac{x}{n}\right)$$

or $$x^2y_2 + xy_1 = -\, n^2y \qquad | \because \text{ of } (i)$$

Differentiating every term n times by Leibnitz's Theorem,

$$\left[y_{n+2} . x^2 + ny_{n+1} . 2x + \frac{n(n-1)}{2!} . y_n . 2\right] + [y_{n+1} . x + ny_n . 1] = -\, n^2y_n$$

or
$$x^2y_{n+2} + (2n + 1)xy_{n+1} + (n^2 - n + n + n^2)y_n = 0$$
$$x^2y_{n+2} + (2n + 1)xy_{n+1} + 2n^2y_n = 0.$$

Example 18. *If* $y = \dfrac{\sin^{-1} x}{\sqrt{1-x^2}}$, *prove that*
$$(1 - x^2)y_{n+2} - (2n + 3)xy_{n+1} - (n + 1)^2y_n = 0.$$

Sol.
$$y = \frac{\sin^{-1} x}{\sqrt{1-x^2}}$$
$$\Rightarrow \quad \sqrt{1-x^2}\,.\,y = \sin^{-1} x$$

Differentiating w.r.t.x
$$\sqrt{1-x^2}\,.\,y_1 + \frac{1}{2\sqrt{1-x^2}}\,.\,(-2x)\,.\,y = \frac{1}{\sqrt{1-x^2}}$$

or
$$(1 - x^2)y_1 - xy = 1$$

Differentiating $(n + 1)$ times by Leibnitz's Theorem,
$$[y_{n+2}\,.\,(1 - x^2) + {}^{n+1}C_1\, y_{n+1}\,.\,(-2x) + {}^{n+1}C_2 y_n\,.\,(-2)] - [y_{n+1}\,.\,x + {}^{n+1}C_1\, y_n\,.\,1] = 0$$

or
$$(1 - x^2)y_{n+2} - 2(n + 1)xy_{n+1} - 2\,.\,\frac{(n+1)n}{2}\,y_n - xy_{n+1} - (n + 1)y_n = 0$$

or
$$(1 - x^2)y_{n+2} - (2n + 3)xy_{n+1} - (n + 1)^2y_n = 0.$$

Example 19. *If* $y = (1 - x^2)^{-1/2} \sin^{-1} x$, *prove that*
$$(1 - x^2)y_{n+1} - (2n + 1)xy_n - n^2y_{n-1} = 0.$$

Sol. Please try yourself.

[**Hint.** Differentiate $(1 - x^2)y_1 - xy = 1$, n times.]

Example 20. *If* $y = e^{\tan^{-1} x}$, *prove that*
$$(1 + x^2)y_{n+2} + \{2(n + 1)x - 1\}y_{n+1} + n(n + 1)y_n = 0.$$

Sol.
$$y = e^{\tan^{-1} x}$$

Differentiating, $$y_1 = e^{\tan^{-1} x}\,.\,\frac{1}{1+x^2} = \frac{y}{1+x^2}$$

$$\Rightarrow \quad (1 + x^2)y_1 = y.$$

Differentiating $(n + 1)$ times by Leibnitz's Theorem,
$$y_{n+2}\,.\,(1 + x^2) + {}^{n+1}C_1\,.\,y_{n+1}\,.\,2x + {}^{n+1}C_2\,.\,y_n\,.\,2 = y_{n+1}$$

or
$$(1 + x^2)\, y_{n+2} + \{2(n + 1)xy_{n+1} + \frac{(n+1)n}{2}\,.\,y_n\,.\,2 = y_{n+1}.$$

or
$$(1 + x^2)\, y_{n+2} + \{2(n + 1)\, x - 1\}\, y_{n+1} + n(n + 1)y_n = 0.$$

Example 21. *If* $y = e^{a \tan^{-1} x}$, *prove that*
$$(1 + x^2)y_{n+1} + (2nx - a)y_n + n(n - 1)y_{n-1} = 0.$$

Sol. Please try yourself.

Example 22. *If* $y = \cos(\log x)$, *prove that*

$$x^2y_{n+2} + (2n+1)xy_{n+1} + (n^2+1)y_n = 0.$$

Sol. $y = \cos(\log x)$

Differentiating $y_1 = -\sin(\log x) \cdot \frac{1}{x}$

$\Rightarrow$ $xy_1 = -\sin(\log x)$

Differentiating again,

$$xy_2 + y_1 = -\cos(\log x) \cdot \frac{1}{x}$$

or $x^2y_2 + xy_1 = -y$

or $x^2y_2 + xy_1 + y = 0$

Now differentiate n times by Leibnitz's Theorem

$$\left[x^2 y_{n+2} + ny_{n+1}(2x) + \frac{n(n-1)}{2} y_n(2)\right] + [xy_{n+1} + ny_n] + y_n = 0.$$

$\Rightarrow$ $x^2y_{n+2} + (2n+1)\, xy_{n+1} + (n^2+1)y_n = 0.$

Example 23. *If* $y = x\cos(\log x)$, *then show that*

$$x^2y_{n+2} + (2n-1)xy_{n+1} + (n^2 - 2n + 2)y_n = 0.$$

Sol. $y = x\cos(\log x)$

Differentiating $y_1 = \cos(\log x) - x\sin(\log x) \cdot \frac{1}{x}$

or $y_1 = \cos(\log x) - \sin(\log x)$ ...(1)

Differentiating again,

$$y_2 = -\sin(\log x) \cdot \frac{1}{x} - \cos(\log x) \cdot \frac{1}{x}$$

or $xy_2 = -\sin(\log x) - \cos(\log x)$

Differentiating again, we get

$$xy_3 + y_2 = -\cos(\log x) \cdot \frac{1}{x} + \sin(\log x) \cdot \frac{1}{x}$$

or $x^2y_3 + xy_2 = -\cos(\log x) + \sin(\log x)$

or $x^2y_3 + xy_2 = -y_1$ [Using (1)]

or $x^2y_3 + xy_2 + y_1 = 0$

Differentiating $(n-1)$ times by Leibnitz's Theorem,

$$\left[y_{n+2} \cdot x^2 + (n-1)\, y_{n+1} \cdot 2x + \frac{(n-1)(n-2)}{2!} y_n \cdot 2\right] + [y_{n+1} \cdot x + (n-1)y_n \cdot 1] + y_n = 0$$

or $x^2y_{n+2} + 2(n-1)xy_{n+1} + (n-1)(n-2)y_n + xy_{n+1} + (n-1)y_n + y_n = 0$

or $x^2y_{n+2} + (2n-1)xy_{n+1} + (n^2 - 3n + 2 + n - 1 + 1)y_n = 0$

or $x^2y_{n+2} + (2n-1)xy_{n+1} + (n^2 - 2n + 2)y_n = 0.$

Example 24. *If* $y = x^{n-1}\log x$, *prove that* $xy = (n-1)\,!$.

Sol. $y = x^{n-1}\log x$ ...(1)

Differentiating $y_1 = x^{n-1} \cdot \frac{1}{x} + (n-1)x^{n-2} \log x$

$\Rightarrow \quad xy_1 = x^{n-1} + (n-1)x^{n-1} \log x$

$\Rightarrow \quad xy_1 = x^{n-1} + (n-1)y$ [Using (1)]

$\Rightarrow \quad xy_1 - (n-1)y = x^{n-1}$

Differentiating $(n-1)$ times by Leibnitz's Theorem,

$$[y_n \cdot x + (n-1)y_{n-1} \cdot 1] - (n-1)y_{n-1} = (n-1)! \qquad \left[\because \frac{d^n}{dx^n}(x^n) = n!\right]$$

$$\Rightarrow \quad xy_n = (n-1)!$$

Example 25. *If* $I_n = \frac{d^n}{dx^n}(x^n \log x)$, *prove that*

$$I_n = nI_{n-1} + (n-1)!$$

Hence prove that

$$I_n = n!\left[\log x + 1 + \frac{1}{2} + \frac{1}{3} + \ldots + \frac{1}{n}\right].$$

Sol.

$$I_n = \frac{d^n}{dx^n}(x^n \log x) = \frac{d^{n-1}}{dx^{n-1}}\left[\frac{d}{dx}(x^n \log x)\right]$$

$$= \frac{d^{n-1}}{dx^{n-1}}\left[x^n \cdot \frac{1}{x} + nx^{n-1} \log x\right]$$

$$= \frac{d^{n-1}}{dx^{n-1}}(x^{n-1}) + n\frac{d^{n-1}}{dx^{n-1}}(x^{n-1} \log x)$$

$= (n-1)! + nI_{n-1}$ which proves the first part.

Now $\quad I_n = (n-1)! + nI_{n-1}$

Dividing both sides by $n!$, we have

$$\frac{I_n}{n!} = \frac{1}{n} + \frac{I_{n-1}}{(n-1)!} \qquad \ldots(1)$$

Replacing n by $n-1, n-2, \ldots\ldots, 3, 2$ in (1), we get

$$\frac{I_{n-1}}{(n-1)!} = \frac{1}{n-1} + \frac{I_{n-2}}{(n-2)!}$$

$$\frac{I_{n-2}}{(n-2)!} = \frac{1}{n-2} + \frac{I_{n-3}}{(n-3)!}$$

....

....

$$\frac{I_3}{3!} = \frac{1}{3} + \frac{I_2}{2!}; \quad \frac{I_2}{2!} = \frac{1}{2} + \frac{I_1}{1!}$$

Adding all the above equations from (1) onwards

$$\frac{I_n}{n!} = \frac{1}{n} + \frac{1}{n-1} + \frac{1}{n-2} + \ldots\ldots + \frac{1}{3} + \frac{1}{2} + I_1 \qquad \ldots(2)$$

Since $\quad I_n = \dfrac{d^n}{dx^n}(x^n \log x)$

$\therefore \quad I_1 = \dfrac{d}{dx}(x \log x) = x \cdot \dfrac{1}{x} + \log x \cdot 1 = 1 + \log x$

$\therefore$ From (2), we have

$$I_n = n!\left[\frac{1}{n} + \frac{1}{n-1} + \frac{1}{n-2} + \ldots\ldots + \frac{1}{3} + \frac{1}{2} + 1 + \log x\right]$$

or $$I_n = n!\left[\log x + 1 + \frac{1}{2} + \frac{1}{3} + \ldots\ldots + \frac{1}{n}\right]$$

which proves the second part.

6.4. DETERMINATION OF THE VALUE OF THE nth DERIVATIVE OF A FUNCTION FOR x = 0

Sometimes, it is required to find the nth derivative of a function for $x = 0$. This is illustrated by the following solved examples.

Example 1. *Find the value of the nth derivative of* $e^{m \sin^{-1} x}$ *for* $x = 0$.

Sol. Let $\quad y = e^{m \sin^{-1} x} \qquad \ldots(i)$

$$y_1 = e^{m \sin^{-1} x} \cdot \frac{m}{\sqrt{1-x^2}} = \frac{my}{\sqrt{1-x^2}} \qquad \ldots(ii)$$

Squaring and cross-multiplying, $\quad (1 - x^2){y_1}^2 = m^2y^2$

Differentiating again, $\quad (1 - x^2)\, 2y_1y_2 + {y_1}^2(-2x) = m^2 \cdot 2yy_1$

Dividing both sides by $2y_1$, $\quad (1 - x^2)\, y_2 - xy_1 = m^2y \qquad \ldots(iii)$

Differentiating every term n times by Leibnitz's Theorem,

$$\left[y_{n+2} \cdot (1-x^2) + ny_{n+1}(-2x) + \frac{n(n-1)}{2!} \cdot y_n(-2)\right] - [y_{n+1} \cdot x + ny_n \cdot 1] = m^2y_n$$

or $$(1 - x^2)y_{n+2} - (2n + 1)xy_{n+1} = (m^2 + n^2)y_n \qquad \ldots(iv)$$

Putting $x = 0$ in (i), (ii), (iii) and (iv)

$$y(0) = e^{m \sin^{-1} 0} = e^0 = 1$$
$$y_1(0) = my(0) = m \cdot 1 = m$$
$$y_2(0) = m^2y(0) = m^2$$
$$y_{n+2}(0) = (m^2 + n^2)y_n(0) \qquad \ldots(v)$$

Putting $n = 1, 2, 3, 4, \ldots\ldots$ in (v)

$$y_3(0) = (m^2 + 1^2)y_1(0) = (m^2 + 1^2)m$$
$$y_4(0) = (m^2 + 2^2)y_2(0) = (m^2 + 2^2)m^2$$
$$y_5(0) = (m^2 + 3^2)y_3(0) = (m^2 + 3^2)\,(m^2 + 1^2)m$$

$$y_6(0) = (m^2 + 4^2)y_4(0) = (m^2 + 4^2)\ (m^2 + 2^2)\ m^2$$

....

....

Hence when n is odd,

$$y_n(0) = m[m^2 + 1^2]\ [m^2 + 3^2]\ ......\ [m^2 + (n-2)^2]$$

and when n is even,

$$y_n(0) = m^2[m^2 + 2^2]\ [m^2 + 4^2]\ ......[m^2 + (n-2)^2].$$

Example 2. *If* $y = e^{m\cos^{-1}x}$, *show that*

$$(1 - x^2)y_{n+2} - (2n + 1)xy_{n+1} - (n^2 + m^2)y_n = 0$$

and calculate $y_n(0)$.

Sol. Let $\quad y = e^{m\cos^{-1}x}$ ...(1)

$$y_1 = e^{m\cos^{-1}x} \cdot \frac{-m}{\sqrt{1-x^2}} = -\frac{my}{\sqrt{1-x^2}} \quad ...(2)$$

Squaring and cross-multiplying, $(1 - x^2)y_1^2 = m^2y^2$

Differentiating again w.r.t.x, we have

$$(1 - x^2) \cdot 2y_1y_2 - 2xy_1^2 = m^2 \cdot 2yy_1$$

Dividing both sides by $2y_1$, we get

$$(1 - x^2)y_2 - xy_1 = m^2y \quad ...(3)$$

Differentiating n times by Leibnitz's Theorem,

$$\left[y_{n+2}(1-x^2) + ny_{n+1}(-2x) + \frac{n(n-1)}{2!}y_n(-2)\right] - [y_{n+1} \cdot x + ny_n \cdot 1] = m^2y_n$$

or $\quad (1 - x^2)y_{n+2} - (2n + 1)\ xy_{n+1} = (m^2 + n^2)y_n$ ...(4)

Putting $x = 0$ in (1), (2), (3) and (4),

$$y(0) = e^{m\cos^{-1}0} = e^{m\frac{\pi}{2}}$$

$$y_1(0) = -\ my(0) = -\ me^{m\frac{\pi}{2}}$$

$$y_2(0) = m^2y(0) = m^2e^{m\frac{\pi}{2}}$$

$$y_{n+2}(0) = (m^2 + n^2)y_n(0) \quad ...(5)$$

Putting $n = 1, 2, 3, 4, ...$ in (5),

$$y_3(0) = (m^2 + 1^2)\ y_1(0) = -\ m(m^2 + 1^2)e^{m\frac{\pi}{2}}$$

$$y_4(0) = (m^2 + 2^2)y_2(0) = -\ m^2(m^2 + 2^2)e^{m\frac{\pi}{2}}$$

$$y_5(0) = (m^2 + 3^2)y_3(0) = -\ m(m^2 + 1^2)\ (m^2 + 3^2)e^{m\frac{\pi}{2}}$$

$$y_6(0) = (m^2 + 4^2)\ y_4(0) = m^2(m^2 + 2^2)\ (m^2 + 4^2)e^{m\frac{\pi}{2}}$$

....

....

Hence, when n is odd,

$$y_n(0) = -m(m^2+1^2)(m^2+3^2) \ldots\ldots [m^2+(n-2)^2]\, e^{m\frac{\pi}{2}}$$

and when n is even,

$$y_n(0) = m^2(m^2+2^2)(m^2+4^2) \ldots\ldots [m^2+(n-2)^2]\, e^{m\frac{\pi}{2}}.$$

Example 3. *If $y = \sin^{-1} x$, prove that $(1-x^2)\, y_{n+2} - (2n+1)\, xy_{n+1} - n^2 y_n = 0$. And find the value of y_n when $x = 0$.*

Sol. $y = \sin^{-1} x$ ...(*i*)

$$y_1 = \frac{1}{\sqrt{1-x^2}} \qquad \text{...}(ii)$$

Squaring and cross-multiplying, $(1-x^2)y_1^2 = 1$

Differentiating again, $(1-x^2) \cdot 2y_1y_2 + y_1^2(-2x) = 0$

Dividing by $2y_1$, $(1-x^2)y_2 - xy_1 = 0$ ...(*iii*)

Differentiating every term n times by Leibnitz's Theorem,

$$\left[y_{n+2} \cdot (1-x^2) + ny_{n+1}(-2x) + \frac{n(n-1)}{2!} \cdot y_n(-2)\right] - [y_{n+1}\, nx + ny_n \cdot 1] = 0$$

or $(1-x^2)y_{n+2} - (2n+1)xy_{n+1} - n^2y_n = 0$ ...(*iv*)

Putting $x = 0$ in (*i*), (*ii*), (*iii*) and (*iv*)

$y(0) = \sin^{-1} 0 = 0,\ y_1(0) = 1,\ y_2(0) = 0,\ y_{n+2}(0) = n^2 y_n(0)$ ...(*v*)

Putting $n = 1, 2, 3, 4, \ldots\ldots$ in (*v*),

$y_3(0) = 1^2 y_1(0) = 1^2 \cdot 1$

$y_4(0) = 2^2 y_2(0) = 2^2 \cdot 0 = 0$

$y_5(0) = 3^2 \cdot y_3(0) = 3^2 \cdot 1^2 \cdot 1$

$y_6(0) = 4^2 \cdot y_4(0) = 4^2 \cdot 0 = 0$

....

....

Hence, when n is odd, $y_n(0) = (n-2)^2 \ldots\ldots 3^2 \cdot 1^2 \cdot 1$ and when n is even, $y_n(0) = 0$.

Example 4. *If $y = (\sin^{-1} x)^2$, prove that*

$$(1-x^2)y_{n+2} - (2n+1)\, xy_{n+1} - n^2y_n = 0.$$

Hence find the value of y_n at $x = 0$.

Sol. $y = (\sin^{-1} x)^2$ ...(*i*)

$$y_1 = 2\sin^{-1} x \cdot \frac{1}{\sqrt{1-x^2}} \qquad \text{...}(ii)$$

Squaring and cross-multiplying,

$$(1-x^2)y_1^2 = 4(\sin^{-1} x)^2 = 4y$$

Differentiating again, $(1 - x^2)\, 2y_1y_2 + y_1^2 (-2x) = 4y_1$

Dividing by $2y_1$, $$(1 - x^2)\, y_2 - xy_1 = 2 \qquad ...(iii)$$

Differentiating every term n times by Leibnitz's Theorem,

$$\left[y_{n+2}\,(1-x^2) + ny_{n+1}\,(-2x) + \frac{n(n-1)}{2!}\,.\,y_n(-2)\right] - [y_{n+1}\,x + n\,.\,y_n\,.\,1] = 0$$

or $$(1 - x^2)y_{n+2} - (2n + 1)\, xy_{n+1} - n^2y_n = 0 \qquad ...(iv)$$

Putting $x = 0$ in (i), (ii), (iii) and (iv),

$$y(0) = 0,\ y_1(0) = 0,\ y_2(0) = 2,\ y_{n+2}(0) = n^2y_n(0) \qquad ...(v)$$

Putting $n = 1, 2, 3, 4, ...$ in (v)

$$y_3(0) = 1^2\,.\,y_1(0) = 1^2\,.\,(0) = 0$$

$$y_4(0) = 2^2\,.\,y_2(0) = 2^2\,.\,2$$

$$y_5(0) = 3^2\,.\,y_3(0) = 3^2.\,0 = 0$$

$$y_6(0) = 4^2.\,y_4(0) = 4^2\,.\,2^2\,.\,2$$

....

....

Hence, when n is odd, $y_n(0) = 0$

When n is even, $y_n(0) = (n - 2)^2\ ......\ 4^2\,.\,2^2\,.\,2.$

Example 5. *If $y = tan^{-1} x$, prove that*

$$(1 + x^2)y_{n+1} + 2nxy_n + n(n - 1)y_{n-1} = 0.$$

Hence determine the values of all the derivatives of y w.r.t. x when x = 0.

Sol. $$y = \tan^{-1} x \qquad ...(i)$$

$$\Rightarrow \quad y_1 = \frac{1}{1+x^2} \quad \text{or} \quad (1 + x^2)y_1 = 1 \qquad ...(ii)$$

Differentiating n times by Leibnitz's Theorem,

$$y_{n+1}(1 + x^2) + ny_n\,.\,2x + \frac{n\,(n-1)}{2!}\,y_{n-1}\,.\,2 = 0$$

or $$(1 + x^2)y_{n+1} + 2nxy_n + n(n - 1)y_{n-1} = 0 \qquad ...(iii)$$

Putting $x = 0$ in (i), (ii) and (iii),

$$y(0) = \tan^{-1} 0 = 0,\ y_1(0) = 1$$

and $$y_{n+1}(0) = -\,n(n - 1)y_{n-1} \qquad ...(iv)$$

Putting $n = 1, 2, 3, 4, ...$ in (iv)

$$y_2(0) = -\,1\,.\,(0)\,.\,y(0) = 0$$

$$y_3(0) = -\,2\,.\,1\,.\,y_1(0) = -\,2\,.\,1\,.\,1 = (-\,1)^1\,.\,2\,!$$

$$y_4(0) = -\,3\,.\,2\,.\,y_2(0) = 0$$

$$y_5(0) = -4 . 3. y_3(0) = -4 . 3 . (-1)^1 2 ! = (-1)^2 . 4 !$$

....

....

Hence when n is odd,

$$y_n(0) = (-1)^{\frac{n-1}{2}} (n-1)!$$

When n is even, $\quad y_n(0) = 0.$

Example 6. *If* $y = \dfrac{\sin^{-1} x}{\sqrt{1-x^2}}$, *prove that*

$$y_n = (n-1)^2 y_{n-2} \text{ for } x = 0.$$

Sol. $\qquad y = \dfrac{\sin^{-1} x}{\sqrt{1-x^2}} \Rightarrow \sqrt{1-x^2}\, y = \sin^{-1} x$

Differentiating w.r.t. x, we get

$$\sqrt{1-x^2}\, y_1 + \frac{1}{2\sqrt{1-x^2}}(-2x)\, y = \frac{1}{\sqrt{1-x^2}}$$

or $$(1-x^2)\, y_1 - xy = 1$$

Differentiating n times by Leibnitz's Theorem,

$$\left[y_{n+1}(1-x^2) + ny_n(-2x) + \frac{n(n-1)}{2!} y_{n-1}(-2)\right] - [y_n . x + ny_{n-1} . 1] = 0$$

or $$(1-x^2)y_{n+1} - 2nxy_n - (n^2-n)\, y_{n-1} - xy_n - ny_{n-1} = 0$$

or $$(1-x^2)\, y_{n+1} - (2n+1)\, xy_n = n^2 y_{n-1}$$

Changing n to $(n-1)$,

$$(1-x^2)y_n - (2n-1)\, xy_{n-1} = (n-1)^2 y_{n-2}$$

Putting $x = 0$, we get

$$y_n(0) = (n-1)^2\, y_{n-2}(0).$$

Example 7. *If* $y = (x + \sqrt{1+x^2})^m$, *find* $y_n(0)$.

Sol. $$y = (x + \sqrt{1+x^2})^m \qquad ...(1)$$

$$\therefore \qquad y_1 = m(x + \sqrt{1+x^2})^{m-1}\left(1 + \frac{1}{2\sqrt{1+x^2}} . 2x\right)$$

$$= m(x + \sqrt{1+x^2})^{m-1}\left(\frac{\sqrt{1+x^2}+x}{\sqrt{1+x^2}}\right)$$

$$= \frac{m(x+\sqrt{1+x^2})^m}{\sqrt{1+x^2}} = \frac{my}{\sqrt{1+x^2}} \qquad ...(2)$$

Squaring and cross-multiplying,

$$(1 + x^2)\, y_1^2 = m^2y^2$$

Differentiating again w.r.t. x,

$$(1 + x^2) \,.\, 2y_1y_2 + 2xy_1^2 = m^2 \,.\, 2yy_1$$

Dividing by $2y_1$,

$$(1 + x^2)y_2 + xy_1 = m^2y \qquad ...(3)$$

Differentiating n times by Leibnitz's Theorem,

$$\left[y_{n+2}\,(1+x^2) + ny_{n+1}\,.\,(2x) + \frac{n(n-1)}{2!}\,y_n.2\right] + [y_{n+1}\,.\,x + ny_n\,.\,1] = m^2y_n$$

or

$$(1 + x^2)y_{n+2} + (2n + 1)\,xy_{n+1} = (m^2 - n^2)\,y_n \qquad ...(4)$$

Putting $n = 0$ in (1), (2), (3) and (4),

$$y(0) = 1,\; y_1(0) = my(0) = m$$

$$y_2(0) = m^2y(0) = m^2$$

$$y_{n+2}(0) = (m^2 - n^2)\,y_n(0) \qquad ...(5)$$

Putting $x = 1, 2, 3, 4, ...$ in (5),

$$y_3(0) = (m^2 - 1^2)\,y_1(0) = m(m^2 - 1^2)$$

$$y_4(0) = (m^2 - 2^2)\,y_2(0) = m^2(m^2 - 2^2)$$

$$y_5(0) = (m^2 - 3^2)\,y_3(0) = m(m^2 - 1^2)\,(m^2 - 3^2)$$

$$y_6(0) = (m^2 - 4^2)\,y_4(0) = m^2(m^2 - 2^2)\,(m^2 - 4^2)$$

....

....

∴ When n is odd,

$$y_n(0) = m(m^2 - 1^2)\,(m^2 - 3^2)\,......\,[m^2 - (n - 2)^2]$$

and when n is even,

$$y_n(0) = m^2\,(m^2 - 2^2)\,(m^2 - 4^2)\,......\,[m^2 - (n - 2)^2].$$

Example 8. (*i*) *Find $y_n(0)$, when $y = \log\,(x + \sqrt{1+x^2})$.*

(*ii*) *Find $y_n(0)$, when $y = \sinh^{-1} x$.*

Sol. (*i*) $\quad y = \log\,(x + \sqrt{1+x^2}) \qquad ...(1)$

$$\therefore \quad y_1 = \frac{1}{x+\sqrt{1+x^2}}\,.\left(1 + \frac{1}{2\sqrt{1+x^2}}\,.\,2x\right)$$

$$= \frac{1}{x+\sqrt{1+x^2}}\left(1 + \frac{x}{\sqrt{1+x^2}}\right)$$

$$= \frac{1}{x+\sqrt{1+x^2}}\left(\frac{\sqrt{1+x^2}+x}{\sqrt{1+x^2}}\right)$$

$\Rightarrow$ $$y_1 = \frac{1}{\sqrt{1+x^2}} \quad ...(2)$$

$\Rightarrow$ $(1 + x^2)\, y_1^{\,2} = 1$

Differentiating again w.r.t. x,

$$(1 + x^2) \,.\, 2y_1y_2 + 2xy_1^{\,2} = 0$$

Dividing by $2y_1$, $\quad (1 + x^2)\, y_2 + xy_1 = 0 \quad ...(3)$

Differentiating n times by Leibnitz's Theorem,

$$\left[y_{n+2}\,(1+x^2) + ny_{n+1}\,.\,2x + \frac{n(n-1)}{2!}\,y_n\,.\,2\right] + (y_{n+1}\,.\,x + ny_n\,.\,1] = 0$$

or $$(1 + x^2)y_{n+2} + (2n + 1)xy_{n+1} + n^2y_n = 0 \quad ...(4)$$

Putting $x = 0$ in (1), (2), (3) and (4), $\;y(0) = 0,\, y_1(0) = 1,\, y_2(0) = 0$

$$y_{n+2}(0) = -\,n^2y_n(0) \quad ...(5)$$

Putting $n = 1, 2, 3, 4,$ ...... in (5),

$$y_3(0) = -1^2y_1(0) = -\,1^2$$
$$y_4(0) = -\,2^2y_2(0) = 0$$
$$y_5(0) = -\,3^2y_3(0) = (-\,3^2)\,(-\,1^2) = (-\,1)^2\,.\,1^2\,.\,3^2$$
$$y_6(0) = -\,4^2y_4(0) = 0$$

....

....

$\therefore$ When n is odd

$$y_n(0) = (-\,1)^{\frac{n-1}{2}}\,1^2\,.\,3^2\,.\,5^2\,......\,(n-2)^2$$

and when n is even,

$$y_n(0) = 0.$$

(*ii*) $$y = \sinh^{-1} x = \log\,(x + \sqrt{1+x^2})$$

Now proceed as in part (*i*).

Example 9. (*i*) *If* $y = [\log\,(x + \sqrt{1+x^2}\,)]^2$, *find all the derivatives of y w.r.t. x, when* $x = 0$.

(*ii*) *If* $y = (\sinh^{-1} x)^2$, *prove that*

$$(1 + x^2)\, y_{n+2} + (2n + 1)\, xy_{n+1} + n^2y_n = 0.$$

Hence find y_n *when* $x = 0$.

Sol. (*i*) $$y = [\log\,(x + \sqrt{1+x^2}\,)]^2 \quad ...(1)$$

$\therefore$ $$y_1 = 2\log\,(x + \sqrt{1+x^2}\,)\,.\,\frac{d}{dx}\,[\log\,(x = \sqrt{1+x^2}\,)]$$

$$= 2\log(x+\sqrt{1+x^2}).\frac{1}{x+\sqrt{1+x^2}}.\left[1+\frac{1}{2\sqrt{1+x^2}}.2x\right]$$

$$= \frac{2\log(x+\sqrt{1+x^2})}{x+\sqrt{1+x^2}}.\frac{\sqrt{1+x^2}+x}{\sqrt{1+x^2}}$$

$$= \frac{2\log(x+\sqrt{1+x^2})}{\sqrt{1+x^2}} \qquad ...(2)$$

Squaring and cross-multiplying,

$$(1+x^2)y_1^{\,2} = 4\,[\log(x+\sqrt{1+x^2})]^2$$

or $$(1+x^2)\,y_1^{\,2} = 4y$$

Differentiating again w.r.t. x,

$$(1+x^2).\,2y_1y_2 + y_1^{\,2}.\,2x = 4y_1$$

Dividing by $2y_1$, $(1+x^2)y_2 + xy_1 = 2$ ...(3)

Differentiating n times by Leibnitz's Theorem,

$$\left[y_{n+2}(1+x^2) + ny_{n+1}.2x + \frac{n(n-1)}{2!}y_n.2\right] + [y_{n+1}.x + ny_n.1] = 0$$

or $$(1+x^2)\,y_{n+2} + (2n+1)\,xy_{n+1} + n^2y_n = 0 \qquad ...(4)$$

Putting $x = 0$ in (1), (2), (3) and (4)

$$y(0) = 0,\ y_1(0) = 0,\ y_2(0) = 2$$

$$y_{n+2}(0) = -\,n^2y_n(0) \qquad ...(5)$$

Putting $n = 1, 2, 3, 4, ...$ in (5), we get

$$y_3(0) = -\,1^2y_1(0) = 0$$
$$y_4(0) = -\,2^2y_2(0) = -\,2.2^2$$
$$y_5(0) = -\,3^2y_3(0) = 0$$
$$y_6(0) = 4^2y_4(0) = (-1)^2.2.2^2.4^2$$

....

....

$\therefore$ When n is odd, $y_n(0) = 0$

and when n is even.

$$y_n(0) = (-1)^{\frac{n-2}{2}}.\,2\,.\,2^2\,.\,4^2\,.\,6^2\,......\,(n-2)^2.$$

(*ii*) Since $\sinh^{-1}x = \log(x+\sqrt{1+x^2})$, part (*ii*) is the same as part (*i*).

Example 10. *If $y \sin(m\sin^{-1}x)$, find y_n at $x = 0$.*

Sol. $$y = \sin(m\sin^{-1}x) \qquad ...(1)$$

$\therefore$ $$y_1 = \cos(m\sin^{-1}x).\frac{m}{\sqrt{1-x^2}} \qquad ...(2)$$

Squaring and cross-multiplying,

$$(1 - x^2)y_1^2 = m^2 \cos^2 (m \sin^{-1} x)$$
$$= m^2 [1 - \sin^2 (m \sin^{-1} x)] = m^2 (1 - y^2)$$

Differentiating w.r.t. x,

$$(1 - x^2) \cdot 2y_1y_2 - 2xy_1^2 = m^2 (-2yy_1)$$

Dividing by $2y_1$,

$$(1 - x^2)y_2 - xy_1 + m^2y = 0 \quad ...(3)$$

Differentiating n times by Leibnitz's Theorem,

$$\left[y_{n+2} (1 - x^2) + ny_{n+1} (-2x) + \frac{n(n-1)}{2!} y_n \cdot (-2) \right] - [y_{n+1} \cdot x + ny_n \cdot 1] + m^2y_n = 0$$

or
$$(1 - x^2)y_{n+2} - (2n + 1)xy_{n+1} + (m^2 - n^2)y_n = 0 \quad ...(4)$$

Putting $x = 0$ in (1), (2), (3) and (4),

$$y(0) = 0, \; y_1(0) = m$$
$$y_2(0) = -m^2y(0) = 0$$
$$y_{n+2}(0) = (n^2 - m^2)y_n(0) \quad ...(5)$$

Putting $n = 1, 2, 3, 4, ...$ in (5),

$$y_3(0) = (1^2 - m^2)y_1(0) = m(1^2 - m^2)$$
$$y_4(0) = (2^2 - m^2)\, y_2(0) = 0$$
$$y_5(0) = (3^2 - m^2)y_3(0) = m(1^2 - m^2)\,(3^2 - m^2)$$
$$y_6(0) = (4^2 - m^2)y_4(0) = 0$$

....

....

$\therefore$ When n is odd.

$$y_n(0) = m(1^2 - m^2)\,(3^2 - m^2)\,(5^2 - m^2) \,......\, [(n - 2)^2 - m^2]$$

and when n is even, $y_n(0) = 0$.

Example 11. *If* $x = \sin\left(\frac{1}{a} \log y\right)$, *find* y_n *when* $x = 0$.

Sol.
$$x = \sin\left(\frac{1}{a} \log y\right) \;\Rightarrow\; \sin^{-1} x = \frac{1}{a} \log y$$

$$\Rightarrow \quad \log y = a \sin^{-1} x \quad \therefore \quad y = e^{a \sin^{-1} x}$$

Now see example 1.

Example 12. *If* $y = x \log\left(\frac{x-1}{x+1}\right)$, *prove that*

$$y_m = (-1)^{m-2} (m-2)! \left[\frac{x - m}{(x-1)^m} - \frac{x + m}{(x+1)^m} \right]$$

Sol. $y = x \log\left(\dfrac{x-1}{x+1}\right) = x\,[\log(x-1) - \log(x+1)]$

or $\quad y = x\log(x-1) - x\log(x+1)$

Differentiating m times by Leibnitz's Theorem, we get

$$y_m = \left[\frac{d^m}{dx^m}\log(x-1)\right].x + m\left[\frac{d^{m-1}}{dx^{m-1}}\log(x-1)\right].1$$

$$-\left[\frac{d^m}{dx^m}\log(x+1)\right].x - m\left[\frac{d^{m-1}}{dx^{m-1}}\log(x+1)\right].1$$

$$= x.\frac{(-1)^{m-1}(m-1)!}{(x-1)^m} + m.\frac{(-1)^{m-2}(m-2)!}{(x-1)^{m-1}} - x.\frac{(-1)^{m-1}(m-1)!}{(x+1)^m} - m.\frac{(-1)^{m-2}(m-2)}{(x+1)^{m-1}}$$

$$= (-1)^{m-2}(m-2)!\left[x.\frac{(-1)(m-1)}{(x-1)^m} + \frac{m}{(x-1)^{m-1}} - x.\frac{(-1)(m-1)}{(x+1)^m} - \frac{m}{(x+1)^{m-1}}\right]$$

$$= (-1)^{m-2}(m-2)!\left[\frac{-mx+x+m(x-1)}{(x-1)^m} - \frac{-mx+x+m(x+1)}{(x+1)^m}\right]$$

$$= (-1)^{m-2}(m-2)!\left[\frac{x-m}{(x-1)^m} - \frac{x+m}{(x+1)^m}\right].$$

Example 13. *If* $u = \dfrac{x^5}{Ax^2 + 2Bx + C}$, *show that*

$$(Ax^2 + 2Bx + C)\,u_n + 2n(Ax + B)\,u_{n-1} + n(n-1)\,A\,u_{n-2} = 0, \text{ if } n > 5.$$

Sol. $(Ax^2 + 2Bx + C)u = x^5$

Differentiating n times by Leibnitz's Theorem, we have

$$u_n.(Ax^2 + 2Bx + C) + nu_{n-1}.(2Ax + 2B) + \frac{n(n-1)}{2!}u_{n-2}.2A = 0$$

$$[\because \text{ if } n > m,\ y = x^m \Rightarrow y_n = 0]$$

$\Rightarrow \quad (Ax^2 + 2Bx + C)\,u_n + 2n\,(Ax + B)\,u_{n-1} + n(n-1)A\,u_{n-2} = 0.$

Example 14. *If f(x) tan x, show that*

$$f^n(0) - {}^nC_2 f^{n-2}(0) + {}^nC_4 f^{n-4}(0) - \ldots\ldots = \sin\frac{n\pi}{2}.$$

Sol. $\quad f(x) = \tan x = \dfrac{\sin x}{\cos x}$

$\Rightarrow \quad f(x).\cos x = \sin x$

Differentiating n times by Leibnitz's Theorem,

$$f^n(x).\cos x + {}^nC_1 f^{n-1}(x).(-\sin x) + {}^nC_2 f^{n-2}(x).(-\cos x)$$

$$+ {}^nC_3 f^{n-3}(x).(\sin x) + {}^nC_4 f^{n-4}(x).\cos x + \ldots\ldots = \sin\left(x + \frac{n\pi}{2}\right)$$

Putting $x = 0$, we get

$$f^n(0) - {}^nC_2 f^{n-2}(0) + {}^nC_4 f^{n-4}(0) - \ldots\ldots = \sin \frac{n\pi}{2}$$

Since $\sin 0 = 0$ and $\cos 0 = 1$.

6.5. TAYLOR'S INFINITE SERIES

If a function $f(x)$ possesses **derivatives of all orders** in the interval $(a, a + h)$, then for every integer n, however large, there corresponds a Taylor's development with Lagrange's form of remainder

$$f(a + h) = f(a) + hf'(a) + \frac{h^2}{2!} f''(a) + \ldots\ldots + \frac{h^{n-1}}{(n-1)!} f^{n-1}(a)$$
$$+ \frac{h^n}{n!} f^n(a + \theta h) \quad \text{where } 0 < \theta < 1$$

or $$f(a + h) = S_n + R_n$$

where $$S_n = f(a) + hf(a) + \frac{h^2}{2!} f''(a) + \ldots\ldots + \frac{h^{n-1}}{(n-1)!} f^{n-1}(a)$$

$$R_n = \frac{h^n}{n!} f^n(a + \theta h)$$

Now, let $R_n \to 0$ as $n \to \infty$, then

$$\underset{n\to\infty}{\text{Lt}}\ S_n = f(a + h)$$

$$= f(a) + hf'(a) + \frac{h^2}{2!} f''(a) + \ldots\ldots + \frac{h^{n-1}}{(n-1)!} f^{n-1}(a) + \ldots\ldots \text{ ad. inf.}$$

Converges to $f(a + h)$.

Thus, if $f(x)$ be a function

(*i*) possessing derivatives of all orders in $(a, a + h)$

(*ii*) $R_n = \frac{h^n}{n!} f^n(a + \theta h)$ where $0 < \theta < 1$ tends to zero as $n \to \infty$ then,

$$f(a + h) = f(a) + hf'(a) + \frac{h^2}{2!} f''(a) + \ldots\ldots + \frac{h^n}{n!} f^n(a) + \ldots\ldots \text{ ad. inf.}$$

☞**Note.** To express $f(x)$ in ascending integral powers of $(x - a)$

$$f(x) = f[a + (x - a)] \qquad | \text{ here } h = x - a$$

$$= f(a) + (x - a) f'(a) + \frac{(x-a)^2}{2!} f''(a) + \ldots\ldots + \frac{(x-a)^n}{n!} f^n(a) + \ldots\ldots \text{ ad. inf.}$$

Application of Taylor's Infinite Series to Expand $f(x + h)$

Method

1. Let $f(x + h)$ = given function

2. Put $h = 0$ and find $f(x)$

3. Differentiate $f(x)$ a number of times and obtain

$$f'(x), f''(x), f'''(x), \ldots\ldots$$

4. Substitute the values of $f(x)$, $f'(x)$, $f''(x)$, ... in

$$f(x+h) = f(x) + hf'(x) + \frac{h^2}{2!} f''(x) + \ldots\ldots$$

Example 1. *Apply Taylor's Theorem to prove that:*

(*a*) $a^{x+h} = a^x \left[1 + h \log a + \frac{h^2}{2!} (\log a)^2 + \frac{h^2}{3!} (\log a)^3 + \ldots\ldots\right]$

(*b*) $e^{x+h} = e^x \left[1 + h + \frac{h^2}{2!} + \frac{h^3}{3!} + \ldots\ldots\right]$

(*c*) $\frac{1}{x+h} = \frac{1}{x}\left[1 - \frac{h}{x} + \frac{h^2}{x^2} - \frac{h^3}{x^3} + \ldots\ldots\right]$

Sol. (*a*) Let $\quad f(x+h) = a^{x+h}$

Putting $h = 0$, $\quad f(x) = a^x$

$$f'(x) = a^x \log a, \; f''(x) = a^x (\log a)^2$$

$$f'''(x) = a^x (\log a)^3 \quad \ldots\ldots\ldots\ldots\ldots\ldots$$

$$\ldots\ldots\ldots\ldots\ldots\ldots\ldots\ldots\ldots\ldots\ldots\ldots\ldots\ldots$$

$$a^{x+h} = f(x+h)$$

$$= f(x) + hf(x) + \frac{h^2}{2!} f''(x) + \frac{h^3}{3!} f'''(x) + \ldots\ldots$$

$$= a^x \left[1 + h \log a + \frac{h^2}{2!} (\log a)^2 + \frac{h^3}{3!} (\log a)^3 + \ldots\ldots\right]$$

(*b*) Please try yourself.

(*c*) Let $\quad f(x+h) = \frac{1}{x+h}$

Putting $h = 0$, $f(x) = \frac{1}{x}$

$$\therefore \quad f'(x) = -\frac{1}{x^2} \; f''(x) = \frac{2}{x^3} = \frac{2!}{x^3}$$

$$f'''(x) = -\frac{6}{x^4} = -\frac{3!}{x^4} \quad \ldots\ldots\ldots\ldots\ldots\ldots$$

$$\ldots\ldots\ldots\ldots\ldots\ldots\ldots\ldots\ldots\ldots\ldots\ldots\ldots\ldots$$

$$\frac{1}{x+h} = f(x+h)$$

$$= f(x) + hf'(x) + \frac{h^2}{2!} f''(x) + \frac{h^2}{3!} f'''(x) + \ldots\ldots$$

$$= \frac{1}{x} - \frac{h}{x^2} + \frac{h^2}{x^3} - \frac{h^3}{x^4} + \ldots\ldots$$

$$= \frac{1}{x}\left[1 - \frac{h}{x} + \frac{h^2}{x^2} - \frac{h^3}{x^3} + \ldots\ldots\right].$$

Example 2. *Apply Taylor's Theorem to prove that:*

(*a*) $\log (x + h) = \log x + \frac{h}{x} - \frac{h^2}{2x^2} + \frac{h^3}{3x^3} - \ldots\ldots$

(*b*) $\log \sin (x + h) = \log \sin x + h \cot x - \frac{h^2}{2} \operatorname{cosec}^2 x + \frac{h^3}{3} \cot x \operatorname{cosec}^2 x + \ldots\ldots$

(*c*) $\log \cos (x + h) = \log \cos x - h \tan x - \frac{h^2}{2} \sec^2 x - \frac{h^3}{3} \sec^2 x \tan x + \ldots\ldots$

Sol. (*a*) Let $f(x + h) = \log (x + h)$

Putting $h = 0$, $f(x) = \log x$

$$f'(x) = \frac{1}{x}, \quad f''(x) = -\frac{1}{x^2}$$

$$f'''(x) = \frac{2}{x^3} \quad \ldots\ldots\ldots\ldots\ldots\ldots$$

....

$\therefore$ $\log (x + h) = f(x + h)$

$$= f(x) + hf'(x) + \frac{h^2}{2!} f''(x) + \frac{h^3}{3!} f'''(x) + \ldots\ldots$$

$$= \log x + \frac{h}{x} - \frac{h^2}{2x^2} + \frac{h^3}{3x^3} - \ldots\ldots$$

(*b*) Let $f(x + h) = \log \sin (x + h)$

Putting $h = 0$, $f(x) = \log \sin x$

$\therefore$ $f'(x) = \frac{\cos x}{\sin x} = \cot x, \quad f''(x) = -\operatorname{cosec}^2 x$

$$f'''(x) = -2 \operatorname{cosec} x (-\operatorname{cosec} x \cot x) = 2 \operatorname{cosec}^2 x \cot x$$

....

$\therefore$ $\log \sin (x + h) = f(x + h)$

$$= f(x) + hf'(x) + \frac{h^2}{2!} f''(x) + \frac{h^3}{3!} f'''(x) + \ldots\ldots$$

$$= \log \sin x + h \cot x - \frac{h^2}{2} \operatorname{cosec}^2 x + \frac{h^3}{3} \operatorname{cosec}^2 x \cot x + \ldots\ldots$$

(*c*) Please try yourself.

Example 3. *Apply Taylor's Theorem to prove that:*

$$(a)\ \sin^{-1}(x+h) = \sin^{-1} x + \frac{h}{\sqrt{1-x^2}} + \frac{x}{(1-x^2)^{3/2}} \cdot \frac{h^2}{2!} + \ldots\ldots$$

$$(b)\ \tan^{-1}(x+h) = \tan^{-1} x + \frac{h}{1+x^2} - \frac{xh^2}{(1+x^2)^2} + \ldots\ldots$$

$$(c)\ \sec^{-1}(x+h) = \sec^{-1} x + \frac{h}{x\sqrt{x^2-1}} - \frac{h^2}{2!} \cdot \frac{2x^2-1}{x^2(x^2-1)^{3/2}} + \ldots\ldots$$

Sol. (*a*) Let $f(x+h) = \sin^{-1}(x+h)$

Putting $h = 0$, $f(x) = \sin^{-1} x$

$$\therefore \quad f'(x) = \frac{1}{\sqrt{1-x^2}}, \quad f''(x) = \frac{d}{dx}[(1-x^2)^{-1/2}]$$

$$= -\frac{1}{2}(1-x^2)^{-3/2}(-2x) = \frac{x}{(1-x^2)^{3/2}}$$

....

$$\sin^{-1}(x+h) = f(x+h)$$

$$= f(x) + hf'(x) + \frac{h^2}{2!}f''(x) + \frac{h^3}{3!}f'''(x) + \ldots\ldots$$

$$= \sin^{-1} x + \frac{h}{\sqrt{1-x^2}} + \frac{x}{(1-x^2)^{3/2}} \cdot \frac{h^2}{2!} + \ldots\ldots$$

(*b*) Let $f(x+h) = \tan^{-1}(x+h)$

Putting $h = 0$, $f(x) = \tan^{-1} x$

$$f'(x) = \frac{1}{1+x^2}$$

$$f''(x) = \frac{d}{dx}[(1+x^2)^{-1}] = -(1+x^2)^{-2} \cdot 2x = -\frac{2x}{(1+x^2)^2}$$

....

$$\therefore \quad \tan^{-1}(x+h) = f(x+h)$$

$$= f(x) + hf'(x) + \frac{h^2}{2!}f''(x) + \ldots\ldots$$

$$= \tan^{-1} x + \frac{h}{1+x^2} - \frac{xh^2}{(1+x^2)^2} + \ldots\ldots$$

(c) Let $f(x+h) = \sec^{-1}(x+h)$

Putting $h = 0$, $f(x) = \sec^{-1} x$

$$f'(x) = \frac{1}{x\sqrt{x^2-1}}, \quad f''(x) = \frac{d}{dx}[x^{-1}(x^2-1)^{-1/2}]$$

$$= -x^{-2}(x^2-1)^{-1/2} + x^{-1} \cdot \left(-\frac{1}{2}\right)(x^2-1)^{-3/2}(2x)$$

$$= -\frac{1}{x^2(x^2-1)^{1/2}} - \frac{1}{(x^2-1)^{3/2}}$$

$$= -\frac{x^2-1+x^2}{x^2(x^2-1)^{3/2}} = -\frac{2x^2-1}{x^2(x^2-1)^{3/2}}$$

....

$\therefore \quad \sec^{-1}(x+h) = f(x+h)$

$$= f(x) + hf'(x) + \frac{h^2}{2!} f''(x) + \ldots\ldots$$

$$= \sec^{-1} x + \frac{h}{x\sqrt{x^2-1}} - \frac{h^2}{2!} \cdot \frac{2x^2-1}{x^2(x^2-1)^{3/2}} + \ldots\ldots$$

Example 4. *Apply Taylor's Theorem to prove that:*

(a) $\tan(x+h) = \tan x + h \sec^2 x + h^2 \sec^2 x \tan x + \frac{h^3}{3} \sec^2 x(1 + 3\tan^2 x) + \ldots\ldots$

(b) $\cos\left(\frac{\pi}{4}+h\right) = \frac{1}{\sqrt{2}}\left(1 - h - \frac{h^2}{2!} + \frac{h^3}{3!} \ldots\ldots\right)$

(c) $\frac{f(x+h)+f(x-h)}{2} = f(x) + \frac{h^2}{2!} f''(x) + \frac{h^4}{4!} f''''(x) + \ldots\ldots$

Sol. (a) Let $f(x+h) = \tan(x+h)$

Putting $h = 0$, $f(x) = \tan x$

$$f'(x) = \sec^2 x,\ f''(x) = 2 \sec x . \sec x \tan x = 2 \sec^2 x \tan x$$

$$f'''(x) = 2(2 \sec x . \sec x \tan x . \tan x + \sec^2 x . \sec^2 x)$$

$$= 2 \sec^2 x (2 \tan^2 x + \sec^2 x)$$

$$= 2 \sec^2 x (2\tan^2 x + 1 + \tan^2 x) = 2 \sec^2 x (1 + 3 \tan^2 x)$$

....

$\therefore \quad \tan(x+h) = f(x+h)$

$$= f(x) + hf'(x) + \frac{h^2}{2!} f''(x) + \frac{h^3}{3!} f'''(x) + \ldots\ldots$$

$$= \tan x + h \sec^2 x + \frac{h^2}{2!} . 2 \sec^2 \tan x + \frac{h^3}{3!} . 2 \sec^4 x (1 + 3 \tan^2 x) + \ldots\ldots$$

$$= \tan x + h \sec^2 x + h^2 \sec^2 x \tan x + \frac{h^3}{3} \sec^2 x (1 + 3 \tan^2 x) + ...$$

(*b*) Let $f(x + h) = \cos(x + h)$ **[Note this step]**

Putting $h = 0, \quad f(x) = \cos x$

$f'(x) = -\sin x, \quad f''(x) = -\cos x$

$f'''(x) = -\sin x$, ...

....

$\therefore \quad \cos(x + h) = f(x + h)$

$$= f(x) + hf'(x) + \frac{h^2}{2!} f''(x) + \frac{h^3}{3!} f'''(x) + ...$$

$$= \cos x - h \sin x - \frac{h^2}{2!} \cos x + \frac{h^3}{3!} \sin x ......$$

Putting $x = \frac{\pi}{4}$, we have

$$\cos\left(\frac{\pi}{4} + h\right) = \frac{1}{\sqrt{2}}\left(1 - h - \frac{h^2}{2!} + \frac{h^3}{3!} ......\right) \qquad \left[\because \cos\frac{\pi}{4} = \sin\frac{\pi}{4} = \frac{1}{\sqrt{2}}\right]$$

(*c*) By Taylor's Theorem, we have

$$f(x + h) = f(x) + hf'(x) + \frac{h^2}{2!} f''(x) + \frac{h^3}{3!} f'''(x) + \frac{h^4}{4!} f^{iv}(x) + ...$$

Changing h to $-h$,

$$f(x - h) = f(x) - hf'(x) + \frac{h^2}{2!} f''(x) - \frac{h^3}{3!} f'''(x) + \frac{h^4}{4!} f^{iv}(x) - ......$$

Adding $f(x + h) + f(x - h) = 2\left[f(x) + \frac{h^2}{2!} f''(x) + \frac{h^4}{4!} f^{iv}(x) + ...\right]$

$$\Rightarrow \quad \frac{f(x+h) - f(x-h)}{2} = f(x) + \frac{h^2}{2!} f''(x) + \frac{h^4}{4!} f^{iv}(x) + ...$$

Example 5. *Expand sin x in ascending powers of* $\left(x - \frac{\pi}{2}\right)$.

Sol. Let $f(x) = \sin x$

then $\sin x = f(x) = f\left[\frac{\pi}{2} + \left(x - \frac{\pi}{2}\right)\right] = f\left(\frac{\pi}{2} + h\right)$ where $h = x - \frac{\pi}{2}$

$$= f\left(\frac{\pi}{2}\right) + hf'\left(\frac{\pi}{2}\right) + \frac{h^2}{2!} f''\left(\frac{\pi}{2}\right) + \frac{h^3}{3!} f'''\left(\frac{\pi}{2}\right) + ... \quad ...(1)$$

Now $\quad f(x) = \sin x \qquad f\left(\frac{\pi}{2}\right) = \sin\frac{\pi}{2} = 1$

$$f'(x) = \cos x \qquad f'\left(\frac{\pi}{2}\right) = \cos\frac{\pi}{2} = 0$$

$$f''(x) = -\sin x \qquad f''\left(\frac{\pi}{2}\right) = -\sin\frac{\pi}{2} = -1$$

$$f'''(x) = -\cos x \qquad f'''\left(\frac{\pi}{2}\right) = -\cos\frac{\pi}{2} = 0$$

$$f^{iv}(x) = \sin x \qquad f^{iv}\left(\frac{\pi}{2}\right) = \sin\frac{\pi}{2} = 1$$

....

$\therefore$ From (1),

$$\sin x = 1 - \frac{h^2}{2!} + \frac{h^4}{4!} - \ldots\ldots$$

$$= 1 - \frac{1}{2!}\left(x - \frac{\pi}{2}\right)^2 + \frac{1}{4!}\left(x - \frac{\pi}{2}\right)^4 - \ldots\ldots$$

Example 6. *Expand sin x in ascending powers of (x – a).*

Sol. Please try yourself.

Hint. $\sin x = f(x) = f[a + (x - a)] = f(a + h)$ where $h = x - a$

Ans. $\sin a + (x - a)\cos a - \frac{(x-a)^2}{2!}\sin a - \frac{(x-a)^3}{3!} - \cos a + \ldots\ldots$

Example 7. *Expand log sin x in powers of (x – 3).*

Sol. Let $\quad f(x) = \log \sin x$

$$\log \sin x = f(x) = f[3 + (x - 3)] = f(3 + yh) \text{ where } h = x - 3$$

$$= f(3) + hf'(3) + \frac{h^2}{2!}f''(3) + \frac{h^3}{3!}f'''(3) + \ldots\ldots \qquad \ldots(1)$$

Now $\quad f(x) = \log \sin x, \qquad f(3) = \log \sin 3$

$$f'(x) = \frac{\cos x}{\sin x} = \cot x, \qquad f'(3) = \cot 3$$

$$f''(x) = -\operatorname{cosec}^2 x \qquad f''(3) = -\operatorname{cosec}^2 3$$

$$f'''(x) = -2\operatorname{cosec} x(-\operatorname{cosec} x \cot x) = 2\operatorname{cosec}^2 x \cot x$$

$$f'''(3) = 2\operatorname{cosec}^2 3 \cot 3$$

....

$\therefore$ From (1)

$$\log \sin x = \log \sin 3 + (x - 3)\cot 3 - \frac{(x-3)^2}{2}\operatorname{cosec}^2 3 + \frac{(x-3)^3}{3}\operatorname{cosec}^2 3 \cot 3 + \ldots\ldots$$

Example 8. *Expand e^x in ascending powers of $(x - 1)$.*

Sol. Let $f(x) = e^x$

$$e^x = f(x) = f[1 + (x - 1)] = f(1 + h) \quad \text{where } h = x - 1$$

$$= f(1) + hf'(1) + \frac{h^2}{2!} f''(1) + \frac{h^3}{3!} f'''(1) + \ldots\ldots \qquad \ldots(1)$$

Now $f(x) = e^x$ $\quad f(1) = e$

$f'(x) = e^x$ $\quad f'(1) = e$

$f''(x) = e^x$ $\quad f''(1) = e$

....

$\therefore$ From (1), $e^x = e\left[1 + h + \frac{h^2}{2!} + \frac{h^3}{3!} + \ldots\ldots\right]$

$$= e\left[1 + (x - 1) + \frac{(x-1)^2}{2!} + \frac{(x-1)^3}{3!} + \ldots\ldots\right]$$

Example 9. *Use Taylor's Theorem to express the polynomial $2x^3 + 7x^2 + x - 6$ in powers of $(x - 2)$.*

Sol. Let $f(x) = 2x^3 + 7x^2 + x - 6$

$\therefore$ $2x^3 + 7x^2 + x - 6 = f(x) = f[2 + (x - 2)] = f(2 + h)$ where $h = x - 2$

$$= f(2) + hf'(2) + \frac{h^2}{2!} f''(2) + \frac{h^3}{3!} f'''(2) \qquad \ldots(1)$$

[as higher derivatives of $f(x)$ vanish]

Now $f(x) = 2x^3 + 7x^2 + x - 6$ $\quad f(2) = 16 + 28 + 2 - 6 = 40$

$f'(x) = 6x^2 + 14x + 1$ $\quad f'(2) = 24 + 28 + 1 = 53$

$f''(x) = 12x + 14$ $\quad f''(2) = 24 + 14 = 38$

$f'''(x) = 12$ $\quad f'''(2) = 12$

$\therefore$ From (1)

$$2x^3 + 7x^2 + x - 6 = 40 + 53h + \frac{38h^2}{2!} + \frac{12h^3}{3!}$$

$$= 40 + 53(x - 2) + 19(x - 2)^2 + 2(x - 2)^2.$$

Example 10. *Expand $2 + x^2 - 3x^5 + 7x^6$ in powers of $(x - 1)$.*

Sol. Let $f(x) = 2 + x^2 - 3x^5 + 7x^6$

$\therefore$ $2 + x^2 - 3x^5 + 7x^6 = f(x) = f[1 + (x - 1) = f(1 + h)$ where $h = x - 1$

$$= f(1) + hf'(1) + \frac{h^2}{2!} f''(1) + \frac{h^3}{3!} f'''(1)$$

$$+ \frac{h^4}{4!} f^{iv}(1) + \frac{h^5}{5!} f^{v}(1) + \frac{h^6}{6!} f^{vi}(1) \qquad \ldots(1)$$

[as higher derivatives of $f(x)$ vanish]

Now $f(x) = 2 + x^2 - 3x^5 + 7x^6$ $\quad f(1) = 7$

$f'(x) = 2x - 15x^4 + 42x^5$ $\quad f'(1) = 29$

$f''(x) = 2 - 60x^3 + 210x^4$ $\quad f''(1) = 152$

$f'''(x) = -180x^2 + 840x^3$ $\quad f'''(1) = 660$

$f^{iv}(x) = -360x + 2520x^2$ $\quad f^{iv}(1) = 2160$

$f^{v}(x) = -360 + 5040x$ $\quad f^{v}(1) = 4680$

$f^{vi}(x) = 5040$ $\quad f^{vi}(1) = 5040$

$\therefore$ From (1)

$$2 + x^2 - 3x^5 + 7x^6 = 7 + 29h + \frac{h^2}{2}(152) + \frac{h^3}{6}(660) + \frac{h^4}{24}(2160) + \frac{h^5}{120}(4680) + \frac{h^6}{720}(5040)$$

$$= 7 + 29(x-1) + 76(x-1)^2 + 110(x-1)^3 + 90(x-1)^4 + 39(x-1)^5 + 7(x-1)^6.$$

Example 11. *Expand tan x in powers of* $\left(x - \frac{\pi}{4}\right)$ *upto first four terms.*

Sol. $f(x) = \tan x$

$$\therefore \quad \tan x = f(x) = f\left[\frac{\pi}{4} + \left(x - \frac{\pi}{4}\right)\right] = f\left(\frac{\pi}{4} + h\right) \quad \text{where } h = x - \frac{\pi}{4}$$

$$= f\left(\frac{\pi}{4}\right) + hf'\left(\frac{\pi}{4}\right) + \frac{h^2}{2!}f''\left(\frac{\pi}{4!}\right) + \frac{h^3}{3!}f'''\left(\frac{\pi}{4}\right) + \ldots \qquad \ldots(1)$$

Now $f(x) = \tan x,$ $\quad f\left(\frac{\pi}{4}\right) = 1$

$f'(x) = \sec^2 x,$ $\quad f\left(\frac{\pi}{4}\right) = (\sqrt{2})^2 = 2$

$f''(x) = 2 \sec x. \sec x \tan x$ $\quad f''\left(\frac{\pi}{4}\right) = 2(\sqrt{2})^2 (1) = 4$

$= 2 \sec^2 x \tan x$

$f''(x) = 2 (2 \sec x. \sec x \tan x \tan x + \sec^2 x. \sec^2 x)$

$= 2 \sec^2 x (2 \tan^2 x \sec^2 x)$

$$f''\left(\frac{\pi}{4}\right) = 2(2)(2 + 2) = 16$$

....

$$\therefore \text{ From (1)} \quad \tan x = 1 + 2h + \frac{h^2}{2}(4) + \frac{h^3}{6}(16) + \ldots\ldots$$

$$= 1 + 2\left(x - \frac{\pi}{4}\right) + 2\left(x - \frac{\pi}{4}\right)^2 + \frac{8}{3}\left(x - \frac{\pi}{4}\right)^3 + \ldots\ldots$$

6.6. MACLAURIN'S INFINITE SERIES

Statement. *If f (x) is a continuous function of x, which is capable of being expanded in ascending powers of x and the expansion be differentiable term by term any number of times, then*

$$f(x) = f(0) + xf'(0) + \frac{x^2}{2!} f''(0) + \frac{x^3}{3!} f'''(0) + \ldots\ldots + \frac{x^n}{n!} f^n(0) + \ldots\ldots ad.\ inf.$$

Proof. Let $f(x) = p_0 + p_1 x + p_2 x^2 + p_3 x^3 + p_4 x^4 + \ldots\ldots$ ...(1)

where p_0, p_1, p_2, p_3, p_4 ...... are all constants.

Differentiating (1) successively w.r.t. x.

$$f'(x) = p_1 + 2p_2 x + 3p_3 x^2 + 4p_4 x^3 + \ldots\ldots \quad \ldots(2)$$

$$f''(x) = 2p_2 + 6p_3 x + 12p_4 x^2 + \ldots\ldots \quad \ldots(3)$$

$$f'''(x) = 6p_3 + 24p_4 x + \ldots\ldots \quad \ldots(4)$$

$$f^{iv}(x) = 24p_4 + \ldots\ldots \quad \ldots(5)$$

....

Putting $x = 0$ in (1), (2), (3), (4), (5), ...

$$f(0) = p_0 \qquad \therefore \quad p_0 = f(0)$$

$$f'(0) = p_1 \qquad \therefore \quad p_1 = f'(0)$$

$$f''(0) = 2p_2 \qquad \therefore \quad p_2 = \frac{1}{2!} f''(0)$$

$$f'''(0) = 6p_3 \qquad \therefore \quad p_3 = \frac{1}{3!} f'''(0)$$

$$f^{iv}(0) = 24p_4 \qquad \therefore \quad p_4 = \frac{1}{4!} f^{iv}(0)$$

....

$\therefore$ From (1) $f(x) = f(0) + xf'(0) + \frac{x^2}{2!} f''(0) + \frac{x^3}{3!} f'''(0) + \ldots\ldots$

Failure of Maclaurin's Theorem

Maclaurin's Theorem fails to expand a function $f(x)$ if

(*i*) **any of the quantities** $f(0), f'(0), f''(0)$, ... is infinite or does not exist. This is the case with log x and cot x.

(*ii*) R_n does not tend to zero as $n \to \infty$.

Example 1. *Expand*

(*a*) e^x (*b*) a^x (*c*) *sin x*

(*d*) *cos x* (*e*) *tan x.* (M.D.U. 1981 S) (*f*) *sec x*

(*g*) *log (sec x + tan x).*

Sol. (*a*) Let $f(x) = e^x$, $\quad f(0) = e^0 = 1$

$f'(x) = e^x$, $\quad f'(0) = 1$

$f''(x) = e^x$, $\quad f''(0) = 1$

$f'''(x) = e^x$, $\quad f'''(0) = 1$

....

$$\therefore \qquad e^x = f(x) = f(0) + xf'(0) + \frac{x^2}{2!} f''(0) + \frac{x^3}{3!} f'''(0) + ...$$

$$= 1 + x + \frac{x^2}{2!} + \frac{x^3}{3!} + ...$$

Note. Putting $x = 1$

$$e = 1 + 1 + \frac{1}{2!} + \frac{1}{3!} + ...$$

(*b*) Let

$f(x) = a^x,$	$f(0) = a^0 = 1$
$f'(x) = a^x \log a,$	$f'(0) = \log a$
$f''(x) = a^x (\log a)^2,$	$f''(0) = (\log a)^2$
$f'''(x) = a^x (\log a)^3,$	$f'''(0) = (\log a)^3$

....

$$\therefore \qquad a^x = f(x)$$

$$= f(0) + xf'(0) + \frac{x^2}{2!} f''(0) + \frac{x^3}{3!} f'''(0) + ...$$

$$= 1 + x \log a + \frac{x^2}{2!} (\log a)^2 + \frac{x^3}{3!} (\log a)^3 + ...$$

(*c*) Let

$f(x) = \sin x$	$f(0) = 0$
$f'(x) = \cos x$	$f'(0) = 1$
$f''(x) = -\sin x$	$f''(0) = 0$
$f'''(x)) = -\cos x$	$f'''(0) = -1$
$f^{iv}(x) = \sin x$	$f^{iv}(0) = 0$
$f^{v}(x) = \cos x$	$f^{v}(0) = 1$

....

$$\therefore \qquad \sin x = f(x)$$

$$= f(0) + xf'(0) + \frac{x^2}{2!} f''(0) + \frac{x^3}{3!} f'''(0) + \frac{x^4}{4!} f^{iv}(0) + \frac{x^5}{5!} f^{v}(0) + ...$$

$$= x - \frac{x^3}{3!} + \frac{x^5}{5!} - ......$$

(*d*)

$f(x) = \cos x$	$f(0) = 1$
$f'(x) = -\sin x$	$f'(0) = 0$
$f''(x) = -\cos x$	$f''(0) = -1$
$f'''(x) = \sin x$	$f'''(0) = 0$
$f^{iv}(x) = \cos x$	$f^{iv}(0) = 1$

....

$$\therefore \qquad \cos x = f(x)$$

$$= f(0) + xf'(0) + \frac{x^2}{2!} f''(0) + \frac{x^3}{3!} f'''(0) + \frac{x^4}{4!} f^{iv}(0) + ...$$

$$= 1 - \frac{x^2}{2!} + \frac{x^4}{4!} - \ldots\ldots$$

(*e*) Let $y = \tan x$ $\quad\therefore\quad y(0) = 0$

$y_1 = \sec^2 x = 1 + \tan^2 x = 1 + y^2 \quad\therefore\quad y_1(0) = 1 + 0 = 1$

$y_2 = 2yy_1 \quad\therefore\quad y_2(0) = 0$

$y_3 = 2y_1y_1 + 2yy_2 + 2y_1^2 + 2yy_2 \quad\therefore\quad y_3(0) = 2 \times 1 + 0 = 2$

$y_4 = 4y_1y_2 + 2y_1y_2 + 2yy_2$

$= 6y_1y_2 + 2yy_3 \quad\therefore\quad y_4(0) = 0 + 0 = 0$

$y_5 = 6y_2^2 + 6y_1y_3 + 2y_1y_3 + 2yy_4$

$= 8y_1y_3 + 6y_2^2 + 2yy_4 \quad\therefore\quad y_5(0) = 8 \times 1 \times 2 + 0 + 0 = 16$

....

$$\therefore \quad \tan x = y(0) + xy_1(0) + \frac{x^2}{2!} y_2(0) + \frac{x^3}{3!} y_3(0) + \frac{x^4}{4!} y_4(0) + \frac{x^5}{5!} y_5(0) + \ldots\ldots$$

$$= x + \frac{x^3}{3!} . 2 + \frac{x^5}{5!} . 16 + \ldots\ldots = x + \frac{x^3}{3} + \frac{2x^5}{15} + \ldots\ldots$$

(*f*) Let $y = \sec x \quad\therefore\quad y(0) = 1$

$y_1 = \sec x \tan x = y \tan x \quad\therefore\quad y_1(0) = 1 \times 0 = 0$

$y_2 = y_1 \tan x + y \sec^2 x$

$= y_1 \tan x + y^3 \quad\therefore\quad y_2(0) = 0 + 1 = 1$

$y_3 = y_2 \tan x + y_1 \sec^2 x + 3y^2y_1 = y_2 \tan x + y^2y_1 + 3y^2y_1$

$= y_2 \tan x + 4y^2y_1$

$\therefore \quad y_3(0) = 1 \times 0 + 4 \times 1 \times 0 = 0$

$y_4 = y_3 \tan x + y_2 \sec^2 x + 8yy_1 . y_1 + 4y^2y_2$

$= y_3 \tan x + y^2y_2 + 8yy_1^2 + 4y^2y_2$

$= y_3 \tan x + 5y^2y_2 + 8yy_1^2$

$\therefore \quad y_4(0) = 0 + 5 \times 1 \times 1 + 8 \times 1 \times 0 = 5$

....

$$\therefore \quad \sec x = y(0) + xy_1(0) + \frac{x^2}{2!} y_2(0) + \frac{x^3}{3!} y_3(0) + \frac{x^4}{4!} y_4(0) + \ldots\ldots$$

$$= 1 + \frac{x^2}{2!} + \frac{5x^4}{5!} + \ldots\ldots$$

(*g*) Let $y = \log (\sec x + \tan x) \quad\therefore\quad y(0) = \log 1 = 0$

$$y_1 = \frac{\sec x \tan x + \sec^2 x}{\sec x + \tan x} = \frac{\sec x (\tan x + \sec x)}{\sec x + \tan x} = \sec x$$

$\therefore \quad y_1(0) = 1$

$y_2 = \sec x \tan x = y_1 \tan \quad\therefore\quad y_2(0) = 1 \times 0 = 0$

$y_3 = y_2 \tan x + y_1 \sec^2 x = y_2 \tan x + y_1 . y_1^2 = y_2 \tan x + y_1^3$

$\therefore \quad y_2(0) = 0 + 1 = 1$

$$y_4 = y_3 \tan x + y_2 \sec^2 x + 3y_1^2 y_2 = y_3 \tan x + y_2 y_1^2 + 3y_1^2 y_2$$
$$= y_3 \tan x + 4y_1^2 y_2$$

$\therefore \quad y_4(0) = 0 + 4 \times 1 \times 0 = 0$

$$y_5 = y_4 \tan x + y_3 \sec^2 x + 8y_1 y_2 . y_2 + 4y_1^2 y_3$$
$$= y_4 \tan x + y_3 . y_1^2 + 8y_1 y_2^2 + 4y_1^2 y_3$$
$$= y_4 \tan x + 5y_1^2 y_3 + 8y_1 y_2^2$$

$\therefore \quad y_5(0) = 0 + 5 \times 1 \times 1 + 8 \times 1 \times 0 = 5$

....

$$\therefore \quad \log(\sec x + \tan x) = y(0) + xy_1(0) + \frac{x^2}{2!} y_2(0) + \frac{x^3}{3!} y_3(0) + \frac{x^4}{4!} y_4(0) + \frac{x^5}{5!} y_5(0) + \ldots\ldots$$

$$= x + \frac{x^3}{3!} + \frac{5x^5}{5!} + \ldots\ldots$$

$$= x + \frac{x^3}{6} + \frac{x^5}{24} + \ldots\ldots$$

☞**Remember :**

1. $e^x = 1 + x + \frac{x^2}{2!} + \frac{x^3}{3!} + \ldots\ldots$
2. $e^{-x} = 1 - x + \frac{x^2}{2!} - \frac{x^3}{3!} + \ldots\ldots$
3. $\sin x = x - \frac{x^3}{3!} + \frac{x^5}{5!} - \ldots\ldots$
4. $\cos x = 1 - \frac{x^2}{2!} + \frac{x^4}{4!} - \ldots\ldots$
5. $\tan x = x + \frac{x^3}{3} + \frac{2x^5}{15} + \ldots\ldots$

Example 2. *Find the expansion of*

(a) log (1 + x) *(b) log (1 – x)*

(c) $\log \sqrt{\frac{1+x}{1-x}}$ *(d) sinh x*

(e) cosh x.

Sol. *(a)* Let $f(x) = \log(1 + x)$ $\quad \therefore \quad f(0) = \log 1 = 0$

$f'(x) = \frac{1}{1+x}$ $\quad \therefore \quad f'(0) = 1$

$f''(x) = -\frac{1}{(1+x)^2}$ $\quad \therefore \quad f''(0) = -1$

$f'''(x) = \frac{2}{(1+x)^3}$ $\quad \therefore \quad f'''(0) = 2$

$f^{iv}(x) = -\frac{6}{(1+x)^4}$ $\quad \therefore \quad f^{iv}(0) = -6$

....

$\therefore \quad \log(1+x) = f(x)$

$$= f(0) + xf'(0) + \frac{x^2}{2!} f''(0) + \frac{x^3}{3!} f'''(0) \frac{x^4}{4!} f^{iv}(0) + \ldots\ldots$$

$$= x - \frac{x^2}{2!} + \frac{2x^3}{3!} - \frac{6x^4}{4!} + \ldots\ldots$$

$$= x - \frac{x^2}{2} + \frac{x^3}{3} - \frac{x^4}{4} + \ldots\ldots.$$

(*b*) Please try yourself. $\left[\textbf{Ans.}\ -x - \frac{x^2}{2} - \frac{x^3}{3} - \frac{x^4}{4} - \ldots\ldots\right]$

☞**Remember : 1.** $\log(1+x) = x - \frac{x^2}{2} + \frac{x^3}{3} - \frac{x^4}{4} + \ldots\ldots$

2. $\log(1-x) = -x - \frac{x^2}{2} - \frac{x^3}{3} - \frac{x^4}{4} - \ldots\ldots$

(*c*) $\quad \log\sqrt{\frac{1+x}{1-x}} = \frac{1}{2}[\log(1+x) - \log(1-x)]$

$$= \frac{1}{2}\left[\left(x - \frac{x^2}{2} + \frac{x^3}{3} - \frac{x^4}{4} + \ldots\ldots\right) - \left(-x - \frac{x^2}{2} - \frac{x^3}{3} - \frac{x^4}{4} - \ldots\ldots\right)\right]$$

$$= \frac{1}{2}\left[2\left(x + \frac{x^3}{3} + \frac{x^5}{5} + \ldots\ldots\right)\right] = x + \frac{x^3}{3} + \frac{x^4}{5} + \ldots\ldots$$

(*d*) $\quad \sinh x = \frac{1}{2}[e^x - e^{-x}]$

$$= \frac{1}{2}\left[\left(1 + x + \frac{x^2}{2!} + \frac{x^3}{3!} + \frac{x^5}{4!} + \ldots\right) - \left(1 - x + \frac{x^2}{2!} - \frac{x^3}{3!} + \frac{x^4}{4!} - \ldots\ldots\right)\right]$$

$$= \frac{1}{2}\left[2\left(x + \frac{x^3}{3!} + \frac{x^5}{5!} + \ldots\ldots\right)\right] = x + \frac{x^3}{3!} + \frac{x^5}{5!} + \ldots\ldots$$

(*e*) Please try yourself.

$\left[\textbf{Hint.}\ \cosh x = \frac{1}{2}(e^x + e^{-x})\right]$ $\left[\textbf{Ans.}\ 1 + \frac{x^2}{2!} + \frac{x^4}{4!} + \ldots\ldots\right]$

Example 3. *Assuming the possibility of expansion, prove that*

(*a*) $e^{\sin x} = 1 + x + \frac{1}{2}x^2 - \frac{1}{8}x^4 + \ldots\ldots$ [CUSAT 2009]

(*b*) $e^{x\cos x} = 1 + x + \frac{1}{2}x^2 - \frac{1}{2}x^3 + \ldots\ldots$ (*c*) $\frac{e^x}{\cos x} = 1 + x + \frac{2x^2}{2!} + \frac{4x^3}{3!} + \ldots\ldots$

(d) $e^x \log(1+x) = x + \frac{x^2}{2} + \frac{x^3}{3} + \frac{3x^5}{40} + \ldots\ldots$

(e) $\log \sec x = \frac{x^2}{2} + \frac{x^4}{12} + \frac{x^6}{45} + \ldots\ldots$ (f) $\log \cos x = -\frac{x^2}{2} - \frac{x^4}{12} - \frac{x^6}{45} + \ldots\ldots$

Sol. (a) Let $y = e^{\sin x}$ $\therefore\ y(0) = 1$

$y_1 = e^{\sin x} \cos x = y \cos x$ $\therefore\ y_1(0) = 1 \times 1 = 1$

$y_2 = y_1 \cos x - y \sin x$ $\therefore\ y_2(0) = 1 \times 1 - 0 = 1$

$$y_3 = y_2 \cos x - y_1 \sin x - (y_1 \sin x + y \cos x)$$
$$= y_2 \cos x - 2y_1 \sin x - y \cos x$$

$\therefore$ $y_3(0) = 1 \times 1 - 0 - 1 \times 1 = 0$

$$y_4 = y_3 \cos x - y_2 \sin x - 2(y_2 \sin x + y_1 \cos x) - (y_1 \cos x - y \sin x)$$
$$= y_3 \cos x - 3y_2 \sin x - 3y_1 \cos x + y \sin x$$

and $y_4(0) = 0 - 0 - 3 \times 1 \times 1 + 0 = -3$

....

$\therefore$ $$e^{\sin x} = y(0) = xy_1(0) + \frac{x^2}{2!} y_2(0) + \frac{x^3}{3!} y_3(0) + \frac{x^4}{4!} y_4(0)$$

$$= 1 + x + \frac{x^2}{2} - \frac{x^4}{8} + \ldots\ldots$$

(b) Let $y = e^{x \cos x}$ $\therefore\ y(0) = e^0 = 1$

$$y_1 = e^{x \cos x}(\cos x - x \sin x) = y(\cos x - x \sin x)$$

$\therefore$ $y_1(0) = 1\,(1 - 0) = 1$

$$y_2 = y_1 (\cos x - x \sin x) + y\,(-\sin x - (\sin x + x \sin x)$$
$$= y_1(\cos x - x \sin x) - y(2 \sin x + x \cos x)$$

$y_2(0) = 1(1 - 0) - 1(0 + 0) = 1.$

$$y_3 = y_2(\cos x - x \sin x) + y_1[-\sin x - (\sin x + x \cos x)]$$
$$- y_1(2 \sin x + x \cos x) - y[2 \cos x + \cos x - x \sin x]$$
$$= y_2(\cos x - x \sin x) - y_1(4 \sin x + 2x \cos x) - y(3 \cos x - x \sin x)$$

and $y_3(0) = 1(1 - 0) - 1(0 + 0) - 1(3 - 0) = 1 - 3 = -2.$

....

$\therefore$ $$e^{x \cos x} = y(0) + xy_1(0) + \frac{x^2}{2!} y_2(0) + \frac{x^3}{3!} y_3(0) + \ldots\ldots$$

$$= 1 + x + \frac{x^2}{2!} - \frac{2x^3}{3!} + \ldots\ldots$$

$$= 1 + x + \frac{x^2}{2} - \frac{x^3}{3} + \ldots\ldots$$

(*c*) Let $y = \dfrac{e^x}{\cos x} = e^x \sec x \qquad \therefore \quad y(0) = 1$

$$y_1 = e^x \sec x + e^x \sec x \tan x = y \tan x$$

$\therefore \qquad y_1(0) = 1 + 0 = 1.$

$$y_2 = y_1 + y_1 \tan x + y \sec^2 x \qquad \therefore \quad y_2(0) = 1 + 0 + 1 = 2$$

$$y_3 = y_2 + y_2 \tan x + y_1 \sec^2 x + y_1 \sec^2 x + y \,.\, 2 \sec x \,.\, \sec x \tan x$$

$$= y_2 + y_2 \tan x + 2y_1 \sec^2 x + 2y \sec^2 x \tan x$$

and $\qquad y_3(0) = 2 + 0 + 2 \times 1 \times 1 + 0 = 4.$

....

$$\therefore \qquad \frac{e^x}{\cos x} = y(0) + xy_1(0) + \frac{x^2}{2!} y_2(0) + \frac{x^3}{3!} y_3(0) + \ldots\ldots$$

$$= 1 + x + \frac{2x^2}{2!} + \frac{4x^3}{3!} + \ldots\ldots$$

(*d*) We know that $e^x = 1 + x + \dfrac{x^2}{2!} + \dfrac{x^3}{3!} + \dfrac{x^4}{4!} + \dfrac{x^5}{5!} + \ldots\ldots$

$$= 1 + x + \frac{x^2}{2} + \frac{x^3}{6} + \frac{x^4}{24} + \frac{x^5}{120} + \ldots\ldots$$

and $\qquad \log(1 + x) = x - \dfrac{x^2}{2} + \dfrac{x^3}{3} - \dfrac{x^4}{4} + \dfrac{x^5}{5} - \ldots\ldots$

....

Multiplying

$$e^x \log(1 + x) = x + x^2 + \frac{x^3}{2} + \frac{x^4}{6} + \frac{x^5}{24} + \ldots\ldots$$

$$- \frac{x^2}{2} - \frac{x^3}{2} - \frac{x^4}{4} - \frac{x^5}{12} - \ldots\ldots$$

$$+ \frac{x^3}{3} + \frac{x^4}{3} + \frac{x^5}{6} + \ldots\ldots$$

$$- \frac{x^4}{4} - \frac{x^5}{4} - \ldots\ldots$$

$$+ \frac{x^5}{5} + \ldots\ldots$$

....

$$= x + \frac{x^2}{2} + \frac{x^3}{3} + \frac{3x^5}{40} + \ldots\ldots$$

(*e*) Let $y = \log \sec x$ $\therefore\ y(0) = \log 1 = 0$

$$y_1 = \frac{\sec x \tan x}{\sec x} = \tan x \qquad y_1(0) = 0$$

$$y_2 = \sec^2 x = 1 + \tan^2 x = 1 + y_1^2 \qquad y_2(0) = 1$$

$$y_3 = 2y_1y_2 \qquad \therefore\ y_3(0) = 0$$

$$y_4 = 2y_2^2 + 2y_1y_3 \qquad \therefore\ y_4(0) = 2$$

$$y_5 = 4y_2y_3 + 2y_2y_3 + 2y_1y_4 = 6y_2y_3 + 2y_1y_4 \ \therefore\ y_5(0) = 0$$

$$y_6 = 6y_3^2 + 6y_2y_4 + 2y_2y_4 + 2y_1y_5 = 6y_3^2 + 8y_2y_4 + 2y_1y_5$$

$$\therefore \quad y_6(0) = 0 + 8 \times 1 \times 2 + 0 = 16.$$

and
$$\log \sec x = y(0) + x.y_1(0) = \frac{x^2}{2!}\, y_2(0) + \frac{x^3}{3!}\, y_3(0) + \ldots\ldots$$

$$= \frac{x^2}{2!} + \frac{2x^4}{4!} + \frac{16x^6}{6!} + \ldots\ldots$$

$$= \frac{x^2}{2} + \frac{x^4}{12} + \frac{x^6}{45} + \ldots\ldots$$

(*f*) Let $y = \log \cos x$ $\quad y(0) = \log 1 = 0$

$$y_1 = \frac{-\sin x}{\cos x} = -\tan x \qquad y_1(0) = 0$$

Proceed further as in part (*e*).

Example 4. *Obtain by Maclaurin's Theorem the first five terms in the expansion of log (1 + sin x).*

Sol. Let $y = \log (1 + \sin x)$ $\quad \therefore\ y(0) = \log 1 = 0$

$$y_1 = \frac{\cos x}{1 + \sin x} = \frac{\sin\left(\frac{\pi}{2} - x\right)}{1 + \cos\left(\frac{\pi}{2} - x\right)} \qquad \text{| Note carefully}$$

$$= \frac{2 \sin\left(\frac{\pi}{4} - \frac{x}{2}\right) \cos\left(\frac{\pi}{4} - \frac{x}{2}\right)}{2 \cos^2\left(\frac{\pi}{4} - \frac{x}{2}\right)}$$

$$\Big|\ \because\ \sin \theta = 2 \sin \frac{\theta}{2} \cos \frac{\theta}{2},\ 1 + \cos \theta = 2 \cos^2 \frac{\theta}{2}$$

$$= \tan\left(\frac{\pi}{4} - \frac{x}{2}\right) \qquad \therefore\ y_1(0) = \tan \frac{\pi}{4} = 1$$

$$y_2 = -\frac{1}{2} \sec^2\left(\frac{\pi}{4} - \frac{x}{2}\right)$$

$$= -\frac{1}{2}\left[1 + \tan^2\left(\frac{\pi}{4} - \frac{x}{2}\right)\right] = -\frac{1}{2} - \frac{1}{2}\, y_1^2$$

$$\therefore \quad y_2(0) = -\frac{1}{2} - \frac{1}{2} . 1 = -1$$

$$y_3 = -y_1y_2 \quad \therefore \quad y_3(0) = -1(-1) = 1$$

$$y_4 = -y_3^2 - y_1y_3 \quad \therefore \quad y_4(0) = -(-1)^2 - 1 \times 1 = -2$$

$$y_5 = -2y_2y_3 - y_2y_3 - y_1y_4$$

$$\therefore \quad y_5(0) = -2(-1) . 1 - (-1) . 1 - 1(-2) = 2 + 1 + 2 = 5$$

....

$$\therefore \log(1 + \sin x) = y(0) + xy_1(0) + \frac{x^2}{2!} y_2(0) + \frac{x^3}{3!} y_3(0) + \ldots\ldots$$

$$= x - \frac{x^2}{2!} + \frac{x^3}{3!} - \frac{2x^4}{4!} + \frac{5x^5}{5!} + \ldots\ldots$$

$$= x - \frac{x^2}{2} + \frac{x^3}{6} - \frac{x^4}{12} + \frac{x^5}{24} + \ldots\ldots$$

Example 5. *Use Maclaurin's Theorem to find the expansion of log (1 + e^x) to the term containing x^4.*

Sol. Let $\quad y = \log(1 + e^x) \qquad \therefore \quad y(0) = \log(1 + 1) = \log 2$

$$y_1 = \frac{e^x}{1+e^x} = 1 - \frac{1}{1+e^x} \qquad \therefore \quad y_1(0) = 1 - \frac{1}{2} = \frac{1}{2}$$

$$y_2 = \frac{e^x}{(1+e^x)^2} = \frac{e^x}{1+e^x} \cdot \frac{1}{1+e^x} = y_1(1 - y_1) \qquad |\text{Note carefully}$$

$$\therefore \quad y_2(0) = \frac{1}{2}\left(1 - \frac{1}{2}\right) = \frac{1}{4}$$

$$y_3 = y_2(1 - y_1) + y_1(-y_2) = y_2 - 2y_1y_2 \qquad \therefore \quad y_3(0) = \frac{1}{4} - 2 . \frac{1}{2} . \frac{1}{4} = 0$$

$$y_4 = y_3 - 2y_2^2 - 2y_1y_3 \qquad \therefore \quad y_4(0) = 0 - 2 . \frac{1}{16} - 0 = -\frac{1}{8}$$

....

$$\log(1 + e^x) = y(0) + xy_1(0) + \frac{x^2}{2!} y_2(0) + \ldots\ldots$$

$$= \log 2 + \frac{1}{2}x + \frac{1}{4} . \frac{x^2}{2!} - \frac{1}{8} . \frac{x^4}{4!} + \ldots\ldots$$

$$= \log 2 + \frac{1}{2}x + \frac{x^2}{8} - \frac{x^4}{192} + \ldots\ldots$$

Example 6. *Expand by Maclaurin's Theorem* $\dfrac{e^x}{1+e^x}$ *as far as the term containing x^3.*

Sol. Let $y = \dfrac{e^x}{1+e^x} = 1 - \dfrac{1}{1+e^x}$ $\therefore\ y(0) = 1 - \dfrac{1}{2} = \dfrac{1}{2}$

Proceed further yourself as in example 5. $\left[\textbf{Ans.}\ \dfrac{1}{2} + \dfrac{1}{4}x - \dfrac{1}{48}x^3 + \ldots\ldots\right]$

Example 7. *Find the first four terms in the expansion of log (1 + tan x).*

Sol. Since $\log(1+z) = z - \dfrac{z^2}{2} + \dfrac{z^3}{3} - \dfrac{z^4}{4} + \ldots\ldots$

$$\therefore \quad \log(1+\tan x) = \tan x - \frac{1}{2}\tan^2 x + \frac{1}{3}\tan^3 x - \frac{1}{4}\tan^4 x + \ldots\ldots$$

$$= \left(x + \frac{x^3}{3} + \frac{2x^5}{15} + \ldots\ldots\right) - \frac{1}{2}\left(x + \frac{x^3}{3} + \frac{2x^5}{15} + \ldots\ldots\right)^2$$

$$+ \frac{1}{3}\left(x + \frac{x^3}{3} + \frac{2x^5}{15} + \ldots\ldots\right)^3 - \frac{1}{4}\left(x + \frac{x^3}{3} + \frac{2x^3}{15} + \ldots\ldots\right)^4 + \ldots\ldots$$

$$\left[\because\ \tan x = x + \frac{x^3}{3} + \frac{2x^5}{15} + \ldots\ldots\right]$$

$$= \left(x + \frac{x^3}{3} + \frac{2x^5}{15} + \ldots\ldots\right) - \frac{1}{2}\left[x + \left(\frac{x^3}{3} + \frac{2x^5}{15} + \ldots\ldots\right)\right]^2$$

$$+ \frac{1}{3}\left[x + \left(\frac{x^3}{3} + \frac{2x^5}{15} + \ldots\ldots\right)\right]^3 - \frac{1}{4}\left[x + \left(\frac{x^3}{3} + \frac{2x^5}{15} + \ldots\ldots\right)\right]^4 + \ldots\ldots$$

$$= \left(x + \frac{x^3}{3} + \ldots\ldots\right) - \frac{1}{2}\left[x^2 + 2x\left(\frac{x^3}{3} + \ldots\ldots\right)\right]$$

$$+ \frac{1}{3}[x^3 + \ldots\ldots] - \frac{1}{4}[x^4 + \ldots\ldots] + \ldots\ldots$$

$$= \left(x + \frac{x^3}{3} + \ldots\ldots\right) - \frac{1}{2}\left(x^2 + \frac{2x^4}{3} + \ldots\ldots\right) + \frac{1}{3}\left(x^3 + \ldots\ldots\right) - \frac{1}{4}\left(x^4 + \ldots\ldots\right) + \ldots\ldots$$

$$= x - \frac{1}{2}x^2 + \frac{2}{3}x^3 - \frac{7}{12}x^4 + \ldots\ldots$$

Example 8. *Apply Maclaurin's Theorem to obtain terms upto x^4 in the expansion of log (1 + $\sin^2 x$).*

Sol. Since $\log(1+z) = z - \dfrac{z^2}{2} + \dfrac{z^3}{3} - \dfrac{z^4}{4} \ldots\ldots$

$$\therefore \quad \log(1+\sin^2 x) = \sin^2 x - \frac{1}{2}\sin^4 x + \ldots\ldots$$

$$= \left(x - \frac{x^3}{3!} + ...\right)^2 - \frac{1}{2}\left(x - \frac{x^3}{3!} + ...\right)^4 + ... \quad \left[\because \sin x = x - \frac{x^3}{3!} + \frac{x^5}{5!} - ......\right]$$

$$= \left[x - \left(\frac{x^3}{6} ......\right)\right]^2 - \frac{1}{2}\left[x - \left(\frac{x^3}{6} ......\right)\right]^4 + ...$$

$$= \left[x^2 - 2x\left(\frac{x^3}{6} - ......\right)\right] - \frac{1}{2}\left[x^4 - ......\right] + ...$$

$$= \left(x^2 - \frac{x^4}{3} - ......\right) - \frac{1}{2}\left(x^4 - ......\right) + ... = x^2 - \frac{5}{6}x^4 + ...$$

Example 9. *Obtain, by Maclaurin's Theorem, the coefficient of x^4 in the expansion of log cos x and verify your result algebraically by the use of well known expansions of log (1 + x) and cos x.*

Sol. Let $y = \log \cos x \quad \therefore \quad y(0) = \log 1 = 0$

$y_1 = -\tan x \quad\quad y_1(0) = 0$

$y_2 = -\sec^2 x \quad\quad y_2(0) = -1$

$y_3 = -2 \sec x \,.\, \sec x \tan x = -2 \sec^2 x \tan x$

$= -2\,(-y_2(\,(-y_1)$

$= -2y_1y_2 \quad \therefore \quad y_3(0) = 0$

$y_4 = -2{y_2}^2 - 2y_1y_3 \quad \therefore \quad y_4(0) = -2$

....

coefficient of $x^4 = \frac{1}{4!}\, y_4(0) = \frac{1}{24}(-2) = -\frac{1}{12}$.

Algebraic verification

$$\log \cos x = \log\left[1 - \frac{x^2}{2!} + \frac{x^4}{4!} - ......\right] \quad \left|\because \cos x = 1 - \frac{x^2}{2!} + \frac{x^4}{4!} - ......\right.$$

$$= \log\left[1 - \left(\frac{x^2}{2} - \frac{x^4}{24} + ...\right)\right]$$

$$= \log(1 - z) \quad \text{where } z = \frac{x^2}{2} - \frac{x^4}{24} + ...$$

$$= -z - \frac{1}{2}z^2 - \frac{1}{3}z^2 - \frac{1}{4}z^4 - ......$$

$$= -\left(\frac{x^2}{2} - \frac{x^4}{24} + ...\right) - \frac{1}{2}\left(\frac{x^2}{2} - \frac{x^4}{24} + ...\right)^2 - ......$$

coefficient of $x^4 = \frac{1}{24} - \frac{1}{2} \cdot \left(\frac{1}{2}\right)^2 = \frac{1}{24} - \frac{1}{8} = -\frac{1}{12}$

Hence the result is verified.

☞**Remember**

1. If $y = e^{ax} \sin bx$ then $y_n = r^n e^{ax} \sin (bx + n\phi)$ where $r = \sqrt{a^2 + b^2}$ and $\phi = \tan^{-1} \dfrac{b}{a}$

2. If $y = e^{ax} \cos bx$ then $y_n = r^n e^{ax} \cos (bx + n\phi)$ where $r = \sqrt{a^2 + b^2}$ and $\phi = \tan^{-1} \dfrac{b}{a}$

3. If $\tan \phi = \dfrac{b}{a}$, then $\dfrac{\sin \phi}{\cos \phi} = \dfrac{b}{a} \Rightarrow \dfrac{\sin \phi}{b} = \dfrac{\cos \phi}{a} = \dfrac{\sqrt{\sin^2 \phi + \cos^2 \phi}}{\sqrt{b^2 + a^2}} = \dfrac{1}{\sqrt{a^2 + b^2}}$

$$\therefore \quad \cos \phi = \frac{a}{\sqrt{a^2 + b^2}}, \sin \phi = \frac{b}{\sqrt{a^2 + b^2}}.$$

Example 10. *Expand $e^{ax} \cos bx$ as an infinite series of ascending powers of x. Give also the (n + 1)th term of the above series.*
Hence deduce that

$$e^{x \cos \alpha} \cos (x \sin \alpha) = 1 + x \cos \alpha + \frac{x^2}{2!} \cos 2\alpha + \frac{x^3}{3!} \cos 3\alpha + \ldots\ldots$$

Sol. Let $\quad y = e^{ax} \cos bx \qquad \therefore \quad y(0) = 1$

$$y_n = r^n e^{ax} \cos (bx + n\phi)$$

where $\quad r = \sqrt{a^2 + b^2}, \phi = \tan^{-1} \dfrac{b}{a}$

so that $\quad y_n(0) = r^n \cos n\phi$

Putting $n = 1, 2, 3 \ldots\ldots$, we have

$$y_1(0) = r \cos \phi = \sqrt{a^2 + b^2} \cdot \frac{a}{\sqrt{a^2 + b^2}} = a$$

$$y_2(0) = r^2 \cos 2\phi = (a^2 + b^2) \cdot \frac{1 - \tan^2 \phi}{1 + \tan^2 \phi}$$

$$= (a^2 + b^2) \cdot \frac{1 - \dfrac{b^2}{a^2}}{1 + \dfrac{b^2}{a^2}} = a^2 - b^2$$

$$y_3(0) = r^3 \cos 3\phi = (a^2 + b^2)^{3/2} (4 \cos^3 \phi - 3 \cos \phi)$$

$$= (a^2 + b^2)^{3/2} \cdot \cos \phi \, [4 \cos^2 \phi - 3]$$

$$= (a^2 + b^2)^{3/2} \cdot \frac{a}{\sqrt{a^2 + b^2}} \left[\frac{4a^2}{a^2 + b^2} - 3\right]$$

$$= a(a^2 + b^2) \cdot \frac{a^2 - 3b^2}{a^2 + b^2} = a (a^2 - 3b^2)$$

....

Also $\qquad y_n(0) = (a^2 + b^2)^{n/2} \cdot \cos\left(n \tan^{-1} \dfrac{b}{a}\right)$

$\therefore$ By Maclaurin's theorem, we have

$$e^{ax} \cos bx = y(0) + xy_1(0) + \frac{x^2}{2!} y_2(0) + \frac{x^3}{3!} y_3(0) + \ldots\ldots + \frac{x^n}{n!} y_n(0) + \ldots\ldots$$

$$= 1 + ax + \frac{a^2 - b^2}{2!} x^2 + \frac{a(a^2 - 3b^2)}{3!} x^3 + \ldots\ldots$$

$$+ \frac{(a^2 + b^2)^{n/2}}{n!} x^n \cos\left(n \tan^{-1} \frac{b}{a}\right) + \ldots\ldots$$

$$(n + 1)\text{th term} = \frac{x^n}{n!} y_n(0) = \frac{(a^2 + b^2)^{n/2}}{n!} x^n \cos\left(n \tan^{-1} \frac{b}{a}\right)$$

Second part : Deduction

Here $\qquad a = \cos \alpha, \qquad b = \sin \alpha$

$$\therefore \qquad r = \sqrt{\cos^2 \alpha + \sin^2 \alpha} = 1 \text{ and } \phi = \tan^{-1} \frac{\sin \alpha}{\cos \alpha} = \tan^{-1} (\tan \alpha) = \alpha.$$

$$\therefore \qquad y_n = r^n e^{ax} \cos (bx + n\phi) = e^{x \cos x} \cos (bx + n\alpha)$$

$$\therefore \qquad y_n(0) = \cos n\alpha$$

Putting $n = 1, 2, 3\ldots\ldots$

$$y_1(0) = \cos \alpha, \qquad y_2(0) = \cos 2\alpha \quad y_3(0) = \cos 3\alpha$$

....

$$\therefore \qquad e^{x \cos x} \cos (x \sin \alpha) = 1 + x \cos \alpha + \frac{x^2}{2!} \cos 2\alpha + \frac{x^3}{3!} \cos 3\alpha + \ldots\ldots$$

Example 11. *Assuming the possibility of expansion, show that*

$$e^{ax} \sin bx = bx + abx^2 + \frac{b(3a^2 - b^2)}{3!} x^3 + \ldots\ldots$$

Sol. Let $\qquad y = e^{ax} \sin bx \qquad \therefore \quad y(0) = 0$

$$y_n = r^n e^{ax} \sin (bx + n\phi)$$

where $\qquad r = \sqrt{a^2 + b^2},\ \phi = \tan^{-1} \dfrac{b}{a}$ so that $y_n(0) = r^n \sin n\phi.$

Putting $n = 1, 2, 3\ldots\ldots$

$$y_1(0) = r \sin \phi = \sqrt{a^2 + b^2} \cdot \frac{b}{\sqrt{a^2 + b^2}} = b$$

$$y_2(0) = r^2 \sin 2\phi = (a^2 + b^2) \cdot 2 \sin \phi \cos \phi$$

$$= 2(a^2 + b^2) \cdot \frac{b}{\sqrt{a^2 + b^2}} \cdot \frac{a}{\sqrt{a^2 + b^2}} = 2ab$$

$$y_3(0) = r^3 \sin 3\phi = (a^2 + b^2)^{3/2} (3 \sin \phi - 4 \sin^3 \phi)$$
$$= (a^2 + b^2)^{3/2}. \sin \phi\, [3 - 4 \sin^2 \phi]$$
$$= (a^2 + b^2)^{3/2} . \frac{b}{\sqrt{a^2 + b^2}}\left[3 - 4\frac{b^2}{a^2 + b^2}\right]$$
$$= b(a^2 + b^2) . \frac{3a^2 - b^2}{a^2 + b^2} = b(3a^2 - b^2)$$

$$\therefore \quad e^{ax} \sin bx = y(0) + xy_1(0) + \frac{x^2}{2!} y_2(0) + \frac{x^3}{3!}(0) + \ldots\ldots$$
$$= bx + abx^2 + \frac{b(3a^2 - b^2)}{3!} x^3 + \ldots\ldots$$

Example 12. *Obtain expansion for $e^x \cos x$ in ascending powers of x.*

Sol. Let $y = e^x \cos x \qquad \therefore \quad y(0) = 1$

$$y_n = r^n e^x \cos (x + n\phi)$$

where $r = \sqrt{1+1} = \sqrt{2} \qquad \phi = \tan^{-1}\frac{1}{1} = \frac{\pi}{4}$

$$y_n(0) = r^n \cos n\phi$$

Putting $n = 1, 2, 3\ldots\ldots$

$$y_1(0) = r \cos \phi = \sqrt{2} \cos \frac{\pi}{4} = \sqrt{2} . \frac{1}{\sqrt{2}} = 1$$
$$y_2(0) = r^2 \cos 2\phi = 2 \cos \frac{\pi}{2} = 2 \times 0 = 0$$
$$y_3(0) = r^3 \cos 3\phi = 2\sqrt{2} \cos \frac{3\pi}{4} = 2\sqrt{2}\left(-\frac{1}{\sqrt{2}}\right) = -2$$

....

$$\therefore \quad e^x \cos x = y(0) + xy_1(0) + \frac{x^2}{2!} y_2(0) + \frac{x^3}{3!} y_3(0) + \ldots\ldots$$
$$= 1 + x - \frac{x^3}{3} + \ldots\ldots$$

Example 13. *Obtain the expansion for $e^x \sin x$ in ascending powers of x.*

Sol. Please try yourself. $\left[\textbf{Ans. } x + x^2 + \frac{x^3}{3} + \ldots\ldots\right]$

Example 14. *Prove that*

$$tan^{-1} x = x - \frac{1}{3} x^3 + \frac{1}{5} x^5 - \ldots\ldots$$

Hence find the value of π.

Sol. Let $y = \tan^{-1} x \qquad \therefore \quad y(0) = \tan^{-1} 0 = 0$

$$y_1 = \frac{1}{1+x^2} \qquad \therefore \; y_1(0) = 1$$

$$y_1 = (1 + x^2)^{-1}$$

$$= 1 - x^2 + x^4 - x^6 + \ldots\ldots \qquad \text{(By Binomial Theorem)}$$

$$y_2 = -2x + 4x^3 - 6x^5 + \ldots\ldots \qquad \therefore \; y_2(0) = 0$$

$$y_3 = -2 + 12x^2 - 30x^4 + \ldots\ldots \qquad \therefore \; y_3(0) = -2$$

$$y_4 = 24x - 120x^3 + \ldots\ldots \qquad \therefore \; y_4(0) = 0$$

$$y_5 = 24 - 360x^2 + \ldots\ldots \qquad \therefore \; y_5(0) = 24$$

$$\therefore \qquad \tan^{-1} x = y(0) + xy_1(0) = \frac{x^2}{2!} y_2(0) + \frac{x^3}{3!} y_3(0) + \ldots\ldots$$

$$= x - \frac{2x^3}{3!} + \frac{24x^5}{5!} - \ldots\ldots = x - \frac{x^3}{3} + \frac{x^5}{5} - \ldots\ldots$$

Putting $x = 1$ in the above expansion

$$\tan^{-1} 1 = 1 - \frac{1}{3} + \frac{1}{5} - \frac{1}{7} + \ldots\ldots$$

$$\frac{\pi}{4} = 1 - \frac{1}{3} + \frac{1}{5} - \frac{1}{7} + \ldots\ldots$$

$$\pi = 4\left[1 - \frac{1}{3} + \frac{1}{5} - \frac{1}{7} + \ldots\ldots\right]$$

whence the value of π can be obtained to any degree of approximation.

Example 15. *Expand* $\sin^{-1} x$ *upto four terms.*

Sol. Let $\quad y = \sin^{-1} x \qquad y(0) = \sin^{-1} 0 = 0$

$$y_1 = \frac{1}{\sqrt{1-x^2}} \qquad y_1(0) = 1$$

$$= (1 - x^2)^{-1/2}$$

$$= 1 + \frac{1}{2}x^2 + \frac{\frac{1}{2}\left(\frac{1}{2}+1\right)}{2!}x^4 + \frac{\frac{1}{2}\left(\frac{1}{2}+1\right)\left(\frac{1}{2}+2\right)}{3!}x^6 + \ldots\ldots$$

| By Binomial Theorem

$$= 1 + \frac{1}{2}x^2 + \frac{3}{8}x^4 + \frac{5}{16}x^6 + \ldots\ldots$$

$$y_2 = x + \frac{3}{2}x^3 + \frac{15}{8}x^5 + \ldots\ldots \qquad \therefore \; y_2(0) = 0$$

$$y_3 = 1 + \frac{9}{2}x^2 + \frac{75}{8}x^4 + \ldots\ldots \qquad \therefore \; y_3(0) = 1$$

$$y_4 = 9x \frac{75}{2}x^3 + \ldots\ldots \qquad \therefore \; y_4(0) = 0$$

$$y_5 = 9 + \frac{25}{2}x^2 + \ldots\ldots \qquad \therefore \quad y_5(0) = 9$$

$$y_6 = 225x + \ldots\ldots \qquad \therefore \quad y_6(0) = 0$$

$$y_7 = 225 + \text{terms containing powers of } x \quad \therefore \quad y_7(0) = 225$$

$$\therefore \qquad \sin^{-1} x = y(0) + xy_1(0) + \frac{x^2}{2!}y_2(0) + \ldots\ldots$$

$$= x + \frac{x^3}{3!} + \frac{9x^5}{5!} + \frac{225x^7}{7!} + \ldots\ldots$$

$$= x + \frac{1}{2}.\frac{x^3}{3} + \frac{1.3}{2.4}.\frac{x^5}{5} + \frac{1.3.5}{2.4.6}.\frac{x^7}{7} + \ldots\ldots$$

Example 16. *Assuming the possibility of expansion, show that*

(a) $\sin(e^x - 1) = x + \frac{1}{2}x^2 - \frac{5}{24}x^4 + \ldots\ldots$

(b) $\log(1 - x + x^2) = -x + \frac{1}{2}x^2 + \frac{2}{3}x^3 + \frac{1}{4}x^4 + \ldots\ldots$

(c) $\log[1 - \log(1 - x)] = x + \frac{1}{6}x^3 + \ldots\ldots$

(d) $(1 + x)^{1+x} = 1 + x + x^2 + \frac{1}{2}x^3 + \ldots\ldots$

Sol. *(a)*

$$e^x = 1 + x + \frac{x^2}{2!} + \frac{x^3}{3!} + \frac{x^4}{4!} + \ldots\ldots$$

$$\Rightarrow \qquad e^x - 1 = x + \frac{x^2}{2} + \frac{x^3}{6} + \frac{x^4}{24} + \ldots\ldots = z \text{ (say)}$$

$$\therefore \qquad \sin(e^x - 1) = \sin z = z - \frac{z^3}{3!} + \frac{z^5}{5!} - \ldots\ldots$$

$$= \left(x + \frac{x^2}{2} + \frac{x^3}{6} + \frac{x^4}{24}\ldots\ldots\right) - \frac{1}{6}\left(x + \frac{x^2}{2} + \frac{x^3}{6} + \frac{x^4}{24} + \ldots\ldots\right)^3 + \ldots\ldots$$

$$= \left(x + \frac{x^2}{2} + \frac{x^3}{6} + \frac{x^4}{24} + \ldots\ldots\right) - \frac{1}{6}\left[x^3 + 3x^2\left(\frac{x^2}{2} + \frac{x^3}{6} + \ldots\ldots\right)\right] + \ldots\ldots$$

$$= x + \frac{1}{2}x^2 - \frac{5x^4}{24} + \ldots\ldots$$

(b)

$$\log(1 - x + x^2) = \log\frac{(1+x)(1-x+x^2)}{1+x} = \log\frac{1+x^3}{1+x}$$

$$= \log(1 + x^3) - \log(1 + x)$$

$$= \left(x^3 - \frac{x^6}{2} + \ldots\ldots\right) - \left(x - \frac{x^2}{2} + \frac{x^3}{3} - \frac{x^4}{4} + \ldots\ldots\right)$$

$$= -x + \frac{1}{2}x^2 + \frac{2}{3}x^3 + \frac{1}{4}x^4 + \ldots\ldots$$

(*c*) $$\log(1 - x) = -x - \frac{x^2}{2} - \frac{x^3}{3} - \ldots\ldots$$

$$\Rightarrow \quad -\log(1 - x) = x + \frac{x^2}{2} + \frac{x^3}{3} + \ldots\ldots = z \text{ (say)}$$

$$\therefore \quad \log[1 - \log(1 - x)] = \log(1 + z) = z - \frac{z^2}{2} + \frac{z^3}{3} - \ldots\ldots$$

$$= \left(x + \frac{x^2}{2} + \frac{x^3}{3} + \ldots\ldots\right) - \frac{1}{2}\left(x + \frac{x^2}{2} + \frac{x^3}{3} + \ldots\ldots\right)^2 + \frac{1}{3}\left(x + \frac{x^2}{2} + \frac{x^3}{3} + \ldots\ldots\right)^3 - \ldots\ldots$$

$$= \left(x + \frac{1}{2}x^2 + \frac{1}{3}x^3 + \ldots\ldots\right) - \frac{1}{2}\left[x^2 + 2x\left(\frac{x^2}{2} + \frac{x^3}{3} + \ldots\ldots\right)\right]$$

$$+ \frac{1}{3}\left[x^3 + 3x^2\left(\frac{x^2}{2} + \frac{x^3}{3} + \ldots\ldots\right)\right] - \ldots\ldots$$

$$= \left(x + \frac{1}{2}x^2 + \frac{1}{3}x^3 + \ldots\ldots\right) - \left(\frac{1}{2}x^2 + \frac{1}{2}x^3 + \ldots\ldots\right) + \frac{1}{3}(x^3 + \ldots\ldots) - \ldots\ldots$$

$$= x + \frac{1}{6}x^3 + \ldots\ldots$$

(*d*) Let $\quad y = (1 + x)^{1+x}$

Taking logarithms of both sides

$$\log y = (1 + x)\log(1 + x)$$

$$= (1 + x)\left(x - \frac{x^2}{2} + \frac{x^3}{3} - \ldots\ldots\right)$$

$$= \left(x - \frac{x^2}{2} + \frac{x^3}{3} - \ldots\ldots\right) + \left(x^2 - \frac{x^3}{2} + \frac{x^4}{3} - \ldots\ldots\right)$$

$$= x + \frac{1}{2}x^2 - \frac{1}{6}x^3 + \ldots\ldots$$

$$= z \quad \text{(say)}$$

$$\therefore \quad y = e^z = 1 + z + \frac{z^2}{2!} + \frac{z^3}{3!} + \ldots\ldots$$

$$= 1 + \left(x + \frac{1}{2}x^2 - \frac{1}{6}x^3 + ...\right) + \frac{1}{2}\left(x + \frac{1}{2}x^2 - \frac{1}{6}x^3 - ......\right)^2$$

$$+ \frac{1}{6}\left(x + \frac{1}{2}x^2 - \frac{1}{6}x^3 + ...\right)^3 + ...$$

$$= 1 + \left(x + \frac{1}{2}x^2 - \frac{1}{6}x^3 + ...\right) + \frac{1}{2}\left[x^2 + 2x\left(\frac{1}{2}x^2 + ...\right)\right] + \frac{1}{6}[x^3 + ...] + ...$$

$\therefore \quad (1 + x)^{1+x} = 1 + x + x^2 + \frac{1}{2}x^3 + ...$

Example 17. *Show that*

(a) $\log [1 + \log (1 + x)] = x - x^2 + \frac{7}{6}x^3 + ...$

(b) $\log \frac{\tan x}{x} = \frac{x^2}{3} + \frac{7}{90}x^4 + ...$

(c) $\tan^{-1} (1 + x) = \frac{\pi}{4} + \frac{x}{2} - \frac{x^2}{4} + \frac{x^3}{12} - ......$

Sol. *(a)* $\quad \log (1 + x) = x - \frac{x^2}{2} + \frac{x^3}{3} - ...... = z$ (say)

$\therefore \quad \log [1 + \log (1 + x)] = \log (1 + z)$

$$= z - \frac{z^2}{2} + \frac{z^3}{3} - ......$$

$$= \left(x - \frac{x^2}{2} + \frac{x^3}{3} - ......\right) - \frac{1}{2}\left(x - \frac{x^2}{2} + \frac{x^3}{3} - ......\right)^2 + \frac{1}{3}\left(x - \frac{x^2}{2} + \frac{x^3}{3} - ......\right)^3 - ......$$

$$= \left(x - \frac{x^2}{2} + \frac{x^3}{3} - ......\right) - \frac{1}{2}\left[x^2 - 2x\left(\frac{x^2}{2} - \frac{x^3}{3} + ...\right)\right] + \frac{1}{3}(x^3 + ...) - ......$$

$$= x - x^2 + \frac{7}{6}x^3 + ...$$

(b) $\quad \log \frac{\tan x}{x} = \log \left[\frac{x + \frac{x^3}{3} + \frac{2x^5}{15} + ...}{x}\right]$

$$= \log \left[1 + \frac{x^2}{3} + \frac{2x^4}{15} + ...\right] = \log (1 + z) \qquad | \text{ (say)}$$

$$= z - \frac{z^2}{2} + \frac{z^3}{3} - \frac{z^4}{4} + \ldots\ldots$$

$$= \left(\frac{x^2}{3} + \frac{2x^4}{15} + \ldots\ldots\right) - \frac{1}{2}\left(\frac{x^2}{3} + \frac{2x^4}{15} + \ldots\ldots\right)^2 + \ldots\ldots$$

$$= \left(\frac{x^2}{3} + \frac{2x^4}{15} + \ldots\ldots\right) - \frac{1}{2}\left[\frac{x^4}{9} + \frac{2x^2}{3}\left(\frac{2x^4}{15} + \ldots\ldots\right)\right] + \ldots\ldots$$

$$= \frac{1}{3}x^2 + \frac{7}{90}x^4 + \ldots\ldots$$

(*c*) Let $y = \tan^{-1}(1 + x)$ $\therefore\ y(0) = \tan^{-1} 1 = \frac{\pi}{4}$

$$y_1 = \frac{1}{1 + (1 + x)^2} \qquad \therefore\ y_1(0) = \frac{1}{2}$$

$$= \frac{1}{2 + 2x + x^2} = \frac{1}{2\left(1 + x + \frac{x^2}{2}\right)} = \frac{1}{2}\left[1 + \left(x + \frac{x^2}{2}\right)\right]^{-1}$$

$$= \frac{1}{2}\left[1 - \left(x + \frac{x^2}{2}\right) + \left(x + \frac{x^2}{2}\right)^2 - \left(x + \frac{x^2}{2}\right)^3 + \ldots\ldots\right]$$

$$= \frac{1}{2}\left[1 - x - \frac{x^2}{2} + x^2 + x^3 + \frac{x^4}{4} - x^3 - \ldots\ldots\right]$$

$$= \frac{1}{2}\left[1 - x + \frac{1}{2}x^2 + \ldots\ldots\right]$$

$$y_2 = \frac{1}{2}[-1 + x + \ldots\ldots] \qquad \therefore\ y_2(0) = -\frac{1}{2}$$

$$y_3 = \frac{1}{2}\ [1 + \text{terms containing } x \text{ and higher powers of } x]$$

$$\therefore \qquad y_3(0) = \frac{1}{2}$$

....

$$\tan^{-1}(1 + x) = y(0) + xy_1(0) + \frac{x^2}{2!}y_2(0) + \ldots\ldots$$

$$= \frac{\pi}{4} + \frac{1}{2}x - \frac{1}{4}x^2 + \frac{1}{12}x^3 + \ldots\ldots$$

Example 18. *Show that*

(*i*) $tan^{-1} \dfrac{2x}{1-x^2} = 2\left(x - \dfrac{x^3}{3} + \dfrac{x^5}{5} \ldots\ldots\right)$

(*ii*) $sin^{-1}\left(\dfrac{2x^2}{1+x^4}\right) = 2\left(x^2 - \dfrac{x^6}{3} + \dfrac{x^{10}}{5} - \ldots\ldots\right)$

(*iii*) $tan^{-1} \dfrac{\sqrt{1+x^2}-1}{x} = \dfrac{x}{2} - \dfrac{x^3}{6} + \dfrac{x^5}{10} - \dfrac{x^7}{14} + \ldots\ldots$

(*iv*) $tan^{-1} \dfrac{x}{\sqrt{1-x^2}} = x + \dfrac{1^2}{3\,!}x^3 + \dfrac{1^2.3^2}{5\,!}x^5 + \ldots\ldots$

(*v*) $tan^{-1} \dfrac{\sqrt{1+x^2}-\sqrt{1-x^2}}{\sqrt{1+x^2}+\sqrt{1-x^2}} = \dfrac{x^2}{2} + \dfrac{x^6}{12} + \ldots\ldots$.

Sol. (*i*) Let $\quad y = \tan^{-1} \dfrac{2x}{1-x^2}$ Put $x = \tan\theta$ so that $\theta = \tan^{-1} x$

Then $\quad y = \tan^{-1}\left(\dfrac{2\tan\theta}{1-\tan^2\theta}\right) = \tan^{-1}(\tan 2\theta) = 2\theta = 2\tan^{-1} x$

$y(0) = 0$

$y_1 = \dfrac{2}{1+x^2} = 2\,(1+x^2)^{-1}$

$= 2(1 - x^2 + x^4 - x^6 + \ldots\ldots) \qquad y_1(0) = 2$

$y_2 = 2\,(-2x + 4x^3 - 6x^5 + \ldots\ldots) \qquad y_2(0) = 0$

$y_3 = 2(-2 + 12x^2 - 30x^4 + \ldots\ldots) \qquad y_3(0) = -4$

$y_4 = 2(24x - 120x^3 + \ldots\ldots) \qquad y_4(0) = 0$

$y_5 = 2(24 - 360x^2 + \ldots\ldots) \qquad y_5(0) = 48$

....

$\therefore \quad \tan^{-1}\left(\dfrac{2x}{1-x^2}\right) = y(0) + xy_1(0) + \dfrac{x^2}{2\,!}y_2(0) + \dfrac{x^3}{3\,!}y_3(0) + \ldots\ldots$

$= 2x - \dfrac{4x^3}{3!} + \dfrac{48x^5}{5!} \ldots\ldots$

$= 2\left(x - \dfrac{x^3}{3} + \dfrac{x^5}{5} \ldots\ldots\right)$

(*ii*) Please try yourself. Put $x^2 = \tan\theta$.

$y_1 = 4x\,(1 + x^4)^{-1} = 4x(1 - x^4 + x^8 \ldots\ldots)$

$= 4(x - x^5 + \ldots\ldots)$

(*iii*) Please try yourself. Put $x = \tan \theta$.

(*iv*) Please try yourself. **[Hint.** Put $x = \sin \theta$]

(*v*) Let $$y = \tan^{-1} \frac{\sqrt{1+x^2} - \sqrt{1-x^2}}{\sqrt{1+x^2} + \sqrt{1-x^2}}$$

Put $x^2 = \cos \theta$ so that $\theta = \cos^{-1} x^2$

Then $$y = \tan^{-1} \frac{\sqrt{2\cos^2 \frac{\theta}{2}} - \sqrt{2\sin^2 \frac{\theta}{2}}}{\sqrt{2\cos^2 \frac{\theta}{2}} + \sqrt{2\sin^2 \frac{\theta}{2}}} = \tan^{-1} \left(\frac{\cos \frac{\theta}{2} - \sin \frac{\theta}{2}}{\cos \frac{\theta}{2} + \sin \frac{\theta}{2}} \right)$$

$$= \tan^{-1} \left(\frac{1 - \tan \frac{\theta}{2}}{1 + \tan \frac{\theta}{2}} \right) = \tan^{-1} \left[\tan \left(\frac{\pi}{4} - \frac{\theta}{2} \right) \right]$$

$$= \frac{\pi}{4} - \frac{\theta}{2} = \frac{\pi}{4} - \frac{1}{2} \cos^{-1} x^2$$

$\therefore$ $$y_1 = 0 - \frac{1}{2} \cdot \frac{-1}{\sqrt{1-x^4}} \cdot 2x = x(1-x^4)^{-\frac{1}{2}}$$

$$= x \left[1 + \frac{1}{2} x^4 + \frac{\frac{1}{2} \cdot \frac{3}{2}}{2!} x^8 + \ldots \right] = x + \frac{1}{2} x^5 + \frac{3}{8} x^9 + \ldots\ldots$$

$$y_2 = 1 + \frac{5}{2} x^4 + \frac{27}{8} x^8 + \ldots\ldots$$
$$y_3 = 10x^3 + 27x^7 + \ldots\ldots$$
$$y_4 = 30x^2 + 189x^6 + \ldots\ldots$$
$$y_5 = 60x + 1134x^5 + \ldots\ldots$$
$$y_6 = 60 + 5670x^4 + \ldots\ldots$$
....

$\therefore$ $$y(0) = 0,\ y_1(0) = 0,\ y_2(0) = 1,\ y_3(0) = y_4(0) = y_5(0) = 0,\ y_6(0) = 60, \ldots\ldots$$

$\therefore$ $$\tan^{-1} \frac{\sqrt{1+x^2} - \sqrt{1-x^2}}{\sqrt{1+x^2} + \sqrt{1-x^2}} = y(0) + xy_1(0) + \frac{x^2}{2!} y_2(0) + \frac{x^3}{3!} y_3(0)$$

$$+ \frac{x^4}{4!} y_4(0) + \frac{x^5}{5!} y_5(0) + \frac{x^6}{6!} y_6(0) + \ldots\ldots$$

$$= \frac{x^2}{2} + \frac{x^6}{12} + \ldots\ldots$$

6.7. EXPANSIONS USING LEIBNITZ THEOREM

Differentiating $(1-x^2)y_2$ *i.e.*, $y_2(1-x^2)$, n times by Leibnitz's Theorem, we get

$$y_{n+2}(1-x^2) + ny_{n+1}(-2x) + \frac{n(n-1)}{2!} y_n (-2)$$

Differentiating xy_1 *i.e.*, y_1x, n times by Leibnitz's Theorem, we have

$$y_{n+1}(x) + ny_n \cdot 1$$

Example 1. *Expand* $e^{a \sin^{-1} x}$ *by Maclaurin's Theorem.*

Hence show that

$$e^\theta = 1 + \sin\theta + \frac{1}{2!}\sin^2\theta + \frac{2}{3!}\sin^3\theta + \ldots\ldots$$

Sol. Let $\quad y = e^{a \sin^{-1} x} \qquad \therefore \quad y(0) = e^0 = 1$

Diff. $y_1 = e^{a \sin^{-1} x} \cdot \frac{a}{\sqrt{1-x^2}} = \frac{ay}{\sqrt{1-x^2}} \qquad \therefore \quad \mathbf{y_1(0) = a}$

Squaring and cross multiplying

$$(1-x^2)\,y_1^2 = a^2y^2$$

Diff. $(1-x^2)\,2y_1y_2 - 2xy_1^2 = 2a^2yy_1$

or $\quad (1-x^2)y_2 - xy_1 = a^2y \quad \therefore \quad \mathbf{y_2(0) = a^2}$

Differentiating n times by Leibnitz's Theorem, we get

$$y_{n+2}(1-x^2) + ny_{n+1}(-2x) + \frac{n(n-1)}{2!} y_n (-2)$$

$$-[y_{n+1}x + ny_n] = a^2y_n$$

or $\quad (1-x^2)\,y_{n+2} - (2n+1)xy_{n+1} - (n^2+a^2)\,y_n = 0 \qquad \ldots(i)$

Putting $x = 0$, $y_{n+2}(0) = (n^2+a^2)\,y_n(0)$

Putting $n = 1, 2, 3, \ldots\ldots$ in (i)

$$y_3(0) = (1^2+a^2)y_1(0) = a(1^2+a^2)$$

$$y_4(0) = (2^2+a^2)y_2(0) = a^2(2^2+a^2)$$

$$y_5(0) = (3^2+a^2)y_3(0) = a(1^2+a^2)\,(3^2+a^2).$$

....

$$\therefore \quad e^{a \sin^{-1} x} = y(0) + xy_1(0) + \frac{x^2}{2!} y_2(0) + \ldots\ldots$$

$$= 1 + ax + \frac{a^2x^2}{2!} + \frac{a(1^2+a^2)}{3!}x^3 + \frac{a^2(2^2+a^2)}{4!}x^4 + \ldots\ldots$$

Second Part: Putting $\sin^{-1} x = \theta$ and $a = 1$, we have

$$e^\theta = 1 + \sin\theta + \frac{1}{2!}\sin^2\theta + \frac{2}{3!}\sin^3\theta + \frac{5}{4!}\sin^4\theta + \ldots\ldots$$

Example 2. *Expand sin (m sin^{-1} x) by Maclaurin's Theorem, as far as x^5.*

Sol. Let $\quad y = \sin(m \sin^{-1} x) \qquad \therefore \quad$ **y(0) = sin 0 = 0**

$$\text{Diff. } y_1 = \cos(m \sin^{-1} x) \cdot \frac{m}{\sqrt{1-x^2}} \qquad \therefore \quad \mathbf{y_1(0) = m}$$

Squaring and cross- multiplying

$$(1 - x^2)\, y_1^{\,2} = m^2 \cos^2 (m \sin^{-1} x) = m^2 [1 - \sin^2 (m \sin^{-1} x)]$$

$$\therefore \qquad (1 - x^2) y_1^{\,2} = m^2(1 - y^2)$$

Differentiating,

$$- 2x \cdot y_1^{\,2} + (1 - x^2) \cdot 2y_1y_2 = m^2(- 2yy_1)$$

or $$(1 - x^2)\, y_2 - xy_1 + m^2y = 0 \qquad \therefore \quad y_2(0) = 0$$

Differentiating n times by Leibnitz's Theorem, we get

$$y_{n+2}\,(1 - x^2) + ny_{n+1}\,(- 2x) + \frac{n(n-1)}{2\,!}\, y_n(- 2) - [y_{n+1}x + ny_n \,.\, 1] + m^2y_n = 0$$

or $$(1 - x^2)y_{n+2} - (2n + 1)\, xy_{n+1} - (n^2 - m^2)y_n = 0$$

Putting $x = 0$, $\quad y_{n+2}(0) = (n^2 - m^2)\, y_n(0)$

Putting $n = 1, 2, 3, \ldots\ldots$

$$y_3(0) = (1^2 - m^2)y_1(0) = m(1^2 - m^2)$$

$$y_4(0) = (2^2 - m^2)y_2(0) = 0$$

$$y_5(0) = (3^2 - m^2)y_3(0) = m(1^2 - m^2)\,(3^2 - m^2)$$

....

$$\therefore \quad \sin(m \sin^{-1} x) = y(0) + xy_1(0) + \frac{x^2}{2\,!}\, y_2(0) + \ldots\ldots$$

$$= mx + \frac{m(1^2 - m^2)}{3\,!}\, x^2 + \frac{m(1^2 - m^2)\,(3^2 - m^2)}{5\,!}\, x^5 + \ldots\ldots$$

Example 3. *Assuming the validity of expansion, show that*

(a) $(\sin^{-1} x)^2 = \dfrac{2}{2\,!}\, x^2 + \dfrac{2 \,.\, 2^2}{4\,!}\, x^4 + \dfrac{2 \,.\, 2^2 \,.\, 4^2}{6\,!}\, x^6 + \ldots\ldots$

(b) $\log(x + \sqrt{1 + x^2}\,) = x - \dfrac{1}{2} \cdot \dfrac{x^3}{3} \; \dfrac{1}{2} \cdot \dfrac{3}{4} \cdot \dfrac{x^5}{5} - \ldots\ldots$

Sol. (*a*) Let $\quad y = (\sin^{-1} x)^2 \qquad \therefore \quad$ **y(0) = 0**

$$\text{Diff.} \quad y_1 = 2 \sin^{-1} x \cdot \frac{1}{\sqrt{1-x^2}} \qquad \therefore \quad \mathbf{y_1(0) = 0}$$

Squaring and cross-multiplying

$$(1 + x^2)y_1^{\,2} = 4(\sin^{-1} x)^2 = 4y$$

Diff. $(1 - x^2) \,.\, 2y_1y_2 - 2xy_1^{\,2} = 4y_1$

or $$(1 - x^2)y_2 - xy_1 = 2 \quad \therefore \quad \mathbf{y_2(0) = 2}$$

Differentiating n times by Leibnitz's Theorem,

$$y_{n+2}(1-x^2) + ny_{n+1}(-2x) + \frac{n(n-1)}{2!} y_n(-2) - [y_{n+1} \cdot x + ny_n \cdot 1] = 0$$

or
$$(1-x^2)y_{n+2} - (2n+1)xy_{n+1} - n^2y_n = 0$$

Putting $x = 0$, $\quad y_{n+2}(0) = n^2 y_n(0)$

Putting $n = 1, 2, 3, ...$

$$y_3(0) = 1^2 \cdot y_1(0) = 0$$
$$y_4(0) = 2^2 \cdot y_2(0) = 2 \cdot 2^2$$
$$y_5(0) = 3^2 \cdot y_3(0) = 0$$
$$y_6(0) = 4^2 \cdot y_4(0) = 2 \cdot 2^2 \cdot 4^2$$

....

....

$$\therefore \quad (\sin^{-1} x)^2 = y(0) + xy_1(0) + \frac{x^2}{2!} y_2(0) + ...$$

$$= \frac{2}{2!}x^2 + \frac{2 \cdot 2^2}{4!}x^4 + \frac{2 \cdot 2^2 \cdot 4^2}{6!}x^6 + ...$$

$$= 2\left(\frac{1}{2!}x^2 + \frac{2^2}{4!}x^4 + \frac{2^2 \cdot 4^2}{6!}x^6 + ...\right)$$

(*b*) Let $\quad y = \log(x + \sqrt{1+x^2}) \qquad \because \ \mathbf{y(0) = \log 1 = 0}$

$$\text{Diff. } y_1 = \frac{1 + \dfrac{2x}{2\sqrt{1+x^2}}}{x + \sqrt{1+x^2}} = \frac{\sqrt{1+x^2} + x}{(x + \sqrt{1+x^2})\sqrt{1+x^2}} = \frac{1}{\sqrt{1+x^2}}$$

$\therefore \quad \mathbf{y_1(0) = 1}$

Squaring and cross-multiplying

$$(1+x^2)y_1^2 = 1$$

$$\text{Diff. } (1+x^2)\,2y_1y_2 + 2xy_1^2 = 0$$

or
$$(1+x^2)y_2 + xy_1 = 0 \qquad \therefore \ \mathbf{y_2(0) = 0}$$

Differentiating n times by Leibnitz's Theorem

$$y_{n+2}(1+x^2) + ny_{n+1} \cdot 2x + \frac{n(n-1)}{2!} y_n \cdot 2 + y_{n+1}x + ny_n \cdot 1 = 0$$

or
$$(1+x^2)y_{n+2} + (2n+1)xy_{n+1} + n^2y_n = 0$$

Putting $x = 0$, $y_{n+2}(0) = -n^2y_n(0)$

Putting $n = 1, 2, 3, ...$

$$y_3(0) = -1^2y_1(0) = -1$$
$$y_4(0) = -2^2y_2(0) = 0$$
$$y_5(0) = -3^2y_3(0) = 3^2 = 9$$

....

$$\therefore \quad \log(x + \sqrt{1+x^2}) = y(0) + xy_1(0) + \frac{x^2}{2!} y^2(0) + ...$$

$$= x - \frac{x^3}{3!} + \frac{9x^5}{5!} - \ldots\ldots$$

$$= x - \frac{1}{2} \cdot \frac{x^3}{3} + \frac{1}{2} \cdot \frac{3}{4} \cdot \frac{x^5}{5} \ldots\ldots$$

Example 4. *If $y = \cos (m \sin^{-1} x)$, prove that*

$$y_{n+2}(0) = (n^2 - m^2)\, y_n(0).$$

Hence evaluate $y_n(0)$ and expand $y = \cos (m \sin^{-1} x)$.

Sol. $y = \cos (m \sin^{-1} x)$ $\quad \therefore$ $\mathbf{y(0) = \cos 0 = 1}$

$$y_1 = -\sin (m \sin^{-1} x) \cdot \frac{m}{\sqrt{1-x^2}} \quad \therefore \quad \mathbf{y_1(0) = 0}$$

Squaring and cross-multiplying

$$(1 - x^2)y_1^2 = m^2 \sin^2 (m \sin^{-1} x)$$

$$= m^2[1 - \cos^2 (m \sin^{-1} x)]$$

or $(1 - x^2)y_1^2 = m^2 (1 - y^2)$

Diff. $(1 - x^2) \cdot 2y_1y_2 - 2xy_1^2 = m^2 (-2yy_1)$

or $(1 - x^2)y_2 - xy_1 + m^2y = 0$ $\quad \therefore$ $\mathbf{y_2(0) = -m^2}$

Differentiating n times by Leibnitz's Theorem, we get

$$(1 - x^2)y_{n+2} - (2n + 1)\, xy_{n+1} - (n^2 - m^2)y_n = 0 \qquad \text{[As in Ex. 2]}$$

Putting $x = 0$, $\quad y_{n+2}(0) = (n^2 - m^2)\, y_n(0)$ ...(I)

Putting $n = 1, 2, 3, \ldots\ldots$

$$y_3(0) = (1^2 - m^2)y_1(0) = 0$$

$$y_4(0) = (2^2 - m^2)y_2(0) = -m^2(2^2 - m^2)$$

$$y_5(0) = (3^2 - m^2)y_3(0) = 0$$

$$y_6(0) = (4^2 - m^2)y_4(0) = -m^2(2^2 - m^2)(4^2 - m^2)$$

....

$\therefore$ When n is odd, $\quad y_n(0) = 0$

When n is even,

$$y_n(0) = -m^2(2^2 - m^2)(4^2 - m^2) \ldots\ldots [(n-2)^2 - m^2] \qquad \text{...(II)}$$

Also $\cos (m \sin^{-1} x) = y(0) + xy_1(0) + \frac{x^2}{2!} y_2(0) + \frac{x^3}{3!} y_3(0) + \ldots\ldots$

$$= 1 - \frac{m^2}{2!} x^2 - \frac{m^2 (2^2 - m^2)}{4!} x^4 - \frac{m^2 (2^2 - m^2)(4^2 - m^2)}{6!} x^6 \ldots\ldots$$

Example 5. *Prove that*

$$(x + \sqrt{1 + x^2})^n = 1 + nx + \frac{n^2x^2}{2!} + \frac{n(n^2 - 1^2)}{3!} x^3 + \frac{n^2(n^2 - 2^2)}{4!} x^4 + \ldots\ldots$$

Sol. Let $\quad y = (x + \sqrt{1+x^2})^n$ $\quad \therefore$ $\mathbf{y(0) = 1}$

Diff. $y_1 = n(x + \sqrt{1+x^2})^{n-1} \cdot \left[1 + \frac{1}{2}(1+x^2)^{-\frac{1}{2}} \cdot 2x\right]$

$$= n\left(x+\sqrt{1+x^2}\right)^{n-1} \cdot \left(1+\frac{x}{\sqrt{1+x^2}}\right)$$

$$= n\left(x+\sqrt{1+x^2}\right)^{n-1} \cdot \frac{\sqrt{1+x^2}+x}{\sqrt{1+x^2}}$$

$$= n\frac{(x+\sqrt{1+x^2})^n}{\sqrt{1+x^2}} = \frac{ny}{\sqrt{1+x^2}} \quad \therefore \quad \mathbf{y_1(0) = n}$$

Squaring and cross-multiplying

$$(1 + x^2)y_1^{\,2} = n^2y^2$$

Diff. $(1 + x^2) \cdot 2y_1y_2 + 2xy_1^{\,2} = n^2.2yy_1$

or $$(1 + x^2)y_2 = xy_1 = n^2y \quad \therefore \quad \mathbf{y_2(0) = n^2}$$

Differentiating *m times* (and not *n* times, since *n* is a given constant) by Leibnitz's Theorem, we get

$$y_{n+2}\,(1 + x^2) + my_{m+1}\,(2x) + \frac{m(m-1)}{2!}\,y_m \cdot (2) + [y_{m+1}\,x + my_m \cdot 1] = n^2y_m$$

or $$(1 + x^2)y_{m+2} + (2m + 1)xy_{m+1} + (m^2 - n^2)y_m = 0$$

Putting $x = 0$, $y_{m+2}\,(0) = (n^2 - m^2)y_m(0)$

Putting $m = 1, 2, 3, \ldots\ldots$

$$y_3(0) = (n^2 - 1^2)\,y_1(0) = n\,(n^2 - 1^2)$$

$$y_4(0) = (n^2 - 2^2)y_2(0) = n^2\,(n^2 - 2^2)$$

....

$$\therefore \quad (x+\sqrt{1+x^2})^n = y(0) + xy_1(0) + \frac{x^2}{2!}\,y_2(0) + \frac{x^3}{3!}\,y_3(0) + \ldots\ldots$$

$$= 1 + nx + \frac{n^2x^2}{2!} + \frac{n(n^2-1^2)}{3!} + x^3 + \frac{n(n^2-2^2)}{4!}\,x^4 + \ldots\ldots$$

MISCELLANEOUS EXAMPLES

Example 1. *(a) Expand sin (x + h) in powers of h and deduce that*

sin (x + h) = sin x cos h + cos x sin h.

(b) Given sin 30° = $\frac{1}{2}$, apply Taylor's Theorem to calculate the value of sin 31°.

Sol. (*a*) Let $f(x + h) = \sin(x + h)$

Putting $h = 0$ $\quad f(x) = \sin x$

$$f'(x) = \cos x \qquad f''(x) = -\sin x$$

$$f'''(x) = -\cos x \qquad f^{iv}(x) = \sin x$$

$$f^{v}(x) = \cos x \qquad f^{vi}(x) = -\sin x$$

$$\therefore \quad \sin(x + h) = f(x + h)$$

$$= f(x) + hf'(x) + \frac{h^2}{2!} f''(x) + \frac{h^3}{3!} f'''(x) + ...$$

$$= \sin x + h \cos x - \frac{h^2}{2!} \sin x - \frac{h^3}{3!} \cos x + \frac{h^4}{4!} \sin x$$

$$+ \frac{h^5}{5!} \cos x - \frac{h^6}{6!} \sin x - ......$$

$$= \sin x \left[1 - \frac{h^2}{2!} + \frac{h^4}{4!} - \frac{h^6}{6!} + ...\right] + \cos x \left[h - \frac{h^3}{3!} + \frac{h^5}{5!} - ......\right]$$

$$= \sin x \cos h + \cos x \sin h.$$

(*b*) $\sin(x + h) = \sin x + h \cos x - \frac{h^2}{2!} \sin x - \frac{h^3}{3!} \cos x + ...$ | Proved in part (*a*)

Put $x = 30°$, $h = 1°$

then $\sin x = \sin 30° = \frac{1}{2}$, $\cos x = \cos 30° = \frac{\sqrt{3}}{2} = \frac{1.732}{2} = .866$

Also $h = 1° = \frac{\pi}{180}$ radians $= .01745$

$h^2 = .000304$, $h^3 = .000005$ ......

$$\therefore \quad \sin 31° = \sin 30° \left[1 - \frac{1}{2}(.01745)^2 + ...\right] + \cos 30° \left[.01745 - \frac{1}{6}(.01745)^3 + ...\right]$$

$$= .5\left[1 - \frac{1}{2} \times .000304 + ...\right] + .866\left[.01745 - \frac{1}{6} \times .000005 + ...\right]$$

$$= .5 - .000076 + .015111 = .515035 = .5150 \text{ approximately.}$$

Example 2. (*a*) *If* $f(x) = x^3 + 2x^2 - 5x + 11$, *calculate the value of* $f\left(\frac{9}{10}\right)$ *by the application of Taylor's series for* $f(x + h)$.

(*b*) *If* $f(x) = x^3 - 2x + 5$, *find the value of* $f(2.001)$ *with the help of Taylor's series for* $f(x + h)$.

(*c*) *If* $f(x) = x^3 + 8x^2 + 15x - 24$, *calculate the value of* $f\left(\frac{11}{10}\right)$ *by the application of Taylor's series.*

Sol. (*a*) By Taylor's series

$$f(x + h) = f(x) + hf'(x) + \frac{h^2}{2!} f''(x) + ...$$

Put $x = 1$ and $h = -\frac{1}{10}$

$$f\left(\frac{9}{10}\right) = f(1) - \frac{1}{10} f'(1) + \frac{1}{2!} \cdot \left(\frac{1}{10}\right)^2 f''(1) - \frac{1}{3!} \cdot \left(\frac{1}{10}\right)^3 \cdot f'''(1) + ... \quad ...(A)$$

Here $f(x) = x^3 + 2x^2 - 5x + 11$ $\therefore$ $f(1) = 9$

$f'(x) = 3x^2 + 4x - 5$ $\therefore$ $f'(1) = 2$

$f''(x) = 6x + 4$ $\therefore$ $f''(1) = 10$

$f'''(x) = 6$ $\therefore$ $f'''(1) = 6$

All higher order differential co-efficients vanish.

$\therefore$ From (A), we get

$$f\left(\frac{9}{10}\right) = 9 - \frac{1}{10} \cdot 2 + \frac{1}{2} \cdot \frac{1}{100} \cdot 10 - \frac{1}{6} \cdot \frac{1}{1000} \cdot 6 = 9 - .2 + .05 - .001 = 8.849.$$

(*b*) Put $x = 2$ and $h = .001$ in Taylor's series

$$f(2.001) = f(2) + (.001) f'(2) + \frac{(.001)^2}{2!} f''(2) + \frac{(.001)^3}{3!} f'''(2) + \ldots\ldots \quad \ldots(A)$$

Here $f(x) = x^3 - 2x + 5$ $\therefore$ $f(2) = 9$

$f'(x) = 3x^2 - 2$ $\therefore$ $f'(2) = 10$

$f''(x) = 6x$ $\therefore$ $f''(2) = 12$

$f'''(x) = 6$ $\therefore$ $f'''(2) = 6$

All higher order derivatives vanish.

$\therefore$ From (A)

$$f(2.001) = 9 + (.001)(10) + \frac{1}{2}(.001)^2 (12) + \frac{1}{6}(.001)^3 (6)$$

$$= 9 + .01 + .000006 + .000000001 = 9.010006001.$$

(*c*) Please try yourself. **[Ans.** 3.511]

Example 3. *If* $f(x) = x^3 - 6x^2 + 7$, *find the value of* $f\left(\frac{21}{20}\right)$ *by Taylor's Theorem.*

Sol. By Taylor's Theorem

$$f(x + h) = f(x) + hf'(x) + \frac{h^2}{2!} f''(x) \frac{h^3}{3!} f'''(x) + \ldots\ldots$$

Put $x = 1$ and $h = \frac{1}{20} = .05$

$$f\left(\frac{21}{20}\right) = f(1) + (.05) f'(1) + \frac{(.05)^2}{2!} f''(1) + \frac{(.05)^3}{3!} f'''(1) + \ldots\ldots \quad \ldots(A)$$

Here $f(x) = x^3 - 6x^2 + 7$ $\therefore$ $f(1) = 2$

$f'(x) = 3x^2 - 12x$ $\therefore$ $f'(1) = -9$

$f''(x) = 6x - 12$ $\therefore$ $f''(1) = -6$

$f'''(x) = 6$ $\therefore$ $f'''(1) = 6$

Higher order derivatives vanish.

$\therefore$ From (A)

$$f\left(\frac{21}{20}\right) = 2 + (.05)(-9) + \frac{.0025}{2}(-6) + \frac{.000125}{6}(6)$$

$$= 2 - .45 - .0075 + .000125 = 1.542625.$$

Example 4. *Calculate the approximate value of $\sqrt{10}$ to four decimal places by taking the first four terms of an appropriate taylor expansion.*

Sol. Let $f(x + h) = \sqrt{x+h}$

Put $h = 0$ then $f(x) = \sqrt{x}$

$$f'(x) = \frac{1}{2} x^{-\frac{1}{2}} = \frac{1}{2\sqrt{x}}$$

$$f''(x) = -\frac{1}{4} x^{-\frac{3}{2}} = -\frac{1}{4x\sqrt{x}}$$

$$f'''(x) = \frac{3}{8} x^{-\frac{5}{2}} = \frac{3}{8x^2\sqrt{x}}$$

....

$$\therefore \quad \sqrt{x+h} = f(x+h)$$

$$= f(x) + hf'(x) + \frac{h^2}{2!} f''(x) + \frac{h^3}{3!} f'''(x) + ...$$

$$= \sqrt{x} \frac{h}{2\sqrt{x}} - \frac{h^2}{8x\sqrt{x}} + \frac{h^3}{48x^2\sqrt{x}} + ...$$

Put $x = 9$ and $h = 1$

$$\sqrt{10} = 3 + \frac{1}{2(3)} - \frac{1}{8(9)(3)} + \frac{1}{48(81)(3)} + ...$$

$$= 3 + .16666 - .00463 + .00009 + ...$$

$$= 3.16212 \text{ nearly.}$$

Example 5. *Calculate the approximate value of $\sqrt{17}$ to four decimal places by taking first three terms of a taylor expansion.*

Sol. Please try yourself. [**Ans.** 4.1231]

Example 6. *Prove that*

$$f\left(\frac{x^2}{1+x}\right) = f(x) - \frac{x}{1+x} f'(x) + \left(\frac{x}{1+x}\right)^2 \frac{1}{2!} f''(x) - \left(\frac{x}{1+x}\right)^3 \frac{1}{3!} f'''(x) + ...$$

Sol.
$$\frac{x^2}{1+x} = x - \frac{x}{1+x} = x + h, \text{ where } h = -\frac{x}{1+x}$$

$$\therefore \quad f\left(\frac{x^2}{1+x}\right) = f(x+h) = f(x) + hf'(x) + \frac{h^2}{2!} f'''(x) + \frac{h^3}{3!} f'''(x) + ...$$

$$= f(x) - \frac{x}{1+x} f'(x) + \left(\frac{x}{1+x}\right)^2 \frac{1}{2!} f''(x) - \left(\frac{x}{1+x}\right)^3 \frac{1}{3!} f'''(x) + ...$$

Example 7. *Show that*

$$f(ax) = f(x) + (a-1)\,xf'(x) + \frac{(a-1)^2\,x^2}{2!}\,f''(x) + \dots..$$

Sol.
$$\begin{aligned} f(ax) &= f(x + ax - x) = f[x + (a-1)x] \\ &= f(x+h) \text{ where } h = (a-1)x \\ &= f(x) + hf'(x) + \frac{h^2}{2!}\,f''(x) + \dots\dots \\ &= f(x) + (a-1)\,xf'(x) + \frac{(a-1)^2\,x^2}{2!}\,f''(x) + \dots\dots \end{aligned}$$

Example 8. *If $0 < x \le 2$, then prove that*

$$\log x = (x-1) - \frac{(x-1)^2}{2} + \frac{(x-1)^3}{3} - \frac{(x-1)^4}{4} + \dots\dots$$

Sol. Here $f(x) = \log x$

But $f(x) = f[1 + (x-1)] = f(a+h)$, where $a = 1$ and $h = x - 1$

$$= f(a) + hf'(a) + \frac{h^2}{2!}\,f''(a) + \frac{h^3}{3!}\,f'''(a) + \frac{h^4}{4!}\,f^{iv}(a) + \dots\dots$$

Putting the values of a and h, we get

$$\log x = f(1) + (x-1)\,f'(1) + \frac{(x-1)^2}{2!}\,f''(1)$$

$$+ \frac{(x-1)^3}{3!}\,f'''(1) + \frac{(x-1)^4}{4!}\,f^{iv}(1) + \dots\dots \qquad \dots(1)$$

Now $f(x) = \log x$ Putting $x = 1$, $f(1) = 0$

$$f'(x) = \frac{1}{x} \qquad f'(1) = 1$$

$$f''(x) = -\frac{1}{x^2} \qquad f''(1) = -1$$

$$f'''(x) = \frac{2}{x^3} \qquad f'''(1) = 2$$

$$f^{iv}(x) = -\frac{6}{x^4} \qquad f^{iv}(1) = -6$$

$\therefore$ From (1), we have

$$\begin{aligned} \log x &= 0 + (x-1) - \frac{(x-1)^2}{2} + \frac{2\,(x-1)^3}{6} - \frac{6\,(x-1)^4}{24} + \dots\dots \\ &= (x-1) - \frac{(x-1)^2}{2} + \frac{(x-1)^3}{3} - \frac{(x-1)^4}{4} + \dots\dots \end{aligned}$$

MODULE–III

PARTIAL DIFFERENTIATION AND CO-ORDINATE SYSTEMS

7

Partial Differentiation and Applications

7.1. FUNCTIONS OF TWO VARIABLES

If three variables x, y, z are so related that the value of z depends upon the values of x and y, then z is called a function of two variables x and y, and this is denoted by $z = f(x, y)$.

z is called the dependent variable while x and y are called independent variables.

For example, the area of a triangle is determined when its base and altitude are known. Thus, area of a triangle is a function of two variables, base and altitude.

In a similar way, a function of more than two variables can be defined.

Geometrical Interpretation :

Let $z = f(x, y)$ be a function of two independent variables x and y defined for all pairs of values of x and y which belong to an area A of the xy-plane. Then to each point (x, y) of this area corresponds a value of z given by the relation $z = f(x, y)$. Representing all these values (x, y, z) by points in space, we get a surface.

Hence the function z = f(x, y) represents a surface.

7.2. CONTINUITY

A function $f(x, y)$ is said to be continuous at a point (a, b) if, for any arbitrarily chosen positive number ε, however small, we can find a corresponding number δ such that $| f(x, y) - f(a, b) | < \varepsilon$ for every point (x, y) within the circle with its centre at (a, b) and radius δ.

i.e., whenever $0 < (x - a)^2 + (y - b)^2 < \delta^2$.

Alternatively, $f(x, y)$ is said to be continuous at (a, b) if $\lim\limits_{\substack{x \to a \\ y \to b}} f(x, y) = f(a, b)$

irrespective of the path along which $x \to a, y \to b$.

It should not be assumed that the path along which the point (x, y) tends to (a, b) is immaterial, because $\lim\limits_{x \to a} \{\lim\limits_{y \to b} f(x, y)\}$ is not always equal to $\lim\limits_{y \to b} \{\lim\limits_{x \to a} f(x, y)\}$.

However, *usually,* the limit is the same irrespective of the path of approach.

In what follows, we shall assume that the functions considered are continuous and their partial differential coefficients as defined in the next section, exist.

7.3. PARTIAL DERIVATIVES OF FIRST ORDER

Let $z = f(x, y)$ be a function of two independent variables x and y. If y is kept constant and x alone is allowed to vary, then z becomes a function of x only. The derivative of z, with respect to x, treating y as constant, is called partial derivative of z w.r.t. x and is denoted by

$$\frac{\partial z}{\partial x} \quad \text{or} \quad \frac{\partial f}{\partial x} \quad \text{or} \quad f_x.$$

Thus, $$\frac{\partial z}{\partial x} = \lim_{h \to 0} \frac{f(x+h, y) - f(x, y)}{h}$$

Similarly, the derivative of z, with respect to y, treating x as constant, is called partial derivative of z w.r.t. y and is denoted by $\frac{\partial z}{\partial y}$ or $\frac{\partial f}{\partial y}$ or f_y

Thus, $$\frac{\partial z}{\partial y} = \lim_{k \to 0} \frac{f(x, y+k) - f(x, y)}{k}$$

$\frac{\partial z}{\partial x}$ and $\frac{\partial z}{\partial y}$ are called **first order partial derivatives of z.**

[**In general,** if z is a function of two or more independent variables, then the **partial derivative of z** w.r.t. **any one** of the independent variables is the **ordinary derivative of z** w.r.t. that variable, **treating all other variables as constant.**]

Geometrical Interpretation :

Let $z = f(x, y)$ be a function of two variables x and y. Then, it represents a surface S. If $y = k$, a constant, then $y = k$ represents a plane parallel to the zx-plane.

$\therefore$ $z = f(x, y)$ and $y = k$ represent a plane curve C which is the section of S by $y = k$.

$\frac{\partial z}{\partial x}$ represents the slope of tangent to C at (x, k, z).

Thus, $\frac{\partial z}{\partial x}$ gives the slope of the tangent drawn to the curve of intersection of the surface $z = f(x, y)$ and a plane parallel to zx-plane.

Similarly, $\frac{\partial z}{\partial y}$ gives the slope of the tangent drawn to the curve of intersection of the surface $z = f(x, y)$ and a plane parallel to yz-plane.

7.4. PARTIAL DERIVATIVES OF HIGHER ORDER

Since the first order partial derivatives $\frac{\partial z}{\partial x}$ and $\frac{\partial z}{\partial y}$ are themselves functions of x and y, they can be further differentiated partially w.r.t. x as well as y. These are called second order partial derivatives of z. The usual notations for these second order partial derivatives are :

$$\frac{\partial}{\partial x}\left(\frac{\partial z}{\partial x}\right) = \frac{\partial^2 z}{\partial x^2} \text{ or } f_{xx} ; \frac{\partial}{\partial y}\left(\frac{\partial z}{\partial y}\right) = \frac{\partial^2 z}{\partial y^2} \text{ or } f_{yy}$$

$$\frac{\partial}{\partial x}\left(\frac{\partial z}{\partial y}\right) = \frac{\partial^2 z}{\partial x \partial y} \text{ or } f_{xy} ; \frac{\partial}{\partial y}\left(\frac{\partial z}{\partial x}\right) = \frac{\partial^2 z}{\partial y \partial x} \text{ or } f_{yx}$$

In general, $$\frac{\partial^2 z}{\partial x \partial y} = \frac{\partial^2 z}{\partial y \partial x} \text{ or } f_{xy} = f_{yx}.$$

Note 1. If $z = f(x)$, a function of single independent variable x, we get $\frac{dz}{dx}$.

If $z = f(x, y)$, a function of two independent variables x and y, we get $\frac{\partial z}{\partial x}$ and $\frac{\partial z}{\partial y}$.

Similarly, for a function of more than two independent variables $x_1, x_2, \ldots\ldots, x_n$, we get $\frac{\partial z}{\partial x_1}, \frac{\partial z}{\partial x_2}, \ldots\ldots, \frac{\partial z}{\partial x_n}$.

Note 2. (*i*) If $z = u + v$, where $u = f(x, y)$, $v = \phi(x, y)$ then z is a function of x and y.

$$\therefore \quad \frac{\partial z}{\partial x} = \frac{\partial u}{\partial x} + \frac{\partial v}{\partial x}; \quad \frac{\partial z}{\partial y} = \frac{\partial u}{\partial y} + \frac{\partial v}{\partial y}.$$

(*ii*) If $z = uv$, where $u = f(x, y)$, $v = \phi(x, y)$ then

$$\frac{\partial z}{\partial x} = \frac{\partial}{\partial x}(uv) = u\frac{\partial v}{\partial x} + v\frac{\partial u}{\partial x}\ ; \quad \frac{\partial z}{\partial y} = \frac{\partial}{\partial y}(uv) = u\frac{\partial v}{\partial y} + v\frac{\partial u}{\partial y}$$

(*iii*) If $z = \dfrac{u}{v}$, where $u = f(x, y)$, $v = \phi(x, y)$ then

$$\frac{\partial z}{\partial x} = \frac{\partial}{\partial x}\left(\frac{u}{v}\right) = \frac{v\dfrac{\partial u}{\partial x} - u\dfrac{\partial v}{\partial x}}{v^2}\ ; \quad \frac{\partial z}{\partial y} = \frac{\partial}{\partial y}\left(\frac{u}{v}\right) = \frac{v\dfrac{\partial u}{\partial y} - u\dfrac{\partial v}{\partial y}}{v^2}$$

(*iv*) If $z = f(u)$, where $u = \phi(x, y)$ then $\dfrac{\partial z}{\partial x} = \dfrac{dz}{du}\cdot\dfrac{\partial u}{\partial x}$; $\dfrac{\partial z}{\partial y} = \dfrac{dz}{du}\cdot\dfrac{\partial u}{\partial y}$.

ILLUSTRATIVE EXAMPLES

Example 1. *Find the first order partial derivatives of the following :*

(*i*) $u = tan^{-1}\dfrac{x^2 + y^2}{x + y}$ (*ii*) $u = cos^{-1}\left(\dfrac{x}{y}\right)$.

Sol. (*i*) $$u = \tan^{-1}\frac{x^2 + y^2}{x + y}$$

$$\frac{\partial u}{\partial x} = \frac{1}{1 + \left(\dfrac{x^2 + y^2}{x + y}\right)^2}\cdot\frac{\partial}{\partial x}\left(\frac{x^2 + y^2}{x + y}\right)$$

$$= \frac{(x + y)^2}{(x + y^2) + (x^2 + y^2)^2}\cdot\frac{(x + y)\dfrac{\partial}{\partial x}(x^2 + y^2) - (x^2 + y^2)\dfrac{\partial}{\partial x}(x + y)}{(x + y)^2}$$

$$= \frac{(x + y)\,.\,2x - (x^2 + y^2)\,.\,1}{(x + y)^2 + (x^2 + y^2)^2}$$

$$\Rightarrow \quad \frac{\partial u}{\partial x} = \frac{x^2 + 2xy - y^2}{(x + y)^2 + (x^2 + y^2)^2} \quad \ldots(1)$$

Since u remains the same if we interchange x and y, u is symmetrical w.r.t. x and y. Interchanging x and y in (1), we have

Similarly, $$\frac{\partial u}{\partial y} = \frac{y^2 + 2xy - x^2}{(x + y)^2 + (x^2 + y^2)^2}$$

(*ii*) $$u = \cos^{-1}\left(\frac{x}{y}\right)$$

$$\frac{\partial u}{\partial x} = \frac{-1}{\sqrt{1 - \left(\dfrac{x}{y}\right)^2}}\cdot\frac{\partial}{\partial x}\left(\frac{x}{y}\right) = \frac{-y}{\sqrt{y^2 - x^2}}\cdot\frac{1}{y} = \frac{-1}{\sqrt{y^2 - x^2}}$$

$$\frac{\partial u}{\partial y} = \frac{-1}{\sqrt{1-\left(\frac{x}{y}\right)^2}} \cdot \frac{\partial}{\partial y}\left(\frac{x}{y}\right) = \frac{-y}{\sqrt{y^2 - x^2}}\left(-\frac{x}{y^2}\right) = \frac{x}{y\sqrt{y^2 - x^2}}.$$

Example 2. *If* $z(x + y) = x^2 + y^2$, *show that* $\left(\frac{\partial z}{\partial x} - \frac{\partial z}{\partial y}\right)^2 = 4\left(1 - \frac{\partial z}{\partial x} - \frac{\partial z}{\partial y}\right)$.

Sol. $$z = \frac{x^2 + y^2}{x + y}$$ [z is symmetrical w.r.t. x and y]

$$\frac{\partial z}{\partial x} = \frac{(x+y)\frac{\partial}{\partial x}(x^2 + y^2) - (x^2 + y^2)\frac{\partial}{\partial x}(x + y)}{(x+y)^2}$$

$$= \frac{(x+y) \,.\, 2x - (x^2 + y^2) \,.\, 1}{(x+y)^2} = \frac{x^2 + 2xy - y^2}{(x+y)^2}$$

Similarly, $$\frac{\partial z}{\partial y} = \frac{y^2 + 2xy - x^2}{(x+y)^2}$$

Now, $$\left(\frac{\partial z}{\partial x} - \frac{\partial z}{\partial y}\right)^2 = \left[\frac{2x^2 - 2y^2}{(x+y)^2}\right]^2 = \frac{4(x+y)^2\,(x-y)^2}{(x+y)^4} = \frac{4(x-y)^2}{(x+y)^2}$$

$$4\left(1 - \frac{\partial z}{\partial x} - \frac{\partial z}{\partial y}\right) = 4\left[1 - \frac{x^2 + 2xy - y^2}{(x+y)^2} - \frac{y^2 + 2xy - x^2}{(x+y)^2}\right]$$

$$= 4\left[\frac{x^2 + 2xy + y^2 - x^2 - 2xy + y^2 - y^2 - 2xy + x^2}{(x+y)^2}\right]$$

$$= \frac{4(x^2 - 2xy + y^2)}{(x+y)^2} = \frac{4(x-y)^2}{(x+y)^2}$$

$$\therefore \quad \left(\frac{\partial z}{\partial x} - \frac{\partial z}{\partial y}\right)^2 = 4\left(1 - \frac{\partial z}{\partial x} - \frac{\partial z}{\partial y}\right).$$

Example 3. *Prove that if* $f(x, y) = \frac{1}{\sqrt{y}} \,.\, e^{-\frac{(x-a)^2}{4y}}$, *then* $f_{xy} = f_{yx}$.

Sol. $$f(x, y) = \frac{1}{\sqrt{y}} \,.\, e^{-\frac{(x-a)^2}{4y}} = y^{-\frac{1}{2}}\, e^{-\frac{(x-a)^2}{4y}}$$

$$f_x = \frac{\partial f}{\partial x} = y^{-\frac{1}{2}} \,.\, e^{-\frac{(x-a)^2}{4y}} \,.\, \frac{\partial}{\partial x}\left[-\frac{(x-a)^2}{4y}\right]$$

$$= y^{-\frac{1}{2}} \,.\, e^{-\frac{(x-a)^2}{4y}} \,.\left[-\frac{2(x-a)}{4y}\right] = -\frac{1}{2}\, y^{-\frac{3}{2}}\,(x-a)\, e^{-\frac{(x-a)^2}{4y}}$$

$$f_y = \frac{\partial f}{\partial y} = -\frac{1}{2} y^{-\frac{3}{2}} . e^{-\frac{(x-a)^2}{4y}} + y^{-\frac{1}{2}} . e^{-\frac{(x-a)^2}{4y}} . \frac{\partial}{\partial y}\left[-\frac{(x-a)^2}{4y}\right]$$

$$= e^{-\frac{(x-a)^2}{4y}} \left[-\frac{1}{2} y^{-\frac{3}{2}} + y^{-\frac{1}{2}} . \frac{(x-a)^2}{4y^2}\right] = \frac{1}{4} y^{-\frac{3}{2}} e^{-\frac{(x-a)^2}{4y}} [-2 + y^{-1}(x-a)^2]$$

$$f_{xy} = \frac{\partial}{\partial x}\left(\frac{\partial f}{\partial y}\right)$$

$$= \frac{1}{4} y^{-\frac{3}{2}} \left\{ e^{-\frac{(x-a)^2}{4y}} . \frac{\partial}{\partial x}\left[-\frac{(x-a)^2}{4y}\right] . [-2 + y^{-1}(x-a)^2] + e^{-\frac{(x-a)^2}{4y}} . 2y^{-1}(x-a) \right\}$$

$$= \frac{1}{4} y^{-\frac{3}{2}} . e^{-\frac{(x-a)^2}{4y}} \left\{ -\frac{2(x-a)}{4y} [-2 + y^{-1}(x-a)^2] + 2y^{-1}(x-a) \right\}$$

$$= \frac{1}{4} y^{-\frac{3}{2}} . e^{-\frac{(x-a)^2}{4y}} . \frac{x-a}{y} \left\{ -\frac{1}{2} [-2 + y^{-1}(x-a)^2] + 2 \right\}$$

$$= \frac{1}{4} y^{-\frac{5}{2}} (x-a) e^{-\frac{(x-a)^2}{4y}} \left[3 - \frac{(x-a)^2}{2y}\right]$$

$$f_{yx} = \frac{\partial}{\partial y}\left(\frac{\partial f}{\partial x}\right) = -\frac{1}{2}(x-a)\left[-\frac{3}{2} y^{-\frac{5}{2}} . e^{-\frac{(x-a)^2}{4y}} + y^{-\frac{3}{2}} . e^{-\frac{(x-a)^2}{4y}} . \frac{(x-a)^2}{4y^2}\right]$$

$$= -\frac{1}{4}(x-a)\, y^{-\frac{5}{2}} . e^{-\frac{(x-a)^2}{4y}} \left[-3 + \frac{(x-a)^2}{2y}\right]$$

$$= \frac{1}{4} y^{-\frac{5}{2}} (x-a) e^{-\frac{(x-a)^2}{4y}} \left[3 - \frac{(x-a)^2}{2y}\right]$$

$\therefore \quad f_{xy} = f_{yx}.$

Example 4. *If $u = x^y$, show that* $\dfrac{\partial^3 u}{\partial x^2 \partial y} = \dfrac{\partial^3 u}{\partial x \partial y \partial x}$.

Sol.

$$u = x^y$$

$$\frac{\partial u}{\partial y} = x^y \log x$$

$$\frac{\partial^2 u}{\partial x \partial y} = \frac{\partial}{\partial x}\left(\frac{\partial u}{\partial y}\right) = yx^{y-1} \log x + x^y . \frac{1}{x} = x^{y-1}(y \log x + 1)$$

$$\frac{\partial^3 u}{\partial x^2 \partial y} = \frac{\partial}{\partial x}\left(\frac{\partial^2 u}{\partial x \partial y}\right) = \frac{\partial}{\partial x}[x^{y-1}(y \log x + 1)] \qquad ...(1)$$

$$\frac{\partial u}{\partial x} = yx^{y-1}$$

$$\frac{\partial^2 u}{\partial y \partial x} = \frac{\partial}{\partial y}\left(\frac{\partial u}{\partial x}\right) = x^{y-1} + yx^{y-1} \log x = x^{y-1}(y \log x + 1)$$

$$\frac{\partial^3 u}{\partial x \partial y \partial x} = \frac{\partial}{\partial x}\left(\frac{\partial^2 u}{\partial y \partial x}\right) = \frac{\partial}{\partial x}\,[x^{y-1}\,(y \log x + 1)] \qquad ...(2)$$

From (1) and (2), $\dfrac{\partial^3 u}{\partial x^2 \partial y} = \dfrac{\partial^3 u}{\partial x \partial y \partial x}$.

Example 5. *If* $u = x^2 \tan^{-1}(y/x) - y^2 \tan^{-1}(x/y)$, *prove that* $\dfrac{\partial^2 u}{\partial x \partial y} = \dfrac{x^2 - y^2}{x^2 + y^2}$.

[CUSAT 2007]

Sol. $u = x^2 \tan^{-1}(y/x) - y^2 \tan^{-1}(x/y)$

$$\frac{\partial u}{\partial x} = 2x \tan^{-1}(y/x) + x^2 \cdot \frac{1}{1 + y^2/x^2}(-y/x^2) - y^2 \cdot \frac{1}{1 + x^2/y^2}(1/y).$$

$$= 2x \tan^{-1}(y/x) - \frac{x^2 y}{x^2 + y^2} - \frac{y^3}{x^2 + y^2} = 2x \tan^{-1}(y/x) - y$$

$$\frac{\partial^2 u}{\partial x \partial y} = 2x \cdot \frac{1}{1 + y^2/x^2}(1/x) - 1 = \frac{2x^2}{x^2 + y^2} - 1 = \frac{x^2 - y^2}{x^2 + y^2}.$$

Example 6. *If* $\theta = t^n e^{-\frac{r^2}{4t}}$, *find the value of n which will make* $\dfrac{1}{r^2}\dfrac{\partial}{\partial r}\left(r^2 \dfrac{\partial \theta}{\partial r}\right) = \dfrac{\partial \theta}{\partial t}$.

Sol. $\theta = t^n e^{-\frac{r^2}{4t}}$

$$\frac{\partial \theta}{\partial r} = t^n \cdot e^{-\frac{r^2}{4t}} \cdot \left(-\frac{2r}{4t}\right) = -\frac{1}{2} r t^{n-1} e^{-\frac{r^2}{4t}}$$

$$\therefore \quad r^2 \frac{\partial \theta}{\partial r} = -\frac{1}{2} r^3 \cdot t^{n-1} e^{-\frac{r^2}{4t}}$$

$$\frac{\partial}{\partial r}\left(r^2 \frac{\partial \theta}{\partial r}\right) = -\frac{1}{2} t^{n-1}\left[3r^2 e^{-\frac{r^2}{4t}} + r^3 e^{-\frac{r^2}{4t}}\left(-\frac{2r}{4t}\right)\right] = -\frac{1}{2} t^{n-1} r^2 e^{-\frac{r^2}{4t}}\left[3 - \frac{r^2}{2t}\right]$$

$$\therefore \quad \frac{1}{r^2}\frac{\partial}{\partial r}\left(r^2 \frac{\partial \theta}{\partial r}\right) = \frac{1}{2} t^{n-1} e^{-\frac{r^2}{4t}}\left(\frac{r^2}{2t} - 3\right)$$

Also
$$\frac{\partial \theta}{\partial t} = n t^{n-1} e^{-\frac{r^2}{4t}} + t^n e^{-\frac{r^2}{4t}} \cdot \left(\frac{r^2}{4t^2}\right) = t^{n-1} e^{-\frac{r^2}{4t}}\left(n + \frac{r^2}{4t}\right)$$

Since
$$\frac{1}{r^2}\frac{\partial}{\partial r}\left(r^2 \frac{\partial \theta}{\partial r}\right) = \frac{\partial \theta}{\partial t} \qquad \text{(given)}$$

$$\therefore \quad \frac{1}{2} t^{n-1} e^{-\frac{r^2}{4t}}\left(\frac{r^2}{2t} - 3\right) = t^{n-1} e^{-\frac{r^2}{4t}}\left(n + \frac{r^2}{4t}\right)$$

$$\Rightarrow \quad \frac{r^2}{4t} - \frac{3}{2} = n + \frac{r^2}{4t} \quad \therefore \quad n = -\frac{3}{2}.$$

Example 7. *If $u = (1 - 2xy + y^2)^{-1/2}$, prove that*

$$\frac{\partial}{\partial x}\left\{(1-x^2)\frac{\partial u}{\partial x}\right\} + \frac{\partial}{\partial y}\left\{y^2 \frac{\partial u}{\partial y}\right\} = 0.$$

Sol. $u = (1 - 2xy + y^2)^{-1/2} = V^{-1/2}$, where $V = 1 - 2xy + y^2$

$$\frac{\partial u}{\partial x} = -\frac{1}{2} V^{-3/2} . \frac{\partial V}{\partial x} = -\frac{1}{2} V^{-3/2} (-2y) = yV^{-3/2}$$

$$\frac{\partial^2 u}{\partial x^2} = y . \frac{\partial}{\partial x} (V^{-3/2}) = y . \left(-\frac{3}{2}\right) V^{-5/2} . \frac{\partial V}{\partial x} = -\frac{3}{2} yV^{-5/2} (-2y) = 3y^2 V^{-5/2}$$

$$\therefore \quad \frac{\partial}{\partial x}\left\{(1-x^2)\frac{\partial u}{\partial x}\right\} = (1 - x^2) . \frac{\partial^2 u}{\partial x^2} + \frac{\partial u}{\partial x} . \frac{\partial}{\partial x} (1 - x^2)$$

$$= (1 - x^2) . 3y^2V^{-5/2} + yV^{-3/2} (-2x) = yV^{-3/2} [3yV^{-1} (1 - x^2) - 2x] \qquad ...(1)$$

Also
$$\frac{\partial u}{\partial y} = -\frac{1}{2} V^{-3/2} \frac{\partial V}{\partial y} = -\frac{1}{2} V^{-3/2} . (-2x + 2y) = V^{-3/2} . (x - y)$$

$$\frac{\partial^2 u}{\partial y^2} = V^{-3/2} . \frac{\partial}{\partial y} (x - y) + (x - y) . \frac{y}{\partial y} (V^{-3/2})$$

$$= V^{-3/2} . (-1) + (x - y). \left(-\frac{3}{2} V^{-5/2}\right). \frac{\partial V}{\partial y}$$

$$= -V^{-3/2} - \frac{3}{2} (x - y) V^{-5/2} . (-2x + 2y) = -V^{-3/2} + 3(x - y)^2 V^{-5/2}$$

$$\therefore \quad \frac{\partial}{\partial y}\left\{y^2 \frac{\partial u}{\partial y}\right\} = y^2 . \frac{\partial^2 u}{\partial y^2} + \frac{\partial u}{\partial y} . \frac{\partial}{\partial y} (y^2)$$

$$= y^2[-V^{-3/2} + 3(x - y)^2 V^{-5/2}] + V^{-3/2} (x - y) . 2y$$

$$= y V^{-3/2} [-y + 3y(x - y)^2 V^{-1} + 2(x - y)]$$

$$= y V^{-3/2} [3y (x - y)^2 V^{-1} + (2x - 3y)] \qquad ...(2)$$

Adding (1) and (2), we have

$$\frac{\partial}{\partial x}\left\{(1-x^2)\frac{\partial u}{\partial x}\right\} + \frac{\partial}{\partial y}\left\{y^2 \frac{\partial u}{\partial y}\right\} = yV^{-3/2} [3yV^{-1}(1 - x^2) - 2x + 3y (x - y)^2 V^{-1} + 2x - 3y]$$

$$= y V^{-3/2}[3yV^{-1} (1 - x^2 + x^2 - 2xy + y^2) - 3y]$$

$$= yV^{-3/2}[3yV^{-1}(1 - 2xy + y^2) - 3y]$$

$$= yV^{-3/2} [3y - 3y] \qquad | \because \; V = 1 - 2xy + y^2$$

$$= 0.$$

Example 8. *If $u = \log (x^3 + y^3 + z^3 - 3xyz)$, show that*

(*i*) $\left(\frac{\partial}{\partial x} + \frac{\partial}{\partial y} + \frac{\partial}{\partial z}\right)^2 u = -\frac{9}{(x + y + z)^2}$

(*ii*) $\frac{\partial^2 u}{\partial x^2} + \frac{\partial^2 u}{\partial y^2} + \frac{\partial^2 u}{\partial z^2} + 2\frac{\partial^2 u}{\partial y \partial z} + 2\frac{\partial^2 u}{\partial z \partial x} + 2\frac{\partial^2 u}{\partial x \partial y} = \frac{-9}{(x + y + z)^2}.$

Sol. (*i*) $u = \log (x^3 + y^3 + z^3 - 3xyz)$

$$\frac{\partial u}{\partial x} = \frac{3x^2 - 3yz}{x^3 + y^3 + z^3 - 3xyz}; \quad \frac{\partial u}{\partial y} = \frac{3y^2 - 3zx}{x^3 + y^3 + z^3 - 3xyz};$$

$$\frac{\partial u}{\partial z} = \frac{3z^2 - 3xy}{x^3 + y^3 + z^3 - 3xyz}$$

Adding, $\frac{\partial u}{\partial x} + \frac{\partial u}{\partial y} + \frac{\partial u}{\partial z} = \frac{3(x^2 + y^2 + z^2 - xy - yz - zx)}{x^3 + y^3 + z^3 - 3xyz} = \frac{3}{x + y + z}$

$[\because \ x^3 + y^3 + z^3 - 3xyz = (x + y + z)(x^2 + y^2 + z^2 - xy - yz - zx)]$

Now, $\left(\frac{\partial}{\partial x} + \frac{\partial}{\partial y} + \frac{\partial}{\partial z}\right)^2 u = \left(\frac{\partial}{\partial x} + \frac{\partial}{\partial y} + \frac{\partial}{\partial z}\right)\left(\frac{\partial}{\partial x} + \frac{\partial}{\partial y} + \frac{\partial}{\partial z}\right) u$

$$= \left(\frac{\partial}{\partial x} + \frac{\partial}{\partial y} + \frac{\partial}{\partial z}\right)\left(\frac{\partial u}{\partial x} + \frac{\partial u}{\partial y} + \frac{\partial u}{\partial z}\right) = \left(\frac{\partial}{\partial x} + \frac{\partial}{\partial y} + \frac{\partial}{\partial z}\right)\left(\frac{3}{x + y + z}\right)$$

$$= -\frac{3}{(x + y + z)^2} - \frac{3}{(x + y + z)^2} - \frac{3}{(x + y + z)^2} = -\frac{9}{(x + y + z)^2} \qquad ...(1)$$

(*ii*) $\left(\frac{\partial}{\partial x} + \frac{\partial}{\partial y} + \frac{\partial}{\partial z}\right)^2 u = \left(\frac{\partial}{\partial x} + \frac{\partial}{\partial y} + \frac{\partial}{\partial z}\right)\left(\frac{\partial}{\partial x} + \frac{\partial}{\partial y} + \frac{\partial}{\partial z}\right) u$

$$= \left(\frac{\partial}{\partial x} + \frac{\partial}{\partial y} + \frac{\partial}{\partial z}\right)\left(\frac{\partial u}{\partial x} + \frac{\partial u}{\partial y} + \frac{\partial u}{\partial z}\right)$$

$$= \frac{\partial}{\partial x}\left(\frac{\partial u}{\partial x} + \frac{\partial u}{\partial y} + \frac{\partial u}{\partial z}\right) + \frac{\partial}{\partial y}\left(\frac{\partial u}{\partial x} + \frac{\partial u}{\partial y} + \frac{\partial u}{\partial z}\right) + \frac{\partial}{\partial z}\left(\frac{\partial u}{\partial x} + \frac{\partial u}{\partial y} + \frac{\partial u}{\partial z}\right)$$

$$= \frac{\partial^2 u}{\partial x^2} + \frac{\partial^2 u}{\partial x \partial y} + \frac{\partial^2 u}{\partial x \partial z} + \frac{\partial^2 u}{\partial y \partial x} + \frac{\partial^2 u}{\partial y^2} + \frac{\partial^2 u}{\partial y \partial z} + \frac{\partial^2 u}{\partial z \partial x} + \frac{\partial^2 u}{\partial z \partial y} + \frac{\partial^2 u}{\partial z^2}$$

$$= \frac{\partial^2 u}{\partial x^2} + \frac{\partial^2 u}{\partial y^2} + \frac{\partial^2 u}{\partial z^2} + 2\frac{\partial^2 u}{\partial y \partial z} + 2\frac{\partial^2 u}{\partial z \partial x} + 2\frac{\partial^2 u}{\partial x \partial y}$$

$$\left[\because \ \frac{\partial^2 u}{\partial y \partial x} = \frac{\partial^2 u}{\partial x \partial y}, \frac{\partial^2 u}{\partial z \partial y} = \frac{\partial^2 u}{\partial y \partial z}, \frac{\partial^2 u}{\partial x \partial z} = \frac{\partial^2 u}{\partial z \partial x}\right]$$

$$\therefore \quad \frac{\partial^2 u}{\partial x^2} + \frac{\partial^2 u}{\partial y^2} + \frac{\partial^2 u}{\partial z^2} + 2\frac{\partial^2 u}{\partial y \partial z} + 2\frac{\partial^2 u}{\partial z \partial x} + 2\frac{\partial^2 u}{\partial x \partial y} = \left(\frac{\partial}{\partial x} + \frac{\partial}{\partial y} + \frac{\partial}{\partial z}\right)^2 u = \frac{-9}{(x + y + z)^2}.$$

[from (1)]

Example 9. *If* $x^x y^y z^z = c$, *show that at* $x = y = z$, $\frac{\partial^2 z}{\partial x \partial y} = -(x \log ex)^{-1}$.

[CUSAT 2008]

Sol. $x^x y^y z^z = c$ defines z as a function of x and y.

Taking log, $x \log x + y \log y + z \log z = \log c$

Differentiating partially w.r.t. y, we have

$$y \cdot \frac{1}{y} + 1 \cdot \log y + z \cdot \frac{1}{z}\frac{\partial z}{\partial y} + \log z \cdot \frac{\partial z}{\partial y} = 0$$

or $$1 + \log y + (1 + \log z) \frac{\partial z}{\partial y} = 0 \quad ...(1)$$

or $$\left.\begin{aligned} \frac{\partial z}{\partial y} &= -\frac{1+\log y}{1+\log z} \\ \text{Similarly,} \quad \frac{\partial z}{\partial x} &= -\frac{1+\log x}{1+\log z} \end{aligned}\right\} \quad ...(2)$$

Differentiating (1) partially w.r.t. x, we have

$$\left(\frac{1}{z}\frac{\partial z}{\partial x}\right)\frac{\partial z}{\partial y} + (1+\log z)\frac{\partial^2 z}{\partial x \partial y} = 0 \quad \text{or} \quad \frac{\partial^2 z}{\partial x \partial y} = -\frac{1}{z(1+\log z)}\frac{\partial z}{\partial x}\cdot\frac{\partial z}{\partial y} \quad ...(3)$$

When $x = y = z$

From (2), $\frac{\partial z}{\partial y} = -1, \frac{\partial z}{\partial x} = -1$

From (3), $$\frac{\partial^2 z}{\partial x \partial y} = -\frac{1}{x(1+\log x)}(-1)(-1) = -\frac{1}{x(\log e + \log x)} = -\frac{1}{x(\log ex)} = -(x \log ex)^{-1}.$$

Example 10. *If $u = f(r)$, where $r^2 = x^2 + y^2$, prove that* $\frac{\partial^2 u}{\partial x^2} + \frac{\partial^2 u}{\partial y^2} = f''(r) + \frac{1}{r} f'(r)$.

Sol. $$r^2 = x^2 + y^2 \quad ...(1)$$

Differentiating partially w.r.t. x, we get $2r\frac{\partial r}{\partial x} = 2x$ or $\frac{\partial r}{\partial x} = \frac{x}{r}$

Similarly, $\frac{\partial r}{\partial y} = \frac{y}{r}$

Now, $u = f(r)$

$$\therefore \quad \frac{\partial u}{\partial x} = f'(r)\cdot\frac{\partial r}{\partial x} = \frac{x}{r} f'(r)$$

Differentiating again w.r.t. x, we get

$$\frac{\partial^2 u}{\partial x^2} = \frac{1}{r}f'(r) + x\cdot\left(-\frac{1}{r^2}\frac{\partial r}{\partial x}\right)f'(r) + \frac{x}{r}f''(r)\cdot\frac{\partial r}{\partial x}$$

$$\left[\because \frac{\partial}{\partial x}(uvw) = vw\frac{\partial}{\partial x}(u) + uw\frac{\partial}{\partial x}(v) + uv\frac{\partial}{\partial x}(w)\right]$$

$$= \frac{1}{r}f'(r) - \frac{x}{r^2}\cdot\frac{x}{r}f'(r) + \frac{x}{r}\cdot f''(r)\cdot\frac{x}{r} = \frac{1}{r}f'(r) - \frac{x^2}{r^3}f'(r) + \frac{x^2}{r^2}f''(r)$$

$$= \frac{r^2 - x^2}{r^3}f'(r) + \frac{x^2}{r^2}f''(r) = \frac{y^2}{r^3}f'(r) + \frac{x^2}{r^2}f''(r) \quad | \text{ Using (1)}$$

Similarly, $$\frac{\partial^2 u}{\partial y^2} = \frac{x^2}{r^3}f'(r) + \frac{y^2}{r^2}f''(r)$$

$$\therefore \quad \frac{\partial^2 u}{\partial x^2} + \frac{\partial^2 u}{\partial y^2} = \frac{x^2+y^2}{r^3}f'(r) + \frac{x^2+y^2}{r^2}f''(r) = \frac{r^2}{r^3}f'(r) + \frac{r^2}{r^2}f''(r) = f''(r) + \frac{1}{r}f'(r).$$

Example 11. *If $x = r \cos\theta$, $y = r \sin\theta$, prove that*

(*i*) $\dfrac{\partial r}{\partial x} = \dfrac{\partial x}{\partial r}$ (*ii*) $\dfrac{1}{r} \cdot \dfrac{\partial x}{\partial \theta} = r \dfrac{\partial \theta}{\partial x}$ (*iii*) $\dfrac{\partial^2 \theta}{\partial x^2} + \dfrac{\partial^2 \theta}{\partial y^2} = 0.$

Sol. (*i*) $\dfrac{\partial r}{\partial x}$ means $\left(\dfrac{\partial r}{\partial x}\right)_y$ = The partial derivative of r w.r.t. x, treating y as constant.

$\therefore$ We express r in terms of x and y.

Squaring and adding the given relations, $r^2 = x^2 + y^2$

Differentiating partially w.r.t. x, we get $2r \dfrac{\partial r}{\partial x} = 2x$ or $\dfrac{\partial r}{\partial x} = \dfrac{x}{r}$

$\dfrac{\partial x}{\partial r}$ means $\left(\dfrac{\partial x}{\partial r}\right)_\theta$ = The partial derivative of x w.r.t. r treating θ as constant.

$\therefore$ We express x in terms of r and θ.

Thus, $x = r\cos\theta$ (given)

$$\frac{\partial x}{\partial r} = \cos\theta = \frac{x}{r} \qquad \left(\because \cos\theta = \frac{x}{r}\right)$$

$$\therefore \quad \frac{\partial r}{\partial x} = \frac{\partial x}{\partial r}$$

(*ii*) Expressing x in terms of r and θ, we have $x = r\cos\theta$

$$\Rightarrow \quad \frac{\partial x}{\partial \theta} = -r\sin\theta = -y \quad \Rightarrow \quad \frac{1}{r}\frac{\partial x}{\partial \theta} = -\frac{y}{r}$$

Expressing θ in terms of x and y, we have $\tan\theta = \dfrac{y}{x}$ or $\theta = \tan^{-1}\dfrac{y}{x}$

$$\frac{\partial \theta}{\partial x} = \frac{1}{1+\dfrac{y^2}{x^2}} \cdot \left(-\frac{y}{x^2}\right) = -\frac{y}{x^2+y^2} = \frac{-y}{r^2(\cos^2\theta + \sin^2\theta)} = -\frac{y}{r^2}$$

$$\Rightarrow \quad r\frac{\partial \theta}{\partial x} = -\frac{y}{r} \quad \therefore \quad \frac{1}{r} \cdot \frac{\partial x}{\partial \theta} = r\frac{\partial \theta}{\partial x}.$$

(*iii*) Expressing θ in terms of x and y, we have $\tan\theta = \dfrac{y}{x}$ or $\theta = \tan^{-1}\dfrac{y}{x}$

$$\frac{\partial \theta}{\partial x} = \frac{1}{1+\dfrac{y^2}{x^2}} \cdot \left(-\frac{y}{x^2}\right) = -\frac{y}{x^2+y^2} = -y(x^2+y^2)^{-1}$$

$$\frac{\partial^2 \theta}{\partial x^2} = y(x^2+y^2)^{-2} \,.\, 2x = \frac{2xy}{(x^2+y^2)^2}$$

$$\frac{\partial \theta}{\partial y} = \frac{1}{1+\dfrac{y^2}{x^2}} \cdot \frac{1}{x} = \frac{x}{x^2+y^2} = x(x^2+y^2)^{-1}$$

$$\frac{\partial^2 \theta}{\partial y^2} = -x(x^2+y^2)^{-2} \,.\, 2y = -\frac{2xy}{(x^2+y^2)^2}$$

$$\therefore \quad \frac{\partial^2 \theta}{\partial x^2} + \frac{\partial^2 \theta}{\partial y^2} = 0.$$

Example 12. *If* $x = e^{r\cos\theta}\cos(r\sin\theta)$ *and* $y = e^{r\cos\theta}\sin(r\sin\theta)$*, prove that*

$$\frac{\partial x}{\partial r} = \frac{1}{r}\cdot\frac{\partial y}{\partial \theta},\ \frac{\partial y}{\partial r} = -\frac{1}{r}\cdot\frac{\partial x}{\partial \theta}.$$

Hence deduce that $\dfrac{\partial^2 x}{\partial r^2} + \dfrac{1}{r}\dfrac{\partial x}{\partial r} + \dfrac{1}{r^2}\dfrac{\partial^2 x}{\partial \theta^2} = 0.$

Sol. $x = e^{r\cos\theta}\cos(r\sin\theta)$

$$\therefore \quad \frac{\partial x}{\partial r} = e^{r\cos\theta}.\cos\theta.\cos(r\sin\theta) - e^{r\cos\theta}\sin(r\sin\theta).\sin\theta$$

$$= e^{r\cos\theta}[\cos\theta\cos(r\sin\theta) - \sin\theta\sin(r\sin\theta)]$$

$$= e^{r\cos\theta}\cos(\theta + r\sin\theta) \quad \text{...(1)}$$

$$\frac{\partial x}{\partial \theta} = e^{r\cos\theta}.(-r\sin\theta).\cos(r\sin\theta) - e^{r\cos\theta}\sin(r\sin\theta).r\cos\theta$$

$$= -re^{r\cos\theta}[\sin\theta\cos(r\sin\theta) + \cos\theta\sin(r\sin\theta)]$$

$$= -re^{r\cos\theta}\sin(\theta + r\sin\theta) \quad \text{...(2)}$$

Also $y = e^{r\cos\theta}\sin(r\sin\theta)$

$$\therefore \quad \frac{\partial y}{\partial r} = e^{r\cos\theta}.\cos\theta.\sin(r\sin\theta) + e^{r\cos\theta}.\cos(r\sin\theta)\sin\theta$$

$$= e^{r\cos\theta}[\sin\theta\cos(r\sin\theta) + \cos\theta\sin(r\sin\theta)]$$

$$= e^{r\cos\theta}\sin(\theta + r\sin\theta) \quad \text{...(3)}$$

$$\frac{\partial y}{\partial \theta} = e^{r\cos\theta}(-r\sin\theta)\sin(r\sin\theta) + e^{r\cos\theta}\cos(r\sin\theta)\times r\cos\theta$$

$$= re^{r\cos\theta}[\cos\theta\cos(r\sin\theta) - \sin\theta\sin(r\sin\theta)]$$

$$= re^{r\cos\theta}\cos(\theta + r\sin\theta) \quad \text{...(4)}$$

From (1) and (4), $\dfrac{\partial x}{\partial r} = \dfrac{1}{r}.\dfrac{\partial y}{\partial \theta}$ ...(5)

From (2) and (3), $\dfrac{\partial y}{\partial r} = -\dfrac{1}{r}.\dfrac{\partial x}{\partial \theta}$ ...(6)

From (5), $\dfrac{\partial^2 x}{\partial r^2} = -\dfrac{1}{r^2}.\dfrac{\partial y}{\partial \theta} + \dfrac{1}{r}.\dfrac{\partial^2 y}{\partial r\partial \theta}$

From (6), $\dfrac{\partial x}{\partial \theta} = -r\dfrac{\partial y}{\partial r}$

$$\therefore \quad \frac{\partial^2 x}{\partial \theta^2} = -r\frac{\partial^2 y}{\partial \theta\partial r} = -r\frac{\partial^2 y}{\partial r\partial \theta}$$

$$\therefore \quad \frac{\partial^2 x}{\partial r^2} + \frac{1}{r}.\frac{\partial x}{\partial r} + \frac{1}{r^2}.\frac{\partial^2 x}{\partial \theta^2} = -\frac{1}{r^2}.\frac{\partial y}{\partial \theta} + \frac{1}{r}.\frac{\partial^2 y}{\partial r\partial \theta} + \frac{1}{r^2}.\frac{\partial y}{\partial \theta} - \frac{1}{r}.\frac{\partial^2 y}{\partial r\partial \theta} = 0.$$

Example 13. *If* $\dfrac{x^2}{a^2+u} + \dfrac{y^2}{b^2+u} + \dfrac{z^2}{c^2+u} = 1$*, prove that*

$$\left(\frac{\partial u}{\partial x}\right)^2 + \left(\frac{\partial u}{\partial y}\right)^2 + \left(\frac{\partial u}{\partial z}\right)^2 = 2\left(x\frac{\partial u}{\partial x} + y\frac{\partial u}{\partial y} + z\frac{\partial u}{\partial z}\right).$$

Sol. Given $$\frac{x^2}{a^2+u}+\frac{y^2}{b^2+u}+\frac{z^2}{c^2+u}=1 \quad ...(1)$$

or $$x^2(a^2+u)^{-1}+y^2(b^2+u)^{-1}+z^2(c^2+u)^{-1}=1$$

Differentiating partially w.r.t. x, we have

$$2x(a^2+u)^{-1}-x^2(a^2+u)^{-2}\cdot\frac{\partial u}{\partial x}-y^2(b^2+u)^{-2}\cdot\frac{\partial u}{\partial x}-z^2(c^2+u)^{-2}\cdot\frac{\partial u}{\partial x}=0$$

or $$\frac{2x}{a^2+u}=\left[\frac{x^2}{(a^2+u)^2}+\frac{y^2}{(b^2+u)^2}+\frac{z^2}{(c^2+u)^2}\right]\frac{\partial u}{\partial x}$$

or $$\frac{2x}{a^2+u}=V\frac{\partial u}{\partial x} \text{ where } V=\frac{x^2}{(a^2+u)^2}+\frac{y^2}{(b^2+u)^2}+\frac{z^2}{(c^2+u)^2}$$

or $$\frac{\partial u}{\partial x}=\frac{2x}{V(a^2+u)}$$

Similarly, $$\frac{\partial u}{\partial y}=\frac{2y}{V(b^2+u)} \quad\text{and}\quad \frac{\partial u}{\partial z}=\frac{2z}{V(c^2+u)}$$

$$\therefore \left(\frac{\partial u}{\partial x}\right)^2+\left(\frac{\partial u}{\partial y}\right)^2+\left(\frac{\partial u}{\partial z}\right)^2=\frac{4}{V^2}\left[\frac{x^2}{(a^2+u)^2}+\frac{y^2}{(b^2+u)^2}+\frac{z^2}{(c^2+u)^2}\right]$$

$$=\frac{4}{V^2}(V)=\frac{4}{V} \quad ...(2)$$

Now, $$2\left(x\frac{\partial u}{\partial x}+y\frac{\partial u}{\partial y}+z\frac{\partial u}{\partial z}\right)=2\left[\frac{2x^2}{V(a^2+u)}+\frac{2y^2}{V(b^2+u)}+\frac{2z^2}{V(c^2+u)}\right]$$

$$=\frac{4}{V}\left[\frac{x^2}{a^2+u}+\frac{y^2}{b^2+u}+\frac{z^2}{c^2+u}\right]$$

$$=\frac{4}{V}(1) \quad \text{[Using (1)]}$$

$$=\left(\frac{\partial u}{\partial x}\right)^2+\left(\frac{\partial u}{\partial y}\right)^2+\left(\frac{\partial u}{\partial z}\right)^2 \quad \text{[Using (2)]}$$

Example 14. *If $u = lx + my$, $v = mx - ly$, show that*

$$\left(\frac{\partial u}{\partial x}\right)_y\left(\frac{\partial x}{\partial u}\right)_v=\frac{l^2}{l^2+m^2},\ \left(\frac{\partial y}{\partial v}\right)_x\left(\frac{\partial v}{\partial y}\right)_u=\frac{l^2+m^2}{l^2}.$$

Sol. Given $$u = lx + my \quad ...(1)$$

$$v = mx - ly \quad ...(2)$$

(*i*) $\left(\frac{\partial u}{\partial x}\right)_y$ = The partial derivative of u w.r.t. x keeping y constant.

$\therefore$ We need a relation expressing u as a function of x and y.

From (1), $$\left(\frac{\partial u}{\partial x}\right)_y = l$$

Also $\left(\dfrac{\partial x}{\partial u}\right)_v$ = The partial derivative of x w.r.t. u keeping v constant.

$\therefore$ We need a relation expressing x as a function of u and v.

Eliminating y between (1) and (2) by multiplying (1) by l, (2) by m and adding the products, we have

$$lu + mv = (l^2 + m^2)x \quad \text{or} \quad x = \frac{lu + mv}{l^2 + m^2}$$

$$\therefore \qquad \left(\frac{\partial x}{\partial u}\right)_v = \frac{l}{l^2 + m^2}$$

Hence $$\left(\frac{\partial u}{\partial x}\right)_y \left(\frac{\partial x}{\partial u}\right)_v = \frac{l^2}{l^2 + m^2}$$

(*ii*) $\left(\dfrac{\partial y}{\partial v}\right)_x$ = The partial derivative of y w.r.t. v keeping x constant.

$\therefore$ We need a relation expressing y as a function of v and x.

From (2), $$y = \frac{mx - v}{l} \qquad \therefore \quad \left(\frac{\partial y}{\partial v}\right)_x = -\frac{1}{l}$$

Also $\left(\dfrac{\partial v}{\partial y}\right)_u$ = Partial derivative of v w.r.t. y keeping u constant

$\therefore$ We need a relation expressing v as a function of y and u.

Eliminating x between (1) and (2), we have $v = \dfrac{mu - (l^2 + m^2)y}{l}$

$$\therefore \qquad \left(\frac{\partial v}{\partial y}\right)_u = -\frac{l^2 + m^2}{l}$$

Hence $$\left(\frac{\partial y}{\partial v}\right)_x \left(\frac{\partial v}{\partial y}\right)_u = \left(-\frac{1}{l}\right)\left(-\frac{l^2 + m^2}{l}\right) = \frac{l^2 + m^2}{l^2}.$$

TEST YOUR KNOWLEDGE

1. Find the first order partial derivatives of the following functions :
   (*i*) $u = y^x$ (*ii*) $u = \log (x^2 + y^2)$
   (*iii*) $u = x^2 \sin \dfrac{y}{x}$ (*iv*) $u = \dfrac{x}{y} \tan^{-1}\left(\dfrac{y}{x}\right)$.
2. If $u = x^2 + y^2 + z^2$, prove that $xu_x + yu_y + zu_z = 2u$.
3. If $z = \log (x^2 + xy + y^2)$, prove that $x\dfrac{\partial z}{\partial x} + y\dfrac{\partial z}{\partial y} = 2$.
4. If $u = x^2y + y^2z + z^2x$, prove that $u_x + u_y + u_z = (x + y + z)^2$.
5. If $u = \dfrac{y}{z} + \dfrac{z}{x} + \dfrac{x}{y}$, show that $x\dfrac{\partial u}{\partial x} + y\dfrac{\partial u}{\partial y} + z\dfrac{\partial u}{\partial z} = 0$.

6. If $f(x, y) = x^3y - xy^3$, find $\left[\dfrac{1}{\dfrac{\partial f}{\partial x}} + \dfrac{1}{\dfrac{\partial f}{\partial y}}\right]_{\substack{x=1\\y=2}}$.

7. If $u = \log(\tan x + \tan y)$, prove that $\sin 2x \dfrac{\partial u}{\partial x} + \sin 2y \dfrac{\partial u}{\partial y} = 2$.

8. If $f(x, y, z) = \begin{vmatrix} x^2 & y^2 & z^2 \\ x & y & z \\ 1 & 1 & 1 \end{vmatrix}$, prove that $f_x + f_y + f_z = 0$.

9. Verify that $\dfrac{\partial^2 u}{\partial x \partial y} = \dfrac{\partial^2 u}{\partial y \partial x}$ for the following functions :

(i) $u = ax^2 + 2hxy + by^2$ (ii) $u = \tan^{-1}\left(\dfrac{x}{y}\right)$ (iii) $u = \log\left(\dfrac{x^2 + y^2}{xy}\right)$

(iv) $u = e^{ax} \sin by$ (v) $u = \log(x \sin y + y \sin x)$.

10. If $z = \log(e^x + e^y)$, show that $rt - s^2 = 0$; where $r = \dfrac{\partial^2 z}{\partial x^2}$, $t = \dfrac{\partial^2 z}{\partial y^2}$, $s = \dfrac{\partial^2 z}{\partial x \partial y}$.

11. If $u = \tan^{-1} \dfrac{xy}{\sqrt{1 + x^2 + y^2}}$, show that $\dfrac{\partial^2 u}{\partial x \partial y} = (1 + x^2 + y^2)^{-3/2}$.

12. If $u = e^{xyz}$, prove that $\dfrac{\partial^3 u}{\partial x \partial y \partial z} = (1 + 3xyz + x^2y^2z^2)e^{xyz}$.

13. If $u = \log(x^2 + y^2) + \tan^{-1}\dfrac{y}{x}$, prove that $\dfrac{\partial^2 u}{\partial x^2} + \dfrac{\partial^2 u}{\partial y^2} = 0$.

14. If $u = \tan^{-1}\left(\dfrac{2xy}{x^2 - y^2}\right)$, prove that $\dfrac{\partial^2 u}{\partial x^2} + \dfrac{\partial^2 u}{\partial y^2} = 0$.

15. Verify that $f_{xy} = f_{yx}$ when f is equal to

(i) $\sin^{-1}\left(\dfrac{y}{x}\right)$ (ii) $\log x \tan^{-1}(x^2 + y^2)$.

16. Find the value of n so that the equation $V = r^n (3 \cos^2 \theta - 1)$ satisfies the relation

$$\frac{\partial}{\partial r}\left(r^2 \frac{\partial V}{\partial r}\right) + \frac{1}{\sin\theta}\frac{\partial}{\partial \theta}\left(\sin\theta \frac{\partial V}{\partial \theta}\right) = 0.$$

17. If $z = \tan(y + ax) - (y - ax)^{3/2}$, show that $\dfrac{\partial^2 z}{\partial x^2} = a^2 \dfrac{\partial^2 z}{\partial y^2}$.

18. If $V = (x^2 + y^2 + z^2)^{-1/2}$, prove that $\dfrac{\partial^2 V}{\partial x^2} + \dfrac{\partial^2 V}{\partial y^2} + \dfrac{\partial^2 V}{\partial z^2} = 0$.

19. If $V = r^m$, where $r^2 = x^2 + y^2 + z^2$, show that $V_{xx} + V_{yy} + V_{zz} = m(m + 1) r^{m-2}$.

20. If $u = \log(x^2 + y^2 + z^2)$, prove that $x \dfrac{\partial^2 u}{\partial y \partial z} = y \dfrac{\partial^2 u}{\partial z \partial x} = z \dfrac{\partial^2 u}{\partial x \partial y}$.

21. If $u = \log \sqrt{x^2 + y^2 + z^2}$, prove that $(x^2 + y^2 + z^2)\left(\frac{\partial^2 u}{\partial x^2} + \frac{\partial^2 u}{\partial y^2} + \frac{\partial^2 u}{\partial z^2}\right) = 1.$

22. If $u = x^2 \tan^{-1} \frac{y}{x} - y^2 \tan^{-1} \frac{x}{y}$, find the value of $\frac{\partial^2 u}{\partial x \partial y}$.

23. If $x^2 + y^2 + z^2 = \frac{1}{u^2}$, prove that $\frac{\partial^2 u}{\partial x^2} + \frac{\partial^2 u}{\partial y^2} + \frac{\partial^2 u}{\partial z^2} = 0.$

24. If $u = \sqrt{x^2 + y^2 + z^2}$, show that

(*i*) $\left(\frac{\partial u}{\partial x}\right)^2 + \left(\frac{\partial u}{\partial y}\right)^2 + \left(\frac{\partial u}{\partial z}\right)^2 = 1$ (*ii*) $\frac{\partial^2 u}{\partial x^2} + \frac{\partial^2 u}{\partial y^2} + \frac{\partial^2 u}{\partial z^2} = \frac{2}{u}.$

25. If $u = e^{x-at} \cos (x - at)$, show that $\frac{\partial^2 u}{\partial t^2} = a^2 \frac{\partial^2 u}{\partial x^2}.$

26. If $v = \frac{1}{\sqrt{t}} e^{\frac{-x^2}{4a^2 t}}$, prove that $\frac{\partial v}{\partial t} = a^2 \frac{\partial^2 v}{\partial x^2}.$

27. If $u = (1 - 2xy + y^2)^{-1/2}$, prove that $x \frac{\partial u}{\partial x} - y \frac{\partial u}{\partial y} = y^2 u^3.$

28. If $u = e^x(x \cos y - y \sin y)$, show that $\frac{\partial^2 u}{\partial x^2} + \frac{\partial^2 u}{\partial y^2} = 0.$

29. If $x = r \cos \theta$, $y = r \sin \theta$, prove that

(*i*) $\frac{\partial^2 r}{\partial x^2} \cdot \frac{\partial^2 r}{\partial y^2} = \left(\frac{\partial^2 r}{\partial x \partial y}\right)^2$ (*ii*) $\frac{\partial^2 r}{\partial x^2} + \frac{\partial^2 r}{\partial y^2} = \frac{1}{r}\left[\left(\frac{\partial r}{\partial x}\right)^2 + \left(\frac{\partial r}{\partial y}\right)^2\right].$

30. If $x^2 = au + bv$, $y^2 = au - bv$, prove that $\left(\frac{\partial u}{\partial x}\right)_y \cdot \left(\frac{\partial x}{\partial u}\right)_v = \frac{1}{2} = \left(\frac{\partial v}{\partial y}\right)_x \cdot \left(\frac{\partial y}{\partial v}\right)_u.$

31. Show that $\frac{\partial^2 z}{\partial x^2} - 2 \frac{\partial^2 z}{\partial x \partial y} + \frac{\partial^2 z}{\partial y^2} = 0$, where $z = xf(x + y) + yg(x + y)$.

32. If $u = f(ax^2 + 2hxy + by^2)$, $v = \phi(ax^2 + 2hxy + by^2)$, prove that $\frac{\partial}{\partial y}\left(u \frac{\partial v}{\partial x}\right) = \frac{\partial}{\partial x}\left(u \frac{\partial v}{\partial y}\right).$

[**Hint.** Given $u = f(z)$, $v = \phi(z)$, where $z = ax^2 + 2hxy + by^2$.

We have to prove that $\frac{\partial u}{\partial y} \cdot \frac{\partial v}{\partial x} + u \frac{\partial^2 v}{\partial y \partial x} = \frac{\partial u}{\partial x} \cdot \frac{\partial v}{\partial y} + u \frac{\partial^2 v}{\partial x \partial y}$ or $\frac{\partial u}{\partial y} \cdot \frac{\partial v}{\partial x} = \frac{\partial u}{\partial x} \cdot \frac{\partial v}{\partial y}$]

$\left(\because \frac{\partial^2 v}{\partial y \partial x} = \frac{\partial^2 v}{\partial x \partial y}\right)$

33. If $u = \log (x^3 + y^3 - x^2 y - xy^2)$, show that $\frac{\partial^2 u}{\partial x^2} + 2 \frac{\partial^2 u}{\partial x \partial y} + \frac{\partial^2 u}{\partial y^2} = -4(x + y)^{-2}.$

[**Hint.** $u = \log \{x^2(x - y) - y^2(x - y)\} = \log (x - y)(x^2 - y^2) = \log (x - y)^2(x + y) = 2 \log (x - y) + \log (x + y)$].

34. If $V = f(r)$ and $r^2 = x^2 + y^2 + z^2$, prove that $V_{xx} + V_{yy} + V_{zz} = f''(r) + \frac{2}{r} f'(r).$

35. If $z = f(x + ay) + \phi(x - ay)$, prove that $\dfrac{\partial^2 z}{\partial y^2} = a^2 \dfrac{\partial^2 z}{\partial x^2}$.

36. Find p and q, if $x = \sqrt{a}\,(\sin u + \cos v)$, $y = \sqrt{a}\,(\cos u - \sin v)$, $z = 1 + \sin(u - v)$ where p and q mean $\dfrac{\partial z}{\partial x}$ and $\dfrac{\partial z}{\partial y}$ respectively.

$\left[\textbf{Hint. } x^2 + y^2 = 2az, \ \therefore\ z = \dfrac{x^2 + y^2}{2a}\right]$

Answers

1. (i) $y^x \log y,\ xy^{x-1}$; (ii) $\dfrac{2x}{x^2 + y^2}, \dfrac{2y}{x^2 + y^2}$ (iii) $2x \sin \dfrac{y}{x} - y \cos \dfrac{y}{x},\ x \cos \dfrac{y}{x}$

(iv) $\dfrac{-x}{x^2 + y^2} + \dfrac{1}{y} \tan^{-1} \dfrac{y}{x},\ \dfrac{x^2}{y(x^2 + y^2)} - \dfrac{x}{y^2} \tan^{-1} \dfrac{y}{x}$ **6.** $-\dfrac{13}{22}$

16. $n = 2, -3$ **22.** $\dfrac{x^2 - y^2}{x^2 + y^2}$ **36.** $p = \dfrac{x}{a},\ q = \dfrac{y}{a}$.

7.5. HOMOGENEOUS FUNCTIONS

A function $f(x, y)$ is said to be homogeneous of degree (or order) n in the variables x and y if it can be expressed in the form $x^n \phi\left(\dfrac{y}{x}\right)$ or $y^n \phi\left(\dfrac{x}{y}\right)$.

An alternative test for a function $f(x, y)$ to be homogeneous of degree (or order) n is that

$$f(tx, ty) = t^n f(x, y).$$

For example, if $f(x, y) = \dfrac{x + y}{\sqrt{x} + \sqrt{y}}$, then

$$\text{(i) } f(x, y) = \frac{x\left(1 + \dfrac{y}{x}\right)}{\sqrt{x}\left(1 + \sqrt{\dfrac{y}{x}}\right)} = x^{1/2} \phi\left(\frac{y}{x}\right)$$

$\Rightarrow$ $f(x, y)$ is a homogeneous function of degree $\frac{1}{2}$ in x and y.

$$\text{(ii) } f(x, y) = \frac{y\left(\dfrac{x}{y} + 1\right)}{\sqrt{y}\left(\sqrt{\dfrac{x}{y}} + 1\right)} = y^{1/2} \phi\left(\frac{x}{y}\right)$$

$\Rightarrow$ $f(x, y)$ is a homogeneous function of degree $\frac{1}{2}$ in x and y.

$$\text{(iii) } f(tx, ty) = \frac{tx + ty}{\sqrt{tx} + \sqrt{ty}} = \frac{t(x + y)}{\sqrt{t}\,(\sqrt{x} + \sqrt{y})} = t^{1/2} f(x, y)$$

$\Rightarrow$ $f(x, y)$ is a homogeneous function of degree $\frac{1}{2}$ in x and y.

Similarly, a function $f(x, y, z)$ is said to be homogeneous of degree (or order) n in the variables x, y, z if

$$f(x, y, z) = x^n \phi\left(\frac{y}{x}, \frac{z}{x}\right) \quad \text{or} \quad y^n\phi\left(\frac{x}{y}, \frac{z}{y}\right) \quad \text{or} \quad z^n \phi\left(\frac{x}{z}, \frac{y}{z}\right).$$

Alternative test is $f(tx, ty, tz) = t^n f(x, y, z)$.

7.6. EULER'S THEOREM ON HOMOGENEOUS FUNCTIONS

If u is a homogeneous function of degree n in x and y, then $\mathbf{x \dfrac{\partial u}{\partial x} + y \dfrac{\partial u}{\partial y} = nu}$.

Since u is a homogeneous function of degree n in x and y, it can be expressed as

$$u = x^n f\left(\frac{y}{x}\right)$$

$$\therefore \quad \frac{\partial u}{\partial x} = nx^{n-1} f\left(\frac{y}{x}\right) + x^n f'\left(\frac{y}{x}\right).\left(-\frac{y}{x^2}\right)$$

$$\Rightarrow \quad x\frac{\partial u}{\partial x} = nx^n f\left(\frac{y}{x}\right) - x^{n-1} yf'\left(\frac{y}{x}\right) \qquad ...(1)$$

Also
$$\frac{\partial u}{\partial y} = x^n f'\left(\frac{y}{x}\right).\frac{1}{x} = x^{n-1} f'\left(\frac{y}{x}\right)$$

$$\Rightarrow \quad y\frac{\partial u}{\partial y} = x^{n-1} yf'\left(\frac{y}{x}\right) \qquad ...(2)$$

Adding (1) and (2), we get $x\dfrac{\partial u}{\partial x} + y\dfrac{\partial u}{\partial y} = nx^n f\left(\dfrac{y}{x}\right) = nu.$

Note. Euler's theorem can be extended to a homogeneous function of any number of variables. Thus, if u is a homogeneous function of degree n in x, y and z, then $x\dfrac{\partial u}{\partial x} + y\dfrac{\partial u}{\partial y} + z\dfrac{\partial u}{\partial z} = nu.$

7.7. IF u IS A HOMOGENEOUS FUNCTION OF DEGREE n IN x AND y, THEN

$$\mathbf{x^2 \frac{\partial^2 u}{\partial x^2} + 2xy\frac{\partial^2 u}{\partial x \partial y} + y^2\frac{\partial^2 u}{\partial y^2} = n(n-1)u}$$

Since u is a homogeneous function of degree n in x and y

$\therefore$ By Euler's theorem, we have $\quad x\dfrac{\partial u}{\partial x} + y\dfrac{\partial u}{\partial y} = nu \qquad ...(1)$

Differentiating (1) partially, w.r.t. x, we have $\quad 1.\dfrac{\partial u}{\partial x} + x\dfrac{\partial^2 u}{\partial x^2} + y.\dfrac{\partial^2 u}{\partial x \partial y} = n\dfrac{\partial u}{\partial x} \qquad ...(2)$

Differentiating (1) partially, w.r.t. y, we have $\quad x\dfrac{\partial^2 u}{\partial y \partial x} + 1.\dfrac{\partial u}{\partial y} + y.\dfrac{\partial^2 u}{\partial y^2} = n.\dfrac{\partial u}{\partial y}$

But
$$\frac{\partial^2 u}{\partial y \partial x} = \frac{\partial^2 u}{\partial x \partial y}$$

$$\therefore \quad x\frac{\partial^2 u}{\partial x \partial y} + \frac{\partial u}{\partial y} + y\frac{\partial^2 u}{\partial y^2} = n\frac{\partial u}{\partial y} \qquad ...(3)$$

Multiplying (2) by x, (3) by y and adding,

$$x^2 \frac{\partial^2 u}{\partial x^2} + 2xy \frac{\partial^2 u}{\partial x \partial y} + y^2 \frac{\partial^2 u}{\partial y^2} + \left(x \frac{\partial u}{\partial x} + y \frac{\partial u}{\partial y} \right) = n \left(x \frac{\partial u}{\partial x} + y \frac{\partial u}{\partial y} \right)$$

or $$x^2 \frac{\partial^2 u}{\partial x^2} + 2xy \frac{\partial^2 u}{\partial x \partial y} + y^2 \frac{\partial^2 u}{\partial y^2} + nu = n \cdot nu \qquad \text{[Using (1)]}$$

$$\Rightarrow \quad x^2 \frac{\partial^2 u}{\partial x^2} + 2xy \frac{\partial^2 u}{\partial x \partial y} + y^2 \frac{\partial^2 u}{\partial y^2} = n^2 u - nu = n(n-1)u.$$

ILLUSTRATIVE EXAMPLES

Example 1. *Verify Euler's theorem for the functions :*

(*i*) $u = (x^{1/2} + y^{1/2})(x^n + y^n)$ (*ii*) $u = \sin^{-1} \frac{x}{y} + \tan^{-1} \frac{y}{x}$.

Sol. (*i*) $u = (x^{1/2} + y^{1/2})(x^n + y^n)$ ...(1)

$$= x^{1/2}\left(1 + \frac{y^{1/2}}{x^{1/2}}\right) x^n \left(1 + \frac{y^n}{x^n}\right) = x^{n+1/2}\left[1 + \left(\frac{y}{x}\right)^{1/2}\right]\left[1 + \left(\frac{y}{x}\right)^n\right] = x^{n+1/2} f\left(\frac{y}{x}\right)$$

[OR $f(tx, ty) = t^{n+1/2} f(x, y)$]

$\Rightarrow$ u is a homogeneous function of degree $(n + \frac{1}{2})$ in x and y

$\therefore$ By Euler's theorem, we should have $x \frac{\partial u}{\partial x} + y \frac{\partial u}{\partial y} = \left(n + \frac{1}{2}\right) u$ ...(2)

From (1), $$\frac{\partial u}{\partial x} = \frac{1}{2} x^{-1/2} (x^n + y^n) + nx^{n-1}(x^{1/2} + y^{1/2})$$

$$x \frac{\partial u}{\partial x} = \frac{1}{2} x^{1/2} (x^n + y^n) + nx^n(x^{1/2} + y^{1/2})$$

$$\frac{\partial u}{\partial y} = \frac{1}{2} y^{-1/2} (x^n + y^n) + ny^{n-1} (x^{1/2} + y^{1/2})$$

$$y \frac{\partial u}{\partial y} = \frac{1}{2} y^{1/2}(x^n + y^n) + ny^n(x^{1/2} + y^{1/2})$$

$$\therefore \quad x \frac{\partial u}{\partial x} + y \frac{\partial u}{\partial y} = \frac{1}{2}(x^{1/2} + y^{1/2})(x^n + y^n) + n(x^n + y^n)(x^{1/2} + y^{1/2})$$

$= \frac{1}{2} u + nu = (n + \frac{1}{2}) u$ which is the same as (2). Hence the verification.

(*ii*) $$u = \sin^{-1} \frac{x}{y} + \tan^{-1} \frac{y}{x} \qquad \text{...(1)}$$

$$= \operatorname{cosec}^{-1} \frac{y}{x} + \tan^{-1} \frac{y}{x} = x^0 f\left(\frac{y}{x}\right)$$

[OR $f(tx, ty) = f(x, y) = t^0 f(x, y)$]

$\Rightarrow$ u is a homogeneous function of degree 0 in x and y.

$\therefore$ By Euler's theorem, we should have $x \frac{\partial u}{\partial x} + y \frac{\partial u}{\partial y} = 0 \times u = 0$ ...(2)

From (1),
$$\frac{\partial u}{\partial x}=\frac{1}{\sqrt{1-\frac{x^2}{y^2}}}\cdot\frac{1}{y}+\frac{1}{1+\frac{y^2}{x^2}}\left(-\frac{y}{x^2}\right)=\frac{1}{\sqrt{y^2-x^2}}-\frac{y}{x^2+y^2}$$

$$x\frac{\partial u}{\partial x}=\frac{x}{\sqrt{y^2-x^2}}-\frac{xy}{x^2+y^2}$$

$$\frac{\partial u}{\partial y}=\frac{1}{\sqrt{1-\frac{x^2}{y^2}}}\left(-\frac{x}{y^2}\right)+\frac{1}{1+\frac{y^2}{x^2}}\cdot\frac{1}{x}=-\frac{x}{y\sqrt{y^2-x^2}}+\frac{x}{x^2+y^2}$$

$$y\frac{\partial u}{\partial y}=-\frac{x}{\sqrt{y^2-x^2}}+\frac{xy}{x^2+y^2}$$

$\therefore$ $x\dfrac{\partial u}{\partial x}+y\dfrac{\partial u}{\partial y}=0$ which is the same as (2). Hence the verification.

Example 2. *If* $u=\tan^{-1}\dfrac{x^3+y^3}{x-y}$, *prove that* $x\dfrac{\partial u}{\partial x}+y\dfrac{\partial u}{\partial y}=\sin 2u.$

Sol. Here u is not a homogeneous function but $\tan u=\dfrac{x^3+y^3}{x-y}=\dfrac{x^3\left[1+\left(\dfrac{y}{x}\right)^3\right]}{x\left[1-\dfrac{y}{x}\right]}=x^2 f\left(\dfrac{y}{x}\right)$

is a homogeneous function of degree 2 in x and y.

$\therefore$ By Euler's theorem, we have

$$x\frac{\partial}{\partial x}(\tan u)+y\frac{\partial}{\partial y}(\tan u)=2\tan u \quad\text{or}\quad x\sec^2 u\frac{\partial u}{\partial x}+y\sec^2 u\frac{\partial u}{\partial y}=2\tan u$$

or
$$x\frac{\partial u}{\partial x}+y\frac{\partial u}{\partial y}=\frac{2\sin u}{\cos u}\cdot\cos^2 u=2\sin u\cos u=\sin 2u.$$

Example 3. *If* $u=\sin^{-1}\left(\dfrac{x+2y+3z}{\sqrt{x^8+y^8+z^8}}\right)$, *show that* $x\dfrac{\partial u}{\partial x}+y\dfrac{\partial u}{\partial y}+z\dfrac{\partial u}{\partial z}+3\tan u=0.$

Sol. Here u is not a homogeneous function.

$$\sin u=f(x,y,z)=\frac{x+2y+3z}{\sqrt{x^8+y^8+z^8}}$$

$$f(tx,ty,tz)=\frac{t(x+2y+3z)}{t^4\sqrt{x^8+y^8+z^8}}=t^{-3}f(x,y,z)$$

$\Rightarrow$ $\sin u$ is a homogeneous function of degree -3 in x, y, z.

$\therefore$ By Euler's theorem, we have

$$x\frac{\partial}{\partial x}(\sin u)+y\frac{\partial}{\partial y}(\sin u)+z\frac{\partial}{\partial z}(\sin u)=-3\sin u$$

or
$$x\cos u\frac{\partial u}{\partial x}+y\cos u\frac{\partial u}{\partial y}+z\cos u\frac{\partial u}{\partial z}+3\sin u=0$$

or $$x\frac{\partial u}{\partial x}+y\frac{\partial u}{\partial y}+z\frac{\partial u}{\partial z}+3\tan u=0.$$

Example 4. *If* $u=\log\frac{x^4+y^4}{x+y}$, *show that* $x\frac{\partial u}{\partial x}+y\frac{\partial u}{\partial y}=3.$

Sol. Here u is not a homogeneous function

$\therefore$ $$u=\log_e\left(\frac{x^4+y^4}{x+y}\right)\Rightarrow e^u=\frac{x^4+y^4}{x+y}$$

which is a homogeneous function of degree 3 in x, y.

$\therefore$ By Euler's theorem, we have $x\frac{\partial}{\partial x}(e^u)+y\frac{\partial}{\partial y}(e^u)=3e^u$

or $$xe^u\frac{\partial u}{\partial x}+ye^u\frac{\partial u}{\partial y}=3e^u \text{ or } x\frac{\partial u}{\partial x}+y\frac{\partial u}{\partial y}=3.$$

Example 5. *If* $u=\sin^{-1}\frac{x+y}{\sqrt{x}+\sqrt{y}}$, *prove that* $x^2\frac{\partial^2 u}{\partial x^2}+2xy\frac{\partial^2 u}{\partial x\partial y}+y^2\frac{\partial^2 u}{\partial y^2}=-\frac{\sin u\cos 2u}{4\cos^3 u}$.

Sol. $u=\sin^{-1}\frac{x+y}{\sqrt{x}+\sqrt{y}}$ is not a homogeneous function but

$\sin u=\frac{x+y}{\sqrt{x}+\sqrt{y}}$ is a homogeneous function of degree $\frac{1}{2}$ in x and y.

$\therefore$ By Euler's theorem, we have

$$x\frac{\partial}{\partial x}(\sin u)+y\frac{\partial}{\partial y}(\sin u)=\frac{1}{2}\sin u$$

or $$x\cos u\frac{\partial u}{\partial x}+y\cos u\frac{\partial u}{\partial y}=\frac{1}{2}\sin u \text{ or } x\frac{\partial u}{\partial x}+y\frac{\partial u}{\partial y}=\frac{1}{2}\tan u \quad \text{...(1)}$$

Differentiating (1) partially w.r.t. x,

$$x\frac{\partial^2 u}{\partial x^2}+1.\frac{\partial u}{\partial x}+y\frac{\partial^2 u}{\partial x\partial y}=\frac{1}{2}\sec^2 u\frac{\partial u}{\partial x}$$

or $$x\frac{\partial^2 u}{\partial x^2}+y\frac{\partial^2 u}{\partial x\partial y}=\left(\frac{1}{2}\sec^2 u-1\right)\frac{\partial u}{\partial x} \quad \text{...(2)}$$

Differentiating (1) partially w.r.t. y,

$$x\frac{\partial^2 u}{\partial y\partial x}+y\frac{\partial^2 u}{\partial y^2}+1.\frac{\partial u}{\partial y}=\frac{1}{2}\sec^2 u\frac{\partial u}{\partial y}$$

or $$x\frac{\partial^2 u}{\partial x\partial y}+y\frac{\partial^2 u}{\partial y^2}=\left(\frac{1}{2}\sec^2 u-1\right)\frac{\partial u}{\partial y} \quad \text{...(3)} \quad \left[\because \frac{\partial^2 u}{\partial y\partial x}=\frac{\partial^2 u}{\partial x\partial y}\right]$$

Multiplying (2) by x, (3) by y and adding

$$x^2 \frac{\partial^2 u}{\partial x^2} + 2xy \frac{\partial^2 u}{\partial x \partial y} + y^2 \frac{\partial^2 u}{\partial y^2} = \left(\frac{1}{2} \sec^2 u - 1\right)\left(x \frac{\partial u}{\partial x} + y \frac{\partial u}{\partial y}\right)$$

$$= \left(\frac{1}{2 \cos^2 u} - 1\right) \cdot \frac{1}{2} \tan u \qquad \text{[Using (1)]}$$

$$= -\frac{2 \cos^2 u - 1}{2 \cos^2 u} \cdot \frac{1}{2} \frac{\sin u}{\cos u} = -\frac{\sin u \cos 2u}{4 \cos^3 u}. \qquad [\because\ 2 \cos^2 u - 1 = \cos 2u]$$

Example 6. *If* $z = xf\left(\frac{y}{x}\right) + g\left(\frac{y}{x}\right)$, *show that* $x^2 \frac{\partial^2 z}{\partial x^2} + 2xy \frac{\partial^2 z}{\partial x \partial y} + y^2 \frac{\partial^2 z}{\partial y^2} = 0.$

[CUSAT 2008]

Sol. Let $\qquad u = xf\left(\frac{y}{x}\right)$ and $v = g\left(\frac{y}{x}\right) + x^0 g\left(\frac{y}{x}\right)$

so that $\qquad z = u + v \qquad$ ...(1)

Since u is a homogeneous function of degree $\qquad n = 1$ in x, y

$$\therefore \qquad x^2 \frac{\partial^2 u}{\partial x^2} + 2xy \frac{\partial^2 u}{\partial x \partial y} + y^2 \frac{\partial^2 u}{\partial y^2} = n(n-1)u = 0 \qquad \text{...(2)}$$

Since v is a homogeneous function of degree $\qquad n = 0$ in x, y

$$\therefore \qquad x^2 \frac{\partial^2 v}{\partial x^2} + 2xy \frac{\partial^2 v}{\partial x \partial y} + y^2 \frac{\partial^2 v}{\partial y^2} = n(n-1)v = 0 \qquad \text{...(3)}$$

Adding (2) and (3), we have

$$x^2 \frac{\partial^2}{\partial x^2}(u+v) + 2xy \frac{\partial^2}{\partial x \partial y}(u+v) + y^2 \frac{\partial^2}{\partial y^2}(u+v) = 0$$

or

$$x^2 \frac{\partial^2 z}{\partial x^2} + 2xy \frac{\partial^2 z}{\partial x \partial y} + y^2 \frac{\partial^2 z}{\partial y^2} = 0. \qquad \text{[Using (1)]}$$

TEST YOUR KNOWLEDGE

1. Verify Euler's theorem for the functions :

   (*i*) $f(x, y) = ax^2 + 2hxy + by^2$ $\qquad$ (*ii*) $u = \dfrac{x^{1/4} + y^{1/4}}{x^{1/5} + y^{1/5}}$ $\qquad$ (*iii*) $f(x, y) = \dfrac{x^2 (x^2 - y^2)^3}{(x^2 + y^2)^3}$.

2. (*i*) If $u = f\left(\frac{y}{x}\right)$, show that $x \frac{\partial u}{\partial x} + y \frac{\partial u}{\partial y} = 0$ (*ii*) If $u = xf\left(\frac{y}{x}\right)$, prove that $x \frac{\partial u}{\partial x} + y \frac{\partial u}{\partial y} = u.$

3. If $\mathrm{V} = \dfrac{x^3 y^3}{x^3 + y^3}$, show that $x \frac{\partial \mathrm{V}}{\partial x} + y \frac{\partial \mathrm{V}}{\partial y} = 3\mathrm{V}.$

4. If $u = \sin^{-1} \dfrac{x}{y} + \tan^{-1} \dfrac{y}{x}$, then find the value of $x \dfrac{\partial u}{\partial x} + y \dfrac{\partial u}{\partial y}$.

5. If $f(x, y) = \sqrt{x^2 - y^2}\ \sin^{-1}\dfrac{y}{x}$, prove that $x\dfrac{\partial f}{\partial x} + y\dfrac{\partial f}{\partial y} = f(x, y)$.

6. If $f(x, y) = \sqrt{y^2 - x^2}\ \sin^{-1}\dfrac{x}{y} + \dfrac{x^2 - y^2}{\sqrt{x^2 + y^2}}$, show that $x\dfrac{\partial f}{\partial x} + y\dfrac{\partial f}{\partial y} - f(x, y) = 0$.

7. If $f(x, y) = \dfrac{1}{x^2} + \dfrac{1}{xy} + \dfrac{\log x - \log y}{x^2 + y^2}$, show that $x\dfrac{\partial f}{\partial x} + y\dfrac{\partial f}{\partial y} + 2f(x, y) = 0$.

8. If $u = \cos\left(\dfrac{xy + yz + zx}{x^2 + y^2 + z^2}\right)$, prove that $x\dfrac{\partial u}{\partial x} + y\dfrac{\partial u}{\partial y} + z\dfrac{\partial u}{\partial z} = 0$.

9. If $u = \cos^{-1}\left(\dfrac{x + y}{\sqrt{x} + \sqrt{y}}\right)$, show that $x\dfrac{\partial u}{\partial x} + y\dfrac{\partial u}{\partial y} + \dfrac{1}{2}\cot u = 0$.

10. If $u = \sin^{-1}\left(\dfrac{x^2 + y^2}{x + y}\right)$, prove that $x\dfrac{\partial u}{\partial x} + y\dfrac{\partial u}{\partial y} = \tan u$.

11. (*a*) If $u = \sin^{-1}\dfrac{\sqrt{x} - \sqrt{y}}{\sqrt{x} + \sqrt{y}}$, show that $\dfrac{\partial u}{\partial x} = -\dfrac{y}{x}\dfrac{\partial u}{\partial y}$.

(*b*) If $\sin u = \dfrac{x^2y^2}{x + y}$, show that $x\dfrac{\partial u}{\partial x} + y\dfrac{\partial u}{\partial y} = 3\tan u$.

12. (*a*) If $u = \log\left(\dfrac{x^5 + y^5 + z^5}{x^2 + y^2 + z^2}\right)$, show that $x\dfrac{\partial u}{\partial x} + y\dfrac{\partial u}{\partial y} + z\dfrac{\partial u}{\partial z} = 3u$.

(*b*) Show that $xu_x + yu_y + zu_z = 2\tan u$, where $u = \sin^{-1}\left(\dfrac{x^3 + y^3 + z^3}{ax + by + cz}\right)$.

13. If $u = \dfrac{x^2y^2}{x + y}$, show that

(*i*) $x\dfrac{\partial^2 u}{\partial x^2} + y\dfrac{\partial^2 u}{\partial x \partial y} = 2\dfrac{\partial u}{\partial x}$ (*ii*) $x\dfrac{\partial^2 u}{\partial x \partial y} + y\dfrac{\partial^2 u}{\partial y^2} = 2\dfrac{\partial u}{\partial y}$.

14. Given $z = x^n f_1\left(\dfrac{y}{x}\right) + y^{-n} f_2\left(\dfrac{x}{y}\right)$, prove that $x^2\dfrac{\partial^2 z}{\partial x^2} + 2xy\dfrac{\partial^2 z}{\partial x \partial y} + y^2\dfrac{\partial^2 z}{\partial y^2} + x\dfrac{\partial z}{\partial x} + y\dfrac{\partial z}{\partial y} = n^2 z$.

15. If $u = (x^2 + y^2)^{1/3}$, show that $x^2\dfrac{\partial^2 u}{\partial x^2} + 2xy\dfrac{\partial^2 u}{\partial x \partial y} + y^2\dfrac{\partial^2 u}{\partial y^2} = -\dfrac{2u}{9}$.

16. If $u = \tan^{-1}\dfrac{x^3 + y^3}{x - y}$, prove that $x^2\dfrac{\partial^2 u}{\partial x^2} + 2xy\dfrac{\partial^2 u}{\partial x \partial y} + y^2\dfrac{\partial^2 u}{\partial y^2} = \sin 4u - \sin 2u = 2\cos 3u \sin u$.

17. If $u = \tan^{-1}\left(\dfrac{y^2}{x}\right)$, show that $x^2\dfrac{\partial^2 u}{\partial x^2} + 2xy\dfrac{\partial^2 u}{\partial x \partial y} + y^2\dfrac{\partial^2 u}{\partial y^2} = -\sin 2u \sin^2 u$.

18. If $u = x^2\tan^{-1}\left(\dfrac{y}{x}\right) - y^2\tan^{-1}\left(\dfrac{x}{y}\right)$, evaluate $x^2\dfrac{\partial^2 u}{\partial x^2} + 2xy\dfrac{\partial^2 u}{\partial x \partial y} + y^2\dfrac{\partial^2 u}{\partial y^2}$.

19. If $u = \operatorname{cosec}^{-1}\left(\dfrac{x^{1/2} + y^{1/2}}{x^{1/3} + y^{1/3}}\right)^{1/2}$, prove that $x^2\dfrac{\partial^2 u}{\partial x^2} + 2xy\dfrac{\partial^2 u}{\partial x \partial y} + y^2\dfrac{\partial^2 u}{\partial y^2} = \dfrac{\tan u}{144}(13 + \tan^2 u)$.

20. If $u = \sin^{-1}\left[\dfrac{x^3 + y^3}{\sqrt{x} + \sqrt{y}}\right]$, prove that $x\dfrac{\partial u}{\partial x} + y\dfrac{\partial u}{\partial y} = \dfrac{5}{2}\tan u$. [*CUSAT 2004*]

Answers

4. 0 **18.** $2u$.

7.8. COMPOSITE FUNCTIONS

(*i*) If $u = f(x, y)$ where $x = \phi(t)$, $y = \psi(t)$

then u is called a composite function of (the **single variable**) t and we can find $\frac{du}{dt}$.

(*ii*) If $z = f(x, y)$ where $x = \phi(u, v)$, $y = \psi(u, v)$

then z is called a composite function of (**two variables**) u and v so that we can find $\frac{\partial z}{\partial u}$ and $\frac{\partial z}{\partial v}$.

7.9. DIFFERENTIATION OF COMPOSITE FUNCTIONS

If u is composite function of t, defined by the relations $u = f(x, y)$; $x = \phi(t)$, $y = \psi(t)$, then

$$\frac{du}{dt} = \frac{\partial u}{\partial x} \cdot \frac{dx}{dt} + \frac{\partial u}{\partial y} \cdot \frac{dy}{dt}.$$

Proof. Here $\qquad u = f(x, y) \qquad \ldots(1)$

Let δt be an increment in t and δx, δy, δu the corresponding increments in x, y and u respectively. Then, we have

$$u + \delta u = f(x + \delta x, y + \delta y) \qquad \ldots(2)$$

Subtracting (1) from (2), we get

$$\delta u = f(x + \delta x, y + \delta y) - f(x, y)$$

$$= f(x + \delta x, y + \delta y) - f(x, y + \delta y) + f(x, y + \delta y) - f(x, y)$$

$$\frac{\delta u}{\delta t} = \frac{f(x + \delta x, y + \delta y) - f(x, y + \delta y)}{\delta t} + \frac{f(x, y + \delta y) - f(x, y)}{\delta t}$$

$$= \frac{f(x + \delta x, y + \delta y) - f(x, y + \delta y)}{\delta x} \cdot \frac{\delta x}{\delta t} + \frac{f(x, y + \delta y) - f(x, y)}{\delta y} \cdot \frac{\delta y}{\delta t} \qquad \ldots(3)$$

As $\delta t \to 0$, δx and δy both $\to 0$, so that

$$\lim_{\delta t \to 0} \frac{\delta u}{\delta t} = \frac{du}{dt}, \quad \lim_{\delta t \to 0} \frac{\delta x}{\delta t} = \frac{dx}{dt}, \quad \lim_{\delta t \to 0} \frac{\delta y}{\delta t} = \frac{dy}{dt}$$

$$\lim_{\delta x \to 0} \frac{f(x + \delta x, y + \delta y) - f(x, y + \delta y)}{\delta y} = \frac{\partial f}{\partial x} = \frac{\partial u}{\partial x}$$

$$\lim_{\delta y \to 0} \frac{f(x, y + \delta y) - f(x, y)}{\delta y} = \frac{\partial f}{\partial y} = \frac{\partial u}{\partial y}$$

$\therefore$ From (1), $$\frac{du}{dt} = \frac{\partial u}{\partial x} \cdot \frac{dx}{dt} + \frac{\partial u}{\partial y} \cdot \frac{dy}{dt}$$

$\frac{du}{dt}$ is called the **total derivative** of u to distinguish it from the partial derivatives $\frac{\partial u}{\partial x}$ and $\frac{\partial u}{\partial y}$.

Cor. 1. If $u = f(x, y, z)$ and x, y, z are function of t, then y is a composite function of t and

$$\frac{du}{dt} = \frac{\partial u}{\partial x}\cdot\frac{dx}{dt} + \frac{\partial u}{\partial y}\cdot\frac{dy}{dt} + \frac{\partial u}{\partial z}\cdot\frac{dz}{dt}.$$

Cor. 2. If $z = f(x, y)$ and x, y are functions of u and v, then

$$\frac{\partial z}{\partial u} = \frac{\partial z}{\partial x}\cdot\frac{\partial x}{\partial u} + \frac{\partial z}{\partial y}\cdot\frac{\partial y}{\partial u}; \quad \frac{\partial z}{\partial v} = \frac{\partial z}{\partial x}\cdot\frac{\partial x}{\partial v} + \frac{\partial z}{\partial y}\cdot\frac{\partial y}{\partial v}.$$

Cor. 3. If $u = f(x, y)$ where $y = \phi(x)$ then since $x = \psi(x)$, u is a composite function of x.

$$\therefore \quad \frac{du}{dx} = \frac{\partial u}{\partial x}\cdot\frac{dx}{dx} + \frac{\partial u}{\partial y}\cdot\frac{dy}{dx} \quad \Rightarrow \quad \frac{du}{dx} = \frac{\partial u}{\partial x} + \frac{\partial u}{\partial y}\cdot\frac{dy}{dx}.$$

Cor. 4. If we are given an implicit function $f(x, y) = c$, then $u = f(x, y)$ where $u = c$

Using Cor. 3, we have $\frac{du}{dx} = \frac{\partial u}{\partial x} + \frac{\partial u}{\partial y}\cdot\frac{dy}{dx}$

But $\frac{du}{dx} = 0 \quad \therefore \quad \frac{\partial u}{\partial x} + \frac{\partial u}{\partial y}\cdot\frac{dy}{dx} = 0$ or $\frac{\partial f}{\partial x} + \frac{\partial f}{\partial y}\cdot\frac{dy}{dx} = 0$

$$\therefore \quad \frac{dy}{dx} = -\frac{\frac{\partial f}{\partial x}}{\frac{\partial f}{\partial y}} = -\frac{f_x}{f_y}.$$

Hence the differential coefficient of $f(x, y)$ w.r.t. x is $\frac{df}{dx} = \frac{\partial f}{\partial x} + \frac{\partial f}{\partial y}\cdot\frac{dy}{dx}$.

Cor. 5. If $f(x, y) = c$, then by Cor. 4, we have $\frac{dy}{dx} = -\frac{f_x}{f_y}$

Differentiating again w.r.t. x, we get

$$\frac{d^2y}{dx^2} = -\frac{f_y\frac{d}{dx}(f_x) - f_x\frac{d}{dx}(f_y)}{f_y^{\,2}} = -\frac{f_y\left[\frac{\partial f_x}{\partial x} + \frac{\partial f_x}{\partial y}\cdot\frac{dy}{dx}\right] - f_x\left[\frac{\partial f_y}{\partial x} + \frac{\partial f_y}{\partial y}\cdot\frac{dy}{dx}\right]}{f_y^{\,2}}$$

$$= -\frac{f_y\left[f_{xx} - f_{yx}\cdot\frac{f_x}{f_y}\right] - f_x\left[f_{xy} - f_{yy}\cdot\frac{f_x}{f_y}\right]}{f_y^{\,2}} = -\frac{f_{xx}f_y^{\,2} - f_xf_yf_{xy} - f_xf_yf_{xy} + f_{yy}f_x^{\,2}}{f_y^{\,3}}$$

Hence $\quad \mathbf{\frac{d^2y}{dx^2} = -\frac{f_{xx}f_y^{\,2} - 2f_xf_yf_{xy} + f_{yy}f_x^{\,2}}{f_y^{\,3}}}$.

ILLUSTRATIVE EXAMPLES

Example 1. *If* $u = \sin^{-1}(x - y)$, $x = 3t$, $y = 4t^3$, *show that* $\dfrac{du}{dt} = \dfrac{3}{\sqrt{1-t^2}}$.

Sol. The given equations define u as a composite function of t.

$$\therefore \quad \frac{du}{dt} = \frac{\partial u}{\partial x}\cdot\frac{dx}{dt} + \frac{\partial u}{\partial y}\cdot\frac{dy}{dt}$$

$$= \frac{1}{\sqrt{1-(x-y)^2}}\,.\,3 + \frac{1}{\sqrt{1-(x-y)^2}}(-1)\,.\,12t^2$$

$$= \frac{3(1-4t^2)}{\sqrt{1-(x-y)^2}} = \frac{3(1-4t^2)}{\sqrt{1-(3t-4t^3)^2}} = \frac{3(1-4t^2)}{\sqrt{1-9t^2+24t^4-16t^6}}$$

$$= \frac{3(1-4t^2)}{\sqrt{(1-t^2)(1-8t^2+16t^4)}} = \frac{3(1-4t^2)}{\sqrt{(1-t^2)(1-4t^2)^2}} = \frac{3}{\sqrt{1-t^2}}.$$

Example 2. *If z is a function of x and y, where* $x = e^u + e^{-v}$ *and* $y = e^{-u} - e^v$, *show that*

$$\frac{\partial z}{\partial u} - \frac{\partial z}{\partial v} = x\frac{\partial z}{\partial x} - y\frac{\partial z}{\partial y}.$$

Sol. Here z is a composite function of u and v.

$$\therefore \quad \frac{\partial z}{\partial u} = \frac{\partial z}{\partial x}\cdot\frac{\partial x}{\partial u} + \frac{\partial z}{\partial y}\cdot\frac{\partial y}{\partial u} = \frac{\partial z}{\partial x}\cdot e^u + \frac{\partial z}{\partial y}(-e^{-u})$$

and
$$\frac{\partial z}{\partial v} = \frac{\partial z}{\partial x}\cdot\frac{\partial x}{\partial v} + \frac{\partial z}{\partial y}\cdot\frac{\partial y}{\partial v} = \frac{\partial z}{\partial x}(-e^{-v}) + \frac{\partial z}{\partial y}(-e^v)$$

Subtracting,
$$\frac{\partial z}{\partial u} - \frac{\partial z}{\partial v} = (e^u + e^{-v})\frac{\partial z}{\partial x} - (e^{-u} - e^v)\frac{\partial z}{\partial y} = x\frac{\partial z}{\partial x} - y\frac{\partial z}{\partial y}.$$

Example 3. *If* $u = f(y - z, z - x, x - y)$, *prove that* $\dfrac{\partial u}{\partial x} + \dfrac{\partial u}{\partial y} + \dfrac{\partial u}{\partial z} = 0.$ [CUSAT 2010]

Sol. Here $u = f(X, Y, Z)$ where $X = y - z$, $Y = z - x$, $Z = x - y$

$\therefore$ u is a composite function of x, y and z.

$$\frac{\partial u}{\partial x} = \frac{\partial u}{\partial X}\cdot\frac{\partial X}{\partial x} + \frac{\partial u}{\partial Y}\cdot\frac{\partial Y}{\partial x} + \frac{\partial u}{\partial Z}\cdot\frac{\partial Z}{\partial x} = \frac{\partial u}{\partial X}(0) + \frac{\partial u}{\partial Y}(-1) + \frac{\partial u}{\partial Z}(1)$$

$$\frac{\partial u}{\partial y} = \frac{\partial u}{\partial X}\cdot\frac{\partial X}{\partial y} + \frac{\partial u}{\partial Y}\cdot\frac{\partial Y}{\partial y} + \frac{\partial u}{\partial Z}\cdot\frac{\partial Z}{\partial y} = \frac{\partial u}{\partial X}(1) + \frac{\partial u}{\partial Y}(0) + \frac{\partial u}{\partial Z}(-1)$$

$$\frac{\partial u}{\partial z} = \frac{\partial u}{\partial X}\cdot\frac{\partial X}{\partial z} + \frac{\partial u}{\partial Y}\cdot\frac{\partial Y}{\partial z} + \frac{\partial u}{\partial Z}\cdot\frac{\partial Z}{\partial z} = \frac{\partial u}{\partial X}(-1) + \frac{\partial u}{\partial Y}(1) + \frac{\partial u}{\partial Z}(0)$$

Adding, $\dfrac{\partial u}{\partial x} + \dfrac{\partial u}{\partial y} + \dfrac{\partial u}{\partial z} = 0.$

Example 4. *If* $w = f(x, y)$, $x = r\cos\theta$, $y = r\sin\theta$, *show that*

$$\left(\frac{\partial w}{\partial r}\right)^2 + \frac{1}{r^2}\left(\frac{\partial w}{\partial \theta}\right)^2 = \left(\frac{\partial f}{\partial x}\right)^2 + \left(\frac{\partial f}{\partial y}\right)^2.$$

Sol. The given equations define w as a composite function of r and θ.

$$\frac{\partial w}{\partial r}=\frac{\partial w}{\partial x}\cdot\frac{\partial x}{\partial r}+\frac{\partial w}{\partial y}\cdot\frac{\partial y}{\partial r}=\frac{\partial w}{\partial x}\cdot\cos\theta+\frac{\partial w}{\partial y}\cdot\sin\theta$$

or $$\frac{\partial w}{\partial r}=\frac{\partial f}{\partial x}\cos\theta+\frac{\partial f}{\partial y}\sin\theta \quad \ldots(1) \quad [\because\ w=f(x,y)]$$

Also $$\frac{\partial w}{\partial \theta}=\frac{\partial w}{\partial x}\cdot\frac{\partial x}{\partial \theta}+\frac{\partial w}{\partial y}\cdot\frac{\partial y}{\partial \theta}=\frac{\partial w}{\partial x}(-r\sin\theta)+\frac{\partial w}{\partial y}(r\cos\theta)$$

or $$\frac{1}{r}\frac{\partial w}{\partial \theta}=-\frac{\partial f}{\partial x}\sin\theta+\frac{\partial f}{\partial y}\cos\theta \quad \ldots(2)$$

Squaring and adding (1) and (2), we get

$$\left(\frac{\partial w}{\partial r}\right)^2+\frac{1}{r^2}\left(\frac{\partial w}{\partial \theta}\right)^2=\left(\frac{\partial f}{\partial x}\right)^2+\left(\frac{\partial f}{\partial y}\right)^2.$$

Example 5. *If* $z=\sqrt{x^2+y^2}$ *and* $x^3+y^3+3axy=5a^2$, *find the value of* $\frac{dz}{dx}$, *when* $x=y=a$.

Sol. The given equations are of the form $z=f(x,y)$ and $\phi(x,y)=c$

$\therefore$ z is composite function of x.

$$\Rightarrow \quad \frac{dz}{dx}=\frac{\partial z}{\partial x}\cdot\frac{dx}{dx}+\frac{\partial z}{\partial y}\cdot\frac{dy}{dx}=\frac{\partial z}{\partial x}+\frac{\partial z}{\partial y}\cdot\frac{dy}{dx} \quad \ldots(1)$$

Now, $$\frac{\partial z}{\partial x}=\frac{1}{2}(x^2+y^2)^{-1/2}\cdot 2x=\frac{x}{\sqrt{x^2+y^2}}$$

Similarly, $$\frac{\partial z}{\partial y}=\frac{y}{\sqrt{x^2+y^2}}$$

Also, differentiating $x^3+y^3+3axy=5a^2$ w.r.t. x, we have

$$3x^2+3y^2\cdot\frac{dy}{dx}+3ay+3ax\cdot\frac{dy}{dx}=0 \quad \text{or} \quad (y^2+ax)\frac{dy}{dx}=-(x^2+ay)$$

$$\therefore \quad \frac{dy}{dx}=-\frac{x^2+ay}{y^2+ax}$$

$\therefore$ From (1), $$\frac{dz}{dx}=\frac{x}{\sqrt{x^2+y^2}}+\frac{y}{\sqrt{x^2+y^2}}\left(-\frac{x^2+ay}{y^2+ax}\right)$$

$$\left[\frac{dz}{dx}\right]_{\substack{x=a\\y=a}}=\frac{a}{\sqrt{a^2+a^2}}+\frac{a}{\sqrt{a^2+a^2}}\cdot\left(-\frac{a^2+a^2}{a^2+a^2}\right)=0.$$

Example 6. *If* $u=xe^yz$, *where* $y=\sqrt{a^2-x^2}$, $z=\sin^2 x$, *find* $\frac{du}{dx}$.

Sol. Here u is a function of x, y and z while y and z are functions of x.

$$\therefore \quad \frac{du}{dx}=\frac{\partial u}{dx}\cdot\frac{dx}{dx}+\frac{\partial u}{\partial y}\cdot\frac{dy}{dx}+\frac{\partial u}{\partial z}\cdot\frac{dz}{dx}$$

$$=e^yz\cdot 1+xe^yz\cdot\tfrac{1}{2}(a^2-x^2)^{-1/2}(-2x)+xe^y\cdot 2\sin x\cos x$$

$$= e^y \left[z - \frac{x^2 z}{\sqrt{a^2 - x^2}} + x \sin 2x \right].$$

Example 7. *Find* $\dfrac{du}{dx}$ *if* $u = \sin (x^2 + y^2)$, *where* $a^2x^2 + b^2y^2 = c^2$.

Sol. The given equations are of the form $u = f(x, y)$ and $\phi(x, y) = k$

$\therefore$ u is a composite function of x.

$$\frac{du}{dx} = \frac{\partial u}{\partial x} \cdot \frac{dx}{dx} + \frac{\partial u}{\partial y} \cdot \frac{dy}{dx} = \frac{\partial u}{\partial x} + \frac{\partial u}{\partial y} \cdot \frac{dy}{dx} \qquad \ldots(1)$$

Now, $\dfrac{\partial u}{\partial x} = 2x \cos (x^2 + y^2)$, $\dfrac{\partial u}{\partial y} = 2y \cos (x^2 + y^2)$

Also, differentiating $a^2x^2 + b^2y^2 = c^2$ w.r.t. x, we have

$$2a^2x + 2b^2y \cdot \frac{dy}{dx} = 0 \quad \text{or} \quad \frac{dy}{dx} = -\frac{a^2x}{b^2y}$$

$\therefore$ From (1),

$$\frac{du}{dx} = 2x \cos (x^2 + y^2) + 2y \cos (x^2 + y^2) \cdot \left[-\frac{a^2x}{b^2y} \right]$$

$$= 2\left[x - \frac{a^2x}{b^2} \right] \cos (x^2 + y^2) = \frac{2(b^2 - a^2)x}{b^2} \cdot \cos (x^2 + y^2).$$

Example 8. *Find* $\dfrac{dy}{dx}$, *when*

(*i*) $x^y + y^x = c$ (*ii*) $(\cos x)^y = (\sin y)^x$.

Sol. (*i*) Let $f(x, y) = x^y + y^x$, then $f(x, y) = c$

[Using Cor. 4] $\therefore$ $\dfrac{dy}{dx} = -\dfrac{f_x}{f_y} \Rightarrow \dfrac{dy}{dx} = -\dfrac{yx^{y-1} + y^x \log y}{x^y \log x + xy^{x-1}}$.

(*ii*) Let $f(x, y) = (\cos x)^y - (\sin y)^x = 0$

$$\frac{dy}{dx} = -\frac{f_x}{f_y} = -\frac{y(\cos x)^{y-1} \cdot (-\sin x) - (\sin y)^x \log (\sin y)}{(\cos x)^y \log (\cos x) - x (\sin y)^{x-1} \cdot \cos y}$$

$$= \frac{y(\cos x)^{y-1} \sin x + (\cos x)^y \log (\sin y)}{(\cos x)^y \log (\cos x) - x (\cos x)^y (\sin y)^{-1} \cos y}$$

$[\because \ (\sin y)^x = (\cos x)^y]$

$$= \frac{(\cos x)^y \left[y \cdot \dfrac{\sin x}{\cos x} + \log \sin y \right]}{(\cos x)^y [\log \cos x - x \cot y]} = \frac{y \tan x + \log \sin y}{\log \cos x - x \cot y}.$$

Example 9. *If* $f(x, y) = 0$, $\phi (y, z) = 0$, *show that* $\dfrac{\partial f}{\partial y} \cdot \dfrac{\partial \phi}{\partial z} \cdot \dfrac{dz}{dx} = \dfrac{\partial f}{\partial x} \cdot \dfrac{\partial \phi}{\partial y}$.

Sol.

$$\frac{dz}{dx} = \frac{dz}{dy} \cdot \frac{dy}{dx} \qquad \ldots(1)$$

$f(x, y) = 0$ gives $\quad \dfrac{dy}{dx} = -\dfrac{\dfrac{\partial f}{\partial x}}{\dfrac{\partial f}{\partial y}}$; $\quad \phi(y, z) = 0$ gives $\dfrac{dz}{dy} = -\dfrac{\dfrac{\partial \phi}{\partial y}}{\dfrac{\partial \phi}{\partial z}}$

$\therefore$ From (1), $\quad \dfrac{dz}{dx} = \dfrac{\dfrac{\partial f}{\partial x} \cdot \dfrac{\partial \phi}{\partial y}}{\dfrac{\partial f}{\partial y} \cdot \dfrac{\partial \phi}{\partial z}} \quad \Rightarrow \quad \dfrac{\partial f}{\partial y} \cdot \dfrac{\partial \phi}{\partial z} \cdot \dfrac{dz}{dx} = \dfrac{\partial f}{\partial x} \cdot \dfrac{\partial \phi}{\partial y}$.

Example 10. *If $\phi(x, y, z) = 0$, show that* $\left(\dfrac{\partial y}{\partial z}\right)_x \left(\dfrac{\partial z}{\partial x}\right)_y \left(\dfrac{\partial x}{\partial y}\right)_z = -1.$

Sol. The given relation defines y as a function of x and z. Treating x as constant

$$\left(\frac{\partial y}{\partial z}\right)_x = -\frac{\dfrac{\partial \phi}{\partial z}}{\dfrac{\partial \phi}{\partial y}}$$

The given relation defines z as a function of x and y. Treating y as constant $\left(\dfrac{\partial z}{\partial x}\right)_y = -\dfrac{\dfrac{\partial \phi}{\partial x}}{\dfrac{\partial \phi}{\partial z}}$

Similarly, $\quad \left(\dfrac{\partial x}{\partial y}\right)_z = -\dfrac{\dfrac{\partial \phi}{\partial y}}{\dfrac{\partial \phi}{\partial x}}$

Multiplying, we get the desired result.

Example 11. *Prove that* $\dfrac{\partial^2 z}{\partial x^2} + \dfrac{\partial^2 z}{\partial y^2} = \dfrac{\partial^2 z}{\partial u^2} + \dfrac{\partial^2 z}{\partial v^2}$ *where* $x = u \cos \alpha - v \sin \alpha$, $y = u \sin \alpha + v \cos \alpha$

Or

By changing the independent variables u and v to x and y by means of the relations $x = u \cos \alpha - v \sin \alpha$, $y = u \sin \alpha + v \cos \alpha$, *show that* $\dfrac{\partial^2 z}{\partial u^2} + \dfrac{\partial^2 z}{\partial v^2}$ *transforms into* $\dfrac{\partial^2 z}{\partial x^2} + \dfrac{\partial^2 z}{\partial y^2}$.

Sol. Here z is a composite function of u and v

$$\frac{\partial z}{\partial u} = \frac{\partial z}{\partial x} \cdot \frac{\partial x}{\partial u} + \frac{\partial z}{\partial y} \cdot \frac{\partial y}{\partial u} = \cos \alpha \frac{\partial z}{\partial x} + \sin \alpha \frac{\partial z}{\partial y}$$

or $\quad \dfrac{\partial}{\partial u}(z) = \left(\cos \alpha \dfrac{\partial}{\partial x} + \sin \alpha \dfrac{\partial}{\partial y}\right) z \Rightarrow \dfrac{\partial}{\partial u} \equiv \cos \alpha \dfrac{\partial}{\partial x} + \sin \alpha \dfrac{\partial}{\partial y} \quad$ ...(1)

Also $\quad \dfrac{\partial z}{\partial v} = \dfrac{\partial z}{\partial x} \cdot \dfrac{\partial x}{\partial v} + \dfrac{\partial z}{\partial y} \cdot \dfrac{\partial y}{\partial v} = -\sin \alpha \dfrac{\partial z}{\partial x} + \cos \alpha \dfrac{\partial z}{\partial y}$

or $\quad \dfrac{\partial}{\partial v}(z) = \left(-\sin \alpha \dfrac{\partial}{\partial x} + \cos \alpha \dfrac{\partial}{\partial y}\right) z$

$\Rightarrow \quad \dfrac{\partial}{\partial v} \equiv -\sin \alpha \dfrac{\partial}{\partial x} + \cos \alpha \dfrac{\partial}{\partial y} \quad$ ...(2)

Now, we shall make use of the equivalence of operators as given by (1) and (2).

$$\frac{\partial^2 z}{\partial u^2} = \frac{\partial}{\partial u}\left(\frac{\partial z}{\partial u}\right) = \left(\cos\alpha\frac{\partial}{\partial x} + \sin\alpha\frac{\partial}{\partial y}\right)\left(\cos\alpha\frac{\partial z}{\partial x} + \sin\alpha\frac{\partial z}{\partial y}\right)$$

$$= \cos^2\alpha\frac{\partial^2 z}{\partial x^2} + \cos\alpha\sin\alpha\frac{\partial^2 z}{\partial x\partial y} + \sin\alpha\cos\alpha\frac{\partial^2 z}{\partial y\partial x} + \sin^2\alpha\frac{\partial^2 z}{\partial y^2}$$

$$= \cos^2\alpha\frac{\partial^2 z}{\partial x^2} + 2\cos\alpha\sin\alpha\frac{\partial^2 z}{\partial x\partial y} + \sin^2\alpha\frac{\partial^2 z}{\partial y^2} \qquad ...(3)$$

$$\frac{\partial^2 z}{\partial v^2} = \frac{\partial}{\partial v}\left(\frac{\partial z}{\partial v}\right) = \left(-\sin\alpha\frac{\partial}{\partial x} + \cos\alpha\frac{\partial}{\partial y}\right)\left(-\sin\alpha\frac{\partial z}{\partial x} + \cos\alpha\frac{\partial z}{\partial y}\right)$$

$$= \sin^2\alpha\frac{\partial^2 z}{\partial x^2} - \sin\alpha\cos\alpha\frac{\partial^2 z}{\partial x\partial y} - \cos\alpha\sin\alpha\frac{\partial^2 z}{\partial y\partial x} + \cos^2\alpha\frac{\partial^2 z}{\partial y^2}$$

$$= \sin^2\alpha\frac{\partial^2 z}{\partial x^2} - 2\cos\alpha\sin\alpha\frac{\partial^2 z}{\partial x\partial y} + \cos^2\alpha\frac{\partial^2 z}{\partial y^2} \qquad ...(4)$$

Adding (3) and (4), $\dfrac{\partial^2 z}{\partial u^2} + \dfrac{\partial^2 z}{\partial v^2} = \dfrac{\partial^2 z}{\partial x^2} + \dfrac{\partial^2 z}{\partial y^2}$.

Example 12. *Transform the equation* $\dfrac{\partial^2 u}{\partial x^2} + \dfrac{\partial^2 u}{\partial y^2} = 0$ *into polar coordinates.*

Sol. The relations connecting cartesian coordinates, (x, y) with polar coordinates (r, θ) are

$$x = r\cos\theta, \quad y = r\sin\theta$$

Squaring and adding, $r^2 = x^2 + y^2$

Dividing, $\tan\theta = \dfrac{y}{x}$

$$\therefore \quad r = \sqrt{x^2 + y^2}, \quad \theta = \tan^{-1}\left(\frac{y}{x}\right)$$

$$\frac{\partial r}{\partial x} = \frac{x}{\sqrt{x^2 + y^2}} = \frac{r\cos\theta}{r} = \cos\theta$$

$$\frac{\partial r}{\partial y} = \frac{y}{\sqrt{x^2 + y^2}} = \frac{r\sin\theta}{r} = \sin\theta$$

$$\frac{\partial \theta}{\partial x} = \frac{1}{1 + \frac{y^2}{x^2}}\cdot\left(-\frac{y}{x^2}\right) = -\frac{y}{x^2 + y^2} = -\frac{r\sin\theta}{r^2} = -\frac{\sin\theta}{r}$$

$$\frac{\partial \theta}{\partial y} = \frac{1}{1 + \frac{y^2}{x^2}}\cdot\frac{1}{x} = \frac{x}{x^2 + y^2} = \frac{r\cos\theta}{r^2} = \frac{\cos\theta}{r}$$

Here u is a composite function of x and y

$$\frac{\partial u}{\partial x} = \frac{\partial u}{\partial r}\cdot\frac{\partial r}{\partial x} + \frac{\partial u}{\partial \theta}\cdot\frac{\partial \theta}{\partial x} = \cos\theta\frac{\partial u}{\partial r} - \frac{\sin\theta}{r}\frac{\partial u}{\partial \theta}$$

or $$\frac{\partial}{\partial x}(u) = \left(\cos\theta\frac{\partial}{\partial r} - \frac{\sin\theta}{r}\frac{\partial}{\partial \theta}\right)u$$

$$\Rightarrow \qquad \frac{\partial}{\partial x} \equiv \cos\theta \frac{\partial}{\partial r} - \frac{\sin\theta}{r}\frac{\partial}{\partial\theta} \qquad ...(1)$$

Also $$\frac{\partial u}{\partial y} = \frac{\partial u}{\partial r}\cdot\frac{\partial r}{\partial y} + \frac{\partial u}{\partial\theta}\cdot\frac{\partial\theta}{\partial y} = \sin\theta\frac{\partial u}{\partial r} + \frac{\cos\theta}{r}\frac{\partial u}{\partial\theta}$$

or $$\frac{\partial}{\partial y}(u) = \left(\sin\theta\frac{\partial}{\partial r} + \frac{\cos\theta}{r}\frac{\partial}{\partial\theta}\right)u \;\Rightarrow\; \frac{\partial}{\partial y} \equiv \sin\theta\frac{\partial}{\partial r} + \frac{\cos\theta}{r}\frac{\partial}{\partial\theta} \qquad ...(2)$$

Now, we shall make use of the equivalence of cartesian and polar operators as given by (1) and (2),

$$\frac{\partial^2 u}{\partial x^2} = \frac{\partial}{\partial x}\left(\frac{\partial u}{\partial x}\right) = \left(\cos\theta\frac{\partial}{\partial r} - \frac{\sin\theta}{r}\frac{\partial}{\partial\theta}\right)\left(\cos\theta\frac{\partial u}{\partial r} - \frac{\sin\theta}{r}\frac{\partial u}{\partial\theta}\right)$$

$$= \cos\theta\frac{\partial}{\partial r}\left(\cos\theta\frac{\partial u}{\partial r} - \frac{\sin\theta}{r}\frac{\partial u}{\partial\theta}\right) - \frac{\sin\theta}{r}\frac{\partial}{\partial\theta}\left(\cos\theta\frac{\partial u}{\partial r} - \frac{\sin\theta}{r}\frac{\partial u}{\partial\theta}\right)$$

$$= \cos\theta\left[\cos\theta\frac{\partial^2 u}{\partial r^2} - \sin\theta\frac{\partial u}{\partial\theta}\left(-\frac{1}{r^2}\right) - \frac{\sin\theta}{r}\cdot\frac{\partial^2 u}{\partial r\partial\theta}\right]$$

$$- \frac{\sin\theta}{r}\left[-\sin\theta\frac{\partial u}{\partial r} + \cos\theta\frac{\partial^2 u}{\partial\theta\partial r} - \frac{\cos\theta}{r}\frac{\partial u}{\partial\theta} - \frac{\sin\theta}{r}\frac{\partial^2 u}{\partial\theta^2}\right]$$

$$= \cos^2\theta\frac{\partial^2 u}{\partial r^2} + \frac{2\cos\theta\sin\theta}{r^2}\frac{\partial u}{\partial\theta} + \frac{\sin^2\theta}{r}\frac{\partial u}{\partial r} - \frac{2\cos\theta\sin\theta}{r}\frac{\partial^2 u}{\partial r\partial\theta} + \frac{\sin^2\theta}{r^2}\frac{\partial^2 u}{\partial\theta^2} \qquad ...(3)$$

$$\frac{\partial^2 u}{\partial y^2} = \frac{\partial}{\partial y}\left(\frac{\partial u}{\partial y}\right) = \left(\sin\theta\frac{\partial}{\partial r} + \frac{\cos\theta}{r}\frac{\partial}{\partial\theta}\right)\left(\sin\theta\frac{\partial u}{\partial r} + \frac{\cos\theta}{r}\frac{\partial u}{\partial\theta}\right)$$

$$= \sin\theta\frac{\partial}{\partial r}\left(\sin\theta\frac{\partial u}{\partial r} + \frac{\cos\theta}{r}\frac{\partial u}{\partial\theta}\right) + \frac{\cos\theta}{r}\frac{\partial}{\partial\theta}\left(\sin\theta\frac{\partial u}{\partial r} + \frac{\cos\theta}{r}\frac{\partial u}{\partial\theta}\right)$$

$$= \sin\theta\left[\sin\theta\frac{\partial^2 u}{\partial r^2} - \frac{\cos\theta}{r^2}\frac{\partial u}{\partial\theta} + \frac{\cos\theta}{r}\frac{\partial^2 u}{\partial r\partial\theta}\right]$$

$$+ \frac{\cos\theta}{r}\left[\cos\theta\frac{\partial u}{\partial r} + \sin\theta\frac{\partial^2 u}{\partial\theta\partial r} - \frac{\sin\theta}{r}\frac{\partial u}{\partial\theta} + \frac{\cos\theta}{r}\frac{\partial^2 u}{\partial\theta^2}\right]$$

$$= \sin^2\theta\frac{\partial^2 u}{\partial r^2} - \frac{2\cos\theta\sin\theta}{r^2}\frac{\partial u}{\partial\theta} + \frac{\cos^2\theta}{r}\frac{\partial u}{\partial r} + \frac{2\cos\theta\sin\theta}{r}\frac{\partial^2 u}{\partial r\partial\theta} + \frac{\cos^2\theta}{r^2}\frac{\partial^2 u}{\partial\theta^2} \qquad ...(4)$$

Adding (3) and (4), $$\frac{\partial^2 u}{\partial x^2} + \frac{\partial^2 u}{\partial y^2} = \frac{\partial^2 u}{\partial r^2} + \frac{1}{r}\frac{\partial u}{\partial r} + \frac{1}{r^2}\frac{\partial^2 u}{\partial\theta^2}$$

$$\therefore \quad \frac{\partial^2 u}{\partial x^2} + \frac{\partial^2 u}{\partial y^2} = 0 \text{ transforms into } \frac{\partial^2 u}{\partial r^2} + \frac{1}{r}\frac{\partial u}{\partial r} + \frac{1}{r^2}\frac{\partial^2 u}{\partial\theta^2} = 0.$$

Example 13. *If $z = f(u, v)$, $u = x^2 - y^2$, $v = 2xy$, then prove that*

$$x\frac{\partial z}{\partial x} - y\frac{\partial z}{\partial y} = 2(x^2 + y^2)\frac{\partial z}{\partial u}$$ [CUSAT 2007]

Sol. Here z is a composite function of x and y

$$\frac{\partial z}{\partial x} = \frac{\partial z}{\partial u}.\frac{\partial u}{\partial x} + \frac{\partial z}{\partial v}.\frac{\partial v}{\partial x} = \frac{\partial z}{\partial u}(2x) + \frac{\partial z}{\partial v}(2y)$$

$$\frac{\partial z}{\partial y} = \frac{\partial z}{\partial u}.\frac{\partial u}{\partial y} + \frac{\partial z}{\partial v}.\frac{\partial v}{\partial y} = \frac{\partial z}{\partial u}(-2y) + \frac{\partial z}{\partial v}(2x)$$

$$\therefore \quad x\frac{\partial z}{\partial x} - y\frac{\partial z}{\partial y} = x\left[\frac{\partial z}{\partial u}(2x) + \frac{\partial z}{\partial v}(2y)\right] - y\left[\frac{\partial z}{\partial u}(-2y) + \frac{\partial z}{\partial v}(2x)\right]$$

$$= 2x^2\frac{\partial z}{\partial u} + 2xy\frac{\partial z}{\partial v} + 2y^2\frac{\partial z}{\partial u} - 2xy\frac{\partial z}{\partial v}$$

$$= 2x^2\frac{\partial z}{\partial u} + 2y^2\frac{\partial z}{\partial u} = 2(x^2 + y^2)\frac{\partial z}{\partial u}.$$

Example 14. *Find the total differential coefficient of x^2y with respect to x when x and y are connected by $x^2 + xy + y^2 = 1$.* [CUSAT 2008]

Sol. Let $\quad u = x^2y$

The given equations are of the form

$$u = f(x, y) \text{ and } Q(x, y) = k$$

$\therefore$ u is a composite function of x

$$\frac{du}{dx} = \frac{\partial u}{\partial x}\frac{dx}{dx} + \frac{\partial u}{\partial y}\frac{dy}{dx} = \frac{\partial u}{\partial x} + \frac{\partial u}{\partial y}.\frac{dy}{dx} \quad \text{...(1)}$$

Now, $\quad \frac{\partial u}{\partial x} = 2xy, \; \frac{\partial u}{\partial y} = x^2$

Also differentiating $x^2 + xy + y^2 = 1$, w.r.t. x, we have

$$2x + y + x\frac{dy}{dx} + 2y\frac{dy}{dx} = 0$$

or

$$\frac{dy}{dx} = -\frac{(2x + y)}{x + 2y}$$

From (1)

$$\frac{du}{dx} = 2xy + x^2\left(\frac{-(2x + y)}{x + 2y}\right)$$

$$= \frac{2xy(x + 2y) - x^2(2x + y)}{x + 2y}$$

$$= \frac{2x^2y + 4xy^2 - 2x^3 - x^2y}{x + 2y}$$

$$= \frac{x^2y + 4xy^2 - 2x^3}{x + 2y}.$$

TEST YOUR KNOWLEDGE

1. (a) Find $\frac{du}{dt}$ when $u = x^2 + y^2$, $x = at^2$, $y = 2at$. Also verify by direct substitution.

   (b) If $u = x^2 + y^2 + z^2$ and $x = e^{2t}$, $y = e^{2t} \cos 3t$, $z = e^{2t} \sin 3t$; find $\frac{du}{dt}$ as a total derivative and verify the result by direct substitution.

2. If $u = \sin \frac{x}{y}$, $x = e^t$, $y = t^2$, find $\frac{du}{dt}$.

3. If $u = x^3 + y^3$, where $x = a \cos t$, $y = b \sin t$, find $\frac{du}{dt}$ and verify the result.

4. (a) If $z = u^2 + v^2$, $u = r \cos \theta$, $v = r \sin \theta$, find $\frac{\partial z}{\partial r}$ and $\frac{\partial z}{\partial \theta}$.

   (b) If $z = \log (u^2 + v)$, $u = e^{x^2 + y^2}$, $v = x^2 + y$, find $\frac{\partial z}{\partial x}$ and $\frac{\partial z}{\partial y}$.

5. If $u = f(r, s)$, $r = x + y$, $s = x - y$, show that $\frac{\partial u}{\partial x} + \frac{\partial u}{\partial y} = 2\frac{\partial u}{\partial r}$.

6. If $z = e^{ax+by} f(ax - by)$, prove that $b \frac{\partial z}{\partial x} + a \frac{\partial z}{\partial y} = 2abz$.

7. If $x = u + v$, $y = uv$ and z is a function of x, y ; show that $u \frac{\partial z}{\partial u} + v\frac{\partial z}{\partial v} = x\frac{\partial z}{\partial x} + 2y\frac{\partial z}{\partial y}$.

8. If $u = f(r, s)$, $r = x + at$, $s = y + bt$, show that $\frac{\partial u}{\partial t} = a\frac{\partial u}{\partial x} + b\frac{\partial u}{\partial y}$.

9. If $u = x \log (xy)$, where $x^3 + y^3 + 3xy = 1$, find $\frac{du}{dx}$.

10. (a) If $u = f(r, s, t)$ and $r = \frac{x}{y}$, $s = \frac{y}{z}$, $t = \frac{z}{x}$, prove that $x \frac{\partial u}{\partial x} + y\frac{\partial u}{\partial y} + z\frac{\partial u}{\partial z} = 0$.

    (b) If $x = u + v + w$, $y = vw + wu + uv$, $z = uvw$ and F is a function of x, y, z, show that

$$u \frac{\partial F}{\partial u} + v\frac{\partial F}{\partial v} + w\frac{\partial F}{\partial w} = x\frac{\partial F}{\partial x} + 2y\frac{\partial F}{\partial y} + 3z\frac{\partial F}{\partial z}.$$

    (c) If $u = f(2x - 3y, 3y - 4z, 4z - 2x)$, prove that $\frac{1}{2}\frac{\partial u}{\partial x} + \frac{1}{3}\frac{\partial u}{\partial y} + \frac{1}{4}\frac{\partial u}{\partial z} = 0$. [*CUSAT 2008*]

    (d) If $u = f(x^2 + 2yz, y^2 + 2zx)$, prove that $(y^2 - zx) \frac{\partial u}{\partial x} + (x^2 - yz) \frac{\partial u}{\partial y} + (z^2 - xy) \frac{\partial u}{\partial z} = 0$.

11. If $z = x^2y$ and $x^2 + xy + y^2 = 1$, show that $\frac{dz}{dx} = 2xy - \frac{x^2(2x + y)}{x + 2y}$.

12. If $x^3 + 3x^2y + 6xy^2 + y^3 = 1$, find $\frac{dy}{dx}$.

13. If $x^y = y^x$, show that $\frac{dy}{dx} = \frac{y(y - x \log y)}{x(x - y \log x)}$ using partial derivative method.

14. If $x^3 + y^3 - 3axy = 0$, find $\frac{dy}{dx}$ and $\frac{d^2y}{dx^2}$ using partial derivative method.

15. Prove that if $y^3 - 3ax^2 + x^3 = 0$, then $\frac{d^2y}{dx^2} + \frac{2a^2x^2}{y^5} = 0$.

16. If $z = xyf\left(\frac{y}{x}\right)$ and z is a constant, show that $\frac{f'\left(\frac{y}{x}\right)}{f\left(\frac{y}{x}\right)} = \frac{x\left[y + x\frac{dy}{dx}\right]}{y\left[y - x\frac{dy}{dx}\right]}$.

17. By changing the independent variables x and y to u and v by means of the relations $u = x - ay$, $v = x + ay$, show that $a^2\frac{\partial^2 z}{\partial x^2} - \frac{\partial^2 z}{\partial y^2}$ transforms into $4a^2\frac{\partial^2 z}{\partial u \partial v}$.

18. If z is a function of x and y, and u and v be two other variables such that $u = lx + my$, $v = ly - mx$, show that

$$\frac{\partial^2 z}{\partial x^2} + \frac{\partial^2 z}{\partial y^2} = (l^2 + m^2)\left(\frac{\partial^2 z}{\partial u^2} + \frac{\partial^2 z}{\partial v^2}\right).$$

19. If $u = u(x, y)$ and $x = e^r \cos\theta$, $y = e^r \sin\theta$, show that $\frac{\partial^2 u}{\partial x^2} + \frac{\partial^2 u}{\partial y^2} = e^{-2r}\left(\frac{\partial^2 u}{\partial r^2} + \frac{\partial^2 u}{\partial \theta^2}\right)$.

20. If $x = \rho\cos\phi$, $y = \rho\sin\phi$, show that $\left(\frac{\partial v}{\partial x}\right)^2 + \left(\frac{\partial v}{\partial y}\right)^2 = \left(\frac{\partial v}{\partial \rho}\right)^2 + \frac{1}{\rho^2}\left(\frac{\partial v}{\partial \phi}\right)^2$.

21. Prove that $\frac{\partial^2 u}{\partial x^2} + \frac{\partial^2 u}{\partial y^2} = \frac{\partial^2 u}{\partial s^2} + \frac{\partial^2 u}{\partial t^2}$, where $x = s\cos\alpha - t\sin\alpha$ and $y = s\sin\alpha + t\cos\alpha$.

22. If by the substitution $u = x^2 - y^2$, $v = 2xy$, $f(x, y) = \phi(u, v)$, show that

$$\frac{\partial^2 f}{\partial x^2} + \frac{\partial^2 f}{\partial y^2} = 4(x^2 + y^2)\left(\frac{\partial^2 \phi}{\partial u^2} + \frac{\partial^2 \phi}{\partial v^2}\right).$$

Answers

1. (a) $4a^2t(t^2 + 2)$ (b) $8e^{4t}$ **2.** $\frac{e^t(t-2)}{t^3}\cos\left(\frac{e^t}{t^2}\right)$

3. $-3a^3\cos^2 t\sin t + 3b^3\sin^2 t\cos t$

4. (a) $2r, 0$ (b) $\frac{2x(2u^2+1)}{u^2+v}, \frac{4yu^2+1}{u^2+v}$

9. $1 + \log(xy) - \frac{x(x^2 + y)}{y(y^2 + x)}$ **12.** $-\frac{x^2 + 2xy + 2y^2}{x^2 + 4xy + y^2}$ **14.** $\frac{x^2 - ay}{ax - y^2}, \frac{2a^3xy}{(ax - y^2)^3}$.

7.10. JACOBIANS

If u and v are functions of two independent variables x and y, then the determinant

$\begin{vmatrix} \frac{\partial u}{\partial x} & \frac{\partial u}{\partial y} \\ \frac{\partial v}{\partial x} & \frac{\partial v}{\partial y} \end{vmatrix}$ is called Jacobian of u, v with respect to x, y and is denoted by the symbol $J\left(\frac{u, v}{x, y}\right)$

or $\frac{\partial(u, v)}{\partial(x, y)}$.

Similarly, if u, v, w be functions of x, y, z, then the Jacobian of u, v, w with respect to x, y, z is

$$J\left(\frac{u, v, w}{x, y, z}\right) \quad \text{or} \quad \frac{\partial(u, v, w)}{\partial(x, y, z)} = \begin{vmatrix} \frac{\partial u}{\partial x} & \frac{\partial u}{\partial y} & \frac{\partial u}{\partial z} \\ \frac{\partial v}{\partial x} & \frac{\partial v}{\partial y} & \frac{\partial v}{\partial z} \\ \frac{\partial w}{\partial x} & \frac{\partial w}{\partial y} & \frac{\partial w}{\partial z} \end{vmatrix}.$$

7.11. PROPERTIES OF JACOBIANS

I. *If u, v are functions of r, s where r, s are functions of x, y, then*

$$\frac{\partial(u, v)}{\partial(x, y)} = \frac{\partial(u, v)}{\partial(r, s)} \times \frac{\partial(r, s)}{\partial(x, y)}.$$

Proof. Since u, v are composite functions of x, y

$$\therefore \quad \left.\begin{aligned} \frac{\partial u}{\partial x} &= \frac{\partial u}{\partial r}\cdot\frac{\partial r}{\partial x} + \frac{\partial u}{\partial s}\cdot\frac{\partial s}{\partial x} = u_r r_x + u_s s_x \\ \frac{\partial u}{\partial y} &= \frac{\partial u}{\partial r}\cdot\frac{\partial r}{\partial y} + \frac{\partial u}{\partial s}\cdot\frac{\partial s}{\partial y} = u_r r_y + u_s s_y \\ \frac{\partial v}{\partial x} &= \frac{\partial v}{\partial r}\cdot\frac{\partial r}{\partial x} + \frac{\partial v}{\partial s}\cdot\frac{\partial s}{\partial x} = v_r r_x + v_s s_x \\ \frac{\partial v}{\partial y} &= \frac{\partial v}{\partial r}\cdot\frac{\partial r}{\partial y} + \frac{\partial v}{\partial s}\cdot\frac{\partial s}{\partial y} = v_r r_y + v_s s_y \end{aligned}\right] \quad \text{...(A)}$$

Now, $$\frac{\partial(u, v)}{\partial(r, s)} \times \frac{\partial(r, s)}{\partial(x, y)} = \begin{vmatrix} u_r & u_s \\ v_r & v_s \end{vmatrix} \cdot \begin{vmatrix} r_x & r_y \\ s_x & s_y \end{vmatrix}$$

Interchanging rows and columns in the second determinant

$$= \begin{vmatrix} u_r & u_s \\ v_r & v_s \end{vmatrix} \cdot \begin{vmatrix} r_x & s_x \\ r_y & s_y \end{vmatrix} = \begin{vmatrix} u_r r_x + u_s s_x & u_r r_y + u_s s_y \\ v_r r_x + v_s s_x & v_r r_y + v_s s_y \end{vmatrix} = \begin{vmatrix} \frac{\partial u}{\partial x} & \frac{\partial u}{\partial y} \\ \frac{\partial v}{\partial x} & \frac{\partial v}{\partial y} \end{vmatrix} \quad \text{[Using (A)]}$$

$$= \frac{\partial(u, v)}{\partial(x, y)}.$$

II. *If J_1 is the Jacobian of u, v with respect to x, y and J_2 is the Jacobian of x, y, with respect to u, v, then $J_1 J_2 = 1$ i.e., $\frac{\partial(u, v)}{\partial(x, y)} \cdot \frac{\partial(x, y)}{\partial(u, v)} = 1$.*

Proof. Let $u = u(x, y)$ and $v = v(x, y)$, so that u and v are functions of x, y.

Differentiating partially w.r.t. u and v, we get

$$\left.\begin{aligned} 1 &= \frac{\partial u}{\partial x}\cdot\frac{\partial x}{\partial u} + \frac{\partial u}{\partial y}\cdot\frac{\partial y}{\partial u} = u_x x_u + u_y y_u \\ 0 &= \frac{\partial u}{\partial x}\cdot\frac{\partial x}{\partial v} + \frac{\partial u}{\partial y}\cdot\frac{\partial y}{\partial v} = u_x x_v + u_y y_v \\ 0 &= \frac{\partial v}{\partial x}\cdot\frac{\partial x}{\partial u} + \frac{\partial v}{\partial y}\cdot\frac{\partial y}{\partial u} = v_x x_u + v_y y_u \\ 1 &= \frac{\partial v}{\partial x}\cdot\frac{\partial x}{\partial v} + \frac{\partial v}{\partial y}\cdot\frac{\partial y}{\partial v} = v_x x_v + v_y y_v \end{aligned}\right] \quad \text{...(A)}$$

Now, $$\frac{\partial(u, v)}{\partial(x, y)}\cdot\frac{\partial(x, y)}{\partial(u, v)} = \begin{vmatrix} u_x & u_y \\ v_x & v_y \end{vmatrix}\begin{vmatrix} x_u & x_v \\ y_u & y_v \end{vmatrix}$$

Interchanging rows and columns in the second determinant

$$= \begin{vmatrix} u_x & u_y \\ v_x & v_y \end{vmatrix}\cdot\begin{vmatrix} x_u & y_u \\ x_v & y_v \end{vmatrix} = \begin{vmatrix} u_x x_u + u_y y_u & u_x x_v + u_y y_v \\ v_x x_u + v_y y_u & v_x x_v + v_y y_v \end{vmatrix}$$

$$= \begin{vmatrix} 1 & 0 \\ 0 & 1 \end{vmatrix} = 1. \qquad \text{[Using (A)]}$$

ILLUSTRATIVE EXAMPLES

Example 1. *If* $r = \sqrt{x^2 + y^2}$, $\theta = \tan^{-1}\frac{y}{x}$, *evaluate* $\frac{\partial(r, \theta)}{\partial(x, y)}$.

Sol. $$r = \sqrt{x^2 + y^2}, \quad \theta = \tan^{-1}\frac{y}{x}$$

$$\frac{\partial r}{\partial x} = \frac{x}{\sqrt{x^2 + y^2}}, \qquad \frac{\partial r}{\partial y} = \frac{y}{\sqrt{x^2 + y^2}}$$

$$\frac{\partial \theta}{\partial x} = \frac{1}{1 + \frac{y^2}{x^2}}\left(-\frac{y}{x^2}\right) = -\frac{y}{x^2 + y^2}, \qquad \frac{\partial \theta}{\partial y} = \frac{1}{1 + \frac{y^2}{x^2}}\left(\frac{1}{x}\right) = \frac{x}{x^2 + y^2}$$

$$\frac{\partial(r, \theta)}{\partial(x, y)} = \begin{vmatrix} \frac{\partial r}{\partial x} & \frac{\partial r}{\partial y} \\ \frac{\partial \theta}{\partial x} & \frac{\partial \theta}{\partial y} \end{vmatrix} = \begin{vmatrix} \frac{x}{\sqrt{x^2 + y^2}} & \frac{y}{\sqrt{x^2 + y^2}} \\ -\frac{y}{x^2 + y^2} & \frac{x}{x^2 + y^2} \end{vmatrix}$$

$$= \frac{x^2}{(x^2 + y^2)^{3/2}} + \frac{y^2}{(x^2 + y^2)^{3/2}} = \frac{x^2 + y^2}{(x^2 + y^2)^{3/2}} = \frac{1}{\sqrt{x^2 + y^2}}.$$

Example 2. *If $x = r \sin\theta \cos\phi$, $y = r \sin\theta \sin\phi$, $z = r\cos\theta$, show that* $\dfrac{\partial(x, y, z)}{\partial(r, \theta, \phi)} = r^2 \sin\theta$.

Sol.
$$\frac{\partial(x, y, z)}{\partial(r, \theta, \phi)} = \begin{vmatrix} \dfrac{\partial x}{\partial r} & \dfrac{\partial x}{\partial \theta} & \dfrac{\partial x}{\partial \phi} \\ \dfrac{\partial y}{\partial r} & \dfrac{\partial y}{\partial \theta} & \dfrac{\partial y}{\partial \phi} \\ \dfrac{\partial z}{\partial r} & \dfrac{\partial z}{\partial \theta} & \dfrac{\partial z}{\partial \phi} \end{vmatrix} = \begin{vmatrix} \sin\theta\cos\phi & r\cos\theta\cos\phi & -r\sin\theta\sin\phi \\ \sin\theta\sin\phi & r\cos\theta\cos\phi & r\sin\theta\sin\phi \\ \cos\theta & -r\sin\theta & 0 \end{vmatrix}$$

Taking out common factors (r from second column and $r \sin\theta$ from third column)

$$= r^2 \sin\theta \begin{vmatrix} \sin\theta\cos\phi & \cos\theta\cos\phi & -\sin\phi \\ \sin\theta\sin\phi & \cos\theta\sin\phi & \cos\phi \\ \cos\theta & -\sin\theta & 0 \end{vmatrix}$$

Expanding by third row

$$= r^2 \sin\theta \left\{\cos\theta \begin{vmatrix} \cos\theta\cos\phi & -\sin\phi \\ \cos\theta\sin\phi & \cos\phi \end{vmatrix} + \sin\theta \begin{vmatrix} \sin\theta\cos\phi & -\sin\phi \\ \sin\theta\sin\phi & \cos\phi \end{vmatrix}\right\}$$

$$= r^2 \sin\theta\,[\cos\theta\,(\cos\theta\cos^2\phi + \cos\theta\sin^2\phi) + \sin\theta\,(\sin\theta\cos^2\phi + \sin\theta\sin^2\phi)]$$

$$= r^2 \sin\theta\,(\cos^2\theta + \sin^2\theta) = r^2 \sin\theta.$$

Example 3. *Show that $u = e^x \sin y$, $v = e^x \cos y$ are functionally independent.*

[CUSAT 2009]

Sol. u and v are functionally independent if $\dfrac{\partial(u, v)}{\partial(x, y)} \neq 0$

$$u = e^x \sin y, \qquad v = e^x \cos y$$

$$\frac{\partial u}{\partial x} = e^x \sin y \qquad \frac{\partial v}{\partial x} = e^x \cos y$$

$$\frac{\partial u}{\partial y} = e^x \cos y \qquad \frac{\partial v}{\partial y} = -e^x \sin y$$

$$\frac{\partial(u, v)}{\partial(x, y)} = \begin{vmatrix} \dfrac{\partial u}{\partial x} & \dfrac{\partial u}{\partial y} \\ \dfrac{\partial v}{\partial x} & \dfrac{\partial v}{\partial y} \end{vmatrix} = \begin{vmatrix} e^x \sin y & e^x \cos y \\ e^x \cos y & -e^x \sin y \end{vmatrix}$$

$$= -e^{2x} \sin^2 y - e^{2x} \cos^2 y$$

$$= -e^{2x} (\sin^2 y + \cos^2 y) = -e^{2x} \neq 0$$

$\therefore$ u and v are functionally independent.

TEST YOUR KNOWLEDGE

1. If $u = x^2 - 2y$, $v = x + y$, prove that $\dfrac{\partial(u, v)}{\partial(x, y)} = 2x + 2$.

2. If $u = x(1 - y)$, $v = xy$; prove that $JJ' = 1$.

3. If $x = r\cos\theta$, $y = r\sin\theta$, verify that $\dfrac{\partial(x, y)}{\partial(r, \theta)} \cdot \dfrac{\partial(r, \theta)}{\partial(x, y)} = 1$.

$\left[\textbf{Hint. } r = \sqrt{x^2 + y^2}, \theta = \tan^{-1}\dfrac{y}{x}.\right]$

4. If $x = r\cos\theta$, $y = r\sin\theta$, $z = z$, evaluate $\dfrac{\partial(x, y, z)}{\partial(r, \theta, z)}$.

5. If $u = \dfrac{yz}{x}$, $v = \dfrac{zx}{y}$ and $w = \dfrac{xy}{z}$, show that $\dfrac{\partial(u, v, w)}{\partial(x, y, z)} = 4$. [*CUSAT 2007*]

6. If $u = \dfrac{y^2}{2x}$, $v = \dfrac{x^2 + y^2}{2x}$, find $\dfrac{\partial(u, v)}{\partial(x, y)}$.

7. If $x = a\cos\xi\cosh\eta$, $y = a\sinh\xi\sin\eta$, show that $\dfrac{\partial(x, y)}{\partial(\xi, \eta)} = \dfrac{a^2}{2}(\cosh 2\xi - \cos 2\eta)$.

$\left[\textbf{Hint. } \cosh^2 x = \dfrac{\cosh 2x + 1}{2}, \sinh^2 x = \dfrac{\cosh 2x - 1}{2}\right]$

8. If $F = xu + v - y$, $G = u^2 + vy + w$, $H = zu - v + vw$, compute $\dfrac{\partial(F, G, H)}{\partial(u, w, v)}$.

9. If $u = \dfrac{x + y}{1 - xy}$ and $v = \tan^{-1} x + \tan^{-1} y$, find $\dfrac{\partial(u, v)}{\partial(x, y)}$.

10. If $u = x^2 - 2y$, $v = x + y + z$, $w = x - 2y + 3z$, find $\dfrac{\partial(u, v, w)}{\partial(x, y, z)}$.

11. If $u = x + y + z$, $uv = y + z$, $uvw = z$, show that $\dfrac{\partial(x, y, z)}{\partial(u, v, w)} = u^2 v$.

12. If $u = xyz$, $v = xy + yz + zx$, $w = x + y + z$, show that $\dfrac{\partial(u, v, w)}{\partial(x, y, z)} = (x - y)(y - z)(z - x)$.

13. If $u = \dfrac{x}{y - z}$, $v = \dfrac{y}{z - x}$, $w = \dfrac{z}{x - y}$, show that $\dfrac{\partial(u, v, w)}{\partial(x, y, z)} = 0$.

Answers

4. r 6. $-\dfrac{y}{2x}$ 8. $xw - x - xyv + 2uv - z$

9. 0 10. $10x + 4$.

7.12. MAXIMA AND MINIMA OF FUNCTIONS OF TWO VARIABLES

A function $f(x, y)$ is said to have a **maximum value at x = a, y = b if f(a, b) > f(a + h, b + k),** for small and independent values of h and k, positive or negative.

A function $f(x, y)$ is said to have a **minimum value at x = a, y = b** if $f(a, b) < f(a + h, b + k)$, for small and independent values of h and k, positive or negative.

Thus $f(x, y)$ has a maximum or minimum value at a point (a, b) according as

$$\Delta f = f(a + h, b + k) - f(a, b) < \text{or} > 0.$$

Geometrically, the surface $z = f(x, y)$ has a maximum at the point (a, b) if $f(a, b)$ is the greatest in the small neighbourhood of the point (a, b). As such, the surface descends in all directions at this point. The point of maxima may well be compared with the highest point of a dome. The surface has a minimum at the point (a, b) if $f(a, b)$ is the smallest in the small neighbourhood of the point (a, b). As such, the surface ascends in all directions at this point. The point of minima may well be compared with the lowest point of a bowl. However, sometimes, at the point (a, b), the tangent plane is horizontal and the surface descends in certain directions and ascends in other directions. Such a point is called a *saddle point*.

A maximum or a minimum value of a function is called its *extreme value*.

7.13. CONDITIONS FOR f(x, y) TO BE MAXIMUM OR MINIMUM

By Taylor's theorem, we have

$$\Delta f = f(a + h, b + k) - f(a, b)$$

$$= [hf_x(a, b) + kf_y(a, b)] + \frac{1}{2!}[h^2 f_{xx}(a, b) + 2hkf_{xy}(a, b) + k^2 f_{yy}(a, b)] + \cdots \quad \ldots(1)$$

For small values of h and k, the second and higher order terms are still smaller and may be neglected.

Thus, sign of Δf = sign of $[hf_x(a, b) + kf_y(a, b)]$

Taking $h = 0$, the sign of Δf changes with the sign of k. Similarly taking $k = 0$, the sign of Δf changes with the sign of h. Since Δf changes sign with h and k, $f(x, y)$ cannot have a maximum or a minimum value at (a, b) unless $f_x(a, b) = 0 = f_y(a, b)$.

Hence the necessary conditions for (x, y) to have a maximum or a minimum value at (a, b) are

$$f_x(a, b) = 0, f_y(a, b) = 0$$

If these conditions are satisfied, then for small values of h and k, we have, from (1),

$$\Delta f = \frac{1}{2!}[h^2 f_{xx}(a, b) + 2hkf_{xy}(a, b) + k^2 f_{yy}(a, b)] + \cdots$$

$$= \frac{1}{2!}[h^2 r + 2hks + k^2 t] + \cdots \text{ where } r = f_{xx}(a, b),\ s = f_{xy}(a, b),\ t = f_{yy}(a, b)$$

or

$$\Delta f = \frac{1}{2r}[h^2 r^2 + 2hkrs + k^2 rt] + \cdots = \frac{1}{2r}[(h^2 r^2 + 2hkrs + k^2 s^2) + k^2 rt - k^2 s^2)] + \cdots$$

$$= \frac{1}{2r}[(hr + ks)^2 + k^2 (rt - s^2)] + \cdots \quad \ldots(2)$$

Now, $(hr + ks)^2$ is always positive and $k^2(rt - s^2)$ will be positive if $rt - s^2 > 0$. In this case, Δf will have the same sign as that of r for all values of h and k.

Hence if $rt - s^2 > 0$, then $f(x, y)$ has a maximum or a minimum at (a, b) according as $r < 0$ or $r > 0$.

If $rt - s^2 < 0$, then Δf changes sign with h and k. Hence there is neither a maximum nor a minimum value at (a, b). The point (a, b) is a saddle point in this case.

If $rt - s^2 = 0$, no conclusion can be drawn about a maximum or a minimum value at (a, b) and hence further investigation is required.

Note. The point (a, b) is called a *stationary point* if $f_x(a, b) = 0$, $f_y(a, b) = 0$. The value $f(a, b)$ is called a *stationary value*. Thus every extreme value is a stationary value but the converse may not be true.

7.14. RULE TO FIND THE EXTREME VALUES OF A FUNCTION z = f(x, y)

(*i*) Find $\frac{\partial z}{\partial x}$ and $\frac{\partial z}{\partial y}$.

(*ii*) Solve $\frac{\partial z}{\partial x} = 0$ and $\frac{\partial z}{\partial y} = 0$ simultaneously.

Let (a, b); (c, d) ... be the solutions of these equations.

(*iii*) For each solution in step (*ii*), find $r = \frac{\partial^2 f}{\partial x^2}$, $s = \frac{\partial^2 f}{\partial x \partial y}$, $t = \frac{\partial^2 f}{\partial y^2}$.

(*iv*) (*a*) If $rt - s^2 > 0$ and $r < 0$ for a particular solution (a, b) of step (*ii*), then z has a maximum value at (a, b).

(*b*) If $rt - s^2 > 0$ and $r > 0$ for a particular solution (a, b) of step (*ii*), then z has a minimum value at (a, b).

(*c*) If $rt - s^2 < 0$ for a particular solution (a, b) of step (*ii*), then z has no extreme value at (a, b).

(*d*) If $rt - s^2 = 0$, the case is doubtful and requires further investigation.

ILLUSTRATIVE EXAMPLES

Example 1. *Examine the function $x^3 + y^3 - 3axy$ for maxima and minima.*

Sol. Here $f(x, y) = x^3 + y^3 - 3axy$

$$f_x = 3x^2 - 3ay, \quad f_y = 3y^2 - 3ax, \quad r = f_{xx} = 6x, \quad s = f_{xy} = -3a, \quad t = f_{yy} = 6y$$

Now, $f_x = 0$ and $f_y = 0$

$\Rightarrow$ $x^2 - ay = 0$ ...(1) $\quad y^2 - ax = 0$ ...(2)

From (1), $y = \frac{x^2}{a}$

$\therefore$ From (2), $\frac{x^4}{a^2} - ax = 0$ or $x(x^3 - a^3) = 0$ or $x = 0, a$

when $x = 0$, $y = 0$; when $x = a$, $y = a$

$\therefore$ There are two stationary points $(0, 0)$ and (a, a).

Now, $rt - s^2 = 36xy - 9a^2$

At (0, 0), $rt - s^2 = -9a^2 < 0$

$\Rightarrow$ There is no extreme value at $(0, 0)$.

At (a, a), $rt - s^2 = 36a^2 - 9a^2 = 27a^2 > 0$

$\Rightarrow$ $f(x, y)$ has extreme value at (a, a)

Now, $r = 6a$

If $a > 0$, $r > 0$ so that $f(x, y)$ has a minimum value at (a, a).

Minimum value $= f(-a, -a) = -a^3 - a^3 + 3a^3 = a^3$

If $a < 0$, $r < 0$ so that $f(x, y)$ has a maximum value at (a, a).

Maximum value $= f(a, a) = a^3 + a^3 - 3a^3 = -a^3$.

Example 2. *Discuss the maxima and minima of $x^3y^2 (1 - x - y)$.*

Sol. Here $f(x, y) = x^3y^2 (1 - x - y) = x^3y^2 - x^4y^2 - x^3y^3$

$$f_x = 3x^2y^2 - 4x^3y^2 - 3x^2y^3,\ f_y = 2x^3y - 2x^4y - 3x^3y^2$$

$$r = f_{xx} = 6xy^2 - 12x^2y^2 - 6xy^3\ ;\ s = f_{xy} = 6x^2y - 8x^3y - 9x^2y^2\ ;\ t = f_{yy} = 2x^3 - 2x^4 - 6x^3y$$

Now, $f_x = 0$ and $f_y = 0$

$\Rightarrow$ $x^2y^2(3 - 4x - 3y) = 0$ ...(1) $x^3y(2 - 2x - 3y) = 0$ ...(2)

Solving (1) and (2), the stationary points are (0, 0) and $(\frac{1}{2}, \frac{1}{3})$.

Now, $rt - s^2 = 6xy^2(1 - 2x - y)\ 2x^3(1 - x - 3y) - [x^2y(6 - 8x - 9y)]^2$

$= 12x^4y^2(1 - 2x - y)(1 - x - 3y) - x^4y^2(6 - 8x - 9y)^2$

At (0, 0), $rt - s^2 = 0$

$\therefore$ At (0, 0) further investigation is needed.

Consider $f(a + h, b + k) - f(a, b) = f(h, k) - f(0, 0) = h^3k^2(1 - h - k)$

$= h^3k^2$ (neglecting $h^4k^2 + h^3k^3$ which is very small as compared to h^3k^2)

which is > 0 if $h > 0$ and < 0 if $h < 0$

$\therefore$ $f(a + h, b + k) - f(a, b)$ does not keep same sign for all small values of h and k (positive or negative).

$\therefore$ There is no extreme value at (0, 0).

At $\left(\frac{1}{2}, \frac{1}{3}\right)$, $rt - s^2 = 12 \cdot \frac{1}{16} \cdot \frac{1}{9}\left(-\frac{1}{3}\right)\left(-\frac{1}{2}\right) - \frac{1}{16} \cdot \frac{1}{9}(1) = \frac{1}{144}(2 - 1) = \frac{1}{14} > 0.$

Also $r = 6 \cdot \frac{1}{2} \cdot \frac{1}{9} - 12 \cdot \frac{1}{4} \cdot \frac{1}{9} - 6 \cdot \frac{1}{2} \cdot \frac{1}{8} = \frac{1}{3} - \frac{1}{3} - \frac{3}{8} = -\frac{3}{8} < 0.$

$\therefore$ $f(x, y)$ has a maximum value at $(\frac{1}{2}, \frac{1}{3})$

Maximum value $= f\left(\frac{1}{2}, \frac{1}{3}\right) = \frac{1}{8} \cdot \frac{1}{9}\left(1 - \frac{1}{2} - \frac{1}{3}\right) = \frac{1}{432}$.

Example 3. *Locate the stationary points of $x^4 + y^4 - 2x^2 + 4xy - 2y^2$ and determine their nature.*

Sol. Here $f(x, y) = x^4 + y^4 - 2x^2 + 4xy - 2y^2$

$$f_x = 4x^3 - 4x + 4y,\quad f_y = 4y^3 + 4x - 4y$$

$$r = f_{xx} = 12x^2 - 4,\quad s = f_{xy} = 4,\ t = f_{yy} = 12y^2 - 4$$

Now, $f_x = 0$ and $f_y = 0$

$\Rightarrow$ $x^3 - x + y = 0$ ...(1) $y^3 + x - y = 0$ ...(2)

Adding (1) and (2), $x^3 + y^3 = 0 \Rightarrow y = -x$.

Putting $y = -x$ in (1), we get $x^3 - 2x = 0$ or $x(x^2 - 2) = 0$ $\therefore$ $x = 0, \pm\sqrt{2}$.

Since $y = -x$, the stationary points are $(0, 0)$, $(\sqrt{2}, -\sqrt{2})$ and $(-\sqrt{2}, \sqrt{2})$.

Now $\qquad rt - s^2 = (12x^2 - 4)(12y^2 - 4) - 16$

At (0, 0), $\qquad rt - s^2 = 16 - 16 = 0.$

$\therefore$ At (0, 0), further investigation is needed.

Now, $\qquad f(0, 0) = 0$

$$f(x, y) = x^4 + y^4 - 2(x^2 - 2xy + y^2) = x^4 + y^4 - 2(x - y)^2$$

When h and k are small, $f(h, k) = 2h^4 > 0$ for $h = k$

and $\qquad f(h, k) = -2(h - k)^2 < 0$ for $h \neq k$

(neglecting $h^4 + k^4$ which is very small)

$\Rightarrow \qquad f(0, 0) < f(h, k)$ for $h = k$

$\qquad f(0, 0) > f(h, k)$ for $h \neq k$

$\Rightarrow$ There is no extreme value at (0, 0).

At $(\sqrt{2}, -\sqrt{2})$, $\quad rt - s^2 = (12 \times 2 - 4)^2 - 16 = 384 > 0$

$$r = 12 \times 2 - 4 = 20 > 0$$

$\therefore$ $f(x, y)$ has a minimum value at $(\sqrt{2}, -\sqrt{2})$.

$$\text{Minimum value} = (\sqrt{2})^4 + (-\sqrt{2})^4 - 2(\sqrt{2})^2 + 4\sqrt{2}(-\sqrt{2}) - 2(-\sqrt{2})^2$$
$$= 4 + 4 - 4 - 8 - 4 = -8.$$

Similarly, $f(x, y)$ has a minimum value -8 at $(-\sqrt{2}, \sqrt{2})$.

Example 4. *Examine for minimum and maximum values: sin x + sin y + sin (x + y).*

Sol. Here $\qquad f(x, y) = \sin x + \sin y + \sin (x + y)$

$$f_x = \cos x + \cos (x + y),\ f_y = \cos y + \cos (x + y)$$
$$r = f_{xx} = -\sin x - \sin (x + y),\ s = f_{xy} = -\sin (x + y),$$
$$t = f_{yy} = -\sin y - \sin (x + y)$$

Now, $\qquad f_x = 0$ and $f_y = 0$

$\Rightarrow \qquad \cos x + \cos (x + y) = 0 \quad \ldots(1) \quad \cos y + \cos (x + y) = 0 \quad \ldots(2)$

Subtracting (2) from (1), $\cos x - \cos y = 0$ or $\cos x = \cos y$ $\therefore$ $x = y$

From (1), $\cos x + \cos 2x = 0$ or $\cos 2x = -\cos x = \cos (\pi - x)$ or $2x = \pi - x$ $\therefore$ $x = \dfrac{\pi}{3}$

$\therefore$ $x = y = \dfrac{\pi}{3}$ is a stationary point.

At $\left(\dfrac{\pi}{3}, \dfrac{\pi}{3}\right)$, $\quad r = -\dfrac{\sqrt{3}}{2} - \dfrac{\sqrt{3}}{2} = -\sqrt{3} < 0,\quad s = \dfrac{\sqrt{3}}{2},\quad t = -\dfrac{\sqrt{3}}{2} - \dfrac{\sqrt{3}}{2} = -\sqrt{3}$

or $\qquad rt - s^2 = 3 - \dfrac{3}{4} = \dfrac{9}{4} > 0$

Also $r < 0$

$\therefore$ $f(x, y)$ has a maximum value at $\left(\frac{\pi}{3}, \frac{\pi}{3}\right)$.

Maximum value $= f\left(\frac{\pi}{3}, \frac{\pi}{3}\right) = \sin\frac{\pi}{3} + \sin\frac{\pi}{3} + \sin\frac{2\pi}{3} = \frac{\sqrt{3}}{2} + \frac{\sqrt{3}}{2} + \frac{\sqrt{3}}{2} = \frac{3\sqrt{3}}{2}$.

Example 5. *A rectangular box, open at the top, is to have a given capacity. Find the dimensions of the box requiring least material for its construction.*

Sol. Let x, y and z be the length, breadth and height respectively. Let V be the given capacity and S, the surface.

V is given $\Rightarrow$ V is constant

$$V = xyz \quad \text{or} \quad z = \frac{V}{xy} \qquad \text{...(1)}$$

$$S = xy + 2xz + 2yz = xy + \frac{2V}{y} + \frac{2V}{x} = f(x, y) \qquad \text{[Using (1)]}$$

$$f_x = y - \frac{2V}{x^2}, \; f_y = x - \frac{2V}{y^2}$$

$$r = f_{xx} = \frac{4V}{x^3}, \; s = f_{xy} = 1, \; t = f_{yy} = \frac{4V}{y^3}$$

Now, $f_x = 0$ and $f_y = 0$

$$\Rightarrow \quad y - \frac{2V}{x^2} = 0 \quad \text{...(1)} \qquad x - \frac{2V}{y^2} = 0 \quad \text{...(2)}$$

From (1), $y = \frac{2V}{x^2}$

$\therefore$ From (2), $x - 2V \cdot \frac{x^4}{4V^2} = 0$ or $x\left(1 - \frac{x^3}{2V}\right) = 0$ or $x = (2V)^{1/3}$ $\quad | \because \; x \neq 0$

and $y = \frac{2V}{x^2} = \frac{2V}{(2V)^{3/2}} = (2V)^{1/3}$

$\therefore$ $x = y = (2V)^{1/3}$ is a stationary point.

At this point, $r = \frac{4V}{2V} = 2 > 0, \; s = 1, \; t = \frac{4V}{2V} = 2$

so that $rt - s^2 = 4 - 1 = 3 > 0$ and $r > 0$

$\Rightarrow$ S is minimum when $x = y = (2V)^{1/3}$

Also $z = \frac{V}{xy} = \frac{V}{(2V)^{2/3}} = \frac{V^{1/3}}{2^{2/3}} = \frac{(2V)^{1/3}}{2} = \frac{y}{2}$

Hence S is minimum when $x = y = 2z = (2V)^{1/3}$.

Example 6. *Prove that if the perimeter of a triangle is constant, its area is maximum when the triangle is equilateral.*

Sol. Let a, b, c be the sides of a triangle whose perimeter $2s$ is constant.

Then $2s = a + b + c$ or $c = 2s - a - b$ $\qquad$ ...(1)

$$\Delta = \sqrt{s(s-a)(s-b)(s-c)} = \sqrt{s(s-a)(s-b)(a+b-s)} \qquad | \because \text{Using (1)}$$

Let $\quad z = \Delta^2 = s(s-a)(s-b)(a+b-s) = f(a, b)$

$$f_a = s(s-b)\frac{\partial}{\partial a}[(s-a)(a+b-s)]$$
$$= s(s-b)[-(a+b-s)+(s-a)] = s(s-b)(2s-2a-b)$$
$$f_b = s(s-a)\frac{\partial}{\partial b}[(s-b)(a+b-s)]$$
$$= s(s-a)[-(a+b-s)+(s-b)] = s(s-a)(2s-a-2b)$$
$$r = f_{aa} = -2s(s-b),\ s = f_{ab} = s[-(2s-a-2b)-(s-a)] = s(2a+2b-3s)$$
$$t = f_{bb} = -2s(s-a)$$

Now, $\quad f_a = 0 \quad$ and $\quad f_b = 0$

$\Rightarrow \quad s(s-b)(2s-2a-b) = 0$ and $s(s-a)(2s-a-2b) = 0$

$\Rightarrow \quad (s-b)(2s-2a-b) = 0 \quad \ldots(1)$

$(s-a)(2x-a-2b) = 0 \quad \ldots(2) \quad | \because \quad s \neq 0$

From (1), $\quad s = b \quad$ or $\quad 2s = 2a+b$

When $s = b$, from (2), $\quad (b-a)(-a) = 0 \quad$ or $\quad b = a \quad | \because \quad a \neq 0$

When $2s = 2a+b$, from (2), $\quad \frac{b}{2}(a-b) = 0 \quad$ or $\quad a = b \quad | \because \quad b \neq 0$

If we express z as a function of b and c, we similarly get $b = c$

$$\therefore \quad a = b = c = \frac{2s}{3}$$
$$r = -2s\left(\frac{s}{3}\right) = -\frac{2s^2}{3} < 0$$
$$\text{'}s\text{'} = s\left(\frac{4s}{3}+\frac{4s}{3}-3s\right) = s\left(-\frac{s}{3}\right) = -\frac{s^2}{3}$$
$$t = -2s\left(\frac{s}{3}\right) = -\frac{2s^2}{3}$$
$$\text{'}rt - s^2\text{'} = \frac{4s^4}{9} - \frac{s^4}{9} = \frac{s^4}{3} > 0. \text{ Also } r < 0$$

$\therefore \quad z$ and hence Δ is maximum when $a = b = c = \frac{2s}{3}$ *i.e.*, when the triangle is equilateral.

Example 7. *Show that the rectangular solid of maximum volume that can be inscribed in a given sphere is a cube.*

Sol. Let x, y, z be the length, breadth and height of the rectangular solid. If V is the volume of the solid, then

$$V = xyz \quad \ldots(1)$$

Since each diagonal of the solid passes through the centre of the sphere

$\therefore \quad$ each diagonal = diameter of sphere = d (say)

$$\Rightarrow \quad \sqrt{x^2+y^2+z^2} = d \quad \text{or} \quad x^2+y^2+z^2 = d^2 \quad \text{or} \quad z = \sqrt{d^2-x^2-y^2}$$

$\therefore \quad$ From (1), $\quad V = xy\sqrt{d^2-x^2-y^2}$

or $\quad V^2 = x^2y^2(d^2-x^2-y^2) = d^2x^2y^2 - x^4y^2 - x^2y^4 = f(x, y)$

$$f_x = 2d^2xy^2 - 4x^3y^2 - 2xy^4 = 2xy^2(d^2-2x^2-y^2)$$

$$f_y = 2d^2x^2y - 2x^4y - 4x^2y^3 = 2x^2y(d^2 - x^2 - 2y^2)$$
$$r = f_{xx} = 2d^2y^2 - 12x^2y^2 - 2y^4,\ s = f_{xy} = 4d^2xy - 8x^3y - 8xy^3$$
$$t = f_{yy} = 2d^2x^2 - 2x^4 - 12x^2y^2$$

Now, $f_x = 0$ and $f_y = 0$

$\Rightarrow$ $d^2 - 2x^2 - y^2 = 0$ ...(1) $[\because\ x \neq 0, y \neq 0]$

$d^2 - x^2 - 2y^2 = 0$ ...(2)

Subtracting (2) from (1), $-x^2 + y^2 = 0 \Rightarrow y = x$

$\therefore$ From (1), $x = y = \dfrac{d}{\sqrt{3}}$

$$z = \sqrt{d^2 - \frac{d^2}{3} - \frac{d^2}{3}} = \frac{d}{\sqrt{3}}$$

At $x = y = \dfrac{d}{\sqrt{3}}$,

$$r = 2d^2 \cdot \frac{d^2}{3} - 12 \cdot \frac{d^2}{3} \cdot \frac{d^2}{3} - 2 \cdot \frac{d^4}{9} = -\frac{8d^4}{9} < 0$$
$$s = 4d^2 \cdot \frac{d^2}{3} - 8 \cdot \frac{d^3}{3\sqrt{3}} \cdot \frac{d}{\sqrt{3}} - 8 \cdot \frac{d}{\sqrt{3}} \cdot \frac{d^3}{3\sqrt{3}} = -\frac{4d^4}{9}$$
$$t = 2d^2 \cdot \frac{d^2}{3} - 2 \cdot \frac{d^4}{9} - 12 \cdot \frac{d^2}{3} \cdot \frac{d^2}{3} = -\frac{8d^4}{9}$$
$$rt - s^2 = \frac{64d^8}{81} - \frac{16d^8}{81} = \frac{16d^8}{27} > 0, \text{ Also } r < 0.$$

$\therefore$ V^2 and hence, V is maximum when $x = y = z$ *i.e.*, when the rectangular solid is a cube.

Example 8. *Find the volume of the largest rectangular parallelopiped that can be inscribed in the ellipsoid,* $\dfrac{x^2}{a^2} + \dfrac{y^2}{b^2} + \dfrac{z^2}{c^2} = 1$.

Sol. Let (x, y, z) be a vertex of the parallelopiped, then it lies on the ellipsoid

$$\frac{x^2}{a^2} + \frac{y^2}{b^2} + \frac{z^2}{c^2} = 1 \quad \text{...(1)}$$

Also its dimensions are $2x$, $2y$, $2z$ so that the volume V is given by $V = 2x \cdot 2y \cdot 2z = 8xyz$

$$\Rightarrow \quad V^2 = 64x^2y^2z^2 = 64x^2y^2c^2\left(1 - \frac{x^2}{a^2} - \frac{y^2}{b^2}\right) = 64c^2\left(x^2y^2 - \frac{x^4y^2}{a^2} - \frac{x^2y^4}{b^2}\right) = f(x, y)$$

$$f_x = 64c^2\left(2xy^2 - \frac{4x^3y^2}{a^2} - \frac{2xy^4}{b^2}\right);\quad f_y = 64c^2\left(2x^2y - \frac{2x^4y}{a^2} - \frac{4x^2y^3}{b^2}\right)$$

$$r = f_{xx} = 64c^2\left(2y^2 - \frac{12x^2y^2}{a^2} - \frac{2y^4}{b^2}\right);\quad s = f_{xy} = 64c^2\left(4xy - \frac{8x^3y}{a^2} - \frac{8xy^3}{b^2}\right)$$

$$t = f_{yy} = 64c^2\left(2x^2 - \frac{2x^4}{a^2} - \frac{12x^2y^2}{b^2}\right)$$

Now, $f_x = 0$ and $f_y = 0$

$$\Rightarrow \quad 128c^2xy^2\left(1-\frac{2x^2}{a^2}-\frac{y^2}{b^2}\right)=0 \quad \text{and} \quad 128c^2x^2y\left(1-\frac{x^2}{a^2}-\frac{2y^2}{b^2}\right)=0$$

$$\Rightarrow \quad 1-\frac{2x^2}{a^2}-\frac{y^2}{b^2}=0 \qquad \ldots(1) \qquad [\because \; x \neq 0, y \neq 0]$$

$$1-\frac{x^2}{a^2}-\frac{2y^2}{b^2}=0 \qquad \ldots(2)$$

Subtracting (2) from (1), $-\dfrac{x^2}{a^2}+\dfrac{y^2}{b^2}=0$ or $y=\dfrac{bx}{a}$

$\therefore$ From (1), $\quad 1-\dfrac{2x^2}{a^2}-\dfrac{x^2}{a^2}=0$ or $x^2=\dfrac{a^2}{3}$ or $x=\dfrac{a}{\sqrt{3}}$

$$\therefore \quad y=\frac{b}{a}\cdot\frac{a}{\sqrt{3}}=\frac{b}{\sqrt{3}}$$

and

$$z^2=c^2\left(1-\frac{x^2}{a^2}-\frac{y^2}{b^2}\right)=c^2\left(1-\frac{1}{3}-\frac{1}{3}\right)=\frac{c^2}{3}$$

$$\therefore \quad z=\frac{c}{\sqrt{3}}$$

Thus $x=\dfrac{a}{\sqrt{3}}$, $y=\dfrac{b}{\sqrt{3}}$ is a stationary point.

At this point, $\quad r=64c^2\left[\dfrac{2b^2}{3}-\dfrac{12}{a^2}\cdot\dfrac{a^2}{3}\cdot\dfrac{b^2}{3}-\dfrac{2}{b^2}\cdot\dfrac{b^4}{9}\right]=-\dfrac{512}{9}\,b^2c^2<0$

$$s=64c^2\left[4\cdot\frac{a}{\sqrt{3}}\cdot\frac{b}{\sqrt{3}}-\frac{8}{a^2}\cdot\frac{a^3}{3\sqrt{3}}\cdot\frac{b}{\sqrt{3}}-\frac{8}{b^2}\cdot\frac{a}{\sqrt{3}}\cdot\frac{b^3}{3\sqrt{3}}\right]=-\frac{256}{9}\,abc^2$$

$$t=64c^2\left[2\cdot\frac{a^2}{3}-\frac{2}{a^2}\cdot\frac{a^4}{9}-\frac{12}{b^2}\cdot\frac{a^2}{3}\cdot\frac{b^2}{3}\right]=-\frac{512}{9}\,a^2c^2$$

$$rt-s^2=\left(\frac{512}{9}\right)^2 a^2b^2c^4-\left(\frac{256}{9}\right)^2 a^2b^2c^4=\left(\frac{256}{9}\right)^2 a^2b^2c^4\,(4-1)>0.$$

Also $\qquad r<0$

$\therefore$ V^2 and hence, V is maximum

When $\quad x=\dfrac{a}{\sqrt{3}}, y=\dfrac{b}{\sqrt{3}}, z=\dfrac{c}{\sqrt{3}}$

Maximum volume $=8\cdot\dfrac{a}{\sqrt{3}}\cdot\dfrac{b}{\sqrt{3}}\cdot\dfrac{c}{\sqrt{3}}=\dfrac{8abc}{3\sqrt{3}}$.

Example 9. *In a plane triangle ABC, find the maximum value of cos A cos B cos C.*

[CUSAT 2008, 2010]

Sol. $\cos A \cos B \cos C=\cos A \cos B \cos[\pi-(A+B)] \qquad | \because \; A+B+C=\pi$

$$=-\cos A \cos B \cos(A+B)=f(A, B)$$

$$\frac{\partial f}{\partial A} = -\cos B[-\sin A\cos(A+B) - \cos A\sin(A+B)]$$
$$= \cos B\sin[A+(A+B)] = \cos B\sin(2A+B)$$
$$\frac{\partial f}{\partial B} = -\cos A[-\sin B\cos(A+B) - \cos B\sin(A+B)]$$
$$= \cos A\sin[B+(A+B)] = \cos A\sin(A+2B)$$
$$r = 2\cos B\cos(2A+B)$$
$$s = -\sin A\sin(A+2B) + \cos A\cos(A+2B)$$
$$= \cos[A+(A+2B)] = \cos(2A+2B)$$
$$t = 2\cos A\cos(A+2B)$$

Now, $\frac{\partial f}{\partial A} = 0$ and $\frac{\partial f}{\partial B} = 0$

$\Rightarrow$ $\cos B\sin(2A+B) = 0$ ...(1)

$\cos A\sin(A+2B) = 0$ ...(2)

If $\cos B = 0$, then $B = \frac{\pi}{2}$

From (2), $\cos A\sin(A+\pi) = 0$ or $\cos A(-\sin A) = 0$

$\Rightarrow$ either $\cos A = 0$ *i.e.*, $A = \frac{\pi}{4}$ which is not possible ($\because$ $A+B+C=\pi \Rightarrow C=0$)

or $\sin A = 0$ *i.e.*, $A = 0$ or π which is not possible.

$\therefore$ $\cos B \neq 0$. Similarly $\cos A \neq 0$

$\therefore$ From (1), $\sin(2A+B) = 0$ or $2A+B=\pi$

From (2), $\sin(A+2B) = 0$ or $A+2B=\pi$

Solving these equations, $A = B = \frac{\pi}{3}$

At $A = B = \frac{\pi}{3}$

$$r = 2\cos\frac{\pi}{3}\cos\pi = -1,\quad s = \cos\frac{4\pi}{3} = \cos\left(\pi+\frac{\pi}{3}\right) = -\cos\frac{\pi}{3} = -\frac{1}{2}$$
$$t = 2\cos\frac{\pi}{3}\cos\pi = -1$$
$$rt - s^2 = 1 - \frac{1}{4} = \frac{3}{4} > 0$$

Also $r = -1 < 0$

$\Rightarrow$ $f(A, B)$ is maximum at $A = B = \frac{\pi}{3}$

Maximum value $= f\left(\frac{\pi}{3}, \frac{\pi}{3}\right) = -\cos\frac{\pi}{3}\cos\frac{\pi}{3}\cos\frac{2\pi}{3} = -\left(\frac{1}{2}\right)\left(\frac{1}{2}\right)\left(-\frac{1}{2}\right) = \frac{1}{8}$.

7.15. LAGRANGE'S METHOD OF UNDETERMINED MULTIPLIERS

To find the maximum or minimum values of a function of three (or more) variables, when the variables are not independent but are connected by some given relation, we try to convert the given function to the one, having least number of independent variables with the help of the given relation. [See examples with 3.14]

When this procedure is not practicable, we use Lagrange's method.

Let $f(x, y, z)$ be a function of x, y, z which is to be examined for maximum or minimum value.

Let the variables x, y, z be connected by the relation $\phi(x, y, z) = 0$ ...(1)

For $f(x, y, z)$ to have a maximum or minimum value, the necessary condition is

$$\frac{\partial f}{\partial x} = 0, \frac{\partial f}{\partial y} = 0, \frac{\partial f}{\partial z} = 0$$

$$\therefore \quad \frac{\partial f}{\partial x} dx + \frac{\partial f}{\partial y} dy + \frac{\partial f}{\partial z} dz = 0 \quad ...(2)$$

Also, from (1), taking differentials, we get $\frac{\partial \phi}{\partial x} dx + \frac{\partial \phi}{\partial y} dy + \frac{\partial \phi}{\partial z} dz = 0$ ...(3)

Multiplying (3) by a parameter λ and adding to (2), we get

$$\left(\frac{\partial f}{\partial x} + \lambda \frac{\partial \phi}{\partial x}\right) dx + \left(\frac{\partial f}{\partial y} + \lambda \frac{\partial \phi}{\partial y}\right) dy + \left(\frac{\partial f}{\partial z} + \lambda \frac{\partial \phi}{\partial z}\right) dz = 0$$

This equation will hold good if $\frac{\partial f}{\partial x} + \lambda \frac{\partial \phi}{\partial x} = 0, \frac{\partial f}{\partial y} + \lambda \frac{\partial \phi}{\partial y} = 0, \frac{\partial f}{\partial z} + \lambda \frac{\partial \phi}{\partial z} = 0$

These equations together with equation (1), give the values of x, y, z and λ for a maximum or minimum.

Lagrange's method does not enable us to find whether there is a maximum or minimum. This fact is determined from the physical considerations of the problem.

Note. The above equations can be easily obtained by considering Lagrange's function

$$F(x, y, z) = f(x, y, z) + \lambda\phi(x, y, z)$$

and considering the stationary values of $F(x, y, z)$. For stationary values of $F(x, y, z)$, $dF = 0$

$$\Rightarrow \quad \left(\frac{\partial f}{\partial x} + \lambda \frac{\partial \phi}{\partial x}\right) dx + \left(\frac{\partial f}{\partial y} + \lambda \frac{\partial \phi}{\partial y}\right) dy + \left(\frac{\partial f}{\partial z} + \lambda \frac{\partial \phi}{\partial z}\right) dz = 0$$

$$\Rightarrow \quad \frac{\partial f}{\partial x} + \lambda \frac{\partial \phi}{\partial x} = 0, \frac{\partial f}{\partial y} + \lambda \frac{\partial \phi}{\partial y} = 0, \frac{\partial f}{\partial z} + \lambda \frac{\partial \phi}{\partial z} = 0.$$

Example 10. *Find the maximum and minimum distances of the point (3, 4, 12) from the sphere*

$$x^2 + y^2 + z^2 = 1.$$

Sol. Let (x, y, z) be any point on the sphere.

Distance of the point A(3, 4, 12) from (x, y, z) is given by $\sqrt{(x-3)^2 + (y-4)^2 + (z-12)^2}$

If the distance is maximum or minimum, so will be the square of the distance.

Let $\quad f(x, y, z) = (x - 3)^2 + (y - 4)^2 + (z - 12)^2$ ...(1)

subject to the condition that $\phi(x, y, z) \equiv x^2 + y^2 + z^2 - 1 = 0$ ...(2)

Consider Lagrange's function $F(x, y, z) = f(x, y, z) + \lambda\phi(x, y, z)$

$$= (x - 3)^2 + (y - 4)^2 + (z - 12)^2 + \lambda(x^2 + y^2 + z^2 - 1)$$

For stationary values, $dF = 0$.

$$\Rightarrow \quad [2(x - 3) + 2\lambda x]dx + [2(y - 4) + 2\lambda y]dy + [2(z - 12) + 2\lambda z]dz = 0$$

$$\Rightarrow \quad 2(x - 3) + 2\lambda x = 0 \quad ...(3)$$

$$2(y - 4) + 2\lambda y = 0 \quad ...(4)$$

$$2(z - 12) + 2\lambda z = 0 \quad ...(5)$$

Multiplying (3) by x, (4) by y, (5) by z and adding, we get

$$2(x^2 + y^2 + z^2) - 6x - 8y - 24z + 2\lambda(x^2 + y^2 + z^2) = 0$$

or $$2 - 6x - 8y - 24z + 2\lambda = 0 \qquad |\because \text{ Using (2)}$$

or $$3x + 4y + 12z = 1 + \lambda \qquad \text{...(6)}$$

From (3), (4) and (5), $x = \dfrac{3}{1+\lambda}, y = \dfrac{4}{1+\lambda}, z = \dfrac{12}{1+\lambda}$.

Putting these values of x, y, z in (6), we have

$$\frac{9}{1+\lambda} + \frac{16}{1+\lambda} + \frac{144}{1+\lambda} = 1 + \lambda \quad \text{or} \quad (1 + \lambda)^2 = 169 \quad \text{or} \quad 1 + \lambda = \pm 13$$

$$\therefore \qquad \lambda = 12 \quad \text{or} \quad -14$$

When $\lambda = 12$, $\qquad x = \dfrac{3}{13}, \quad y = \dfrac{4}{13}, \quad z = \dfrac{12}{13}$

When $\lambda = -14$, $\qquad x = -\dfrac{3}{13}, \quad y = -\dfrac{4}{13}, \quad z = -\dfrac{12}{13}$

Thus, we get two points P $\left(\dfrac{3}{13}, \dfrac{4}{13}, \dfrac{12}{13}\right)$ and Q $\left(-\dfrac{3}{13}, -\dfrac{4}{13}, -\dfrac{12}{13}\right)$ on the sphere which are at a *maximum or minimum* distance from the given point A.

Now, $$\text{AP} = \sqrt{\left(3 - \frac{3}{13}\right)^2 + \left(4 - \frac{4}{13}\right)^2 + \left(12 - \frac{12}{13}\right)^2} = 12$$

$$\text{AQ} = \sqrt{\left(3 + \frac{3}{13}\right)^2 + \left(4 + \frac{4}{13}\right)^2 + \left(12 + \frac{12}{13}\right)^2} = 14$$

$\therefore$ P $\left(\dfrac{3}{13}, \dfrac{4}{13}, \dfrac{12}{13}\right)$ is at a minimum distance from A and the minimum distance = 12.

Q $\left(-\dfrac{3}{13}, -\dfrac{4}{13}, -\dfrac{12}{13}\right)$ is at a maximum distance from A and the maximum distance = 14.

Example 11. *Find the minimum value of $x^2 + y^2 + z^2$, given that $ax + by + cz = p$.*

Sol. Let $$u = x^2 + y^2 + z^2 \qquad \text{...(1)}$$

where $$\phi(x, y, z) \equiv ax + by + cz - p = 0 \qquad \text{...(2)}$$

Consider Lagrange's function, $F(x, y, z) = (x^2 + y^2 + z^2) + \lambda(ax + by + cz - p)$

For stationary values, $dF = 0$

$$\Rightarrow \quad (2x + \lambda a)dx + (2y + \lambda b)dy + (2z + \lambda c)dz = 0$$

$$\Rightarrow \quad 2x + \lambda a = 0 \quad \text{...(3)} \qquad 2y + \lambda b = 0 \quad \text{...(4)} \qquad 2z + \lambda c = 0 \quad \text{...(5)}$$

Multiplying (3) by x, (4) by y, (5) by z and adding, we get

$$2(x^2 + y^2 + z^2) + \lambda(ax + by + cz) = 0$$

or $$2u + \lambda p = 0 \qquad |\because \text{ Using (1) and (2)}$$

$$\therefore \qquad \lambda = -\frac{2u}{p}$$

From (3), (4) and (5), $\qquad x = \dfrac{au}{p}, y = \dfrac{bu}{p}, z = \dfrac{cu}{p}$

$\therefore$ From (1), $\quad u = \dfrac{(a^2+b^2+c^2)u^2}{p^2}$ or $u = \dfrac{p^2}{a^2+b^2+c^2}$.

This is the **maximum or minimum** value of u. Now u is the square of the distance of any point P(x, y, z) on the plane (2) from the origin. Also, the length of perpendicular from O on the plane is $\dfrac{p}{\sqrt{a^2+b^2+c^2}}$.

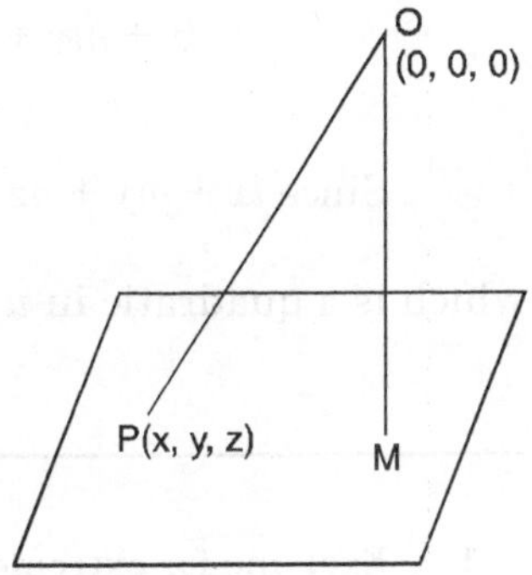

Clearly, OP is least when P coincides with M, the foot of the perpendicular from O on the plane.

Hence the minimum value of $u = \dfrac{p^2}{a^2+b^2+c^2}$.

Example 12. *Prove that the stationary values of* $u = \dfrac{x^2}{a^4}+\dfrac{y^2}{b^4}+\dfrac{z^2}{c^4}$ *where* $lx + my + nz = 0$ *and* $\dfrac{x^2}{a^2}+\dfrac{y^2}{b^2}+\dfrac{z^2}{c^2} = 1$ *are the roots of the equation*

$$\frac{l^2a^4}{1-a^2u}+\frac{m^2b^4}{1-b^2u}+\frac{n^2c^4}{1-c^2u} = 0.$$

Sol. Consider Lagrange's function,

$$F(x, y, z) = \left(\frac{x^2}{a^4}+\frac{y^2}{b^4}+\frac{z^2}{c^4}\right) + \lambda\,(lx+my+nz) + \mu\left(\frac{x^2}{a^2}+\frac{y^2}{b^2}+\frac{z^2}{c^2}-1\right)$$

For stationary values, $dF = 0$

$$\Rightarrow \quad \left(\frac{2x}{a^4}+\lambda l+\frac{2\mu x}{a^2}\right)dx+\left(\frac{2y}{b^4}+\lambda m+\frac{2\mu y}{b^2}\right)dy+\left(\frac{2z}{c^4}+\lambda n+\frac{2\mu z}{c^2}\right)dz = 0$$

$$\Rightarrow \quad \frac{2x}{a^4}+\lambda l+\frac{2\mu x}{a^2} = 0 \qquad ...(1)$$

$$\Rightarrow \quad \frac{2y}{b^4}+\lambda m+\frac{2\mu y}{b^2} = 0 \qquad ...(2)$$

and
$$\frac{2z}{c^4}+\lambda n+\frac{2\mu z}{c^2} = 0 \qquad ...(3)$$

Multiplying (1), (2), (3) by x, y, z respectively and adding, we get

$$2\left(\frac{x^2}{a^4}+\frac{y^2}{b^4}+\frac{z^2}{c^4}\right)+\lambda(lx+my+nz)+2\mu\left(\frac{x^2}{a^2}+\frac{y^2}{b^2}+\frac{z^2}{c^2}\right) = 0$$

$$\Rightarrow \quad 2u+\lambda(0)+2\mu\,(1) = 0 \qquad | \text{ from given relations}$$

$$\Rightarrow \quad \mu = -u$$

$\therefore$ Equation (1) becomes $\dfrac{2x}{a^4}+\lambda l-\dfrac{2ux}{a^2} = 0$ or $\dfrac{2x}{a^4}(1-a^2u) = -\lambda l$ or $x = -\dfrac{\lambda l a^4}{2(1-a^2u)}$

Similarly, $\quad y = -\dfrac{\lambda m b^4}{2(1-b^2u)}, \; z = -\dfrac{\lambda n c^4}{2(1-c^2u)}$

To eliminate λ between them, multiply these values of x, y, z by l, m, n respectively and add. Then

$$lx + my + nz = -\frac{\lambda}{2}\left[\frac{l^2a^4}{1-a^2u} + \frac{m^2b^4}{1-b^2u} + \frac{n^2c^4}{1-c^2u}\right]$$

Since $lx + my + nz = 0$, we have $\dfrac{l^2a^4}{1-a^2u} + \dfrac{m^2b^4}{1-b^2u} + \dfrac{n^2c^4}{1-c^2u} = 0$

which is a quadratic in u and gives two stationary values of u.

TEST YOUR KNOWLEDGE

1. Examine for extreme values:
   (*i*) $x^2 + y^2 + 6x + 12$ (*ii*) $x^3 + y^3 + 3xy$ (*iii*) $3x^2 - y^2 + x^3$
   (*iv*) $x^2y^2 - 5x^2 - 8xy - 5y^2$ (*v*) $x^3 + 3xy^2 - 15x^2 - 15y^2 + 72x$
   (*vi*) $x^2 - xy + y^2 + 3x - 2y + 1$ (*vii*) $x^3 + y^3 - 63(x + y) + 12xy$ (*viii*) $xy + \dfrac{a^3}{x} + \dfrac{a^3}{y}$
   (*ix*) $\sin x \sin y \sin (x + y)$ (*x*) $\cos x + \cos y + \cos (x + y)$.
2. Find the stationary points of the function $z = x^3y^2 (12 - x - y)$ satisfying the condition $x > 0, y > 0$ and examine their nature.
3. Find the minimum values of $x^2 + y^2 + z^2$ when
   (*i*) $x + y + z = 3a$ (*ii*) $xyz = a^3$.
4. Find the minimum value of the function $x + y + z$ subject to the condition $\dfrac{a}{x} + \dfrac{b}{y} + \dfrac{c}{z} = 1$.
5. Find the minimum value of $ax + by + cz$ subject to the condition $\dfrac{a^2}{x} + \dfrac{b^2}{y} + \dfrac{c^2}{z} = 1$.
6. Find the minimum value of the function $x^2 + y^2 + z^2$ subject to the condition
   (*i*) $ax + by + cz = a + b + c$ (*ii*) $xy + yz + zx = 3a^2$.
7. Find the points on the surface $z^2 = xy + 1$ nearest to the origin.
8. Given $f(x, y, z) = \dfrac{5xyz}{x + 2y + 4z}$, find the values of x, y, z for which $f(x, y, z)$ is a maximum, subject to the condition $xyz = 8$.
9. A rectangular box, open at the top, is to have a volume of 32 c.c. Find the dimensions of the box requiring least material for its construction.
10. Find the dimensions of the rectangular box, open at the top, of maximum capacity whose surface is 432 sq. cm.
11. A tent on a square base of side x, has its sides vertical of height y and the top is a regular pyramid of height h. Find x and y in terms of h, if the canvas required for its construction is to be minimum for the tent to have a given capacity.
12. The temperature T at any point (x, y, z) in space is $T = 400\, xyz^2$. Find the highest temperature on the surface of the unit sphere $x^2 + y^2 + z^2 = 1$.
13. Find the shortest and the longest distances from the point $(1, 2, -1)$ to the sphere
    $x^2 + y^2 + z^2 = 24$.

14. Find the maximum and minimum distances from the origin to the curve $5x^2 + 6xy + 5y^2 - 8 = 0$.
15. Prove that of all the rectangular parallelopipeds of the same volume, the cube has the least surface.
16. Prove that of all the rectangular parallelopipeds of given surface, cube has the maximum volume.
17. Divide 24 into three parts such that the continued product of the first, square of the second and the cube of the third may be maximum.
18. Find the stationary values of $x^2 + y^2 + z^2$ subject to $ax^2 + by^2 + cz^2 = 1$ and $lx + my + nz = 0$.
19. Find the maximum value of $x^m y^n z^p$ when $x + y + z = a$.
20. If $u = a^3x^2 + b^3y^2 + c^3z^2$ where $x^{-1} + y^{-1} + z^{-1} = 1$, show that the stationary value of u is given by

$$x = \frac{\Sigma a}{a},\ y = \frac{\Sigma a}{b},\ z = \frac{\Sigma a}{c}.$$

Answers

1. (*i*) Min. value $= 3$ at $(-3, 0)$ (*ii*) Max. value $= 1$ at $(-1, -1)$
(*iii*) Max. value $= 4$ at $(-2, 0)$ (*iv*) Max. value $= 0$ at $(0, 0)$
(*v*) Max. value $= 112$ at $(4, 0)$; Min. value $= 108$ at $(6, 0)$
(*vi*) Min. value $= -\frac{4}{3}$ at $\left(-\frac{4}{3}, \frac{1}{3}\right)$
(*vii*) Max. value $= 784$ at $(-7, -7)$; Min. value $= -216$ at $(3, 3)$
(*viii*) Min. value $= 3a^2$ at (a, a)
(*ix*) Max. value $= \frac{3\sqrt{3}}{8}$ at $\left(\frac{\pi}{3}, \frac{\pi}{3}\right)$; Min. value $= -\frac{3\sqrt{3}}{8}$ at $\left(\frac{2\pi}{3}, \frac{2\pi}{3}\right)$
(*x*) Min. value $= -\frac{3}{2}$ at $\left(\frac{2\pi}{3}, \frac{2\pi}{3}\right)$
2. Maximum at $(6, 4)$ 3. (*i*) $3a^2$ (*ii*) $3a^2$
4. $(\sqrt{a} + \sqrt{b} + \sqrt{c})^2$ at $(\sqrt{a}(\sqrt{a} + \sqrt{b} + \sqrt{c}), \sqrt{b}(\sqrt{a} + \sqrt{b} + \sqrt{c}), \sqrt{c}(\sqrt{a} + \sqrt{b} + \sqrt{c}))$
5. $(a^{3/2} + b^{3/2} + c^{3/2})^2$
6. (*i*) $\frac{(a+b+c)^2}{a^2+b^2+c^2}$ (*ii*) $3a^2$ at (a, a, a) and $(-a, -a, -a)$
7. $(0, 0, \pm 1)$ 8. $(4, 2, 1)$ 9. 4 cm × 4 cm × 2 cm
10. Length = Breadth = 12 cm, Height = 6 cm
11. $x = \sqrt{5}h,\ y = \frac{1}{2}h$ 12. 50 13. $\sqrt{6}, 3\sqrt{6}$
14. 4, 1 17. 4, 8, 12
18. Two stationary values of u are roots of $\frac{l^2}{au-1} + \frac{m^2}{bu-1} + \frac{n^2}{cu-1} = 0$.
19. $m^m n^n p^p a^{m+n+p}/(m+n+p)^{m+n+p}$.

7.16. ERRORS AND APPROXIMATIONS

Let $$z = f(x, y) \quad ...(1)$$

If δx, δy are small increments in x and y respectively and δz, the corresponding increment in z, then

$$z + \delta z = f(x + \delta x, y + \delta y) \quad ...(2)$$

Subtracting (1) from (2), we get

$$\delta z = f(x+\delta x, y+\delta y) - f(x, y) = f(x, y) + \delta x \frac{\partial f}{\partial x} + \delta y \frac{\partial f}{\partial y} + \cdots - f(x, y)$$

$$= \delta x \frac{\partial f}{\partial x} + \delta y \frac{\partial f}{\partial y} \text{ (approximately)} \mid \text{Neglecting higher powers of } \delta x, \delta y$$

$$\therefore \quad \delta z = \frac{\partial f}{\partial x} \cdot \delta x + \frac{\partial f}{\partial y} \cdot \delta y \text{ (approximately)}$$

⇒ If δx and δy are small changes (or errors) in x and y respectively, then an approximate change (or error) in z is δz.

Replacing δx, δy, δz by dx, dy, dz respectively, we have $\mathbf{dz} = \frac{\partial \mathbf{f}}{\partial \mathbf{x}} \mathbf{dx} + \frac{\partial \mathbf{f}}{\partial \mathbf{y}} \mathbf{dy}$.

Note 1. If δx is the error in x, then **relative error** $= \frac{\delta x}{x}$

percentage error $= \frac{\delta x}{x} \times 100$.

Note 2. $f'(x)\, dx$ is called the **differential of** $f(x)$.

ILLUSTRATIVE EXAMPLES

Example 1. *Find the percentage error in the area of an ellipse when an error of + 1 percent is made in measuring the major and minor axes.* [CUSAT 2007]

Sol. If x and y are semi-major and semi-minor axes of the ellipse, then its area A is given by $A = \pi xy$

Taking logs, $\log A = \log \pi + \log x + \log y$

Taking differentials $\frac{dA}{A} = 0 + \frac{dx}{x} + \frac{dy}{y}$

or $$\frac{dA}{A} \times 100 = \frac{dx}{x} \times 100 + \frac{dy}{y} \times 100 = 1 + 1 = 2.$$

∴ Error in the area = 2%.

Example 2. *The work that must be done to propel a ship of displacement D for a distance 'S' in time 't' is proportional to* $\frac{S^2 D^{2/3}}{t^2}$. *Find approximately the increase of work necessary when the displacement is increased by 1%, the time diminished by 1% and the distance diminished by 2%.* [CUSAT 2010]

Sol. Let the work done be W, then $W \propto \frac{S^2 D^{2/3}}{t^2}$

$$\Rightarrow \quad W = k \cdot \frac{S^2 D^{2/3}}{t^2}, \text{ where } k \text{ is a constant.}$$

Taking logs, $\log W = \log k + 2 \log S + \frac{2}{3} \log D - 2 \log t$

Taking differentials,

$$\frac{dW}{W} = 0 + 2\frac{dS}{S} + \frac{2}{3}\frac{dD}{D} - 2\frac{dt}{t}$$

or

$$\frac{dW}{W} \times 100 = 2\frac{dS}{S} \times 100 + \frac{2}{3}\frac{dD}{D} \times 100 - 2\frac{dt}{t} \times 100$$

$$= 2(-2) + \frac{2}{3}(1) - 2(-1) = -\frac{4}{3}$$

$\therefore$ Approximate increase of work $= -\frac{4}{3}\%$.

Example 3. *The diameter and altitude of a can in the shape of a right circular cylinder are measured as 4 cm and 6 cm respectively. The possible error in each measurement is 0.1 cm. Find approximately the maximum possible error in the value computed for the volume and lateral surface.*

Sol. Let the diameter and altitude of the can be denoted by D and H respectively.

Then radius $= \frac{D}{2}$

(*i*) The volume V of the can is given by

$$V = \pi r^2 h = \frac{\pi}{4} D^2 H \qquad [\because \; V = f(D, H)]$$

$$\therefore \quad dV = \frac{\partial V}{\partial D} dD + \frac{\partial V}{\partial H} dH = \frac{\pi}{4}[2DHdD + D^2 dH]$$

$$= \frac{\pi}{4}[2 \times 4 \times 6 \times 0.1 + 4^2 \times 0.1] = \frac{\pi}{4}(6.4) = 1.6\,\pi \text{ cu cm.}$$

(*ii*) The lateral surface S of the can is given by

$$S = 2\pi rh = \pi DH \qquad [\because \; S = f(D, H)]$$

$$\therefore \quad dS = \frac{\partial S}{\partial D} dD + \frac{\partial S}{\partial H} dH$$

$$= \pi[HdD + DdH] = \pi[6 \times 0.1 + 4 \times 0.1] = \pi \text{ sq. cm.}$$

Example 4. *The height of a tower is determined by observing the elevation θ and φ of its summit from two points in a direct line with the foot of the tower and at a distance 'a' apart. Show that the error in the calculated height due to small errors dθ and dφ is approximately*

$$a(\sin^2 \theta \, d\phi - \sin^2 \phi \, d\theta) \operatorname{cosec}^2 (\theta - \phi).$$

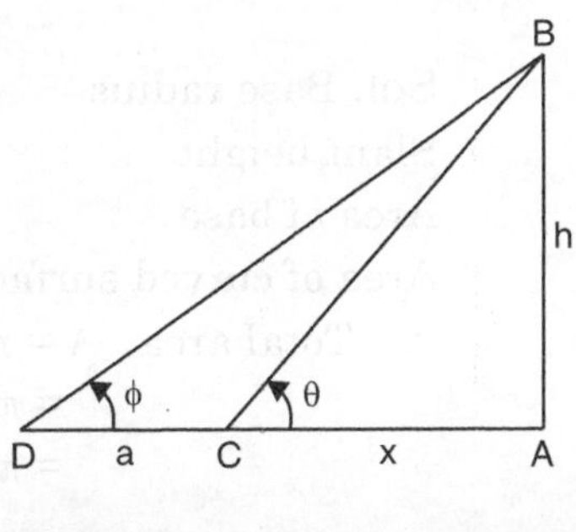

Sol. Let h be the height of the tower AB and C and D, the two points of observation so that

$$CD = a, \quad \angle ACB = \theta,$$
$$\angle ADB = \phi. \text{ Let } AC = x$$

From rt. $\angle d$. ΔBAC, $\quad x = h \cot \theta \qquad \ldots(1)$

From rt. $\angle d$. ΔBAD, $\quad x + a = h \cot \phi \qquad \ldots(2)$

Subtracting (1) from (2),

$$a = h(\cot\phi - \cot\theta) = h\left(\frac{\cos\phi}{\sin\phi} - \frac{\cos\theta}{\sin\theta}\right)$$

$$= h\left[\frac{\sin\theta\cos\phi - \cos\theta\sin\phi}{\sin\theta\sin\phi}\right] = \frac{h\sin(\theta-\phi)}{\sin\theta\sin\phi}$$

$$\Rightarrow \qquad h = \frac{a\sin\theta\sin\phi}{\sin(\theta-\phi)} = f(\theta,\phi) \qquad \ldots(1)$$

Taking logs, $\log h = \log a + \log\sin\theta + \log\sin\phi - \log\sin(\theta-\phi)$

Taking differentials

$$\frac{dh}{h} = 0 + \frac{\cos\theta}{\sin\theta}d\theta + \frac{\cos\phi}{\sin\phi}d\phi - \frac{\cos(\theta-\phi)}{\sin(\theta-\phi)}(d\theta - d\phi)$$

or

$$\frac{dh}{h} = \left[\frac{\cos\theta}{\sin\theta} - \frac{\cos(\theta-\phi)}{\sin(\theta-\phi)}\right]d\theta + \left[\frac{\cos\phi}{\sin\phi} + \frac{\cos(\theta-\phi)}{\sin(\theta-\phi)}\right]d\phi$$

$$= \frac{\sin(\theta-\phi)\cos\theta - \cos(\theta-\phi)\sin\theta}{\sin\theta\sin(\theta-\phi)}d\theta + \frac{\sin(\theta-\phi)\cos\phi + \cos(\theta-\phi)\sin\phi}{\sin\phi\sin(\theta-\phi)}d\phi$$

$$= \frac{\sin[(\theta-\phi)-\theta]}{\sin\theta\sin(\theta-\phi)}d\phi + \frac{\sin[(\theta-\phi)+\phi]}{\sin\phi\sin(\theta-\phi)}d\phi$$

$$= \frac{\sin(-\phi)}{\sin\theta\sin(\theta-\phi)}d\theta + \frac{\sin\theta}{\sin\phi\sin(\theta-\phi)}d\phi$$

$$= \frac{\sin^2\theta\,d\phi - \sin^2\phi\,d\theta}{\sin\theta\sin\phi\sin(\theta-\phi)} \qquad [\because \sin(-\phi) = -\sin\phi]$$

$$\therefore \quad dh = h\cdot\frac{\sin^2\theta\,d\phi - \sin^2\phi\,d\theta}{\sin\theta\sin\phi\sin(\theta-\phi)} = \frac{a\sin\theta\sin\phi}{\sin(\theta-\phi)}\cdot\frac{\sin^2\theta\,d\phi - \sin^2\phi\,d\theta}{\sin\theta\sin\phi\sin(\theta-\phi)} \qquad [\because \text{Using (1)}]$$

$$= a(\sin^2\theta\,d\phi - \sin^2\phi\,d\theta)\operatorname{cosec}^2(\theta-\phi).$$

Example 5. *The height h and semi-vertical angle α of a cone are measured, and from them A, the total area of the cone, including the base is calculated. If A and α are in error by small quantities δA and δα respectively, find the corresponding error in the area. Show further that, if* $\alpha = \frac{\pi}{6}$, *an error of + 1% in h will be approximately compensated by an error of – 0.33° in α.*

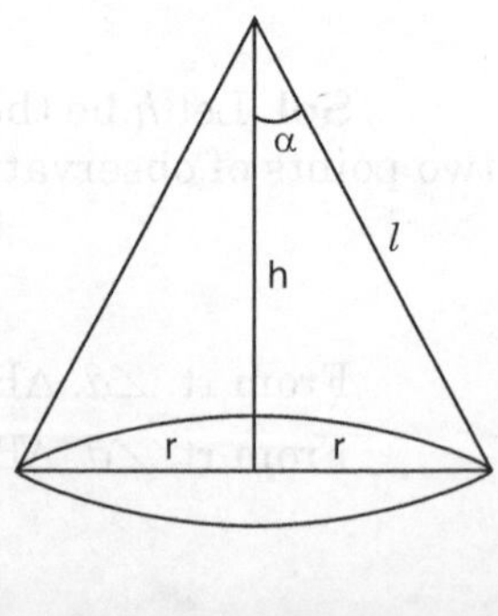

Sol. Base radius $r = h\tan\alpha$

Slant height $l = h\sec\alpha$

Area of base $= \pi r^2$

Area of curved surface $= \pi r l$

$\therefore$ Total area $A = \pi r^2 + \pi r l = \pi r(r + l)$

$= \pi h\tan\alpha(h\tan\alpha + h\sec\alpha)$

$= \pi h^2(\tan^2\alpha + \tan\alpha\sec\alpha)$

$$\delta A = \frac{\partial A}{\partial h}\delta h + \frac{\partial A}{\partial \alpha}\delta\alpha$$
$$= 2\pi h\,(\tan^2\alpha + \tan\alpha\sec\alpha)\,\delta h + \pi h^2\,(2\tan\alpha\sec^2\alpha + \sec^3\alpha + \tan\alpha\sec\alpha\tan\alpha)\,\delta\alpha$$
$$= 2\pi h\tan\alpha(\tan\alpha + \sec\alpha)\,\delta h + \pi h^2\sec\alpha(2\tan\alpha\sec\alpha + \sec^2\alpha + \tan^2\alpha)\,\delta\alpha \quad ...(1)$$

which gives the error in A.

Putting $\alpha = \frac{\pi}{6}$ and $\delta h = 1\%$ of $h = \frac{h}{100}$ in (1), we have

$$\delta A = 2\pi h\cdot\frac{1}{\sqrt{3}}\left(\frac{1}{\sqrt{3}}+\frac{2}{\sqrt{3}}\right)\cdot\frac{h}{100} + \pi h^2\cdot\frac{2}{\sqrt{3}}\left(\frac{2}{\sqrt{3}}\cdot\frac{2}{\sqrt{3}}+\frac{4}{4}+\frac{1}{3}\right)\delta\alpha$$
$$= \frac{2\pi h^2}{100} + 2\sqrt{3}\pi h^2\delta\alpha$$

Since the error in h is to be compensated by the error in α, $\delta A = 0$

$$\Rightarrow \quad \frac{1}{100} + \sqrt{3}\,\delta\alpha = 0 \quad \text{or} \quad \delta\alpha = -\frac{1}{100\sqrt{3}} \text{ radians}$$

or
$$\delta\alpha = -\frac{.01}{1.732}\times 57.3 \text{ degrees} \qquad [\because\ 1 \text{ radian} = 57.3° \text{ nearly}]$$
$$= -0.33 \text{ degrees.}$$

Example 6. *If the sides and angles of a triangle ABC vary in such a way that its circum-radius remains constant, prove that* $\frac{da}{\cos A}+\frac{db}{\cos B}+\frac{dc}{\cos C} = 0.$

Sol. The circum-radius R of a ΔABC is given by

$$R = \frac{a}{2\sin A} = \frac{b}{2\sin B} = \frac{c}{2\sin C}$$

Now, $a = 2R\sin A$ (R is constant)

Taking differentials, $da = 2R\cos A\,dA$ or $\frac{da}{\cos A} = 2R\,dA$

Similarly, $\frac{db}{\cos B} = 2R\,dB$; $\frac{dc}{\cos C} = 2R\,dC$

Adding, $$\frac{da}{\cos A}+\frac{db}{\cos B}+\frac{dc}{\cos C} = 2R\,(dA + dB + dC) \quad ...(1)$$

Also $A + B + C = \pi$

Taking differentials, $dA + dB + dC = 0$

$\therefore$ From (1), $$\frac{da}{\cos A}+\frac{db}{\cos B}+\frac{dc}{\cos C} = 0.$$

Example 7. *In a plane triangle, if the sides a, b be constant, prove that the variations of its angles are given by the relations* $\frac{dA}{\sqrt{a^2 - b^2\sin^2 A}} = \frac{dB}{\sqrt{b^2 - a^2\sin^2 B}} = -\frac{dC}{c}$*, the letters having their usual significance.*

Sol. By sine formula, $$\frac{a}{\sin A} = \frac{b}{\sin B}$$

$$a \sin B = b \sin A \qquad ...(1)$$

Taking differentials $a \cos B \cdot dB = b \cos A \cdot dA$

or $$\frac{dA}{a \cos B} = \frac{dB}{b \cos A} = \frac{dA + dB}{a \cos B + b \cos A} \qquad ...(2)$$

$$\left[\because \text{ If } \frac{a}{b} = \frac{c}{d}, \text{ then each} = \frac{a+c}{b+d}\right]$$

Now, $$a \cos B = a\sqrt{1 - \sin^2 B} = \sqrt{a^2 - a^2 \sin^2 B} = \sqrt{a^2 - b^2 \sin^2 A} \qquad \text{[Using (1)]}$$

$$b \cos A = b\sqrt{1 - \sin^2 A} = \sqrt{b^2 - b^2 \sin^2 A} = \sqrt{b^2 - a^2 \sin^2 B} \qquad \text{[Using (1)]}$$

$a \cos B + b \cos A = c$ [By projection formula]

Also $A + B + C = \pi \Rightarrow A + B = \pi - C$ so that $dA + dB = -\,dC$

$\therefore$ From (2), $$\frac{dA}{\sqrt{a^2 - b^2 \sin^2 A}} = \frac{dB}{\sqrt{b^2 - a^2 \sin^2 B}} = \frac{-\,dC}{c}.$$

Example 8. *If $x^2 + y^2 + z^2 - 2xyz = 1$, show that* $\dfrac{dx}{\sqrt{1-x^2}} + \dfrac{dy}{\sqrt{1-y^2}} + \dfrac{dz}{\sqrt{1-z^2}} = 0$.

Sol. Given $$x^2 + y^2 + z^2 - 2xyz = 1 \qquad ...(1)$$

Taking differentials, we get

$$2xdx + 2ydy + 2zdz - 2yzdx - 2zxdy - 2xydz = 0$$

or $$(x - yz)\,dx + (y - zx)\,dy + (z - xy)\,dz = 0 \qquad ...(2)$$

Now, $$(x - yz)^2 = x^2 - 2xyz + y^2z^2 = 1 - y^2 - z^2 + y^2z^2 \qquad |\text{Using (1)}$$

$$= (1 - y^2) - z^2(1 - y^2) = (1 - y^2)(1 - z^2)$$

$\Rightarrow$ $$x - yz = \sqrt{(1-y^2)(1-z^2)}$$

Similarly, $$y - zx = \sqrt{(1-z^2)(1-x^2)} \quad \text{and} \quad z - xy = \sqrt{(1-x^2)(1-y^2)}$$

$\therefore$ From (2), $$\sqrt{(1-y^2)(1-z^2)}\,dx + \sqrt{(1-z^2)(1-x^2)}\,dy + \sqrt{(1-x^2)(1-y^2)}\,dz = 0$$

Dividing by $\sqrt{(1-x^2)(1-y^2)(1-z^2)}$, we get $$\frac{dx}{\sqrt{1-x^2}} + \frac{dy}{\sqrt{1-y^2}} + \frac{dz}{\sqrt{1-z^2}} = 0.$$

Example 9. *If $u = xyz$, find d^2u.*

Sol. We know that if $u = f(x, y, z)$, then

$$du = \frac{\partial u}{\partial x}dx + \frac{\partial u}{\partial y}dy + \frac{\partial u}{\partial z}dz = \left(dx\frac{\partial}{\partial x} + dy\frac{\partial}{\partial y} + dz\frac{\partial}{\partial z}\right)u$$

$\therefore$ $d^2u = d(du)$

$$= \left(dx\frac{\partial}{\partial x} + dy\frac{\partial}{\partial y} + dz\frac{\partial}{\partial z}\right)\left(dx\frac{\partial}{\partial x} + dy\frac{\partial}{\partial y} + dz\frac{\partial}{\partial z}\right)u = \left(dx\frac{\partial}{\partial x} + dy\frac{\partial}{\partial y} + dz\frac{\partial}{\partial z}\right)^2 u$$

$$= \left[(dx)^2\frac{\partial^2}{\partial x^2} + (dy)^2\frac{\partial^2}{\partial y^2} + (dz^2)^2\frac{\partial^2}{\partial z^2} + 2dxdy\frac{\partial^2}{\partial x\partial y} + 2dydz\frac{\partial^2}{\partial y\partial z} + 2dzdx\frac{\partial^2}{\partial z\partial x}\right]u$$

$$= \frac{\partial^2 u}{\partial x^2}(dx)^2 + \frac{\partial^2 u}{\partial y^2}(dy)^2 + \frac{\partial^2 u}{\partial z^2}(dz)^2 + 2\frac{\partial^2 u}{\partial x\partial y}dxdy + 2\frac{\partial^2 u}{\partial y\partial z}dydz + 2\frac{\partial^2 u}{\partial z\partial x}dzdx \quad ...(1)$$

Here $u = xyz$

$$\frac{\partial u}{\partial x} = yz, \ \frac{\partial u}{\partial y} = zx, \ \frac{\partial u}{\partial z} = xy$$

$$\frac{\partial^2 u}{\partial x^2} = \frac{\partial^2 u}{\partial y^2} = \frac{\partial^2 u}{\partial z^2} = 0$$

$$\frac{\partial^2 u}{\partial x\partial y} = z, \ \frac{\partial^2 u}{\partial y\partial z} = x, \ \frac{\partial^2 u}{\partial z\partial x} = y$$

∴ From (1), we have $d^2u = 2zdxdy + 2xdydz + 2ydzdx$.

Example 10. *The focal length of a mirror is given by the formula* $\frac{1}{v} - \frac{1}{u} = \frac{2}{f}$. *If equal errors δ and made in the determination of u and v, show that the relative error in the focal length is given by* $\left(\frac{1}{u} + \frac{1}{v}\right)\delta$. (CUSAT 2008)

Sol. $$\frac{1}{v} - \frac{1}{u} = \frac{2}{f} \quad .... (1)$$

Taking differentials, we get,

$$-\frac{1}{v^2}dv + \frac{1}{u^2}du = \frac{-2}{f^2}df$$

or $$-\frac{1}{v^2}\delta + \frac{1}{u^2}\delta = \frac{-2}{f^2}df$$

$$\Rightarrow \quad -\left(\frac{1}{v^2} - \frac{1}{u^2}\right)\delta = -\frac{2}{f^2}df$$

$$\Rightarrow \quad -\left(\frac{1}{v} - \frac{1}{u}\right)\left(\frac{1}{v} + \frac{1}{u}\right) = \frac{-2}{f}\frac{df}{f}$$

∴ From equation (1), $\frac{1}{v} + \frac{1}{u} = \frac{df}{f}$

or the relative error in $f = \frac{1}{v} + \frac{1}{u}$.

TEST YOUR KNOWLEDGE

1. Find the percentage error in the area of a rectangle when an error of +1 percent is made in measuring its length and breadth.
2. The radius of a circle is found to be 100 cm. Find the relative error in the area of the circle due to an error of 1 mm in measuring the radius.
3. The radius of a sphere is found to be 10 cm with a possible error of 0.02 cm. What is the relative error in computing the volume?
4. What error in the common logarithm of a number will be produced by an error of 1% in the number?
5. The density of the atmosphere at height z metres is given by $\rho = \rho_0 e^{\frac{-z}{800}}$, where ρ_0 is the density at sea level. Find the approximate increase in height corresponding to a fall of 1% in the density.
6. If the kinetic energy $k = \dfrac{wv^2}{2g}$, find approximately the change in the kinetic energy as w changes from 49 to 49.5 and v changes from 1600 to 1590.
7. If the H.P. required to propel a steamer varies as the cube of the velocity and the square of its length, prove that a 3% increase in velocity and 4% increase in length will require an increase of about 17% in H.P.

   [**Hint.** If H denotes the horse-power, v the velocity and l, the length, then $H \propto v^3 l^2 \Rightarrow H = kv^3 l^2$, where k is a constant.]
8. The indicated horse-power I of an engine is calculated from the formula $I = \dfrac{PLAN}{33000}$, where $A = \dfrac{\pi}{4} d^2$. Assuming that errors of r percent may have been made in measuring P, L, N and d, find the greatest possible error in I.
9. The period T of a simple pendulum is $T = 2\pi\sqrt{\dfrac{l}{g}}$. Find the maximum error in T due to possible errors upto 1% in l and 2.5% in g.
10. The quantity Q of water flowing over a notch is given by $Q = \dfrac{8}{15} \times 0.64 \times \sqrt{2g} H^{5/2}$, where H is the head of the notch. What is the percentage error in Q caused by measuring H as 0.198 instead of 0.2?
11. The period T of a simple pendulum of length l is given by $T = 2\pi\sqrt{\dfrac{l}{g}}$. Find the (*i*) error and (*ii*) percent error (relative error) made in computing T by using $l = 2$ ft and $g = 32$ ft/sec^2, if the true values are $l = 1.95$ ft and $g = 32.2$ ft/sec^2.
12. The voltage V across a resistor is measured with an error h and the resistance R is measured with an error k. Show that the error in calculating the power $W = \dfrac{V^2}{R}$ generated in the resistor is $\dfrac{V}{R^2}(2Rh - Vk)$.

    [**Hint.** Here $dV = h$, $dR = k$, $dW = ?$]

13. The torsional rigidity of a length of a wire is obtained from the formula $N = \frac{8\pi Il}{t^2 r^4}$. If l is decreased by 2%, r increased by 2%, t increased by 1.5%, show that the value of N is diminished by 13% approximately.

14. If a triangle ABC be slightly varied but so as to remain inscribed in the same circle, prove that

$$\frac{\delta a}{\cos A} + \frac{\delta b}{\cos B} + \frac{\delta c}{\cos C} = 0.$$

15. In determining specific gravity by the formula $S = \frac{A}{A - w}$, where A is the weight in air and w is the weight in water; A can be read within 0.01 lb and w within 0.02 lb. Find approximately the maximum error in S, if the readings are A = 1.1 lb, w = 0.6 lb. Find also the maximum relative error.

16. The range R of a projectile which starts with a velocity v at an elevation α is given by $R = \frac{v^2 \sin 2\alpha}{g}$. Find the percentage error in R due to an error of 1% in v and an error of $\frac{1}{2}$ % in α.

17. If $ax^2 + by^2 + cz^2 = 1$ and $lx + my + nz = 0$, prove that $\frac{dx}{bny - cmz} = \frac{dy}{clz - anx} = \frac{dz}{amx - bly}$.

18. If Δ be the area of a triangle, prove that the error in Δ resulting from a small error in c is given by

$$\delta\Delta = \frac{1}{4}\Delta[s^{-1} + (s - a)^{-1} + (s - b)^{-1} - (s - c)^{-1}]\delta c.$$

19. Two sides a, b of a triangle and included angle C are measured. Show that the error δc in the computed length of third side c due to a small error in the angle C is given by $\delta c = a \sin B \cdot \delta C$.

20. If $z = xy(x + y)$, find d^2z, when x and y are the independent variables.

21. The deflection at the centre of a rod of length l and diameter d supported at its ends and loaded at the centre with a weight w varies as wl^3d^{-4}. What is the increase in the deflection corresponding to p% increase in w, q% decrease in l and r% increase in d?

22. At a distance of 50 metres from the foot of the tower the elevation of its top is 30°. If the possible errors in measuring the distance and elevation are 2 cm and 0.05 degrees, find the approximate error in calculating the height.

23. In estimating the cost of a pile bricks measured as 6 m × 50 m × 4 m, the tape is stretched 1% beyond the standard length. If the count is 12 bricks to 1 cu. m and bricks cost ₹ 100 per 1000, find the approximate error in the cost.

24. If $f(x, y) = \tan^{-1} xy$, compute $f(0.9, + 1.2)$ approximately by the use of differentials.

Answers

1. 2%	**2.** 0.002 sq. cm.	**3.** 0.006 cc	**4.** 0.0043429
5. 8 m	**6.** – 4500 units	**8.** $5r$%	**9.** 1.75%
10. – 2.5%	**11.** (*i*) 0.049	(*ii*) 1.585%	**15.** 0.112, 0.051
16. $2 + \alpha \cot 2\alpha$	**20.** $2y(dx)^2 + 2x(dy)^2 + 4(x + y)\,dxdy$		**21.** $(p - 3q - 4r)$%
22. 7 cm	**23.** ₹ 43.20	**24.** 0.8239.	

8

Coordinate Systems

8.1. COORDINATE SYSTEMS

There are several coordinate systems that are convenient to use with the various physical applications. The most common are the two dimensional and three dimensional rectangular coordinate system. The other coordinate systems that have very much applications are Polar coordinate system, cylindrical coordinate system and spherical coordinate system.

We can relate the rectangular coordinate system to polar coordinate system, cylindrical coordinate system and to spherical coordinate system.

8.2. POLAR COORDINATE SYSTEM

Polar coordinate system is another method for determining the position of a point in the plane. It also gives the direction.

Suppose O be a fixed point, called the Origin or Pole and OX a fixed line, called the initial line or Polar axis. Let P be any point in the plane and join OP and it is called radius vector. Let angle XOP = θ measured in anticlockwise direction and OP = r. The position of P is denoted by r, θ). A negative angle would be measured clockwise. Each point is determined by a distance r and an angle θ.

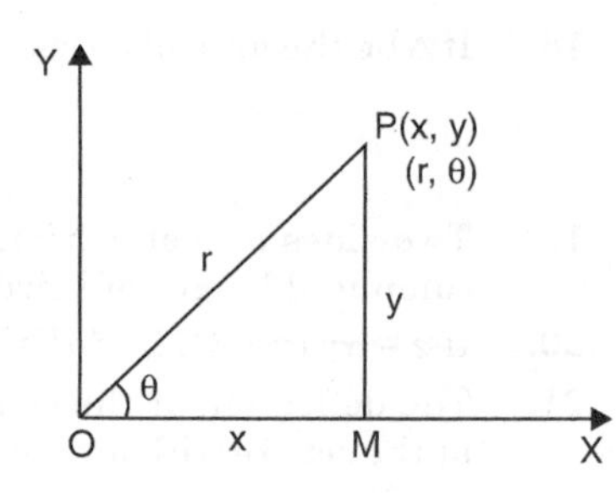

The relation between cartesian and polar coordinates is given by

$$x = r\cos\theta \qquad r^2 = \sqrt{(x^2 + y^2)}$$

$$y = r\sin\theta \text{ and } \theta = \tan^{-1}\left(\frac{y}{x}\right)$$

Using these relations one can easily transform between the cartesian and polar coordinates.

Note: *(i)* 1. A curve consists of all points (r, θ) which satisfies $r = f(\theta)$ is called Polar curve.

Example: $a\,(1 + \cos\theta)$, $r = 2$, $r = 2\cos\theta$

Note:

(i) $r = a$, constant represents circle centered at the origin and radius a.

(ii) $\theta = \theta_0$, constant represents the line through the origin.

(iii) The Jacobian of cartesian to polar coordinates is $J\left(\frac{x, y}{r, \theta}\right) = r$.

ILLUSTRATIVE EXAMPLES

Example 1. *Transform Cartesian into Polar coordinates :*

(i) (5, 4) *(ii) (2, 0).*

Sol. (*i*) Here $x = 5$, $y = 4$

Relation connecting Cartesian and Polar coordinates (r, θ) are

$$r = \sqrt{x^2 + y^2} \quad ...(1)$$

$$\theta = \tan^{-1} (y/x) \quad ...(2)$$

Putting the values of x and y in equations (1) and (2), we get

$$r = \sqrt{5^2 + 4^2} = 6.4$$

$$\theta = \tan^{-1} (4/5) = 38.66°$$

∴ The transformed Polar coordinate (6.4, 38.66°)

(*ii*) Here $x = 2$, $y = 0$

$$r = \sqrt{x^2 + y^2} \quad ...(1)$$

$$\theta = \tan^{-1} (y/x) \quad ...(2)$$

Substituting the values of x and y, we get

$$r = \sqrt{2^2 + 0^2} = 2$$

$$\theta = \tan^{-1} (0/2) = 0$$

∴ The polar coordinate is (2, 0).

Example 2. *Transform polar coordinate into cartesian coordinates (10, 30°).*

Sol. Here $v = 10$, $\theta = 30°$

$$x = r \cos \theta \quad ...(1)$$

$$y = r \sin \theta \quad ...(2)$$

Putting the values of r and θ in equation (1) and (2), we get

$$x = 10 \cos 30° = 8.66$$

$$y = 10 \sin 30° = 5$$

∴ The cartesian coordinate is (8.66, 5).

Example 3. *Convert the curve $x^2 + y^2 = 4x$ into polar curve.*

Sol. Here the given curve is

$$x^2 + y^2 = 4x \quad ...(1)$$

The relation between Cartesian and Polar coordinate is

$$x = r \cos \theta, \ y = r \cos \theta$$

Putting the values of x and y in equation (1), we get

$$r^2 \cos^2 \theta + r^2 \sin^2 \theta = 4r \cos \theta$$

$$\therefore \quad r^2 = 4r \cos \theta$$

or $$r = 4 \cos \theta.$$

8.3. CYLINDRICAL POLAR COORDINATE SYSTEM *[CUSAT 2007]*

Different coordinate system exists for representing three coordinate. Usually the geometry of the application suggests an appropriate system. Cylindrical polar coordinate system is suitable for problems involving cylinders. Cylindrical polar coordinate system is an extension of polar coordinate system by adding a third coordinate z, which gives the height of appoint above the plane. The notation for this system is P (r, θ, z), where

r-the distance from the origin to P′, the orthogonal projection of P onto XY-plane.

θ-the angle between positive X-axis and the line joining origin to P′, the projection of P

z-the distance PP′.

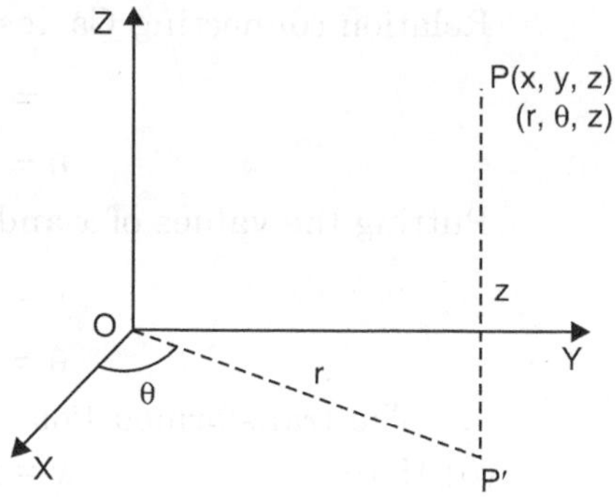

If (x, y, z) is the cartesian coordinate of P referred to OX, OY, OZ as axis, the following formula give the relation between cartesian coordinate and cylindrical polar coordinate.

$$x = r\cos\theta$$
$$y = r\sin\theta, \quad 0 \le r < \infty,\ 0 \le \theta \le 2\pi$$
$$z = z, \quad -\infty < z < \infty$$

The relation between Cartesian Coordinate to cylindrical polar coordinate is given below:

$$r^2 = \sqrt{(x^2 + y^2)}$$
$$\theta = \tan^{-1}(y/x)$$
$$z = z$$

Note:

(*i*) $r = a$, (a constant) represents right circular cylinder of radius a whose axis is the Z-axis

(*ii*) $\theta = \theta_0$, (a constant) represents half planes adjoining Z-axis.

(*iii*) $z = z_0$ (a constant) represents a plane parallel to XY plane.

(*iv*) The Jacobean of cartesian coordinate to cylindrical polar coordinate $J\left(\frac{x, y, z}{r, \theta, z}\right) = r$.

Example 1. *Express the point $(x, y, z) = (1, -\sqrt{3}, 2)$ in cylindrical coordinates.*

Sol. Here $x = 1, y = -\sqrt{3}, z = 2$

The relation connecting cartesian and cylindrical coordinates is

$$r = \sqrt{x^2 + y^2} \qquad \text{...(1)}$$
$$\theta = \tan^{-1}(y/x) \qquad \text{...(2)}$$
$$z = z \qquad \text{...(3)}$$

Putting the values of x, y, z in equations (1), (2) and (3), we get

$$r = \sqrt{1^2 + (\sqrt{3})^2} = 2$$
$$\theta = \tan^{-1}(-\sqrt{3}/1) = -\frac{\pi}{3}$$
$$z = 2$$

$\therefore$ The cylindrical coordinate is $\left(2, -\frac{\pi}{3}, 2\right)$.

Example 2. *Identify the surface for each of the following equations :*

(*i*) $r = 4$ (*ii*) $r^2 + z^2 = 25.$

Sol. (*i*) Here $r = 4$

Therefore $x^2 + y^2 = 4^2$, which is a circle, and $z = z$,

By allowing z to vary, we get a cylinder of radius 4.

Henec $r = 4$ represents a cylinder.

(*ii*) Here $r^2 + z^2 = 25$

Putting $r^2 = x^2 + y^2$, we get

$x^2 + y^2 + z^2 = 5^2$, which is a sphere centered at the origin with a radius of 5.

8.4. SPHERICAL POLAR COORDINATE SYSTEM [*CUSAT 2010*]

Spherical coordinates are a system of curvilinear coordinates that are natural for describing positions on a sphere or spheroid. Let P be any point in the space and M′ be the foot of the perpendicular from P on the XY-plane. Let (x, y, z) be the Cartesian Coordinate of P with respect to OX, OY, OZ axes.

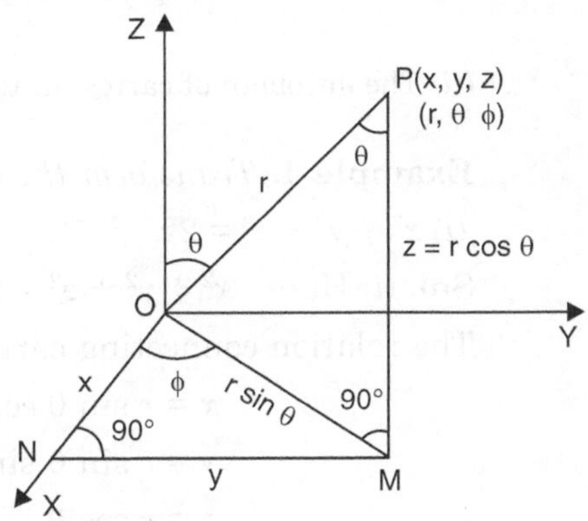

The spherical polar coordinate of P is given by (r, θ, φ), where r, $0 \le r < \infty$ is the distance from the origin to a given point P φ, $0 \le \phi \le 2\pi$ is the angle between the positive X-axis and the line from the origin to the point M, projection of P on the XY-plane. θ, $0 \le \theta \le \pi$ is the angle between the positive Z-axis and the line OP.

Relation Connecting Cartesian and Spherical Coordinates

From right angled triangle OMP,

$$\sin \theta = \frac{OM}{OP}$$

$$OM = OP \sin \theta = r \sin \theta$$

$$\cos \theta = \frac{PM}{OP}$$

$$PM = r \cos \theta$$

$$\therefore \quad z = r \cos \theta$$

Similarly, from right angled triangle ONM

$$x = ON = OM \cos \phi = r \sin \theta \cos \phi$$

$$y = MN = OM \sin \phi = r \sin \theta \sin \phi$$

Hence, we have the following formulae of transformations from cartesian to spherical polar coordinates

$$x = r \sin \theta \cos \phi \qquad \ldots(1)$$

$$y = r \sin \theta \sin \phi \qquad \ldots(2)$$

$$z = r\cos\theta,\ 0 \le r < \infty,\ 0 \le \phi \le 2\pi,\ 0 \le \theta \le \pi \quad ...(3)$$

From equations 1, 2 and 3 we have

$$r^2 = x^2 + y^2 + z^2$$

$$\sin\phi = \frac{y}{\sqrt{x^2 + y^2}}$$

$$\cos\phi = \frac{x}{\sqrt{x^2 + y^2}}$$

$$\cos\theta = \frac{z}{r}$$

Note:

(*i*) $r = a$ constant represents sphere of radius a with centre at origin.

(*ii*) The Jacobian of cartesian to spherical polar coordinates is $J\left(\frac{x, y, z}{r, \theta, \phi}\right) = r^2 \sin\theta$.

Example 1. *Transform the equations into spherical polar coordinates*

(*i*) $x^2 + y^2 + z^2 = 2z$ (*ii*) $x^2 + y^2 - 3z^2 = 0$.

Sol. (*i*) Here $x^2 + y^2 + z^2 = 2z$ ...(1)

The relation connecting cartesian and spherical polar coordinates is

$$x = r\sin\theta\cos\phi$$

$$y = r\sin\theta\sin\phi,$$

$$z = r\cos\theta$$

Putting the values of x, y and z in equation (1), we get

$$(r\sin\theta\cos\phi)^2 + (r\sin\phi\sin\phi)^2 + (r\cos\theta)^2 = 2r\cos\theta$$

or $$r^2\sin^2\theta(\cos^2\phi + \sin^2\phi) + r^2\cos^2\theta = 2r\cos\theta$$

or $$r^2 = 2r\cos\theta$$

$\therefore$ $$r = 2\cos\theta$$

(*ii*) Here $x^2 + y^2 - 3z^2 = 0$ ...(1)

Equation (1) can be written as $x^2 + y^2 + z^2 - 4z^2 = 0$ ...(2)

$$x = r\sin\theta\cos\phi$$

$$y = r\sin\theta\sin\phi,$$

$$z = r\cos\theta$$

Putting these values of x, y, z in equation (2), we get

$$r^2 - 4r^2\cos\theta = 0$$

$$r^2(1 - 4\cos^2\theta) = 0.$$

TEST YOUR KNOWLEDGE

Change to spherical polar coordinates and to cylindrical coordinates

(*i*) $x^2 + y^2 = 3$

(*ii*) $x^2 - z^2 = 4$

(*iii*) $x^2 + y^2 + 2z^2 = 4$.

Answers

(*i*) $r^2 \sin^2 \theta = 3$, $r^2 = 3$, $z = 0$

(*ii*) $r^2 (\sin^2 \theta \cos^2 \phi - \cos^2 \theta) = 4$, $r^2 \cos^2 \theta - z^2 = 4$

(*iii*) $r^2 (1 + \cos^2 \theta) = 4$, $r^2 + 2z^2 = 4$.

MODULE–IV

INTEGRAL CALCULUS

9

Application of Definite Integrals—Rectification

The process of finding the length of an arc of a curve between two given points on it is called *rectification.*

9.1. ARC FORMULAE FOR CARTESIAN EQUATIONS

To find the length of the curve y = f(x) between the points, where x = a, x = b.

Or

To show that the length of the arc of the curve y = f(x) between the points, where x = a, x = b is given by

$$\int_a^b \sqrt{1+\left(\frac{dy}{dx}\right)^2}\, dx$$

where y and $\frac{dy}{dx}$ *are continuous and single-valued functions in the interval* $a \le x \le b$ *and the integrand does not change sign in this interval.*

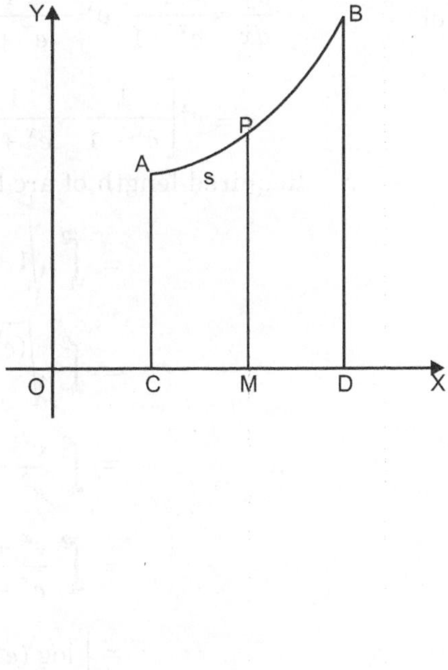

Let AB be the curve $y = f(x)$ where A, B are the points whose abscissae are a and b respectively and CA, DB are their ordinates.

Let P(x, y) be any point on the curve and draw MP $\perp$ x-axis.

If s denotes the length of the arc AP measured from the fixed point A to the variable point P, then s is clearly a function of x.

From Differential Calculus, we know that

$$\frac{ds}{dx} = \sqrt{1+\left(\frac{dy}{dx}\right)^2} \qquad ...(i)$$

$$\therefore \quad \int_a^b \sqrt{1+\left(\frac{dy}{dx}\right)^2}\, dx = \int_a^b \frac{ds}{dx}\, dx = \left[s\right]_a^b$$

= (value of s at $x = b$) – (value of s at $x = a$)

= arc AB – 0 = arc AB.

Hence $\quad$ **arc AB =** $\int_a^b \sqrt{1+\left(\frac{dy}{dx}\right)^2}\, dx.$

Cor. 1. From (1), we have $\mathbf{s} = \int \sqrt{1+\left(\frac{dy}{dx}\right)^2}\,dx.$

Cor. 2. *If the equation of the curve is of the form $x = f(y)$, then the length of the arc between the points, where $y = c$, $y = d$ is given by*

$$\int_c^d \sqrt{1+\left(\frac{dx}{dy}\right)^2}\,dy.$$

ILLUSTRATIVE EXAMPLES

Example 1. *Find the length of the arc of the curve*

(*a*) $y = \log \frac{e^x - 1}{e^x + 1}$ *from* $x = 1$ *to* $x = 2$.

(*b*) $y = \log \sec x$ *from* $x = 0$ *to* $x = \frac{\pi}{3}$.

Sol. (*a*) The equation of the curve is $y = \log \frac{e^x - 1}{e^x + 1}$

$\therefore \quad y = \log (e^x - 1) - \log(e^x + 1)$

or $\quad \frac{dy}{dx} = \frac{1}{e^x - 1} \cdot e^x - \frac{1}{e^x + 1} \cdot e^x$

$$= e^x\left[\frac{1}{e^x - 1} - \frac{1}{e^x + 1}\right] = e^x\left[\frac{e^x + 1 - e^x + 1}{e^{2x} - 1}\right] = \frac{2e^x}{e^{2x} - 1}$$

$\therefore$ Required length of arc from $x = 1$ to $x = 2$ is

$$= \int_1^2 \sqrt{1+\left(\frac{dy}{dx}\right)^2}\,dx = \int_1^2 \sqrt{1+\frac{4e^{2x}}{(e^{2x}-1)^2}}\,dx$$

$$= \int_1^2 \sqrt{\frac{(e^{2x}-1)^2 + 4e^{4x}}{(e^{2x}-1)^2}}\,dx = \int_1^2 \sqrt{\frac{(e^{2x}+1)^2}{(e^{2x}-1)^2}}\,dx$$

$$= \int_1^2 \frac{e^{2x}+1}{e^{2x}-1}\,dx \qquad \text{| Divide the num. and den. by } e^x$$

$$= \int_1^2 \frac{e^x + e^{-x}}{e^x - e^{-x}}\,dx$$

$$= \left| \log (e^x - e^{-x}) \right|_1^2 = \log (e^2 - e^{-2}) - \log (e - e^{-1})$$

$$= \log\left(e^2 - \frac{1}{e^2}\right) - \log\left(e - \frac{1}{e}\right) = \log\left(\frac{e^4 - 1}{e^2}\right) - \log\left(\frac{e^2 - 1}{e}\right)$$

$$= \log\left(\frac{e^4 - 1}{e^2} \cdot \frac{e}{e^2 - 1}\right) = \log\left(\frac{e^2 + 1}{e}\right) = \log(e + e^{-1}).$$

(*b*) The equation of the curve is $y = \log \sec x$.

$$\therefore \quad \frac{dy}{dx} = \frac{1}{\sec x} \cdot \sec x \tan x = \tan x.$$

$\therefore$ Required length of the arc from $x = 0$ to $x = \frac{\pi}{3}$ is

$$= \int_0^{\pi/3} \sqrt{1 + \left(\frac{dy}{dx}\right)^2}\, dx = \int_0^{\pi/3} \sqrt{1 + \tan^2 x}\, dx.$$

$$= \int_0^{\pi/3} \sec x\, dx = \Big| \log (\sec x + \tan x) \Big|_0^{\pi/3}.$$

$$\Big| \because \int \sec x\, dx = \log (\sec x + \tan x)$$

$$= \log\left(\sec\frac{\pi}{3} + \tan\frac{\pi}{3}\right) - \log (\sec 0 + \tan 0)$$

$$= \log (2 + \sqrt{3}) - \log (1 + 0) = \log (2 + \sqrt{3}) - 0 = \log (2 + \sqrt{3}).$$

Example 2. *Show that in the catenary* $y = c \cosh \frac{x}{c}\left[\text{or } y = \frac{c}{2}(e^{x/c} + e^{-x/c})\right]$ *the length of the arc from the vertex (where* $x = 0$*) to any point* (x, y) *is given by*

(*i*) $s = c \sinh \frac{x}{c}$ (*ii*) $s^2 = y^2 - c^2$

Sol. (*i*) The equation of the catenary is $y = c \cosh \frac{x}{c}$ ...(1)

$$\therefore \quad \frac{dy}{dx} = c \cdot \sinh \frac{x}{c} \cdot \frac{1}{c} = \sinh \frac{x}{c}.$$

$\therefore$ Required length of the arc from $x = 0$ to $x = x$ is

$$s = \int_0^x \sqrt{1 + \left(\frac{dy}{dx}\right)^2}\, dx = \int_0^x \sqrt{1 + \sinh^2 \frac{x}{c}}\, dx$$

$$= \int_0^x \cosh \frac{x}{c}\, dx = \Big| c \sinh \frac{x}{c} \Big|_0^x$$

$$= \left(c \sinh \frac{x}{c} - c \sinh 0\right) = c \sinh \frac{x}{c} - 0 \qquad [\because \sinh 0 = 0]$$

$$= c \sinh \frac{x}{c}, \text{ which proves part } (i).$$

(*ii*) Now

$$s^2 = c^2 \sinh^2 \frac{x}{c}$$

$$= c^2 \left(\cosh^2 \frac{x}{c} - 1\right) \qquad \Big| \begin{array}{l} \because \cosh^2\theta - \sinh^2\theta = 1 \\ \therefore \sinh^2\theta = \cosh^2\theta - 1 \end{array}$$

$$= c^2 \cosh^2 \frac{x}{c} - c^2 = y^2 - c^2 \qquad \Big| \text{ Using (1)}$$

Hence the result

Example 3. (*i*) *Find the length of the arc of the semi-cubical parabola $ay^2 = x^3$ from the vertex to the point (a, a).*

(*ii*) *Find the length of the curve $3ay^2 = 2x^3$ between the limits (0, 0) and (x, y).*

Sol. (*i*) The given curve is $ay^2 = x^3$

$$\therefore \quad y = \frac{1}{\sqrt{a}} . x^{3/2} \qquad | \text{ Taking +ve sign with the square root.}$$

$$\therefore \quad \frac{dy}{dx} = \frac{1}{\sqrt{a}} . \frac{3}{2} . x^{1/2} = \frac{3\sqrt{x}}{2\sqrt{a}}.$$

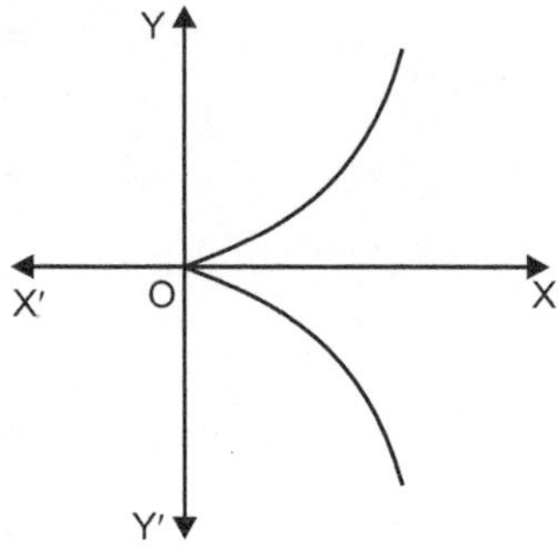

The vertex of the semi-cubical parabola $ay^2 = x^3$ is at the origin (0, 0).

Hence the limits of integration are $x = 0$ to $x = a$.

$\therefore$ Required length of the arc from the vertex (0, 0) to the point (a, a) is

$$= \int_0^a \sqrt{1 + \left(\frac{dy}{dx}\right)^2}\, dx$$

$$= \int_0^a \sqrt{1 + \frac{9x}{4a}}\, dx = \frac{1}{2\sqrt{a}} \int_0^a (9x + 4x)^{1/2}\, dx$$

$$= \frac{1}{2\sqrt{a}} . \left| \frac{(9x + 4x)^{3/2}}{\frac{3}{2}.9} \right|_0^a = \frac{1}{2\sqrt{a}} . \frac{2}{27} \left| (9x + 4a)^{3/2} \right|_0^a$$

$$= \frac{1}{27\sqrt{a}} [(13a)^{3/2} - (4a)^{3/2}] = \frac{a^{3/2}}{27\sqrt{a}} [13\sqrt{13} - 8] = \frac{a}{27} (13\sqrt{13} - 8).$$

(*ii*) The given curve is $3ay^2 = 2x^3$ or $y^2 = \frac{2x^3}{3a}$ $\therefore$ $y = \sqrt{\frac{2}{3a}} . x^{3/2}$

$$\therefore \quad \frac{dy}{dx} = \sqrt{\frac{2}{3a}} . \frac{3}{2} . x^{1/2} = \sqrt{\frac{3x}{2a}}\, x^{1/2} = \sqrt{\frac{3x}{2a}}.$$

$\therefore$ Required length of the arc from (0, 0) to (x, y) is

$$= \int_0^x \sqrt{1 + \left(\frac{dy}{dx}\right)^2}\, dx = \int_0^x \sqrt{1 + \frac{3x}{2a}}\, dx$$

$$= \frac{1}{\sqrt{2a}} \int_0^x (3x + 2a)^{1/2}\, dx = \frac{1}{\sqrt{2a}} \left| \frac{(3x + 2a)^{3/2}}{\frac{3}{2}.3} \right|_0^x$$

$$= \frac{1}{\sqrt{2a}} . \frac{2}{9} \left| (3x + 2a)^{3/2} \right|_0^x = \frac{2}{9\sqrt{2a}} [(3x + 2a)^{3/2} - (2a)^{3/2}]$$

$$= \frac{2}{9\sqrt{2a}}(2a)^{3/2}\left[\left(\frac{3x}{2a}+1\right)^{3/2}-1\right]$$

$$= \frac{4a}{9}\left[\left(1+\frac{3x}{2a}\right)^{3/2}-1\right].$$

Example 4. *Show that the length an arc of the curve $x^2 = a^2(1 - e^{y/a})$ measured from (0, 0) to (x, y) is* $a \log \dfrac{a+x}{a-x} - x$.

Sol. The given curve is $x^2 = a^2(1 - e^{y/a})$ or $x^2 = a^2 - a^2e^{y/a}$

or $$a^2e^{y/a} = a^2 - x^2 \quad \therefore \quad e^{y/a} = \frac{a^2-x^2}{a^2}$$

Taking logarithms, we get $\dfrac{y}{a} = \log\left(\dfrac{a^2-x^2}{a^2}\right)$

$$\therefore \quad y = a\log\left(\frac{a^2-x^2}{a^2}\right) = a\,[\log(a^2-x^2) - \log a^2]$$

$$\therefore \quad \frac{dy}{dx} = a\left[\frac{1}{a^2-x^2}.(-2x) - 0\right] = \frac{-2ax}{a^2-x^2}.$$

$\therefore$ Required length of the arc from (0, 0) to (x, y) is

$$= \int_0^x \sqrt{1+\left(\frac{dy}{dx}\right)^2}\,dx = \int_0^x \sqrt{1+\frac{4a^2x^2}{(a^2-x^2)^2}}\,dx$$

$$= \int_0^x \sqrt{\frac{(a^2-x^2)^2+4a^2x^2}{(a^2-x^2)^2}}\,dx = \int_0^x \frac{a^2+x^2}{a^2-x^2}\,dx$$

$$= \int_0^x \left(\frac{2a^2}{a^2-x^2}-1\right)dx = \left|\, 2a^2.\frac{1}{2a}\log\frac{a+x}{a-x} - x \,\right|_0^x$$

$$\left[\because \int\frac{1}{a^2-x^2}dx = \frac{1}{2a}\log\frac{a+x}{a-x}\right]$$

$$= \left(a\log\frac{a+x}{a-x} - x\right) - (a\log 1 - 0) = a\log\frac{a+x}{a-x} - x. \qquad |\because \log 1 = 0$$

Example 5. *Find the length of the arc of the parabola $y^2 = 4ax$*

(i) from the vertex to an extremity of the latus rectum.

(ii) cut off by the latus rectum.

(iii) cut off by the line $3y = 8x$.

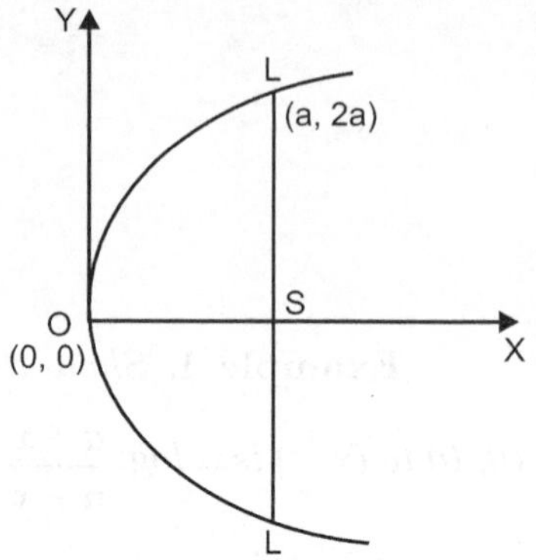

Sol. Let O be the vertex and L be an extremity of the latus rectum of the parabola $y^2 = 4ax$

$\therefore$ $$x = \frac{1}{4a} y^2$$

or $$\frac{dx}{dy} = \frac{1}{4a} . 2y = \frac{y}{2a}$$

(*i*) For the length of the arc from the vertex (0, 0) to an end L(a, $2a$) of latus rectum, y varies from 0 to $2a$.

$\therefore$ Required length of arc OL

$$= \int_0^{2a} \sqrt{1+\left(\frac{dx}{dy}\right)^2}\, dy = \int_0^{2a} \sqrt{1+\frac{y^2}{4a^2}}\, dy$$

$$= \frac{1}{2a} \int_0^{2a} \sqrt{y^2+4a^2}\, dy$$

$$= \frac{1}{2a} . \left[\frac{y}{2} \sqrt{y^2+4a^2} + \frac{4a^2}{2a} \sinh^{-1} \frac{y}{2a} \right]_0^{2a}$$

$$\left[\because \int \sqrt{x^2+a^2}\, dx = \frac{x}{2}\sqrt{a^2+x^2} + \frac{a^2}{a} \sinh^{-1} \frac{x}{a} \right]$$

$$= \frac{1}{2a} . \left[\frac{2a}{2} \sqrt{4a^2+4a^2} + 2a^2 \sinh^{-1} 1 - 0 \right]$$

$$= \frac{1}{2a} [a . 2\sqrt{2}a + 2a^2 \sinh^{-1} 1] = a[\sqrt{2} + \sinh^{-1} 1]$$

$$= a[\sqrt{2} + \log (1 + \sqrt{1+1})] \qquad \Big| \because \sinh^{-1} x = \log (x + \sqrt{x^2+1})$$

$$= a[\sqrt{2} + \log (\sqrt{2}+1)].$$

(*ii*) Arc cut off by the latus rectum

$= 2 \times$ arc cut off by semi-latus rectum $= 2 \times$ arc OL

$= 2a[\sqrt{2} + \log (\sqrt{2}+1)].$ | By part (*i*) above.

(*iii*) The parabola is $y^2 = 4ax$ ...(1)

and the line is $3y = 8x$ ...(2)

Solving (1) and (2) for y [eliminate x from the two equations], we get

$$y^2 = 4a\left(\frac{3y}{8}\right) = \frac{3ay}{2} \quad \text{or} \quad 2y^2 - 3ay = 0$$

or $y(2y - 3a) = 0$ which gives $y = 0$ and $y = \dfrac{3a}{2}$.

Hence for the arc intercepted by the line (2) on the curve (1), the limits of integration are $y = 0$ to $y = \dfrac{3a}{2}$.

$\therefore$ Required length of arc of parabola

$$= \int_0^{3a/2} \sqrt{1+\left(\frac{dx}{dy}\right)^2}\, dy = \int_0^{3a/2} \sqrt{1+\frac{y^2}{4a^2}}\, dy \qquad \Big|\ \because\ \frac{dx}{dy} = \frac{y}{2a}$$

$$= \frac{1}{2a}\int_0^{3a/2} \sqrt{y^2+4a^2}\, dy = \frac{1}{2a}\left[\frac{y}{2}\sqrt{y^2+4a^2} + \frac{4a^2}{2}\sinh^{-1}\frac{y}{2a}\right]_0^{3a/2}$$

$$= \frac{1}{2a}\left[\frac{3a}{4}\sqrt{\frac{9a^2}{4}+4a^2} + 2a^2\sinh^{-1}\frac{3}{4} - 0\right]$$

$$= \frac{1}{2a}\left[\frac{3a}{4}\cdot\frac{5a}{2} + 2a^2\log\left(\frac{3}{4}+\sqrt{\frac{9}{16}+1}\right)\right] \qquad [\because\ \sinh^{-1}x = \log(x+\sqrt{x^2+1})]$$

$$= \frac{1}{2a}\left[\frac{15a^2}{8} + 2a^2\log\left(\frac{3}{4}+\frac{5}{4}\right)\right] = a\left[\frac{15}{16}+\log 2\right].$$

Example 6. *Find the length of the arc of the parabola $x^2 = 4ay$ from the vertex to an extremity of the latus rectum.*

Sol. Let O be the vertex and L, an extremity of the latus rectum of the parabola

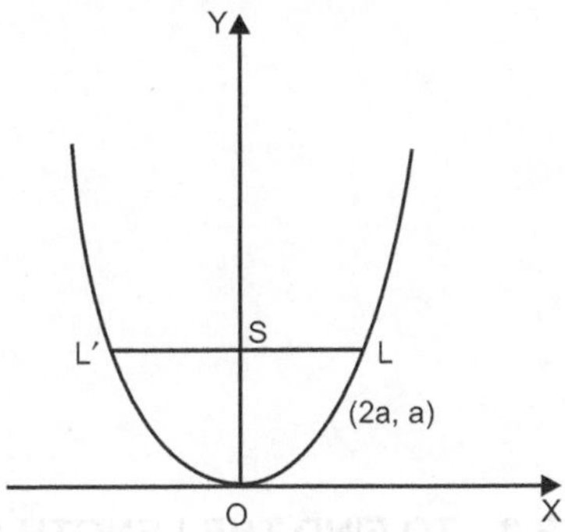

$$x^2 = 4ay \text{ or } y = \frac{1}{4a}x^2$$

$$\therefore \qquad \frac{dy}{dx} = \frac{2x}{4a} = \frac{x}{2a}.$$

For the length of the arc from O(0, 0) to L(2a, a), x varies from 0 to 2a.

$\therefore$ Required length of arc OL

$$= \int_0^{2a}\sqrt{1+\left(\frac{dy}{dx}\right)^2}\, dx = \int_0^{2a}\sqrt{1+\frac{x^2}{4a^2}}\, dx$$

$$= \frac{1}{2a}\int_0^{2a}\sqrt{x^2+4a^2}\, dx = \frac{1}{2a}\left[\frac{x}{2}\sqrt{x^2+4a^2} + \frac{4a^2}{2}\sinh^{-1}\frac{x}{2a}\right]_0^{2a}$$

$$= \frac{1}{2a}\left[\frac{2a}{2}\cdot\sqrt{4a^2+4a^2} + 2a^2\sinh^{-1}1 - 0\right]$$

$$= \frac{1}{2a}\,[a\,.2\sqrt{2}a + 2a^2\log(1+\sqrt{1+1})] \qquad \Big|\ \because\ \sinh^{-1}x = \log(x+\sqrt{x^2+1})$$

$$= a\,[\sqrt{2}+\log(\sqrt{2}+1)].$$

Example 7. *In the evolute of parabola, viz. $4(x-2a)^3 = 27ay^2$, show that the length of the curve from its cusp ($x = 2a$) to the point where it meets the parabola is $2a\,(3\sqrt{3}-1)$.*

Sol. The equation of the parabola is $y^2 = 4ax$ ...(1)

and that of the evolute is $27ay^2 = 4(x-2a)^3$ ...(2)

To find the points of intersection, we have to solve (1) and (2). Thus putting the value of y^2 from (1) in (2), we get

$$27a(4ax) = 4(x-2a)^3 \quad \text{or} \quad 27a^2x = (x-2a)^3$$

or $$27a^2x = x^3 - 6ax^2 + 12a^2x - 8a^3$$

or $\quad x^3 - 6x^2a - 15xa^2 - 8a^3 = 0 \quad$ ...(3)

or $\quad (x + a)^2 (x - 8a) = 0 \quad$ [$\because$ $x = -a$ satisfies (3)]

$\therefore \quad x = -a, -a$ or $8a$.

If $x = -a$, then from (1), y is imaginary, $\therefore$ $x = -a$ is rejected and hence $x = 8a$.

Thus the limits of integration are $x = 2a$ and $x = 8a$.

$\therefore$ From (2), $\quad y^2 = \dfrac{4}{27a}(x - 2a)^3$

$$\therefore \quad y = \frac{2}{3\sqrt{3a}}.(x - 2a)^{3/2}$$

$$\therefore \quad \frac{dy}{dx} = \frac{2}{3\sqrt{3a}}.\frac{3}{2}(x - 2a)^{1/2} = \frac{1}{\sqrt{3a}}(x - 2a)^{1/2}$$

$\therefore$ Required length of evolute from its cups to its point of intersection with the parabola

$$= \int_{2a}^{8a} \sqrt{1 + \left(\frac{dy}{dx}\right)^2}\, dx = \int_{2a}^{8a} \sqrt{1 + \frac{x - 2a}{3a}}\, dx$$

$$= \frac{1}{\sqrt{3a}} \int_{2a}^{8a} \sqrt{x + a}\, dx = \frac{1}{\sqrt{3a}} . \left| \frac{(x + a)^{3/2}}{3/2} \right|_{2a}^{8a}$$

$$= \frac{2}{3\sqrt{3a}} \left| (x + a)^{3/2} \right|_{2a}^{8a} = \frac{2}{3\sqrt{3a}}[(8a + a)^{3/2} - (2a + a)^{3/2}]$$

$$= \frac{2}{3\sqrt{3a}}.[27a\sqrt{a} - 3\sqrt{3}a^{3/2}] = \frac{6\sqrt{3}a\sqrt{a}}{3\sqrt{3}\sqrt{a}}(3.\sqrt{3} - 1) = 2a(3\sqrt{3} - 1).$$

9.2. TO FIND THE LENGTH OF THE CURVE WHEN THE LIMITS OF INTEGRATION ARE NOT GIVEN

When limits of integration are not given, first of all we have to find it by tracing the curve roughly.

(*i*) **Symmetry.** If the given equation of the curve contains only even powers of x [Replace x by $-x$ no chanse] then the curve is symmetrical about Y-axis. If the equation contains only even powers of y, then the curve is symmetrical about X-axis.

(*ii*) Check whether the curve passes through the origin or not.

(*iii*) Find the points where the curve meets X-axis by putting $y = 0$ in the given equation.

(*iv*) Find the points where the curve meets Y-axis by putting $x = 0$.

ILLUSTRATIVE EXAMPLES

Example 1. *Find the perimeter of the loop of the curve*

(*a*) $9ay^2 = (x - 2a)(x - 5a)^2$

(*b*) $3ay^2 = x(x - a)^2$. [CUSAT 2010]

Sol. (*a*) The given curve is $\quad 9ay^2 = (x - 2a)(x - 5a)^2 \quad$ ...(1)

To trace the curve roughly to find the limits of integration :

(*i*) The curve is symmetrical about the X-axis.

(*ii*) The curve does not pass through the origin.

(*iii*) The curve meets X-axis, where putting $y = 0$, we get

$$(x - 2a)(x - 5a)^2 = 0 \quad \text{or} \quad x = 2a, 5a$$

i.e., at the points $(2a, 0)$ and $(5a, 0)$.

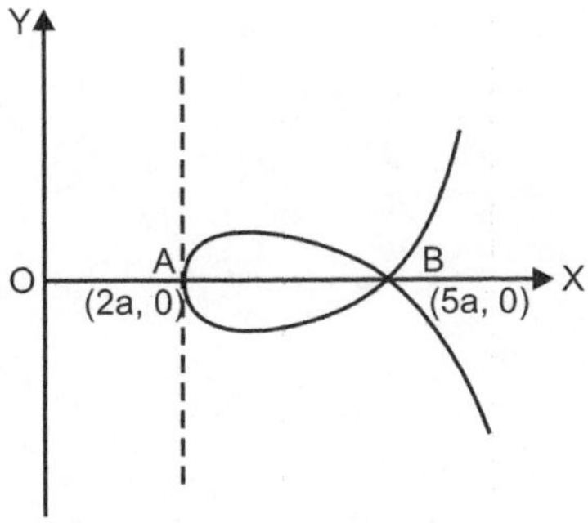

Shifting the origin to $(2a, 0)$, the equation (1) transforms to $9ay^2 = x(x - 3a)^2$ and the tangent at the new origin is $x = 0$ *i.e.*, new y-axis.

Thus at $(2a, 0)$, the tangent is || to Y-axis.

Again shifting the origin to $(5a, 0)$, the given equation (1) transforms to $9ay^2 = (x + 3a)x^2$.

The tangents at the new origin are given by

$$3ax^2 = 9ay^2 \quad \text{or} \quad y^2 = \frac{1}{3}x^2.$$

$\therefore$ $y = \pm\frac{1}{\sqrt{3}}x$, showing that the new origin *i.e.*, $(5a, 0)$ is a node.

Thus there is a loop between $x = 2a$ and $x = 5a$.

(*iv*) The curve has no asymptotes.

(*v*) From (1), $\quad y^2 = \frac{1}{9a}(x - 2a)(x - 5a)^2$ ...(2)

For $x < 2a$, y^2 is –ve,

$\therefore$ y is imaginary.

Thus no portion of the curve lies to the left of the line $x = 2a$.

Again when x increases from $5a$ to ∞, y also increases from 0 to ∞.

Thus the shape of the curve is as shown in the figure.

Now for the loop, the limits of integration are

$$x = 2a \text{ to } x = 5a,$$

and the perimeter of the loop = 2 × arc of loop above X-axis.

Now from (2), $\quad y = \frac{1}{3\sqrt{a}}(x - 5a)\sqrt{x - 2a}$

$$\therefore \quad \frac{dy}{dx} = \frac{1}{3\sqrt{a}}\left[\sqrt{x-2a} + \frac{x-5a}{2\sqrt{x-2a}}\right] = \frac{1}{3\sqrt{a}}\left[\frac{2(x-2a)+x-5a}{2\sqrt{x-2a}}\right]$$

$$= \frac{3x - 9a}{6\sqrt{a}\sqrt{x-2a}} = \frac{x - 3a}{2\sqrt{a}\sqrt{x-2a}}.$$

Required perimeter = 2 × arc of loop above X-axis

$$= 2\int_{2a}^{5a}\sqrt{1+\left(\frac{dy}{dx}\right)^2}\,dx = 2\int_{2a}^{5a}\sqrt{1+\frac{(x-3a)^2}{4a(x-2a)}}\,dx$$

$$= 2\int_{2a}^{5a}\sqrt{\frac{4ax - 8a^2 + x^2 + 9a^2 - 6ax}{4a(x-2a)}}\,dx$$

$$= 2\int_{2a}^{5a} \frac{x-a}{2\sqrt{a}\sqrt{x-2a}}\,dx = \frac{1}{\sqrt{a}}\int_{2a}^{5a} \frac{x-2a+a}{\sqrt{x-2a}}\,dx$$

$$= \frac{1}{\sqrt{a}}\int_{2a}^{5a} [(x-2a)^{1/2} + a(x-2a)^{-1/2}]\,dx$$

$$= \frac{1}{\sqrt{a}}\left[\frac{2}{3}(x-2a)^{3/2} + 2a\,(x-2a)^{1/2}\right]_{2a}^{5a}$$

$$= \frac{1}{\sqrt{a}}\left[\frac{2}{3}(3a)^{3/2} + 2a\,(3a)^{1/2} - 0\right] = \frac{1}{\sqrt{a}}\,.\,2\sqrt{3a}\left[\frac{3a}{3} + a\right] = 4\sqrt{3}\,a.$$

(*b*) The equation of the given curve is $3ay^2 = x(x-a)^2$ ...(1)

To trace the curve roughly to find the limits of integration.

(*i*) The curve is symmetrical about X-axis.

(*ii*) The curve passes through the origin. The tangent at the origin is given by $a^2x = 0$ or $x = 0$ *i.e.*, the Y-axis.

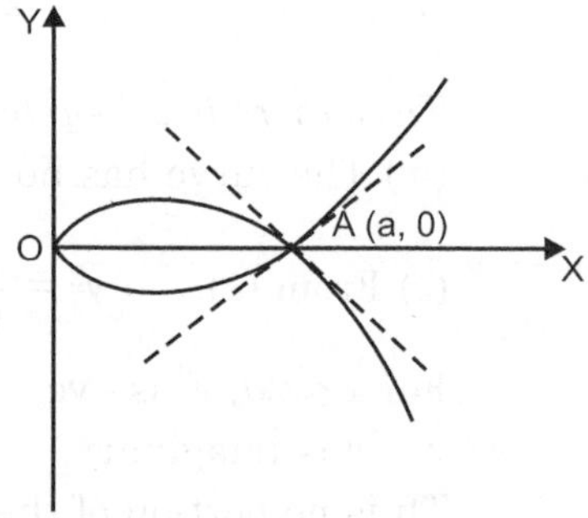

(*iii*) The curve meets X-axis, where putting $y = 0$ in (1), we get

$x(x-a)^2 = 0$ or $x = 0, a$, *i.e.*, at the points (0, 0) and (a, 0).

The curve meets Y-axis, where putting $x = 0$, we get

$y^2 = 0$ or $y = 0$ *i.e.*, at the origin (0, 0).

Shifting the origin to (a, 0), the equation (1) transforms to $3ay^2 = (x+a)x^2$.

The tangents at the origin are given by $3ay^2 = ax^2$ or $y = \pm\frac{1}{\sqrt{3}}x$, showing that the new origin *i.e.*, the point (a, 0) is a node.

Thus *there is a loop between x = 0 and x = a.*

(*iv*) The curve has no asymptote.

(*v*) From (1), $y^2 = \frac{1}{3a}x(x-a)^2$ ...(2)

when $x < 0$, y^2 is –ve, $\therefore$ y is imaginary.

Thus no portion of the curve lies to the left of y-axis ($x = 0$)

Again as x increases from a to ∞, y also increases from 0 to ∞.

Hence the shape of the curve is as shown in the figure.

Now for half the loop, x varies from 0 to a and perimeter of loop = 2 × arc of loop above X-axis.

From (2), $$y = \frac{1}{\sqrt{3a}}\sqrt{x}(x-a)$$

$$\therefore \quad \frac{dy}{dx} = \frac{1}{\sqrt{3a}}\left[\sqrt{x} + \frac{x-a}{2\sqrt{x}}\right] = \frac{1}{\sqrt{3a}}\left[\frac{2x+x-a}{2\sqrt{x}}\right] = \frac{3x-a}{2\sqrt{3ax}}$$

or $$1+\left(\frac{dy}{dx}\right)^2 = 1+\frac{(3x-a)^2}{12ax} = \frac{12ax+9x^2-6ax+a^2}{12ax} = \frac{9x^2+6ax+a^2}{12ax} = \frac{(3x+a)^2}{12ax}$$

∴ Required perimeter of the loop

$$= 2\int_0^a \sqrt{1+\left(\frac{dy}{dx}\right)^2}\,dx = 2\int_0^a \frac{3x+a}{2\sqrt{3a}\sqrt{x}}\,dx$$

$$= \frac{1}{\sqrt{3a}}\int_0^a (3x^{1/2}+ax^{-1/2})\,dx = \frac{1}{\sqrt{3a}}\left[3\,.\,\frac{x^{3/2}}{3/2}+a\,.\,\frac{x^{1/2}}{1/2}\right]_0^a$$

$$= \frac{1}{\sqrt{3a}}\left[2x\sqrt{x}+2a\sqrt{x}\right]_0^a = \frac{1}{\sqrt{3a}}\left[2a\sqrt{a}+2a\sqrt{a}-0\right] = \frac{4a}{\sqrt{3}}.$$

Example 2. *Show that the whole length of the curve* $x^2(a^2-x^2) = 8a^2y^2$ *is* $\pi a\sqrt{2}$.

Sol. *To trace the curve roughly to find the limits of integration*

(*i*) The curve is symmetrical about the axes.

(*ii*) The curve passes through the origin and the tangents at the origin are given by

$$a^2x^2 = 8a^2y^2 \quad \text{or} \quad y = \pm\frac{1}{2\sqrt{2}}x, \text{ showing that origin is a node.}$$

(*iii*) The curve meets the X-axis, where putting $y = 0$, in the given equation,

$$x^2(a^2-x^2) = 0 \quad \text{or} \quad x = 0, \pm a.$$

Hence the curve meets the X-axis at (0, 0), (a, 0), ($-a$, 0)

The curve meets Y-axis, where putting $x = 0$, get

$$8a^2y^2 = 0 \quad i.e., \quad y = 0 \quad i.e., \quad \text{at the origin } (0, 0).$$

Shifting the origin to (a, 0), the given equation transforms to

$$(x+a)^2[a^2-(x+a)^2] = 8a^2y^2 \quad \text{or} \quad 8a^2y^2 = (x+a)^2(-x^2-2ax) = -x(x+2a)(x+a)^2$$

and the tangent at the new origin is $-2a^3x = 0$ or $x = 0$ *i.e.*, the new Y-axis.

Thus the tangent at (a, 0) is || to Y-axis.

(*iv*) The curve has no asymptotes.

(*v*) The given curve is $y^2 = \frac{1}{8a^2}x^2(a^2-x^2)$ ...(1)

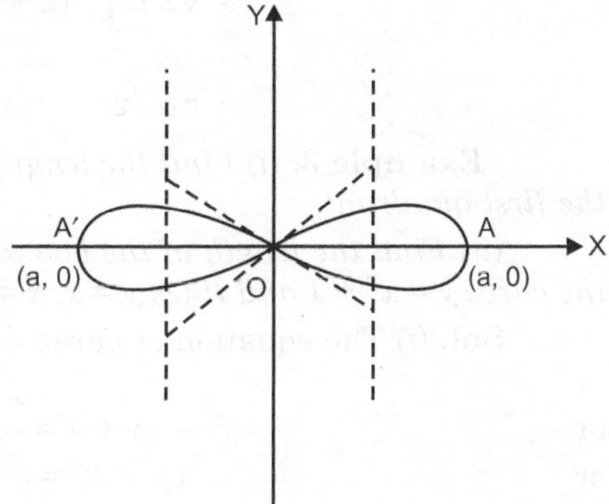

When $x > a$ numerically, y^2 is –ve,

∴ y is imaginary.

Thus no portion of the curve lies beyond the lines $x = a$, $x = -a$.

The shape of the curve is as shown in figure. The curve consists of symmetrical loops.

∴ whole length of the curve

= 2 × length of a loop

= 4 × arc of loop in first quadrant.

For half the loop in the first quadrant x varies from 0 to a.

From (1), $$y = \frac{1}{2\sqrt{2}a} x\sqrt{a^2 - x^2}$$

$$\therefore \quad \frac{dy}{dx} = \frac{1}{2\sqrt{2}a}\left[\sqrt{a^2-x^2} + x.\frac{-2x}{2\sqrt{a^2-x^2}}\right] = \frac{1}{2a\sqrt{2}}\left[\frac{a^2-x^2-x^2}{\sqrt{a^2-x^2}}\right] = \frac{a^2-2x^2}{2a\sqrt{2}\sqrt{a^2-x^2}}.$$

and $$1+\left(\frac{dy}{dx}\right)^2 = 1+\frac{(a^2-2x^2)^2}{8a^2(a^2-x^2)} = \frac{8a^4-8a^2x^2+a^4+4x^4-4a^2x^2}{8a^2(a^2-x^2)}$$

$$= \frac{4x^4-12a^2x^2+9a^4}{8a^2(a^2-x^2)} = \frac{(3a^2-2x^2)^2}{8a^2(a^2-x^2)}.$$

$\therefore$ Required whole length of the curve

$$= 4 \text{ arc OA} = 4\int_0^a \sqrt{1+\left(\frac{dy}{dx}\right)^2}\,dx$$

$$= 4\int_0^a \frac{3a^2-2x^2}{2\sqrt{2}a\sqrt{a^2-x^2}}\,dx \qquad \Big| \begin{array}{l}\text{Put } x = a\sin\theta \\ \therefore \ dx = a\cos\theta\, d\theta\end{array}$$

$$= \frac{4}{2\sqrt{2}a}\int_0^{\pi/2} \frac{3a^2-2a^2\sin^2\theta}{a\cos\theta}.\,a\cos\theta\,d\theta = \sqrt{2}\,a\int_0^{\pi/2}(3-2\sin^2\theta)\,d\theta$$

$$= \sqrt{2}\,a\int_0^{\pi/2}[3-(1-\cos 2\theta)]\,d\theta \qquad \Big| \begin{array}{l}\because \ \cos 2\theta = 1-2\sin^2\theta \\ \therefore \ 2\sin^2\theta = 1-\cos 2\theta\end{array}$$

$$= \sqrt{2}\,a\int_0^{\pi/2}(2+\cos 2\theta)\,d\theta = \sqrt{2}a\left[2\theta+\frac{\sin 2\theta}{2}\right]_0^{\pi/2} = \sqrt{2}a.\left[\pi+\frac{1}{2}\sin\pi - 0\right]$$

$$= \pi a\sqrt{2}. \qquad \Big| \because \ \sin\pi = 0$$

Example 3. (*i*) *Find the length of the arc of the parabola $y^2 - 4y + 2x = 0$ which lies in the first quadrant.*

(*ii*) *Find the length of the boundary of the region by the curve $y = x^2 + 1$ and lines $y = x$, $x = 0$ and $y = 2$.*

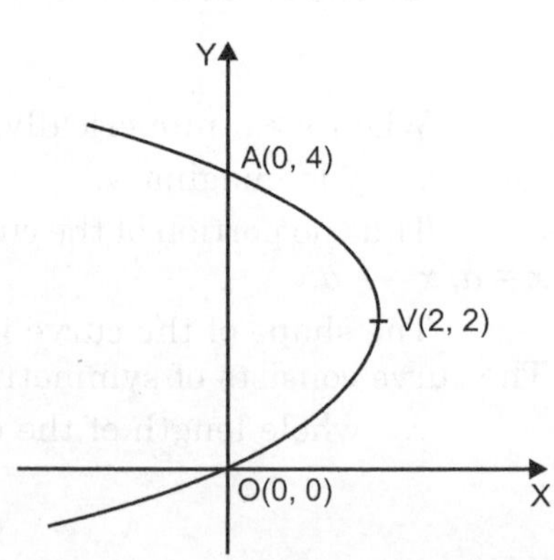

Sol. (*i*) The equation of parabola is $y^2 - 4y + 2x = 0$ ...(1)

or $$y^2 - 4y + 4 = -2x + 4$$

or $$(y-2)^2 = -2(x-2)$$

It is left-handed parabola with vertex at (2, 2). It meets y-axis *i.e.*, $x = 0$, where, from (1),

$$y^2 - 4y = 0 \quad \text{or} \quad y = 0, 4$$

$\therefore$ The parabola meets the y-axis at (0, 0) and (0, 4). Length of the arc OVA which lies in the first quadrant

$$= \int_0^4 \sqrt{1 + \left(\frac{dx}{dy}\right)^2}\, dy \qquad ...(2)$$

From(1), $x = \dfrac{4y - y^2}{2}$

$$\frac{dx}{dy} = \frac{4 - 2y}{2} = 2 - y$$

$$1 + \left(\frac{dx}{dy}\right)^2 = 1 + (2 - y)^2 = (y - 2)^2 + 1$$

From (2), length of arc OVA

$$= \int_0^4 \sqrt{(y-2)^2 + 1}\, dy = \left[\frac{(y-2)\sqrt{(y-2)^2+1}}{2} + \frac{1}{2}\sinh^{-1}\frac{y-2}{1}\right]_0^4$$

$$= \left[\frac{2\sqrt{5}}{2} + \frac{1}{2}\sinh^{-1} 2\right] - \left[\frac{-2\sqrt{5}}{2} + \frac{1}{2}\sinh^{-1}(-2)\right]$$

$$= 2\sqrt{5} + \frac{1}{2}\sinh^{-1} 2 - \frac{1}{2}\sinh^{-1}(-2)$$

$$= 2\sqrt{5} + \frac{1}{2}\log(2 + \sqrt{2^2+1}) - \frac{1}{2}\log(-2 + \sqrt{(-2)^2+1})$$

$$\left|\ \because \sinh^{-1} x = \log(x + \sqrt{x^2+1})\right.$$

$$= 2\sqrt{5} + \frac{1}{2}\log\frac{\sqrt{5}+2}{\sqrt{5}-2}.$$

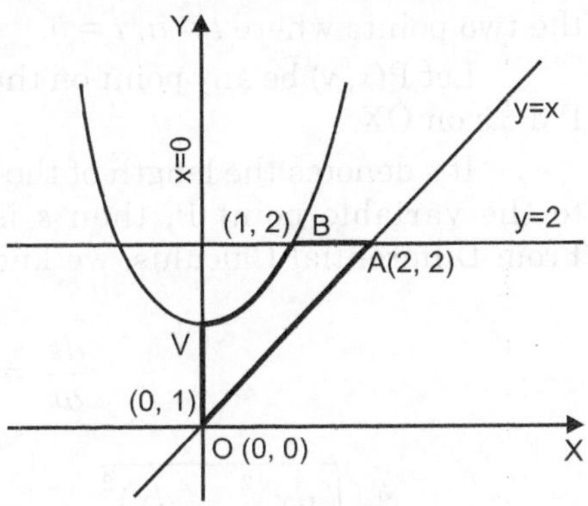

(*ii*) The equation $y = x^2 + 1$ represents an upward parabola with vertex at (0, 1). It meets the line $y = 2$ where $2 = x^2 + 1$

or $x = \pm 1$ *i.e.*, in the points (1, 2) and (– 1, 2).

The line $y = x$ meets the line $y = 2$ at (2, 2).

We have to find the length of OABVO

Now OV = 1, AB = 1

$$OA = \sqrt{(2-0)^2 + (2-0)^2} = \sqrt{8} = 2\sqrt{2}$$

Length of arc VB of parabola

$$= \int_0^1 \sqrt{1 + \left(\frac{dy}{dx}\right)^2}\, dx = \int_0^1 \sqrt{1 + (2x)^2}\, dx$$

$$= \int_0^1 \sqrt{4\left(x^2 + \frac{1}{4}\right)}dx = 2\int_0^1 \sqrt{x^2 + \left(\frac{1}{2}\right)^2}\, dx$$

$$= 2\left[\frac{x\sqrt{x^2 + \left(\frac{1}{2}\right)^2}}{2} + \frac{\left(\frac{1}{2}\right)^2}{2}\sinh^{-1}\left(\frac{x}{1/2}\right)\right]_0^1$$

$$= 2\left[\frac{\sqrt{5}}{4} + \frac{1}{8}\sinh^{-1} 2\right] = \frac{\sqrt{5}}{2} + \frac{1}{4}\log\,(2+\sqrt{2^2+1}) = \frac{\sqrt{5}}{2} + \frac{1}{4}\log\,(2+\sqrt{5})$$

$\therefore$ Length OABVO = OA + AB + arc VB + OV

$$= 2\sqrt{2} + 1 + \frac{\sqrt{5}}{2} + \frac{1}{4}\log\,(2+\sqrt{5}) + 1$$

$$= 2 + 2\sqrt{2} + \frac{\sqrt{5}}{2} + \frac{1}{4}\log\,(2+\sqrt{5}).$$

9.3. ARC FORMULA FOR PARAMETRIC EQUATIONS

To show that the length of the arc of the curve x = f(t), y = φ(t) between the points where t = a, t = b is given by

$$\int_a^b \sqrt{\left(\frac{dx}{dt}\right)^2 + \left(\frac{dy}{dt}\right)^2}\, dt$$

where x and y are continuous and single-valued functions of t in the interval a ≤ t ≤ b.

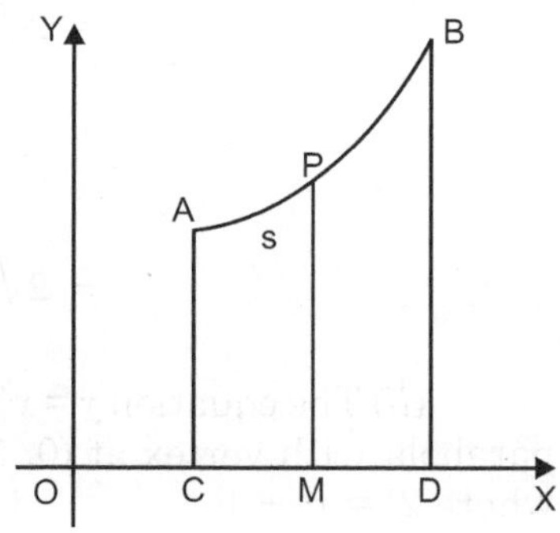

Let AB be the curve $x = f(t)$, $y = \phi(t)$, where A,B are the two points where $t = a$, $t = b$.

Let P(x, y) be any point on the curve. Draw AC, BD and PM ⊥*s* on OX.

If s denotes the length of the arc from the fixed point A to the variable point P, then s is clearly a function of t. From Differential Calculus, we know that

$$\frac{ds}{dt} = \sqrt{\left(\frac{dx}{dt}\right)^2 + \left(\frac{dy}{dt}\right)^2}$$

$$\therefore \int_a^b \sqrt{\left(\frac{dx}{dt}\right)^2 + \left(\frac{dy}{dt}\right)^2}\, dt = \int_a^b \frac{ds}{dt}\,.\,dt = \Big[\, s\,\Big]_a^b$$

= (value of s when $t = b$) – (value of s when $t = a$)

= arc AB – 0 = arc AB.

Hence $$\text{arc AB} = \int_a^b \sqrt{\left(\frac{dx}{dt}\right)^2 + \left(\frac{dy}{dt}\right)^2}\, dt \quad \text{...(1)}$$

Note. Here again it is assumed that the integrand does not change sign in the range of integration $a \le t \le b$.

Cor. From (1), $$s = \int \sqrt{\left(\frac{dx}{dt}\right)^2 + \left(\frac{dy}{dt}\right)^2}\, dt$$

ILLUSTRATIVE EXAMPLES

Example 1. *Find the length of the arc of the curve*

(a) $x = e^{\theta} \sin \theta,\ y = e^{\theta} \cos \theta$ *from* $\theta = 0$ *to* $\theta = \frac{\pi}{2}$.

(b) $x = e^{\theta}\left(\sin \frac{\theta}{2} + 2 \cos \frac{\theta}{2}\right),\ y = e^{\theta}\left(\cos \frac{\theta}{2} - 2 \sin \frac{\theta}{2}\right)$ *measured from* $\theta = 0$ *to* $\theta = \pi$.

(c) $x = a \sin 2\theta (1 + \cos 2\theta),\ y = a \cos 2\theta (1 - \cos 2\theta)$ *from* $(0,0)$ *to* (x, y).

Sol. (a) The given curve is $x = e^{\theta} \sin \theta,\ y = e^{\theta} \cos \theta$

Now $$\frac{dx}{d\theta} = e^{\theta} . \cos \theta + \sin \theta . e^{\theta} = e^{\theta} (\cos \theta + \sin \theta)$$

and $$\frac{dy}{d\theta} = e^{\theta} (- \sin \theta) + \cos \theta . e^{\theta} = e^{\theta}. (\cos \theta - \sin \theta)$$

$$\therefore \quad \left(\frac{dx}{d\theta}\right)^2 + \left(\frac{dy}{d\theta}\right)^2 = e^{2\theta} [(\cos \theta + \sin \theta)^2 + (\cos \theta - \sin \theta)^2]$$

$$= e^{2\theta} . 2(\cos^2 \theta + \sin^2 \theta) = 2e^{2\theta} \qquad \Big| \because \cos^2 \theta + \sin^2 \theta = 1$$

Hence the required length of arc $$= \int_0^{\pi/2} \sqrt{\left(\frac{dx}{d\theta}\right)^2 + \left(\frac{dy}{d\theta}\right)^2}\, d\theta = \int_0^{\pi/2} \sqrt{2e^{2\theta}} . d\theta$$

$$= \sqrt{2} \int_0^{\pi/2} e^{\theta}\, d\theta = \sqrt{2} \left| e^{\theta} \right|_0^{\pi/2} = \sqrt{2} (e^{\pi/2} - e^0) = \sqrt{2} (e^{\pi/2} - 1).$$

$$\int_0^{\pi/2} e^{\theta} d\theta = \sqrt{2} = \sqrt{2} \left| e^{\theta} \right|_0^{\pi/2} = \sqrt{2} (e^{\pi/2} - e^0) = \sqrt{2} (e^{\pi/2} - 1).$$

(b) The given curve is $x = e^{\theta}\left(\sin \frac{\theta}{2} + \cos \frac{\theta}{2}\right),\ y = e^{\theta}\left(\cos \frac{\theta}{2} - 2 \sin \frac{\theta}{2}\right)$

$$\therefore \quad \frac{dx}{d\theta} = e^{\theta}\left(\frac{1}{2} \cos \frac{\theta}{2} - \sin \frac{\theta}{2}\right) + e^{\theta}\left(\sin \frac{\theta}{2} + 2 \cos \frac{\theta}{2}\right) = \frac{5}{2} e^{\theta} \cos \frac{\theta}{2}.$$

and $$\frac{dy}{d\theta} = e^{\theta}\left(-\frac{1}{2} \sin \frac{\theta}{2} - \cos \frac{\theta}{2}\right) + e^{\theta}\left(\cos \frac{\theta}{2} - 2 \sin \frac{\theta}{2}\right) = -\frac{5}{2} e^{\theta} \sin \frac{\theta}{2}$$

$$\therefore \quad \left(\frac{dx}{d\theta}\right)^2 + \left(\frac{dy}{d\theta}\right)^2 = \frac{25}{4} e^{2\theta}\left(\cos^2\frac{\theta}{2} + \sin^2\frac{\theta}{2}\right) = \frac{25}{4} e^{2\theta}$$

Hence the required length $= \int_0^{\pi} \sqrt{\left(\frac{dx}{d\theta}\right)^2 + \left(\frac{dy}{d\theta}\right)^2}\, d\theta$

$$= \int_0^{\pi} \frac{5}{2} e^{\theta}\, d\theta = \frac{5}{2} \left| e^{\theta} \right|_0^{\pi} = \frac{5}{2}(e^{\pi} - e^0) = \frac{5}{2}(e^{\pi} - 1).$$

(*c*) The given curve is $x = a \sin 2\theta\, (1 + \cos 2\theta)$, $y = a \cos 2\theta\, (1 - \cos 2\theta)$

$$\therefore \quad \frac{dx}{d\theta} = 2a \cos 2\theta\, (1 + \cos 2\theta) + a \sin 2\theta\, (-2 \sin 2\theta)$$

$$= 2a\, (\cos 2\theta + \cos^2 2\theta - \sin^2 2\theta)$$

$$= 2a.\, (\cos 2\theta + \cos 4\theta) \qquad \Big|\ \because\ \cos^2 A - \sin^2 A = \cos 2A$$

$$= 2a\,.\,2\cos\frac{4\theta + 2\theta}{2}\,.\,\cos\frac{4\theta - 2\theta}{2}$$

$$\Big|\ \because\ \cos C + \cos D = 2\cos\frac{C+D}{2}\cos\frac{C-D}{2}$$

$$= 4a \cos 3\theta \cos \theta.$$

and
$$\frac{dy}{d\theta} = -\,2a \sin 2\theta\, (1 - \cos 2\theta) + a \cos 2\theta\, (2 \sin 2\theta)$$

$$= 2a \sin 2\theta\, (2 \cos 2\theta - 1) = 2a\, (2 \sin 2\theta \cos 2\theta - \sin 2\theta)$$

$$= 2a\, (\sin 4\theta - \sin 2\theta) \qquad \Big|\ \because\ 2 \sin A \cos A = \sin 2A$$

$$= 2a\,.\, 2 \cos 3\theta \sin \theta = 4a \cos 3\theta \sin \theta.$$

$$\Big|\ \because\ \cos C - \sin D = 2\cos\frac{C+D}{2}\sin\frac{C-D}{2}$$

When $x = 0,\ a \sin 2\theta\, (1 + \cos 2\theta) = 0$
and $y = 0,\ a \cos 2\theta\, (1 - \cos 2\theta) = 0$ $\Big\}$ which give $\theta = 0$

Hence for the arc from (0, 0) to (*x*, *y*) *x* varies from 0 to *x*, *y* varies from 0 to *y* or θ varies from 0 to θ.

$$\therefore \quad \text{Required length} = \int_0^{\theta} \sqrt{\left(\frac{dx}{d\theta}\right)^2 + \left(\frac{dy}{d\theta}\right)^2}\, d\theta$$

$$= \int_0^{\theta} \sqrt{16a^2 \cos^2 3\theta \cos^2 \theta + 16a^2 \cos^2 3\theta \sin^2 \theta}\; d\theta$$

$$= \int_0^{\theta} 4a \cos 3\theta \sqrt{\cos^2 \theta + \sin^2 \theta}\; d\theta = \int_0^{\theta} 4a \cos 3\theta\, d\theta$$

$$= 4a. \left| \frac{\sin 3\theta}{3} \right|_0^{\theta} = \frac{4a}{3} (\sin 3\theta - 0) = \frac{4a}{3} \sin 3\theta.$$

Example 2. (*a*) *Show that the length of an arc of the curve* $x \sin \theta + y \cos \theta = f'(\theta)$, $x \cos \theta - y \sin \theta = f''(\theta)$ *is given by* $s = f(\theta) + f''(\theta) + c$.

(*b*) *Show that in the ellipse* $x = a \cos \theta$, $y = b \sin \theta$, *the perimeter may be expressed as*

$$2a\pi\left[1-\left(\frac{1}{2}\right)^2 \cdot \frac{e^2}{1}-\left(\frac{1.3}{2.4}\right)^2 \cdot \frac{e^4}{3}-\left(\frac{1.3.5}{2.4.6}\right)^2 \cdot \frac{e^6}{5}-\ldots\right]$$

where e is the eccentricity of ellipse.

Sol. (*a*) The curve is given by $x \sin \theta + y \cos \theta = f'(\theta)$ ...(1)

$$x \cos \theta - y \sin \theta = f''(\theta) \qquad ...(2)$$

Multiplying (1) by $\sin \theta$ and (2) by $\cos \theta$, and adding,

$$x(\sin^2 \theta + \cos^2 \theta) = \sin \theta\, f'(\theta) + \cos \theta\, f''(\theta)$$

or

$$x = \sin \theta\, f'(\theta) + \cos \theta\, f''(\theta) \qquad ...(3)$$

Again multiplying (1) by $\cos \theta$ and (2) by $\sin \theta$, and subtracting,

$$y(\cos^2 \theta + \sin^2 \theta) = \cos \theta\, f'(\theta) - \sin \theta\, f''(\theta)$$

or

$$y = \cos \theta\, f'(\theta) - \sin \theta\, f''(\theta) \qquad ...(4)$$

From (3),
$$\frac{dx}{d\theta} = \sin \theta.\, f''(\theta). + f'(\theta).\cos \theta + \cos \theta f'''(\theta) - \sin \theta.\, f''(\theta)$$
$$= [f'(\theta) + f'''(\theta)] \cos \theta$$

From (4),
$$\frac{dy}{d\theta} = \cos \theta.\, f''(\theta) - \sin \theta.\, f'(\theta) - \cos \theta\, f''(\theta) - \sin \theta.\, f'''(\theta)$$
$$= -[f'(\theta) + f'''(\theta)] \sin \theta.$$

$$\therefore \quad \left(\frac{dx}{d\theta}\right)^2 + \left(\frac{dy}{d\theta}\right)^2 = [f'(\theta) + f'''(\theta)]^2(\cos^2 \theta + \sin^2 \theta)$$
$$= [f'(\theta) + f'''(\theta)]^2. \qquad ...(5)$$

Hence the length of the arc is given by

$$s = \int \sqrt{\left(\frac{dx}{d\theta}\right)^2 + \left(\frac{dy}{d\theta}\right)^2}\, d\theta = \int [f'(\theta) + f'''(\theta)]\, d\theta \qquad | \text{Using (5)}$$
$$= f(\theta) + f''(\theta) + c$$

where c is an arbitrary constant.

(*b*) The equations of the ellipse are

$$x = a \cos \theta,\ y = b \sin \theta$$

$$\therefore \quad \frac{dx}{d\theta} = -a \sin \theta,\ \frac{dy}{d\theta} = b \cos \theta.$$

The ellipse is symmetrical about both the axes, and in the first quadrand, θ varies from 0 to $\frac{\pi}{2}$.

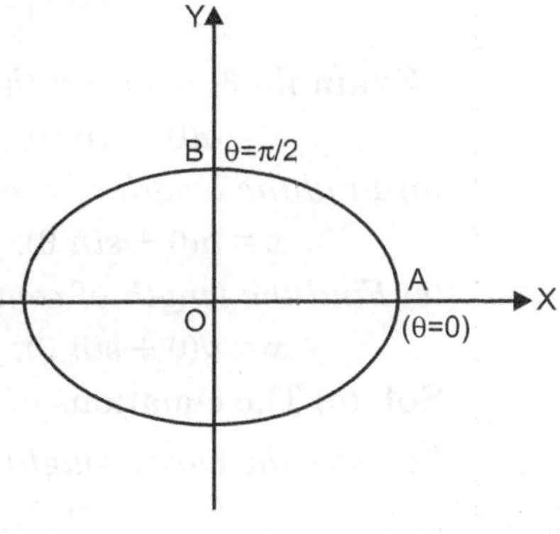

$\therefore$ Perimeter of the ellipse

$$= 4 \times \text{length of arc in first quadrand,}$$

$$= 4\int_0^{\pi/2}\sqrt{\left(\frac{dx}{d\theta}\right)^2+\left(\frac{dy}{d\theta}\right)^2}\,d\theta = 4\int_0^{\pi/2}\sqrt{a^2\sin^2\theta+b^2\cos^2\theta}\,d\theta$$

$$= 4\int_0^{\pi/2}\sqrt{a^2\sin^2\theta+a^2(1-e^2)\cos^2\theta}\,d\theta \qquad \Big|\because\ b^2=a^2(1-e^2)$$

$$= 4\int_0^{\pi/2}\sqrt{a^2(\sin^2\theta+\cos^2\theta)-a^2e^2\cos^2\theta}\,d\theta$$

$$= 4a\int_0^{\pi/2}\sqrt{1-e^2\cos^2\theta}\,d\theta = 4a\int_0^{\pi/2}(1-e^2\cos^2\theta)^{1/2}\,d\theta$$

$$= 4a\int_0^{\pi/2}\left[1+\frac{1}{2}(-e^2\cos^2\theta)+\frac{\frac{1}{2}\left(\frac{1}{2}-1\right)}{2!}(-e^2\cos^2\theta)^2\right.$$

$$\left.+\frac{\frac{1}{2}(\frac{1}{2}-1)(\frac{1}{2}-2)}{3!}.(-e^2\cos^2\theta)^3+\ldots\ldots\right]d\theta$$

[On expanding by Binomial Theorem]

$$= 4a\int_0^{\pi/2}\left[1-\frac{1}{2}e^2\cos^2\theta-\frac{1}{2.4}e^4\cos^4\theta-\frac{1.3}{2.4.6}e^6\cos^6\theta\ldots\right]d\theta$$

$$= 4a\left[\int_0^{\pi/2}d\theta-\frac{1}{2}e^2\int_0^{\pi/2}\cos^2\theta\,d\theta-\frac{1}{2.4}e^4.\int_0^{\pi/2}\cos^4\theta\,d\theta\right.$$

$$\left.-\frac{1.3}{2.4.6}e^6\int_0^{\pi/2}\cos^6\theta\,d\theta-\ldots\right]$$

$$= 4a\left[\frac{\pi}{2}-\frac{1}{2}e^2.\frac{1}{2}.\frac{\pi}{2}-\frac{1}{2.4}e^4.\frac{3.1}{4.2}.\frac{\pi}{2}-\frac{1.3}{2.4.6}.e^6\frac{5.3.1}{6.4.2}.\frac{\pi}{2}-\ldots\right]$$

$$= 4a.\frac{\pi}{2}\left[1-\frac{1}{2^2}.e^2-\frac{1.3}{2^2.4^2}e^4-\frac{1^2.3^2.5^2}{2^2.4^2.6^2}.e^6-\ldots\right]$$

$$= 2a\pi\left[1-\left(\frac{1}{2}\right)^2e^2-\left(\frac{1.3}{2.4}\right)^2.\frac{e^4}{3}-\left(\frac{1.3.5}{2.4.6}\right)^2.\frac{e^6}{5}-\ldots\right].$$

Example 3. (*a*) *Find the length of one arch of the cycloid*

$x = a(\theta - \sin\theta),\ y = a(1 - \cos\theta)$

(*b*) *Find the length of complete cycloid given by*

$x = a(\theta + \sin\theta),\ y = a(1 - \cos\theta),\ a > 0$

(*c*) *Find the length of complete cycloid given by*

$x = a(\theta + \sin\theta),\ y = a(1 + \cos\theta).$

Sol. (*a*) The equations of the cycloid are $x = a(\theta - \sin\theta),\ y = a(1 - \cos\theta)$

To trace the curve roughly to find the limits of integration :

(*i*) Since x is an odd function of θ ; and y, an even function of θ, $\therefore$ the curve is symmetrical about Y-axis.

(*ii*) Putting $x = 0$, we get $\theta - \sin\theta = 0$ which gives $\theta = 0$, Now $\theta = 0$ also makes $y = 0$. Thus the curve passes through the origin.

(*iii*) $$\frac{dx}{d\theta} = a(1 - \cos\theta) \text{ and } \frac{dy}{d\theta} = a\sin\theta$$

$$\therefore \quad \frac{dy}{dx} = \frac{a\sin\theta}{a(1-\cos\theta)} = \frac{2\sin\frac{\theta}{2}\cos\frac{\theta}{2}}{2\sin^2\frac{\theta}{2}} = \cot\frac{\theta}{2}$$

(*iv*) The various values are as below :

$\theta = 0$	$\frac{\pi}{2}$	π	$\frac{3\pi}{2}$	2π
$x = 0$	$a\left(\frac{\pi}{2} - 1\right)$	$a\pi$	$a\left(\frac{3\pi}{2} + 1\right)$	$2a\pi$
$y = 0$	a	$2a$	a	0
$\frac{dy}{dx} = \infty$	1	0	-1	∞.

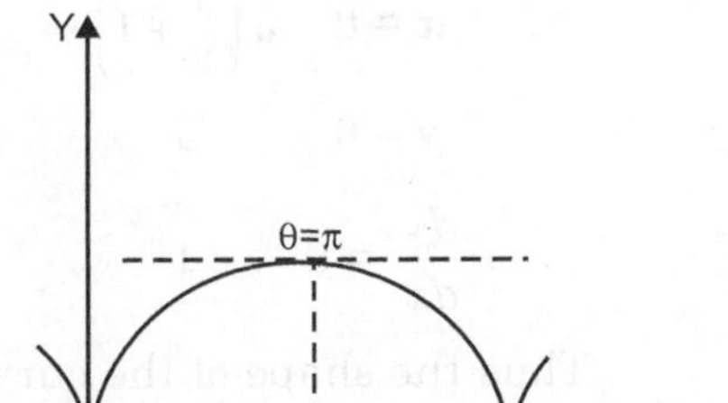

Thus one arch of the curve is the arc OAB and, we get similar other arches extending on both sides of Y-axis.

Thus the limits for one complete arch are from $\theta = 0$ to $\theta = 2\pi$.

$\therefore$ Required length of one each

$$= \int_0^{2\pi} \sqrt{\left(\frac{dx}{d\theta}\right)^2 + \left(\frac{dy}{d\theta}\right)^2}\, d\theta$$

$$= \int_0^{2\pi} \sqrt{a^2(1-\cos\theta)^2 + a^2\sin\theta}\, d\theta$$

$$= a\int_0^{2\pi} \sqrt{1 + \cos^2\theta - 2\cos\theta + \sin^2\theta}\, d\theta$$

$$= a\int_0^{2\pi} \sqrt{2(1-\cos\theta)}\, d\theta = a\int_0^{2\pi} \sqrt{4\sin^2\frac{\theta}{2}}\, d\theta \qquad \Big|\ \because\ 1 - \cos\theta = 2\sin^2\frac{\theta}{2}$$

$$= 2a\int_0^{2\pi} \sin\frac{\theta}{2}\, d\theta = 2a\left|\frac{-\cos\frac{\theta}{2}}{\frac{1}{2}}\right|_0^{2\pi}.$$

$$= -4a[\cos\pi - \cos 0] = -4a(-1-1) = 8a.$$

Note. The arch is clearly symmetrical about the line $\theta = \pi$. Hence we could easily calculate the length of half the arch from $\theta = 0$ to $\theta = \pi$ and then double it to get the full length. **Make use of symmetry wherever possible.**

(*b*) The equations of the cycloid are $x = a(\theta + \sin\theta)$, $y = a(1 - \cos\theta)$.

Note. The '*complete cycloid*' means '*one arch of the cycloid*' *i.e.*, Portion between two successive cusps.

To trace the curve roughly to find the limits of integration :

(*i*) Since x is an odd function of θ, and y an even function of θ,

$\therefore$ The cycloid is symmetrical about the Y-axis.

(*ii*) Putting $x = 0$, we get $\theta + \sin\theta = 0$ which gives $\theta = 0$. Now when $\theta = 0$, $y = 0$.

$\therefore$ The cycloid passes through the origin.

(*iii*) $\dfrac{dx}{d\theta} = a(1 + \cos\theta)$, $\dfrac{dy}{d\theta} = a\sin\theta$

$$\therefore \quad \frac{dy}{dx} = \frac{a\sin\theta}{a(1+\cos\theta)} = \frac{2\sin\frac{\theta}{2}\cos\frac{\theta}{2}}{2\cos^2\frac{\theta}{2}} = \tan\frac{\theta}{2}.$$

(*iv*) The various values are as below :

$\theta = 0$	$\frac{\pi}{2}$	θ	
$x = 0$	$a\left(\frac{\pi}{2} + 1\right)$	$a\pi$	$\because$ of symm. about Y-axis trace its image in Y-axis
$y = 0$	a	$2a$	
$\frac{dy}{dx} = 0$	1	∞	

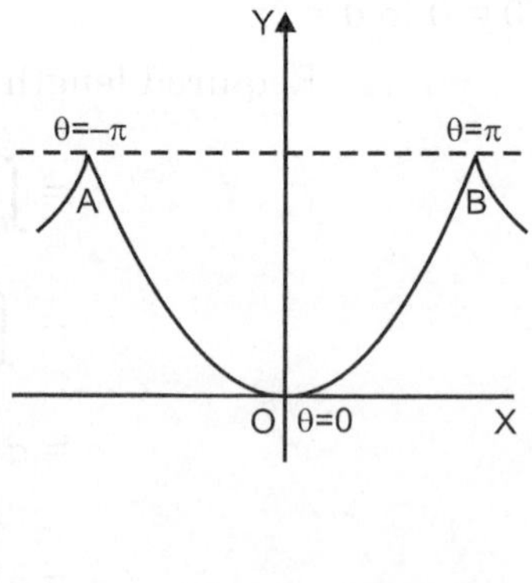

Thus the shape of the curve is as shown is the figure.

The complete arch AOB, is symmetrical about the Y-axis,

$\therefore$ Arc AOB = 2 arc OB and for arc OB, θ varies from 0 to π.

$\therefore$ Length of complete cycloid = Arc AOB = 2 arc OB

$$= 2\int_0^{\pi}\sqrt{\left(\frac{dx}{d\theta}\right)^2 + \left(\frac{dy}{d\theta}\right)^2}\, d\theta$$

$$= 2\int_0^{\pi}\sqrt{a^2(1+\cos\theta)^2 + a^2\sin^2\theta}\, d\theta$$

$$= 2a\int_0^{\pi}\sqrt{1+\cos^2\theta + 2\cos\theta + \sin^2\theta}\, d\theta = 2a\int_0^{\pi}\sqrt{2(1+\cos\theta)}\, d\theta$$

$$= 2a\int_0^{\pi}\sqrt{2.2\cos^2\frac{\theta}{2}}\, d\theta \qquad \because\ 1+\cos\theta = 2\cos^2\frac{\theta}{2}$$

$$= 4a \int_0^{\pi} \cos\frac{\theta}{2}\, d\theta = 4a \left| \frac{\sin\frac{\theta}{2}}{\frac{1}{2}} \right|_0^{\pi}$$

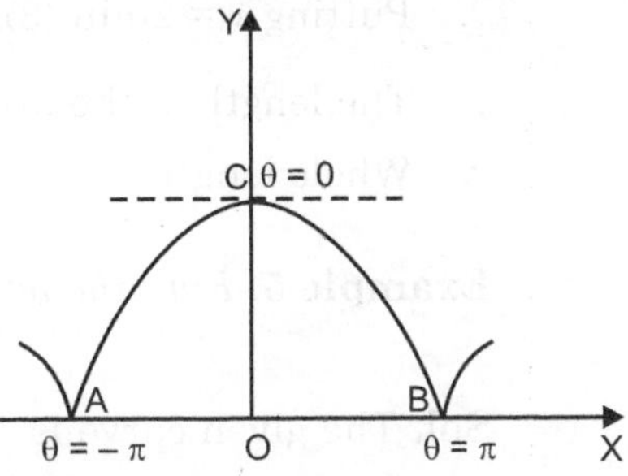

$$= 8a \left[\sin\frac{\pi}{2} - \sin 0 \right]$$

$$= 8a\,(1 - 0) = 8a.$$

(c) *Please try yourself.* **[Ans. 89]**

[**Hint.** Shape of the curve is as shown in the following figure].

Example 4. *Prove that in the cycloid $x = a(\theta + \sin\theta)$, $y = a(1 - \cos\theta)$, the length of the arc from the vertex to point (x, y) is $\sqrt{8ay}$. Hence find the whole length of the curve.*

Sol. The equation of the curve is $x = a(\theta + \sin\theta)$, $y = a(1 - \cos\theta)$

The shape of the curve is as shown in the figure [See. Ex. 3 (*b*)]

The vertex of the cycloid is at origin, which corresponds to $\theta = 0$.

Now $$\frac{dx}{d\theta} = a(1 + \cos\theta), \frac{dy}{d\theta} = a\sin\theta$$

$$\therefore \quad \left(\frac{dx}{d\theta}\right)^2 + \left(\frac{dy}{d\theta}\right)^2 = a^2(1+\cos\theta)^2 + a^2\sin^2\theta = a^2(1 + \cos^2\theta + 2\cos\theta + \sin^2\theta)$$

$$= 2a^2(1 + \cos\theta) = 2a^2 . 2\cos^2\frac{\theta}{2} = 4a^2\cos^2\frac{\theta}{2} \qquad ...(1)$$

$\therefore$ Required length of arc from the vertex, (0, 0) to (x, y) *i.e.*, from $\theta = 0$ to $\theta = \theta$ is

$$= \int_0^{\theta} \sqrt{\left(\frac{dx}{d\theta}\right)^2 + \left(\frac{dy}{d\theta}\right)^2}\, d\theta = \int_0^{\theta} 2a\cos\frac{\theta}{2}\, d\theta \qquad | \text{ Using (1)}$$

$$= 2a \left| \frac{\sin\frac{\theta}{2}}{\frac{1}{2}} \right|_0^{\theta} = 4a\left(\sin\frac{\theta}{2} - 0\right) = 4a\sin\frac{\theta}{2} \qquad ...(2)$$

But $$y = a(1 - \cos\theta) = 2a\sin^2\frac{\theta}{2}$$

$$\therefore \quad \sin\frac{\theta}{2} = \sqrt{\frac{y}{2a}}.$$

$\therefore$ From (2), required length $= 4a . \sqrt{\dfrac{y}{2a}} = \sqrt{8ay}$ ...(3)

For the length of the arc OB, y varies from 0 to $2a$.

$\therefore$ Putting $y = 2a$ in (3),

$\therefore$ The length of the arc OB $= \sqrt{8a \times 2a} = 4a$

$\therefore$ Whole length AOB $= 2$ arc OB $= 2 \times 4a = 8a$.

Example 5. *Find the length of the loop of the curve* $x = t^2,\ y = t - \frac{1}{3}t^3$.

Sol. The given curve is $x = t^2,\ y = t - \frac{1}{3}t^3$. ...(1)

The trace the curve roughly of find the limits of integration :

(*i*) Since x is an even function of t and y and odd function of t, curve is symmetrical about the X-axis.

(*ii*) From (1), $x = 0$ when $t = 0$. Also $t = 0$ makes $y = 0$.

$\therefore$ the curve passes through the origin.

(*iii*) The curve meets X-axis, where putting $y = 0$ in (1),

$$t - \frac{1}{3}t^3 = 0 \quad \text{or} \quad t\left(1 - \frac{1}{3}t^2\right) = 0$$

$$\therefore \qquad t = 0 \quad \text{or} \quad t = \pm\sqrt{3}$$

When $t = 0$, $x = 0$ and when $t = \pm\sqrt{3}$, $x = 3$

$\therefore$ The curve meets X-axis at (0, 0), (3, 0)

The curve meets Y-axis at the origin (0, 0).

(*iv*) When $x \to \infty$, t also $\to \infty$ and then $y \to -\infty$,

and
$$\underset{x \to \infty}{\text{Lt}} \frac{y}{x} = \underset{t \to \infty}{\text{Lt}} \frac{t - \frac{1}{3}t^3}{t^2} = \underset{x \to \infty}{\text{Lt}} \left(\frac{1}{t} - \frac{1}{3}t\right) = -\infty$$

$\therefore$ There are no asymptotes, but the curve extends to ∞.

(*v*) From (1), $\dfrac{dx}{dt} = 2t$, and $\dfrac{dy}{dt} = 1 - \dfrac{1}{3}.\,3t^2 = 1 - t^2$

$$\therefore \qquad \frac{dy}{dx} = \frac{\frac{dy}{dt}}{\frac{dx}{dt}} = \frac{1 - t^2}{2t}$$

$$\therefore \qquad \frac{dy}{dt} = 0 \text{ when } 1 - t^2 = 0 \text{ or } t = \pm 1,$$

When $t = 1$, $x = 1$, $y = 1 - \dfrac{1}{3} = \dfrac{2}{3}$

When $t = -1$, $x = 1$, $y = -1 + \dfrac{1}{3} = -\dfrac{2}{3}$

$\therefore$ At the points $\left(1, \frac{2}{3}\right), \left(1 - \frac{2}{3}\right)$, the tangent is || to X-axis.

Also $\frac{dy}{dx} \to \infty$ as $t \to 0$ and then $x \to 0, y \to 0$ *i.e.*, the tangent is $\perp$ to X-axis at the origin (0, 0),

When $t = \pm 3, \frac{dv}{dx} = \pm \frac{1}{\sqrt{3}}$ *i.e.*, at the point (3, 0) there are two tangents inclined at an angle $\mp$ 30° with the X-axis.

$\therefore$ The point (3, 0) is a node.

From (1), $t = \sqrt{x}$ $\therefore$ When x is –ve, t is imaginary.

Hence no portion of the curve lies to the left to Y-axis,

When $t \to -\infty, x \to \infty, y \to \infty$; and when $t \to \infty, x \to \infty, y \to -\infty$.

As t increases from $-\infty$ to ∞, passing through the values $-\sqrt{3}, -1, 0, 1, \sqrt{3}$ found above, the curve, PBCOABQ, is described as shown in figure.

For the upper half of the loop, t varies from 0 to $\sqrt{3}$.

$\therefore$ Length of the loop = 2 × length of upper half of loop

$$= 2\int_0^{\sqrt{3}} \sqrt{\left(\frac{dx}{dt}\right)^2 + \left(\frac{dy}{dt}\right)^2}\, dt$$

$$= 2\int_0^{\sqrt{3}} \sqrt{(2t)^2 + (1 - t^2)^2}\, dt$$

$$= 2\int_0^{\sqrt{3}} \sqrt{4t^2 + 1 + t^4 - 2t^2}\, dt$$

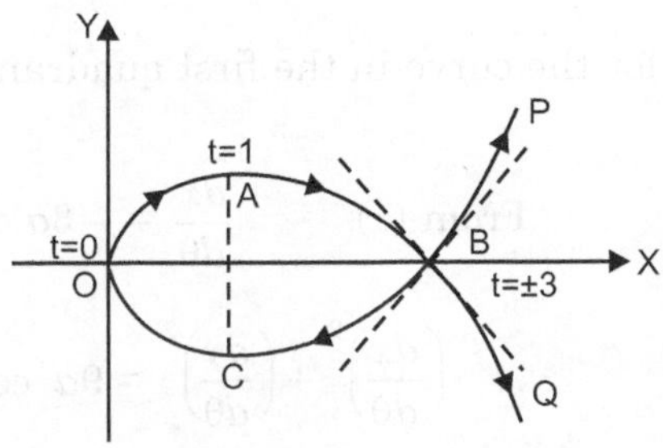

$$= 2\int_0^{\sqrt{3}} (1 + t^2)\, dt = 2\left| t + \frac{t^3}{3} \right|_0^{\sqrt{3}} = 2\left[\sqrt{3} + \frac{3\sqrt{3}}{3} - 0\right] = 2(\sqrt{3} + \sqrt{3}) = 4\sqrt{3}.$$

Example 6. *(a) Show that the length of the curve*
$x = a \cos^3 \theta, y = a \sin^3 \theta$ *is 6a.*
Also find the portion of the length which lies in the positive quadrant.
(b) Find the whole length of the curve (Astroied)
$$x^{2/3} + y^{2/3} = a^{2/3}.$$
(c) Find the length of one quadrant of the curve $\left(\frac{x}{a}\right)^{2/3} + \left(\frac{y}{b}\right)^{2/3} = 1.$

Also find the whole length.

Sol. (*a*) The equations of the curve are $x = a\cos^3\theta,\ y = a\sin^3\theta$ ...(1)

(*i*) The curve is symmetrical about both the axes.

When $|x| < a$, $|y| < a$ also.

Hence the curve lies within the square bounded by the lines

$$x = \pm a,\ y = \pm a$$

$$x = 0, \text{ gives } \cos^2\theta = 0 \text{ i.e., } \cos\theta = 0 \text{ or } \theta = \frac{\pi}{2},$$

When $\theta = \frac{\pi}{2}$, $y = a$.

Hence the curve does not pass through the origin.

The various values of x, y for some of θ are

$\theta = 0$	$\frac{\pi}{2}$	π	$\frac{3\pi}{2}$	2π
$x = a$	0	$-a$	0	a
$y = 0$	a	0	$-a$	0

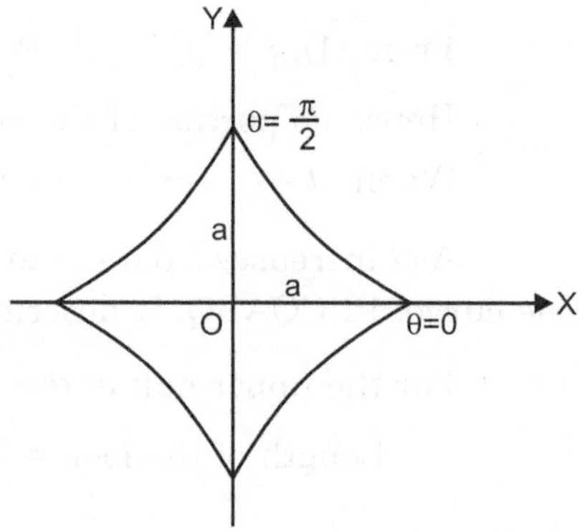

As θ increases from 0 to $\frac{\pi}{2}$, x decreases from a to 0.

The shape of the curve is as shown in the figure and for the curve in the first quadrant, θ varies from 0 to $\frac{\pi}{2}$.

From (1) $\frac{dx}{d\theta} = -3a\cos^2\theta\sin\theta$ and $\frac{dy}{d\theta} = 3a\sin^2\theta\cos\theta$

$$\therefore \left(\frac{dx}{d\theta}\right)^2 + \left(\frac{dy}{d\theta}\right)^2 = 9a^2\cos^4\theta\sin^2\theta + 9a^2\sin^4\theta\cos^2\theta = 9a^2\sin^2\theta\cos^2\theta(\cos^2\theta + \sin^2\theta)$$

$$= 9a^2\sin^2\theta\cos^2\theta \qquad |\because \sin^2\theta + \cos^2\theta = 1$$

$\therefore$ Required whole length of the curve

$$= 4 \times \text{length of the curve in the first quadrant}$$

$$= 4\int_0^{\pi/2}\sqrt{\left(\frac{dx}{d\theta}\right)^2 + \left(\frac{dy}{d\theta}\right)^2}\,d\theta = 4\int_0^{\pi/2} 3a\sin\theta\cos\theta\,d\theta$$

$$= 12a\int_0^{\pi/2}\sin\theta\cos\theta\,d\theta = 12a\left|\frac{\sin^2\theta}{2}\right|_0^{\pi/2} = 6a\left[\sin^2\frac{\pi}{2} - \sin^2 0\right]$$

$$= 6a\,[1 - 0] = 6a.$$

Second part :

Portion of the curve in the positive (first) quadrant

$$= \frac{1}{4} \times \text{whole length of curve} = \frac{1}{4} \times 6a = \frac{3a}{2}.$$

(*b*) The equation of the curve is $x^{2/3} + y^{2/3} = a^{2/3}$

or $$\left(\frac{x}{a}\right)^{2/3} + \left(\frac{y}{a}\right)^{2/3} = 1 \qquad ...(1)$$

Also $\cos^2\theta + \sin^2\theta = 1$ ...(2)

From(1) and (2), we have

$$\left(\frac{x}{a}\right)^{2/3} = \cos^2\theta \quad \therefore \quad x = a\cos^3\theta$$

and $$\left(\frac{y}{a}\right)^{2/3} = \sin^2\theta \quad \therefore \quad y = a\sin^3\theta$$

Thus $x = a\cos^3\theta,\ y = a\sin^3\theta$ ...(3)

are the parametric equations of the curve.

Now proceed as in part (*a*) above of find the required length.

(*c*) The given curve is $\left(\frac{x}{a}\right)^{2/3} + \left(\frac{y}{b}\right)^{2/3} = 1$ ...(1)

Comparing it with $\cos^2\theta + \sin^2\theta = 1$, we get

$$\left(\frac{x}{a}\right)^{2/3} = \cos^2\theta \text{ and } \left(\frac{y}{b}\right)^{2/3} = \sin^2\theta$$

or $$x = a\cos^3\theta,\ y = b\sin^3\theta \qquad ...(2)$$

which are the parametric equations of the given curve.

(*i*) The curve is symmetrical about both the axes.

Since $|x| \le a,\ |y| \le b$,

$\therefore$ the curve lies entirely within a rectangle bounded by the lines $x = \pm a,\ y = \pm b$.

(*ii*) The curve does not pass through the origin, since no value of θ makes *both* x and y zero.

(*iii*) The various values of x, y for some values of θ are

$\theta = 0$	$\frac{\pi}{2}$	π	$\frac{3\pi}{2}$	2π
$x = a$	0	$-a$	0	a
$y = 0$	b	0	$-b$	0

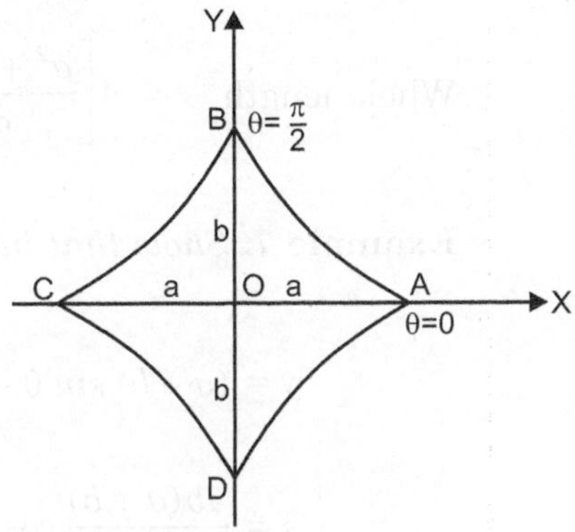

As θ increases from 0 to $\frac{\pi}{2}$, x decreases from a to 0 any y increase from 0 to b.

Thus the shape of the curve is as shown in the figure.

For the portion of the curve in the first quadrant, θ various from 0 to $\frac{\pi}{2}$.

From (2), $\frac{dx}{d\theta} = -3a\cos^2\theta\sin\theta$, and $\frac{dy}{d\theta} = 3b\sin^2\theta\cos\theta$,

$$\therefore \left(\frac{dx}{d\theta}\right)^2 + \left(\frac{dy}{d\theta}\right)^2 = 9a^2\cos^4\theta\sin^2\theta + 9b^2\sin^4\theta\cos^2\theta = 9\sin^2\theta\cos^2\theta\,(a^2\cos^2\theta + b^2\sin^2\theta)$$

Hence the required length of arc in the first quadrant

$$= \int_0^{\pi/2} \sqrt{\left(\frac{dx}{d\theta}\right)^2 + \left(\frac{dy}{d\theta}\right)^2}\, d\theta$$

$$= \int_0^{\pi/2} \sqrt{9\sin^2\theta\cos^2\theta(a^2\cos^2\theta + b^2\sin^2\theta)}\, d\theta$$

Put $a^2\cos^2\theta + b^2\sin^2\theta = t^2$ **[Remember]**

$\therefore\ 2(b^2 - a^2)\cos\theta\sin\theta\, d\theta = 2t\, dt$

or $\sin\theta\cos\theta\, d\theta = \frac{t\, dt}{b^2 - a^2}$

When $\theta = 0, t = a$, when $\theta = \frac{\pi}{2}, t = b$

$$= 3\int_0^{\pi/2} (a^2\cos^2\theta + b^2\sin^2\theta)^{1/2}\sin\theta\cos\theta\, d\theta$$

$$= 3\int_a^b t\,.\frac{t\, dt}{b^2 - a^2} = \frac{3}{b^2 - a^2}\int_a^b t^2\, dt = \frac{3}{b^2 - a^2}\cdot\left|\frac{t^3}{3}\right|_a^b$$

$$= \frac{b^3 - a^3}{b^2 - a^2} = \frac{(b-a)(b^2 + ab + a^2)}{(b-a)(b+a)} = \frac{a^2 + ab + b^2}{a+b}.$$

Whole length $= 4\left[\frac{a^2 + ab + b^2}{a+b}\right]$.

Example 7. *Show that in the epicycloid* $x = (a+b)\cos\theta - b\cos\left(\frac{a+b}{b}\theta\right)$,

$$y = (a+b)\sin\theta - b\sin\left(\frac{a+b}{b}\theta\right),$$

$$s = \frac{4b(a+b)}{a}\cos\left(\frac{a}{2b}\theta\right),$$ *where s is measured from the vertex* $\left(\text{where } \theta = \frac{\pi b}{a}\right)$.

Sol. The equations of the given curve are

$$x = (a+b)\cos\theta - b\cos\left(\frac{a+b}{b}\theta\right)$$

$$y = (a+b)\sin\theta - b\sin\left(\frac{a+b}{b}\theta\right)$$

$$\therefore \quad \frac{dx}{d\theta} = -(a+b)\sin\theta + b\,.\sin\left(\frac{a+b}{b}\theta\right).\frac{a+b}{b} = (a+b)\left[\sin\left(\frac{a+b}{b}\theta\right) - \sin\theta\right]$$

$$= 2(a+b)\cos\left(\frac{a+2b}{2b}\theta\right)\sin\frac{a\theta}{2b}$$

$$\because \sin C - \sin D = 2\cos\frac{C+D}{2}\sin\frac{C-D}{2}.$$

and

$$\frac{dy}{d\theta} = (a+b)\cos\theta - b\cos\left(\frac{a+b}{b}\theta\right).\frac{a+b}{b}$$

$$= (a+b)\left[\cos\theta - \cos\left(\frac{a+b}{b}\theta\right)\right]$$

$$= 2(a+b)\sin\left(\frac{a+2b}{2b}\theta\right)\sin\frac{a\theta}{2b}$$

$$\because \cos C - \cos D = 2\sin\frac{C+D}{2}\sin\frac{D-C}{2}.$$

$$\therefore \quad \left(\frac{dx}{d\theta}\right)^2 + \left(\frac{dy}{d\theta}\right)^2 = 4(a+b)^2\cos^2\left(\frac{a+2b}{2b}\theta\right)\sin^2\frac{a\theta}{2b} + 4(a+b)^2\sin^2\left(\frac{a+2b}{2b}\theta\right)\sin^2\frac{a\theta}{2b}$$

$$= 4(a+b)^2\sin^2\frac{a\theta}{2b}\left[\cos^2\left(\frac{a+2b}{2b}\theta\right) + \sin^2\left(\frac{a+2b}{2b}\theta\right)\right]$$

$$= 4(a+b)^2\sin^2\frac{a\theta}{2b}(1) = 4(a+b)^2\sin^2\frac{a\theta}{2b}.$$

$\therefore$ Required length of arc from vertex $\left(\theta = \frac{\pi b}{a}\right)$ to any point is

$$= \int_{\pi b/a}^{\theta}\sqrt{\left(\frac{dx}{d\theta}\right)^2 + \left(\frac{dy}{d\theta}\right)^2}\,d\theta = \int_{\pi b/a}^{\theta} 2(a+b)\sin\frac{a\theta}{2b}\,d\theta$$

$$= 2(a+b) \cdot \left| \frac{-\cos\left(\frac{a\theta}{2b}\right)}{\frac{a}{2b}} \right|_{\pi b/a}^{\theta} = 2(a+b) \cdot \frac{2b}{a} \left[\cos\frac{\pi}{2} - \cos\frac{a\theta}{2b} \right]$$

$$= \frac{-4b(a+b)}{a} \cos\frac{a\theta}{2b} = \frac{4b(a+b)}{a} \cos\frac{a\theta}{2b} \text{ (Numerically).}$$

Example 8. *Show that the length of the arc of the curve given by $x = a\,(3 \sin \theta - \sin^3 \theta)$ and $y = a \cos^3 \theta$, measured from (0, a) to any point (x, y) is $\frac{3a}{2}$ (θ + sin θ cos θ).*

Sol. The equations of the given curve are

$$x = a(3 \sin \theta - \sin^3 \theta) \quad \text{and} \quad y = a \cos^3 \theta$$

$$\therefore \quad \frac{dx}{d\theta} = a(3 \cos \theta - 3 \sin^2 \theta \cos \theta) = 3a \cos \theta\,(1 - \sin^2 \theta) = 3a \cos^3 \theta$$

and

$$\frac{dy}{d\theta} = -3a \cos^2 \theta \sin \theta$$

$$\therefore \quad \left(\frac{dx}{d\theta}\right)^2 + \left(\frac{dy}{d\theta}\right)^2 = 9a^2 \cos^6 \theta + 9a^2 \cos^4 \theta \sin^2 \theta = 9a^2 \cos^4 \theta\,(1) = 9a^2 \cos^4 \theta.$$

$\therefore$ Required length of arc measured from (θ, a) to (x, y) is

$$= \int_0^\theta \sqrt{\left(\frac{dx}{d\theta}\right)^2 + \left(\frac{dy}{d\theta}\right)^2}\, d\theta = \int_0^\theta 3a \cos^2 \theta\, d\theta$$

$$= 3a \int_0^\theta \frac{1 + \cos 2\theta}{2}\, d\theta = \frac{3a}{2} \left| \theta + \frac{\sin 2\theta}{2} \right|_0^\theta$$

$$= \frac{3a}{2} (\theta + \sin \theta \cos\theta - 0) = \frac{3a}{2} (\theta + \sin \theta \cos \theta).$$

9.4. ARC FORMULA FOR POLAR EQUATIONS

Show that the length of the arc of the curve r = f(θ) between the points, where $\theta = \alpha$, $\theta = \beta$ is

$$= \int_\alpha^\beta \sqrt{\mathbf{r}^2 + \left(\frac{\mathbf{dr}}{\mathbf{d\theta}}\right)^2}\, \mathbf{d\theta}.$$

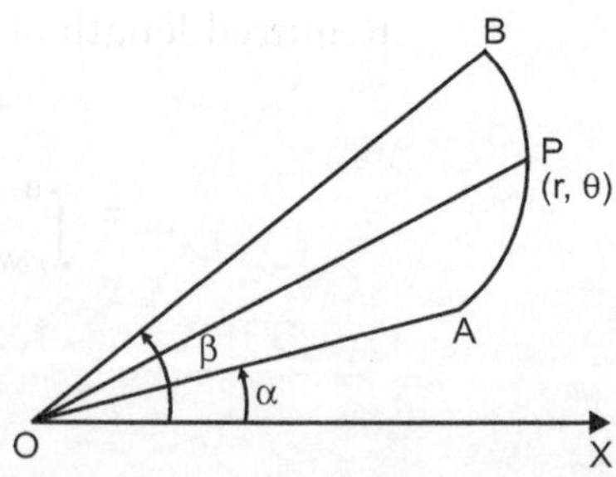

Let AB be the curve $r = f(\theta)$ and A, B the points where $\theta = \alpha$, $\theta = \beta$.

Let P (r, θ) be any point on the curve. If s denotes the length of the arc measured from the fixed point A to the variable point P, then clearly s is a function of θ.

From Differential Calculus, we know that

$$\frac{ds}{d\theta} = \sqrt{r^2 + \left(\frac{dr}{d\theta}\right)^2} \qquad ...(1)$$

$$\therefore \quad \int_\alpha^\beta \sqrt{r^2 + \left(\frac{dr}{d\theta}\right)^2}\, d\theta = \int_\alpha^\beta \frac{ds}{d\theta}\,.\, d\theta = [s]_\alpha^\beta$$

$$= (\text{value of } s \text{ when } \theta = \beta) - (\text{value of } s \text{ when } \theta = \alpha)$$

$$= \text{arc AB} - 0 = \text{arc AB}.$$

Hence $\quad \textbf{arc AB} = \int_\alpha^\beta \sqrt{\mathbf{r}^2 + \left(\frac{\mathbf{dr}}{\mathbf{d\theta}}\right)^2}\, \mathbf{d\theta}.$

Cor. 1. From (1), $\quad \mathbf{s} = \int \sqrt{\mathbf{r}^2 + \left(\frac{\mathbf{dr}}{\mathbf{d\theta}}\right)^2}\, \mathbf{d\theta}$

Cor.2. When the equation of the curve is of the form $\theta = \mathbf{f(r)}$, the length of the arc from r_1 to r_2 is given by

$$\int_{r_1}^{r_2} \sqrt{1 + \mathbf{r}^2 \left(\frac{\mathbf{d\theta}}{\mathbf{dr}}\right)^2}\, \mathbf{dr}.$$

ILLUSTRATIVE EXAMPLES

Example 1. (*a*) *Find the length of the arc of the equiangular spiral* $r = ae^{\theta \cot \alpha}$ *between the points for which the radii vectors are* r_1 *and* r_2.

(*b*) *Find the length of the arc of the spiral* $r = a\theta$ *between the points whose radii vectors are* r_1 *and* r_2.

(*c*) *Show that the length of the arc of the hyperbolic spiral* $r\theta = a$ *taken from the point* $r = a$ *to* $r = 2a$ *is* $a\left[\sqrt{5} - \sqrt{2} + \log\left(\frac{2 + \sqrt{8}}{1 + \sqrt{5}}\right)\right]$

Sol. (*a*) The equation of the curve is $r = ae^{\theta \cot \alpha}$

$$\therefore \quad \frac{dr}{d\theta} = ae^{\theta \cot \alpha}\,.\,\cot \alpha = r \cot \alpha. \qquad | \because \; r = ae^{\theta \cot \alpha}$$

$\therefore$ Required length of arc from $r = r_1$ to $r = r_2$ is

$$= \int_{r_1}^{r_2} \sqrt{1 + r^2 \left(\frac{d\theta}{dr}\right)^2}\, dr = \int_{r_1}^{r_2} \sqrt{1 + r^2 \frac{1}{r^2 \cot^2 \alpha}}\, dr$$

$$= \int_{r_1}^{r_2} \sqrt{1 + \tan^2 \alpha}\, dr = \int_{r_1}^{r_2} \sec \alpha\, dr = \sec \alpha\,.[r]_{r_1}^{r_2} = (r_2 - r_1) \sec \alpha.$$

(*b*) The given curve is $r = a\theta$ or $\theta = \dfrac{r}{a}$ $\therefore$ $\dfrac{d\theta}{dr} = \dfrac{1}{a}$.

$\therefore$ Required length of arc from $r = r_1$ to $r = r_2$ is

$$= \int_{r_1}^{r_2} \sqrt{1 + r^2 \left(\frac{d\theta}{dr}\right)^2}\, dr = \int_{r_1}^{r_2} \sqrt{1 + r^2 \cdot \frac{1}{a^2}}\, dr$$

$$= \frac{1}{a} \int_{r_1}^{r_2} \sqrt{r^2 + a^2}\, dr = \frac{1}{a} \left[\frac{r}{2} \sqrt{r^2 + a^2} + \frac{a^2}{2} \sinh^{-1} \frac{r}{a} \right]_{r_1}^{r_2}$$

$$\left[\because \int \sqrt{x^2 + a^2}\, dx = \frac{x}{2} \sqrt{x^2 + a^2} + \frac{a^2}{2} \sinh^{-1} \frac{x}{a} \right]$$

$$= \frac{1}{2a} [f(r)]_{r_1}^{r_2} \quad \text{where } f(r) = r\sqrt{r^2 + a^2} + a^2 \sinh^{-1} \frac{r}{a} = \frac{1}{2a}[f(r_2) - f(r_1)].$$

(*c*) The equation of the given curve is $r\theta = a$ or $\theta = \dfrac{a}{r}$

$$\therefore \quad \frac{d\theta}{dr} = -\frac{a}{r^2}.$$

Hence the required length from $r = a$ to $r = 2a$ is

$$= \int_a^{2a} \sqrt{1 + r^2 \left(\frac{d\theta}{dr}\right)^2}\, dr = \int_a^{2a} \sqrt{1 + r^2 \cdot \frac{a^2}{r^2}}\, dr$$

$$= \int_a^{2a} \sqrt{\frac{r^2 + a^2}{r}}\, dr = \int_a^{2a} \frac{\sqrt{r^2 + a^2}}{r^2} \cdot r\,dr$$

Put $r^2 + a^2 = t^2$ $\therefore$ $2r\,dr = 2t\,dt$ or $r\,dr = t\,dt$
when $r = a$, $t = \sqrt{2}\,a$, and when $r = 2a$, $t = \sqrt{5}a$

$$= \int_{a\sqrt{2}}^{a\sqrt{5}} \frac{t}{t^2 - a^2} \cdot t\,dt = \int_{a\sqrt{2}}^{a\sqrt{5}} \frac{t^2 - a^2 + a^2}{t^2 - a^2}\, dt$$

$$= \int_{a\sqrt{2}}^{a\sqrt{5}} \left(1 + \frac{a^2}{t^2 - a^2}\right) dt = \left[t + a^2 \cdot \frac{1}{2a} \log \frac{t - a}{t + a} \right]_{a\sqrt{2}}^{a\sqrt{5}}$$

$$= a(\sqrt{5} - \sqrt{2}) + \frac{a}{2} \left[\log \frac{\sqrt{5} - 1}{\sqrt{5} + 1} - \log \frac{\sqrt{2} - 1}{\sqrt{2} - 1} \right]$$

$$= a(\sqrt{5} - \sqrt{2}) + \frac{a}{2} \log \left(\frac{\sqrt{5} - 1}{\sqrt{5} + 1} \times \frac{\sqrt{2} + 1}{\sqrt{2} - 1} \right)$$

$$= a\,(\sqrt{5}-\sqrt{2}) + \frac{a}{2}\log\left[\frac{5-1}{(\sqrt{5}+1)^2} \times \frac{(\sqrt{2}+1)^2}{2-1}\right]$$

$$= a\,(\sqrt{5}-\sqrt{2}) + \frac{a}{2}\log\frac{4(\sqrt{2}+1)^2}{(\sqrt{5}+1)^2} = a\,(\sqrt{5}-\sqrt{2}) + \frac{a}{2}\log\left[\frac{2(\sqrt{2}+1)}{\sqrt{5}+1}\right]^2$$

$$= a(\sqrt{5}-\sqrt{2}) + \frac{a}{2}\,.\,2\log\frac{2(\sqrt{2}+1)}{\sqrt{5}+1} = a\left(\sqrt{5}-\sqrt{2}+\log\frac{\sqrt{8}+2}{\sqrt{5}+1}\right).$$

Example 2. (*a*) *Find the length of the arc of the cardioid $r = a(1 - \cos\theta)$ between the points whose vectorial angles are α and β.*

(*b*) *Find the perimeter of the curve $r = a(1 - \cos\theta)$.*

(*c*) *Show that the arc of the upper half of the curve $r = a\,(1 - \cos\theta)$ is bisected by $\theta = \frac{2\pi}{3}$*

Sol. (*a*) The given curve is $r = a(1 - \cos\theta)$

$\therefore \quad \frac{dr}{d\theta} = a\sin\theta$

$\therefore$ Required length of arc between $\theta = \alpha$, $\theta = \beta$ is

$$= \int_\beta^\alpha \sqrt{r^2 + \left(\frac{dr}{d\theta}\right)^2}\,d\theta = \int_\alpha^\beta \sqrt{a(1-\cos\theta)^2 + a^2\sin^2\theta}\,d\theta$$

$$= a\int_\alpha^\beta \sqrt{1+\cos^2\theta - 2\cos\theta + \sin^2\theta}\,d\theta$$

$$= a\int_\alpha^\beta \sqrt{2-2\cos\theta}\,d\theta = a\int_\alpha^\beta \sqrt{2\,.\,2\sin^2\frac{\theta}{2}}\,d\theta \qquad \because\ 1-\cos\theta = 2\sin^2\frac{\theta}{2}$$

$$= 2a\int_\alpha^\beta \sin\frac{\theta}{2}\,d\theta = 2a.\left|\frac{-\cos\frac{\theta}{2}}{\frac{1}{2}}\right|_\alpha^\beta$$

$$= 4a\left[-\cos\frac{\beta}{2} + \cos\frac{\alpha}{2}\right] = 4a\left(\cos\frac{\alpha}{2} - \cos\frac{\beta}{2}\right).$$

[**Note.** *When the limits of integration are given, we need* **not** *trace curve.*]

(*b*) The curve is $r = a\,(1 - \cos\theta)$ ...(1)

To trace the curve roughly to find the limits of integration:

(*i*) The curve is symmetrical about initial line,

$\therefore$ The equation of the curve remains unaltered on changing θ to $-\theta$.

(*ii*) Since $|\cos\theta| \le 1$.

$\therefore$ $|r| \le 2a$, hence the curve lies within the circle

$$r = 2a.$$

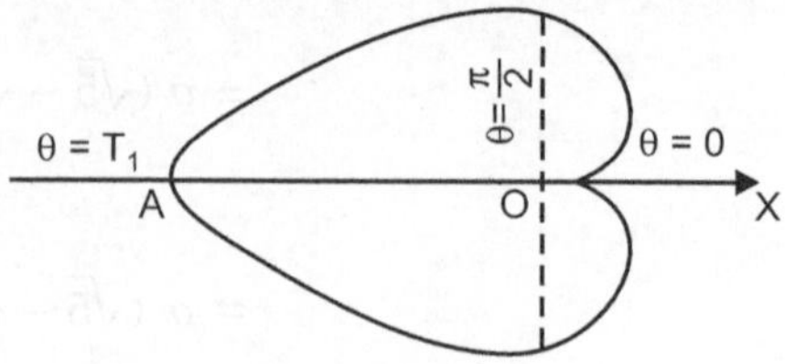

(*iii*) When $r = 0$, $\theta = 0$.

Hence the curve passes through the pole and $\theta = 0$ is the tangent to the curve at the pole.

(*iv*) The various values of r for some particular values of θ are :

$\theta = 0$	$\frac{\pi}{2}$	π
$r = 0$	a	$2a$.

We need not take the values from π to 2π, since the curve is symmetrical about the initial line.

Hence the shape of the curve is as shown in the figure.

For upper half of the curve, θ varies from 0 to π.

Now from (1), $\dfrac{dr}{d\theta} = a\sin\theta$

$$\therefore \quad r^2 + \left(\frac{dr}{d\theta}\right)^2 = a^2(1-\cos\theta)^2 + a^2\sin^2\theta = a^2(1+\cos^2\theta - 2\cos\theta + \sin^2\theta)$$

$$= a^2(2 - 2\cos\theta) = 2a^2(1-\cos\theta) = 2a^2 \,.\, 2\sin^2\frac{\theta}{2}$$

$$= 4a^2\sin^2\frac{\theta}{2} \qquad \ldots(2)$$

$\therefore$ Required perimeter of the curve

$= 2 \times$ length of arc of the upper half of the curve

$$= 2\int_0^{\pi} \sqrt{r^2 + \left(\frac{dr}{d\theta}\right)^2}\, d\theta = 2\int_0^{\pi} 2a\sin\frac{\theta}{2}\, d\theta \qquad | \text{ Using (2)}$$

$$= 4a.\left|\frac{-\cos\frac{\theta}{2}}{\frac{1}{2}}\right|_0^{\pi} = 8a\left[-\cos\frac{\pi}{2} + \cos 0\right] = 8a[-0+1] = 8a.$$

(*c*) Length of the arc from $\theta = 0$ to $\theta = \dfrac{2\pi}{3}$ is

$$= \int_0^{2\pi/3} \sqrt{r^2 + \left(\frac{dr}{d\theta}\right)^2}\, d\theta = \int_0^{2\pi/3} 2a\sin\frac{\theta}{2}\, d\theta \qquad | \text{ Using (2)}$$

$$= 2a.\left|\frac{-\cos\frac{\theta}{2}}{\frac{1}{2}}\right|_0^{2\pi/3} = -4a\left[\cos\frac{\theta}{2}\right]_0^{2\pi/3}$$

$$= -4a\left[\cos\frac{\pi}{3} - \cos 0\right] = -4a\left(\frac{1}{2} - 1\right) = 2a$$

Also entire length of the curve = $8a$ | Provied in (*b*)

$\therefore$ Length of the upper half of the curve = $4a$.

$\therefore$ The arc of the upper half of the curve is bisected by $\theta = \frac{2\pi}{3}$.

Example 3. *Find the entire length of the cardioid $r = a\,(1 + \cos\theta)$ and show that the arc of the upper half is bisected by $\theta = \frac{\pi}{3}$.*

Sol. The equation of the curve is

$$r = a\,(1 + \cos\theta)$$

(*i*) Since the equation remains unchanged on changing θ to $-\theta$,

$\therefore$ The curve is symmetrical about the initial line.

(*ii*) Since $|\cos\theta| \le 1$. $\therefore$ $|\theta| \le 2a$.

Thus the curve lies entirely within the circle $r = 2a$.

(*iii*) Putting $r = 0$, we get $1 + \cos\theta = 0$ or $\cos\theta = -1$

$\therefore$ $\theta = \pi$.

Hence the curve passes through the pole and the tangent at the pole is $\theta = \pi$.

(*iv*) The various values of r for some particular values of θ are :

$\theta = 0$	$\frac{\pi}{3}$	$\frac{\pi}{2}$	$\frac{2\pi}{3}$	π
$r = 2a$	$\frac{3a}{2}$	a	$\frac{a}{2}$	0

Thus the shape of the curve is as shown in the figure. For the upper half of the curve, θ varies from 0 to π.

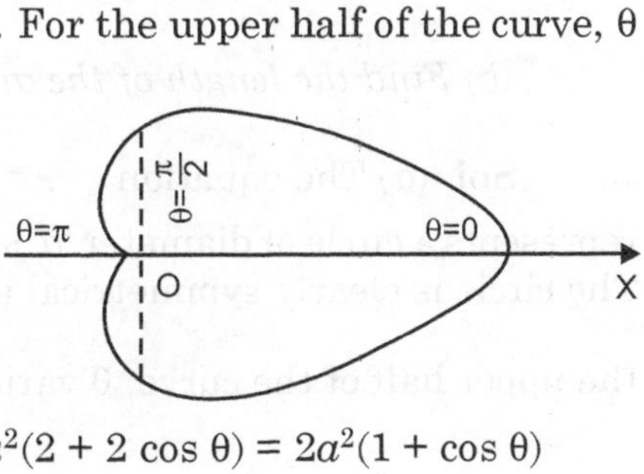

From (1), $\frac{dr}{d\theta} = -a\sin\theta$

$$\therefore \quad r^2 + \left(\frac{dr}{d\theta}\right)^2 = a^2(1+\cos\theta)^2 + a^2\sin^2\theta$$

$$= a^2(1 + \cos^2\theta + 2\cos\theta + \sin^2\theta) = a^2(2 + 2\cos\theta) = 2a^2(1 + \cos\theta)$$

$$= 2a^2 \,.\, 2\cos^2\frac{\theta}{2} = 4a^2\cos^2\frac{\theta}{2} \qquad ...(2) \qquad \Big|\ \because\ 1 + \cos\theta = 2\cos^2\frac{\theta}{2}$$

$\therefore$ Required entire length of the curve

= 2 × length of the upper half of the curve

$$= 2\int_0^{\pi}\sqrt{r^2 + \left(\frac{dr}{d\theta}\right)^2}\,d\theta = 2\int_0^{\pi} 2a\cos\frac{\theta}{2}\,d\theta. \qquad |\text{ Using (2)}$$

$$= 4a\left|\frac{\sin\frac{\theta}{2}}{\frac{1}{2}}\right|_0^\pi = 8a\left[\sin\frac{\pi}{2} - \sin 0\right] = 8a\,(1-0) = 8a.$$

Second part :

Now length of the upper half of the curve $= \frac{1}{2} \times 8a = 4a.$

Also length of the arc from $\theta = 0$ to $\theta = \frac{\pi}{3}$

$$= \int_0^{\pi/3} \sqrt{r^2 + \left(\frac{dr}{d\theta}\right)^2}\, d\theta = \int_0^{\pi/3} 2a\cos\frac{\theta}{2}\, d\theta \qquad |\text{ Using (2)}$$

$$= 2a\left|\frac{\sin\frac{\theta}{2}}{\frac{1}{2}}\right|_0^{\pi/3} = 4a\left(\sin\frac{\pi}{6} - \sin 0\right) = 4a\left(\frac{1}{2} - 0\right) = 2a = \frac{1}{2} \times 4a$$

$$= \frac{1}{2}\text{ . length of the upper half of the curve.}$$

$\therefore$ The length of the upper half of the curve is bisected by $\theta = \frac{\pi}{3}$.

Example 4. (*a*) *Find the perimeter of the curve $r = a \cos\theta$.*

(*b*) *Find the length of the arc of the parabola $\frac{2a}{r} = 1 + \cos\theta$ cut off by its latus rectum.*

Sol. (*a*) The equation $\quad r = a\cos\theta \qquad \ldots(1)$

represents a circle of diameter '*a*' and passing through the pole. The circle is clearly symmetrical about the initial line and for the upper half of the curve, θ varies from 0 to $\frac{\pi}{2}$.

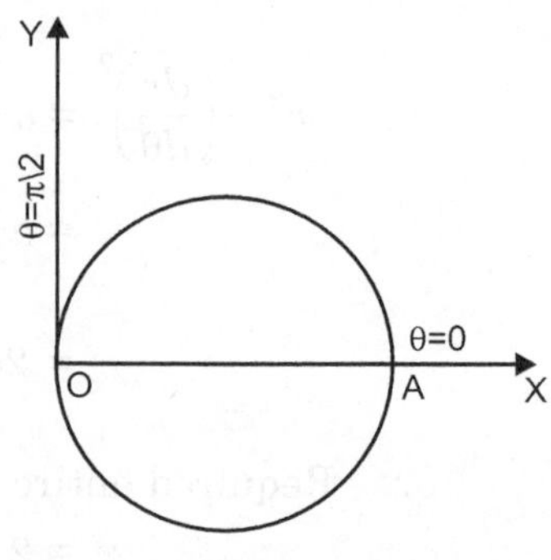

From (1), $\quad \frac{dr}{d\theta} = -\,a\sin\theta.$

$\therefore$ The whole perimeter

$$= 2\int_0^{\pi/2} \sqrt{r^2 + \left(\frac{dr}{d\theta}\right)^2}\, d\theta$$

$$= 2\int_0^{\pi/2} \sqrt{a^2\cos^2\theta + a^2\sin^2\theta}\, d\theta$$

$$= 2a \int_0^{\pi/2} \sqrt{\cos^2\theta + \sin^2\theta}\, d\theta = 2a \int_0^{\pi/2} 1\, d\theta = 2a\,[\theta]_0^{\pi/2} = 2a\left(\frac{\pi}{2} - 0\right) = \pi a.$$

(*b*) The equation of parabola is $\dfrac{2a}{r} = 1 + \cos\theta$

or $$r = \frac{2a}{1+\cos\theta} = \frac{2a}{2\cos^2\frac{\theta}{2}} = a\sec^2\frac{\theta}{2} \qquad ...(1)$$

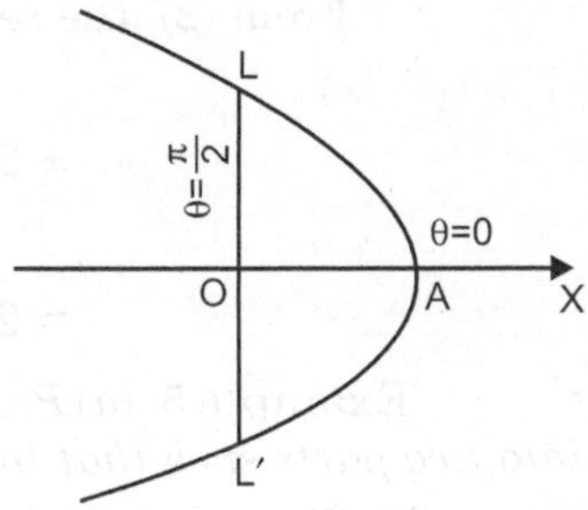

The given equation represents a parabola with focus at the pole and the axis of parabola as the initial line.

The curve is symmetrical about the initial line and for the upper half of the curve AL interepted by the latus rectum LOL′, θ varies from 0 to $\frac{\pi}{2}$.

From (1), $$\frac{dr}{d\theta} = 2a\sec\frac{\theta}{2}.\sec\frac{\theta}{2}\tan\frac{\theta}{2}.\frac{1}{2} = a\sec^2\frac{\theta}{2}\tan\frac{\theta}{2}.$$

∴ Required length of curve intercepted by the latus rectum *i.e.*, the length L′AL
= 2 × length of the arc AL

$$= 2\int_0^{\pi/2} \sqrt{r^2 + \left(\frac{dr}{d\theta}\right)^2}\, d\theta = 2\int_0^{\pi/2} \sqrt{a^2\sec^4\frac{\theta}{2} + a^2\sec^4\frac{\theta}{2}\tan^2\frac{\theta}{2}}\, d\theta$$

$$= 2a\int_0^{\pi/2} \sec^2\frac{\theta}{2}\sqrt{1+\tan^2\frac{\theta}{2}}\, d\theta$$

$$= 2a\int_0^{\pi/2} \sec^2\frac{\theta}{2}.\sec\frac{\theta}{2}\, d\theta = 2a\int_0^{\pi/2} \sec^3\frac{\theta}{2}\, d\theta \qquad ...(2)$$

Now, $$\int \sec^3\frac{\theta}{2}\, d\theta = \int \underset{\text{I}}{\sec\frac{\theta}{2}}.\underset{\text{II}}{\sec^2\frac{\theta}{2}}\, d\theta \qquad \text{| Integrate by parts}$$

$$= \sec\frac{\theta}{2}.\,2\tan\frac{\theta}{2} - \int \frac{1}{2}\sec\frac{\theta}{2}\tan\frac{\theta}{2}.\,2\tan\frac{\theta}{2}\, d\theta$$

$$= 2\sec\frac{\theta}{2}\tan\frac{\theta}{2} - \int \sec\frac{\theta}{2}\tan^2\frac{\theta}{2}\, d\theta = 2\sec\frac{\theta}{2}\tan\frac{\theta}{2} - \int \sec\frac{\theta}{2}\left(\sec^2\frac{\theta}{2} - 1\right) d\theta$$

$$= 2\sec\frac{\theta}{2}\tan\frac{\theta}{2} - \int \sec^3\frac{\theta}{2}\, d\theta + \int \sec\frac{\theta}{2}\, d\theta$$

or $$2\int \sec^3\frac{\theta}{2}\, d\theta = 2\sec\frac{\theta}{2}\tan\frac{\theta}{2} + \int \sec\frac{\theta}{2}\, d\theta$$

$$= 2\sec\frac{\theta}{2}\tan\frac{\theta}{2} + 2\log\left(\sec\frac{\theta}{2} + \tan\frac{\theta}{2}\right) \quad \Big| \because \int \sec x\,dx = \log(\sec x + \tan x)$$

$$\therefore \quad \int \sec^3\frac{\theta}{2}\,d\theta = \sec\frac{\theta}{2}\tan\frac{\theta}{2} + \log\left(\sec\frac{\theta}{2} + \tan\frac{\theta}{2}\right)$$

$\therefore$ From (2), the required length

$$= 2a\left[\sec\frac{\theta}{2}.\tan\frac{\theta}{2} + \log\left(\sec\frac{\theta}{2} + \tan\frac{\theta}{2}\right)\right]_0^{\pi/2}$$

$$= 2a[(\sqrt{2}\,.1 + \log(\sqrt{2}+1) - 0 - \log(1+0)] = 2a\,[\sqrt{2} + \log(\sqrt{2}+1)].$$

Example 5. (*a*) *Prove that the cardioid* $r = a\,(1 + \cos\theta)$ *is divided by the line* $4r\cos\theta = 3a$ *into two parts such that the lengths of the arcs on either side of this line are equal.*

(*b*) *Show that the length of the arc of that part of the cardioid* $r = a\,(1 + \cos\theta)$, *which lies on the side of the line* $4r = 3a\sec\theta$ *remote from the pole, is equal to* $4a$.

Sol. (*a*) The equation of the curve is $\qquad r = a(1 + \cos\theta)$ $\qquad$ ...(1)

and that of the line is $\qquad 4r\cos\theta = 3a$ $\qquad$ ...(2)

Solving (1) and (2),

[Putting the value of r from (1) in (2)], we get

$$4a(1 + \cos\theta)\cos\theta = 3a \quad \text{or} \quad 4\cos^2\theta + 4\cos\theta - 3 = 0$$

or $\qquad (2\cos\theta - 1)(2\cos\theta + 3) = 0$

which gives $\cos\theta = \frac{1}{2}$ or $-\frac{3}{2}$.

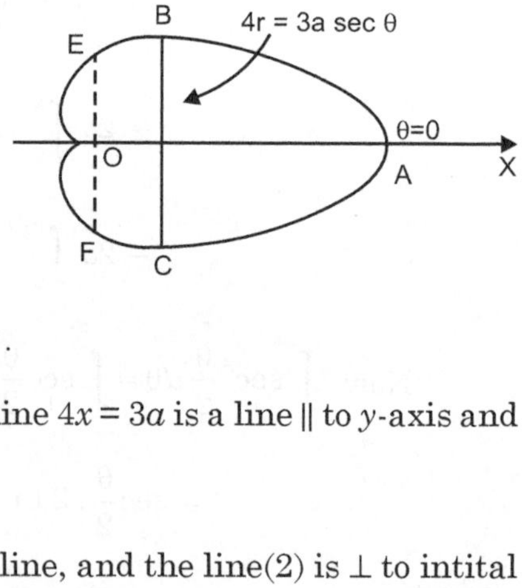

But $\quad \cos\theta = -\frac{3}{2}$ is impossible, since $|\cos\theta| \le 1$.

$\therefore \quad \cos\theta = \frac{1}{2}$ which gives $\theta = \frac{\pi}{3}$.

$\therefore$ The line (2) meets the curve (1) at B where $\theta = \frac{\pi}{3}$.

The line $4r\cos\theta = 3a$(on *changing to cartesians*) *i.e.*, line $4x = 3a$ is a line || to y-axis and at a distance $\frac{3a}{4}$ from the origin.

Since the cardioid (1) is symmetrical about the initial line, and the line(2) is $\perp$ to intital line,

$\therefore \qquad$ arc AB = arc AC.

For the upper half of the curve AB, θ varies from 0 to $\frac{\pi}{3}$ and for the upper half of the curve BEO, θ varies from $\frac{\pi}{3}$ to π.

We are to prove that arc AB = arc BEO

Now from (1), $\dfrac{dr}{d\theta} = -\, a \sin \theta$

$$\therefore \quad r^2 + \left(\frac{dr}{d\theta}\right)^2 = a^2 (1 + \cos \theta)^2 + a^2 \sin^2 \theta = a^2 (1 + \cos^2 \theta + 2 \cos \theta + \sin^2 \theta)$$

$$= a^2 (2 + 2 \cos \theta) = 2a^2 (1 + \cos \theta)$$

$$= 2a^2 \,.\, 2 \cos^2 \frac{\theta}{2} = 4a^2 \cos^2 \frac{\theta}{2} \qquad \ldots(3)$$

$$\therefore \quad \text{Length of arc} \quad AB = \int_0^{\pi/3} \sqrt{r^2 + \left(\frac{dr}{d\theta}\right)^2}\, d\theta = \int_0^{\pi/3} 2a \cos \frac{\theta}{2}\, d\theta \qquad | \text{ Using (3)}$$

$$= 2a \,.\left[\frac{\sin \frac{\theta}{2}}{\frac{1}{2}}\right]_0^{\pi/3} = 4a\left(\sin \frac{\pi}{6} - \sin 0\right) = 4a\left(\frac{1}{2} - 0\right) = 2a$$

and

$$\text{length of arc BEO} = \int_{\pi/3}^{\pi} \sqrt{r^2 + \left(\frac{dr}{d\theta}\right)^2}\, d\theta = \int_{\pi/3}^{\pi} 2a \cos \frac{\theta}{2}\, d\theta \qquad | \text{Using (3)}$$

$$= 4a.\left[\sin \frac{\theta}{2}\right]_{\pi/3}^{\pi} = 4a\left(\sin \frac{\pi}{2} - \sin \frac{\pi}{6}\right) = 4a\left(1 - \frac{1}{2}\right) = 2a$$

$\therefore$ arc AB = arc BEO.

$\therefore$ The line (2), *i.e.*, BC divides the curve (1) into two parts such that the lengths of the arcs on either side of the line are equal.

(*b*) Length of arc intercepted by the line (2) which is remote from the pole = arc CAB

$= 2 \times$ arc AB | By symmetry

$= 2 \times 2a = 4a$ | $\because$ arc AB $= 2a$, found above in part (*a*)

Hence the result.

Example 6. (*a*) *Show that the whole length of the limicon $r = a + b \cos \theta$ $(a > b)$ is equal to that of an ellipse whose semi-axes are equal in length to the maximum and minimum radii vectors of the limicon.*

(*b*) *Prove that the perimeter of the limicon $r = a + b \cos \theta$, if $\dfrac{b}{a}$ be small, is approximately*

$$2\pi a \left(1 + \frac{1}{4}\frac{b^2}{a^2}\right).$$

Sol. (*a*) The equation of the limicon is $r = a + b \cos \theta$, $(a > b)$ ...(1)

(*i*) Since the equation remains unaltered when θ changes to $-\theta$, $\therefore$ The curve is *symmetrical* about the initial line.

(*ii*) Putting $r = 0$, we get $a + b \cos \theta = 0$

or $$\cos \theta = -\frac{a}{b} > 1 \text{ numerically, which is impossible, } \because |\cos \theta| \le 1.$$

$\therefore$ The curve does not pass through the pole.

(*iii*) The values of r for some values of θ are

$\theta = 0$	$\frac{\pi}{3}$	$\frac{\pi}{2}$	$\frac{2\pi}{3}$	π
$r = a + b$	$a + \frac{b}{2}$	a	$a - \frac{b}{2}$	$a - b$

Thus the shape of the curve is as shown in the figure, and for the upper half of the curve, θ varies from 0 to π.

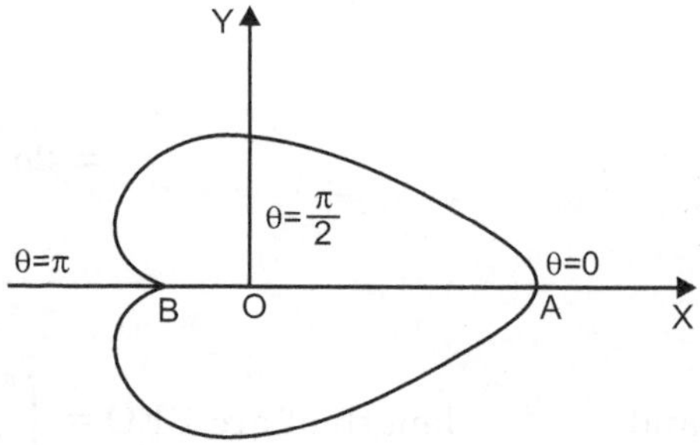

From (1), $$\frac{dr}{d\theta} = -b \sin \theta.$$

$\therefore$ Whole length of the limicon

$$= 2\int_0^{\pi} \sqrt{r^2 + \left(\frac{dr}{d\theta}\right)^2}\, d\theta$$

$$= 2\int_0^{\pi} \sqrt{(a + b \cos \theta)^2 + b^2 \sin^2 \theta}\, d\theta$$

$$= 2\int_0^{\pi} \sqrt{a^2 + b^2 + 2ab \cos \theta}\, d\theta \qquad ...(2)$$

The maximum value of $\cos \theta$ is 1 and its minimum value is -1

$\therefore$ From (1),

$\therefore$ maximum value of $r = a + b \,.\, 1 = a + b$.

and minimum value of $r = a + b(-1) = a - b$.

$\therefore$ The parametric equations of the ellipse whose semi-axes are the maximum and minimum radii vectors of (1), *i.e.*, $a + b$ and $a - b$ are

$$x = (a + b) \cos t, \quad y = (a - b) \sin t$$

$$\therefore \quad \frac{dx}{dt} = -(a + b) \sin t, \quad \frac{dy}{dt} = (a - b) \cos t$$

$$\therefore \quad \left(\frac{dx}{dt}\right)^2 + \left(\frac{dy}{dt}\right)^2 = (a + b)^2 \sin^2 t + (a - b)^2 \cos^2 t$$

$$= a^2 (\sin^2 t + \cos^2 t) + b^2(\sin^2 t + \cos^2 t) - 2ab (\cos^2 t - \sin^2 t)$$

$$= a^2 + b^2 - 2ab \cos 2t \qquad ...(3)$$

The ellipse is symetrical about both the axes, and in the first quadrant, t varies from 0 to $\frac{\pi}{2}$.

$\therefore$ Whole length of the ellipse

$$= 4\int_0^{\pi/2} \sqrt{\left(\frac{dx}{dt}\right)^2 + \left(\frac{dy}{dt}\right)^2}\, dt$$

$$= 4\int_0^{\pi/2} \sqrt{a^2 + b^2 - 2ab \cos 2t}\, dt \qquad \text{| Using (3)}$$

| Put $2t = \pi - \theta \quad \therefore \quad 2dt = -d\theta$
when $t = 0, \theta = \pi$, when $t = \frac{\pi}{2}, \theta = 0$

$$= 4\int_\pi^0 \sqrt{a^2 + b^2 - 2ab \cos (\pi - \theta)} \cdot \left(-\frac{d\theta}{2}\right)$$

$$= 2\int_0^\pi \sqrt{a^2 + b^2 - 2ab \cos \theta}\, d\theta \qquad \left|\ \because \int_a^b f(\theta)\, d\theta = -\int_b^a f(\theta)\, d\theta\right.$$

= whole length of the limicon. | from (2)

(*b*) From (2) above, the whole length of the limicon

$$= 2\int_0^\pi \sqrt{a^2 + b^2 + 2ab \cos \theta}\, d\theta = 2a\int_0^\pi \left[1 + \left(2\frac{b}{a}\cos\theta + \frac{b^2}{a^2}\right)\right]^{1/2} d\theta$$

(Expand by Binomial theorem)

$$= 2a\int_0^\pi \left[1 + \frac{1}{2}\left(2\frac{b}{a}\cos\theta + \frac{b^2}{a^2}\right) + \frac{\frac{1}{2}\left(-\frac{1}{2}\right)}{2\,!} \times \left(\frac{2b}{a}\cos\theta + \frac{b^2}{a^2}\right)^2 + ...\right] d\theta$$

$$= 2a\int_0^\pi \left[1 + \frac{b}{a}\cos\theta + \frac{b^2}{2a^2} - \frac{1}{8}\cdot\frac{4b^2}{a^2}\cos^2\theta + ...\right] d\theta$$

$$\left[\text{Neglecting } \frac{b^3}{a^3} \text{ and higher powers of } \frac{b}{a}, \text{ as } \frac{b}{a} \text{ is small}\right]$$

$$= 2a\int_0^\pi \left[1 + \frac{b}{a}\cos\theta + \frac{b^2}{2a^2}(1 - \cos^2\theta)\right] d\theta = 2a\int_0^\pi \left[1 + \frac{b}{a}\cos\theta + \frac{b^2}{2a^2}\sin^2\theta\right] d\theta.$$

$$= 2a\int_0^\pi \left[1 + \frac{b}{a}\cos\theta + \frac{b^2}{2a^2}\cdot\frac{1 - \cos 2\theta}{2}\right] d\theta$$

$$= 2a\int_0^{\pi}\left[1+\frac{b}{a}\cos\theta+\frac{b^2}{4a^2}-\frac{b^2}{4a^2}\cos 2\theta\right]d\theta$$

$$= 2a\left[\theta+\frac{b}{a}\sin\theta+\frac{b^2}{4a^2}\theta-\frac{b^2}{4a^2}\frac{\sin 2\theta}{2}\right]_0^{\pi}$$

$$= 2a\left[\pi+\frac{b^2}{4a^2}\pi-0\right]=2a\pi\left(1+\frac{b^2}{4a^2}\right)\text{approx.}$$

Example 7. *Show that the length of a loop of the curve*

$$3x^2y - y^3 = (x^2 + y^2)^3 \text{ is } 2\int_0^1 \frac{dr}{\sqrt{1-r^6}}.$$

Sol. The given curve is $3x^2y - y^3 = (x^2 + y^2)^3$ ...(1)

Changing to polars [by putting $x = r\cos\theta$, $y = r\sin\theta$],

(1) becomes, $3r^3\cos^2\theta\sin\theta - r^3\sin^3\theta - r^6$

or $r^3 = 3\cos^2\theta\sin\theta - \sin^3\theta = 3\sin\theta(1-\sin^2\theta) - \sin^3\theta$

$= 3\sin\theta - 4\sin^3\theta = \sin 3\theta$ ...(2)

[**Remember** : *To find the limits of integration for a loop of a polar curve, put $r = 0$ and find the two consecutive values of θ*].

$\therefore$ For a loop, putting $r = 0$ in (2), we get

$\sin 3\theta = 0 \quad \therefore \quad 3\theta = 0, \pi$ (Two consecutive values)

$\therefore \quad \theta = 0, \dfrac{\pi}{3}$

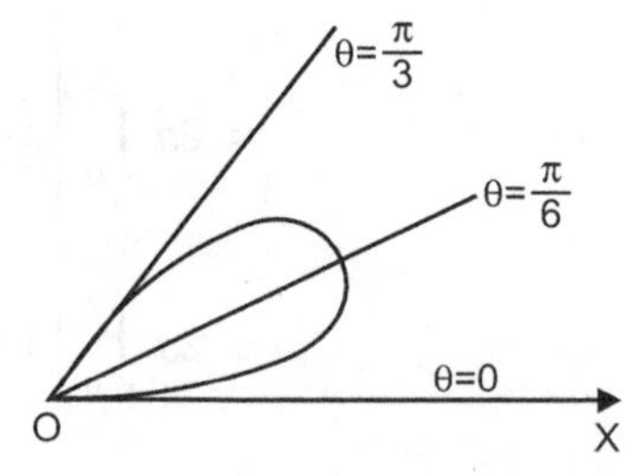

For a loop θ varies from 0 to $\dfrac{\pi}{3}$

For half of the loop, θ varies from 0 to $\dfrac{\pi}{6}$

(By symmetry)

Form (2) when $\theta - 0$, $r = 0$ when $\theta = \dfrac{\pi}{6}$, $r = 1$.

$\therefore$ For the lower half of the curve r varies from 0 to 1.

Now from (2), $3\theta = \sin^{-1} r^3$ or $\theta = \dfrac{1}{3}\sin^{-1} r^3$

$$\therefore \quad \frac{d\theta}{dr} = \frac{1}{3}.\frac{1}{\sqrt{1-r^6}}.3r^2 = \frac{r^2}{\sqrt{1-r^6}}$$

$$\therefore \quad 1 + r^2 \left(\frac{d\theta}{dr}\right)^2 = 1 + r^2 \cdot \frac{r^4}{1 - r^6} \quad \frac{1 - r^6 + r^6}{1 - r^6} = \frac{1}{1 - r^6}$$

$$\therefore \quad \text{Required length of loop} = 2\int_0^1 \sqrt{1 + r^2\left(\frac{d\theta}{dr}\right)^2}\, dr = 2\int_0^1 \frac{1}{\sqrt{1 - r^6}}\, dr.$$

TEST YOUR KNOWLEDGE

1. Find the perimeter of the circle $x^2 + y^2 = a^2$.
2. Find the length of the catenary $y = a \cosh \dfrac{x}{a}$ from $x = 0$ to $x = b$.
3. Find the arc length of the curve $3y^2 = x(x - 1)^2$, $y \geq 0$, between the points where the curve meets the x-axis.
4. Find the whole length of the loop of the curve
   (*i*) $3ax^2 = y^2 (a - y)$ (*ii*) $3ax^2 = y(y - a)^2$
   (*iii*) $3ay^2 = x^2 (a - x)$ (*iv*) $9ay^2 = x(x - 3a)^2$ (*v*) $9y^2 = (x - 4)(x - 7)^2$.
5. Find the whole length of the arc of the curve $8ay^2 = x^2(a^2 - 2x^2)$.
6. Find the length of the arc of the curve $x^2 + y^2 - 2ax = 0$ in first quadrant.
7. Show that the length of the arc of the curve $x = 3 - 4t^2$, $y = t^3 - 3t$ from $t = 1$ to $t = 3$ is 32.
8. Rectify the cycloid $= x = a(\theta - \sin\theta)$, $y = a(1 + \cos\theta)$.
9. Show that $8a$ is the length of the arc of the cycloid whose equations are $x = a(t - \sin t)$, $y = a(1 - \cos t)$ lying between two successive cusps.
10. Find the length of the arc of the cruve
    $x = a(\cos t + t \sin t)$, $y = a(\sin t - t \cos t)$ from $t = 0$ to $t = 2$.
11. A curve is given by the equations $x = a(\cos\theta + \theta \sin\theta)$, $y = a(\sin\theta = \theta\cos\theta)$.
    Find the length of the arc from $\theta = 0$ to $\theta = \alpha$.
12. Find the length of an arch of the cycloid given by $x = a(\theta + \sin\theta)$, $y = a(1 + \cos\theta)$.
13. Find the entire length of the astroid $x = a\cos^3\theta$, $y = b\sin^3\theta$.
14. Show that the lenth of the arc of the curve $r = a(1 + \cos\theta)$ between the points $\theta = 0$ to the point where $\theta = \dfrac{\pi}{2}$ is $2a\sqrt{2}$.
15. Find the perimeter of the curve $r = a(1 + \sin\theta)$.
16. Find the intrinsic equation of $y^3 = ax^2$.
17. Find the intrinsic equation of the curve $x = a(t + \sin t)$,$y = a(1 + \cos t)$.
18. Show that the intrinsic equation of the equi angular spiral $r = ae^{m\theta}$, S being measured from the point $(a, 0)$ is

$$S = \frac{a\sqrt{1 + m^2}}{m}(e^{m\psi} - 1).$$

[**Hint.** See Ex. 2 with Article 8. Here $\cot\alpha = m$ so that $\sec\alpha = \dfrac{\sqrt{1 + m^2}}{m}$]

Answers

1. $2\pi a$ **2.** $a \sinh \dfrac{b}{a}$ **3.** $\dfrac{2}{\sqrt{3}}$ **4.** (*i*) $\dfrac{4a}{\sqrt{3}}$ (*ii*) $\dfrac{4a}{\sqrt{3}}$ (*iii*) $\dfrac{4a}{\sqrt{3}}$

(*iv*) $4\sqrt{3}\,a$ (*v*) $4\sqrt{3}$ **5.** πa **6.** $\dfrac{1}{2}\pi a$ **8.** $8a$ **10.** $2a$

11. $\dfrac{1}{2}a\,\alpha^2$ **12.** $8a$ **13.** $4\left(\dfrac{a^2+ab+b^2}{a+b}\right)$ **15.** $8a$

16. $27S = 8a\,(\operatorname{cosec}^2 \psi - 1)$ **17.** $S + 4a \sin \psi = 0.$

10
Quadrature

Quadrature. *The process of finding the area bounded by a given curve is called* **quadrature.**

10.1. AREA FORMULAE FOR CARTESIAN EQUATIONS

To prove that the area bounded by the curve y = f(x), the x-axis and ordinates x = a, x = b is

$$\int_a^b y\,dx,$$

where y = f(x) is a continuous single-valued function and y does not change sign in the interval a ≤ x ≤ b.

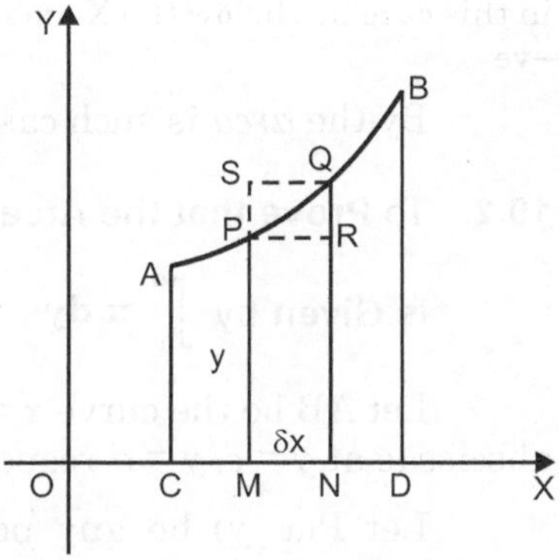

Let AB be the curve $y = f(x)$ and CA, DB be the two ordinates at $x = a$ and $x = b$ respectively.

Suppose y is an increasing function of x in the interval $a \le x \le b$.

Let P(x, y) be any point on the curve and Q$(x + \delta x, y + \delta y)$ a neighbouring point on it. Draw their ordinates PM, QN.

As x changes, the area ACMP also changes and hence is clearly a function of x.

Let A denote the area ACMP, then the area ACNQ = A + δA, so that

area PMNQ = δA.

Complete the rectangle PRQS. Then the area PMNQ lies between the area of rectangles PMNR and SMNQ.

i.e., δQ lies between $y\,\delta x$ and $(y + \delta y)\,\delta x$.

Dividing throughout by δx, $\dfrac{\delta A}{\delta x}$ lies between y and $y + \delta y$.

In the limiting position when Q → P, $\delta x \to 0$ and $\delta y \to 0$.

$\therefore \quad \underset{\delta x \to 0}{\text{Lt}} \dfrac{\delta A}{\delta x}$ lies between y and $\underset{\delta x \to 0}{\text{Lt}} (y + \delta y)$

$\therefore \quad \dfrac{dA}{dx}$ lies between y and a quantity which tends to y.

$$\therefore \qquad \frac{dA}{dx} = y$$

Integrating both sides w.r.t. x, from $x = a$ to $x = b$, we have

$$\int_a^b y\,dx = \int_a^b \frac{dA}{dx}\,.\,dx = [A]_a^b$$

$$= (\text{value of A when } x = b) - (\text{value of A when } x = a)$$

$$= \text{area ACDB} - 0.$$

Hence area ACDB = $\int_a^b$ **ydx.**

Note. 1. The area bounded by the curve AB, the ordinates at A and B and the x-axis is often called '*the area under the curve AB*'.

Note. 2. Sign of the area. We know that if $y = f(x) > 0$ over the range $a \le x \le b$, then the area $\int_a^b f(x)\,dx$ *is* +ve and if y is < 0 in the range $a \le x \le b$, then the area $\int_a^b f(x)\,dx$ is –ve. The curve $y = f(x)$ in this case lies below the X-axis over the range $a \le x \le b$. Thus we consider the areas below the x-axis as –ve.

By the *area* is such cases, we mean the numerical value of the area.

10.2. To Prove that the Area of the Curve x = f(y) between y-axis and the Lines y = c, y = d is Given by $\int_c^d$ x dy

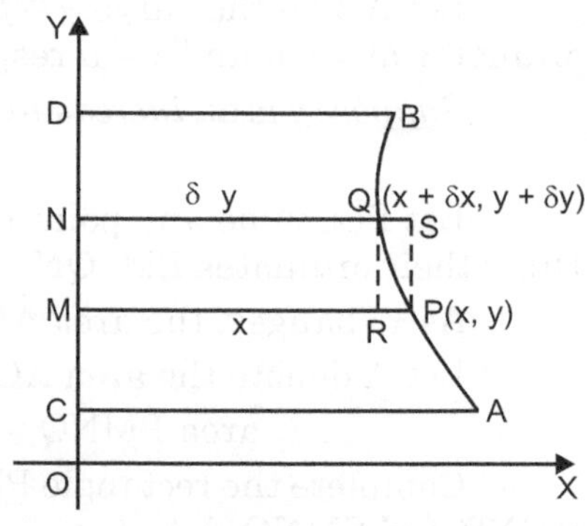

Let AB be the curve $x = f(y)$ and let CA, DB be the abscissae at $y = c$, $y = d$ respectively.

Let P(x, y) be any point on the curve and let Q($x + \delta x$, $y + \delta y$) be a neighbouring point on it.

Draw PM, QN ⊥'s on Y-axis from P and Q.

As y changes, the area ACMP also changes and hence is clearly a function of y. Let A denote the area ACMP then the area ACNQ = A + δA, so that area PMNQ = δA.

Complete the rectangle PRQS.

Then the area PMNQ, lies between the areas of the rectangles RMNQ and PMNS,

i.e., δA lies between $x\,\delta y$, and $(x + \delta x)\,\delta y$.

Dividing throughout by δy, we have $\dfrac{\delta A}{\delta y}$ lies between x and $x + \delta x$.

In the limiting position, when Q → P, then $\delta x \to 0$, $\delta y \to 0$

$$\therefore \quad \underset{\delta y \to 0}{\text{Lt}} \frac{\delta A}{\delta y} \text{ lies between } x \text{ and } \underset{\delta x \to 0}{\text{Lt}} (x + \delta x)$$

or $\dfrac{dA}{dy}$ lies between x and a quantity which tends to x.

$$\therefore \quad \frac{dA}{dy} = x.$$

Integrating both sides, between the limits c to d, we have

$$\int_c^d x\,dy = \int_c^d \frac{dA}{dy}\,dy = [A]_c^d$$

$$= (\text{value of A when } y = d) - (\text{value of A when } y = c)$$

$$= \text{Area ACDB} - 0 = \text{Area ACDB}.$$

Hence area $\quad$ **ACDB** $= \int_c^d \mathbf{x\,dy}.$

ILLUSTRATIVE EXAMPLES

Example 1. (*a*) *Find the area enclosed by the curve $y = 3x - x^2$, x-axis and the lines $x = 0$ and $x = 3$.*

(*b*) *Find the area bounded by the x-axis, the curve $y = c\cosh\frac{x}{c}$ and the ordinates $x = 0$, $x = a$.*

(*c*) *Find the area bounded by the parabola $y = 2x - x^2$ and x-axis.*

Sol. (*a*) The given curve is $y = 3x - x^2 = x(3 - x) \geq 0$ for $0 \leq x \leq 3$.

$\therefore$ Required area bounded by the curve, x-axis and the ordinates $x = 0$, $x = 3$ is

$$= \int_0^3 y\,dx = \int_0^3 (3x - x^2)\,dx = \left[\frac{3x^2}{2} - \frac{x^3}{3}\right]_0^3 = \frac{27}{2} - \frac{27}{3} - 0 = \frac{27}{6} = \frac{9}{2}.$$

(*b*) The given curve is $y = c\cosh\frac{x}{c}$.

$$\therefore \text{ Required area} = \int_0^a y\,dx = \int_0^a c\cosh\frac{x}{c}\,dx = c^2\left[\sinh\frac{x}{c}\right]_0^a$$

$$= c^2\left[\sinh\frac{a}{c} - \sinh 0\right] = c^2\sinh\frac{a}{c}.$$

(*c*) The given parabola is $y = 2x + x^2$ or $x^2 - 2x + 1 = 1 - y$ or $(x - 1)^2 = -(y - 1)$

It is a downward parabola with vertex (1, 1).

It meets the x-axis

i.e., $\quad y = 0$ where $0 = 2x - x^2$ or $x(2 - x) = 0$

$\therefore \quad x = 0, 2$

$\therefore$ The parabola meets the x-axis at O(0, 0) and A(2, 0).

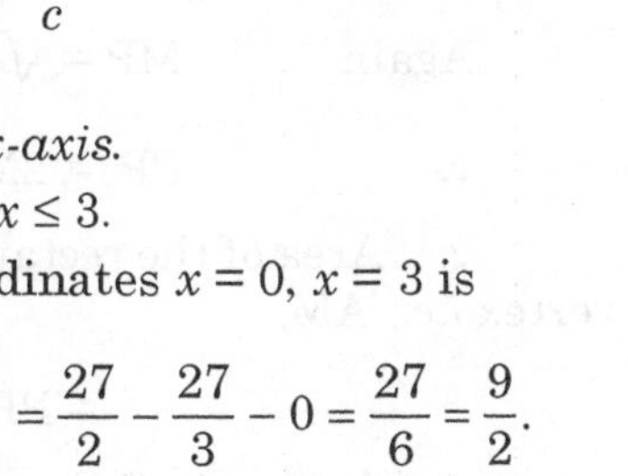

Required area (shown shaded)

$$= \int_0^2 y\,dx = \int_0^2 (2x - x^2)\,dx$$

$$= \left[2 \cdot \frac{x^2}{2} - \frac{x^3}{3}\right]_0^2 = 4 - \frac{8}{3} = \frac{4}{3}.$$

Example 2. *Show that the area cut-off a parabola by any double ordinate is two-thirds of the corresponding rectangle contained by the double ordinate and its distance from the vertex.*

Sol. Let the equation of the parabola be $y^2 = 4ax$ ...(1)

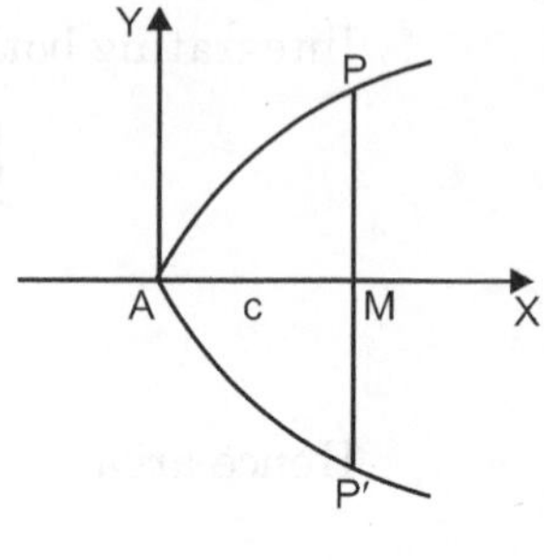

and that of the double ordinate PP′ be $x = c$.

The curve is symmetrical about the x-axis and for the part of the curve above the x-axis (taking +ve sign before the square root), from (1), we have

$$y = \sqrt{4ax}$$

∴ Area cut-off the parabola by the double ordinate PMP′

$$= \text{area PAP} = 2 \text{ area PAM}$$

$$= 2\int_0^c y\,dx = 2\int_0^c \sqrt{4ax}\,dx = 2\int_0^c 2\sqrt{a}\,\sqrt{x}\,dx$$

$$= 4\sqrt{a}\left[\frac{x^{3/2}}{3/2}\right]_0^c = \frac{8}{3}\sqrt{a}\,(c^{3/2} - 0) = \frac{8}{3}\sqrt{a}\,c^{3/2} \qquad \text{...(2)}$$

Again $\quad \text{MP} = \sqrt{4ax} = \sqrt{4ac} = 2\sqrt{a}\sqrt{c}$ $\quad | \because \text{AM} = c$

∴ $\quad \text{PP}' = 2\text{MP} = 2\sqrt{a}\,\sqrt{c}$.

∴ Area of the rectangle formed by the double ordinate PMP′ and its distance from the vertex *i.e.*, AM,

$$= \text{PP}' \times \text{AM} = 4\sqrt{a}\,\sqrt{c} \times c = 4\sqrt{a}\,c^{3/2} \qquad \text{...(3)}$$

∴ Area cut-off the parabola by the double ordinate PP′

$$= \frac{8}{3}\sqrt{a}\,c^{3/2} = \frac{2}{3}(4\sqrt{a}\,c^{3/2})$$ | *proved above in* (2)

$$= \frac{2}{3} \times \text{area of the rectangle formed by PP}' \text{ and AM.}$$

Hence the result.

Example 3. *Find the whole area of the ellipse* $\dfrac{x^2}{a^2} + \dfrac{y^2}{b^2} = 1$.

Sol. The equation of the ellipse is $\dfrac{x^2}{a^2} + \dfrac{y^2}{b^2} = 1$. ...(1)

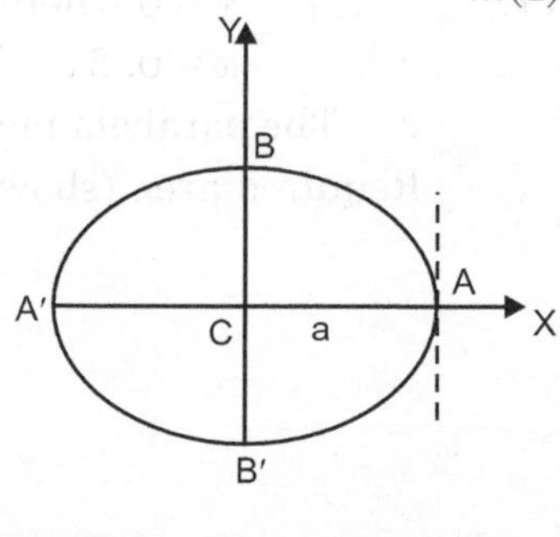

The ellipse is symmetrical about both the axes and so the whole area of the ellipse is four times the area under the curve in the first quadrant.

i.e., Whole area of the ellipse = 4 × area CAB

Now for the area CAB, x varies from 0 to a, and from (1)

$$\frac{y^2}{b^2} = 1 - \frac{x^2}{a^2} = \frac{a^2 - x^2}{a^2}$$

$\therefore$ $$y = \frac{b}{a}\sqrt{a^2 - x^2}$$ | Taking +ve sign before the radical.

$\therefore$ $$\text{Area CAB} = \int_0^a y\,dx = \int_0^a \frac{b}{a}\sqrt{a^2 - x^2}\,dx$$

$$= \frac{b}{a}\left[\frac{x}{2}\sqrt{a^2 - x^2} + \frac{a^2}{2}\sin^{-1}\frac{x}{a}\right]_0^a = \frac{b}{a}\left[0 + \frac{a^2}{2}\sin^{-1}1 - 0 - 0\right]$$

$$= \frac{b}{a}\cdot\left(\frac{a^2}{2}\cdot\frac{\pi}{2}\right) = \frac{ab\pi}{4}$$ $\left|\ \because \sin^{-1}1 = \frac{\pi}{2}\right.$

$\therefore$ Whole area of the ellipse = 4 × area CAB = $4 \times \frac{\pi\, ab}{4} = \pi\, ab.$

Example 4. *Find the whole area of*

(a) the astroid $x^{2/3} + y^{2/3} = a^{2/3}$ *(b) the curve* $a^2y^2 = a^2x^2 - x^4$

(c) the curve $a^2y^2 = x^3(2a - x)$.

Sol. (*a*) The equation of the astroid is $x^{2/3} + y^{2/3} = a^{2/3}$ ...(1)

To trace the curve roughly to find the limits of integration.

(*i*) The curve is symmetrical about both the axes.

(*ii*) The curve does not pass through the origin.

(*iii*) The curve meets X-axis, where putting $y = 0$ in (1), $x^{2/3} = a^{2/3}$ or $x^2 = a^2$

$\therefore$ $x = \pm a.$ *i.e.,* in the points $(a, 0), (-a, 0)$.

Similarly the curve meets the Y-axis in the points $(0, a), (0, -a)$.

(*iv*) From (1), $y^{2/3} = a^{2/3} - x^{2/3}$ ...(2)

Now if $|x| > a$,

$y^{2/3}$ is –ve,

$\therefore$ y^2 is also –ve and hence y is imaginary.

$\therefore$ The curve does not lie beyond the lines $x = \pm a$.

Similarly the curve does not lie beyond the lines $y = \pm a$.

From (2), when $x = 0$, $y = a$, and as x increases from 0 to a, y decreases from a to 0.

Hence the shape of the curve is as shown in the figure.

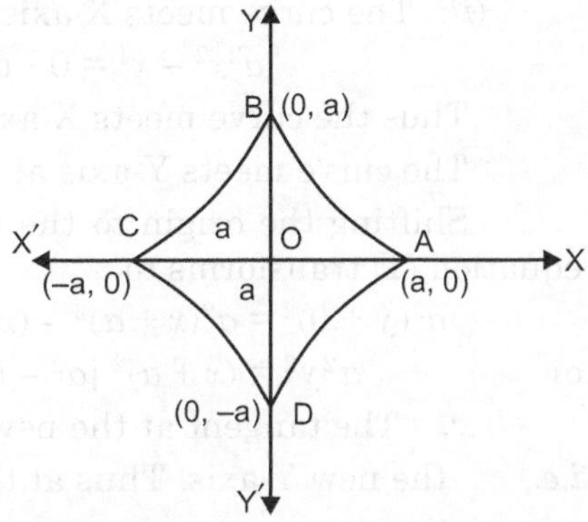

Since the curve is symmetrical about both the axes.

$\therefore$ Whole area of the curve

= 4 × area under the curve in first quadrant.

i.e., Area ACBD = 4 × area OAB and for the area OAB *i.e.,* area under the curve in the first quadrant, x varies from 0 to a.

Now from (1), $y^{2/3} = a^{2/3} - x^{2/3}$

$\therefore$ $y = (a^{2/3} - x^{2/3})^{3/2}$

$\therefore$ Area OAB $= \int_0^a y\,dx = \int_0^a (a^{2/3} - x^{2/3})^{3/2}\,dx$

| Put $x = a\sin^3\theta$, $\therefore$ $dx = 3a\sin^2\theta\cos\theta\,d\theta$
| when $x = 0, \theta = 0$, when $x = a, \theta = \frac{\pi}{2}$

$$= \int_0^{\pi/2} (a^{2/3} - a^{2/3}\sin^2\theta)^{3/2} \,.\, 3a\sin^2\theta\cos\theta\,d\theta$$

$$= \int_0^{\pi/2} a(1-\sin^2\theta)^{3/2} \,.\, 3a\sin^2\theta\cos\theta\,d\theta = 3a^2\int_0^{\pi/2} \cos^3\theta.\sin^2\theta\cos\theta\,d\theta$$

$$= 3a^2\int_0^{\pi/2} \cos^4\theta.\sin^2\theta\,d\theta = 3a^2 \,.\, \frac{3.1.1}{6.4.2} \,.\, \frac{\pi}{2} = \frac{3\pi a^2}{32}$$

| $\because \int_0^{\pi/2} \sin^p x\cos^q x\,dx = \frac{(p-1)(p-3)\ldots(q-1)(q-3)\ldots}{(p+q)(p+q-2)\ldots} \,.\, \frac{\pi}{2}$ if p and q are both even.

$\therefore$ Whole area of the curve $= 4 \times$ area OAB $= 4 \times \frac{3\pi a^2}{32} = \frac{3\pi a^2}{8}$.

(*b*) The equation of the curve is $a^2y^2 = a^2x^2 - x^4$ ...(1)

To trace the curve roughly to find the limits of integration :

(*i*) Since in (1), powers of x and y are both even, $\therefore$ the curve is symmetrical about both the axes.

(*ii*) It clearly passes through the origin and the tangents at the origin [equating the lowest degree terms to zero in (1)] are given by

$$a^2y^2 - a^2x^2 = 0 \quad \text{or} \quad y^2 = x^2 \quad \therefore \quad y = \pm x.$$

These tangents at the origin being real and different, origin is a node.

(*iii*) The curve meets X-axis, where putting $y = 0$ in (1),

$$a^2x^2 - x^4 = 0 \quad \text{or} \quad x^2(a^2 - x^2) = 0 \quad \therefore \quad x = 0, \pm a.$$

Thus the curve meets X-axis in the points O(0, 0), A(a, 0) and A′($-a$, 0).

The curve meets Y-axis at the origin.

Shifting the origin to the point A(a, 0) [by changing x to $x + a$ y to $y + 0$ in (1)], the equation (1) transforms to

$$a^2(y + 0)^2 = a^2(x + a)^2 - (x + a)^4$$

or $$a^2y^2 = (x + a)^2\,[a^2 - (x + a)^2] = (x + a)^2\,(-2ax - x^2) = -x(x + a)^2\,(x + 2a)$$

$\therefore$ The tangent at the new origin is given by $x = 0$

i.e., the new Y-axis. Thus at the point A(a, 0), the tangent is || to Y-axis.

From (1), $y^2 = \frac{x^2}{a^2}(a^2 - x^2)$ ...(2)

(*iv*) If $| x | > a$, y^2 is –ve $\therefore$ y is imaginary. Hence the curve does not lie beyond the lines $x = \pm a$.

(*v*) The curve has no asymptote.

Thus the shape of the curve is as shown in the figure. Since the curve is symmetrical about both the axes, whole area of the curve

= 4 × area of the curve in first quadrant, and for the area of the curve in first quadrant, x varies from 0 to a.

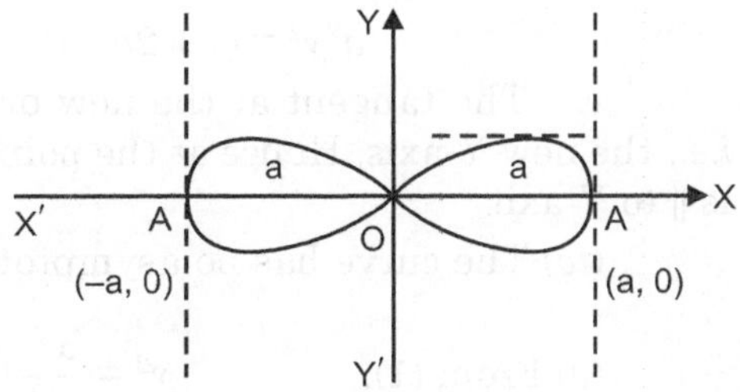

Now from (2), taking +ve sign with the square root, we have

$$y = \frac{x}{a}\sqrt{a^2 - x^2}$$

$\therefore$ Area of the curve in the first quadrant

$$= \int_0^a y\,dx = \int_0^a \frac{x}{a}\sqrt{a^2 - x^2}\,dx$$

$$= \frac{1}{a}\int_0^a x\sqrt{a^2 - x^2}\,dx$$

Put $x = a \sin \theta$, $\therefore$ $dx = a \cos \theta\, d\theta$
When $x = 0, \theta = 0$, when $x = a, \theta = \frac{\pi}{2}$

$$= \frac{1}{a}\int_0^{\pi/2} a \sin \theta \,.\, a \cos \theta \,.\, a \cos \theta\, d\theta$$

$$= a^2 \int_0^{\pi/2} \sin \theta \cos^2 \theta\, d\theta = -a^2 \int_0^{\pi/2} \cos^2 \theta\,(-\sin \theta\, d\theta)$$

$$= -a^2 \left[\frac{\cos^3 \theta}{3}\right]_0^{\pi/2} = -\frac{a^3}{3}\left[\cos^3 \frac{\pi}{2} - \cos^3 0\right] = -\frac{a^2}{3}(0 - 1) = \frac{a^2}{3}.$$

$\therefore$ Whole area under the curve $= 4 \times \dfrac{a^2}{3} = \dfrac{4}{3}a^2.$

(*c*) The equation of the curve is $a^2y^2 = x^3\,(2a - x)$ ...(1)

To tract the curve roughly to find the limits of integration :

(*i*) The curve is symmetrical about X-axis.

(*ii*) The curve passes through the origin and the tangents at the origin are [equating to zero the lowest degree terms in (1)]

$$a^2y^2 = 0 \quad \text{or} \quad y = 0, y = 0$$

Since these two tangents are real and coincident $\therefore$ origin is a cusp and X-axis is the cuspidal tangent.

(*iii*) The curve meets X-axis, where putting $y = 0$ in (1), we have

$$x^3(2a - x) = 0 \quad \therefore \quad x = 0, 2a.$$

Hence the curve meets X-axis in the points (0, 0), (2a, 0).

The curve meets Y-axis only at the origin (0, 0).

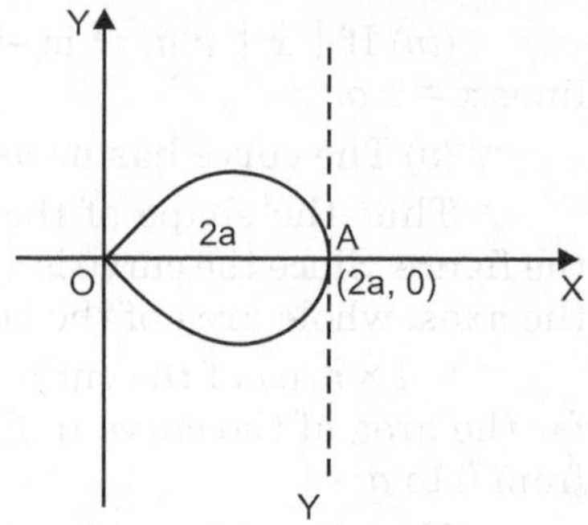

Shifting the origin to the point A $(2a, 0)$ the equation (1) transforms to

$$a^2y^2 = (x + 2a)^3 (2a - x - 2a) = - x(x + 2a)^3$$

$\therefore$ The tangent at the new origin is given by $x = 0$, *i.e.*, the new Y-axis. Hence at the point A$(2a, 0)$, the tangent is || to Y-axis.

(*iv*) The curve has no asymptotes.

(*v*) From (1), $\quad y^2 = \dfrac{x^3}{a^2}(2a - x)$

If x is –ve, y^2 is –ve $\therefore$ y is imaginary.

Again if x is $> 2a$, y^2 is –ve,

$\therefore$ y is imaginary.

Hence the curve does not lie on the left of y-axis and beyond the line $x = 2a$.

Hence the shape of the curve is as shown in the figure.

The whole area of the curve

= 2 × area of the curve in the first quadrant,

and for the curve in the first quadrant x varies from 0 to $2a$.

$\therefore$ Area of the curve in first quadrant

$$= \int_0^{2a} y\,dx = \int_0^{2a} \frac{x^{3/2}}{a}\sqrt{2a - x}\,dx \qquad \left| \text{From (1), } y = \frac{x^{3/2}}{a}\sqrt{2a - x}\right.$$

| Put $x = 2a \sin^2 \theta$, $\therefore$ $dx = 4a \sin\theta \cos\theta\, d\theta$
| As x varies from 0 to $2a$, θ varies from 0 to $\pi/2$

$$= \frac{1}{a}\int_0^{\pi/2} (2a \sin^2\theta)^{3/2} \sqrt{2a - 2a\sin^2\theta}\,.\,4a \sin\theta\cos\theta\, d\theta$$

$$= \frac{1}{a}\int_0^{\pi/2} (2a)^2\,.\,\sin^3\theta\cos\theta\,.\,4a\sin\theta\cos\theta\, d\theta$$

$$= 16a^2 \int_0^{\pi/2} \sin^4\theta\cos^2\theta\, d\theta = 16a^2\,.\,\frac{3\,.\,1\,.\,1}{6\,.\,4\,.\,2}\,.\,\frac{\pi}{2} = \frac{\pi a^2}{2}$$

$\therefore$ Whole area of the curve $= 2 \times \dfrac{\pi a^2}{2} = \pi a^2$.

Example 5. *Find the area of the loop of the following curves :*

(*a*) $y^2(a - x) = x^2(a + x)$ (*b*) $y^2 = (x - 2)(x - 4)^2$

(*c*) $x(x^2 + y^2) = a(x^2 - y^2)$ (*d*) $a^3y^2 = x^4(b + x)$.

Sol. (*a*) The equation of the curve is $y^2(a - x) = x^2(a + x)$ ...(1)

To trace the curve roughly to find the limits of integration :

(*i*) The curve is symmetrical about the X-axis.

(*ii*) The curve passes through the origin and the tangents at the origin are given by

$$y^2 = x^2 \quad \text{or} \quad y = \pm\, x.$$

Since these tangents are real and different, $\therefore$ Origin is a node.

(*iii*) The curve meets X-axis, where putting $y = 0$ in (1),

$$x^2(a + x) = 0 \quad \therefore \quad x = 0, \; -a, \quad i.e., \quad \text{in the points } (0, 0), (-a, 0)$$

The curve meets Y-axis in the origin (0, 0).

(*iv*) The asymptote || to y-axis is given by $a - x = 0$, *i.e.*, $x = a$.

The curve has no other asymptote.

(*v*) From (1), we have (taking +ve root)

$$y = x\sqrt{\frac{a + x}{a - x}} \qquad \ldots(2)$$

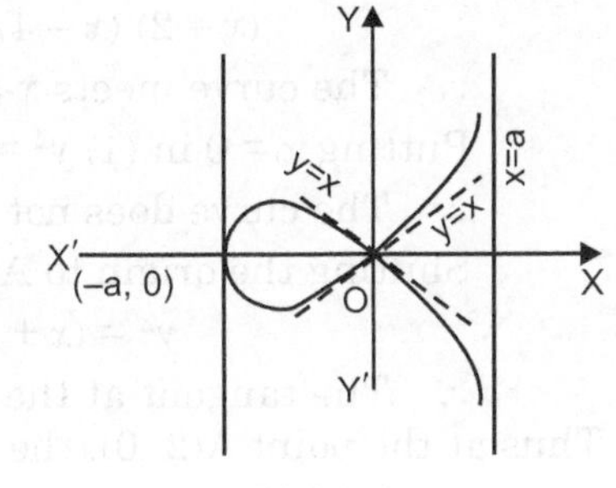

When $x < -a$, or when $x > a$, y is imaginary. Hence no portion of the curve lies beyond the lines $x = \pm a$.

Hence the approximate shape of the curve is as shown in the figure and the loop lies between $x = -a$ and $x = 0$.

$\therefore$ Required area of the loop = 2 × area of the loop above the X-axis

$$= 2 \cdot \int_{-a}^{0} y \, dx = 2 \int_{-a}^{0} x \sqrt{\frac{a + x}{a - x}} \, dx \qquad \Big| \; From \; (2)$$

| Put $x = a \sin \theta$, $\therefore$ $dx = a \cos \theta \, d\theta$
| When $x = 0$, $\theta = 0$, when $x = -a$, $\theta = -\pi/2$

$$= 2 \int_{-\pi/2}^{0} a \sin \theta \sqrt{\frac{a(1 + \sin \theta)}{a(1 - \sin \theta)}} \cdot a \cos \theta \, d\theta$$

$$= 2a^2 \int_{-\pi/2}^{0} \sin \theta \cos \theta \sqrt{\frac{1 + \sin \theta}{1 - \sin \theta}} \cdot \sqrt{\frac{1 + \sin \theta}{1 + \sin \theta}} \, d\theta \qquad \Big| \; \textbf{Note this step}$$

$$= 2a^2 \int_{-\pi/2}^{0} \frac{\sin \theta \cos \theta (1 + \sin \theta)}{\cos \theta} \, d\theta$$

$$= 2a^2 \int_{-\pi/2}^{0} \left[\sin \theta + \frac{1}{2}(1 - \cos 2\theta) \right] d\theta \qquad \Big| \; \because \sin \theta = \frac{1 - \cos 2\theta}{2}$$

$$= 2a^2 \left[-\cos \theta + \frac{1}{2}\theta - \frac{\sin 2\theta}{4} \right]_{-\pi/2}^{0}$$

$$= 2a^2 \left[-\cos 0 - \left\{ -\cos\left(-\frac{\pi}{2}\right) + \frac{1}{2}\left(-\frac{\pi}{2}\right) - \frac{\sin(-\pi)}{4} \right\} \right]$$

$$= 2a^2 \left[-1 + \frac{\pi}{4} \right] = \frac{a^2}{2}(\pi - 4) = \frac{a^2}{2}(4 - \pi) \text{ numerically} \qquad [\because \; \pi < 4].$$

Hence the area of the loop $= \dfrac{a^2}{2}(4 - \pi)$.

(*b*) The equation of the curve is $y^2 = (x-2)(x-4)^2$ ...(1)

To trace the curve roughly to find the limits of integration :

(*i*) The curve is symmetrical about the *x*-axis.

(*ii*) The curve does not pass through the origin.

(*iii*) The curve meets the *x*-axis, where putting $y = 0$ in (1),

$$(x-2)(x-4)^2 = 0, \text{ i.e., } x = 2, 4.$$

$\therefore$ The curve meets *x*-axis in the points A(2, 0), B(4, 0).

Putting $x = 0$ in (1) $y^2 = -32$; *i.e.*, *y* is imaginary.

$\therefore$ The curve does not meet *y*-axis.

Shifting the origin to A(2, 0) the equation (1) transforms to

$$y^2 = (x+2-2)(x+2-4)^2 = x(x-2)^2.$$

$\therefore$ The tangent at the new origin is given by $4x = 0$, *i.e.*, $x = 0$ *i.e.*, the new Y-axis. Thus at the point A(2, 0), the tangent is || to Y-axis.

Again shifting the origin to the point B(4, 0), the equation (1) transforms to

$$y^2 = (x+4-2)(x+4-4)^2 = x^2(x+2).$$

The tangents at the new origin are given by $y^2 = 2x^2$ or $y = \pm 2x$. These tangents being real and distinct, the new origin *i.e.*, the point B(4, 0) is a node.

(*iv*) The curve has no asymptotes.

(*v*) From (1), when $0 < x < 2$, y^2 is –ve, and $\therefore$ *y* is imaginary. As $x > 4$, increased and $\to \infty$, *y* also increases and $\to \infty$.

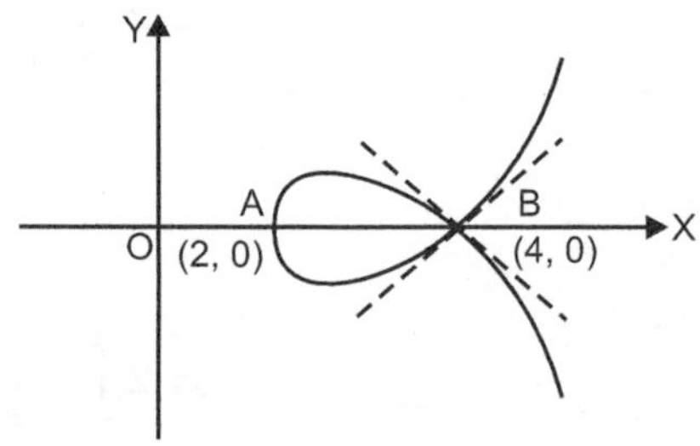

Thus the shape of the curve is as shown in the figure. The loop lies between $x = 2$ to $x = 4$ and for the upper half of the loop, *x* varies from 2 to 4.

Thus the required area of the loop = 2 × area under the upper half of the loop

$$= 2\int_2^4 y\,dx = 2\int_2^4 (x-4)\sqrt{x-2}\,dx$$ Putting +ve value of *y* from (1)

| Put $x - 2 = t^2$ $\therefore$ $dx = 2t\,dt$.
When $x = 2, t = 0$, when $x = 4, t = \sqrt{2}$

$$= 2\int_0^{\sqrt{2}} (t^2-4)\,.\,t\,.\,2t\,dt = 4\int_0^{\sqrt{2}} (t^4 - 2t^2)\,dt$$

$$= 4\left[\frac{t^5}{5} - 2\frac{t^3}{3}\right]_0^{\sqrt{2}} = 4\left[\frac{4\sqrt{2}}{5} - 2\,.\,\frac{2\sqrt{2}}{3}\right]$$

$$= 4\left(-\frac{8\sqrt{2}}{15}\right) = \frac{32\sqrt{2}}{15} \text{ numerically.}$$ | Neglecting –ve sign.

(*c*) The equation of the curve is $x(x^2 + y^2) = a(x^2 - y^2)$ ...(1)

To trace the curve roughly to find the limits of integration :

(*i*) The curve is symmetrical about the *x*-axis.

(*ii*) The curve passes through the origin and the tangents at the origin are given by $x^2 - y^2 = 0$ or $y = \pm x$. Since these tangents at the origin are real and distinct. $\therefore$ the origin is a node.

(*iii*) The curve meets the x-axis, where putting $y = 0$ in (1),

$$x^3 = ax^2 \quad \text{or} \quad x^2(x - a) = 0 \quad \therefore \quad x = 0, a \ \textit{i.e.,} \ \text{at the points } (0, 0), (a, 0).$$

The curve meets the y-axis at the origin.

From (1), $\quad y^2(x + a) = x^2(a - x)$ ...(2)

Shifting the origin to the point $(a, 0)$, the equation (2) transforms to

$$y^2(x + 2a) = (x + a)^2 (a - x - a) = - x(x + a)^2.$$

$\therefore$ The tangents at the new origin are given by $x = 0$ *i.e.,* the new y-axis. Hence at the point $(a, 0)$, the tangent is $\parallel$ to y-axis.

(*iv*) From (2), the asymptote $\parallel$ to y-axis is given by $x + a = 0$. The curve has no other asymptote.

(*v*) From (2), taking +ve square root,

$$y = x\sqrt{\frac{a - x}{a + x}} \qquad ...(3)$$

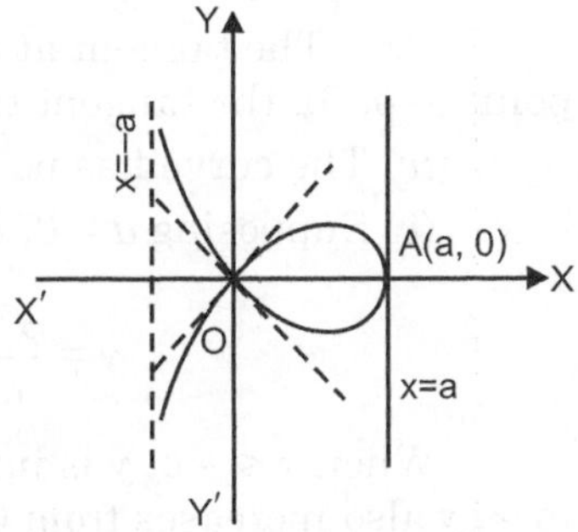

When $x < - a$ or when $x > a$, y is imaginary. $\therefore$ No portion of the curve lies beyond the lines $x = \pm a$.

As x decreases from 0 to $- a$, y is –ve and decreases from 0 to $- \infty$. Again when x increases from 0 to a, y first increases from 0 and then decreases to zero at $x = a$.

Hence the shape of the curve is as shown in the figure.

For the upper half of the loop, x varies from 0 to a.

$\therefore$ Required area of the loop

$$= 2 \times \text{area of upper half of loop}$$

$$= 2\int_0^a y\,dx = 2\int_0^a x\sqrt{\frac{a - x}{a + x}}\,dx \qquad | \text{ From (3)}$$

| Put $x = a \sin \theta \ \therefore \ dx = a \cos \theta \, d\theta$.
When $x = 0, \theta = 0$, when $x = a, \theta = \pi/2$

$$= 2\int_0^{\pi/2} a \sin \theta \, . \sqrt{\frac{a(1 - \sin \theta)}{a(1 + \sin \theta)}} \, . \, a \cos \theta \, d\theta$$

$$= 2a^2 \int_0^{\pi/2} \sin \theta \cos \theta \sqrt{\frac{1 - \sin \theta}{1 + \sin \theta}} \, . \sqrt{\frac{1 - \sin \theta}{1 - \sin \theta}} \, d\theta \qquad | \textbf{ Note this step}$$

$$= 2a^2 \int_0^{\pi/2} \frac{\sin \theta \cos \theta \, (1 - \sin \theta)}{\cos \theta} \, d\theta = 2a^2 \int_0^{\pi/2} (\sin \theta - \sin^2 \theta) \, d\theta$$

$$= 2a^2 \int_0^{\pi/2} \left[\sin \theta - \frac{1}{2}(1 - \cos 2\theta)\right] d\theta = 2a^2 \left[- \cos \theta - \frac{1}{2}\,\theta + \frac{1}{2} . \frac{\sin 2\theta}{2}\right]_0^{\pi/2}$$

$$= 2a^2 \left[- 0 - \frac{1}{2}\frac{\pi}{2} + \frac{1}{4} \sin \pi + \cos 0 - 0\right] = 2a^2 \left(1 - \frac{\pi}{4}\right) = \frac{a^2}{2}(4 - \pi).$$

(*d*) The equation of the curve is $a^2y^2 = x^4\,(b + x)$ ...(1)

To trace the curve roughly to find the limits of integration :

(*i*) The curve is symmetrical about the x-axis.

(*ii*) The curve passes through the origin and the tangents at the origin are given by

$$y^2 = 0 \quad i.e., \quad y = 0,\ y = 0.$$

These tangents being real and coincident, the origin is a cusp and $y = 0$ *i.e.,* x-axis is cuspidal tangent.

(*iii*) The curve meets x-axis, where putting $y = 0$ in (1)

$$x^4(x + b) = 0 \quad \therefore \quad x = 0, -b \text{ i.e., at the points } (0, 0), (-b, 0).$$

The curve meets y-axis only at the origin.

Shifting the origin to the point $(-b, 0)$, the equation (1) transform to

$$a^3y^2 = (x - b)^4\,(b + x - b) = (x - b)^4\,x$$

$\therefore$ The tangent at the new origin is given by $x = 0$ *i.e.,* the new y-axis. Thus at the point $(-b, 0)$, the tangent to the curve is || to y-axis.

(*iv*) The curve has no asymptote.

(*v*) Supposing $a > 0$, $b > 0$, from (1), (taking +ve square root), we have

$$y = \frac{x^2}{a}\sqrt{\frac{b + x}{a}} \qquad ...(2)$$

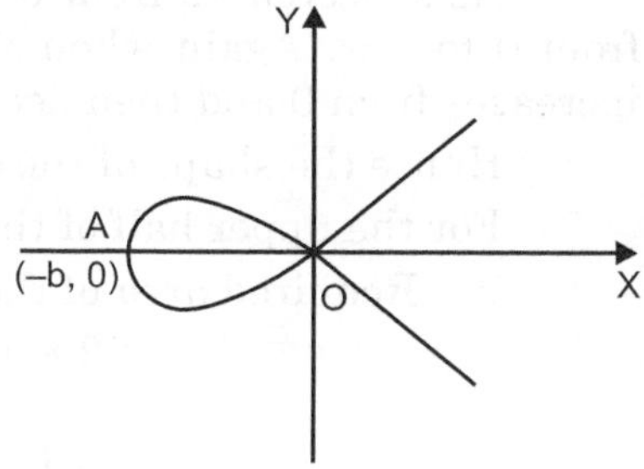

When $x < -b$, y is imaginary. As x increases from 0 to ∞, y also increases from 0 to ∞.

Hence the shape of the curve is as shown in the figure. For the upper half of the loop, x various from $-b$, to 0.

$\therefore$ Required area of the loop

$= 2 \times$ area of upper half of the loop

$$= 2\int_{-b}^{0} y\,dx = \frac{2}{a}\int_{-b}^{0} x^2\sqrt{\frac{b + x}{a}}\,dx \qquad | \text{ Using (2)}$$

$$= \frac{2}{a^{3/2}}\int_{-b}^{0} x^2\sqrt{b + x}\,dx.$$

| Put $b + x = t^2$, $\therefore\ dx = 2t\,dt$ when $x = -b$, $t = 0$ and when $x = 0$, $t = \sqrt{b}$.

$$= \frac{2}{a^{3/2}}\int_{0}^{\sqrt{b}} (t^2 - b)^2 . t . 2t\,dt = \frac{4}{a^{3/2}}\int_{0}^{\sqrt{b}} (t^4 + b^2 - 2bt^2)\,t^2\,dt$$

$$= \frac{4}{a^{3/2}}\left[\frac{t^7}{7} + b^2.\frac{t^3}{3} - 2b.\frac{t^5}{5}\right]_0^{\sqrt{b}} = \frac{4}{a^{3/2}}\left[\frac{b^3\sqrt{b}}{7} + b^2\frac{b\sqrt{b}}{3} - 2b.\frac{b^2\sqrt{b}}{5} - 0\right]$$

$$= \frac{4b^3\sqrt{b}}{a^{3/2}}\left(\frac{1}{7} + \frac{1}{3} - \frac{2}{5}\right) = \frac{4b^3\sqrt{b}}{a\sqrt{a}}\left(\frac{15 + 35 - 42}{105}\right) = \frac{32b^{7/2}}{105a^{3/2}} = \frac{32}{105}b^{7/2}\,a^{-3/2}.$$

Example 6. *Find the areas between the following curves and their asymptotes :*

(*a*) $x^2y^2 = a^2(y^2 - x^2)$ (*b*) $x(x^2 + y^2) = a(x^2 - y^2)$

(*c*) $y^2(2a - x) = x^3$ (*d*) $x^2(x^2 + y^2) = a^2(y^2 - x^2)$

(*e*) $xy^2 = 4a^2(2a - x)$.

Sol. (*a*) The equation of the curve is $x^2y^2 = a^2(y^2 - x^2)$ ...(1)

To trace the curve roughly to find the limits of integration :

(*i*) The curve is symmetrical about both the axes.

(*ii*) The curve passes through the origin and the tangents at the origin are given by

$$y^2 - x^2 = 0 \quad \text{or} \quad y = \pm x.$$

Since the tangents at the origin are real and distinct.

∴ origin is a node.

(*iii*) The curve meets the *x*-axis and *y*-axis both at the origin only.

(*iv*) Equating the co-eff. of y^2 (highest power of *y* in the equation) to zero in (1), the asymptotes ∥ to *y*-axis are given by $a^2 - x^2 = 0$ or $x = \pm a$.

The curve has no other asymptote.

(*v*) From (1), we have $$y^2 = \frac{a^2x^2}{a^2 - x^2}$$

Taking +ve square root, we have $y = \dfrac{ax}{\sqrt{a^2 - x^2}}$ ...(2)

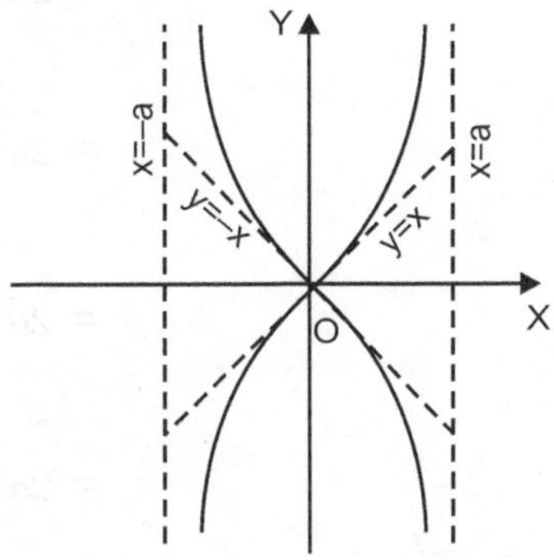

If $|x| > a$, *y* is imaginary. Hence no portion of the curve lies beyond the lines $x = \pm a$.

When $x = 0, y = 0$. As *x* increases from 0 to *a*, *y* increases from 0 to ∞.

Hence the shape of the curve is as shown in the figure. For the portion of the curve in the first quadrant, *x* varies from 0 to *a*.

Since the curve is symmetrical about both the axes,

∴ Required whole area between the curve and its asymptotes

$= 4 \times$ area of the curve in first quadrant and its asymptote

$$= 4\int_0^a y\, dx = 4\int_0^a \frac{ax}{\sqrt{a^2 - x^2}}\, dx \qquad | \text{ Using (2)}$$

| Put $x = a \sin\theta$ ∴ $dx = a\cos\theta\, d\theta$
| When $x = 0, \theta = 0$, when $x = a, \theta = \pi/2$

$$= 4\int_0^{\pi/2} \frac{a \,.\, a\sin\theta}{a\cos\theta} \,.\, a\cos\theta\, d\theta$$

$$= 4a^2\int_0^{\pi/2} \sin\theta\, d\theta = 4a^2\,[-\cos\theta]_0^{\pi/2} = 4a^2(0 + 1) = 4a^2.$$

(*b*) The equation of the curve is $x(x^2 + y^2) = a(x^2 - y^2)$ ...(1)

For the shape of the curve, see Fig. of Ex. 5(c) on page 481. The curve is symmetrical about the x-axis and for the area of the curve above x-axis and its asymptote, x varies from $-a$ to 0.

From (1), $y^2(a + y) = x^2(a - x)$

$$\therefore \quad y = x\sqrt{\frac{a-x}{a+x}} \qquad ...(2)$$

$\therefore$ Required area between the curve and its asymptote = 2 × area of the curve and its asymptote above x-axis.

$$= 2\int_{-a}^{0} y\,dx = 2\int_{-a}^{0} x\sqrt{\frac{a-x}{a+x}}\,dx \qquad \text{| From (2)}$$

| Put $x = a \sin\theta \quad \therefore \quad dx = a\cos\theta\,d\theta$
| When $x = -a, \theta = -\pi/2$; when $x = 0, \theta = 0$

$$= 2\int_{-\pi/2}^{0} a\sin\theta\sqrt{\frac{a(1-\sin\theta)}{a(1+\sin\theta)}}\,.\,a\cos\theta\,d\theta$$

$$= 2a^2\int_{-\pi/2}^{0} \sin\theta\cos\theta\sqrt{\frac{1-\sin\theta}{1+\sin\theta}}\sqrt{\frac{1-\sin\theta}{1-\sin\theta}}\,d\theta$$

$$= 2a^2\int_{-\pi/2}^{0} \frac{\sin\theta\cos\theta\,.\,(1-\sin\theta)}{\cos\theta}\,d\theta$$

$$= 2a^2\int_{-\pi/2}^{0}\left[\sin\theta - \frac{1}{2}(1-\cos 2\theta)\right]d\theta = 2a^2\left[-\cos\theta - \frac{1}{2}\theta + \frac{1}{4}\sin 2\theta\right]_{-\pi/2}^{0}$$

$$= 2a^2\left[-1 - 0 + 0 + \cos\left(-\frac{\pi}{2}\right) + \frac{1}{2}\cdot\left(-\frac{\pi}{2}\right) - \frac{1}{4}\sin(-\pi)\right]$$

$$= 2a^2\left(-1 - \frac{\pi}{4}\right) = -\frac{a^2}{2}(\pi + 4)$$

$$= \frac{a^2}{2}(\pi + 4) \text{ numerically.} \qquad \text{| Neglecting the –ve sign}$$

(*c*) The equation of the curve is $y^2(2a - x) = x^3$. ...(1)

To trace the curve roughly to find the limits of integration.

(*i*) The curve is symmetrical about the x-axis.

(*ii*) The curve passes through the origin and the tangents at the origin are given by

$$2ay^2 = 0 \text{ or } y = 0, y = 0.$$

Since the tangents at the origin are real and coincident, $\therefore$ origin is a cusp and x-axis is the cuspidal tangent.

(*iii*) The curve meets the x-axis and y-axis both at the origin only.

(*iv*) Equating to zero the co-eff. of y^2, the asymptote || to y-axis is $2a - x = 0$ or $x = 2a$. There is no other asymptote of the curve.

(*v*) From (1), $$y^2 = \frac{x^3}{2a-x}$$

Taking +ve, square root, we have $y = x\sqrt{\dfrac{x}{2a-x}}$ ...(2)

When $x > 2a$ or $x < 0$, y is imaginary. Thus the curve does not lie to the left of the line $x = 0$ *i.e.*, y-axis and to the right of $x = 2a$.

Again as x increases from 0 to $2a$, y increases from 0 to ∞. Thus the shape of the curve is roughly as shown in the figure.

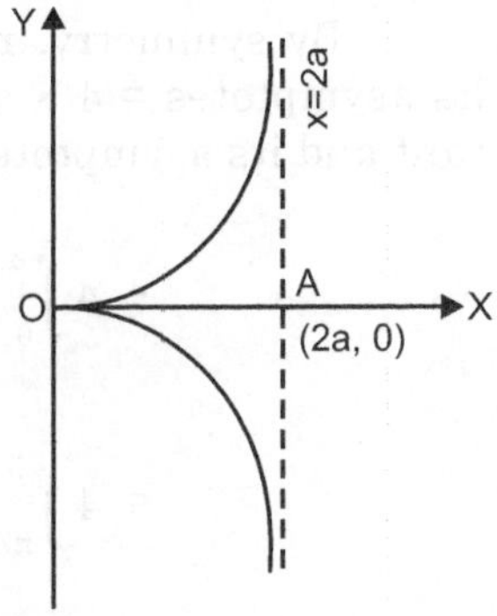

Since the curve is symmetrical about the x-axis, $\therefore$ whole area between the curve and its asymptote

= 2 × area between the curve and its asymptote in the first quadrant.

For the curve in first quadrant, x varies from 0 to $2a$.

$\therefore$ Required area between the curve and its asymptote.

$$= 2\int_0^{2a} y\,dx = 2\int_0^{2a} x\sqrt{\frac{x}{2a-x}}\,dx \quad | \text{ From (2)}$$

| Put $x = 2a\sin^2\theta \quad \therefore \quad dx = 4a\sin\theta\cos\theta\,d\theta$
| when $x = 0, \theta = 0$, when $x = 2a, \theta = \pi/2$

$$= 2\int_0^{\pi/2} 2a\sin^2\theta\sqrt{\frac{2a\sin^2\theta}{2a(1-\sin^2\theta)}}\,.\,4a\sin\theta\cos\theta\,d\theta$$

$$= 16a^2\int_0^{\pi/2}\sin^2\theta\,.\frac{\sin\theta}{\cos\theta}\,.\sin\theta\cos\theta\,d\theta$$

$$= 16a^2\int_0^{\pi/2}\sin^4\theta\,d\theta = 16a^2\,\frac{3\,.\,1}{4\,.\,2}\,.\frac{\pi}{2} = 3\pi a^2.$$

(*d*) The equation of the curve is $x^2(x^2 + y^2) = a^2(y^2 - x^2)$ ...(1)

To trace the curve roughly to find the limits of integration :

(*i*) The curve is symmetrical about both the axes.

(*ii*) The curve passes through the origin and the tangents at the origin are given by $y^2 - x^2 = 0$ or $y = \pm x$. The tangents being real and different, origin is a node.

(*iii*) The curve meets the x-axis and y-axis both only at the origin.

(*iv*) Equating to zero the co-eff. of y^2 in (1), the symptotes || to y-axis are given by $x^2 - a^2 = 0$ or $x = \pm a$.

The curve has no other asymptote.

(*v*) From (1), $y^2(a^2 - x^2) = x^2(x^2 + a^2)$

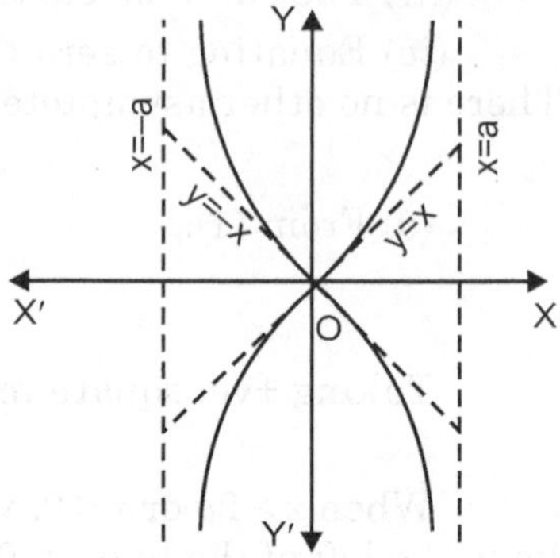

$$\therefore \qquad y = x\sqrt{\frac{a + x^2}{a^2 - x^2}} \qquad \ldots(2)$$

When $|x| > a$, y is imaginary.

$\therefore$ No portion of the curve lies beyond the lines $x = \pm a$.

As x increases from 0 to a, y increases from 0 to ∞. Thus the shape of the curve is as shown in the figure.

By symmetry, required area between the curve and its asymptotes = 4 × area between the curve in first quadrant and its asymptote.

$$= 4\int_0^a y\,dx = 4\int_0^a x\sqrt{\frac{a^2 + x^2}{a^2 - x^2}}\,dx$$

Put $x^2 = a^2 \cos\theta \;\therefore\; 2x\,dx = -a^2 \sin\theta\,d\theta$
when $x = 0, \theta = \pi/2$, when $x = a, \theta = 0$

$$= 4\int_{\pi/2}^0 \sqrt{\frac{a^2(1+\cos\theta)}{a^2(1-\cos\theta)}}\left(\frac{-a^2\sin\theta}{2}\right)d\theta$$

$$= 4a^2\int_0^{\pi/2}\sqrt{\frac{2\cos^2\frac{\theta}{2}}{2\sin^2\frac{\theta}{2}}}\left(\frac{2\sin\frac{\theta}{2}\cos\frac{\theta}{2}}{2}\right)d\theta$$

$\because \int_a^b f(x)\,dx = -\int_b^a f(x)\,dx$ and
$1 + \cos\theta = 2\cos^2\frac{\theta}{2}; 1 - \cos\theta = 2\sin^2\frac{\theta}{2}$

$$= 4a^2\int_0^{\pi/2}\cos^2\frac{\theta}{2}\,d\theta = 2a^2\int_0^{\pi/2}(1+\cos\theta)\,d\theta$$

$$= 2a^2\left[\theta + \sin\theta\right]_0^{\pi/2} = 2a^2\left(\frac{\pi}{2} + 1\right).$$

(*e*) The equation of the curve is $xy^2 = 4a^2(2a - x)$ ...(1)

To trace the curve roughly to find the limits of integration :

(*i*) The curve is symmetrical about the x-axis.

(*ii*) The curve does not pass through the origin.

(*iii*) The curve meets the x-axis, where putting $y = 0$, we have

$$4a^2(2a - x) = 0 \quad \text{or} \quad x = 2a \quad i.e., \quad \text{at the point } (2a, 0)$$

The curve does not meet the y-axis.

Shifting the origin to the point $(2a, 0)$, (1) transforms to

$$(x + 2a)\,y^2 = 4a^2\,(2a - 2a - x) = -4a^2x.$$

The tangent at the new origin is given by $x = 0$ *i.e.*, the new y-axis. Hence at the point $(2a, 0)$ the tangent to the curve is || to y-axis.

(*iv*) The asymptote || to y-axis is $x = 0$ *i.e.*, y-axis itself. The curve has no other asymptote.

(*v*) From (1), $y^2 = 4a^2 \left(\dfrac{2a - x}{x}\right) \quad \therefore \quad y = 2a\sqrt{\dfrac{2a - x}{x}}$ ...(2)

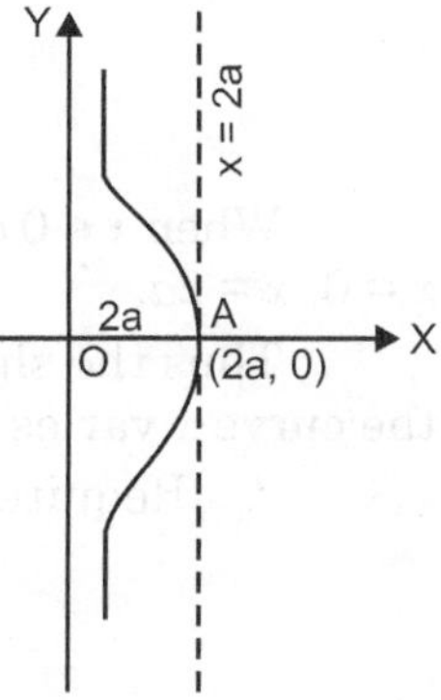

When $x < 0$ or when $x > 2a$, y is imaginary. Hence no portion of the curve lies to the left of y-axis and to the right of the line $x = 2a$. When $x = 2a$, $y = 0$. As x decreases from $2a$ to 0, y increases from 0 to ∞. Hence the shape of the curve is as shown in the figure.

For the area between the curve and its asymptote, x varies from 0 to $2a$.

$\therefore$ Required area between the curve and its asymptote

$= 2 \times$ area between the curve in first quadrant and its asymptote

$$= 2\int_0^{2a} y\,dx = 2\int_0^{2a} 2a\sqrt{\frac{2a - x}{x}}\,dx \qquad | \text{ From (2)}$$

| Put $\mathbf{x = 2a \sin^2 \theta} \quad \therefore \quad dx = 4a \sin\theta \cos\theta\, d\theta$
when $x = 0, \theta = 0$, when $x = 2a$, $\theta = \dfrac{\pi}{2}$

$$= 4a\int_0^{\pi/2} \sqrt{\frac{2a(1 - \sin^2\theta)}{2a\sin^2\theta}}\,.\,4a \sin\theta\cos\theta\,d\theta = 16a^2\int_0^{\pi/2} \frac{\cos\theta}{\sin\theta}\,.\,\sin\theta\cos\theta\,d\theta$$

$$= 16a^2\int_0^{\pi/2} \cos^2\theta\,d\theta = 16a^2\,.\,\frac{1}{2}\,.\,\frac{\pi}{2} = 4\pi a^2.$$

Example 7. (*a*) *Prove that the area of the curve $a^4y^2 = x^5(2a - x)$ is to that of a circle whose radius is a, as 5 to 4.*

(*b*) *Show that the area of the loop of the curve $y^2(a + x) = x^2(3a - x)$ is equal to the area between the curve and its asymptote.*

Sol. (*a*) The equation of the curve is $a^4y^2 = x^5(2a - x)$ ...(1)

To trace the curve roughly to find the limits of integration :

(*i*) The curve is symmetrical about x-axis.

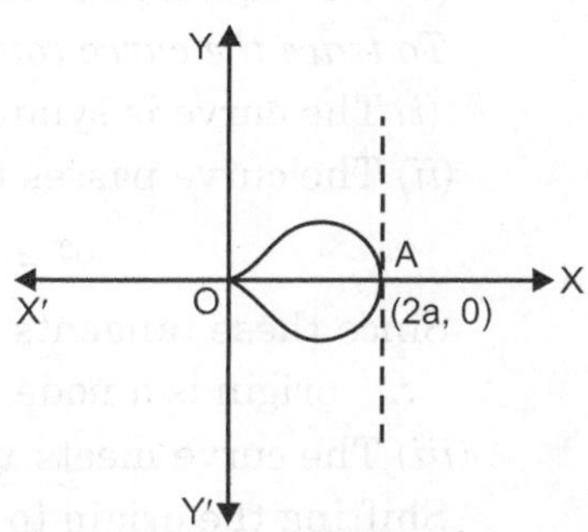

(*ii*) The curves passes thro' the origin and the tangents at the origin are given by $y^2 = 0$.

i.e., $y = 0$, $y = 0$. Since these two tangents are real and coincident, $\therefore$ origin is a cusp and x-axis ($y = 0$) is the cuspidal tangent.

(*iii*) The curve meets x-axis in the points (0, 0) and $(2a, 0)$ and y-axis in (0, 0).

Shifting the origin to the point $(2a, 0)$, the equation (1) transforms to

$$a^4y^2 = (x + 2a)^5 + (2a - x - 2a) = -\,x(x + 2a)^5$$

$\therefore$ The tangent at the new origin is given by $x = 0$ *i.e.,* the new y-axis. Hence at the point $(2a, 0)$, the tangent to the curve is || to y-axis.

(*iv*) The curve has no asymptote.

(*v*) From (1), taking +ve square root, we have

$$y = \frac{x^2}{a^2}\sqrt{x(2a-x)} \qquad \ldots(2)$$

When $x < 0$ or when $x > 2a$, y is imaginary. Hence the curve does not lie beyond the lines $x = 0$, $x = 2a$.

Thus the shape of the curve is roughly as shown in the figure and for the upper half of the curve x varies from 0 to $2a$.

∴ Required whole area of the curve.

= 2 × area under the curve in the first quadrant

$$= 2\int_0^{2a} y\,dx = 2\int_0^{2a} \frac{x^2}{a^2}\sqrt{x(2a-x)}\,dx \qquad \Big|\ \textit{Using } (2)$$

| Put $\mathbf{x = 2a \sin^2 \theta}$ ∴ $dx = 4a \sin\theta \cos\theta\, d\theta$
| when $x = 0, \theta = 0$; when $x = 2a, \theta = \frac{\pi}{2}$

$$= 2\int_0^{\pi/2} \frac{4a^2 \sin^4\theta}{a^2}\sqrt{2a\sin^2\theta \,.\, 2a\,(1-\sin^2\theta)} \times 4a \sin\theta\cos\theta\, d\theta$$

$$= 64a^2 \int_0^{\pi/2} \sin^6\theta \cos^2\theta\, d\theta = 64a^2 \,.\, \frac{5\,.\,3\,.\,1\,.\,1}{8\,.\,6\,.\,4\,.\,2}\,.\,\frac{\pi}{2} = \frac{5\pi a^2}{4}$$

Area of circle of radius 'a' = πa^2.

$$\therefore \qquad \frac{\text{Area of the curve}}{\text{Area of circle whose radius is } a} = \frac{\frac{5\pi a^2}{4}}{\pi a^2} = \frac{5}{4}.$$

Hence the result.

(*b*) The equation of the curve is $y^2\,(a + x) = x^2\,(3a - x)$ ...(1)

To trace the curve roughly to find the limits of integration :

(*i*) The curve is symmetrical about the x-axis.

(*ii*) The curve passes through the origin and the tangents at the origin are given by

$$y^2 = 3x^2 \quad \text{or} \quad y = \pm\sqrt{3}x\,.$$

Since these tangents are real and distinct,

∴ origin is a node.

(*iii*) The curve meets x-axis in the points (0, 0) and (3a, 0) and y-axis at (0, 0).

Shifting the origin to the point (3a, 0) the equation (1) transforms to

$$y^2(a + 3a + x) = (x + 3a)^2\,(3a - x - 3a)$$

or $$y^2(x + 4a) = -\,x(x + 3a)^2$$

∴ The tangent at the new origin is $x = 0$. *i.e.*, the new y-axis.

Hence at the point (3a, 0), the tangent to the curve is || to y-axis.

(*iv*) Equating to zero the co-eff. of y^2 in (1), the asymptote || to y-axis is given by $x + a = 0$. There is no other asymptote of the curve.

(*v*) From (1), $\quad y = x\sqrt{\dfrac{3a - x}{a + x}}$ ...(2) | Taking +ve sign with the radical

When $x > 3a$ or when $x < -a$, y is imaginary.

Thus the curve does not lie beyond the lines $\quad x = -a$, $x = 3a$.

As x increases from 0 to $3a$, y first increases and then decreases to 0.

Thus there is a loop between $x = 0$, $x = 3a$.

Again as x decreases from 0 to $-a$, y also decreases from 0 to $-\infty$.

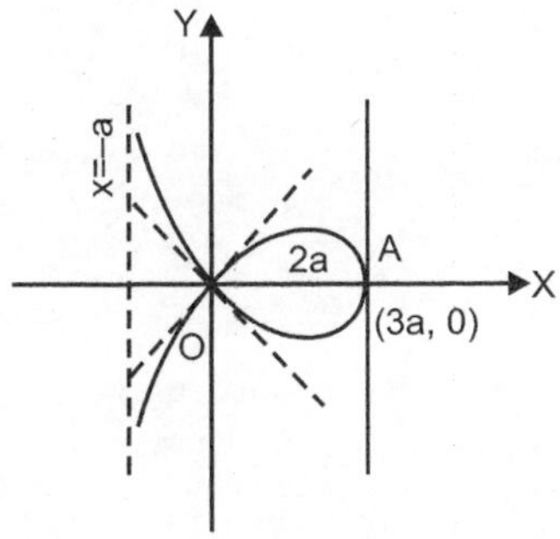

Thus the shape of the curve is as shown in the figure.

For the loop of the curve x varies from 0 to $3a$ and for the area between the curve and the asymptote, x varies from $-a$ to 0.

$\therefore$ Area of the loop

$$= 2 \times \text{area of the upper half of the loop}$$

$$= 2\int_0^{3a} y\, dx = 2\int_0^{3a} x\sqrt{\frac{3a - x}{a + x}}\, dx \qquad \text{| From (2)}$$

$$= 2\int_0^{3a} x\sqrt{\frac{2a + (a - x)}{2a - (a - x)}}\, dx \qquad \text{| Note this step}$$

| Put $a - x = 2a\cos\theta$, $\quad \therefore \quad dx = 2a\sin\theta\, d\theta$
when $\quad x = 0, \theta = \pi/3$; when $x = 3a, \theta = \pi$

$$= 2\int_{\pi/3}^{\pi} (a - 2a\cos\theta)\,.\sqrt{\frac{2a + 2a\cos\theta}{2a - 2a\cos\theta}}\,.\,2a\sin\theta\, d\theta$$

$$= 4a^2\int_{\pi/3}^{\pi} (1 - 2\cos\theta)\sqrt{\frac{1 + \cos\theta}{1 - \cos\theta}}\,.\sin\theta\, d\theta$$

$$= 4a^2\int_{\pi/3}^{\pi} (1 - 2\cos\theta)\,.\frac{1 + \cos\theta}{\sqrt{1 - \cos^2\theta}}\,.\sin\theta\, d\theta$$

$$= 4a^2\int_{\pi/3}^{\pi} (1 - 2\cos\theta)(1 + \cos\theta)d\theta = 4a^2\int_{\pi/3}^{\pi} (1 - \cos\theta - 2\cos^2\theta)d\theta$$

$$= 4a^2\int_{\pi/3}^{\pi} [1 - \cos\theta - (1 + \cos 2\theta)]\, d\theta = -4a^2\int_{\pi/3}^{\pi} (\cos 2\theta + \cos\theta)d\theta$$

$$= -4a^2\left[\frac{\sin 2\theta}{2} + \sin\theta\right]_{\pi/3}^{\pi} = -4a^2\left[0 - \left(\frac{1}{2}\sin\frac{2\pi}{3} + \sin\frac{\pi}{3}\right)\right]$$

$$= 4a^2\left(\frac{1}{2}\cdot\frac{\sqrt{3}}{2}+\frac{\sqrt{3}}{2}\right)=3\sqrt{3}\,a^2 \qquad ...(3)$$

And the area between the curve and its asymptote

$$= 2 \times \text{area between curve and its asymptote above the } x\text{-axis}$$

$$= 2\int_{-a}^{0} y\,dx = 2\int_{-a}^{0} x\sqrt{\frac{3a-x}{a+x}}\,dx \qquad \Big|\ \textit{From } (2)$$

$$\left[\begin{array}{l}\text{Put } a-x = 2a\cos\theta \ \therefore\ dx = 2a\sin\theta\,d\theta \\ \text{when } x=-a, \theta=0\,; \text{ when } x=0, \theta=\pi/3\end{array}\right]$$

$$= 2\int_{0}^{\pi/3}(a-2a\cos\theta)\,.\sqrt{\frac{2a\,(1+\cos\theta)}{2a\,(1-\cos\theta)}}\,.\,2a\sin\theta\,d\theta$$

$$= 4a^2\int_{0}^{\pi/3}(1-2\cos\theta)\,.\,\frac{(1+\cos\theta)}{\sqrt{1-\cos^2\theta}}\,.\,\sin\theta\,d\theta$$

$$= 4a^2\int_{0}^{\pi/3}(1-2\cos\theta)(1+\cos\theta)\,d\theta = 4a^2\int_{0}^{\pi/3}(1-\cos\theta-2\cos^2\theta)\,d\theta$$

$$= 4a^2\int_{0}^{\pi/3}[1-\cos\theta-(1+\cos 2\theta)]\,d\theta$$

$$= -\,4a^2\int_{0}^{\pi/3}(\cos 2\theta+\cos\theta)\,d\theta = -\,4a^2\left[\frac{1}{2}\sin 2\theta+\sin\theta\right]_0^{\pi/3}$$

$$= -\,4a^2\left(\frac{1}{2}\sin\frac{2\pi}{3}+\sin\frac{\pi}{3}-0\right) = -\,4a^2\left(\frac{\sqrt{3}}{4}+\frac{\sqrt{3}}{2}\right)$$

$$= -\,3\sqrt{3}\,a^2 = 3\sqrt{3}\,a^2 \ (\textit{numerically})$$

This is the same as (3). Hence the result.

Example 8. *If A is the vertex, O the centre and P(x, y) any point on the hyperbola* $\frac{x^2}{a^2}-\frac{y^2}{b^2}=1$, *prove that* $x = a\cosh\frac{2S}{ab}$, $y = b\sinh\frac{2S}{ab}$ *where S is the sectorial area OPA.*

Sol. The equation of the hyperbola is

$$\frac{x^2}{a^2}-\frac{y^2}{b^2}=1 \quad\text{or}\quad \frac{y^2}{b^2}=\frac{x^2}{a^2}-1=\frac{x^2-a^2}{a^2}$$

$$\therefore \qquad y = \frac{b}{a}\sqrt{x^2-a^2} \qquad ...(1) \qquad \Big|\ \text{Taking +ve sign with the square root}$$

Now $\quad S =$ sectorial area OPA

$= \text{area OMP} - \text{area AMP} \quad \ldots(2)$

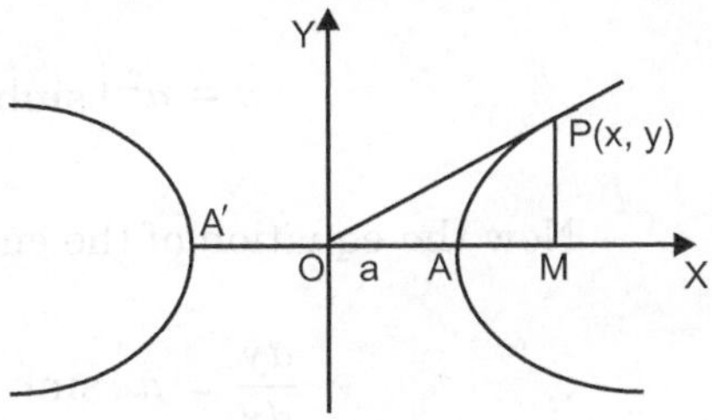

But area $\quad \text{OMP} = \frac{1}{2}\,\text{OM}\,.\,\text{MP}$

$$= \frac{1}{2}\,.\,x\,.\,y$$

$$= \frac{1}{2}\,x\,.\,\frac{b}{a}\sqrt{x^2 - a^2} \qquad \ldots(3)$$

and area AMP $= \int_a^x y\,dx = \int_a^x \frac{b}{a}\sqrt{x^2 - a^2}\,dx \qquad | \because \; OA = a$

$$= \frac{b}{a}\left[\frac{x}{2}\sqrt{x^2 - a^2} - \frac{a^2}{2}\cosh^{-1}\frac{x}{a}\right]_a^x$$

$$\left[\because \int \sqrt{x^2 - a^2}\,dx = \frac{x}{2}\sqrt{x^2 - a^2} - \frac{a^2}{2}\cosh^{-1}\frac{x}{a}\right]$$

$$= \frac{b}{a}\left[\frac{x}{2}\sqrt{x^2 - a^2} - \frac{a^2}{2}\cosh^{-1}\frac{x}{a} - 0 + \frac{a^2}{2}\cosh^{-1}1\right]$$

$$= \frac{b}{a}\left[\frac{x}{2}\sqrt{x^2 - a^2} - \frac{a^2}{2}\cosh^{-1}\frac{x}{a}\right] \qquad \ldots(4) \qquad | \because \; \cosh^{-1}1 = 0$$

$\therefore$ From (2), $\quad S = \frac{1}{2}\frac{b}{a}\,x\sqrt{x^2 - a^2} - \frac{b}{a}\left[\frac{x}{2}\sqrt{x^2 - a^2} - \frac{a^2}{2}\cosh^{-1}\frac{x}{a}\right]$

$$= \frac{ab}{2}\cosh^{-1}\frac{x}{a} \qquad \therefore \quad \cosh^{-1}\frac{x}{a} = \frac{2S}{ab}$$

$\therefore \qquad x = a\cosh\frac{2S}{ab}$

and from (1), $\qquad y = \frac{b}{a}\sqrt{x^2 - a^2} = \frac{b}{a}\sqrt{a^2\left(\cosh^2\frac{2S}{ab} - 1\right)} = \frac{b}{a}\,.\,a\sinh\frac{2S}{ab} = b\sinh\frac{2S}{ab}.$

Hence the result.

Example 9. *In the catenary, $y = a\cosh\frac{x}{a}$, prove that the area between the curve, the x-axis and the ordinates of two points on the curve varies as the length of the intervening arc.*

Sol. Let the two ordinates be at $x = b$, $x = c$.

Then area under the curve between $x = b$, $x = c$ is

$$A = \int_b^c y\,dx = \int_b^c a\cosh\frac{x}{a}\,dx = a^2\left|\,\sinh\frac{x}{a}\right|_b^c$$

$$= a^2 \left(\sinh \frac{c}{a} - \sinh \frac{b}{a} \right) \qquad \ldots(1)$$

Now the equation of the curve is $y = a \cosh \dfrac{x}{a}$.

$$\therefore \qquad \frac{dy}{dx} = a \,.\, \sinh \frac{x}{a} \,.\, \frac{1}{a} = \sinh \frac{x}{a}$$

$\therefore$ Length of intervening arc = length of arc between $x = b$, $x = c$ is

$$\text{L} = \int_b^c \sqrt{1+\left(\frac{dy}{dx}\right)^2}\, dx = \int_b^c \sqrt{1+\sinh^2 \frac{x}{a}}\, dx = \int_b^c \cosh \frac{x}{a} dx = a\left[\sinh \frac{x}{a}\right]_b^c$$

$$= a \left(\sinh \frac{c}{a} - \sinh \frac{b}{a} \right) \qquad \ldots(2)$$

From (1) and (2), $\quad$ A = aL $\quad$ or $\quad$ A varies as L

Area under the curve between two ordinates varies as the length of the intervening arc.

10.3. AREA BETWEEN TWO CURVES

The area bounded by the curves y = f(x), y = φ(x) and the ordinates x = a, x = b is

$\displaystyle\int_{\mathbf{a}}^{\mathbf{b}}$ (**y** *of upper curve* – **y** *of lower curve*) **dx**.

Let AB, CD be the curves $y = f(x)$, $y = \phi(x)$ and MCA, NDB be the ordinates $x = a$, $x = b$.

$\therefore$ Area ACDB = area AMNB – area CMND

$$= \int_a^b f(x)\, dx - \int_a^b \phi(x)\, dx = \int_a^b [f(x) - \phi(x)]\, dx$$

$$= \int_a^b [\mathbf{y} \textit{ of upper curve} - \mathbf{y} \textit{ of lower curve}]\, \mathbf{dx}$$

ILLUSTRATIVE EXAMPLES

Example 1. *Prove that the area common to the two parabolas* $x^2 = 4ay$ *and* $y^2 = 4ax$ *is* $\dfrac{16a^2}{3}$.

Sol. The equations of the two parabolas are

$$x^2 = 4ay \qquad \ldots(1)$$

and

$$y^2 = 4ax \qquad \ldots(2)$$

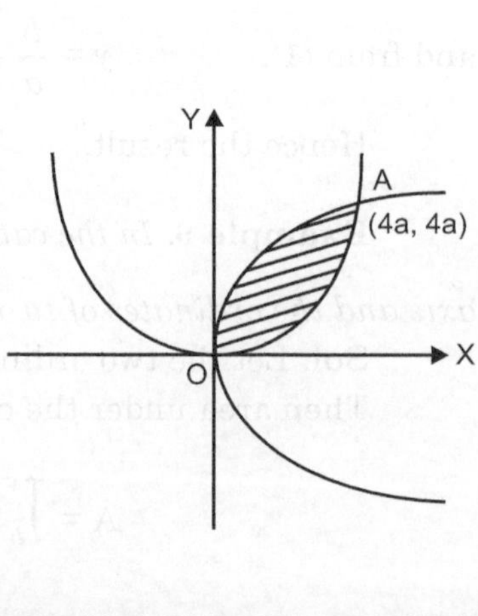

Both the parabolas have their vertices at the origin. Their axes are along OY and OX respectively.

Solving (1) and (2) [eliminating y from these], we have

$$\frac{x^4}{16a^2} = 4ax \quad \text{or} \quad x(x^3 - 64a^3) = 0$$

$\therefore \qquad x = 0 \quad \text{or} \quad 4a.$

When $x = 0$, $y = 0$, when $x = 4a$, from (1), $y = 4a$.

Hence the two parabolas intersect in the points (0, 0), $(4a, 4a)$.

$\therefore$ Required area common to both the parabolas

$$= \text{shaded area}$$

$$= \int_0^{4a} (y \text{ of upper curve} - y \text{ of lower curve})\, dx$$

$$= \int_0^{4a} \left(\sqrt{4ax} - \frac{x^2}{4a}\right) dx \qquad \text{| From (2) and (1)}$$

$$= \left[2\sqrt{a} \cdot \frac{x^{3/2}}{3/2} - \frac{1}{4a} \cdot \frac{x^3}{3}\right]_0^{4a} = \left[\frac{4\sqrt{a}}{3} x^{3/2} - \frac{1}{12a} x^3\right]_0^{4a}$$

$$= \frac{4\sqrt{a}}{3} \cdot (4a)^{3/2} - \frac{1}{12a} \cdot (4a)^3 - 0 = \frac{32a^2}{3} - \frac{16a^2}{3} = \frac{16}{3} a^2.$$

Example 2. *Find the area included between the parabola $x^2 = 4ay$ and the curve $y(x^2 + 4a^2) = 8a^3$.*

Sol. The equation of the parabola is $x^2 = 4ay$ ...(1)

It is an upward parabola with y-axis as its axis and the origin at its vertex.

The equation of the curve is $y = \dfrac{8a^3}{x^2 + 4a^2}$ ...(2)

(*i*) The curve (2) is symmetrical about Y-axis.

(*ii*) The curve (2) does not pass through the origin. The curve does not meet x-axis and it meets y-axis in the point $(0, 2a)$.

(*iii*) $y = 0$, *i.e.*, X-axis is asymptote to curve (2).

(*iv*) From (2), when $x = 0$, $y = 2a$. As x increases from 0 to ∞ y decreases from $2a$ to 0. Thus the shape of the curve is as shown in the figure.

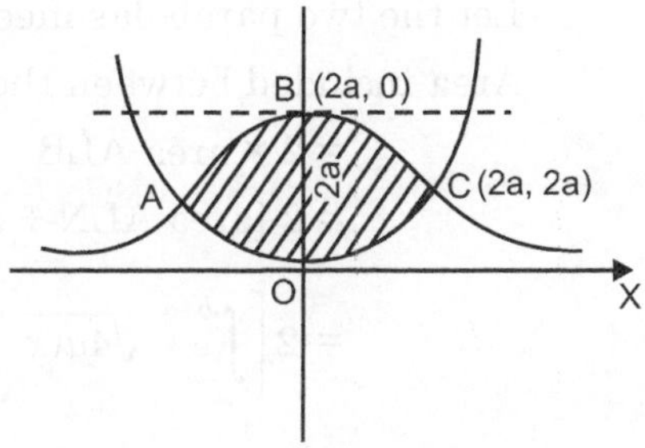

Eliminating y from (1) and (2), we get

$$x^2 = 4ay = 4a \cdot \frac{8a^3}{x^2 + 4a^2}$$

or $\qquad x^4 + 4a^2x^2 - 32a^4 = 0$

or $\qquad (x^2 + 8a^2)(x^2 - 4a^2) = 0$

Rejecting $x^2 + 8a^2 = 0$, which gives imaginary values of x, we have

$$x^2 - 4a^2 = 0 \quad \text{or} \quad x = \pm 2a.$$

Hence the points of intersection area $A(-2a, a)$, $C(2a, a)$.

$\therefore$ Required area OABC $= -2 \times$ area OBC

$$= 2\int_0^{2a} (y \text{ of upper curve} - y \text{ of lower curve})\, dx$$

$$= 2\int_0^{2a} \left(\frac{8a^3}{x^2+4a^2} - \frac{x^2}{4a}\right) dx \qquad \text{| From (2) and (1)}$$

$$= 2\left[8a^3 \cdot \frac{1}{2a} \cdot \tan^{-1}\frac{x}{2a} - \frac{1}{4a} \cdot \frac{x^3}{3}\right]_0^{2a} = 2\left[4a^2 \tan^{-1} 1 - \frac{1}{12a} \cdot (2a)^3 - 0\right]$$

$$= 2\left[4a^2 \cdot \frac{\pi}{4} - \frac{8}{12}a^2\right] = 2\left(\pi a^2 - \frac{2}{3}a^2\right) = \frac{2}{3}a^2(3\pi - 2).$$

Example 3. *Show that the area included between the parabolas* $y^2 = 4a(x + a)$ *and* $y^2 = 4b(b - x)$ *is* $\frac{8}{3}(a + b)\sqrt{ab}$.

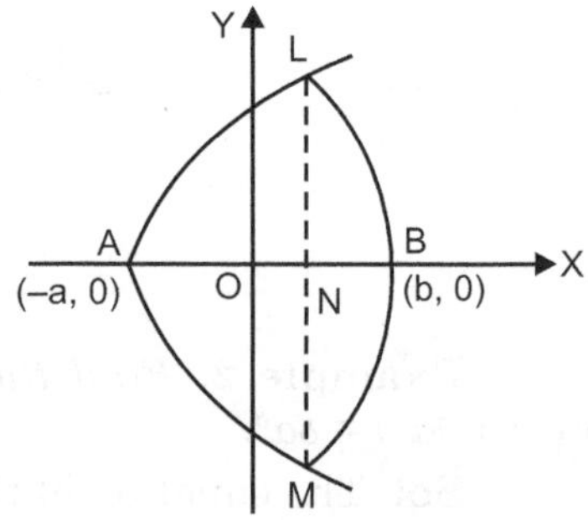

Sol. The equations of the two parabolas are

$$y^2 = 4a(x + a) \qquad ...(1)$$

and
$$y^2 = 4b(b - x) = -4b(x - b) \qquad ...(2)$$

Now (1) represents a parabola to the right whose vertex is at $(-a, 0)$, and latus rectum $4a$, and axis as x-axis.

Also (2) also represents a parabola to the left whose vertex is $(b, 0)$, latus rectum $4b$, and axis as x-axis.

Solving (1) and (2), [Subtracting (2) from (1)]

$$0 = 4a(x + a) + 4b(x - b)$$

or
$$x(a + b) = b^2 - a^2 \quad \therefore \quad x = b - a.$$

Let the two parabolas meet in L and M. The x-co-ordinate of L or M is $b - a$.

Area included between the parabolas = area ALBM

= 2 × area ALB

= 2 [area ALN + area LNB] | *By symmetry*

$$= 2\left[\int_{-a}^{b-a} \sqrt{4a(x+a)}\, dx + \int_{b-a}^{b} \sqrt{4b(b-x)}\, dx\right]$$

$$= 4\sqrt{a} \cdot \left[\frac{(x+a)^{3/2}}{3/2}\right]_{-a}^{b-a} + 4\sqrt{b}\left[\frac{(b-x)^{3/2}}{-3/2}\right]_{b-a}^{b}$$

$$= \frac{8\sqrt{a}}{3}[b^{3/2} - 0] - \frac{8\sqrt{b}}{3}[0 - a^{3/2}] = \frac{8}{3}a^{1/2}b^{3/2} + \frac{8}{3}b^{1/2}a^{3/2} = \frac{8}{3}\sqrt{ab}\,(a + b).$$

Example 4. (*a*) *Find the area common to the circle* $x^2 + y^2 = 4$ *and the ellipse* $x^2 + 4y^2 = 9$.

(*b*) *Find the area common to the two curves* $y^2 = ax$ *and* $x^2 + y^2 = 4ax$.

(*c*) *Prove that the area enclosed by the curves* $x^2 + y^2 = 2ax$ *and* $y^2 = ax$ *is* $a^2(\pi/4 - 2/3)$.

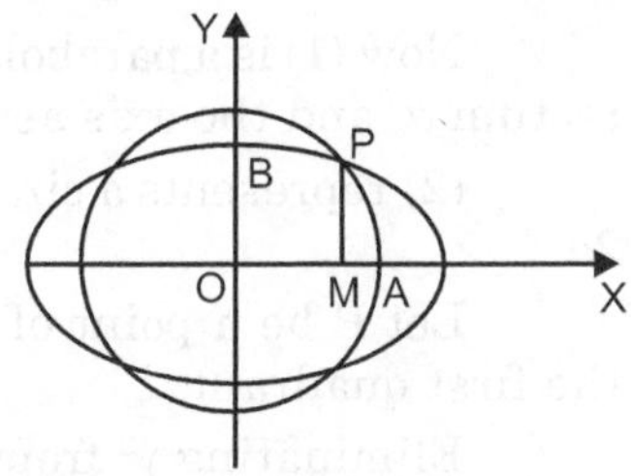

Sol. (*a*) Let P be the point of intersection in the first quadrant, of the circle and the ellipse, as shown in the figure. Draw MP $\perp$ x-axis.

Eliminating y^2 from the two equations of circle $x^2 + y^2 = 4$ and the ellipse $x^2 + 4y^2 = 9$, we have

$$x^2 + 4(4 - x^2) = 9$$

or
$$3x^2 = 7$$

which gives
$$x = \sqrt{\frac{7}{3}} \text{ for P}.$$

Rejecting – ve value, as P is in first quadrant.

Now for the ellipse $\quad y = \frac{1}{2}\sqrt{9 - x^2}$

and $\quad$ for the circle $\quad y = \sqrt{4 - x^2}$.

Since the ellipse and circle are both symmetrical about both the axes.

$\therefore$ Required area common to the circle and ellipse.

= 4 × area common to circle and ellipse in first quadrant.

= 4 × area OAPB = 4 [area OBPM + area MPA]

$$= 4\left[\int_0^{\sqrt{7/3}} (y \text{ of ellipse})\, dx + \int_{\sqrt{7/3}}^{2} (y \text{ of circle})\, dx\right]$$

$$= 4\left[\int_0^{\sqrt{7/3}} \frac{1}{2}\sqrt{9 - x^2}\, dx + \int_{\sqrt{7/3}}^{2} \sqrt{4 - x^2}\, dx\right]$$

$$= 2\left[\frac{x}{2}\sqrt{9 - x^2} + \frac{9}{2}\sin^{-1}\frac{x}{3}\right]_0^{\sqrt{7/3}} + 4\left[\frac{x}{2}\sqrt{4 - x^2} + \frac{4}{2}\sin^{-1}\frac{x}{2}\right]_{\sqrt{7/3}}^{2}$$

$$\left[\because \int \sqrt{a^2 - x^2}\, dx = \frac{x}{2}\sqrt{a^2 - x^2} + \frac{a^2}{2}\sin^{-1}\frac{x}{2}\right]$$

$$= \left[\sqrt{\frac{7}{3}}\sqrt{9 - \frac{7}{3}} + 9\sin^{-1}\frac{\sqrt{7}}{3\sqrt{3}} - 0\right] + 2\left[0 + 4\sin^{-1} 1 - \sqrt{\frac{7}{2}}\sqrt{4 - \frac{7}{3}} - 4\sin^{-1}\frac{\sqrt{7}}{2\sqrt{3}}\right]$$

$$= \sqrt{\frac{7}{3}}\left[\sqrt{\frac{20}{3}} - 2\sqrt{\frac{5}{3}}\right] + 9\sin^{-1}\frac{\sqrt{7}}{3\sqrt{3}} + 8 \cdot \frac{\pi}{2} - 8\sin^{-1}\frac{\sqrt{7}}{2\sqrt{3}}$$

$$= 0 + 9\sin^{-1}\sqrt{\frac{7}{27}} + 4\pi - 8\sin^{-1}\sqrt{\frac{7}{12}} = 4\pi + 9\sin^{-1}\sqrt{\frac{7}{27}} - 8\sin^{-1}\sqrt{\frac{7}{12}}.$$

(*b*) The equations of the curves are

$y^2 = ax$ ...(1) $\quad$ and $\quad x^2 + y^2 - 4ax = 0$ ...(2)

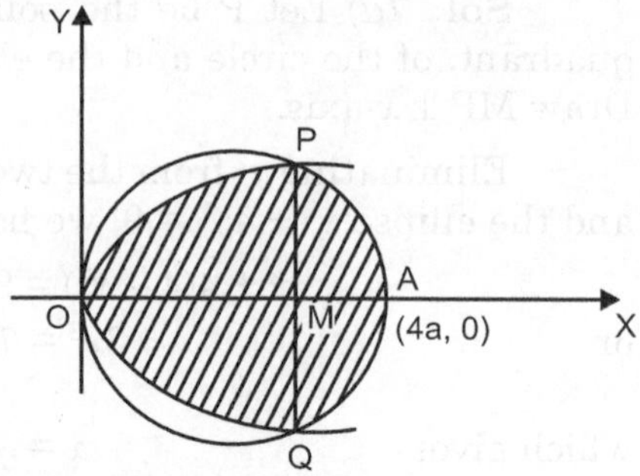

Now (1) is a parabola whose vertex is origin, latus rectum a, and the axis as X-axis.

(2) represents a circle of centre $(2a, 0)$ and radius $2a$.

Let P be a point of intersection of (1) and (2) in the first quadrant.

Eliminating y^2 from (1) and (2), we have

$$x^2 + ax - 4ax = 0$$

or $$x(x - 3a) = 0$$

$\therefore\ x = 0$ or $3a$ $\therefore$ $3a$ is the x-co-ordinate of P.

If the circle meets X-axis in A, its co-ordinates are clearly $(4a, 0)$.

Now y for circle $= \sqrt{4ax - x^2}$ | From (2)

and y for parabola $= \sqrt{ax}$, | From (1)

$\therefore$ Required area common to parabola and circle (shown shaded in figure)

$= 2 \times$ area AOP | *By symmetry*

$= 2$[area OPM + area MPA]

$$= 2\left[\int_0^{3a} (y\ of\ parabola)\, dx + \int_{3a}^{4a} (y\ of\ circle)\, dx\right]$$

$$= 2\left[\int_0^{3a} \sqrt{ax}\, dx + \int_{3a}^{4a} \sqrt{4ax - x^2}\, dx\right]$$

$$= 2\sqrt{a} \int_0^{3a} \sqrt{x}\, dx + 2\int_{3a}^{4a} \sqrt{4a^2 - (x - 2a)^2}\, dx$$

$$= 2\sqrt{a} \cdot \frac{2}{3}[x^{3/2}]_0^{3a} + 2.\left[\frac{1}{2}(x - 2a)\sqrt{4a^2 - (x - 2a)^2} + \frac{4a^2}{2}\sin^{-1}\frac{x - 2a}{2a}\right]_{3a}^{4a}$$

$$= \frac{4}{3}\sqrt{a}\,.\,(3a)^{3/2} + \left[(x - 2a)\sqrt{4ax - x^2} + 4a^2\sin^{-1}\frac{x - 2a}{2a}\right]_{3a}^{4a}$$

$$= 4\sqrt{3}\ a^2 + 0 + 4a^2 \sin^{-1} 1 - a\,.\,\sqrt{3}\,a^2 - 4a^2 \sin^{-1}\frac{1}{2}$$

$$= 4\sqrt{3}\,a^2 + 4a^2\,.\,\frac{\pi}{2} - \sqrt{3}\,a^2 - 4a^2\,.\,\frac{\pi}{6}$$ | $\because \sin^{-1} 1 = \pi/2$, $\sin^{-1} 1/2 = \pi/6$

$$= 3\sqrt{3}\,a^2 + 2a^2\pi - \frac{2a^2\pi}{3} = 3\sqrt{3}\,a^2 + \frac{4\pi a^2}{3} = a^2\left(3\sqrt{3} + \frac{4\pi}{3}\right).$$

Caution. The area *included between the two curves* is not as found above, but the unshaded area in the figure and is given by

$$2\left[\int_0^{3a} (y \text{ of circle} - y \text{ of parabola})\, dx\right] \text{ etc.}$$

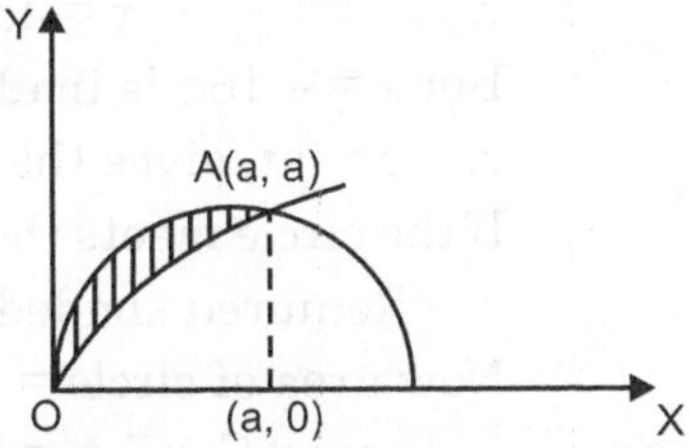

(c) The equation of the circle can be written as

$$(x - a)^2 + y^2 = a^2 \; ; \qquad ...(1)$$

It centre is $(a, 0)$ and radius 'a'. Given parabola is

$$y^2 = ax \qquad ...(2)$$

Solving the given Eqn. (2) and (1) of parabola and circle we find that the two curves intersect at O(0, 0) and A(a, a).

So, limits of integration are from $x = 0$ to $x = a$.

$\therefore$ The required area

$$= \int_0^a (y \text{ of circle} - y \text{ of parabola})\, dx$$

$$= \int_0^a (\sqrt{2ax - x^2} - \sqrt{ax})\, dx = I_1 - I_2 \qquad \text{(say)}$$

Now $\quad I_1 = \int_0^a \sqrt{2ax - x^2}\, dx = \int_0^a \sqrt{a^2 - (a - x)^2}\, dx$

$$= \int_0^{\pi/2} \sqrt{a^2 - a^2 \cos^2 \theta} \,.\, a \sin \theta \, d\theta \qquad [\text{Put } a - x = a \cos \theta]$$

$$= a^2 \int_0^{\pi/2} \sin^2 \theta \, d\theta = a^2 \,.\, \frac{1}{2} \,.\, \frac{\pi}{2} = \frac{\pi a^2}{4}$$

Also $\quad I_2 = \int_0^a \sqrt{ax}\, dx = \sqrt{a} \left| \frac{x^{3/2}}{3/2} \right|_0^a = \frac{2}{3} a^2.$

Hence required area $= I_1 - I_2 = a^2 (\pi/4 - 2/3)$.

Example 5. (a) *Show that the larger of the two areas into which the circle $x^2 + y^2 = 64a^2$ is divided by the parabola $y^2 = 12ax$ is $\frac{16}{3} a^2 (8\pi - \sqrt{3})$.*

(b) *Find the area of the smaller portion enclosed by the curves $y^2 = 8x$ and $x^2 + y^2 = 9$.*

Sol. (a) The equation of the circle is $x^2 + y^2 = 64a^2$ $\qquad$ ...(1)

Its centre is (0, 0) and radius is $8a$.

The equation of the parabola is

$$y^2 = 12ax \qquad ...(2)$$

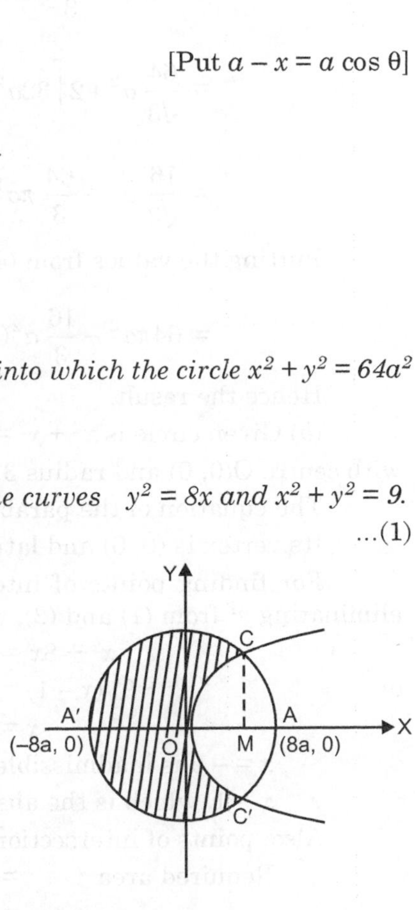

Its vertex is (0, 0) and latus rectum is $12a$.

To find the points of intersection of (1) and (2).

Eliminating y^2 from (1) and (2), we get

$$x^2 + 12ax = 64a^2$$

or $\quad x^2 + 12ax - 64a^2 = 0$

or $\quad (x + 16a)(x - 4a) = 0$

$\therefore \quad x = 4a \quad \text{or} \quad -16a.$

But $x = -16a$ is inadmissible, $\because$ When $x = -16a$, from (2), y is imaginary.

$\therefore$ $x = 4a$, gives the abscissa of C, the point of intersection.

If the circle meets the X-axis in the points A, A′, then these points are A(8a, 0), A′(8a, 0).

$\therefore$ Required shaded area = area of circle – area COC′A ...(3)

Now area of circle $= \pi \cdot (8a)^2 = 64\pi a^2$ ...(4)

and area COC′ A = 2 area COA = 2 (area COM + area MCA)

$$= 2\left[\int_0^{4a} (y \text{ of parabola})\, dx + \int_{4a}^{8a} (y \text{ of circle})\, dx\right]$$

$$= 2\left[\int_0^{4a} \sqrt{12ax}\, dx + \int_{4a}^{8a} \sqrt{64a^2 - x^2}\, dx\right]$$

$$= 2 \cdot \sqrt{12a} \cdot \left[\frac{x^{3/2}}{3/2}\right]_0^{4a} + 2\left[\frac{x}{2}\sqrt{64a^2 - x^2} + \frac{64a^2}{2}\sin^{-1}\frac{x}{8a}\right]_{4a}^{8a}$$

$$= 2.2\sqrt{3a} \cdot \frac{2}{3}(4a)^{3/2} + 2\left[0 + 32a^2 \sin^{-1} 1 - 2a \cdot \sqrt{48a^2} - 32a^2 \sin^{-1}\frac{1}{2}\right]$$

$$= \frac{64}{\sqrt{3}}a^2 + 2\left[32a^2 \cdot \frac{\pi}{2} - 8\sqrt{3} \cdot a^2 - 32a^2 \cdot \frac{\pi}{6}\right] = \left(\frac{64}{\sqrt{3}} - 16\sqrt{3}\right)a^2 + 32a^2\pi - \frac{32\pi a^2}{3}$$

$$= \frac{16}{\sqrt{3}}a^2 + \frac{64}{3}\pi a^2 = \frac{16}{3}a^2(\sqrt{3} + 4\pi) \qquad ...(5)$$

Putting the values from (4) and (5) in (3), we get the required shaded area

$$= 64\pi a^2 - \frac{16}{3}a^2(\sqrt{3} + 4\pi) = \frac{16a^2}{3}(12\pi - \sqrt{3} - 4\pi) = \frac{16}{3}a^2(8\pi - \sqrt{3}).$$

Hence the result.

(b) Given circle is $x^2 + y^2 = 9$ ...(1)

with centre O(0, 0) and radius 3.

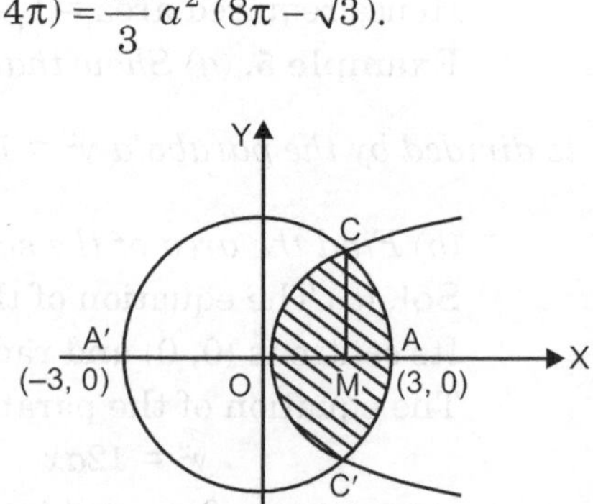

The equation of the parabola is $y^2 = 8x$ ...(2)

Its vertex is (0, 0) and latus rectum 8.

For finding points of intersection of (1) and (2) eliminating y^2 from (1) and (2), we get

$$x^2 + 8x = 9 \quad \text{or} \quad x^2 + 8x - 9 = 0$$

or $\quad (x + 9)(x - 1) = 0$

$$x = -9 \quad \text{or} \quad x = 1$$

$x = -9$ is inadmissible since it makes y imaginary in (2).

$\therefore$ $x = 1$, which is the abscissa of C, the point of intersection.

Also, points of intersection of (1) with x-axis are A(3, 0) and A′(– 3, 0).

$\therefore$ Required area $= 2 \times$ area ACO

$= 2$ [Area OMC + Area MAC]

$$= 2\left[\int_0^1 (y \text{ of parabola})\, dx + \int_1^3 (y \text{ of circle})\, dx\right]$$

$$= 2\left[\int_0^1 \sqrt{8}\,x\, dx + \int_1^3 \sqrt{9-x^2}\, dx\right] = 2\left[\left|\frac{\sqrt{8}\,x^{3/2}}{3/2}\right|_0^1 + \left|\frac{x}{2}\sqrt{9-x^2} + \frac{9}{2}\sin^{-1}\frac{x}{3}\right|_1^3\right]$$

$$= 2\left[\frac{2}{3}\sqrt{8} + \frac{9}{2}\sin^{-1} 1 - \frac{1}{2}\sqrt{8} - \frac{9}{2}\sin^{-1}\frac{1}{3}\right]$$

$$= 2\left[\frac{4\sqrt{2}}{3} - \frac{2\sqrt{2}}{2} + \frac{9}{2}\cdot\frac{\pi}{2} - \frac{9}{2}\sin^{-1}\frac{1}{3}\right] = \frac{2\sqrt{2}}{3} + \frac{9\pi}{2} - 9\sin^{-1}\frac{1}{3}.$$

Example 6. (*a*) *Show that the area of the segment cut-off from the parabola* $y^2 = 2x$ *by the straight line* $y = 4x - 1$ *is* $\frac{9}{32}$.

(*b*) *Find the area of the segment cut-off from the parabola* $y^2 = 4x$ *by the straight line* $2x - 3y + 4 = 0$.

(*c*) *Find the area bounded by the parabola* $x^2 = 8y$ *and the line* $x - 2y + 8 = 0$.

Sol. (*a*) The equation of the parabola is $y^2 = 2x$ ...(1)

and that of the line is $y = 4x - 1$ ...(2)

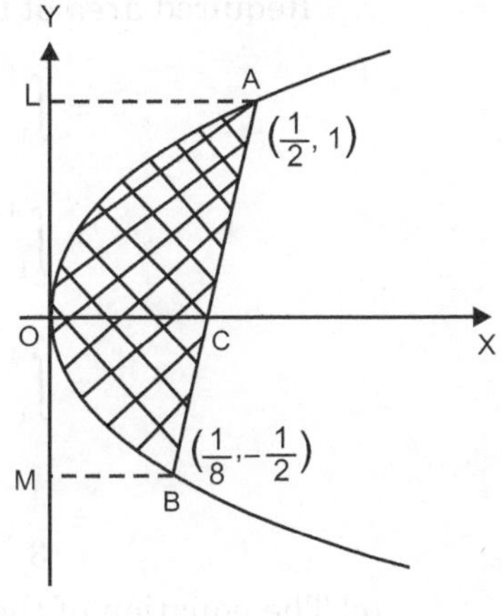

Solving (1) and (2), we have [eliminating y]

$$(4x - 1)^2 = 2x \quad \text{or} \quad 16x^2 - 8x + 1 = 2x$$

or $\quad 16x^2 - 10x + 1 = 0 \quad \text{or} \quad (8x - 1)(2x - 1) = 0$

$\therefore \quad x = \frac{1}{8} \quad \text{or} \quad x = \frac{1}{2}.$

From (2), when $x = \frac{1}{8}$, $y = -\frac{1}{2}$ and when $x = \frac{1}{2}$, $y = 1$.

Hence the line (2) meets the parabola in the points $A\left(\frac{1}{2}, 1\right)$ and $B\left(\frac{1}{8}, -\frac{1}{2}\right)$

Required area (shown shaded)

= area ACBML – area AOBML

= (area between the line segment AB and y-axis)

– (area between the arc of AOB of parabola and y-axis)

$$= \int_{-1/2}^{1} (x \text{ of line} - x \text{ of parabola})\, dy$$

$$= \int_{-1/2}^{1} \left(\frac{y+1}{4} - \frac{y^2}{2}\right) dy = \left[\frac{y^2}{8} + \frac{y}{4} - \frac{y^3}{6}\right]_{-1/2}^{1}$$

$$= \left(\frac{1}{8}+\frac{1}{4}-\frac{1}{6}\right)-\left(\frac{1}{32}-\frac{1}{8}+\frac{1}{48}\right)=\frac{5}{24}+\frac{7}{96}=\frac{27}{96}=\frac{9}{32}.$$

(*b*) The equation of the parabola is $y^2 = 4x$ ...(1)

and that of straight line is

$$2x - 3y + 4 = 0 \quad ...(2)$$

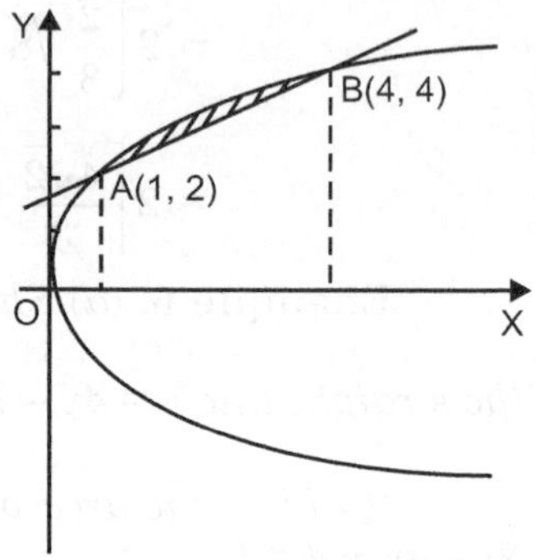

To find the points of intersection of (1) and (2) :

From (2), $$y = \frac{2x+4}{3} = \frac{2(x+2)}{3} \quad ...(3)$$

Putting this value of y in (1), we have

$$\frac{4}{9}(x+2)^2 = 4x \quad \text{or} \quad x^2 + 4x + 4 = 9x$$

$$x^2 - 5x + 4 = 0 \quad \text{or} \quad (x-1)(x-4) = 0$$

$\therefore$ $$x = 1, \quad x = 4.$$

From (3), when $x = 1, y = 2$; when $x = 4, y = 4$.

Thus the line (2) meets parabola (1) in the points A(1, 2) and B(4, 4).

$\therefore$ Required area of the segment cut-off by the line

$$= \int_1^4 (y \text{ of upper curve} - y \text{ of lower curve})\, dx$$

$$= \int_1^4 (y \text{ of parabola} - y \text{ of line})\, dx$$

$$= \int_1^4 \left(\sqrt{4x} - \frac{2x+4}{3}\right) dx = \left[2.\frac{2}{3}x^{3/2} - \frac{1}{3}(x^2+4x)\right]_1^4$$

$$= \frac{4}{3}(4^{3/2}-1) - \frac{1}{3}(16+16-1-4) = \frac{28}{3} - \frac{27}{3} = \frac{1}{3}.$$

(*c*) The equation of the parabola is $x^2 = 8y$ ...(1)

and that of straight line is $x - 2y + 8 = 0$ ...(2)

To find the points of intersection of (1) and (2) :

From (2), $$x = 2y - 8 \quad ...(3)$$

Putting this value of x in (1), we get

$$(2y-8)^2 = 8y \quad \text{or} \quad 4y^2 + 64 - 32y = 8y$$

or $$y^2 - 10y + 16 = 0 \quad \text{or} \quad (y-8)(y-2) = 0$$

$$y = 2, 8.$$

From (3) when $y = 2, x = -4$; when $y = 8, x = 8$.

Thus the line (2) meets parabola (1) in the points A(– 4, 2) and B(8, 8).

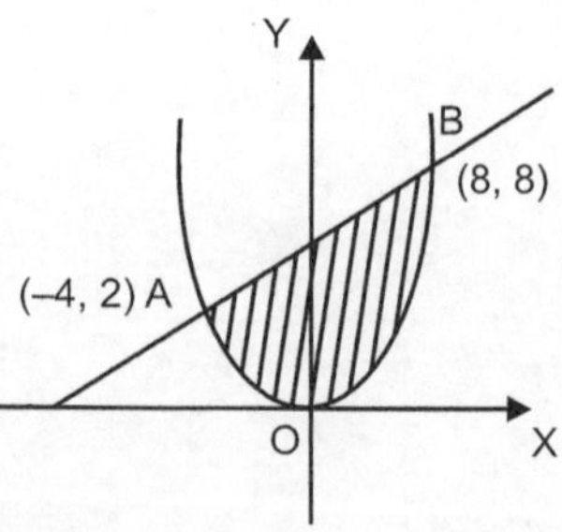

$\therefore$ Required area bounded by the parabola and the line.

$$= \int_{-4}^8 (y \text{ of line} - y \text{ of parabola})\, dx$$

$$= \int_{-4}^{8} \left(\frac{x+8}{2} - \frac{x^2}{8} \right) dx = \left[\frac{x^2}{4} + 4x - \frac{x^3}{24} \right]_{-4}^{8}$$

$$= \left(16 + 32 - \frac{64}{3} \right) - \left(4 - 16 + \frac{8}{3} \right) = \frac{80}{3} + \frac{28}{3} = \frac{108}{3} = 36.$$

10.4. AREA FORMULAE FOR PARAMETRIC EQUATIONS

(*i*) **Prove that the area bounded by the curve $\mathbf{x = f(t), y = \phi(t)}$, the X-axis and the ordinates at the points where $\mathbf{t = a, t = b}$ is $\displaystyle\int_a^b \mathbf{y \frac{dx}{dt} \, dt}$.**

The parametric equations of the curve are $x = f(t)$, $y = \phi(t)$.

From 8.4.1, we know that

$$\text{Required area} = \int_{t=a}^{t=b} y \, dx = \int_a^b \mathbf{y \frac{dx}{dt} \, . \, dt} = \int_a^b \phi(t) \, . \, f'(t) \, dt.$$

(*ii*) **Similarly the area bounded by the curve $\mathbf{x = f(t), y = \phi(t)}$, the y-axis and the abscissae at the points where $\mathbf{t = c, t = d}$ is given by $\displaystyle\int_c^d \mathbf{x \frac{dy}{dt} \, dt}$.**

Example 1. (*i*) *Find the area enclosed between one arch of the cycloid $x = a(\theta - \sin \theta)$, $y = a(1 - \cos \theta)$ and its base.*

(*ii*) *Find the area between the cycloid $x = a(\theta + \sin \theta)$ $y = a(1 - \cos \theta)$ and its base. Also find the area between the curve and the axis of x.*

Sol. (*i*) The parametric equations of the cycloid are

$$x = a(\theta - \sin \theta), \; y = a(1 - \cos \theta) \qquad ...(1)$$

The cycloid is symmetrical about the line $x = a\pi$,

i.e., the line through the point where $\theta = \pi$ and $\perp$ to X-axis, and the base is x-axis.

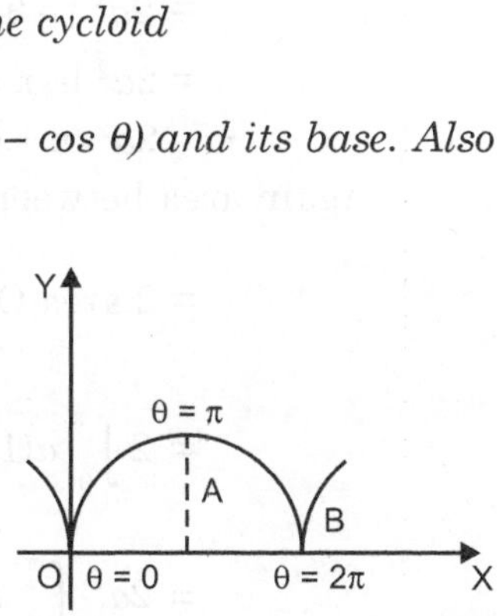

For the first half of the cycloid, θ varies from 0 to π.

$\therefore$ Required area bounded by the cycloid and its base

= area under the cycloid

$$= 2 \int_0^{\pi} y \frac{dx}{d\theta} d\theta = 2 \int_0^{\pi} a(1 - \cos \theta) \, . \, a(1 - \cos \theta) \, d\theta \qquad | \text{ From } (1)$$

$$= 2a^2 \int_0^{\pi} (1 - \cos \theta)^2 \, d\theta = 2a^2 \int_0^{\pi} \left(2 \sin^2 \frac{\theta}{2} \right)^2 d\theta$$

$$= 8a^2 \int_0^{\pi} \sin^4 \frac{\theta}{2} \, d\theta = 8a^2 \int_0^{\pi/2} \sin^4 t \, . \, 2dt$$

$$= 16a^2 \, . \int_0^{\pi/2} \sin^4 t \, dt = 16a^2 \, . \frac{3.1}{4.2} . \frac{\pi}{2} = 3\pi a^2.$$

Put $\dfrac{\theta}{2} = t$

$\therefore$ $d\theta = 2dt$.

when $\theta = 0, t = 0$.

when $\theta = \pi, t = \dfrac{\pi}{2}$

(*ii*) The parametric equations of the cycloid are

$$x = a(\theta + \sin\theta),\ y = a(1 - \cos\theta)$$

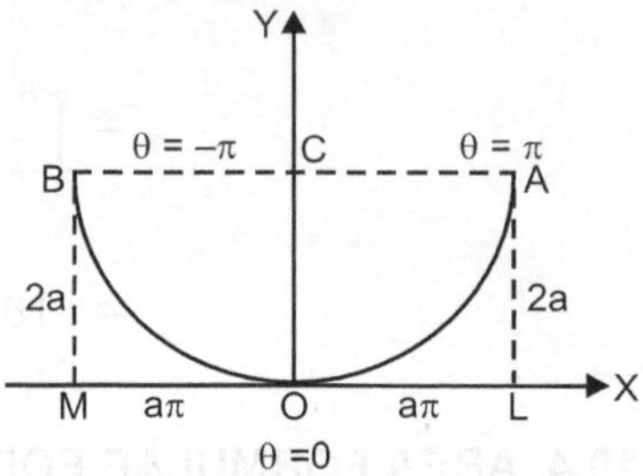

The cycloid is symmetrical about the Y-axis, and the base is the line AB || to X-axis and at a distance $2a$ from it.

For the half of cycloid, θ varies from 0 to π.

The area between the curve and its base

= area BOA = 2 area OCA

$$= 2\int_0^\pi x\frac{dy}{d\theta}\,.\,d\theta = 2\int_0^\pi a\,(\theta + \sin\theta)\,.\,(a\sin\theta)\,d\theta$$

$$= 2a^2\int_0^\pi (\theta\sin\theta + \sin^2\theta)\,d\theta$$

$$= 2a^2\int_0^\pi \theta\sin\theta\,d\theta + 2a^2\int_0^\pi \frac{1-\cos 2\theta}{2}\,d\theta$$

$$= 2a^2\left[\theta(-\cos\theta) - \int 1\,.\,(-\cos\theta)\,d\theta\right]_0^\pi + a^2\left[\theta - \frac{\sin 2\theta}{2}\right]_0^\pi$$

$$= 2a^2\,[-\theta\cos\theta + \sin\theta]_0^\pi + a^2\,(\theta - \sin\theta\cos\theta)_0^\pi$$

$$= 2a^2\,[-\pi\cos\pi + \sin\pi - 0] + a^2\,.\,[\pi - 0 - 0]$$

$$= 2a^2\,(\pi + 0) + a^2\pi = 3a^2\pi.$$

Again area between the curve and X-axis ...(2)

$$= 2 \text{ area OAL} = 2\int_0^\pi y\frac{dx}{d\theta}\,d\theta$$

$$= 2\int_0^\pi a(1-\cos\theta)\,.\,a(1+\cos\theta)\,d\theta = 2a^2\int_0^\pi (1-\cos^2\theta)\,d\theta$$

$$= 2a^2\int_0^\pi \sin^2\theta\,d\theta = 4a^2\,.\int_0^{\pi/2}\sin^2\theta\,d\theta = 4a^2\,.\,\frac{1}{2}\,.\,\frac{\pi}{2} = \pi a^2.$$

Aliter. Area between the curve and X-axis

= area of rectangle ALMB – area between curve and its base

$= \text{AL} \times \text{LM} - 3\pi a^2$ | *Using* (2)

$= 2a \times 2a\pi - 3\pi a^2 = 4\pi a^2 - 3\pi a^2 = \pi a^2.$

Example 2. (*a*) *Find the area of the curve* $x = a\cos^3 t,\ y = b\sin^3 t$.

(*b*) *Find the whole area of the curve* $x = a\,\dfrac{1-t^2}{1+t^2},\ y = \dfrac{2at}{1+t^2}$.

Sol. (*a*) The equations of the curve are

$$x = a\cos^3 t,\ y = b\sin^3 t \qquad ...(1)$$

The curve is symmetrical about both the axes.

$\because$ $\quad |x| \le a$ and $|y| \le b$,

$\therefore$ The curve lies within the rectangle bounded by the lines

$$x = \pm a,\ y = \pm b.$$

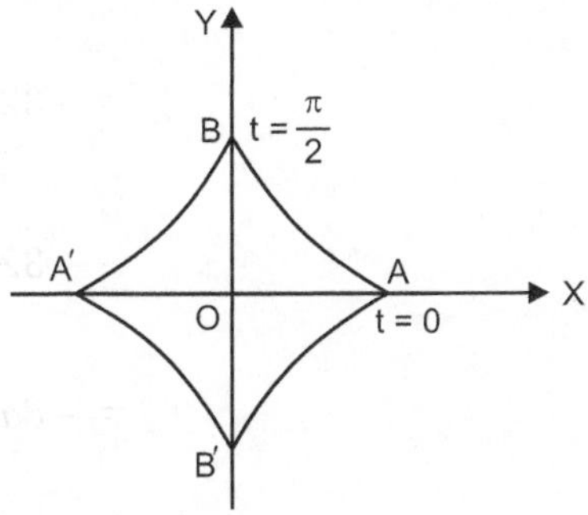

In the first quadrant, t varies from 0 to $\frac{\pi}{2}$.

$\therefore$ Required area = 4 × area OAB

$$= 4\int_0^{\pi/2} y\frac{dx}{dt}\,dt$$

$$= 4\int_0^{\pi/2} b\sin^3 t\,.\,3a\cos^2 t\,(-\sin t)\,dt$$

$$= -12ab\int_0^{\pi/2}\sin^4 t\cos^2 t\,dt$$

$$= -12ab\,.\,\frac{3\,.\,1\,.\,1}{6\,.\,4\,.\,2}\,.\,\frac{\pi}{2} = -\frac{3}{8}\pi ab = \frac{3}{8}\pi ab \text{ numerically.}$$

(*b*) The parametric equations of the curve are

$$x = a\,\frac{1-t^2}{1+t^2},\quad y = \frac{2at}{1+t^2} \qquad \ldots(1)$$

Squaring and adding, we get

$$x^2 + y^2 = a^2\left[\frac{(1-t^2)^2 + 4t^2}{(1+t^2)^2}\right] = a^2\,\frac{(1+t^2)^2}{(1+t^2)^2} = a^2$$ | **Note this step**

Thus the equations (1) represent a circle whose centre is (0, 0) and radius is a.

For the curve in the first quadrant, t varies from 0 to 1.

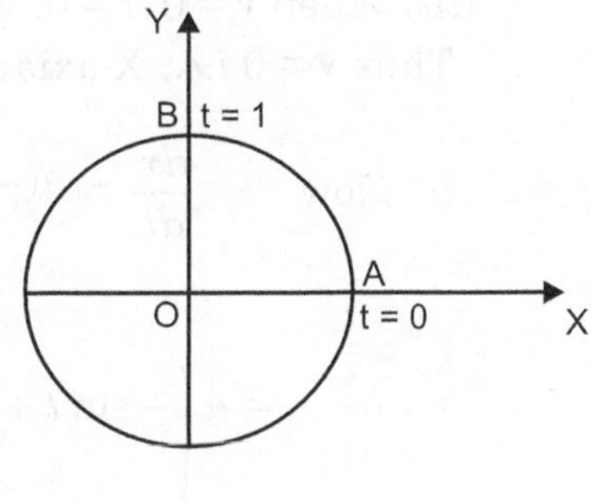

From (1),

$$\frac{dx}{dt} = a\,.\,\frac{(1+t^2)(-2t) - (1-t^2)\,2t}{(1+t^2)^2} = \frac{-4at}{(1+t^2)^2}.$$

$\therefore$ Required whole area

$$= 4 \times \text{area OAB}$$

$$= 4\int_0^1 y\frac{dx}{dt}\,.\,dt = 4\int_0^1 \frac{2at}{1+t^2}\,.\,\frac{-4at}{(1+t^2)^2}\,dt$$

$$= -32a^2\int_0^1 \frac{t^2}{(1+t^2)^3}\,dt$$

Put $t = \tan\theta$

$\therefore$ $dt = \sec^2\theta\,d\theta$

When $t = 0, \theta = 0$;

When $t = 1, \theta = \frac{\pi}{4}$

$$= -32a^2\int_0^{\pi/4}\frac{\tan^2\theta}{(1+\tan^2\theta)^3}\,.\,\sec^2\theta\,d\theta$$

$$= -32a^2 \int_0^{\pi/4} \frac{\sin^2\theta}{\cos^2\theta \,.\, \sec^6\theta} \,.\, \sec^2\theta \, d\theta$$

$$= -32a^2 \int_0^{\pi/4} \sin^2\theta \cos^2\theta \, d\theta = -8a^2 \int_0^{\pi/4} (2\sin\theta\cos\theta)^2 \, d\theta$$

$$= -8a^2 \int_0^{\pi/4} \sin^2 2\theta \, d\theta = -8a^2 \int_0^{\pi/4} \frac{1-\cos 4\theta}{2} \, d\theta$$

$$= -4a^2 \left[\theta - \frac{\sin 4\theta}{4}\right]_0^{\pi/4} = -4a^2 \left[\frac{\pi}{4} - 0 - 0\right] = -\pi a^2$$

$= \pi a^2$ in magnitude.

Example 3. *Prove that the whole area between the four infinite branches of the tractrix*

$$x = a\cos t + \frac{1}{2} a \log \tan^2 \frac{t}{2},\ y = a \sin t \text{ is } \pi a^2 .$$

Sol. The equations of the curve are $x = a\cos t + \frac{a}{2}\log\tan^2\frac{t}{2}$ ...(1)

$y = a \sin t$ ...(2)

(*i*) Since x is an even function of t and y is an odd function of t, the curve is symmetrical about the X-axis.

(*ii*) Putting $y = 0$, we have $a \sin t = 0$ or $t = 0$.

But $t = 0$ does not make $x = 0$.

Hence the curve does not pass through the origin.

$\because \ |\sin t| \le 1, \quad \therefore \ |y| \le a$ *i.e.*, the curve lies between the lines $y = \pm a$.

(*iii*) When $y = 0$, $t = 0$, which makes x infinite. *i.e.*, if $x \to \infty$, $y \to 0$.

Thus $y = 0$ *i.e.*, X-axis is an asymptote of the curve.

(*iv*) Now $\dfrac{dx}{dt} = a(-\sin t) + \dfrac{a}{2} \,.\, \dfrac{1}{\tan^2 \frac{t}{2}} \,.\, 2\tan\dfrac{t}{2}\sec^2\dfrac{t}{2}\,.\,\dfrac{1}{2}$

$$= a\left(-\sin t + \frac{1}{2\sin\frac{t}{2}\cos\frac{t}{2}}\right) = a\left(-\sin t + \frac{1}{\sin t}\right) = a\left(\frac{1-\sin^2 t}{\sin t}\right) = a\,\frac{\cos^2 t}{\sin t}$$

and $\dfrac{dy}{dt} = a\cos t$

$$\therefore \quad \frac{dy}{dx} = \frac{\frac{dy}{dt}}{\frac{dx}{dt}} = a\cos t \,.\, \frac{\sin t}{a\cos^2 t} = \tan t$$

The corresponding values of x, y for different values of t are

$t = -\pi$	$-\dfrac{\pi}{2}$	0	$\dfrac{\pi}{2}$	π
$x = \infty$	0	$-\infty$	0	∞
$y = 0$	$-a$	0	a	0
$\dfrac{dy}{dx} = 0$	$-\infty$	0	∞	0

Thus it follows that when $t = 0$, the corresponding point is $(-\infty, 0)$

i.e., a point at ∞ on the –ve side of X-axis.

When $t = \pi$, the corresponding point is $(\infty, 0)$.

Also at $t = \pm \dfrac{\pi}{2}$ *i.e.,* at the points $(0, -a)$ and $(0, a)$, since $\dfrac{dy}{dx} = \pm \infty$,

$\therefore$ the tangents is || to Y-axis.

In this case the line Y-axis itself is the tangent to the curve.

Thus the shape of the curve is roughly as shown in figure.

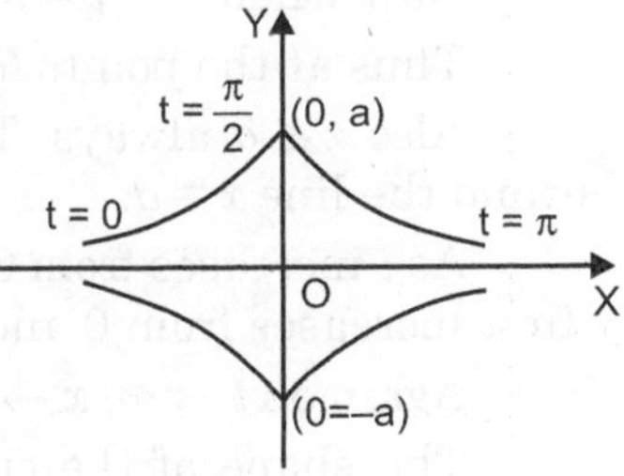

By symmetry,

Required area = 4 × area in the second quadrant

$$= 4 \int_0^{\pi/2} y \frac{dx}{dt}\, dt$$

$$= 4 \int_0^{\pi/2} a \sin t \,.\, a \frac{\cos^2 t}{\sin t}\, dt = 4a^2 \int_0^{\pi/2} \cos^2 t\, dt = 4a^2 \,.\, \frac{1}{2} \cdot \frac{\pi}{2}$$

$= \pi a^2$. Hence the result.

Example 4. *Find the area of the loop of the curve $x = a(1 - t^2)$, $y = at(1 - t^2)$.*

Sol. The parametric equations of the given curve are

$$x = a(1 - t^2), \quad y = at(1 - t^2) \qquad \ldots(1)$$

To trace the curve roughly to find the limits of integration :

(*i*) Since x is an even and y an odd function of t, $\therefore$ the curve is symmetrical about the X-axis.

(*ii*) $x = 0$ gives $t = \pm 1$ which also makes $y = 0$.

$\therefore$ The curve passes through the origin. Since the curve passes through the origin for two different values of t *i.e.,* + 1 and – 1 $\therefore$ origin is a node.

(*iii*) The curve meets X-axis, where putting $y = 0$, we have

$$at(1 - t^2) = 0 \quad \therefore \quad t = 0 \quad \text{or} \quad 1, -1.$$

When $t = 0$, $x = a$; When $t = \pm 1$, $x = 0$.

Thus the curve meets X-axis in the points $(0, 0)$, $(a, 0)$.

(*iv*) The curve has no asymptote *i.e.,* for no value of t, x (or y) tends to ∞ and y(or x) tends to a finite quantity.

(*v*) $\dfrac{dx}{dt} = -2at, \quad \dfrac{dy}{dt} = a(1 = -3t^2)$

$$\therefore \quad \frac{dy}{dx} = \frac{\frac{dy}{dt}}{\frac{dx}{dt}} = \frac{1-3t^2}{-2t} = \frac{3t^2-1}{2t}$$

Now $\frac{dy}{dx} = 0$ when $3t^2 - 1 = 0$ *i.e.,* $t = \pm \frac{1}{\sqrt{3}}$.

When $\quad t = \pm \frac{1}{\sqrt{3}}, x = \frac{2}{3}a, y = \pm \frac{2}{3\sqrt{3}}a.$

Thus at the points $\left(\frac{2}{3}a, \pm \frac{2}{3\sqrt{3}}a\right)$ the tangent is || to X-axis.

Again $\quad \frac{dy}{dx} = \infty, \quad$ when $\quad t = 0.$

Now when $\quad t = 0, x = a, y = 0.$

Thus at the points $(a, 0)$, the tangents is ⊥ to X-axis.

Also $x \le a$ always. Thus no portion of the curve lies beyond the line $x = a$.

As t increases from 0 to 1, x decreases from a to 0 and y first increases from 0 and then decreases to zero.

Again as $t \to \infty$, $x \to -\infty$ and $y \to -\infty$.

The shape of the curve is roughly as shown in the figure and for the upper half of the loop, t varies from 0 to 1.

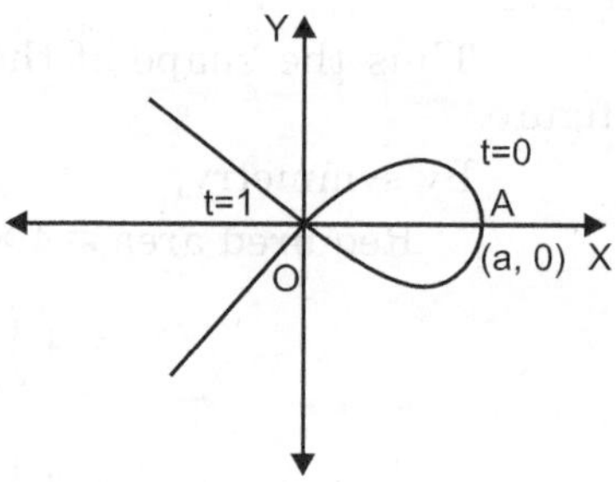

∴ Required area of the loop

$$= 2\int_0^1 y\frac{dx}{dt}\,dt = 2\int_0^1 at\,(1-t^2)\,.\,(-2at)\,dt$$

$$= -4a^2\int_0^1 (t^2 - t^4)\,dt = -4a^2\left[\frac{t^3}{3} - \frac{t^5}{5}\right]_0^1$$

$$= -4a^2\left[\frac{1}{3} - \frac{1}{5} - 0\right] = \frac{-8}{15}a^2 = \frac{8}{15}a^2 \text{ (in magnitude)}.$$

Example 5. *Show that the area bounded by the cissoid*

$$x = a\sin^2 t, \qquad y = a\frac{\sin^3 t}{\cos t} \textit{ and its asymptote is } \frac{3\pi a^2}{4}.$$

Sol. The equation of the curve are

$$x = a\sin^2 t, \quad y = a\frac{\sin^3 t}{\cos t} \qquad \text{...(1)}$$

To trace the curve roughly to find the limits of integration :

(*i*) Since x is an even function of t, and y an odd function of t, the curve is symmetrical about the X-axis.

(*ii*) $x = 0$ gives $\sin^2 t = 0$ or $t = 0$, which also makes $y = 0$,

$\therefore$ The curve passes through the origin.

(*iii*) The curve meets both the axis only at the origin.

(*iv*) $\dfrac{dx}{dt} = 2a \sin t \cos t,$

and $$\frac{dy}{dt} = a \cdot \frac{\cos t \cdot 3\sin^2 t \cos t - \sin^3 t(-\sin t)}{\cos^2 t}$$

$$= a \cdot \frac{\sin^2 t(3\cos^2 t + \sin^2 t)}{\cos^2 t} = a\,\frac{\sin^2 t(1 + 2\cos^2 t)}{\cos^2 t}$$

$$\therefore \quad \frac{dy}{dx} = \frac{\frac{dy}{dt}}{\frac{dx}{dt}} = \frac{a\sin^2 t(1 + 2\cos^2 t)}{\cos^2 t} \cdot \frac{1}{2a \sin t \cos t} = \frac{(1 + \cos^2 t)\sin t}{2\cos^2 t}$$

$$\frac{dy}{dx} = 0 \text{ when } \sin t = 0 \qquad \Big|\ \because\ 1 + 2\cos^2 t \neq 0\ i.e.,\ \cos^2 t \neq -\frac{1}{2}\ (\text{–ve})$$

or when $t = 0$ *i.e.*, $x = 0, y = 0$.

Hence at the origin, the tangents is || to X-axis, *i.e.*, the X-axis itself.

Again $\dfrac{dy}{dx} = \infty$ if $\cos t = 0$ or $t = \dfrac{\pi}{2}$.

When $t = \dfrac{\pi}{2}, x = a, \ y = \infty.$

Thus the line $x = a$ is an asymptote of the curve $[\because$ as $y \to \infty, x \to a]$

(*v*) Since $x = a \sin^2 t$, x can't be –ve.

Also x can't be $> a$, $\because \sin^2 t \le 1$.

Hence no portion of the curve lies to the left of Y-axis and beyond the line $x = a$.

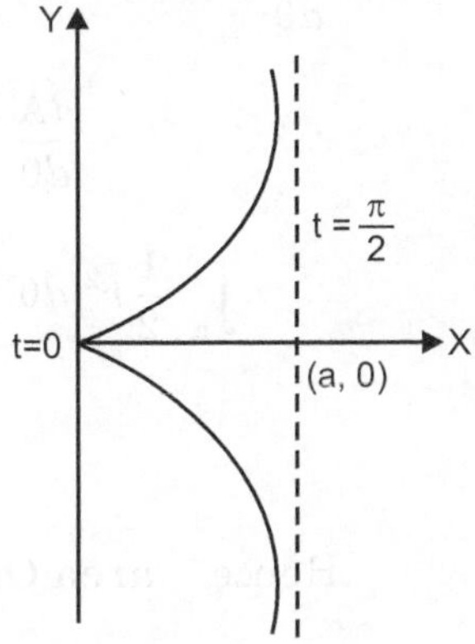

Thus for the upper half of the curve, t varies from 0 to $\dfrac{\pi}{2}$.

$\therefore$ Required area between the curve and its asymptote

$= 2 \times$ area of the upper half of the curve

$$= 2 \times \int_0^{\pi/2} y\,\frac{dx}{dt}\,.\,dt$$

$$= 2\int_0^{\pi/2} a\,\frac{\sin^3 t}{\cos t}\,.\,2a \sin t \cos t\,dt$$

$$= 4a^2 \int_0^{\pi/2} \sin^2 t\,dt = 4a^2 \cdot \frac{3.1}{4.2} \cdot \frac{\pi}{2} = \frac{3\pi a^2}{4}.$$

Hence the result.

10.5. AREA FORMULAE FOR POLAR EQUATIONS

Prove that the area bounded by the curve r = f(θ) and the radii vectors θ = α ; θ = β is

$$\int_{\alpha}^{\beta} \frac{1}{2} r^2 d\theta$$

where $r = f(\theta)$ *is finite, continuous and single-valued function of* θ *in the interval* $\alpha \le \theta \le \beta$.

Let AB be the curve $r = f(q)$ and OA, OB the radii vectors $\theta = \alpha$, $\theta = \beta$.

Let $P(r, \theta)$ be any point on the curve. Let $Q(r + \delta r, \theta + \delta\theta)$ be a neighbouring point on the curve.

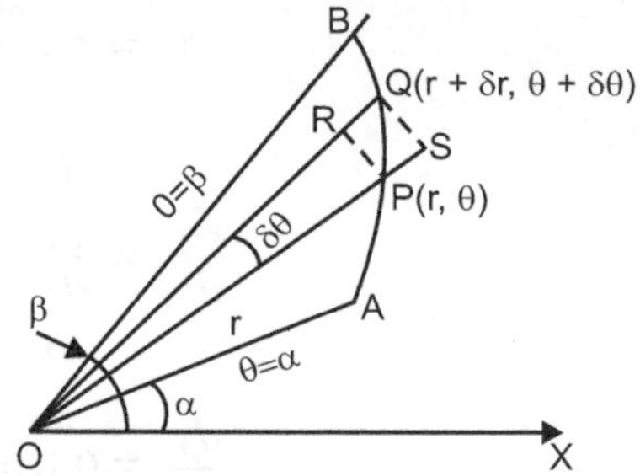

Then if A denotes the *sectorial area* OA P, then clearly A is a function of θ. Also sectorial area OAQ = A + δA, so that sectorial area OPQ = δA.

With O as centre and OP, OQ as radii, draw circular arcs to meet OQ, OP (produced) in R and S respectively.

Then the area OPQ lies between the circular sectorial areas OPR and OSQ.

i.e., δA lies between $\frac{1}{2} r^2 \,\delta\theta$ and $\frac{1}{2}(r + \delta r)^2 \,\delta\theta$

$\left[\because \text{ Area of a circular sector} = \frac{1}{2} \text{ (radius)}^2 \times \text{circular measure of the angle of sector} \right]$

$\therefore \quad \frac{\delta A}{\delta\theta}$ lies between $\frac{1}{2} r^2$ and $\frac{1}{2} (r + \delta r)^2$.

Proceeding to limits as Q → P *i.e.,* as $\delta\theta \to 0 \;\therefore\; \delta r \to 0$, we get

$\frac{dA}{d\theta}$ lies between $\frac{1}{2} r^2$ and a quantity which $\to \frac{1}{2} r^2$

$$\therefore \quad \frac{dA}{d\theta} = \frac{1}{2} r^2$$

$$\therefore \quad \int_{\alpha}^{\beta} \frac{1}{2} r^2 \, d\theta = \int_{\alpha}^{\beta} \frac{dA}{d\theta} \, d\theta = [A]_{\alpha}^{\beta}$$

= (value of A when θ = β) – (value of A when θ = α)

= area OAB – 0.

Hence **area OAB** $= \frac{1}{2} \int_{\alpha}^{\beta} r^2 \, d\theta$.

Note. 1. In some cases, it is more convenient to transform a given cartesian equation into polars than to solve for *y*. In such cases the above formula is applied after changing to polars.

Note. 2. Determination of the limits of integration

(*i*) In case of a **loop,** the limits of integration for finding its area two *successive* values of θ which make $r = 0$.

(*ii*) If the curve is symmetrical about both the axes, the integral must be evaluated from 0 to $\frac{\pi}{2}$ and the whole result multiplied by 4.

(*iii*) If the curve is symmetrical about the x-axis, *i.e.*, initial line only, the integral may be evaluated from 0 to π and the whole result multiplied by 2.

ILLUSTRATIVE EXAMPLES

Example 1. *Find the area between the following curves and the radii vectors :*

(*i*) $r = a\theta^{-1/2}$; $\theta = \alpha$ *to* $\theta = \beta$

(*ii*) $r = a\theta \cos\theta$; $\theta = 0$ *to* $\theta = \frac{\pi}{2}$

(*iii*) $r = ae^{\theta \cot \alpha}$; $\theta = \beta$ *to* $\theta = \beta + \gamma\ (\gamma < 2\pi)$.

Sol. (*i*) The given curve is $r = a\theta^{-1/2}$

$$\therefore \quad \text{Required sectorial area} = \int_\alpha^\beta \frac{1}{2} r^2\, d\theta = \frac{1}{2}\int_\alpha^\beta a^2\, \theta^{-1}\, d\theta = \frac{1}{2} a^2 \int_\alpha^\beta \frac{1}{\theta}\, d\theta$$

$$= \frac{a^2}{2}[\log\theta]_\alpha^\beta = \frac{a^2}{2}(\log\beta - \log\alpha) = \frac{a^2}{2}\log\frac{\beta}{\alpha}.$$

(*ii*) The curve is $r = a\theta\cos\theta$

$\therefore$ Required sectorial area

$$= \int_0^{\pi/2} \frac{1}{2} r^2\, d\theta = \frac{1}{2}\int_0^{\pi/2} a^2\theta^2\cos^2\theta\, d\theta = \frac{a^2}{2}\int_0^{\pi/2} \theta^2 . \frac{1+\cos 2\theta}{2}\, d\theta$$

$$= \frac{a^2}{4}\int_0^{\pi/2} \theta^2\, d\theta + \frac{a^2}{4}\int_0^{\pi/2} \theta^2 \cos 2\theta\, d\theta$$

$$= \frac{a^2}{4}\left[\frac{\theta^3}{3}\right]_0^{\pi/2} + \frac{a^2}{4}\left[\left(\theta . \frac{\sin 2\theta}{2}\right)_0^{\pi/2} - \int_0^{\pi/2} 2\theta . \frac{\sin 2\theta}{2}\, d\theta\right]$$

$$= \frac{a^2}{12}.\frac{\pi^3}{8} + \frac{a^2}{4}\left[0 - \int_0^{\pi/2} \theta\sin 2\theta\, d\theta\right]$$

$$= \frac{a^2\pi^3}{96} - \frac{a^2}{4}\left[\left\{\theta . \left(\frac{-\cos 2\theta}{2}\right)\right\}_0^{\pi/2} - \int_0^{\pi/2} 1 . \frac{-\cos 2\theta}{2}\, d\theta\right]$$

$$= \frac{a^2\pi^3}{96} - \frac{a^2}{4}\left[\frac{\pi}{2}.\frac{1}{2} + \frac{1}{2}\left(\frac{\sin 2\theta}{2}\right)_0^{\pi/2}\right]$$

$$= \frac{\pi^3 a^2}{96} - \frac{\pi a^2}{16} - \frac{1}{16}a^2.(\sin\pi - \sin 0) = \frac{\pi^3 a^2}{96} - \frac{\pi a^2}{16} - 0 = \frac{\pi a^2}{16}\left(\frac{\pi^2}{6} - 1\right).$$

(*iii*) The given curve is $r = ae^{\theta\cot\alpha}$

$$\therefore \quad \text{Required sectorial area} = \int_\beta^{\beta+\gamma} \frac{1}{2} r^2\, d\theta$$

$$= \int_{\beta}^{\beta+\gamma} \frac{1}{2} a^2 e^{2\theta \cot \alpha}\, d\theta = \frac{a^2}{2} \cdot \left| \frac{e^{2\theta \cot \alpha}}{2 \cot \alpha} \right|_{\beta}^{\beta+\gamma}$$

$$= \frac{1}{4} a^2 \tan \alpha \,[e^{2(\beta+\gamma)\cot \alpha} - e^{2\beta \cot \alpha}] = \frac{1}{4} a^2 \tan \alpha . e^{2\beta \cot \alpha} (e^{2\gamma \cot \alpha} - 1).$$

Example 2. *Find the whole area of the*

(*i*) *Circle* $r = 2a \cos \theta$ (*ii*) *Cardioid* $r = a(1 - \cos \theta)$

(*iii*) *Curve* $r^2 = a^2 \cos^2 \theta + b^2 \sin^2 \theta$.

Sol. (*i*) The equation of the given circle is $r = 2a \cos \theta$ ...(1)

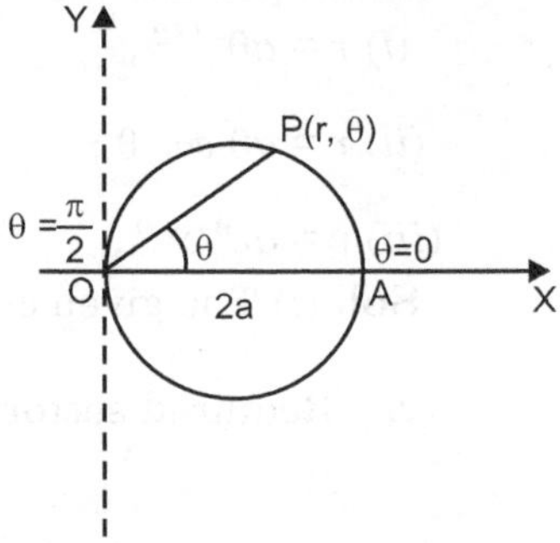

Let OX be the initial line and P(r, θ) be any point on the circle. When $\theta = 0$, $r = 2a$; when $\theta = \pi/2$, $r = 0$.

Since on changing θ to – θ, the equation is unaltered.

∴ The circle is symmetrical about the initial line and for the upper half of the circle, θ varies from 0 to π/2.

$$\therefore \quad \text{Required area} = 2 \int_0^{\pi/2} \frac{1}{2} r^2\, d\theta = 2 \int_0^{\pi/2} (2a \cos \theta)^2\, d\theta$$

$$= 4a^2 \int_0^{\pi/2} \cos^2 \theta\, d\theta = 4a^2 . \frac{1}{2} . \frac{\pi}{2} = \pi a^2.$$

(*ii*) The cardioid is $r = a(1 - \cos \theta)$ ...(1)

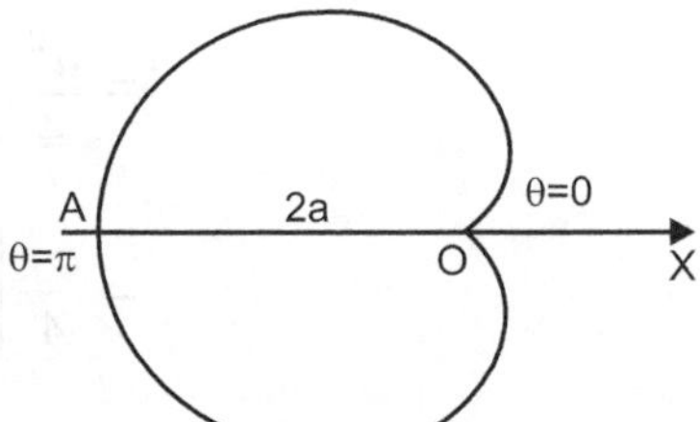

The curve is symmetrical about the initial line (∵ the equation is unchanged on changing θ to – θ). For the upper half of the curve, θ varies from 0 to π.

∴ Required area

$$= 2 \int_0^{\pi} \frac{1}{2} r^2\, d\theta$$

$$= \int_0^{\pi} a^2 (1 - \cos \theta)^2\, d\theta = a^2 \int_0^{\pi} \left(2 \sin^2 \frac{\theta}{2}\right)^2 d\theta$$

$$= 4a^2 \int_0^{\pi} \sin^4 \frac{\theta}{2}\, d\theta$$

$$\left[\text{Put } \frac{\theta}{2} = t \;\therefore\; d\theta = 2dt. \text{ When } \theta = 0, t = 0, \text{ and when } \theta = \pi, t = \frac{\pi}{2}\right]$$

$$= 4a^2 \int_0^{\pi/2} \sin^4 t \,.\, 2\, dt = 8a^2 \int_0^{\pi/2} \sin^4 t\, dt = 8a^2 . \frac{3.1}{4.2} . \frac{\pi}{2} = \frac{3}{2} \pi a^2.$$

(*iii*) The equation of the curve is $r^2 = a^2 \cos^2 \theta + b^2 \sin^2 \theta$ ...(1)

Since on changing θ to – θ, (1) is unaltered, ∴ the curve is symmetrical about the initial line *i.e.,* X-axis. Again on changing θ to π – θ, (1) becomes

$$r^2 = a^2 \cos^2 (\pi - \theta) + b^2 \sin^2 (\pi - \theta)$$

$$= a^2 (-\cos \theta)^2 + b^2 (\sin \theta)^2 = a^2 \cos^2 \theta + b^2 \sin^2 \theta$$

which is the same as (1).

Thus the curve (1) is symmetrical about the line $\theta = \frac{\pi}{2}$ *i.e.,* the Y-axis.

For the curve in the first quadrant, θ varies from 0 to $\frac{\pi}{2}$.

$$\therefore \quad \text{Required area} = 4\int_0^{\pi/2} \frac{1}{2} r^2 \,.\, d\theta = 2\int_0^{\pi/2} (a^2 \cos^2 \theta + b^2 \sin^2 \theta)\, d\theta$$

$$= 2a^2 \,.\, \frac{1}{2}\frac{\pi}{2} + 2b^2 \,.\, \frac{1}{2} \,.\, \frac{\pi}{2} = (a^2 + b^2)\frac{\pi}{2}.$$

Example 3. *Find the area of a loop of the curve $r^2 = a^2 \cos 2\theta$ and hence find its total area.*

Sol. The curve is $r^2 = a^2 \cos 2\theta$ ...(1)

The curve is clearly symmetrical about both the axes [∵ the equation remains unchanged on changing θ to – θ or π – θ].

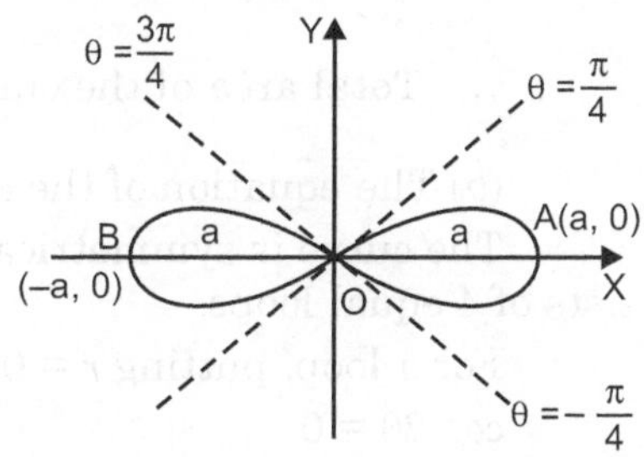

For the loop, putting $r = 0$, we have $\cos 2\theta = 0$.

$$2\theta = \frac{-\pi}{2}, \frac{\pi}{2}$$

[Two consecutive values (**Remember**)]

or $$\theta = \frac{-\pi}{4}, \frac{\pi}{4}.$$

Hence a loop lies between the radii vectors $\theta = \frac{-\pi}{4}$ and $\theta = \frac{\pi}{4}$. The curve consists of two equal loops and for half the loop in first quadrant, θ varies from 0 to $\frac{\pi}{4}$.

$$\therefore \quad \text{Area of a loop} = 2\int_0^{\pi/4} \frac{1}{2} r^2\, d\theta = \int_0^{\pi/4} a^2 \cos 2\theta d\theta = a^2 \,.\left[\frac{\sin 2\theta}{2}\right]_0^{\pi/4} = \frac{a^2}{2}(1 - 0) = \frac{a^2}{2}.$$

$$\therefore \quad \text{Total area of the curve} = 2 \times \frac{a^2}{2} = a^2.$$

Example 4. *Find the area of a loop of the curve*

(*a*) *$r = a \sin 3\theta$ (three leaved rose)*

(*b*) *$r = a \cos 2\theta$ (four leaved rose)*

Find also the total area in each case.

Sol. The equation $r = a \sin n\theta$ or $r = a \cos n\theta$ has n or $2n$ loops according as n is odd or even and for a loop, put $r = 0$ and get two consecutive values of θ.

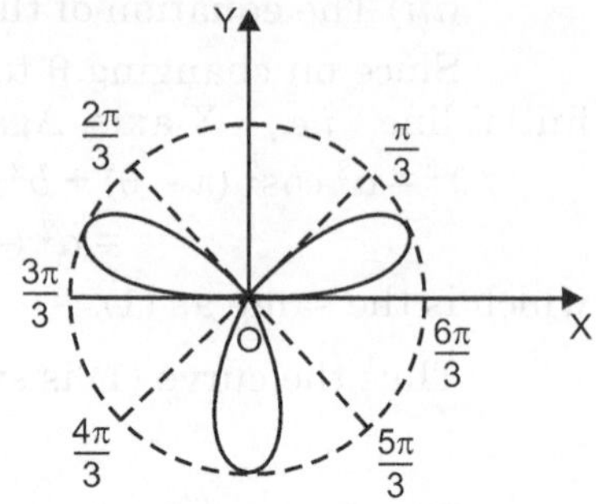

(*a*) The equation of the curve is

$$r = a \sin 3\theta \qquad \ldots(1)$$

For a loop, putting $\theta = 0$, we get $\sin 3\theta = 0$.

$\therefore \quad 3\theta = 0, \pi \qquad$ (*two consecutive values*)

$\therefore \quad \theta = 0, \frac{\pi}{3}$. Hence one loop is obtained when θ varies from 0 to $\frac{\pi}{3}$.

$\therefore$ Area of one loop

$$= \int_0^{\pi/3} \frac{1}{2} r^2 \, d\theta = \frac{1}{2} \int_0^{\pi/3} a^2 \sin^2 3\theta \, d\theta$$

$$= \frac{a^2}{2} \int_0^{\pi/3} \frac{1 - \cos 6\theta}{2} \, d\theta = \frac{a^2}{4} \left[\theta - \frac{\sin 6\theta}{6}\right]_0^{\pi/3} = \frac{a^2}{4}\left[\frac{\pi}{3} + 0 - 0\right] = \frac{\pi a^2}{12}$$

$\therefore$ Total area of the curve = 3 × area of one loop $= 3 \times \frac{\pi a^2}{12} = \frac{1}{4} \pi a^2$.

(*b*) The equation of the curve is $r = a \cos 2\theta \qquad \ldots(2)$

The curve is symmetrical about the initial line and it consists of 4 equal loops.

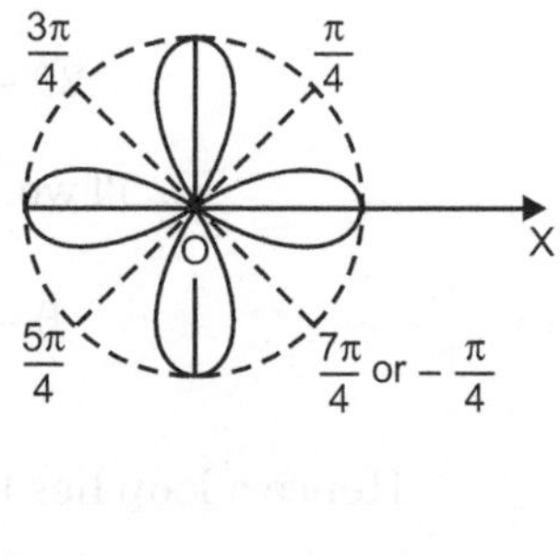

For a loop, putting $r = 0$, we get
$\cos 2\theta = 0$

$\therefore \quad 2\theta = \frac{-\pi}{2}, \frac{\pi}{2} \qquad$ (Two consecutive values)

or $\quad \theta = \frac{-\pi}{4}, \frac{\pi}{4}$

Thus one loop of the curve is obtained as θ varies from $-\frac{\pi}{4}$ to $\frac{\pi}{4}$.

$\therefore$ Area of one loop $= \int_{-\pi/4}^{\pi/4} \frac{1}{2} r^2 \, d\theta = \frac{1}{2} \int_{-\pi/4}^{\pi/4} a^2 \cos^2 2\theta \, d\theta$

$$= \frac{a^2}{2} \cdot \int_{-\pi/4}^{\pi/4} \frac{1 + \cos 4\theta}{2} \, d\theta = \frac{a^2}{4}\left[\theta + \frac{\sin 4\theta}{4}\right]_{-\pi/4}^{\pi/4}$$

$$= \frac{a^2}{4}\left[\frac{\pi}{4} + \frac{1}{4} \sin \pi - \left(\frac{-\pi}{4}\right) - \frac{1}{4} \sin(-\pi)\right] = \frac{a^2}{4} \cdot \left(\frac{\pi}{2} + 0\right) = \frac{\pi a^2}{8}.$$

$\therefore$ Total area of the curve $= 4 \times \frac{\pi a^2}{8} = \frac{\pi a^2}{2}$.

Example 5. *Show that the area of a loop of $r = a \cos n\theta$ is $\frac{\pi a^2}{4n}$, n being integral. Also prove that the whole area is $\frac{\pi a^2}{4}$ or $\frac{\pi a^2}{2}$ according as n is odd or even.*

Sol. The equation of the curve is $r = a \cos n\theta$ ...(1)

For a loop, putting $r = 0$, we get $\cos n\theta = 0$.

$$n\theta = \frac{-\pi}{2}, \frac{\pi}{2} \qquad \text{(Two consecutive values)}$$

or
$$\theta = \frac{-\pi}{2n}, \frac{\pi}{2n}.$$

Thus one loop of the curve is obtained as θ varies from $\frac{-\pi}{2n}$ to $\frac{\pi}{2n}$.

$\therefore$ Area of one loop

$$= \int_{-\pi/2n}^{\pi/2n} \frac{1}{2} r^2 \, d\theta = \frac{1}{2} \int_{-\pi/2n}^{\pi/2n} a^2 \cos^2 n\theta \, d\theta$$

$$= \frac{a^2}{2} \int_{-\pi/2n}^{\pi/2n} \frac{1 + \cos 2n\theta}{2} \, d\theta = \frac{a^2}{4} \left[\theta + \frac{\sin 2n\theta}{2n} \right]_{-\pi/2n}^{\pi/2n}$$

$$= \frac{a^2}{4} \left[\frac{\pi}{2n} - \left(-\frac{\pi}{2n} \right) + \frac{1}{2n} \{\sin \pi - \sin(-\pi)\} \right] = \frac{a^2}{4} \left[\frac{\pi}{n} + \frac{1}{2n}(0) \right] = \frac{\pi a^2}{4n}.$$

Case I. When n is odd, the curve has n loops.

$\therefore$ Total area of the curve $= n \times \frac{\pi a^2}{4n} = \frac{\pi a^2}{4}$.

Case II. When n is even, the curve has $2n$ loops.

$\therefore$ Total area of the curve $= 2n \times \frac{\pi a^2}{4n} = \frac{\pi a^2}{2}$.

Example 6. *Show that the area contained between the circle $r = a$ and the curve $r = a \cos 5\theta$ is equal to three-fourths of the area of the circle.*

Sol. The equation of the curve is $r = a \cos 5\theta$ ...(1)

For one loop of the curve, putting $r = 0$ in (1), we have

$$\cos 5\theta = 0$$

$\therefore$
$$5\theta = -\frac{\pi}{2}, \frac{\pi}{2} \qquad \text{(Two consecutive values)}$$

or
$$\theta = -\frac{\pi}{10}, \frac{\pi}{10}.$$

Thus one loop of the curve is obtained as θ varies from $-\dfrac{\pi}{10}$ or $\dfrac{\pi}{10}$.

∴ Area of one loop

$$= \int_{-\pi/10}^{\pi/10} \frac{1}{2} r^2 \, d\theta = \frac{1}{2} \int_{-\pi/10}^{\pi/10} a^2 \cos^2 5\theta \, d\theta$$

$$= \frac{a^2}{2} \int_{-\pi/10}^{\pi/10} \frac{1 + \cos 10\theta}{2} \, d\theta = \frac{a^2}{4} \left[\theta + \frac{\sin 10\theta}{10} \right]_{-\pi/10}^{\pi/10}$$

$$= \frac{a^2}{4} \left[\frac{\pi}{10} - \left(-\frac{\pi}{10} \right) \right] + \frac{a^2}{40} [\sin \pi - \sin(-\pi)]$$

$$= \frac{a^2}{4} \cdot \frac{\pi}{5} + \frac{a^2}{40} (0) = \frac{\pi a^2}{20}.$$

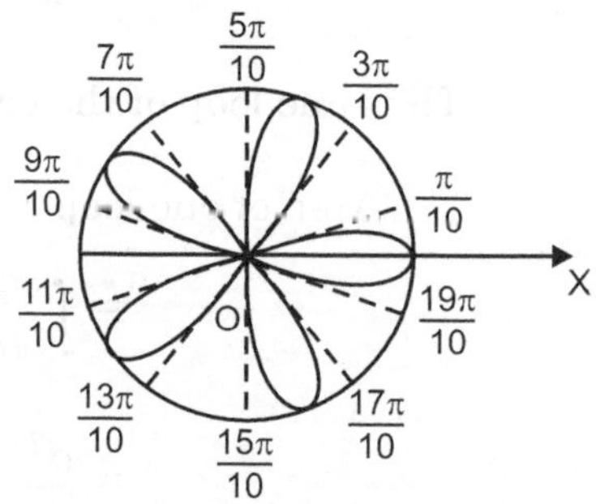

Since this curve consists of five equal loops.

∴ Total area of the curve $= 5 \times \dfrac{\pi a^2}{20} = \dfrac{\pi a^2}{4}$,

Area of the circle $= \pi a^2$.

∴ Area between the circle $r = a$ and the curve (1)

$$= \pi a^2 - \frac{1}{4} \pi a^2 = \frac{3}{4} \pi a^2 = \frac{3}{4} . \text{ area of the circle.}$$

Example 7. *Find the area of a loop of the curve* $r = a \cos 3\theta + b \sin 3\theta$.

Sol. The given curve is $r = a \cos 3\theta + b \sin 3\theta$ ...(1)

Let $a = k \cos \alpha$, $b = k \sin \alpha$,

so that $k = \sqrt{a^2 + b^2}$ and $\alpha = \tan^{-1} \left(\dfrac{b}{a} \right)$.

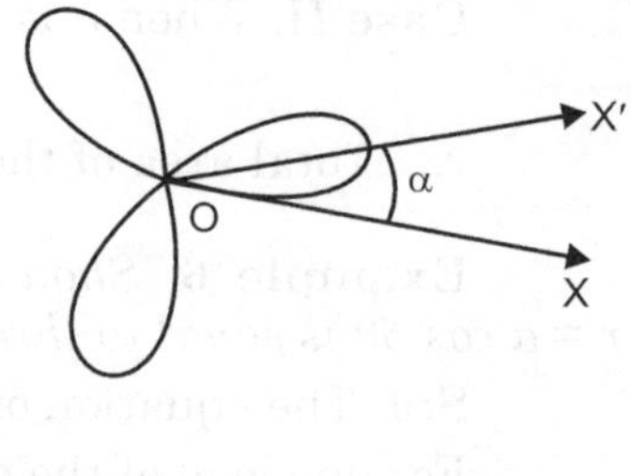

∴ From (1), the given equation of the curve becomes $r = k \cos 3\theta \cos \alpha + k \sin 3\theta \sin \alpha = k \cos (3\theta - \alpha)$.

Turning the initial line through an angle α, this equation becomes

$r = k \cos 3\theta$ ...(2)

For a loop, putting $r = 0$ in (2), we have

$\cos 3\theta = 0 \quad \therefore \quad 3\theta = -\dfrac{\pi}{2}, \dfrac{\pi}{2}$ (Two consecutive values)

or $\theta = -\dfrac{\pi}{6}, \dfrac{\pi}{6}$.

Thus a loop of the curve (2) is obtained as θ varies from $-\dfrac{\pi}{6}$ to $\dfrac{\pi}{6}$.

$\therefore$ Area of one loop of the curve

$$= \int_{-\pi/6}^{\pi/6} \frac{1}{2} r^2 \, d\theta = \frac{1}{2} \int_{-\pi/6}^{\pi/6} k^2 \cos^2 3\theta \, d\theta \qquad | \textit{ Using } (2) \text{ and not } (1)$$

$$= \frac{1}{2} k^2 \int_{-\pi/6}^{\pi/6} \frac{1 + \cos 6\theta}{2} \, d\theta = \frac{k^2}{4} \left[\theta + \frac{\sin 6\theta}{6} \right]_{-\pi/6}^{\pi/6}$$

$$= \frac{1}{4} k^2 \left[\frac{\pi}{6} - \left(-\frac{\pi}{6} \right) \right] + \frac{k^2}{24} [\sin \pi - \sin (-\pi)]$$

$$= \frac{1}{4} k^2 \cdot \frac{\pi}{3} + \frac{k^2}{24} (0) = \frac{k^2 \pi}{12} = \frac{\pi(a^2 + b^2)}{12} \qquad | \because \; k^2 = a^2 + b^2$$

Example 8. (*a*) *Prove that the sum of the areas of the two loops of the limicon* $r = a + b \cos \theta$ $(b > a)$ *is equal to* $\pi \left(a^2 + \frac{1}{2} b^2 \right)$.

(*b*) *Calculate the ratio of the area of the larger to the area of the smaller loop of the curve*

$$r = \frac{1}{2} + \cos 2\theta.$$

Sol. (*a*) The equation of the curve is $r = a + b \cos \theta \; (b > a)$

The curve is symmetrical about the initial line.

Let $\theta = \alpha$ when $r = 0$

$\therefore \quad a + b \cos \alpha = 0$

$\therefore \quad \cos \alpha = -\frac{a}{b},$

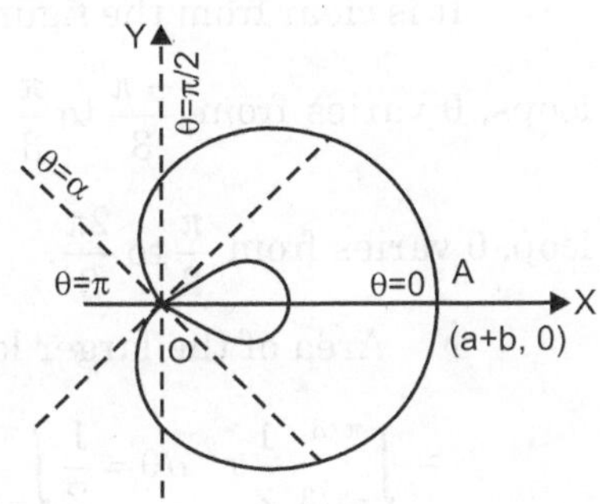

and the line $\theta = \alpha$ is the tangent at the pole.

For the larger loop, θ varies from 0 to α and for the smaller, θ varies from α to π, above the initial line.

$\therefore$ Required sum of areas of the two loops

$$= 2 \left[\int_0^{\alpha} \frac{1}{2} r^2 \, d\theta + \int_{\alpha}^{\pi} \frac{1}{2} r^2 \, d\theta \right]$$

$$= 2 \int_0^{\pi} \frac{1}{2} r^2 \, d\theta \qquad \left| \because \int_a^b f(x) \, dx + \int_b^c f(x) \, dx = \int_a^c f(x) \, dx \right.$$

$$= \int_0^{\pi} (a + b \cos \theta)^2 \, d\theta = \int_0^{\pi} (a^2 + b^2 \cos^2 \theta + 2ab \cos \theta) \, d\theta$$

$$= \int_0^{\pi} \left[a^2 + b^2 \frac{1 + \cos 2\theta}{2} + 2ab \cos \theta \right] d\theta$$

$$= \left[\left(a^2 + \frac{b^2}{2} \right) \theta + \frac{b^2}{4} \sin 2\theta + 2ab \sin \theta \right]_0^{\pi} = \left(a^2 + \frac{b^2}{2} \right) \pi + 0 = \pi \left(a^2 + \frac{b^2}{2} \right).$$

(*b*) The equation of the curve is $r = \frac{1}{2} + \cos 2\theta$.

(*i*) The curve is symmetrical about the initial line and the line $\theta = \frac{\pi}{2}$, *i.e.*, about both the axes.

(*ii*) Putting $r = 0$, $\cos 2\theta = -\frac{1}{2} = \cos \frac{2\pi}{3}$.

$$\therefore \qquad 2\theta = 2n\pi \pm \frac{2\pi}{3} \quad \text{or} \quad \theta = n\pi \pm \frac{\pi}{3}$$

Putting $n = 0, 1$, we get

$$\theta = \pm \frac{\pi}{3} \text{ and } \pi \pm \frac{\pi}{3} \qquad i.e., \quad \theta = -\frac{\pi}{3}, \frac{\pi}{3}, \frac{2\pi}{3}, \frac{4\pi}{3}.$$

(*iii*) The corresponding values of θ and r are given below :

$\theta =$	0	45°	60°	90°	120°	135°	180°
$r =$	$\frac{3}{2}$	$\frac{1}{2}$	0	$-\frac{1}{2}$	0	$\frac{1}{2}$	$\frac{3}{2}$

With the above data, the shape of the curve is roughly as shown in the figure.

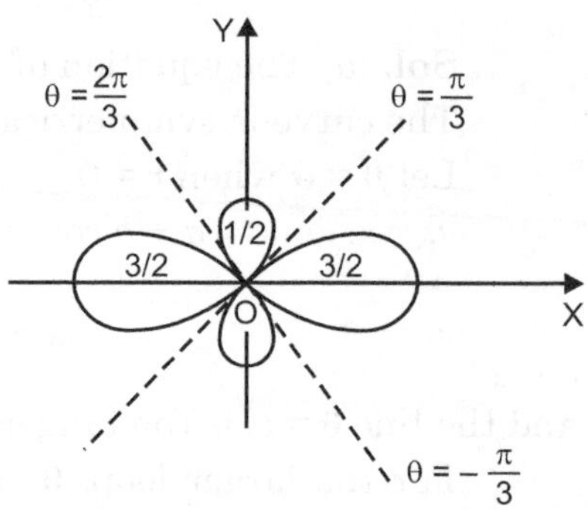

It is clear from the figure, that for one of the larger loops, θ varies from $\frac{-\pi}{3}$ to $\frac{\pi}{3}$ and for one of the smaller loop, θ varies from $\frac{\pi}{3}$ to $\frac{2\pi}{3}$.

$\therefore$ Area of the larger loop

$$= \int_{-\pi/3}^{\pi/3} \frac{1}{2} r^2 \, d\theta = \frac{1}{2} \int_{-\pi/3}^{\pi/3} \left(\frac{1}{2} + \cos 2\theta \right)^2 d\theta$$

$$= \frac{1}{2} \int_{-\pi/3}^{\pi/3} \left(\frac{1}{4} + \cos 2\theta + \cos^2 2\theta \right) d\theta = \frac{1}{2} \int_{-\pi/3}^{\pi/3} \left(\frac{1}{4} + \cos 2\theta + \frac{1 + \cos 4\theta}{2} \right) d\theta$$

$$= \frac{1}{2} \left[\frac{3}{4} \theta + \frac{\sin 2\theta}{2} + \frac{1}{2} \cdot \frac{\sin 4\theta}{4} \right]_{-\pi/3}^{\pi/3}$$

$$= \frac{3}{8} \left[\frac{\pi}{3} - \left(-\frac{\pi}{3} \right) \right] + \frac{1}{4} \left[\sin \frac{2\pi}{3} - \sin \left(-\frac{2\pi}{3} \right) \right] + \frac{1}{16} \left[\sin \frac{4\pi}{3} - \sin \left(-\frac{4\pi}{3} \right) \right]$$

$$= \frac{3}{8} \cdot \frac{2\pi}{3} + \frac{1}{4} \cdot 2 \sin \frac{2\pi}{3} + \frac{1}{16} \cdot 2 \sin \frac{4\pi}{3}$$

$$= \frac{\pi}{4} + \frac{1}{2} \cdot \frac{\sqrt{3}}{2} + \frac{1}{8} \cdot \left(-\frac{\sqrt{3}}{2} \right) = \frac{\pi}{4} + \frac{\sqrt{3}}{4} - \frac{\sqrt{3}}{16} = \frac{1}{16} (4\pi + 3\sqrt{3}).$$

Area of the smaller loop

$$= \frac{1}{2}\int_{\pi/3}^{2\pi/3} \frac{1}{2} r^2 \, d\theta = \frac{1}{2}\int_{\pi/3}^{2\pi/3} \left(\frac{1}{2} + \cos 2\theta\right)^2 d\theta$$

$$= \frac{1}{2}\left[\frac{3}{4}\theta + \frac{\sin 2\theta}{2} + \frac{1}{2}\cdot\frac{\sin 4\theta}{4}\right]_{\pi/3}^{2\pi/3} \qquad | \text{ as above}$$

$$= \frac{3}{8}\left[\frac{2\pi}{3} - \frac{\pi}{3}\right] + \frac{1}{4}\left(\sin\frac{4\pi}{3} - \sin\frac{2\pi}{3}\right) + \frac{1}{16}\left(\sin\frac{8\pi}{3} - \sin\frac{4\pi}{3}\right)$$

$$= \frac{\pi}{8} + \frac{1}{4}\left(-\frac{\sqrt{3}}{2} - \frac{\sqrt{3}}{2}\right) + \frac{1}{16}\left[\frac{\sqrt{3}}{2} - \left(-\frac{\sqrt{3}}{2}\right)\right] = \frac{\pi}{8} - \frac{\sqrt{3}}{4} + \frac{\sqrt{3}}{16} = \frac{1}{16}(2\pi - 3\sqrt{3}).$$

$$\therefore \quad \text{Required ratio} = \frac{\text{area of larger loop}}{\text{area of smaller loop}} = \frac{\frac{1}{16}(4\pi + 3\sqrt{3})}{\frac{1}{16}(2\pi - 3\sqrt{3})} = \frac{4\pi + 3\sqrt{3}}{2\pi - 3\sqrt{3}}.$$

Example 9. *Find the area of the loop of the following curves :*

(*a*) $x^3 + y^3 = 3axy$ (*b*) $(x^2 + y^2)^2 = 4axy^2$

(*c*) $x^4 + y^4 = 4a^2xy$ (*d*) $x^5 + y^5 = 5ax^2y^2$

(*e*) $x^4 + 3x^2y^2 + 2y^4 = a^2xy$.

Sol. (*a*) The given curve is $x^3 + y^3 = 3axy$.

Changing to polars [by putting $x = r\cos\theta$, $y = r\sin\theta$], we have $r^3(\cos^3\theta + \sin^3\theta) = 3ar^2 \sin\theta\cos\theta$

or
$$r = \frac{3a\sin\theta\cos\theta}{\sin^3\theta + \cos^3\theta} \qquad \ldots(1)$$

For a loop, putting r = 0, we get

$\sin\theta\cos\theta = 0$ *i.e.,* $\sin\theta = 0$ and $\cos\theta = 0$

or $\theta = 0$ and $\frac{\pi}{2}$ (*To consecutive values*)

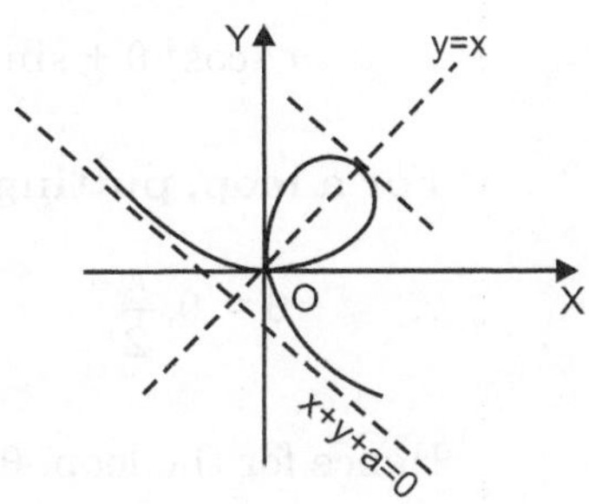

Hence as θ varies from 0 to $\frac{\pi}{2}$, we get the loop.

$$\therefore \qquad \text{Required area} = \int_0^{\pi/2} \frac{1}{2} r^2 \, d\theta$$

$$= \frac{9a^2}{2}\int_0^{\pi/2} \frac{\sin^2\theta\cos^2\theta}{(\sin^3\theta + \cos^3\theta)^2} d\theta$$

[Divide the numerator and denominator by $\cos^6\theta$]

$$= \frac{9a^2}{2}\int_0^{\pi/2} \frac{\tan^2\theta \sec^2\theta\, d\theta}{(1+\tan^3\theta)^2}$$

Put $t = \tan^3\theta$ $\therefore$ $dt = 3\tan^2\theta\sec^2\theta\, d\theta$

when $\theta = 0, t = 0,$ when $\theta \to \frac{\pi}{2}, t \to \infty.$

$$= \frac{9a^2}{2}\int_0^{\infty} \frac{1}{(1+t)^2}\cdot\frac{dt}{3} = \frac{3a^2}{2}\int_0^{\infty}(1+t)^{-2}\, dt$$

$$= \frac{3a^2}{2}\left[\frac{(1+t)^{-1}}{-1}\right]_0^{\infty} = \frac{-3a^2}{2}\left[\frac{1}{1+t}\right]_0^{\infty} = \frac{-3a^2}{2}[0-1] = \frac{3a^2}{2}.$$

(*b*) The equation of the curve is $(x^2 + y^2)^2 = 4a\, xy^2$.

Changing to polars [by putting $x = r\cos\theta, y = r\sin\theta$], we get

$$[r^2(\cos^2\theta + \sin^2\theta)]^2 = 4a\, r^3\cos\theta\sin^2\theta \quad \text{or} \quad r = 4a\cos\theta\sin^2\theta.$$

For a loop, putting r = 0, we get

$$\cos\theta\sin^2\theta = 0 \quad \text{or} \quad \sin\theta = 0 \text{ and } \cos\theta = 0$$

$\therefore$ $\theta = 0, \frac{\pi}{2}$ (Two consecutive values)

Thus the loop is obtained as θ varies from 0 to $\frac{\pi}{2}$

$\therefore$ Required area of loop

$$= \frac{1}{2}\int_0^{\pi/2} r^2\, d\theta = \frac{1}{2}\int_0^{\pi/2} 16a^2\cos^2\theta\sin^4\theta\, d\theta = 8a^2\cdot\frac{1.3.1}{6.4.2}\cdot\frac{\pi}{2} = \frac{\pi a^2}{4}.$$

(*c*) The equation of the curve is $x^4 + y^4 = 4a^2xy$

Changing to polars, [by putting $x = r\cos\theta, y = r\sin\theta$], we get

$$r^4(\cos^4\theta + \sin^4\theta) = 4a^2r^2\sin\theta\cos\theta \quad \text{or} \quad r^2 = \frac{4a^2\sin\theta\cos\theta}{\sin^4\theta + \cos^4\theta}.$$

For a loop, putting r = 0, we get $\sin\theta\cos\theta = 0$ *i.e.*, $\sin\theta = 0$ and $\cos\theta = 0$

$\therefore$ $\theta = 0, \frac{\pi}{2}$ (Two consecutive values)

Hence for the loop, θ varies from 0 to $\frac{\pi}{2}$.

$\therefore$ Required area of the loop $= \int_0^{\pi/2}\frac{1}{2}r^2\, d\theta$

$$= \frac{1}{2}\int_0^{\pi/2}\frac{4a^2\sin\theta\cos\theta}{\sin^4\theta + \cos^4\theta}\, d\theta$$ [Divide the num. and denom. by $\cos^4\theta$]

$$= 2a^2 \int_0^{\pi/2} \frac{\tan\theta \sec^2\theta}{1+\tan^4\theta}\, d\theta$$

Put $\tan^2\theta = t$ $\therefore$ $2\tan\theta\sec^2\theta\, d\theta = dt$
when $\theta = 0, t = 0$; when $\theta \to \frac{\pi}{2}, t \to \infty$

$$= 2a^2 \int_0^{\infty} \frac{\frac{1}{2}dt}{1+t^2} = a^2 \left[\tan^{-1} t\right]_0^{\infty} = a^2\left(\frac{\pi}{2} - 0\right) = \frac{1}{2}\pi a^2.$$

(*d*) The equation of the curve is $x^5 + y^5 = 5ax^2y^2$.
Changing to polars by putting $x = r\cos\theta$, $y = r\sin\theta$, we get

$$r^5(\sin^5\theta + \cos^5\theta) = 5ar^4\sin^2\theta\cos^2\theta \quad \text{or} \quad r = \frac{5a\sin^2\theta\cos^2\theta}{\sin^5\theta+\cos^5\theta}$$

For a loop, putting $r = 0$, we get $\sin^2\theta\cos^2\theta = 0$

or $\sin\theta = 0$ and $\cos\theta = 0$ $\therefore$ $\theta = 0, \frac{\pi}{2}$ (*Two consecutive values*)

Hence for the loop, θ varies from 0 to $\frac{\pi}{2}$.

$\therefore$ Required area of the loop $= \int_0^{\pi/2} \frac{1}{2}r^2\, d\theta$

$$= \frac{25a^2}{2}\int_0^{\pi/2} \frac{\sin^4\theta\cos^4\theta}{(\sin^5\theta+\cos^5\theta)^2}\, d\theta$$

[Divide the num. and denom. by $\cos^{10}\theta$]

$$= \frac{25a^2}{2}\int_0^{\pi/2} \frac{\tan^4\theta\sec^2\theta\, d\theta}{(1+\tan^5\theta)^2}$$

Put $\tan^5\theta = t$ $\therefore$ $5\tan^4\theta\sec^2\theta\, d\theta = dt$
when $\theta = 0, t = 0$; when $\theta \to \frac{\pi}{2}, t \to \infty$

$$= \frac{25a^2}{2}\int_0^{\infty} \frac{1}{(1+t)^2}\cdot\frac{dt}{5} = \frac{5a^2}{2}\cdot\left|-\frac{1}{1+t}\right|_0^{\infty} = \frac{5a^2}{2}(-0+1) = \frac{5}{2}a^2.$$

(*e*) The equation of the curve is $x^4 + 3x^2y^2 + 2y^4 = a^2xy$.
Changing to polars, by putting $x = r\cos\theta$, $y = r\sin\theta$, we get

$$r^4(\cos^4\theta + 3\sin^2\theta\cos^2\theta + 2\sin^4\theta) = a^2r^2\sin\theta\cos\theta$$

or
$$r^2 = \frac{a^2\sin\theta\cos\theta}{\cos^4\theta + 3\sin^2\theta\cos^2\theta + 2\sin^4\theta}$$

For the loop, putting $r = 0$, we get $\sin\theta\cos\theta = 0$

or $\sin\theta = 0$ and $\cos\theta = 0$ $\therefore$ $\theta = 0, \frac{\pi}{2}$ (*Two consecutive values*)

$$\therefore \text{ Required area } = \int_0^{\pi/2}\frac{1}{2}r^2\, d\theta = \frac{a^2}{2}\int_0^{\pi/2}\frac{\sin\theta\cos\theta\, d\theta}{\cos^4\theta + 3\sin^2\theta\cos^2\theta + 2\sin^4\theta}$$

[Divide the num. and denom. by $\cos^4\theta$]

$$= \frac{1}{2} a^2 \int_0^{\pi/2} \frac{\tan\theta \sec^2\theta \, d\theta}{1 + 3\tan^2\theta + 2\tan^4\theta}$$

| Put $\tan^2\theta = t$ $\therefore$ $2\tan\theta\sec^2\theta \, d\theta = dt$
when $\theta = 0, t = 0$; when $\theta \to \frac{\pi}{2}, t \to \infty$

$$= \frac{a^2}{2} \int_0^\infty \frac{\frac{1}{2} dt}{1 + 3t + 2t^2} = \frac{a^2}{2} \int_0^\infty \frac{1}{(1+t)(1+2t)} dt$$

$$= \frac{a^2}{4} \int_0^\infty \left[\frac{2}{1+2t} - \frac{1}{1+t}\right] dt = \frac{a^2}{4} [\log(1+2t) - \log(1+t)]_0^\infty$$

$$= \frac{a^2}{4} \left[\log \frac{1+2t}{1+t}\right]_0^\infty = \frac{a^2}{4} \left[\underset{t\to\infty}{\text{Lt}} \log \frac{1+2t}{1+t} - \log \frac{1+0}{1-0}\right]$$

$$= \frac{a^2}{4} . \underset{t\to\infty}{\text{Lt}} \log \left[\frac{2 + \frac{1}{t}}{1 + \frac{1}{t}}\right] - 0 = \frac{a^2}{4} . \log 2.$$

Example 10. *If s be the length of the curve $r = a \tanh \frac{\theta}{2}$ between the origin and $\theta = 2\pi$ and A be the area under the curve between the same two points, prove that $A = a(s - a\pi)$.*

Sol. The equation of the curve is $r = a \tanh \frac{\theta}{2}$ ...(1)

$$\frac{dr}{d\theta} = \frac{1}{2} a \operatorname{sech}^2 \frac{\theta}{2}$$

$$\therefore \quad r^2 + \left(\frac{dr}{d\theta}\right)^2 = a^2 \tanh^2 \frac{\theta}{2} + \frac{1}{4} a^2 \operatorname{sech}^4 \frac{\theta}{2}$$

$$= \frac{a^2}{4}\left[4\tanh^2 \frac{\theta}{2} + \operatorname{sech}^4 \frac{\theta}{2}\right] = \frac{a^2}{4}\left[\operatorname{sech}^4 \frac{\theta}{2} + 4\left(1 - \operatorname{sech}^2 \frac{\theta}{2}\right)\right]$$

| $\because$ $\operatorname{sech}^2 x + \tanh^2 x = 1$

$$= \frac{a^2}{4}\left[\operatorname{sech}^4 \frac{\theta}{2} - 4\operatorname{sech}^2 \frac{\theta}{2} + 4\right] = \frac{a^2}{4}\left(2 - \operatorname{sech}^2 \frac{\theta}{2}\right)^2$$

$$\therefore \quad s = \int_0^{2\pi} \sqrt{r^2 + \left(\frac{dr}{d\theta}\right)^2} \, d\theta = \int_0^{2\pi} \frac{a}{2}\left(2 - \operatorname{sech}^2 \frac{\theta}{2}\right) d\theta$$

$$= \frac{a}{2}\left[2\theta - 2\tanh\frac{\theta}{2}\right]_0^{2\pi} = \frac{a}{2}[4\pi - 2\tanh\pi - 0 - 2\tanh 0]$$

$$= 2a\pi - a\tanh\pi$$

| $\because$ $\tanh 0 = 0$

or $\quad s - a\pi = a\pi - a \tanh \pi \qquad \ldots(2)$

Also $\quad A = \int_0^{2\pi} \frac{1}{2} r^2 \, d\theta = \int_0^{2\pi} \frac{1}{2} . a^2 \tanh^2 \frac{\theta}{2} \, d\theta$

$$= \frac{a^2}{2} \int_0^{2\pi} \left(1 - \text{sech}^2 \frac{\theta}{2}\right) d\theta \qquad | \because \ \text{sech}^2 x + \tanh^2 x = 1$$

$$= \frac{a^2}{2} \left[\theta - 2 \tanh \frac{\theta}{2}\right]_0^{2\pi} = \frac{a^2}{2} [2\pi - 2 \tanh \pi - 0]$$

$$= a[a\pi - a \tanh \pi] = a(s - a\pi) \qquad | \textit{ Using } (2)$$

Hence the result.

10.6. AREA BETWEEN TWO POLAR CURVES

Prove that the area bounded by the curves r = f(θ) and r = F(θ) and the radii vectors θ = α, θ = β is $\displaystyle\int_\alpha^\beta \frac{1}{2} (r_1^2 - r_2^2) \, d\theta$

where r_1 is the 'r' of the outer curve and r_2 the 'r' of the inner curve.

Let AB, CD be the two given curves $r = f(\theta)$ and $r = F(\theta)$ and OCA, ODB the radii vectors $\theta = \alpha, \theta = \beta$ respectively.

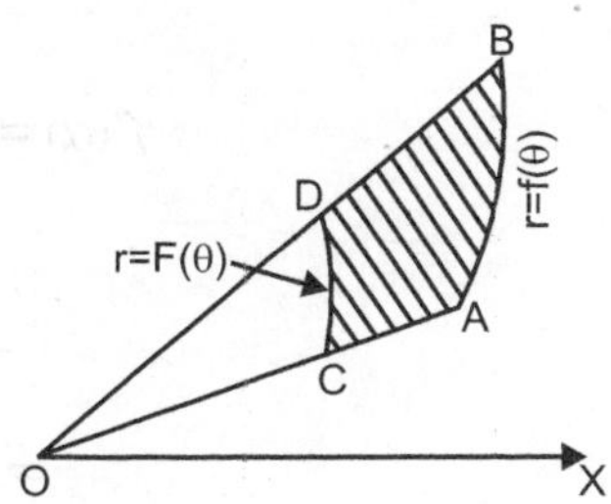

Then the area ABCD

$$= \text{area OAB} - \text{area OCD}$$

$$= \frac{1}{2} \int_\alpha^\beta [f(\theta)]^2 \, d\theta - \frac{1}{2} \int_\alpha^\beta [F(\theta)]^2 \, d\theta$$

$$= \frac{1}{2} \int_\alpha^\beta \{[f(\theta)]^2 - [F(\theta)]^2\} \, d\theta = \frac{1}{2} \int_\alpha^\beta (r_1^2 - r_2^2) \, d\theta$$

where $\quad r_1 \, [= f(\theta)]$ is the value of 'r' of the outer curve

and $\quad r_2 \, [= F(\theta)]$ is the value of 'r' of the inner curve.

ILLUSTRATIVE EXAMPLES

Example 1. *Prove that the area between the cissoid* $r = a \dfrac{\sin^2 \theta}{\cos \theta}$ *and its asymptote is* $\dfrac{3\pi a^2}{4}$.

Sol. The equation of the curve is $r = a \dfrac{\sin^2 \theta}{\cos \theta} \qquad \ldots(1)$

(*i*) Since the equation is unaltered on changing θ to – θ, the curve is symmetrical about the initial line.

(*ii*) Putting $r = 0$, we get $\dfrac{a \sin^2 \theta}{\cos \theta} = 0 \quad \therefore \quad \sin \theta = 0 \quad \text{or} \quad \theta = 0.$

Thus the curve passes through the pole and the tangents at the pole are given by

$$\sin^2\theta = 0 \quad \text{or} \quad \theta = 0,\ \theta = 0.$$

Thus the origin is a cusp and $\theta = 0$ (the initial line) is the cuspidal tangent.

(*iii*) As $\theta \to \frac{\pi}{2}, r \to \infty$. $\therefore$ The curve has an asymptote which is determined as below :

$$\frac{1}{r} = \frac{\cos\theta}{a\sin^2\theta} = f(\theta) \text{ say.}$$

$\therefore$ $f(\theta) = 0$ implies $\cos\theta = 0$ or $\theta = (2n+1)\frac{\pi}{2} = \alpha$ (*say*)

Now $f'(\theta) \frac{1}{a} \cdot \frac{\sin^2\theta(-\sin\theta) - \cos\theta(2\sin\theta\cos\theta)}{\sin^4\theta} = \frac{1}{a} \cdot \frac{-(\sin^2\theta + 2\cos^2\theta)}{\sin^3\theta}$

$$= -\frac{1+\cos^2\theta}{a\sin^3\theta}$$

$$f'(\alpha) = f'\left(\overline{2n+1}\frac{\pi}{2}\right) = -\frac{1+\cos^2(2n+1)\frac{\pi}{2}}{a\sin^3(2n+1)\frac{\pi}{2}} = \frac{1+\cos^2\left(n\pi+\frac{\pi}{2}\right)}{a\sin^3\left(n\pi+\frac{\pi}{2}\right)}$$

$$= -\frac{1+(-1)^{2n}\cos^2\frac{\pi}{2}}{a(-1)^{3n}\sin^3\frac{\pi}{2}} \qquad \left|\ \because \begin{aligned}\sin(n\pi+\theta) &= (-1)^n\sin\theta\\ \cos(n\pi+\theta) &= (-1)^n\cos\theta\end{aligned}\right.$$

$$= -\frac{1+0}{a(-1)^n} = -\frac{1}{a(-1)^n} \qquad \left|\ \because \cos\frac{\pi}{2} = 0, \sin\frac{\pi}{2} = 1, (-1)^{2n} = 1\right.$$

$\therefore$ The asymptotes are given by

$$r\sin(\theta-\alpha) = \frac{1}{f'(\alpha)}$$

or $$r\sin\left(\theta - \overline{2n+1}\frac{\pi}{2}\right) = \frac{(-1)^n a}{-1}$$

or $$-r\sin\left(n\pi + \frac{\pi}{2} - \theta\right) = -(-1)^n a$$

or $$-r.(-1)^n\sin\left(\frac{\pi}{2} - \theta\right) = -(-1)^n a$$

or $$r\cos\theta = a.$$

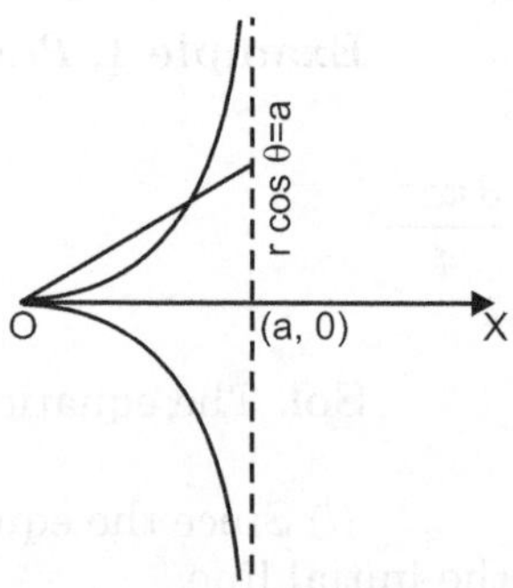

Hence there is only one asymptote $r\cos\theta = a$ *i.e.*, $x = a$.

(*iv*) When $\theta = 0$, $r = 0$. As θ increases from 0 to $\frac{\pi}{2}, r$ increases from 0 to ∞. As θ varies from $\frac{\pi}{2}$ to π, r is negative and increases from $-\infty$ to 0.

Thus the shape of the curve is as shown in the figure.

The required area between the curve and the asymptote

$$= 2\int_0^{\pi/2} \frac{1}{2}(r_1^2 - r_2^2)\, d\theta = \int_0^{\pi/2} \left(a^2 \sec^2 \theta - a^2 \frac{\sin^4 \theta}{\cos^2 \theta}\right) d\theta$$

$$= \int_0^{\pi/2} a^2 \sec^2 \theta\,(1-\sin^4 \theta)\, d\theta = a^2 \int_0^{\pi/2} \sec^2 \theta\,(1-\sin^2 \theta)\,(1+\sin^2 \theta)\, d\theta$$

$$= a^2 \int_0^{\pi/2} \sec^2 \theta \,.\, \cos^2 \theta\,(1 + \sin^2 \theta)\, d\theta$$

$$= a^2 \int_0^{\pi/2} (1 + \sin^2 \theta)\, d\theta = a^2 \int_0^{\pi/2} d\theta + a^2 \int_0^{\pi/2} \sin^2 \theta\, d\theta$$

$$= a^2 \left[\theta\right]_0^{\pi/2} + a^2 \,.\, \frac{1}{2} \,.\, \frac{\pi}{2} = a^2 \,.\, \frac{\pi}{2} + a^2 \,.\, \frac{\pi}{4} = \pi a^2 \left(\frac{1}{2} + \frac{1}{4}\right) = \frac{3}{4}\,\pi a^2.$$

Example 2. *Find the area between the curve $r = a(\sec \theta + \cos \theta)$ and its asymptote.*

Sol. The equation of the curve is $r = a\,(\sec \theta + \cos \theta)$. ...(1)

To trace the curve roughly to find the limits of integration :

(*i*) The curve is symmetrical about the initial line.

(*ii*) Putting $r = 0$, we get $\sec \theta + \cos \theta = 0$ or $\cos^2 \theta + 1 = 0$ or $\cos^2 \theta = -1$ which makes $\cos \theta$ imaginary. Thus the curve does not pass through the pole.

(*iii*) From (1), $r = \dfrac{a(1 + \cos^2 \theta)}{\cos \theta}$. As $\theta \to \frac{\pi}{2}, r \to \infty$.

Thus the curve has an asymptote which is determined as below :

$$\frac{1}{r} = \frac{\cos \theta}{a(1 + \cos^2 \theta)} = f(\theta),\ (say)$$

$$f(\theta) = 0 \text{ gives } \cos \theta = 0 \quad \therefore \quad \theta = (2n + 1)\frac{\pi}{2} = \alpha\ (say).$$

$$f'(\theta) = \frac{1}{a} \,.\, \frac{(1 + \cos^2 \theta)(-\sin \theta) - 2\cos^2 \theta\,(-\sin \theta)}{(1 + \cos^2 \theta)^2}$$

$$= \frac{-\sin \theta\,(1 - \cos^2 \theta)}{a(1 + \cos^2 \theta)^2} = -\frac{\sin^3 \theta}{a(1 + \cos^2 \theta)^2}.$$

$$\therefore \qquad f'(\alpha) = f'\left(\overline{2n+1}\,\frac{\pi}{2}\right) = -\frac{\sin^3\left(n\pi\,\frac{\pi}{2}\right)}{a\left[1+\cos^2\left(n\pi+\frac{\pi}{2}\right)\right]^2}$$

$$= -\frac{(-1)^{3n}\sin^3\frac{\pi}{2}}{a\left[1+(-1)^{2n}\cos^2\frac{\pi}{2}\right]^2} = -\frac{(-1)^n\,.1}{a(1+0)^2} = \frac{(-1)^{n+1}}{a}.$$

Thus the asymptote is given by $\qquad r\sin(\theta-\alpha) = \dfrac{1}{f'(\alpha)}$

or $\qquad r\sin\left(\theta-\overline{2n+1}\,\frac{\pi}{2}\right) = \dfrac{a}{(-1)^{n+1}} \quad$ or $\quad -r\sin\left(n\pi+\frac{\pi}{2}-\theta\right) = \dfrac{a}{(-1)^{n+1}}$

or $\qquad -r\,.(-1)^n\sin\left(\frac{\pi}{2}-\theta\right) = \dfrac{a}{(-1)^{n+1}} \quad$ or $\quad (-1)^{n+1}\,r\cos\theta = \dfrac{a}{(-1)^{n+1}}$

or $\qquad (-1)^{2n+2}\,r\cos\theta = a$

or $\qquad r\cos\theta = a.$

Thus the curve has only one asymptote $r\cos\theta = a$ or $r = a\sec\theta$.

(iv) When $\theta = 0$, $r = 2a$. As θ increases from 0 to $\frac{\pi}{2}$, r increases from $2a$ to ∞ and as θ increases from $\frac{\pi}{2}$ to π, r increases from $-\infty$ to $2a$.

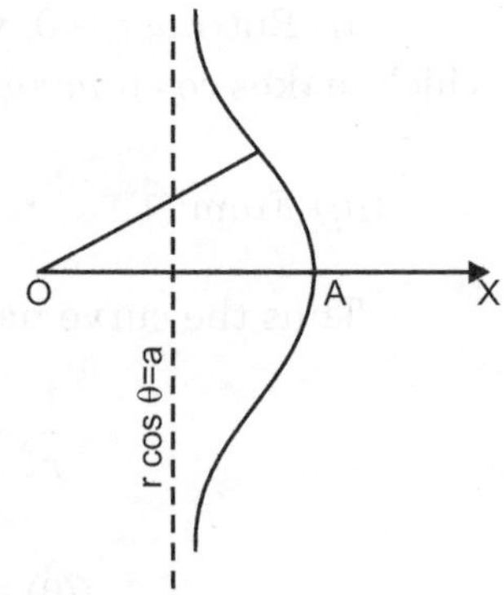

Thus the shape of the curve is roughly as shown in the figure.

$\therefore$ Required area between the curve and the asymptote

$$= 2\int_0^{\pi/2}\frac{1}{2}(r_1^2 - r_2^2)\,d\theta$$

$$= \int_0^{\pi/2}[a^2(\sec\theta+\cos\theta)^2 - a^2\sec^2\theta]\,d\theta$$

$$= a^2.\int_0^{\pi/2}(2\sec\theta\cos\theta+\cos^2\theta)\,d\theta = a^2\int_0^{\pi/2}(2+\cos^2\theta)\,d\theta$$

$\qquad\qquad | \because \sec\theta\cos\theta = 1$

$$= 2a^2\int_0^{\pi/2}d\theta + a^2\int_0^{\pi/2}\cos^2\theta\,d\theta = 2a^2\,.\,[\theta]_0^{\pi/2} + a^2\,\frac{1}{2}\cdot\frac{\pi}{2}$$

$$= 2a^2\left(\frac{\pi}{2}-0\right) + a^2\,.\,\frac{\pi}{4} = \pi a^2 + \frac{\pi a^2}{4} = \frac{5}{4}\,\pi a^2.$$

Example 3. (*a*) *Prove that the area common to the circles* $r = a\sqrt{2}$ *and* $r = 2a \cos \theta$ *is* $a^2 (\pi - 1)$.

(*b*) *Find the area common to the circle* $r = a$ *and the cardioid* $r = a (1 + \cos \theta)$.

(*c*) *Show that the area of the region included between the cardioids*

$$r = a (1 + \cos \theta) \text{ and } r = a(1 - \cos \theta) \text{ is } \frac{a^2}{2}(3\pi - 8).$$

Sol. (*a*) The equations of the circle are

$$r = a\sqrt{2} \qquad ...(1)$$

and

$$r = 2a \cos \theta \qquad ...(2)$$

Solving the two equations (eliminating r), we get

$$a\sqrt{2} = 2a \cos \theta \quad \text{or} \quad \cos \theta = \frac{1}{\sqrt{2}}$$

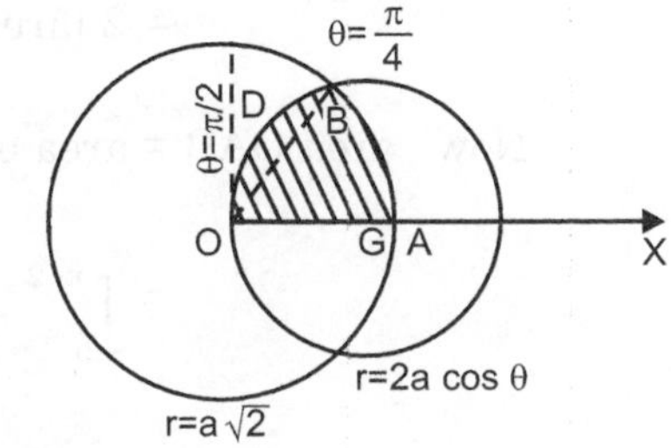

$$\therefore \qquad \theta = \pm \frac{\pi}{4}.$$

$\therefore$ for the point of intersection B, $\theta = \frac{\pi}{4}$.

Now required area common to two circles

$$= 2 \text{ area OABD} = 2 \text{ [area OAB + area OBDO]} \qquad ...(3)$$

Now area OAB = area of circle (1) bounded by $\theta = 0$; $\theta = \frac{\pi}{4}$

$$= \int_0^{\pi/4} \frac{1}{2} r^2 \, d\theta = \int_0^{\pi/4} \frac{1}{2} . 2a^2 \, d\theta = a^2 \, [\theta]_0^{\pi/4} = \frac{\pi a^2}{4}$$

and area OBDO = area of circle (2) bounded by $\theta = \frac{\pi}{4}, \theta = \frac{\pi}{4}$

$$= \int_{\pi/4}^{\pi/2} \frac{1}{2} r^2 \, d\theta = \int_{\pi/4}^{\pi/2} \frac{1}{2} . 4a^2 \cos^2 \theta \, d\theta = a^2 \int_{\pi/4}^{\pi/2} (1 + \cos 2\theta) \, d\theta$$

$$= a^2 \left[\theta + \frac{\sin 2\theta}{2}\right]_{\pi/4}^{\pi/2} = a^2 . \left(\frac{\pi}{2} - \frac{\pi}{4}\right) + \frac{a^2}{2}\left(\sin \pi - \sin \frac{\pi}{2}\right)$$

$$= a^2 . \frac{\pi}{4} + \frac{a^2}{2}(0 - 1) = \frac{a^2}{2}\left(\frac{\pi}{2} - 1\right)$$

$\therefore$ From (3), required area common to two circles

$$= 2\left[\frac{\pi a^2}{4} + \frac{a^2}{2}\left(\frac{\pi}{2} - 1\right)\right] = 2\left(\frac{\pi a^2}{2} - \frac{a^2}{2}\right) = a^2(\pi - 1)$$

(*b*) The circle is $r = a$ ...(1)

and the cardioid is $r = a(1 + \cos\theta)$ ...(2)

Solving the two equations (eliminating r) to find the points of intersection, we get

$$a = a(1 + \cos\theta) \text{ or } \cos\theta = 0$$

$$\therefore \quad \theta = \pm\frac{\pi}{2}$$

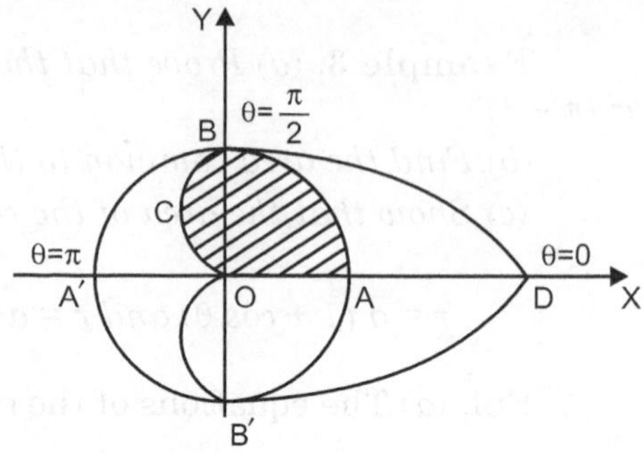

$\therefore$ For the point of intersection B, $\theta = \dfrac{\pi}{2}$

$\therefore$ Required area common to two curves (1) and (2)

= 2 × shaded area OABCO

= 2 [area OAB + area OBCO) ...(3)

Now area OAB = area of the circle (1) bounded by $\theta = 0, \theta = \dfrac{\pi}{2}$

$$= \int_0^{\pi/2} \frac{1}{2} r^2\, d\theta = \int_0^{\pi/2} \frac{1}{2} a^2\, d\theta = \frac{a^2}{2}[\theta]_0^{\pi/2} = \frac{a^2\pi}{4}.$$

and area OBCO = area of cardioid (2) bounded by the radii vectors $\theta = \dfrac{\pi}{2}, \theta = \pi$

$$= \int_{\pi/2}^{\pi} \frac{1}{2} r^2\, d\theta = \frac{1}{2}\int_{\pi/2}^{\pi} a^2(1+\cos\theta)^2\, d\theta = \frac{a^2}{2}\int_{\pi/2}^{\pi} (1+2\cos\theta+\cos^2\theta)d\theta$$

$$= \frac{a^2}{2}\int_{\pi/2}^{\pi}\left(1+2\cos\theta+\frac{1+\cos 2\theta}{2}\right)d\theta = \frac{a^2}{2}\int_{\pi/2}^{\pi}\left(\frac{3}{2}+2\cos\theta+\frac{1}{2}\cos 2\theta\right)d\theta$$

$$= \frac{a^2}{2}\left[\frac{3}{2}\theta + 2\sin\theta + \frac{1}{2}\cdot\frac{\sin 2\theta}{2}\right]_{\pi/2}^{\pi}$$

$$= \frac{3a^2}{4}\left(\pi - \frac{\pi}{2}\right) + a^2\left(\sin\pi - \sin\frac{\pi}{2}\right) + \frac{a^2}{8}(\sin 2\pi - \sin\pi)$$

$$= \frac{3a^2}{4}\cdot\frac{\pi}{2} + a^2(0-1) + \frac{a^2}{8}(0) = \frac{a^2}{8}(3\pi - 8).$$

$\therefore$ From (3), the required area

$$= 2\left[\frac{\pi a^2}{4} + \frac{a^2}{8}(3\pi - 8)\right] = a^2\left(\frac{\pi}{2} + \frac{3\pi}{4} - 2\right) = a^2\left(\frac{5\pi}{4} - 2\right).$$

(*c*) The equations of the two cardioids are

$r = a(1 + \cos\theta)$ ...(1)

and $r = a(1 - \cos\theta)$ ...(2)

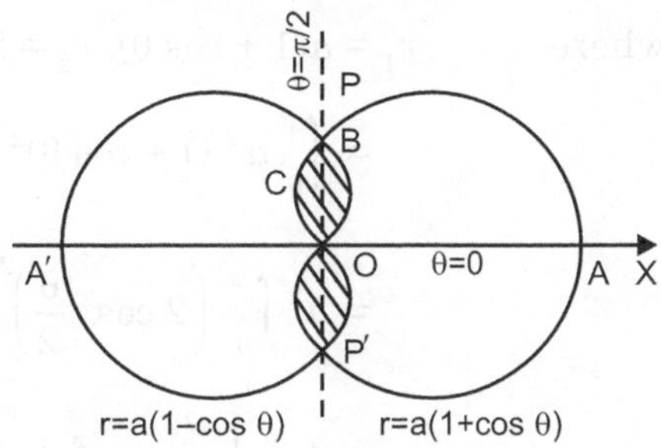

Solving (1) and (2) to find the points of intersection, (eliminating r), we get

$$a(1 + \cos\theta) = a(1 - \cos\theta)$$

or $$2\cos\theta = 0$$

$$\therefore \quad \cos\theta = 0 \quad \text{or} \quad \theta = \pm\frac{\pi}{2}.$$

$\therefore$ For the point of intersection P, $\theta = \dfrac{\pi}{2}$.

Since both the cardioids are symmetrical about the initial line,

$\therefore$ By symmetry

Required area common between two cardioids = 2 area OBPCO = 4 area OBPO

where area OBPO is the area of the curve (2) between the radii vectors $\theta = 0$, $\theta = \dfrac{\pi}{2}$

$$\therefore \quad \text{Required area} = 4\int_0^{\pi/2} \frac{1}{2} r^2\, d\theta = 2\int_0^{\pi/2} a^2(1 - \cos\theta)^2\, d\theta$$

$$= 2a^2 \int_0^{\pi/2} (1 - 2\cos\theta + \cos^2\theta)\, d\theta$$

$$= 2a^2\,[\theta - 2\sin\theta]_0^{\pi/2} + 2a^2 \,.\int_0^{\pi/2} \cos^2\theta\, d\theta = 2a^2\left[\frac{\pi}{2} - 2\right] + 2a^2 \,.\, \frac{1}{2} \,.\, \frac{\pi}{2}$$

$$= 2a^2\left(\frac{\pi}{2} - 2 + \frac{\pi}{4}\right) = 2a^2\left(\frac{3\pi}{4} - 2\right) = \frac{a^2}{2}(3\pi - 8).$$

Example 4. (*a*) *Find the area outside the circle $r = 2a\cos\theta$ and inside the cardioid $r = a(1 + \cos\theta)$.*

(*b*) *Find the area inside the circle $r = \sin\theta$ and outside the cardioid $r = 1 - \cos\theta$.*

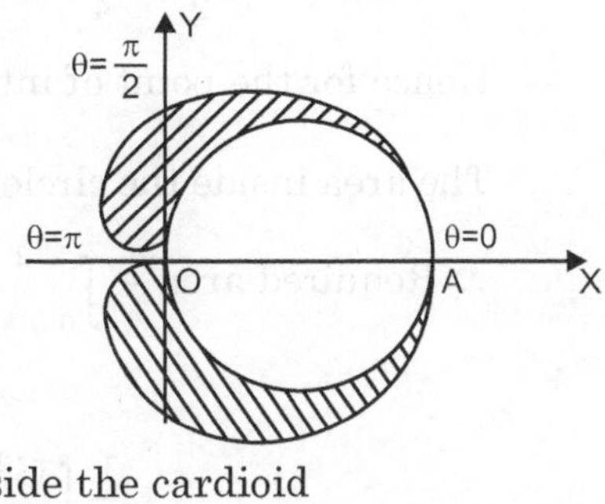

Sol. (*a*) The equation of the circle is

$$r = 2a\cos\theta \qquad \ldots(1)$$

and that of the cardioid is

$$r = a\,(1 + \cos\theta) \qquad \ldots(2)$$

$\because$ Both the curves are symmetrical about the initial line.

$\therefore$ The required shaded area outside the circle and inside the cardioid

= 2 × shaded area above the initial line

= 2 [area of half cardioid – area of half the circle]

$$= 2\left[\int_0^{\pi} \frac{1}{2} r_1^2\, d\theta - \int_0^{\pi/2} \frac{1}{2} r_2^2\, d\theta\right]$$

where $r_1 = a(1 + \cos\theta),\ r_2 = 2a\cos\theta$

$$= \int_0^{\pi} a^2 (1 + \cos\theta)^2\, d\theta - \int_0^{\pi/2} 4a^2 \cos^2\theta\, d\theta$$

$$= a^2 \int_0^{\pi} \left(2\cos^2\frac{\theta}{2}\right)^2 d\theta - 4a^2 . \int_0^{\pi/2} \cos^2\theta\, d\theta = 4a^2 \int_0^{\pi} \cos^4\frac{\theta}{2} d\theta - 4a^2 . \frac{1}{2}.\frac{\pi}{2}$$

$$= 4a^2 \int_0^{\pi/2} \cos^4 t \,.\, 2\, dt - \pi a^2, \text{ where } t = \frac{\theta}{2}$$

$$= 8a^2 . \frac{3.1}{4.2}\frac{\pi}{2} - \pi a^2 = \frac{3\pi}{2} a^2 - \pi a^2 = \frac{1}{2}\pi a^2.$$

(*b*) The equation of the circle is $r = \sin\theta$ ...(1)

and that of the cardioid is $r = 1 - \cos\theta$ ...(2)

To find the points of intersection, solving (1) and (2) (eliminating r), we get

$$1 - \cos\theta = \sin\theta$$

or $$2\sin^2\frac{\theta}{2} = 2\sin\frac{\theta}{2}\cos\frac{\theta}{2} \quad \text{or} \quad \sin\frac{\theta}{2}\left(\sin\frac{\theta}{2} - \cos\frac{\theta}{2}\right) = 0$$

$\therefore$ either $\sin\frac{\theta}{2} = 0$ which gives $\theta = 0$

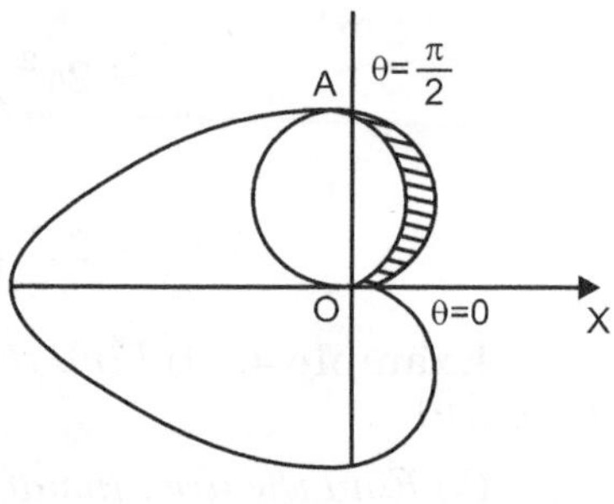

or $$\sin\frac{\theta}{2} - \cos\frac{\theta}{2} = 0$$

i.e., $$\sin\frac{\theta}{2} = \cos\frac{\theta}{2} \quad \text{or} \quad \tan\frac{\theta}{2} = 1$$

$$\frac{\theta}{2} = \frac{\pi}{4} \quad \therefore \quad \theta = \frac{\pi}{2}$$

Hence for the point of intersection O, $\theta = 0$ and for the point of intersection A, $\theta = \frac{\pi}{2}$.

The area inside the circle (1) and outside the cardioid (2) is shown shaded in the figure.

$\therefore$ Required area $= \int_0^{\pi/2} \frac{1}{2}(r_1^2 - r_2^2)\, d\theta$ where $r_1 = \sin\theta,\ r_2 = 1 - \cos\theta$

| From (1) and (2)

$$= \frac{1}{2}\int_0^{\pi/2} [\sin^2\theta - (1 - \cos\theta)^2] d\theta = \frac{1}{2}\int_0^{\pi/2} [\sin^2\theta - (1 + \cos^2\theta - 2\cos\theta)] d\theta$$

$$= \frac{1}{2}\int_0^{\pi/2} (\sin^2\theta - \cos^2\theta + 2\cos\theta - 1)\, d\theta$$

$$= \frac{1}{2}\int_0^{\pi/2} \sin^2\theta\, d\theta - \frac{1}{2}\int_0^{\pi/2} \cos^2\theta\, d\theta + \int_0^{\pi/2} \cos\theta\, d\theta - \frac{1}{2}\int_0^{\pi/2} d\theta$$

$$= \frac{1}{2}\cdot\frac{1}{2}\cdot\frac{\pi}{2} - \frac{1}{2}\cdot\frac{1}{2}\cdot\frac{\pi}{2} + \Big[\sin\theta\Big]_0^{\pi/2} - \frac{1}{2}\Big[\theta\Big]_0^{\pi/2}$$

$$= \frac{\pi}{8}\cdot\frac{\pi}{8} + (1-0) - \frac{1}{2}\left(\frac{\pi}{2} - 0\right) = 1 - \frac{\pi}{4}.$$

Example 5. (*a*) *Find the area common to the ellipses* $x^2 + 2y^2 = a^2$ *and* $2x^2 + y^2 = a^2$.
(*b*) *Show that the area common to the ellipses* $a^2x^2 + b^2y^2 = 1$, $b^2x^2 + a^2y^2 = 1$,

where $0 < a < b$ *is* $\frac{4}{ab}\tan^{-1}\left(\frac{a}{b}\right)$.

Sol. (*a*) The equation of the first ellipse is $x^2 + 2y^2 = a^2$ ...(1)

or $$\frac{x^2}{a^2} + \frac{y^2}{\frac{a^2}{2}} = 1.$$

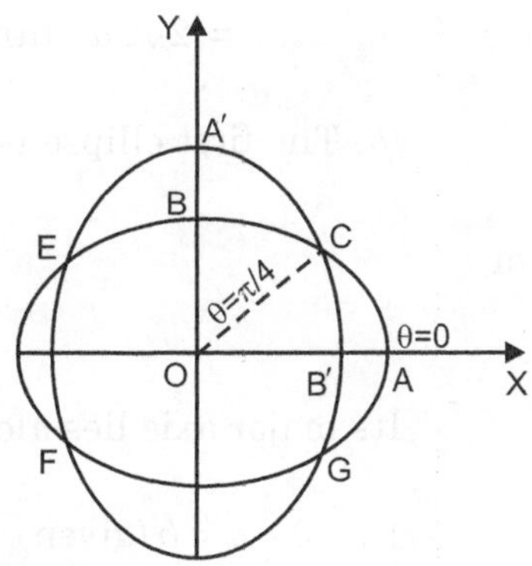

Its major axis lies along X-axis and minor axis along Y-axis.

The equation of the second ellipse is

$$2x^2 + y^2 = a^2 \quad ...(2)$$

or $$\frac{x^2}{\frac{a^2}{2}} + \frac{y^2}{a^2} = 1.$$

Its major axis lies along Y-axis and minor axis along the X-axis.

Changing to polars (by putting $x = r\cos\theta$, $y = r\sin\theta$), the equation (1) and (2) become

$$r^2(\cos^2\theta + 2\sin^2\theta) = a^2 \quad\text{or}\quad r^2 = \frac{a^2}{\cos^2\theta + 2\sin^2\theta} \quad ...(3)$$

and $$r^2(\sin^2\theta + 2\cos^2\theta) = a^2 \quad\text{or}\quad r^2 = \frac{a^2}{2\cos^2\theta + \sin^2\theta} \quad ...(4)$$

To find the points of intersection, solving (3) and (4), we have

$$\frac{a^2}{\cos^2\theta + 2\sin^2\theta} = \frac{a^2}{2\cos^2\theta + 2\sin^2\theta} \quad | \text{ Eliminating } r^2$$

or $$\cos^2\theta + 2\sin^2\theta = 2\cos^2\theta + \sin^2\theta \quad\text{or}\quad \sin^2\theta = \cos^2\theta \quad\text{or}\quad \tan^2\theta = 1$$

$$\therefore \quad \tan\theta = \pm 1 = \tan\left(\pm\frac{\pi}{4}\right)$$

$\therefore \quad \theta = \pm\frac{\pi}{4}$. Hence for the point of intersection C, $\theta = \frac{\pi}{4}$.

The area OB′CB is symmetrical about the line OC ($y = x$)

[$\because$ on changing x to y to x, (1) becomes $y^2 + 2x^2 = a^2$, which is (2)]

$\therefore$ Required common area = 4 area OB′CB

$$= 4 \times 2 \text{ area OB′C} = 8 \int_0^{\pi/4} \frac{1}{2} r^2 d\theta = 4 \int_0^{\pi/4} \frac{a^2}{2\cos^2\theta + \sin^2\theta}$$ | From (4)

Divide the num. and den. by $\cos^2\theta$

$$= 4a^2 \int_0^{\pi/4} \frac{\sec^2\theta d\theta}{2 + \tan^2\theta}$$

Put $\tan\theta = t$ $\therefore$ $\sec^2\theta d\theta = dt$

When $\theta = 0, t = 0$; when $\theta = \frac{\pi}{4}, t = 1$

$$= 4a^2 \int_0^1 \frac{dt}{t^2 + 2} = 4a^2 . \frac{1}{\sqrt{2}} \left[\tan^{-1} \frac{1}{\sqrt{2}} \right]_0^1 = 2\sqrt{2}\, a^2 \left[\tan^{-1} \frac{1}{\sqrt{2}} - 0 \right]$$

$$= 2\sqrt{2}\, a^2 \tan^{-1}\left(\frac{1}{\sqrt{2}}\right).$$

(*b*) The first ellipse is $a^2x^2 + b^2y^2 = 1$

or $$\frac{x^2}{\left(\frac{1}{a^2}\right)} + \frac{y^2}{\left(\frac{1}{b^2}\right)} = 1$$

Its major axis lies along X-axis and minor axis along Y-axis.

$\left(\because \ 0 < a < b \text{ (given)}, \therefore \ \frac{1}{a} > \frac{1}{b}\right)$

The second ellipse is $b^2x^2 + a^2y^2 = 1$ ...(2)

or $$\frac{x^2}{\left(\frac{1}{b^2}\right)} + \frac{y^2}{\left(\frac{1}{a^2}\right)} = 1$$ [See figure with part (*a*)]

$\because \ \frac{1}{a} > \frac{1}{b}$, $\therefore$ its major axis lies along Y-axis and minor axis along X-axis.

Changing to polars (by putting $x = r\cos\theta$, $y = r\sin\theta$), we get

$$r^2(a^2\cos^2\theta + b^2\sin^2\theta) = 1 \quad \text{or} \quad r^2 = \frac{1}{a^2\cos^2\theta + b^2\sin^2\theta} \qquad ...(3)$$

and $$r^2(b^2\cos^2\theta + a^2\sin^2\theta) = 1 \quad \text{or} \quad r^2 = \frac{1}{b^2\cos^2\theta + a^2\sin^2\theta} \qquad ...(4)$$

To find the points of intersection of (3) and (4), eliminating r^2, we get

$$a^2\cos^2\theta + b^2\sin^2\theta = b^2\cos^2\theta + a^2\sin^2\theta$$

or $$(b^2 - a^2)\sin^2\theta = (b^2 - a^2)\cos^2\theta \text{ or } \sin^2\theta = \cos^2\theta \text{ or } \tan^2\theta = 1$$

or $\quad$ at C, $\tan\theta = 1 \quad \therefore \quad \theta = \dfrac{\pi}{4}$.

Since on interchanging x and y in (1), we get (2),

$\therefore$ the area OB′CB is symmetrical about the line OC $\left(y = x \text{ or } \theta = \dfrac{\pi}{4} \right)$

$\therefore$ Required common are = 4, area OB′CB

$$= 4 \times 2 \text{ area OB′C} = 8 \int_0^{\pi/4} \frac{1}{2} r^2 \, d\theta$$

$$= 4 \int_0^{\pi/4} \frac{1}{b^2 \cos^2\theta + a^2 \sin^2\theta} \, d\theta \qquad \text{| From (4)}$$

Divide the numerator and denominator by $\cos^2\theta$

$$= 4 \int_0^{\pi/4} \frac{\sec^2\theta}{b^2 + a^2 \tan^2\theta} \, d\theta$$

| Put $\tan\theta = t \quad \therefore \quad \sec^2\theta \, d\theta = dt$
| When $\theta = 0 \quad t = 0$
| When $\theta = \pi/4, t = 1$

$$= 4 \int_0^1 \frac{dt}{b^2 + a^2 t^2} = \frac{4}{a^2} \int_0^1 \frac{1}{t^2 + \dfrac{b^2}{a^2}} \, dt$$

$$= \frac{4}{a^2} \cdot \frac{1}{\left(\dfrac{b}{a}\right)} \left[\tan^{-1} \left(\frac{t}{b/a} \right) \right]_0^1$$

$$= \frac{4}{ab} \left[\tan^{-1} \frac{a}{b} - \tan^{-1} 0 \right] = \frac{4}{ab} \left[\tan^{-1} \frac{a}{b} - 0 \right]$$

$$= \frac{4}{ab} \tan^{-1} \frac{a}{b}.$$

TEST YOUR KNOWLEDGE

1. Find the area enclosed by the curve $y^2 = x$ and the straight line $x = 4$.
2. Find the area bounded by the parabola $y^2 = 4ax$ and its latus rectum.

   [**Hint**. Reqd. area $= 2\int_0^a \sqrt{4ax}\, dx$, since equation of latus rectum is $x = a$]

3. Find the whole area of the circle $x^2 + y^2 = a^2$.
4. Find the area of the loop of the curve $ay^2 = x^2 (a - x)$.
5. Find the whole area of the curve $a^2x^2 = y^3(2a - y)$.
6. Find the area of a loop of the curve $a^4y^2 = x^4(a^2 - x^2)$.
7. Find the area between the curve $y^2(4 - x) = x^2$ and its asymptote.

8. Find the area of the portion bounded by the curve $y^2(a - x) = x^2(a + x)$ and its asymptote.
9. Find the area included between the curve $x^2 y = 4a^2(2a - y)$ and its asymptote.
10. Find the area of the loop of the curve $y^2 = x^4 (a + x)$.
11. Show that the area of the loop of the curve $ay^2 = (x - a)(x - 5a)^2$ is $\frac{256}{15} a^2$.
12. Find the area of the loop of the curve $3ay^2 = x(x - a)^2$.
13. Find the area enclosed by the curve $xy^2 = a^2(a - x)$ and y-axis.
14. Find the area of a loop of the curve $y^2 = x^2 (4 - x^2)$.
15. In the ellipse $\frac{x^2}{a^2} + \frac{y^2}{b^2} = 1$, prove that $x = a \cos \frac{2S}{ab}$, $y = b \sin \frac{2S}{ab}$
    where S is the sectorial area bounded by the ellipse, x-axis and the line joining (0, 0) to (x, y).
16. Find the area included between the parabolas $y^2 = 4ax$ and $x^2 = 4by$.
17. Show that the area enclosed between the parabolas

$$y^2 = 4a(x + a) \text{ and } y^2 = - 4a(x - a) \text{ is } \frac{16a^2}{3}.$$

18. Find the area of the region bounded by the parabolas $y = x^2$ and $y = 4 - x^2$.
19. Find the area lying between the parabola $y = 4x - x^2$ and the line $y = x$.
20. Find the area of the region enclosed by the circle $x^2 + y^2 = a^2$ and the line $x + y = a$ is the first quadrant.
21. Find the area cut off from the parabola $4y = 3x^2$ by the line $2y = 3x + 12$.
22. Find the area of the region bounded by the curve $y = 2\sqrt{1 - x^2}$ and the x-axis.

    $\left[\textbf{Hint.} \text{ The curve is upper half of the ellipse } \frac{x^2}{1} + \frac{y^2}{4} = 1\right]$
23. Using integration, find the area of the region bounded by the triangle whose vertices are (–1, 1), (0, 5) and (3, 2).
24. Find the area of the region :
    (*i*) $\{(x, y) : x^2 \le y \le x\}$ (*ii*) $\{(x, y) : x^2 \le y \le |x|\}$
    (*iii*) $\{(x, y) : x^2 + y^2 \le 1 \le x + y\}$ (*iv*) $\{(x, y) : y^2 \le 4x, 4x^2 + 4y^2 \le 9\}$.
25. Find the area of the region enclosed between the two circle $x^2 + y^2 = 1$ and $(x - 1)^2 + y^2 = 1$.
26. Find the area of the circle $(x - 2)^2 + (y + 1)^2 = 16$.

    [**Hint.** Shift the origin to the point (2, – 1)]
27. Find the area of the astroid $x^{2/3} + y^{2/3} = a^{2/3}$ or $x = a \cos^3 t$, $y = a \sin^3 t$.
28. Find the area of the cardioid $r = a(1 + \cos \theta)$.
29. Find the area of a loop of the curve $r = a \sin 2\theta$. Also find the total area.
30. Find the area of a loop of the curve $r^2 = a^2 \sin 2\theta$.
31. Find the area of a loop of the curve $r = \sqrt{3} \cos 3\theta + \sin 3\theta$.
32. Find the area of the loop of the curve $x^6 + y^6 = a^2x^2y^2$.
33. Prove that the area of the loop of the curve
    (*i*) $x^3 + y^3 = 3axy$ is three times the area of a loop of the curve $r^2 = a^2 \cos 2\theta$.
    (*ii*) $x^5 + y^5 = 5ax^2y^2$ is five times the area of a loop of the curve $r^2 = a^2 \cos 2\theta$.

34. If P be any point on the ellipse $\frac{x^2}{a^2} + \frac{y^2}{b^2} = 1$, and S be the sectorical area bounded by the curve, x-axis and the line joining the origin to P, show that $x = a \cos \frac{2S}{ab}$, $y = b \sin \frac{2S}{ab}$.

Answers

1. $\frac{32}{3}$ **2.** $\frac{8}{3}a^2$ **3.** πa^2 **4.** $\frac{8}{15}a^2$

5. πa^2 **6.** $\frac{\pi a^2}{16}$ **7.** $\frac{64}{3}$ **8.** $\frac{a^2}{2}(\pi + 4)$

9. $4\pi a^2$ **10.** $\frac{32}{105}a^{7/2}$ **12.** $\frac{8a^2}{15\sqrt{3}}$ **13.** πa^2

14. $\frac{16}{3}$ **16.** $\frac{16ab}{3}$ **18.** $\frac{16\sqrt{2}}{3}$ **19.** $\frac{9}{2}$

20. $\frac{a^2}{4}(\pi - 2)$ **21.** 27 **22.** π **23.** $\frac{15}{2}$

24. *(i)* $\frac{1}{6}$ *(ii)* $\frac{1}{3}$ *(iii)* $\frac{\pi - 2}{4}$

(iv) $\frac{9\pi}{8} - \frac{9}{4}\sin^{-1}\frac{1}{3} + \frac{\sqrt{2}}{6}$ **25.** $\frac{2\pi}{3} - \frac{\sqrt{3}}{2}$ **26.** 16π

27. $\frac{3}{8}\pi a^2$ **28.** $\frac{3}{2}\pi a^2$ **29.** $\frac{1}{8}\pi a^2, \frac{1}{2}\pi a^2$ **30.** $\frac{1}{2}a^2$

31. $\frac{\pi}{3}$ **32.** $\frac{\pi a^2}{12}$.

11

Volumes of Solids of Revolution

11.1. DEFINITIONS

If a plane area is revolved about a fixed straight line lying in its own plane, then the body so generated by the plane area is called the **solid of revolution** and the surface generated by the boundary of the plane area is called the **surface of revolution.** The fixed line about which the plane area rotates is called the **axis of revolution.**

For example ; a right angled triangle when revolved about one of its sides (forming the rt. angle) generates a right circular cone.

Remember. The section of a solid of revolution by a plane perpendicular to the axis of revolution is a circle, having it centre on the axis of revolution.

VOLUME FORMULAE FOR CARTESIAN EQUATIONS

11.2. Prove that the volume of the solid generated by the revolution about the x-axis of the area bounded by the curve y = f(x), the x-axis and the ordinates x = a, x = b is

$$\int_a^b \pi y^2 dx$$

where y = f(x) is a finite, continuous and single-valued function of x in the interval $a \le x \le b$.

Let AB be the curve $y = f(x)$ and CA, DB be the ordinates $x = a$, $x = b$ respectively.

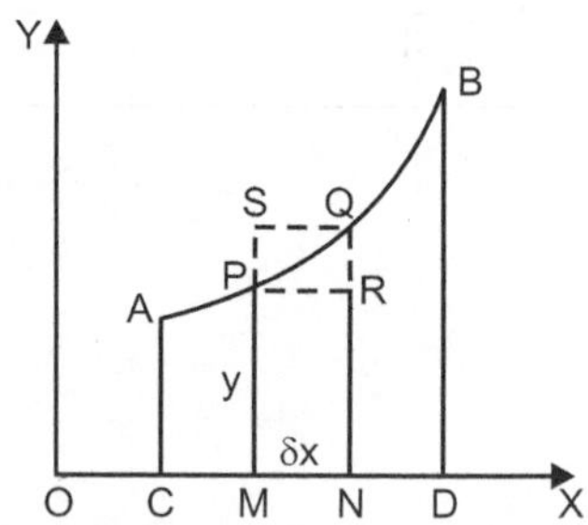

Let P(x, y) be **any point** on the curve AB.

Draw PM ⊥ OX.

∴ OM = x and PM = y.

Let V denote the volume of the solid generated by the revolution about X-axis of the area ACMP.

As x increases, *i.e.,* MP moves towards the right, V also increases.

Thus, the volume V depends on x and is therefore a function of x.

Let Q($x + \delta x$, $y + \delta y$) be a point on the curve in the neighbourhood of P. Then the volume of the solid generated by the revolution about x-axis of the area ACNQ will be V + δV, so that the volume of the solid generated by the revolution about x-axis of the area PMNQ is δV.

Complete the rectangle PRQS.

Then the volume of the solid generated by the revolution about the x-axis of the area PMNQ lies between the right circular cylinders generated by the rectangles PMNR and SMNQ *i.e.,*

δV lies between $\pi y^2\, \delta x$ and $\pi(y + \delta y)^2\, \delta x$

or $$\frac{\delta V}{\delta x} \text{ lies between } \pi y^2 \text{ and } \pi(y + \delta y)^2$$

In the limiting position as $Q \to P$, $\delta x \to 0$ and $\therefore \quad \delta y \to 0$

$$\therefore \quad \underset{\delta x \to 0}{\text{Lt}} \frac{\delta V}{\delta x} \text{ lies between } \pi y^2 \text{ and } \underset{\delta y \to 0}{\text{Lt}} \pi(y + \delta y)^2$$

or $$\frac{\delta V}{dx} \text{ lies between } \pi y^2 \text{ and a quantity which approaches } \pi y^2.$$

$$\therefore \quad \frac{dV}{dx} = \pi y^2$$

$$\therefore \quad \int_a^b \pi y^2 \, dx = \int_a^b \frac{dV}{dx} dx \left[V \right]_a^b$$

$$= (\text{volume V when } x = b) - (\text{volume V when } x = a)$$

$$= \text{volume generated by the area ACDB} - 0.$$

Hence the volume of the solid generated by the area ACDB about the x-axis is $\int_a^b \pi \mathbf{y}^2 \, \mathbf{dx}$.

Cor. Revolution about y-axis. *The volume of the solid generated by revolution about the y-axis of the area bounded by the curve $x = f(y)$, the y-axis and the abscissae $y = a$, $y = b$ is* $\int_a^b \pi \mathbf{x}^2 \, \mathbf{dy}$.

The result follows immediately on interchanging x and y in the above proposition.

An important advice. [Remember] :

(*i*) If the generating curve is **symmetrical about the x-axis** and it is required to find the volume generated by the *revolution of the area about the x-axis,* in such a case we shall revolve only **one** of the symmetrical portions, because the part of the curve above the x-axis generates the same volume as the part of the curve below the x-axis when revolved about the x-axis. Thus the volume generated by the revolution of the area DCOAB about OX = volume generated by the area OABM or OCDM.

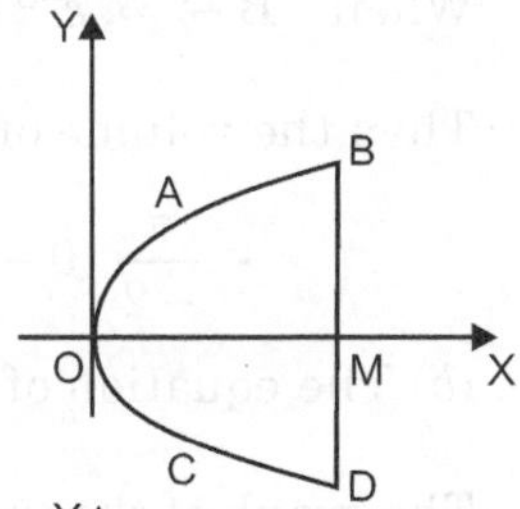

(*ii*) If the curve is *symmetrical about y-axis* and the same curve be made to revolve *about the x-axis* (the curve lying only on one side of the x-axis), then the volume generated = 2 × volume generated by the portion OAB lying to the right of y-axis.

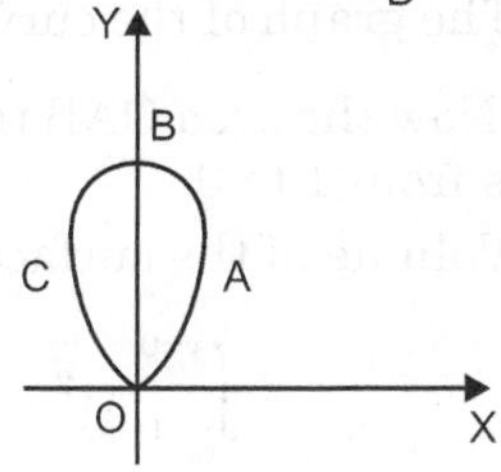

ILLUSTRATIVE EXAMPLES

Example 1. *(a) Let B be a number > 1. What is the volume of solid generated by the area under the curve $y = e^{-x}$ between 1 and B (the axis of revolution being the x-axis) ? Does the volume approach a limit as B becomes large ? If so, what limit ?*

(b) Find the volume generated by rotating about the y-axis the area bounded by the co-ordinate axes and the graph of the curve $y = \cos x$ from $x = 0$ to $x = \frac{\pi}{2}$.

(c) The hyperbola $\frac{x^2}{a^2} - \frac{y^2}{b^2} = 1$ revolves about the axis of x. Show that the volume cut off from one of the two solids thus obtained by a plane perpendicular to x-axis, and distant h from the vertex, is

$$\frac{\pi b^2 h^2 (3a + h)}{3a^2}.$$

Sol. *(a)* The equation of the curve is $y = e^{-x}$.

$\therefore$ Required volume of solid of revolution of the curve about x-axis.

$$= \int_1^B \pi y^2 \, dx = \pi \int_1^B e^{-2x} dx = \pi . \left[\frac{e^{-2x}}{-2}\right]_1^B = \frac{\pi}{-2} [e^{-2B} - e^{-2}] \quad \ldots(1)$$

When $B \to \infty, e^{-B} = \frac{1}{e^B} \to \frac{1}{\infty} = 0.$

Then the volume of the solid in the limit when B is very large (*i.e.*, $B \to \infty$)

$$= \frac{\pi}{-2} (0 - e^{-2}) = \frac{\pi}{2e^2} \quad | \text{ From (1)}$$

(b) The equation of the curve is $y = \cos x$ ...(1)

The graph of the curve from $x = 0$ to $x = \frac{\pi}{2}$ is as shown in the figure. Now the area OAB revolves about y-axis. For the area OAB, y varies from 1 to 0.

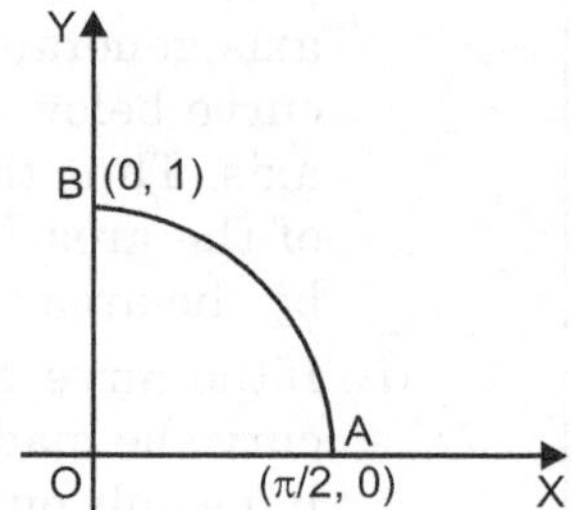

Volume of the surface generated by revolution about y-axis.

$$= \int_{y=1}^{y=0} \pi x^2 \, dy = \int_{x=0}^{x=\pi/2} \pi x^2 \frac{dy}{dx} dx \quad | \textbf{ Note this step}$$

$$\left[\because \text{ when } y = 1, x = 0 \text{ and when } y = 0, x = \frac{\pi}{2}\right] \quad | \text{ from (1)}$$

$$= \int_0^{\pi/2} \pi x^2 (-\sin x) \, dx \quad \Big| \because \text{ from (1), } \frac{dy}{dx} = -\sin x$$

$$= -\pi \int_0^{\pi/2} \underset{\text{I}}{x^2} \underset{\text{II}}{\sin x} \, dx \quad | \text{ Integrate by parts}$$

$$= -\pi\left[\left|\, x^2 . (-\cos x)\right|_0^{\pi/2} - \int_0^{\pi/2} 2x(-\cos x)\, dx\right]$$

$$= -\pi\left[0 + 2\int_0^{\pi/2} \underset{\text{I}}{x}\, \underset{\text{II}}{\cos x}\, dx\right] \qquad \text{| Integrate again by parts}$$

$$= -2\pi\left[\left|\, x . \sin x\right|_0^{\pi/2} - \int_0^{\pi/2} 1 . \sin x\, dx\right]$$

$$= -2\pi\left[\left(\frac{\pi}{2} - 0\right) + \left|\cos x\right|_0^{\pi/2}\right] = -2\pi\left(\frac{\pi}{2} - 0 - 1\right) = -2\pi\left(\frac{\pi}{2} - 1\right)$$

$$= 2\pi\left(\frac{\pi}{2} - 1\right) = \pi(\pi - 2) \qquad \text{Numerically}$$

$[\because\ \pi > 2]$

(c) The given hyperbola is $\dfrac{x^2}{a^2} - \dfrac{y^2}{b^2} = 1$ ...(1)

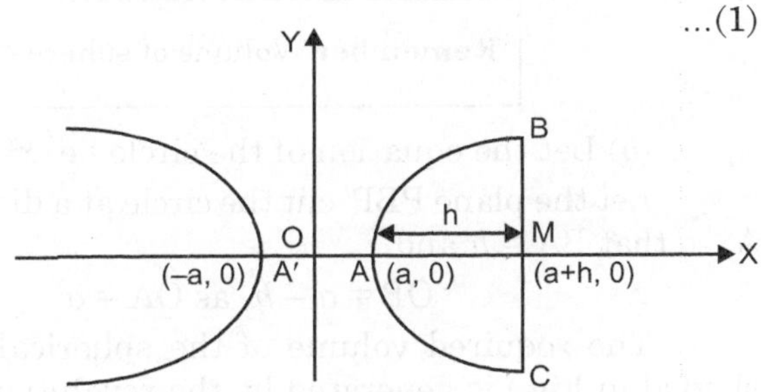

Volume of the solid cut off by a plane ⊥ to x-axis and at a distance h from the vertex x is the same as the volume obtained by rotating the portion of the curve from $x = a$ to $x = a + h$ about the x-axis.

$\therefore$ Required volume $= \displaystyle\int_a^{a+h} \pi y^2\, dx$

$$= \pi\int_h^{a+h} \frac{b^2}{a^2}(x^2 - a^2)\, dx \qquad \left|\ \because \text{ From (1), } y^2 = \frac{b^2}{a^2}(x^2 - a^2)\right.$$

$$= \pi\frac{b^2}{a^2}\left[\frac{x^3}{3} - a^2 x\right]_a^{a+h} = \pi\frac{b^2}{a^2}\left[\frac{(a+h)^3 - (a^3)}{3} - a^2(a + h - a)\right]$$

$$= \pi\frac{b^2}{a^2}\left[\frac{1}{3}(a^3 + 3a^2h + 3ah^2 + h^3 - a^3) - a^2h\right] = \pi\,\frac{b^2}{a^2}\left[\frac{3ah^2 + h^3}{3}\right] = \frac{\pi b^2 h^2 (3a + h)}{3a^2}.$$

Example 2. (a) *Find the volume of a sphere of radius a.*

Or

The circle $x^2 + y^2 = a^2$ is revolved about x-axis. Find the volume of the sphere so formed.
(b) *Find the volume of a spherical cap of height h cut off from a sphere of radius a.*

(c) *A segment is cut off from a sphere of radius a by a plane at a distance $\dfrac{a}{2}$ from the centre. Show that the volume of the segment is $\dfrac{5}{32}$ of the volume of the sphere.*

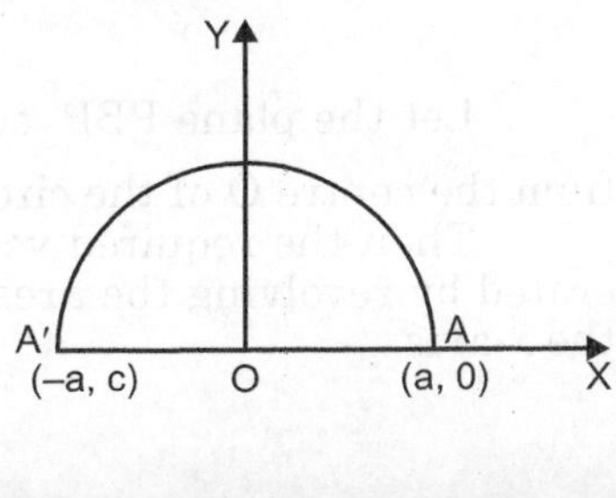

Sol. (a) The sphere is the solid of revolution generated by the revolution of a semi-circular area about its bounding diameter.

Let the equation of the circle of radius a be

$$x^2 + y^2 = a^2 \qquad \ldots(1)$$

whose centre is O.

Now for the semi-circle above the x-axis, x varies from $-a$ to a.

$\therefore$ Required volume of the sphere

$$= \int_{-a}^{a} \pi y^2 \, dx = \pi \int_{-a}^{a} (a^2 - x^2) \, dx \qquad | \text{ From (1)}$$

$$= \pi \left[a^2 x - \frac{x^3}{3} \right]_{-a}^{a} = \pi \left[a^3 - \frac{a^3}{3} - \left(-a^3 + \frac{a^3}{3} \right) \right]$$

$$= \pi \left(2a^3 - 2\frac{a^3}{3} \right) = \frac{4}{3} \pi a^3.$$

Remember. Volume of sphere $= \frac{4}{3} \pi \text{ (radius)}^3$

(*b*) Let the equation of the circle be $x^2 + y^2 = a^2 \qquad \ldots(1)$

Let the plane PBP′ cut the circle at a distance h from A, so that BA $= h$ and

$\therefore$ OB $= a - h$, as OA $= a$.

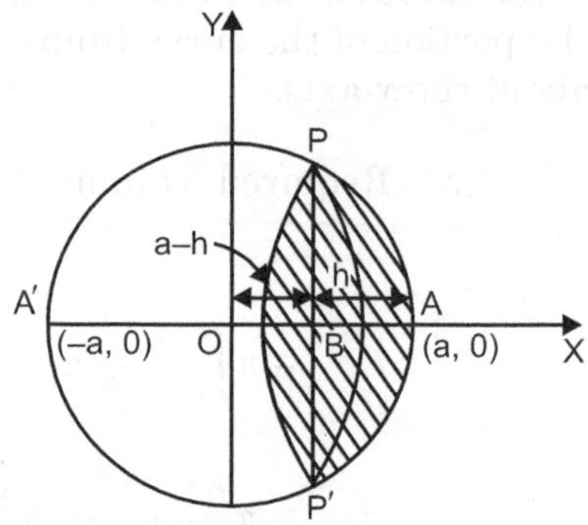

The required volume of the spherical cap (shown shaded in Fig.) is generated by the revolution of the area PBA of the circle (1) about the x-axis.

For the portion of the curve PA, x varies from OB to OA *i.e.*, from $a - h$ to a.

$\therefore$ Required volume of spherical cap

$$= \int_{a-h}^{a} \pi y^2 \, dx = \int_{a-h}^{a} \pi(a^2 - x^2) \, dx \qquad | \text{ From (1)}$$

$$= \pi \left[a^2 x - \frac{x^3}{3} \right]_{a-h}^{a} = \pi a^2 [a - (a - h)] - \frac{\pi}{3} [a^3 - (a - h)^3]$$

$$= \pi a^2 h - \frac{\pi}{3} [3a^2 h - 3ah^2 + h^3]$$

$$= \pi \left(ah^2 - \frac{h^3}{3} \right) = \pi h^2 \cdot \left(a - \frac{h}{3} \right).$$

(*c*) Let the equation of the circle be $x^2 + y^2 = a^2 \qquad \ldots(1)$

Let the plane PBP′ cut the circle at a distance $\frac{a}{2}$ from the centre O of the circle.

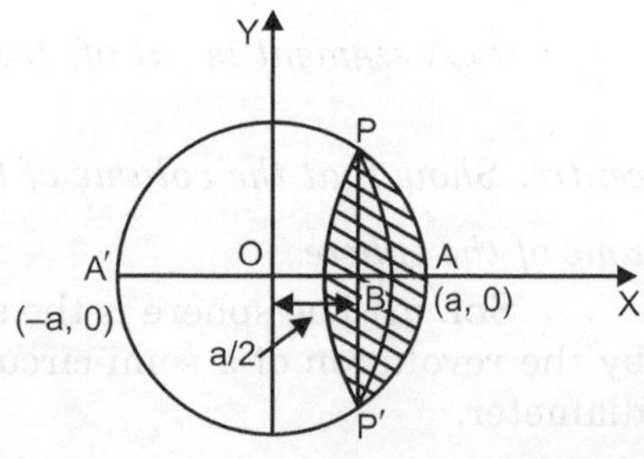

Then the required volume of the segment is generated by revolving the area PBA of the circle (1) about the x-axis.

Now for the arc PA, x varies from $\frac{a}{2}$ to a.

$\therefore$ Required volume

$$= \int_{a/2}^{a} \pi y^2 \, dx = \pi \int_{a/2}^{a} (a^2 - x^2) \, dx \qquad | \text{ Using (1)}$$

$$= \pi \left[a^2 x - \frac{x^3}{3} \right]_{a/2}^{a} = \pi \left[a^3 - \frac{a^3}{3} - \left(\frac{a^3}{2} - \frac{a^3}{24} \right) \right]$$

$$= \pi a^3 \left[1 - \frac{1}{3} - \frac{1}{2} + \frac{1}{24} \right] = \pi a^3 \left(\frac{24 - 8 - 12 + 1}{24} \right)$$

$$= \frac{5\pi a^3}{24} = \frac{5}{32} \left(\frac{4}{3} \pi a^3 \right) = \frac{5}{32} \text{ (volume of the sphere of radius } a).$$

Example 3. *(a) The area of the parabola $y^2 = 4ax$ lying between the vertex and the latus rectum is revolved about the x-axis. Find the volume generated.*

(b) A paraboloid of revolution is generated by rotating the parabola $y^2 = 4ax$ about OX. Find the volume generated by that portion of the curve which lies between $x = 0$ and $x = h$. If R is the area of the cross-section at $x = h$, show that the volume is half that of a cylinder of base area R and length h.

(c) Find the volume of the reel-shaped solid formed by the revolution about the y-axis of the part of the parabola $y^2 = 4ax$ cut off by the latus rectum.

Sol. (*a*) The equation of the parabola is $y^2 = 4ax$. ...(1)

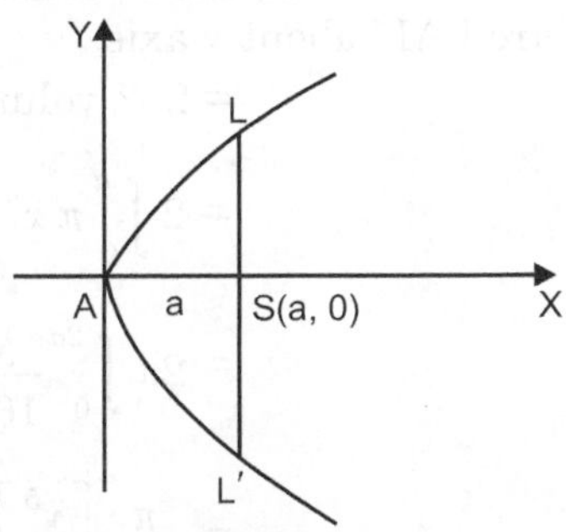

Let A be its vertex and LSL′ be the latus rectum.

The required volume is generated by the revolution of the area ASL about the x-axis and for the arc AL, x varies from 0 to a.

$\therefore$ Required volume

$$= \int_0^a \pi y^2 \, dx = \pi \int_0^a 4ax \, dx \qquad | \text{ From (1)}$$

$$= 4a\pi \left[\frac{x^2}{2} \right]_0^a = 2a\pi(a^2 - 0) = 2\pi a^3.$$

(*b*) The equation of the parabola is $y^2 = 4ax$ ...(1)

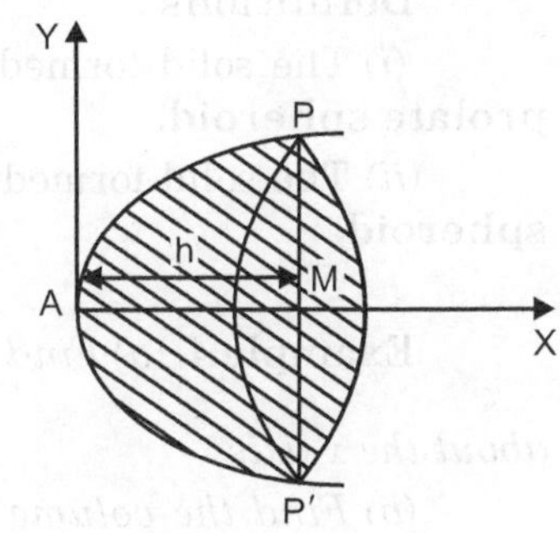

Let A(0, 0) be the vertex of the parabola and PMP′ the ordinate at $x = h$.

Then the required volume of the paraboloid is obtained by revolving the area APM about x-axis and for the AP of the curve x varies from 0 to h.

$\therefore$ Volume of the paraboloid of revolution is

$$V_1 = \int_0^h \pi y^2 \, dx = \pi \int_0^h (4ax) \, dx \qquad | \text{ Using (1)}$$

$$= 4a\pi \left[\frac{x^2}{2} \right]_0^h = 2a\pi(h^2 - 0) = 2\pi a h^2 \qquad ...(2)$$

Now $\quad AM = h \quad \therefore \quad MP = \sqrt{4a \cdot h} = 2\sqrt{ah}$ $\quad \Big| \because y = \sqrt{4ax}$ $\quad$ | From (1)

The section of the paraboloid by a plane PMP′ at a distance h from the vertex is a circle of radius $MP = 2\sqrt{ah}$ and the centre M.

Since R is the area of this cross-section (circle)

$$\therefore \qquad R = \pi \cdot MP^2 = \pi . 4ah = 4\pi\, ah \qquad \ldots(3)$$

Now volume of cylinder of base area R and height h is

$$V_2 = \text{base area} \times \text{height}$$
$$= Rh = 4\pi a h^2 \qquad \ldots(4) \quad | \text{Using (3)}$$

From(2) and (4), $V_1 = \frac{1}{2} V_2$ which proves the required result.

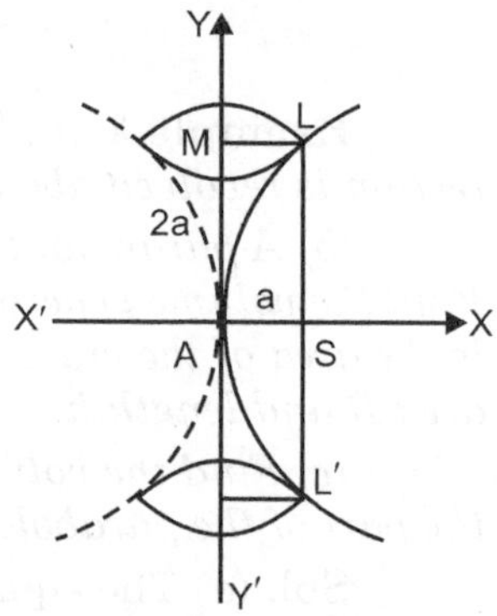

(c) The equation of the parabola is

$$y^2 = 4ax \quad \text{or} \quad x = \frac{y^2}{4a} \qquad \ldots(1)$$

Let LSL′ be the latus rectum of the parabola.

Since the curve is symmetrical about the x-axis and the arc LAL′ revolves about the y-axis (not the axis of symmetry).

$\therefore$ Required volume of the reel formed by revolving the arc LAL′ about y-axis.

= 2 × volume of the solid formed by the revolution of the arc AL about y-axis

$$= 2\int_0^a \pi x^2\, dy \qquad | \because \text{ for the AL, } y \text{ varies from 0 to } 2a$$

$$= 2\pi \int_0^{2a} \frac{y^4}{16a^2} dy = \frac{\pi}{8a^2}\int_0^{2a} y^4\, dy \qquad | \text{ From (1)}$$

$$= \frac{\pi}{8a^2}\left[\frac{y^5}{5}\right]_0^{2a} = \frac{\pi}{40a^2}(2a)^5 = \frac{\pi}{40a^2}.32a^5 = \frac{4}{5}\pi a^3.$$

11.3. PROLATE AND OBLATE SPHEROIDS

Definitions :

(*i*) The solid formed by the revolution of the ellipse about the *major axis* is called a **prolate spheroid.**

(*ii*) The solid formed by revolving the ellipse about the *minor axis* is called an **oblate spheroid.**

Example 4. (*a*) *Find the volume of the solid generated by revolving the ellipse* $\frac{x^2}{a^2} + \frac{y^2}{b^2} = 1$ *about the x-axis.*

(*b*) *Find the volume of the prolate spheroid generated by an ellipse whose major and minor axes are* $(24\pi)^{1/3}$ *and* $(3\pi)^{1/3}$.

(c) *Prove that the volume of the solid generated by the revolution of an ellipse round its minor axis, is a mean proportional between those generated by the revolution of the ellipse and of auxiliary circle round the major axis.*

Sol. (a) The equation of the ellipse is

$$\frac{x^2}{a^2}+\frac{y^2}{b^2}=1 \quad \text{or} \quad \frac{y^2}{b^2}=1-\frac{x^2}{a^2}=\frac{a^2-x^2}{a^2}$$

$$\therefore \quad y^2=\frac{b^2}{a^2}(a^2-x^2) \qquad ...(1)$$

[See Figure with part (c)]

Since the ellipse is symmetrical about y-axis.

$\therefore$ Required volume generated by the ellipse by the revolution about x-axis.

= 2 × volume generated by the arc BA in first quadrant about the x-axis.

Now for the arc BA, x varies from 0 to a.

$$\therefore \quad \text{Required volume} = 2\int_0^a \pi y^2\,dx = 2\pi\int_0^a \frac{b^2}{a^2}(a^2-x^2)dx \qquad | \text{ From (1)}$$

$$= 2\pi\frac{b^2}{a^2}\left[a^2x-\frac{x^3}{3}\right]_0^a = 2\pi\frac{b^2}{a^2}\left[a^3-\frac{a^3}{3}-0\right] = 2\pi\frac{b^2}{a^2}\frac{2a^3}{3}=\frac{4}{3}\pi ab^2.$$

(b) Let the equation of the ellipse be $\frac{x^2}{a^2}+\frac{y^2}{b^2}=1$ ...(1)

where we are given that

$$\left.\begin{array}{l}\text{length of major axis} = 2a = (24\pi)^{1/3}\\ \text{length of minor axis} = 2b = (3\pi)^{1/3}\end{array}\right\} \qquad ...(2)$$

and

Now prolate spheroid is the solid generated by revolving the ellipse (1) about the x-axis.

$$\therefore \quad \text{Required volume} = 2\int_0^a \pi y^2\,dx$$

$$= 2\pi\int_0^a \frac{b^2}{a^2}(a^2-x^2)dx \qquad \left| \because \text{ From (1), } y^2=\frac{b^2}{a^2}(a^2-x^2)\right.$$

$$= 2\pi\frac{b^2}{a^2}\left[a^2x-\frac{x^3}{3}\right]_0^a = 2\pi\frac{b^2}{a^2}\left(a^3-\frac{a^3}{3}\right)=\frac{4}{3}\pi ab^2$$

$$= \frac{1}{6}\pi\,.\,(2a)(2b)^2 = \frac{1}{6}\pi(24\pi)^{1/3}\,.(3\pi)^{2/3}$$

$$= \frac{1}{6}\pi(24\pi)^{1/3}(9\pi^2)^{1/3} = \frac{1}{6}\pi\,.\,(24\pi\times 9\pi^2)^{1/3} = \frac{1}{6}\pi\,.\,6\pi=\pi^2.$$

(c) Let the equation of the ellipse be

$$\frac{x^2}{a^2}+\frac{y^2}{b^2}=1$$

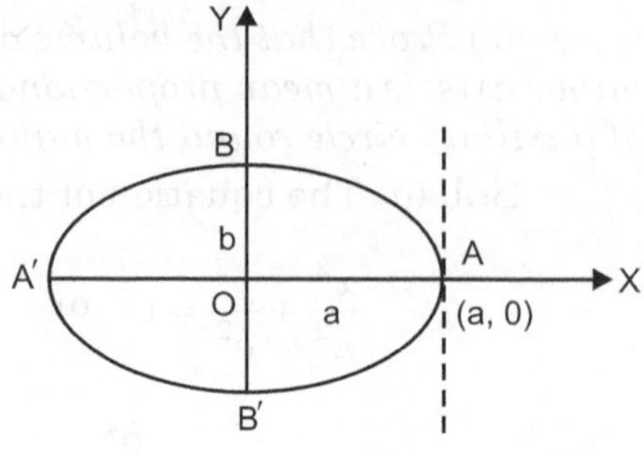

or
$$\frac{x^2}{a^2}=1-\frac{y^2}{b^2}=\frac{1}{b^2}(b^2-y^2)$$

$$x^2=\frac{a^2}{b^2}(b^2-y^2) \qquad \ldots(1)$$

Since the curve is symmetrical about x-axis.

$\therefore$ Volume generated by revolution of the ellipse about y-axis

= 2 × Vol. generated by revolution of the arc AB about y-axis.

But for the arc AB, y varies from 0 to b.

$\therefore$ Required volume is

$$V_1=2\int_0^b \pi x^2\,dy=2\pi\int_0^b \frac{a^2}{b^2}(b^2-y^2)\,dy=2\pi\frac{a^2}{b^2}\left[b^2y-\frac{y^3}{3}\right]_0^b=2\pi\frac{a^2}{b^2}\left(a^3-\frac{b^3}{3}\right)$$

$$=\frac{4}{3}\pi a^2 b \qquad \ldots(2)$$

Also if V_2 is the volume of the solid generated by the ellipse (1) about x-axis, then [interchanging a and b in (2)], we get

$$V_2=\frac{4}{3}\pi b^2 a \qquad \ldots(3)$$

The equation of the auxiliary circle of the ellipse (1) is

$$x^2+y^2=a^2 \quad \text{or} \quad \frac{x^2}{a^2}+\frac{y^2}{a^2}=1$$

$\therefore$ The volume V_3 of the solid generated by the revolution of the auxiliary circle about x-axis is given by [putting $b = a$ in (3)],

$$V_3=\frac{4}{3}\pi a^3 \qquad \ldots(4)$$

Now
$$V_2V_3=\frac{4}{3}\pi b^2 a\,\frac{4}{3}\pi a^3=\frac{16}{9}\pi^2 b^2 a^4=\left(\frac{4}{3}\pi b a^2\right)^2=V_1^2$$

$\therefore$
$$V_1=\sqrt{V_2V_3}.$$

Thus V_1 is the mean proportional (*i.e.*, geometric mean) between V_2 and V_3.

Hence the result.

Example 5. *Find the volume of the solid formed by the revolution of the loop about the x-axis, of the following curves:*

(*a*) $y^2 = \dfrac{x^2(a+x)}{(a-x)}$ (*b*) $y^2(a+x) = x^2(3a-x)$ (*c*) $y^2 = x^2(a-x)$.

Sol. (*a*) The equation of the curve is $y^2 = \dfrac{x^2(a+x)}{(a-x)}$

or $$y^2(a-x) = x^2(a+x) \qquad \ldots(1)$$

To trace the curve roughly to find the limits of integration:

(*i*) The curve is symmetrical about the *x*-axis.

(*ii*) The curve passes through the origin and the tangents at the origin are given by

$$y^2 = x^2 \quad \text{or} \quad y = \pm x.$$

Since these tangents are real and distinct, ∴ origin is a node.

(*iii*) The curve meets the *x*-axis at the point $(-a, 0)$, and shifting the origin to the point $(-a, 0)$, the equation (1) transforms to

$$y^2(a-x+a) = (x-a)^2(a+x-a) \quad \text{or} \quad y^2(2a-x) = x(x-a)^2$$

∴ The tangent at the new origin is $x = 0$ *i.e.*, new *y*-axis.

∴ The tangent at the point $(-a, 0)$ is || to *y*-axis.

The curve meets *y*-axis only at origin.

(*iv*) The curve has an asymptote $a - x = 0$ *i.e.*, $x = a$ || to *y*-axis.

(*v*) From (1), $y = x\sqrt{\dfrac{a+x}{a-x}}$

When $x > a$ or when $x < -a$. *y* is imaginary.

∴ The curve does not lie beyond the lines $x = \pm a$.

Thus the shape of the curve is roughly as shown in the figure.

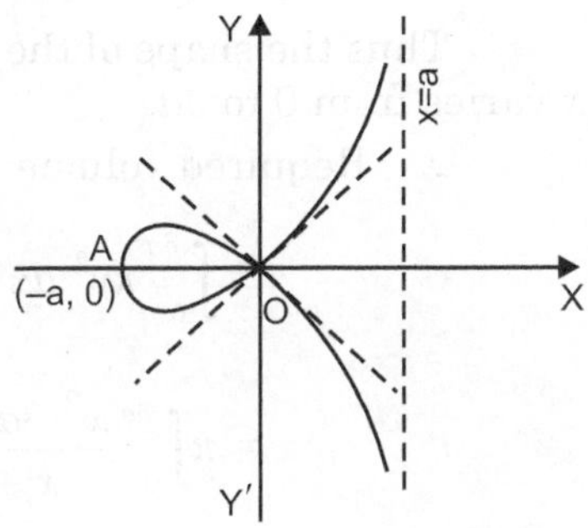

For the upper half of the loop, *x* varies from $-a$ to 0.

∴ Required volume $= \displaystyle\int_{-a}^{0} \pi y^2\,dx$

$$= \pi \int_{-a}^{0} \frac{x^2(a+x)}{a-x}\,dx \qquad | \text{ From (1)}$$

$$= \pi \int_{-a}^{0} \frac{ax^2+x^3}{a-x}\,dx \qquad \text{[Divide the num. by the denominator]}$$

$$= \pi \int_{-a}^{0} \left[-x^2 - 2ax - 2a^2 + \frac{2a^3}{a-x}\right] dx$$

$$= \pi \left[-\frac{x^3}{3} - ax^2 - 2a^2x - 2a^3\log(a-x)\right]_{-a}^{0}$$

$$= \pi\left[-2a^3 \log a - \left(\frac{a^3}{3} - a^3 + 2a^3 - 2a^3 \log 2a\right)\right]$$

$$= \pi\left[-\frac{4}{3}a^3 + 2a^3 (\log 2a - \log a)\right] = 2\pi a^3 \left(\log 2 - \frac{2}{3}\right).$$

(*b*) The equation of the curve is $y^2(a + x) = x^2(3a - x)$ ...(1)

To trace the curve roughly of find the limits of integration :

(*i*) The curve is symmetrical about the x-axis.

(*ii*) It passes through the origin and the tangents at the origin are given by $y^2 = x^2$ or $y = \pm x$.

$\therefore$ Origin is a node.

(*iii*) The curve meets x-axis in (0, 0) and $(3a, 0)$.

The curve meets y-axis at the origin.

(*iv*) The asymptote || to y-axis is $a + x = 0$ or $x = -a$.

The curve has no other asymptote.

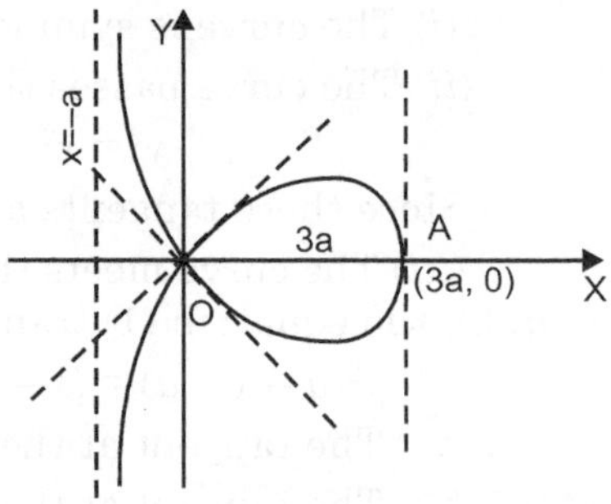

(*v*) Form (1), $y = x\sqrt{\dfrac{3a - x}{a + x}}$

When $x > 3a$ or when $x < -a$, y is imaginary. Thus the curve does not lie beyond the lines $x = -a$, $x = 3a$.

Thus the shape of the curve is as shown in the figure and for the upper half of the loop, x varies from 0 to $3a$.

$\therefore$ Required volume

$$= \int_0^{3a} \pi y^2 \, dx$$

$$= \pi\int_0^{3a} \frac{x^2(3a - x)}{x + a} dx$$

$$= \pi\int_0^{3a} \frac{3ax^2 - x^3}{x + a} dx$$ [Divide the num. by the denom.]

$$= \pi\int_0^{3a}\left[-x^2 + 4ax - 4a^2 + \frac{4a^3}{x+a}\right]dx = \pi\left[-\frac{x^3}{3} + 2ax^2 - 4a^2x + 4a^3 \log(x+a)\right]_0^{3a}$$

$$= \pi[-9a^3 + 18a^3 - 12a^3 + 4a^3 \log 4a - 4a^3 \log a]$$

$$= \pi\left[4a^3(\log 4a - \log a) - 3a^3\right] = \pi a^3 \left[4 \log\left(\frac{4a}{a}\right) - 3\right]$$

$$= \pi a^3[4 \log 2^2 - 3] = \pi a^3[8 \log 2 - 3].$$

(*c*) The equation of the curve is $y^2 = x^2(a - x)$ ...(1)

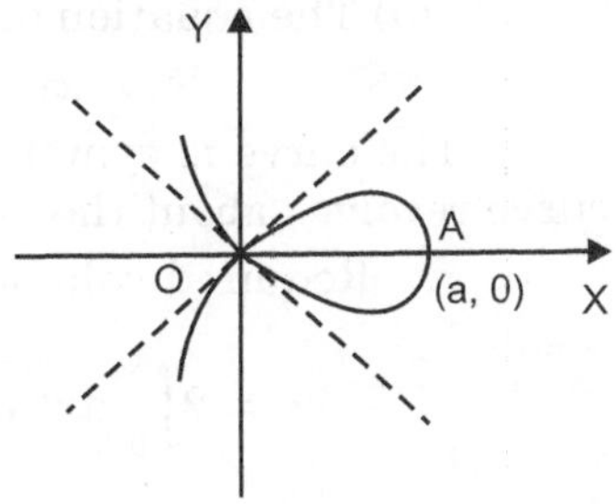

(*i*) The curve is symmetrical about the x-axis and it has a loop between $x = 0$ to $x = a$.

Thus the required volume

$$= \int_0^a \pi y^2\, dx = \pi \int_0^a x^2(a - x)dx$$

$$= \pi\left[a\frac{x^3}{3} - \frac{x^4}{4}\right]_0^a = \pi\left[\frac{a^4}{3} - \frac{a^4}{4}\right] = \frac{\pi}{12}a^4.$$

Example 6. (*a*) *Find the volume of the solid generated by the revolution of the curve $y(a^2 + x^2) = a^3$ about its asymptote.*

(*b*) *Find the volume of the solid generated by the revolution of the area between the curve $xy^2 = 4a^2(2a - x)$ and its asymptote about the asymptote.*

Sol. (*a*) The equation of the curve is $y(x^2 + a^2) = a^3$ ...(1)

To trace the curve roughly to find the limits of integration :

(*i*) The curve is symmetrical about x-axis.

(*ii*) The curve does not pass through the origin.

(*iii*) The curve does not meet x-axis. It meets y-axis in the point $(0, a)$.

(*iv*) The asymptote || to x-axis is $y = 0$ *i.e.*, the x-axis itself. The curve has no other asymptote.

(*v*) From (1), $x^2 = \dfrac{a^2(a - y)}{y}$ $\therefore$ $x = a\sqrt{\dfrac{a - y}{y}}$

When $y < 0$ or $y > a$, x is imaginary. Thus the curve does not lie below x-axis and above the line $y = a$. When $y = a$, $x = 0$. As y decreases from a to 0, x increases from 0 to ∞. Thus the shape of the curve is as shown in figure.

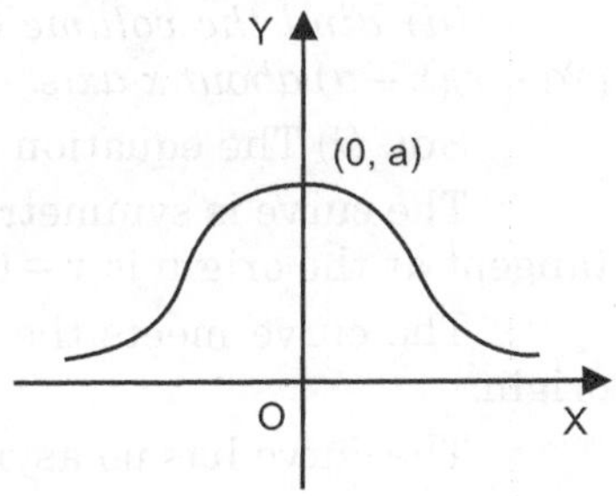

Since the curve is symmetrical about y-axis and the revolution is made about the asymptote (x-axis).

$\therefore$ Required volume = 2 × volume generated by half of the curve in first quadrant.

For the arc in first quadrant, x varies from 0 to ∞.

$$\therefore \quad \text{Required volume} = 2\int_0^\infty \pi y^2\, dx = 2\pi \int_0^\infty \left(\frac{a^3}{a^2 + x^2}\right)^2 dx$$

$$\left[\begin{array}{l}\text{Put } x = a\tan\theta \quad \therefore \quad dx = a\sec^2\theta\, d\theta \\ \text{when } x = 0, \theta = 0\,;\ \text{when } x \to \infty, \theta \to \pi/2\end{array}\right]$$

$$= 2\pi a^6 \int_0^{\pi/2} \frac{a\sec^2\theta\, d\theta}{(a^2 + a^2\tan^2\theta)^2} = 2\pi a^6 \int_0^{\pi/2} \frac{a\sec^2\theta\, d\theta}{a^2\sec^4\theta}$$

$$= 2\pi a^3 \int_0^{\pi/2} \cos^2\theta\, d\theta = 2\pi a^3 \cdot \frac{1}{2} \cdot \frac{\pi}{2} = \frac{1}{2}\pi^2 a^3.$$

(*b*) The equation of the curve is

$$xy^2 = 4a^2(2a - x) \qquad ...(1)$$

The curve is symmetrical about the x-axis and the asymptote is the y-axis. Since the curve revolves about the y-axis *i.e.*, the asymptote.

$\therefore$ Required volume

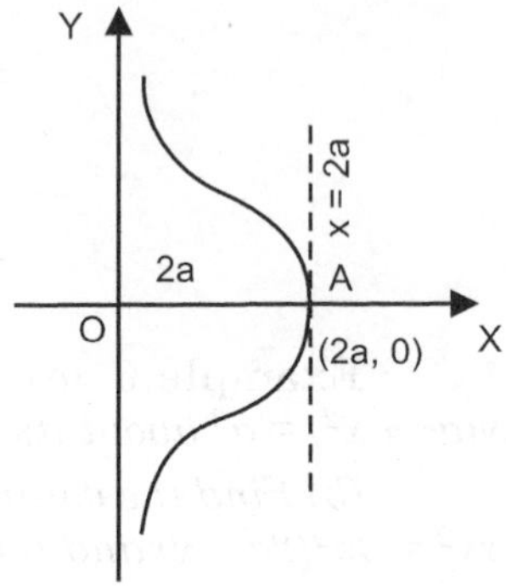

$$= 2\int_0^\infty \pi x^2\, dy \qquad \Big|\ \because \text{ For the upper half of the curve } y \text{ varies from 0 to } \infty$$

$$= 2\pi \int_0^\infty \left(\frac{8a^3}{y^2 + 4a^2}\right)^2 dy \qquad \Big|\ \because \text{ From (1), } x = \frac{8a^3}{y^2 + 4a^2}$$

$$= 128\pi a^6 \int_0^\infty \frac{1}{(y^2 + 4a^2)^2}\, dy$$

| Put $y = 2a \tan\theta \quad \therefore \quad dy = 2a\sec^2\theta\, d\theta$
When $y = 0, \theta = 0$, when $y \to \infty, \theta \to \pi/2$

$$= 128\pi a^6 \int_0^{\pi/2} \frac{2a\sec^2\theta\, d\theta}{(4a^2 + \tan^2\theta + 4a^2)^2} = 128\pi a^6 \int_0^{\pi/2} \frac{2a\sec^2\theta\, d\theta}{16a^4\sec^4\theta}$$

$$= 16\pi a^3 \int_0^{\pi/2} \cos^2\theta\, d\theta = 16\pi a^3 \cdot \frac{1}{2} \cdot \frac{\pi}{2} = 4\pi^2 a^3.$$

Example 7. (*i*) *The loop of the curve* $2ay^2 = x(x - a)^2$ *revolves about x-axis, find the volume of the solid so generated.*

(*ii*) *Find the volume of the solid obtained by revolving the loop of the curve* $a^2y^2 = x^2(2a - x)(x - a)$ *about x-axis.*

Sol. (*i*) The equation of the curve is $\quad 2ay^2 = x(x - a)^2 \qquad ...(1)$

The curve is symmetrical about the x-axis. The curve passes through the origin and the tangent at the origin is $x = 0$ *i.e.*, y-axis.

The curve meets the x-axis in points (0, 0) and (a, 0). It meets the y-axis only at the origin.

The curve has no asymptote. From (1),

$$y = \pm (x - a)\sqrt{\frac{x}{2a}}$$

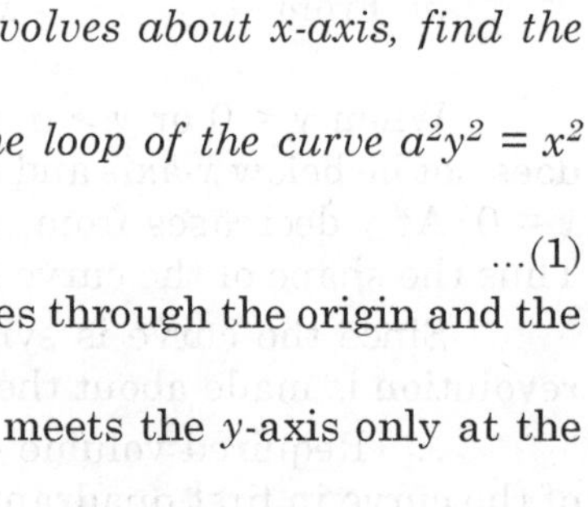

For y to be real, $x \geq 0$.

$\therefore$ No portion of the curve lies to the left of y-axis.

Thus the shape of the curve is as shown in the figure. For the upper half of its loop, x varies from 0 to a.

Required volume formed by the revolution of the loop about x-axis

$$= \int_0^a \pi y^2\, dx = \pi \int_0^a \frac{x(x-a)^2}{2a} dx = \frac{\pi}{2a} \int_0^a (x^3 - 2ax^2 + a^2 x)\, dx$$

$$= \frac{\pi}{2a}\left[\frac{x^4}{4} - 2a.\frac{x^3}{3} + a^2 . \frac{x^2}{2}\right]_0^a = \frac{\pi}{2a}\left[\frac{a^4}{4} - \frac{2a^4}{3} + \frac{a^4}{2}\right] = \frac{\pi}{2a}.\frac{a^4}{12} = \frac{\pi a^3}{24}.$$

(*ii*) The equation of the curve is $a^2y^2 = x^2(2a - x)(x - a)$ ...(1)

The curve is symmetrical about the x-axis. The curve passes through the origin and the tangents at the origin are given by

$$a^2y^2 = -\,2a^2x^2 \quad \text{or} \quad y^2 = -\,2x^2 \quad \text{or} \quad y = \pm i\sqrt{2}x.$$

The tangents being imaginary, the origin is a conjugate point.

The curve meets the x-axis in points (0, 0), (a, 0) and ($2a$, 0). In meets the y-axis only at the origin. The curve has no asymptotes.

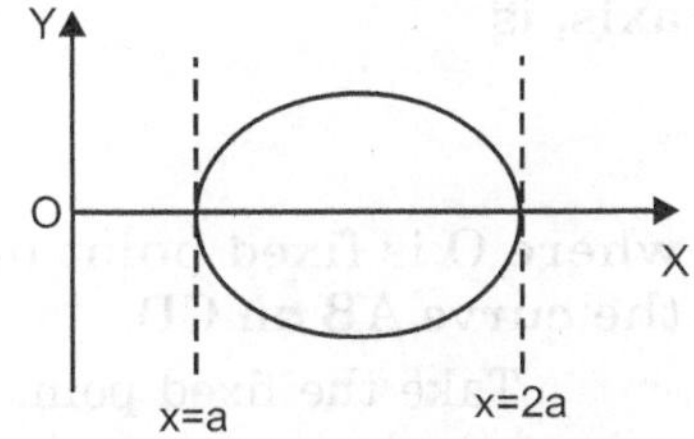

From (1), $\quad y = \pm \frac{x}{a}\sqrt{(2a - x)(x - a)}$

When $x < a$ or $x > 2a$, y is imaginary. For y to be real, $a < x < 2a$.

Thus the shape of the curve is as shown in the figure. For the upper half of its loop, x varies from a to $2a$.

Reqd. volume formed by the revolution of the loop about x-axis

$$= \int_a^{2a} \pi\, y^2 dx = \pi \int_a^{2a} \frac{x^2}{a^2}(2a - x)(x - a) dx$$

$$= \frac{\pi^2}{a^2} \int_a^{2a} (-\,x^4 + 3ax^3 - 2a^2x^2) dx = \frac{\pi}{a^2}\left[-\frac{x^5}{5} + 3a\,.\frac{x^4}{4} - 2a^2\,.\frac{x^3}{3}\right]_a^{2a}$$

$$= \frac{\pi}{a^2}\left[-\frac{32a^5}{5} + 12a^5 - \frac{16a^5}{3} + \frac{a^5}{5} - \frac{3a^5}{4} + \frac{2a^5}{3}\right] = \frac{23\pi a^3}{60}.$$

Example 8. *A basin is formed by the revolution of the curve $x^3 = 64y$ ($y > 0$) about the axis of y. If the depth of the basin is 8 cm, how many cubic cm of water will it hold ?*

Sol. The equation of the generating curve is

$$x^3 = 64y \ (y > 0) \qquad ...(1)$$

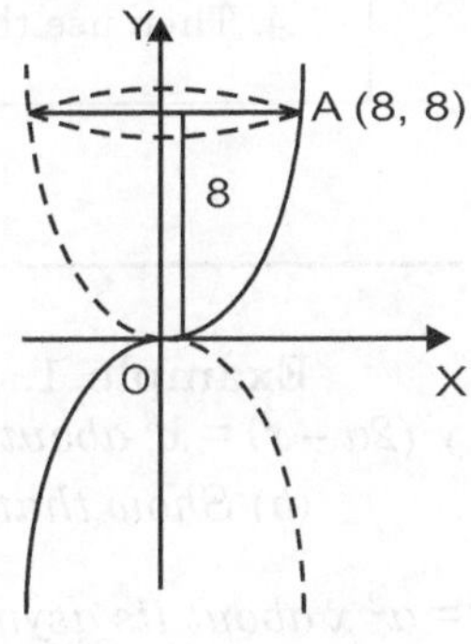

The curve is symmetrical in opposite quadrants.

The shape of the curve is as shown in the figure.

Since the height of the basin is 8 cm, so when $y = 8$, from (1)

$$x^3 = 64 \times 8 \quad \therefore \quad x = 8.$$

Thus A(8, 8) is the point on the curve (1) at a height of 8 cm.

Thus the basin is formed by the revolution of the arc OA about y-axis, where A is the point (8, 8).

$\therefore$ Required volume $= \int_0^8 \pi x^2\, dy = \pi \int_0^8 (64y)^{2/3}\, dy$

$$= 16\pi \int_0^8 y^{2/3}\, dy = 16\pi . \left[\frac{y^{5/3}}{\frac{5}{3}} \right]_0^8$$

$$= \frac{48}{5} \pi[(8)^{5/3} - 0] = \frac{48}{5} \pi(32) = \frac{1536}{5} \pi \text{ cubic cm.}$$

11.4. REVOLUTION ABOUT ANY AXIS

The volume of the solid generated by the revolution about any axis CD of the area bounded by the curve AB, the axis CD and the perpendiculars AC, BD on the axis, is

$$\int_{OC}^{OD} \pi(PM)^2 d(OM)$$

where O is fixed point on the axis CD ad PM is perpendicular from any point P of the curve AB on CD.

Take the fixed point O on CD as origin and CD, the axis of revolution as x-axis.

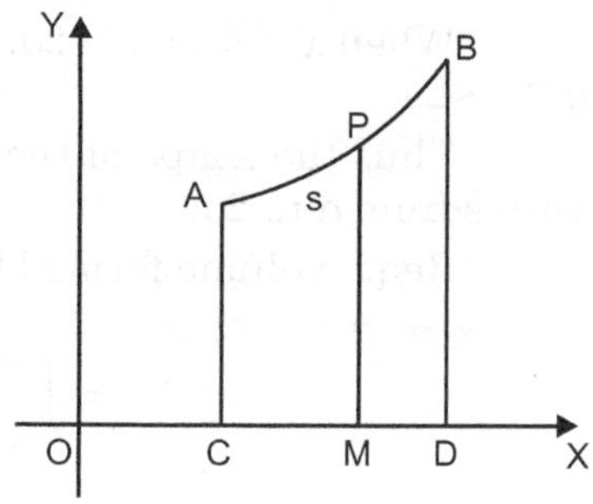

Let OY $\perp$ to CD be taken as as y-axis.

Let P(x, y) be any point on the curve referred to OC and OY as axes.

Draw PM $\perp$ OC, so that

$$OM = x \text{ and } MP = y.$$

If OC $= a$ and OD $= b$, then the required volume

$$= \int_a^b \pi y^2\, dx = \pi \int_a^b (PM)^2 . d(OM)$$

Remember : Practical method for questions :

1. Take any point P(x, y) on the curve.
2. Draw PM $\perp$ on the line about which the curve is to be revolved and find the length of PM.
3. Find the distance OM of the foot of $\perp$ from a fixed point O (say) on the line and take its differential.
4. Then use the formula $\int \pi(PM)^2 d(OM)$ with proper limits for integration.

ILLUSTRATIVE EXAMPLES

Example 1. (*a*) *Find the volume of the solid formed by the revolution of the cissoid* $y^2(2a - x) = x^3$ *about its asymptote.*

(*b*) *Show that the volume of the solid generated by the revolution of the curve* $(a - x)\, y^2 = a^2 x$ *about its asymptote is* $\frac{1}{2} \pi^2 a^3$.

Sol. (*a*) The equation of the curve is $y^2(2a - x) = x^3$ or $y^2 = \dfrac{x^3}{2a - x}$ ...(1)

The curve is symmetrical about the x-axis and the asymptote is the line

$$2a - x = 0 \quad \text{or} \quad x = 2a.$$

If P(x, y) be any point on the curve and PM ⊥ on the asymptote (the axis of revolution), and PN ⊥ OX, then PM = NA = OA – ON = $2a - x$ and AM = NP = y, where A is the pt. of intersection of the asymptote and the x-axis.

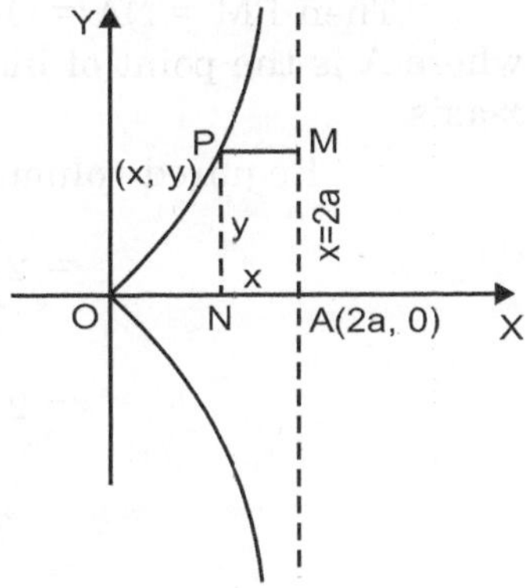

∴ Required volume

$$= 2\int \pi(\text{PM})^2 \,.\, d(\text{AM}) \qquad ...(2)$$

Now $\quad \text{AM} = y = \dfrac{x^{3/2}}{\sqrt{2a - x}}$ | *From* (1)

∴ $\quad d(\text{AM}) = dy$

$$= \frac{(2a - x)^{1/2} \,.\, \frac{3}{2}x^{1/2} - x^{3/2} \,.\, \frac{1}{2}(2a - x)^{-1/2} \,.(-1)}{2a - x} dx$$

$$= \frac{3x^{1/2}(2a - x) + x^{3/2}}{2(2a - x)^{3/2}} dx = \frac{x^{1/2}(6a - 3x + x)}{2(2a - x)^{3/2}} dx = \frac{\sqrt{x}(3a - x)}{(2a - x)^{3/2}} dx$$

From (2),

∴ Required volume

$$= 2\pi\int_0^{2a} (2a - x)^2 \,.\, \frac{\sqrt{x}(3a - x)}{(2a - x)^{3/2}} dx$$

$$= 2\pi\int_0^{2a} (3a - x)\sqrt{x}\,\sqrt{2a - x}\, dx$$

| Put $x = 2a \sin^2 \theta$ ∴ $dx = 4a \sin \theta \cos \theta \, d\theta$
when $x = 0, \theta = 0$, when $x = 2a, \theta = \pi/2$

$$= 2\pi \int_0^{\pi/2} (3a - 2a \sin^2 \theta) \,.\, \sqrt{2a \sin^2 \theta} \,.\, \sqrt{2a\,(1 - \sin^2 \theta)} \,.\, 4a \sin \theta \cos \theta \, d\theta$$

$$= 16\pi a^3 \int_0^{\pi/2} (3 - 2 \sin^2 \theta) \,.\, \sin^2 \theta \cos^2 \theta \, d\theta$$

$$= 16\pi a^3 \int_0^{\pi/2} (3\sin^2 \theta \cos^2 \theta - 2 \sin^4 \theta \cos^2 \theta) \, d\theta$$

$$= 16\pi a^3 \left[3 \,.\, \frac{1.1}{4.2} \,.\, \frac{\pi}{2} - 2 \,.\, \frac{3.1.1}{6.4.2} \,.\, \frac{\pi}{2}\right] = 16\pi a^3 \left[\frac{3\pi}{16} - \frac{\pi}{16}\right] = 2\pi^2 a^3.$$

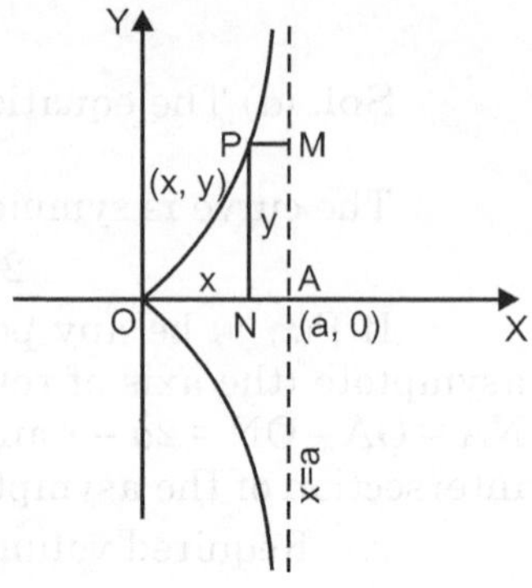

(*b*) The equation of the curve is $y^2(a-x) = a^2x$ ...(1)

The curve is symmetrical about the x-axis and the asymptote is the line $x = a$.

Let P(x, y) be any point on the curve. Draw PM ⊥ on the asymptote (the axis of revolution and PN ⊥ OX.

Then PM = NA = OA − ON = $a - x$ and AM = PN = y where A is the point of intersection of the asymptote and the x-axis.

∴ Required volume

$$= 2\int \pi(\text{PM})^2 \, .d(\text{AM}) \qquad ...(2)$$

$$= 2\int_0^{\infty} \pi(a-x)^2 \, . \, dy$$

$$= 2\pi \int_0^{\infty} \left[a - \frac{ay^2}{a^2+y^2}\right]^2 .dy$$

| ∵ From (1), $x(a^2+y^2) = ay^2$ ∴ $x = \dfrac{ay^2}{a^2+y^2}$

$$= 2\pi \int_0^{\infty} \frac{a^6}{(a^2+y^2)^2} dy$$

| Put $y = a\tan\theta$ ∴ $dy = a\sec^2\theta \, d\theta$ when $y = 0, \theta = 0$, when $y \to \infty, \theta \to \pi/2$

$$= 2\pi a^6 \int_0^{\pi/2} \frac{a\sec^2\theta \, d\theta}{[a^2(1+\tan^2\theta)]^2} = 2\pi a^6 \int_0^{\pi/2} \frac{a\sec^2\theta \, d\theta}{a^4\sec^4\theta}$$

$$= 2\pi a^3 \int_0^{\pi/2} \cos^2\theta \, d\theta = 2\pi a^3 \, . \, \frac{1}{2}\frac{\pi}{2} = \frac{1}{2}\pi^2 a^3.$$

Note. From (2), *we can also proceed as follows :*

We have $\text{AM} = y = \dfrac{a\sqrt{x}}{\sqrt{a-x}}$ | From (1)

$$\therefore \quad d(\text{AM}) = dy = a \, . \, \frac{\sqrt{a-x}\,.\,\dfrac{1}{2\sqrt{x}} - \sqrt{x}\,.\,\dfrac{1}{2\sqrt{a-x}}(-1)}{a-x} dx$$

$$= a.\frac{a-x+x}{2\sqrt{x}(a-x)^{3/2}}dx = \frac{a^2dx}{2\sqrt{x}(a-x)^{3/2}}.$$

∴ Now from (2), the required volume

$$= 2\,.\int \pi(\text{PM})^2 \, .d(\text{AM}) = 2\pi\int_0^a (a-x)^2 \, . \, \frac{a^2dx}{2\sqrt{x}(a-x)^{3/2}}$$

$$= \pi a^2 \int_0^a \sqrt{\frac{a-x}{x}}\, dx \qquad \left| \begin{array}{l} \text{Put } \; x = a\sin^2\theta \quad \therefore\; dx = 2a\sin\theta\cos\theta d\theta \\ \text{when } x = 0, \theta = 0, \text{ when } x = a, \theta = \pi/2 \end{array} \right.$$

$$= \pi a^2 \int_0^{\pi/2} \sqrt{\frac{a(1-\sin^2\theta)}{a\sin^2\theta}}\; 2a\sin\theta\cos\theta\, d\theta$$

$$= 2\pi a^3 \int_0^{\pi/2} \cos^2\theta\, d\theta = 2\pi a^3 . \frac{1}{2} . \frac{\pi}{2} = \frac{1}{2}\pi^2 a^3.$$

Example 2. (*a*) *The ellipse* $b^2x^2 + a^2y^2 = a^2b^2$ *is divided into two parts by the line* $x = \frac{a}{2}$, *and the smaller part is rotated through four right angles about this line. Prove that the volume generated is*

$$\pi a^2 b\left(\frac{3\sqrt{3}}{4} - \frac{\pi}{3}\right).$$

(*b*) *Find the volume of the torus obtained by rotating the area bounded by the circle* $x^2 + y^2 = a^2$ *around the line* $x = b$ $(b > a)$.

Sol. (*a*) The equation of the ellipse is $b^2x^2 + a^2y^2 = a^2b^2$ ...(1)

and CDC' is the line $x = \frac{a}{2}$, about which the area CDC′AC is revolved. Let P(*x*, *y*) be any point on the arc CAC of the ellipse. From P draw PM ⊥ CC′ and PN ⊥ OX.

Then $\quad PM = DN = ON - OD = x - \frac{a}{2}$ and $PN = DM = y$.

Since the arc CAC′ is symmetrical about the *x*-axis.

Required volume $= 2\int \pi(PM)^2 . d(DM)$ ...(2)

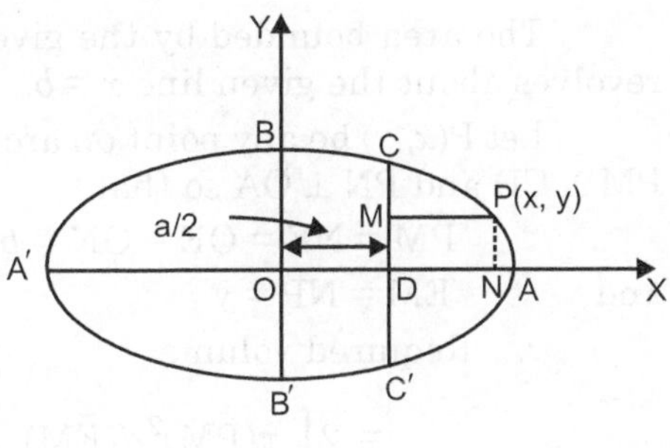

From (1), $\quad a^2y^2 = a^2b^2 - b^2x^2 = b^2(a^2 - x^2)$

$$\therefore \quad DM = y = \frac{b}{a}\sqrt{a^2 - x^2}$$

$$\therefore \quad d(DM) = dy = \frac{b}{a}\frac{1}{2} . \frac{-2x}{\sqrt{a^2 - x^2}}\, dx$$

$$= \frac{-bx}{a\sqrt{a^2 - x^2}}\, dx$$

∴ From (2), Required volume

$$= \int_{a/2}^{a} \left(x - \frac{a}{2}\right)^2 . \frac{-bx}{a\sqrt{a^2 - x^2}} dx \qquad \left| \begin{array}{l} \text{Put } \; x = a\sin\theta \;\therefore\; dx = a\cos\theta\, d\theta \\ \text{when } x = a/2, \theta = \pi/6, \text{ when } x = a, \theta = \pi/2 \end{array} \right.$$

$$= 2\pi \int_{\pi/6}^{\pi/2} \left(a \sin\theta - \frac{a}{2}\right)^2 . \frac{-b . a \sin\theta}{a\sqrt{a^2(1 - \sin^2\theta)}} . a \cos\theta \, d\theta$$

$$= 2\pi \int_{\pi/6}^{\pi/2} -\frac{ba^2}{4}(2\sin\theta - 1)^2 . \sin\theta \, d\theta$$

$$= -\frac{\pi ba^2}{2} \int_{\pi/6}^{\pi/2} (4\sin^3\theta - 4\sin^2\theta + \sin\theta)\, d\theta$$

$$= \frac{-\pi ba^2}{2} \int_{\pi/6}^{\pi/2} \left[3\sin\theta - \sin 3\theta - 4 . \frac{1 - \cos 2\theta}{2} + \sin\theta\right] d\theta$$

$$= \frac{-\pi ba^2}{2} \int_{\pi/6}^{\pi/2} (4\sin\theta - \sin 3\theta + 2\cos 2\theta - 2)\, d\theta$$

$$= \frac{-\pi ba^2}{2} \left[-4\cos\theta + \frac{\cos 3\theta}{3} + \sin 2\theta - 2\theta\right]_{\pi/6}^{\pi/2}$$

$$= \frac{-\pi ba^2}{2} \left[-2 . \frac{\pi}{2} - \left(-4 . \frac{\sqrt{3}}{2} + \frac{\sqrt{3}}{2} - 2 . \frac{\pi}{6}\right)\right]$$

$$= \frac{-\pi ba^2}{2} \left[\frac{-2\pi}{3} + \frac{3\sqrt{3}}{2}\right] = \pi a^2 b \left(\frac{3\sqrt{3}}{2} - \frac{\pi}{3}\right). \qquad \text{(Numerically)}$$

(*b*) The equation of the given circle is

$$x^2 + y^2 = a^2 \qquad \ldots(1)$$

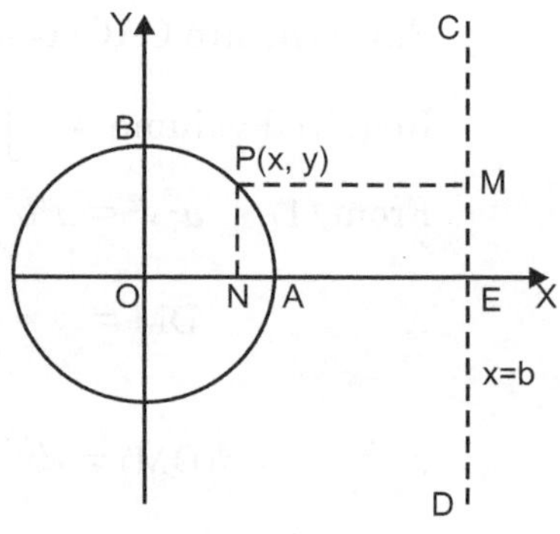

The area bounded by the given circle with centre O revolves about the given line $x = b$.

Let P(x, y) be any point on arc AB of the circle. Draw PM ⊥ CD and PN ⊥ OA so that

$$\text{PM} = \text{NE} = \text{OE} - \text{ON} = b - x$$

and $\quad \text{EM} = \text{NP} = y.$

∴ Required volume

$$= 2\int \pi(\text{PM})^2 d(\text{EM})$$

$$= 2\int_0^a \pi(b - x)^2 dy \qquad | \because \text{ For the arc AB, } y \text{ varies from 0 to } a.$$

$$= 2\pi \int_0^a (b^2 + x^2 - 2bx)\,dy = 2\pi \int_0^a [b^2 + (a^2 - y^2) - 2b\sqrt{a^2 - y^2}]\,dy$$

$$= 2\pi \int_0^a [(b^2 + a^2) - y^2 - 2b\sqrt{a^2 - y^2}]\,dy$$

$$= 2\pi\left[(b^2 + a^2)y - \frac{y^3}{3} - 2b\left(\frac{y}{2}\sqrt{a^2 - y^2} + \frac{a^2}{2}\sin^{-1}\frac{y}{a}\right)\right]_0^a$$

$$= 2\pi\left[(b^2 + a^2)a - \frac{a^3}{3} - a^2b\sin^{-1}1 + 0\right]$$

$$= 2\pi\left[ab^2 + a^3 - \frac{a^3}{3} - a^2b\cdot\frac{\pi}{2}\right] = 2\pi\left[ab^2 + \frac{2}{3}a^3 - a^2b\cdot\frac{\pi}{2}\right]$$

$$= \frac{2\pi}{6}\left[6ab^2 + 4a^3 - 3a^2b\,\pi\right] = \frac{\pi}{3}\left[6ab^2 + 4a^3 - 3a^2b\pi\right].$$

Example 3. (*a*) *Find the volume of the spindle formed by the revolution of a parabolic arc about the line joining the vertex to one extremity of the latus rectum.*

(*b*) *A quadrant of a circle, of radius a, revolves about its chord. Show that the volume of the spindle generated is* $\frac{\pi}{6\sqrt{2}}(10 - 3\pi)a^3$.

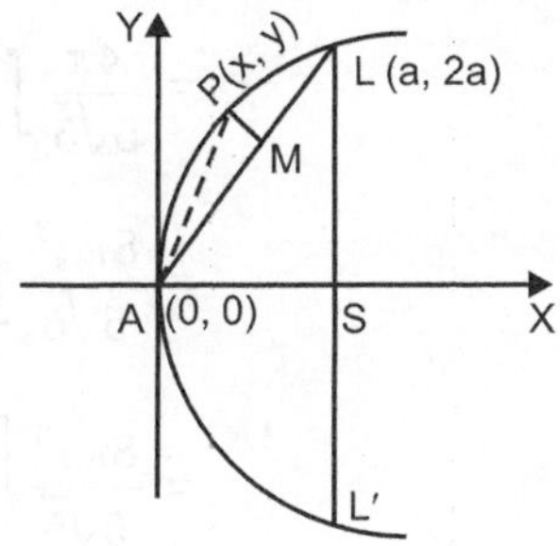

Sol. (*a*) The equation of the parabola is $y^2 = 4ax$.

Let LSL′ be the latus rectum and A the vertex of the parabola.

Then the co-ordinates of L are $(a, 2a)$ and of A are $(0, 0)$.

Now the equation of the line AL is

$$y - 0 = \frac{2a - 0}{a - 0}(x - 0) \quad \text{or} \quad 2x - y = 0$$

Let P(x, y) be any point on the arc AL of the parabola. Draw PM ⊥ AL and join AP.

Then PM = ⊥ distance of P(x, y) form the lime $2x - y = 0$

$$= \frac{2x - y}{\sqrt{4 + 1}} = \frac{2x - y}{\sqrt{5}} = \frac{2x - 2\sqrt{ax}}{\sqrt{5}} \qquad \Big|\ \because\ y^2 = 4ax$$

$$= \frac{2\sqrt{x}(\sqrt{x} - \sqrt{a})}{\sqrt{5}}$$

$$\therefore \qquad PM^2 = \frac{4x(x + a - 2\sqrt{ax})}{5}$$

Now $AM^2 = AP^2 - PM^2 = x^2 + y^2 - \frac{4x(x + a - 2\sqrt{ax})}{5}$

$$= x^2 + 4ax - \frac{4}{5}x(x + a - 2\sqrt{ax}) \qquad |\ \because\ y^2 = 4ax$$

$$= \frac{1}{5}[5x^2 + 20ax - 4x^2 - 4ax + 8\sqrt{a}\,x^{3/2}] = \frac{1}{5}(x^2 + 16ax + 8\sqrt{a}\,x^{3/2}) = \frac{1}{5}(x + 4\sqrt{ax})^2$$

$$\therefore \quad \text{AM} = \frac{1}{\sqrt{5}}(x + 4\sqrt{a}\sqrt{x})$$

$$\therefore \quad d(\text{AM}) = \frac{1}{\sqrt{5}}\left(1 + 4\sqrt{a}\,.\frac{1}{2\sqrt{x}}\right)dx = \frac{1}{\sqrt{5}}\left(\frac{\sqrt{x}+2\sqrt{a}}{\sqrt{x}}\right)dx$$

$\therefore$ Required volume $= \int \pi(\text{PM})^2.d(\text{AM})$

$$= \pi\int_0^a \frac{4x(\sqrt{x}-\sqrt{a})^2}{5}\cdot\frac{1}{\sqrt{5}}\left(\frac{\sqrt{x}+2\sqrt{a}}{\sqrt{x}}\right)dx$$

| $\because$ For the arc AL, x varies from 0 to a.

$$= \frac{4\pi}{5\sqrt{5}}\int_0^a \sqrt{x}(\sqrt{x}-\sqrt{a})^2.(\sqrt{x}+2\sqrt{a})\,dx$$

| Put $x = at^2 \quad \therefore \quad dx = 2at\,dt$
when $x = 0, t = 0$; when $x = a, t = 1$

$$= \frac{4\pi}{5\sqrt{5}}\int_0^1 \sqrt{at}(\sqrt{a}\,t-\sqrt{a})^2.(\sqrt{a}\,t+2\sqrt{a})\,2at\,dt$$

$$= \frac{8\pi a^3}{5\sqrt{5}}\int_0^1 t^2(t+2)(t^2-2t+1)dt = \frac{8\pi a^3}{5\sqrt{5}}\int_0^1 t^2(t^3-3t+2)\,dt$$

$$= \frac{8\pi a^3}{5\sqrt{5}}\left[\frac{t^6}{6} - 3\,.\frac{t^4}{4} + 2\frac{t^3}{3}\right]_0^1 = \frac{8\pi a^3}{5\sqrt{5}}\left[\frac{1}{6}-\frac{3}{4}+\frac{2}{3}\right] = \frac{8\pi a^3}{5\sqrt{5}}\left[\frac{2-9+8}{12}\right] = \frac{2\pi a^3}{15\sqrt{5}}.$$

(*b*) Let the equation of the generating circle be

$$x^2 + y^2 = a^2 \qquad \text{...(1)}$$

Let A and B be the extremities of the arc (in the first quadrant). Then, the co-ordinates of A and B are $(a, 0)$ and $(0, a)$ respectively.

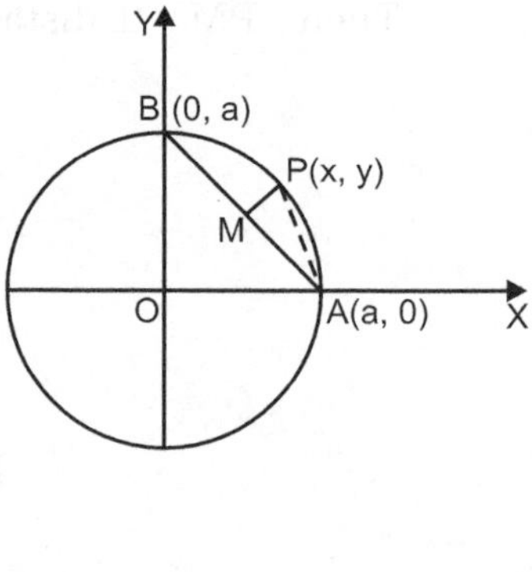

Then the line AB *i.e.*, the line joining the points $(a, 0)$ and $(0, a)$ is

$$y - 0 = \frac{a-0}{0-a}(x-a) = -(x-a)$$

or $$x + y - a = 0$$

Let P(x, y) be any point on the arc AB.

Draw PM ⊥ on chord AB and join AP.

Then $$\text{PM} = \frac{x+y-a}{\sqrt{2}} = \frac{1}{\sqrt{2}}(x - a + \sqrt{a - x^2}) \qquad \text{| From (1)}$$

$$\therefore \quad \text{PM}^2 = \frac{1}{2}.[x^2 + a^2 + (a^2 - x^2) - 2ax + 2x\sqrt{a^2-x^2} - 2a\sqrt{a^2-x^2}]$$

$$= a^2 - ax + \sqrt{a^2-x^2}(x-a) = a(a-x) - (a-x)\sqrt{a^2-x^2}$$

$$= (a - x)(a - \sqrt{a^2 - x^2}) \qquad \ldots(2)$$

Now $AM^2 = AP^2 - PM^2 = (x - a)^2 + y^2 - \left(\dfrac{x + y - a}{\sqrt{2}}\right)^2$

$$= \frac{1}{2}[2(x-a)^2 + 2y^2 - \{(x-a)+y\}^2] = \frac{1}{2}[2(x-a)^2 + 2y^2 - (x-a)^2 - y^2 - 2y(x-a)]$$

$$= \frac{1}{2}[(x-a)^2 + y^2 - 2(x-a)y] = \frac{1}{2}(x - a - y)^2$$

$$\therefore \quad AM = \frac{1}{\sqrt{2}}(x - a - y) = \frac{1}{\sqrt{2}}(x - a - \sqrt{a^2 - x^2}) \qquad | \text{ From (1)}$$

$$\therefore d(AM) = \frac{1}{\sqrt{2}}\left[1 - \frac{1}{2}(a^2 - x^2)^{-1/2}(-2x)\right]dx = \frac{1}{\sqrt{2}}\left[1 + \frac{x}{\sqrt{a^2 - x^2}}\right]dx = \frac{1}{\sqrt{2}}\frac{\sqrt{a^2 - x^2} + x}{\sqrt{a^2 - x^2}}dx$$

For the arc AB, x varies from 0 to a.

$$\therefore \quad \text{Required volume} = \int \pi(PM)^2 \, . \, d(AM)$$

$$= \pi \int_0^a (a - x)(a - \sqrt{a^2 - x^2}) . \frac{1}{\sqrt{2}} . \frac{\sqrt{a^2 - x^2} + x}{\sqrt{a^2 - x^2}} dx$$

| Put $x = a \sin \theta \quad \therefore \quad dx = a \cos \theta \, d\theta$
When $x = 0, \theta = 0, \quad$ when $x = a, \theta = \pi/2$

$$= \frac{\pi}{\sqrt{2}} \int_0^{\pi/2} a(1 - \sin \theta)(a - a \cos \theta) . \frac{a \cos \theta + a \sin \theta}{a \cos \theta} . a \cos \theta \, d\theta$$

$$= \frac{\pi}{\sqrt{2}} a^3 \int_0^{\pi/2} (1 - \sin \theta)(1 - \cos \theta).(\sin \theta + \cos \theta) \, d\theta$$

$$= \frac{\pi a^3}{\sqrt{2}} \int_0^{\pi/2} (\sin \theta + \cos \theta - 2 \sin \theta \cos \theta - 1 + \sin^2 \theta \cos \theta + \cos^2 \theta \sin \theta) \, d\theta$$

$$= \frac{\pi a^3}{\sqrt{2}} \left[- \cos \theta + \sin \theta - \sin^2 \theta - \theta + \frac{\sin^3 \theta}{3} - \frac{\cos^3 \theta}{3} \right]_0^{\pi/2}$$

$$= \frac{\pi a^3}{\sqrt{2}} \left[0 + 1 - 1 - \frac{\pi}{3} + \frac{1}{2} - 0 - \left(-1 - \frac{1}{3} \right) \right] = \frac{\pi a^3}{\sqrt{2}} \left(\frac{5}{3} - \frac{\pi}{2} \right) = \frac{\pi a^3}{6\sqrt{2}} (10 - 3\pi).$$

11.5. VOLUME FORMULA FOR PARAMETRIC EQUATIONS

(*i*) The volume of the solid generated by the revolution about the x-axis, of the area bounded by the curve x = f (t), y = φ(t), the x-axis and the ordinates, where t = a, t = b is

$$\int_a^b \pi y^2 \frac{dx}{dt}\, dt. \qquad \text{| Remember}$$

(*ii*) The volume of the solid generated by the revolution about the y-axis, of the area bounded by the curve x = f(t), y = φ(t), the y-axis and the abscissae at the points, where t = a, t = b is

$$\int_a^b \pi x^2 \frac{dy}{dt}\, dt. \qquad \text{| Remember}$$

Example 1. *Find the volume of the solid formed by the revolution of one arch of the cycloid.*

(*i*) $x = a(\theta - \sin\theta)$, $y = a(1 - \cos\theta)$ *about its base.*

(*ii*) $x = a(\theta + \sin\theta)$, $y = a(1 + \cos\theta)$ *about its base.*

(*iii*) $x = a(\theta + \sin\theta)$, $y = a(1 - \cos\theta)$ *about the tangent at the vertex(i.e., about x-axis).*

(*iv*) *Find the volume of the solid generated by revolving about the y-axis the area between the first arch of the cycloid* $x = \theta + \sin\theta$, $y = 1 - \cos\theta$ *and x-axis.*

Sol. (*i*) The equations of the cycloid are

$$x = a(\theta - \sin\theta),\ y = a(1 - \cos\theta)$$

The cycloid is symmetrical about the line through the point, where θ = π, ⊥ to *x*-axis.

For the first half of the cycloid in the first quadrant, θ varies from 0 to π.

The base is *x*-axis.

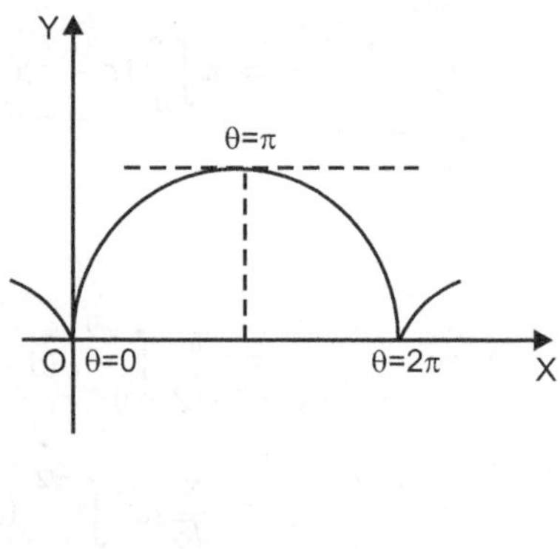

∴ Required volume of the solid generated by revolution about its base (*x*-axis)

$$= 2\int_0^{\pi} \pi y^2 . \frac{dx}{d\theta} . d\theta = 2\pi \int_0^{\pi} a^2(1 - \cos\theta)^2 . a(1 - \cos\theta)\, d\theta \qquad \text{| From (1)}$$

$$= 2\pi a^3 \int_0^{\pi} \left(2\sin^2\frac{\theta}{2}\right)^2 . \left(2\sin^2\frac{\theta}{2}\right) d\theta$$

| Put $\theta/2 = t$ ∴ $d\theta = 2dt$ when $\theta = 0, t = 0$, when $\theta = \pi, t = \pi/2$

$$= 2\pi a^3 \int_0^{\pi/2} (2\sin^2 t)^2 . 2dt = 32\pi a^3 \int_0^{\pi/2} \sin^6 t\, dt = 32\pi a^3 . \frac{5.3.1}{6.4.2} . \frac{\pi}{2} = 5\pi^2 a^3.$$

(*ii*) The equations of the cycloid are

$$x = a(\theta + \sin\theta),\ y = a(1 + \cos\theta)$$

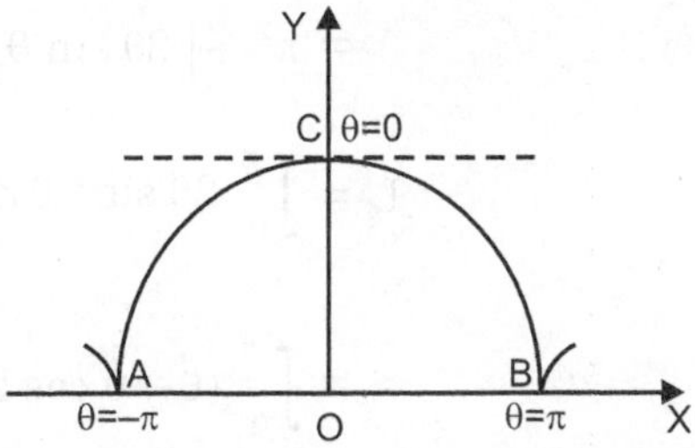

The curve is symmetrical about the y-axis and the base is AB, the x-axis.

For the half of the cycloid, θ varies from 0 to π.

∴ Required volume

$$= 2\int_0^{\pi} \pi y^2 \frac{dx}{d\theta}\, d\theta$$

$$= 2\pi \int_0^{\pi} a^2(1 + \cos\theta)^2 \,.\, a(1 + \cos\theta)\, d\theta$$

$$= 2\pi a^3 \int_0^{\pi} \left(2\cos^2\frac{\theta}{2}\right)^2 . \,2\cos^2\frac{\theta}{2}\, d\theta = 2\pi a^3 \int_0^{\pi/2} 8\cos^6 t \,.2dt \qquad \text{where } \frac{\theta}{2} = t$$

$$= 32\pi a^3 \,.\, \frac{5.3.1}{6.4.2} \,.\, \frac{\pi}{2} = 5\pi^2 a^3.$$

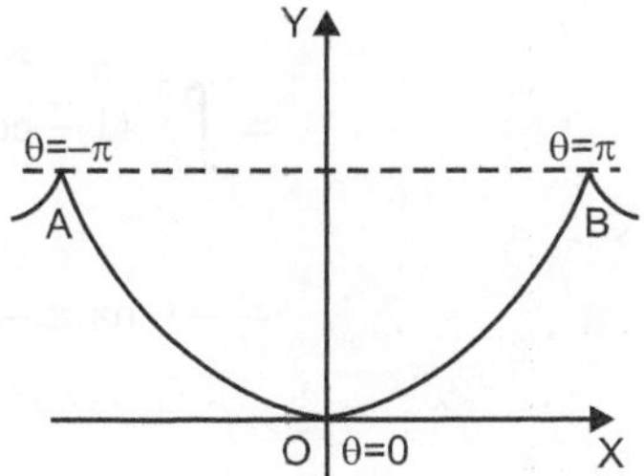

(*iii*) The equations of the cycloid are

$$x = a(\theta + \sin\theta),\ y = a(1 - \cos\theta)$$

The cycloid is symmetrical about the y-axis and the tangent at the vertex is x-axis.

For the half of the curve, θ varies from 0 to π.

∴ Required volume

$$= 2\int_0^{\pi} \pi y^2 \frac{dx}{d\theta}\, d\theta = 2\pi \int_0^{\pi} a^2(1 - \cos\theta)^2 \,.\, a(1 + \cos\theta) d\theta$$

$$= 2\pi a^3 \int_0^{\pi} \left(2\sin^2\frac{\theta}{2}\right)^2 . \left(2\cos^2\frac{\theta}{2}\right) d\theta$$

$$= 2\pi a^3 \int_0^{\pi/2} (2\sin^2 t)^2\, (2\cos^2 t)\,.\, 2dt \qquad \Big|\ \text{where } \frac{\theta}{2} = t$$

$$= 32\pi a^3 \int_0^{\pi/2} \sin^4 t\cos^2 t\, dt = 32\pi a^3 \cdot \frac{3\,.1\,.1}{6\,.4\,.2} \cdot \frac{\pi}{2} = \pi^2 a^3.$$

(*iv*) **For Figure see Ex. 1** (*iii*) with $a = 1$. The required volume

$$V = \int_{y=0}^{y=2} \pi x^2\, dx = \pi \int_{\theta=0}^{\theta=\pi} x^2 \frac{dy}{d\theta}\, d\theta$$

$$= \pi \int_0^{\pi} (\theta + \sin\theta)^2 \sin\theta\, d\theta = \pi \int_0^{\pi} (\theta^2 + \sin^2\theta + 2\theta\sin\theta)\sin\theta\, d\theta$$

$$= \pi \int_0^{\pi} (\theta^2 \sin\theta + 2\theta\sin^2\theta + \sin^3\theta) d\theta = \pi(I_1 + I_2 + I_3)$$

$$I_1 = \int_0^{\pi} \theta^2 \sin\theta\, d\theta = \Big[-\theta^2\cos\theta\Big]_0^{\pi} + \int_0^{\pi} 2\theta\cos\theta\, d\theta$$

$$= \pi^2 + \left| 2\theta \sin\theta \right|_0^\pi - \left| -2\cos\theta \right|_0^\pi = \pi^2 + 2(-1-1) = \pi^2 - 4$$

$$I_2 = \int_0^\pi 2\theta \sin^2\theta \, d\theta = \int_0^\pi 2\theta \left[\frac{1-\cos 2\theta}{2}\right] d\theta$$

$$= \int_0^\pi (\theta - \theta\cos 2\theta)\, d\theta = \left|\frac{\theta^2}{2}\right|_0^\pi - \left|\theta\frac{\sin 2\theta}{2}\right|_0^\pi + \int_0^\pi 1.\frac{\sin 2\theta}{2}\, d\theta$$

$$= \frac{\pi^2}{2} - 0 + \frac{1}{2}\left|\frac{-\cos 2\theta}{2}\right|_0^\pi = \frac{\pi^2}{2} - \frac{1}{4}.0 = \frac{\pi^2}{2}$$

$$I_3 = \int_0^\pi \sin^3\theta\, d\theta = \int_0^\pi \sin^2\theta . \sin\theta\, d\theta$$

$$= \int_0^\pi (1-\cos^2\theta)\sin\theta\, d\theta = \left|-\cos\theta\right|_0^\pi + \left|\frac{\cos^3\theta}{3}\right|_0^\pi$$

$$= -(\cos\pi - \cos 0) + \frac{1}{3}(\cos^3\pi - \cos^3 0) = -(-2) + \frac{1}{3}.[(-1)-(1)] = 2 - \frac{2}{3} = \frac{4}{3}$$

$$\therefore \quad V = \pi(I_1 + I_2 + I_3) = \pi\left[\pi^2 - 4 + \frac{\pi^2}{2} + \frac{4}{3}\right] = \pi\left[\frac{3}{2}\pi^2 - \frac{8}{3}\right]$$

Example 2. (*a*) *Prove that the volume of the solid generated by the revolution about the x-axis of the loop of the curve* $x = t^2,\ y = t - \frac{1}{3}t^3$ *is* $\frac{3\pi}{4}$.

(*b*) *Find the volume of the spindle shaped solid generated by revolving the astroid* $x^{2/3} + y^{2/3} = a^{2/3}$ *about the x-axis.*

(*c*) *Find the volume of the solid generated by revolvtion of the tractrix*

$x = a\cos t + \frac{a}{2}\log\tan^2\frac{t}{2},\ y = a\sin t$ *about its asymptote .*

(*d*) *Prove that the volume of the solid generated by the revolution of the tractrix.*

$x = a\cos t + \frac{a}{2}\log\tan^2\frac{t}{2},\ y = a\sin t,$ *about its asymptote half of a sphere of redius a.*

Sol. (*a*) The equations of the curve are

$$x = t^2,\ y = t - \frac{1}{3}t^3 \qquad \ldots(1)$$

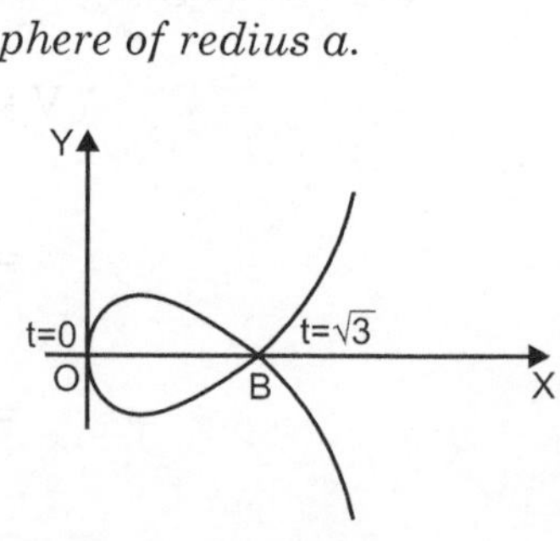

The curve is symmetrical about the x-axis and for the loop, putting $y = 0$, we get

$$t\left(1 - \frac{1}{3}t^2\right) = 0$$

$\therefore \qquad t = 0, \pm\sqrt{3}.$

Thus for the upper half of the loop t-varies from 0 to $\sqrt{3}$.

$$\therefore \qquad \text{Required volume} = \int_0^{\sqrt{3}} \pi y^2 \frac{dx}{dt}\,dt = \int_0^{\sqrt{3}} \pi.\left(t - \frac{1}{3}t^3\right)^2 .\,2t\,dt$$

$$= 2\pi\int_0^{\sqrt{3}} t\left(t^2 - \frac{1}{9}t^6 - \frac{2}{3}t^4\right)dt = 2\pi\int_0^{\sqrt{3}}\left(t^3 + \frac{1}{9}t^7 - \frac{2}{3}t^5\right)dt$$

$$= 2\pi\left[\frac{t^4}{4} + \frac{1}{9}.\frac{t^8}{8} - \frac{2}{3}.\frac{t^6}{6}\right]_0^{\sqrt{3}} = 2\pi\left[\frac{9}{4} + \frac{1}{9}.\frac{81}{8} - \frac{2}{3}.\frac{27}{6}\right]$$

$$= 2\pi\left[\frac{9}{4} + \frac{9}{8} - 3\right] = 2\pi\left[\frac{18 + 9 - 24}{8}\right] = \frac{3\pi}{4}$$

(*b*) The equation of the curve is

$$x^{2/3} + y^{2/3} = a^{2/3}$$

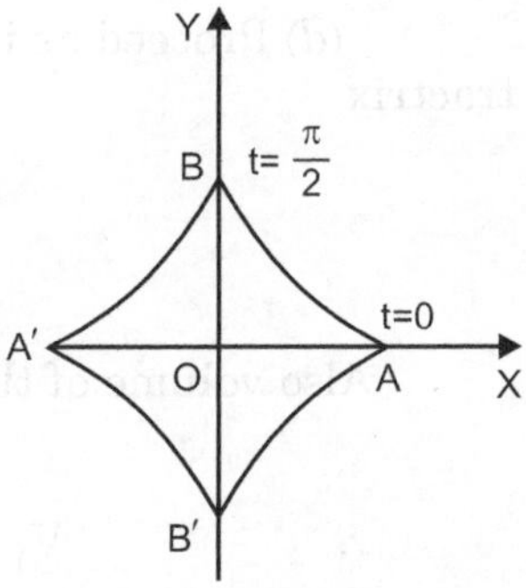

The parametric equations of the curve are

$$x = a\cos^3 t,\ y = a\sin^3 t \qquad \ldots(1)$$

The curve is symmetrical about the axes and for the portion of the curve in the first quadrant, t varies from 0 to $\frac{\pi}{2}$.

$\therefore$ Required volume = 2 × volume generated by the arc in the first quadrant

$$= 2\int_0^{\pi/2} \pi y^2 \frac{dx}{dt}\,dt = 2\pi\int_0^{\pi/2} a^2\sin^6 t.(-\,3a\cos^2 t\sin t)\,dt$$

$$= -\,6\pi a^3\int_0^{\pi/2} \sin^7 t\,.\cos^2 t\,dt = -\,6\pi a^2\,.\,\frac{6.4.2.1}{9.7.5.3.1}$$

$$= \frac{-\,32}{105}\pi a^3 = \frac{32}{105}\pi a^3 \text{ (in magnitude)}.$$

(*c*) The equations of the curve are

$$x = a\cos t + \frac{a}{2}\log\tan^2\frac{t}{2},\ y = a\sin t \qquad \ldots(1)$$

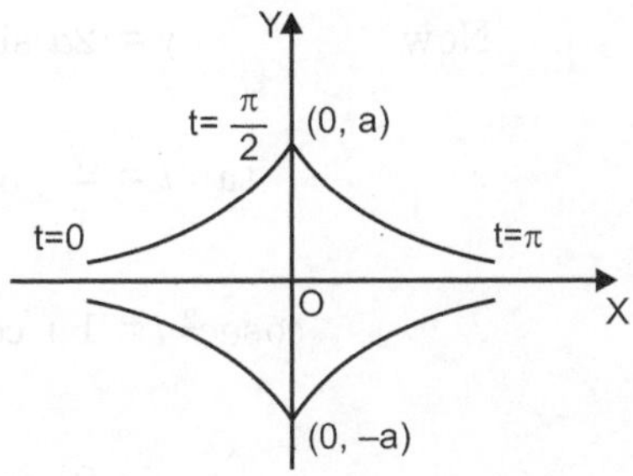

The curve is symmetrical about both the axes and the asymptote is $y = 0$ *i.e.*, x-axis.

From (1),

$$\frac{dx}{dt} = -\,a\sin t + \frac{a}{2}.\frac{1}{\tan^2 t/2}.\,2\tan\frac{t}{2}.\sec^2\frac{t}{2}.\frac{1}{2}$$

$$= - a \sin t + \frac{a}{2 \sin t/2 \cos t/2}$$

$$= a \sin t + \frac{a}{\sin t} = a \frac{(1 - \sin^2 t)}{\sin t} = a \frac{\cos^2 t}{\sin t}$$

For the curve in the second quadrant, t varies from 0 to $\frac{\pi}{2}$

$\therefore$ Required volume = 2 × vol. generated by the arc in the second quadrant

$$= 2 \int_0^{\pi/2} \pi y^2 \frac{dx}{dt} dt = 2\pi \int_0^{\pi/2} a^2 \sin^2 t . \frac{a \cos^2 t}{\sin t} dt$$

$$= 2\pi a^3 \int_0^{\pi/2} \cos^2 t \sin t \, dt = - 2\pi a^3 \left[\frac{\cos^3 t}{3} \right]_0^{\pi/2} = \frac{-2\pi a^3}{3} [0 - 1] = \frac{2}{3} \pi a^3.$$

(*d*) Proceed as in part (*c*) to get volume of the solid generated by the revolution of the tractrix.

$$= \frac{2}{3} \pi a^3 = V_1 (say).$$

Also volume of the sphere of radius $a = \frac{4}{3} \pi a^3 = V_2$(say).

$\therefore \quad V_1 = \frac{1}{2} V_2$. Hence the result.

Example 3. *Find the volume of the solid generated by the arc of the cissiod* $x = 2a \sin^2 t$, $y = \frac{2a \sin^3 t}{\cos t}$ *about its asymptote.*

Sol. The equations of the curve are

$$x = 2a \sin^2 t, \; y = \frac{2a \sin^3 t}{\cos t} \qquad ...(1)$$

Now $\quad y = 2a \sin^2 t . \frac{\sin t}{\cos t} = x \tan t$

$\therefore \quad \tan t = \frac{y}{x}$ or $\cot t = \frac{x}{y}$

$\therefore \quad \operatorname{cosec}^2 t = 1 + \cot^2 t = 1 + \frac{x^2}{y^2} = \frac{x^2 + y^2}{y^2}$ or $\sin^2 t = \frac{y^2}{x^2 + y^2}$.

$\therefore \quad x = 2a \sin^2 t = 2a \frac{y^2}{x^2 + y^2}$ or $x^3 + xy^2 = 2ay^2$ or $y^2(2a - x) = x^3$

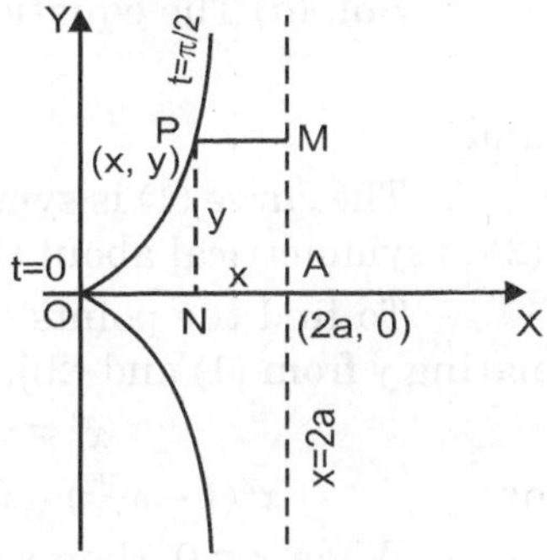

The curve is symmetrical about the x-axis and the asymptote is the line $x = 2a$.

Let P(x, y) be any point on the curve. Draw PM ⊥ on the asymptote and PN ⊥ OX. Then

$$\text{PM} = \text{AN} = \text{OA} - \text{ON} = 2a - x$$
$$= 2a - 2a \sin^2 t$$
$$= 2a(1 - \sin^2 t) = 2a \cos^2 t$$

and $$\text{AM} = \text{PN} = y = 2a \frac{\sin^3 t}{\cos t}$$

$$\therefore \quad d(\text{AM}) = 2a \frac{\cos t \,.\, 3 \sin^2 t \,.\, \cos t - \sin^3 t(-\sin t)}{\cos^2 t} dt$$

$$= 2a \,.\, \frac{\sin^2 t(3\cos^2 t + \sin^2 t)}{\cos^2 t} dt = 2a \,.\, \frac{\sin^2 t(1 + \cos^2 t)}{\cos^2 t} dt$$

∴ For the upper half of the arc, t varies from 0 to $\frac{\pi}{2}$.

∴ Required volume = 2 × vol. generated by the revolution of upper half of the curve

$$= 2\int \pi(\text{PM})^2 . d(\text{AM})$$

$$= 2\int_0^{\pi/2} \pi(2a\cos^2 t)^2 2a \frac{\sin^2 t(1 + 2\cos^2 t)}{\cos^2 t} dt$$

$$= 16\pi a^3 \int_0^{\pi/2} (\sin^2 t \cos^2 t + 2\cos^4 t \sin^2 t)\, dt$$

$$= 16\pi a^3 \left[\frac{1.1}{4.2}.\frac{\pi}{2} + 2.\frac{3.1.1}{6.4.2}.\frac{\pi}{2}\right] = 16\pi a^3 \left[\frac{\pi}{16} + \frac{\pi}{16}\right] = 2\pi^2 a^3.$$

11.6. VOLUME BETWEEN TWO SOLIDS

The volume of the solid generated by the revolution about the x-axis, of the area bounded by the curves y = f(x), y = φ(x), and the ordinates x = a, x = b is

$$\int_b^a \pi(y_1^2 - y_2^2)dx$$ **| Remember**

where y_1 is the 'y' of the upper curve and y_2 that of the lower curve.

Example. (*a*) *Find the volume of the solid of revolution abtained by rotating the area included between the curves $y^2 = x^3$ and $x^2 = y^3$ about the x-axis.*

(*b*) *The figure bounded by a parabola and the tangents at the extremities of its latus rectum revolves about axis of the parabola. Find the volume of the solid thus generated.*

Sol. (*a*) The equations of the curves are

$$y^2 = x^3 \quad ...(1)$$

and $$x^2 = y^3 \quad ...(2)$$

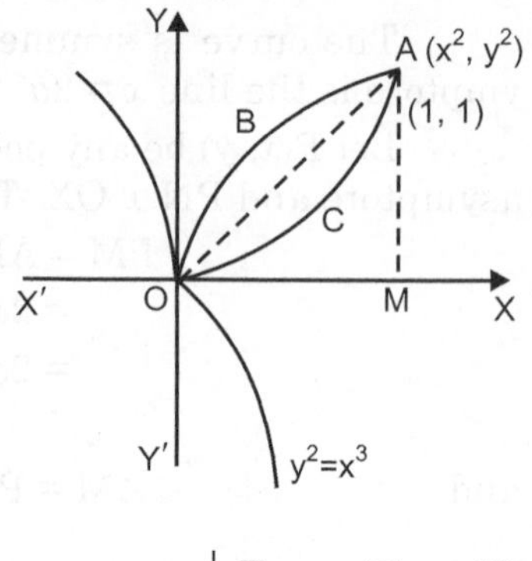

The curve (1) is symmetrical about x-axis and the curve (2) is symmetrical about the y-axis.

To find the points of intersection of (1) and (2), [eliminating y from (1) and (2)], we have

$$x^2 = y^3 = (y^2)^{3/2} = (x^3)^{3/2} = x^{9/2} \quad | \text{ From (1)}$$

or $$x^2(1 - x^{5/2}) = 0 \quad \therefore \quad x = 0, 1$$

When $x = 0$, then $y = 0$; when $x = 1$, $y = 1$ | From (1) or (2)

Hence the two curves meet in the points O (0, 0) and A(1, 1).

$$\therefore \quad \text{Required volume} = \pi \int_0^1 (y_1^2 - y_2^2) dx$$

where y_1 is the y of upper curve (2) and y_2 is the y of the lower curve (1)

$$= \pi \int_0^1 (x^{4/3} - x^3) dx \qquad \Big| \because \text{ From (2), } y_1^2 = y^2 = x^{4/3} \text{ and from (1), } y_2^2 = y^2 = x^3$$

$$= \pi \left[\frac{3}{7} x^{7/3} - \frac{x^4}{4} \right]_0^1 = \pi \left(\frac{3}{7} - \frac{1}{4} \right) = \frac{5}{28} \pi.$$

(*b*) Let the equation of the parabola be $y^2 = 4ax$ ...(1)

Let LSL′ be its latus rectum and LT, L′T be the tangents at the extremities L,L′. The equation of the tangent at L(a, $2a$) is

$$y.\ 2a = 2a(x + a) \quad | \text{ Using } yy_1 = 2a(x + x_1)$$

or $$y = x + a \quad ...(2)$$

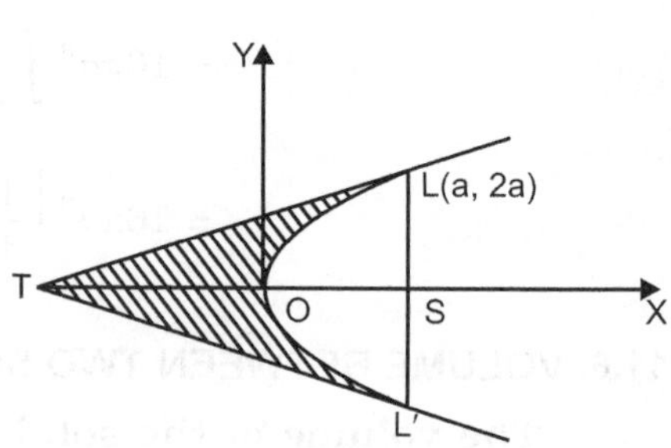

This meets the x-axis, where putting $y = 0$, we get

$$x = -a.$$

Thus the point T is $(-a, 0)$. The curve and the two tangents are symmertical about the x-axis.

$\therefore$ Required volume

$$= \int_{-a}^{a} \pi y^2 dx - \int_0^a \pi y^2 dx$$

(for tangent) (for parabola)

$$= \pi \int_{-a}^{a} (x + a)^2 \, dx - \pi \int_0^a 4ax \, dx$$

$$= \pi \left| \frac{(x + a)^3}{3} \right|_{-a}^{a} - 2a\pi \left| x^2 \right|_0^a = \frac{\pi}{3}(8a^3 - 0) - 2a\pi(a^2 - 0) = \frac{2}{3} \pi a^3.$$

11.7. THE VOLUME FORMULAE FOR POLAR CURVES

The volume of the solid generated by the revolution of the area bounded by the curve $r = f(\theta)$ and the radii vectors $\theta = \alpha, \theta = \beta$

(*i*) about the initial line OX ($\theta = 0$) is

$$\int_{\alpha}^{\beta} \frac{2}{3}\pi r^3 \sin\theta\, d\theta$$ | **Remember**

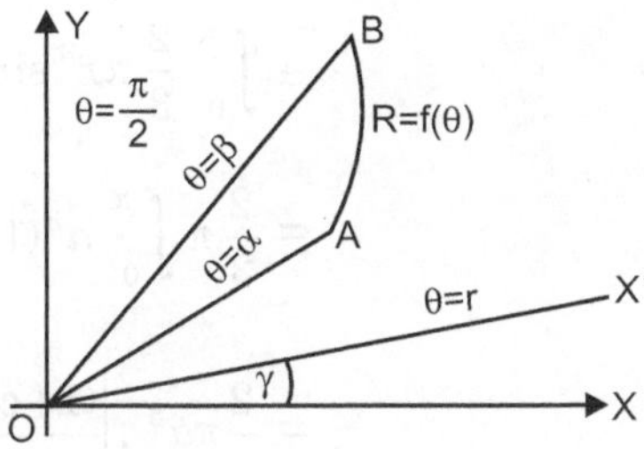

(*ii*) about the line OY $\left(\theta = \frac{\pi}{2}\right)$ is

$$\int_{\alpha}^{\beta} \frac{2}{3}\pi r^3 \cos\theta\, d\theta$$ | **Remember**

(*iii*) about any line OX′ ($\theta = \gamma$) is

$$\int_{\alpha}^{\beta} \frac{2}{3}\pi r^3 \sin(\theta - \gamma)\, d\theta$$ | **Remember**

The proofs of these are beyond the scope of the book and so they are omitted.

Example 1. (*a*) *Find the volume of the solid generated by the revolution of $r = 2a^2 \cos\theta$ about the intial line.*

(*b*) *The cardioid $r = a(1 + \cos\theta)$, revolves about the initial line, find the volume of the solid generated.* [CUSAT 2010]

(*c*) *The arc of cardioid $r = a(1 + \cos\theta)$, specified by $-\frac{\pi}{2} \le \theta \le \frac{\pi}{2}$ is rotated about the line $\theta = 0$, prove that the volume generated is $\frac{5}{2}\pi a^3$.*

Sol. (*a*) The equations of the curve is

$$r = 2a\cos\theta \qquad \ldots(1)$$

(1) is clearly a circle passing through the pole. The curve is symmetrical about the initial line and for the upper half of the circle θ varies from 0 of $\frac{\pi}{2}$.

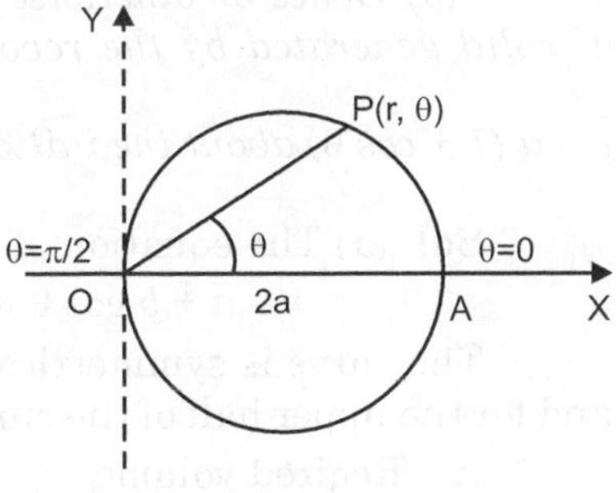

$$\left[\because \begin{array}{l} \text{when } \theta = 0, r = 2a\,; \\ \text{when } \theta = \frac{\pi}{2}, r = 0 \end{array}\right]$$

$$\therefore \quad \text{Required volume} = \int_0^{\pi/2} \frac{2}{3}\pi r^3 \sin\theta\, d\theta = \frac{2}{3}\pi\int_0^{\pi/2} (2a\cos\theta)^3 \sin\theta\, d\theta$$

$$= \frac{16}{3}\pi a^3 \int_0^{\pi/2} \cos^3\theta \sin\theta\, d\theta = -\frac{16}{3}\pi a^3 . \left[\frac{\cos^4\theta}{4}\right]_0^{\pi/2} = -\frac{4}{3}\pi a^3 (0 - 1) = \frac{4}{3}\pi a^3.$$

(*b*) The equation of the cardioid is $r = a\,(1 + \cos\theta)$.

The cardioid is symmetrical about the initial line and for the upper half of the curve, θ varies from 0 to π.

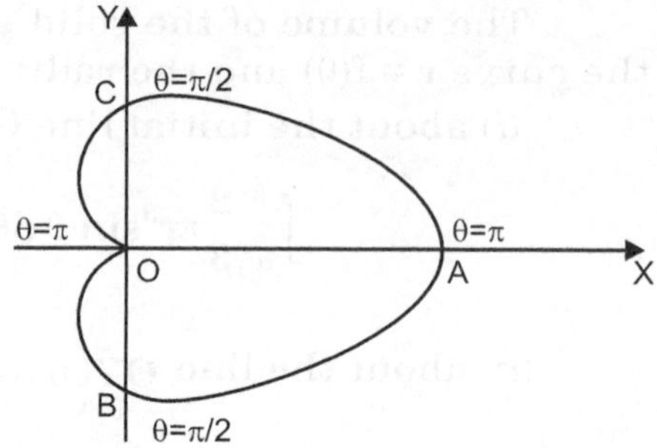

∴ Required volume

$$= \int_0^{\pi} \frac{2}{3}\pi r^3 \sin\theta\, d\theta$$

$$= \frac{2}{3}\pi \int_0^{\pi} a^3(1 + \cos\theta)^3 .\sin\theta\, d\theta$$

$$= \frac{2}{3}\pi a^3 . \left|\frac{(1+\cos\theta)^4}{-4}\right|_0^{\pi} = -\frac{1}{6}\pi a^3 \Big| (1+\cos\theta)^4 \Big|_0^{\pi} = -\frac{1}{6}\pi a^3 (0-16) = \frac{8}{3}\pi a^3.$$

(*c*) The cardioid $r = a(1 + \cos\theta)$ is symmetrical about the initial line and the arc BAC of the curve revolves about the initial line.

For the upper half of the arc AC, θ varies from 0 to $\frac{\pi}{2}$

$$\therefore \quad \text{Reqired volume} = \int_0^{\pi/2} \frac{2}{3}\pi\, r^3 \sin\theta\, d\theta = \frac{2}{3}\pi a^3 \int_0^{\pi/2} (1 + \cos\theta)^3 \sin\theta\, d\theta$$

$$= -\frac{2}{3}\pi a^3 . \left[\frac{(1 + \cos\theta)^4}{4}\right]_0^{\pi/2} = -\frac{1}{6}\pi a^3\,(1 - 16) = \frac{15}{6}\pi a^3 = \frac{5}{2}\pi a^3.$$

Example 2. (*a*) *Show that the volume of the solid formed by the revolution of the curve* $r = a + b\cos\theta\ (a > b)$ *about the initial line is* $\frac{4}{3}\pi a(a^2 + b^2)$.

(*b*) *Hence or otherwise show that the volume of solid generated by the revolution of the cardioid* $r = a\,(1 + \cos\theta)$ *about the initial line is* $\frac{8}{3}\pi a^3$.

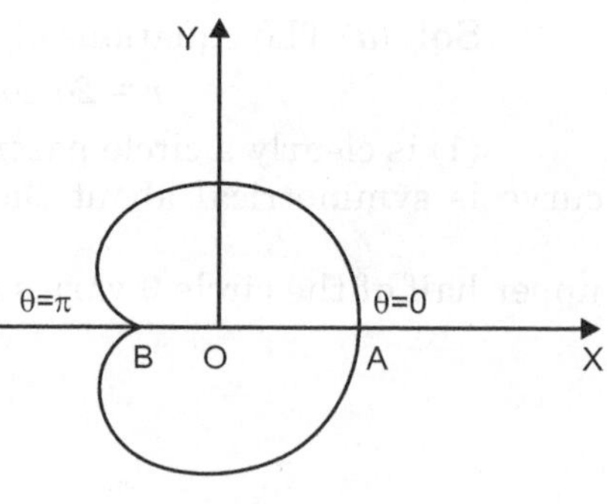

Sol. (*a*) The equation of the curve is

$$r = a + b\cos\theta\ (a > b) \qquad \text{...(1)}$$

The curve is symmetrical about the initial line and for the upper half of the curve θ varies from 0 to π.

∴ Reqired volume

$$= \int_0^{\pi} \frac{2}{3}\pi\, r^3 \sin\theta\, d\theta = \frac{2}{3}\pi \int_0^{\pi} (a + b\cos\theta)^3 \sin\theta\, d\theta$$

$$= -\frac{2}{3}\frac{\pi}{b} \int_0^{\pi} (a + b\cos\theta)^3(-\,b\sin\theta\, d\theta)$$

$$= -\frac{2}{3}\frac{\pi}{b}\left[\frac{(a+b\cos\theta)^4}{4}\right]_0^{\pi} = -\frac{2\pi}{3b}\left[\frac{(a-b)^4}{4} - \frac{(a+b)^4}{4}\right]$$

$$= \frac{\pi}{6b}[(a+b)^4 - (a-b)^4] = \frac{\pi}{6b}[(a+b)^2 + (a-b)^2][(a+b)^2 - (a-b)^2]$$

$$= \frac{\pi}{6b}.\,2(a^2+b^2).\,4ab = \frac{4}{3}\pi a(a^2+b^2) \qquad \ldots(2)$$

(*b*) Putting $b = a$, in (1), the curve becomes

$r = a\,(1 + \cos\theta)$ which is the given cardioid.

∴ Required volume by the revolution of the cardioid about the initial line [Putting $b = a$ in (2)] is

$$\frac{4}{3}\pi a(a^2+a^2) = \frac{8}{3}\pi a^3.$$

Example 3. (*a*) *Find the volume of the solid generated by revolving one loop of the lemniscate* $r^2 = a^2\cos 2\theta$ *about the line* $\theta = \frac{\pi}{2}$.

(*b*) *The area of the inner loop of the curve* $r = 1 + 2\cos\theta$ *is rotated through two right angles about the initial line. Show that the volume of the solid so formed is* $\frac{\pi}{12}$.

Sol. (*a*) The given curve is $r^2 = a^2\cos 2\theta$ ...(1)

The curve is symmetrical about the initial line.

For a loop, putting $r = 0$ in (1), we get $\cos 2\theta = 0$.

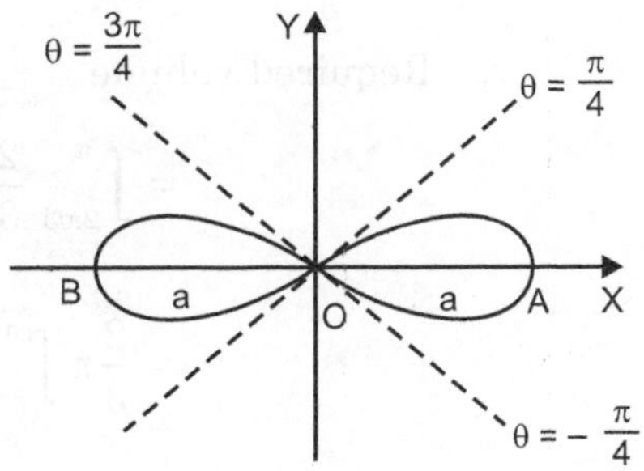

∴ $2\theta = \pm\frac{\pi}{2}$

or $\theta = \pm\frac{\pi}{4}$ [Two consecutive values]

Thus as θ varies from $-\frac{\pi}{4}$ to $\frac{\pi}{4}$, we get a loop.

For the upper half of this loop, θ varies from 0 to $\frac{\pi}{4}$.

∴ Required volume obtained by revolution of the loop about OY $\left(\theta = \frac{\pi}{2}\right)$

$$= 2\int_0^{\pi/4}\frac{2}{3}\pi r^3\cos\theta\,d\theta = \frac{4}{3}\pi\int_0^{\pi/4} a^3(\cos 2\theta)^{3/2}\cos\theta\,d\theta$$

$$= \frac{4}{3}\pi a^3 \int_0^{\pi/4} (1 - 2\sin^2\theta)^{3/2} \cos\theta \, d\theta \qquad \Big| \because \cos 2\theta = 1 - 2\sin^2\theta$$

Put $\sqrt{2}\sin\theta = \sin\phi$ **[Note this]**
$\therefore \sqrt{2}\cos\theta \, d\theta = \cos\phi \, d\phi$.
when $\theta = 0, \phi = 0$ when $\theta = \pi/4, \phi = \pi/2$

$$= \frac{4}{3}\pi a^3 \int_0^{\pi/2} (1 - \sin^2\phi)^{3/2} \frac{1}{\sqrt{2}} \cos\phi \, d\phi$$

$$= \frac{4}{3\sqrt{2}}\pi a^3 \int_0^{\pi/2} \cos^4\phi \, d\phi = \frac{4}{3\sqrt{2}}\pi a^3 \cdot \frac{3.1}{4.2}\frac{\pi}{2} = \frac{\pi^2 a^3}{4\sqrt{2}}.$$

(*b*) The equation of the curve is $r = 1 + 2\cos\theta$.
The curve is symmetrical about the initial line.

Also	θ =	0	$\frac{\pi}{3}$	$\frac{\pi}{2}$	$\frac{2\pi}{3}$	π	$\frac{4\pi}{3}$
	r =	3	2	1	0	-1	0

Thus the inner loop is obtained as θ varies from $\frac{2\pi}{3}$ to $\frac{4\pi}{3}$. Since the loop (*or* curve) is symmetrical about the initial line, $\therefore$ for lower half of the inner loop, θ varies from $\frac{2\pi}{3}$ to π.

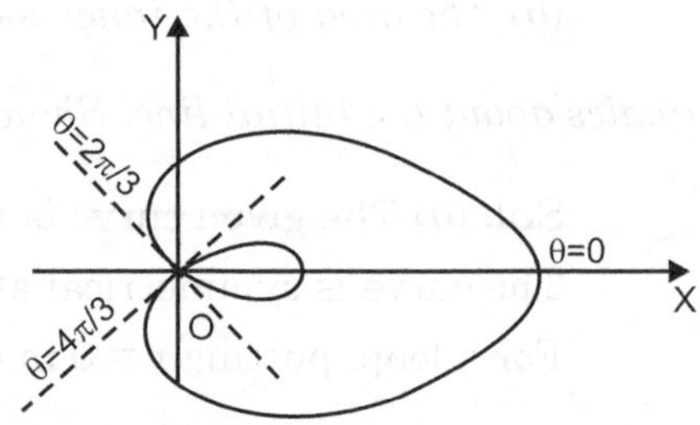

$\therefore$ Required volume

$$= \int_{2\pi/3}^{\pi} \frac{2}{3}\pi r^3 \sin\theta \, d\theta$$

$$= \frac{2}{3}\pi \int_{2\pi/3}^{\pi} (1 + 2\cos\theta)^3 \sin\theta \, d\theta = \frac{2}{3}\pi . \left[\frac{(1 + 2\cos\theta)^4}{-8}\right]_{2\pi/3}^{\pi}$$

$$= -\frac{\pi}{12}[(-1)^4 - (1-1)^4] = -\frac{\pi}{12}(1 - 0) = -\frac{\pi}{12} = \frac{\pi}{12} \text{ (in magnitude)}$$

Example 4. *Show that if the area lying within the cardioid $r = 2a(1 + \cos\theta)$ and without the parabola $r(1 + \cos\theta) = 2a$, revolves about the initial line, the volume generated is $18\pi a^3$.*

Sol. The equation of the cardioid is $r = 2a(1 + \cos\theta)$ ...(1)

and that of the parabola is $r = \dfrac{2a}{1 + \cos\theta}$ ...(2)

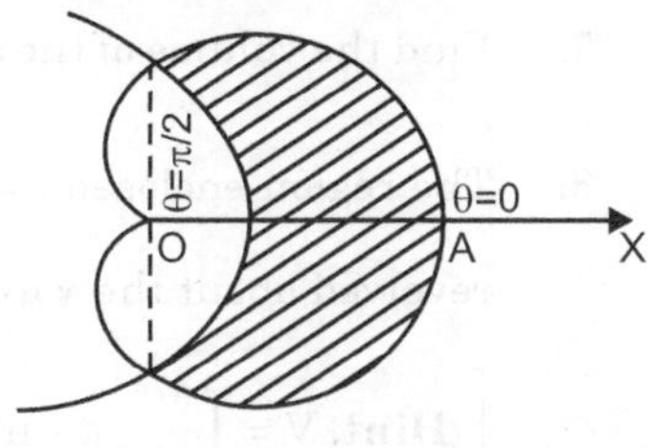

Both the curve are symmetrical about the initial line. The upper half of the shaded area revolves about the initial line and it will generate the required volume.

The two curve intersect where solving (1) and (2) [eliminating r],

$$2a(1 + \cos\theta) = \frac{2a}{1 + \cos\theta}$$

or $\quad (1 + \cos\theta)^2 = 1 \quad$ or $\quad 1 + \cos\theta = \pm 1$

Now either $\quad 1 + \cos\theta = 1 \quad$ *i.e.,* $\quad \cos\theta = 0 \quad$ or $\quad \theta = \pm\dfrac{\pi}{2}$

or $\quad 1 + \cos\theta = -1 \quad$ *i.e.,* $\quad \cos\theta = -2$, which is impossible.

Thus the two curves intersect at the points where $\theta = \dfrac{\pi}{2}.\ \theta = -\dfrac{\pi}{2}$.

For the upper half, θ varies from 0 to $\dfrac{\pi}{2}$.

$\therefore$ Required volume $= \displaystyle\int_0^{\pi/2} \frac{2}{3}\pi[(r \text{ of outer curve})^3 - (r \text{ of inner curve})^3 \sin\theta\, d\theta$

$$= \frac{2}{3}\pi \int_0^{\pi} \left[8a^3(1+\cos\theta)^3 - \frac{8a^3}{(1+\cos\theta)^3}\right] \sin\theta\, d\theta$$

$$= \frac{16}{3}\pi a^3 . \left[\frac{(1+\cos\theta)^4}{-4} - \frac{(1+\cos\theta)^{-2}}{(-2)(-1)}\right]_0^{\pi/2}$$

$$= \frac{16}{3}\pi a^3\left[-\frac{1}{4}(1-16) - \frac{1}{2}\left(1 - \frac{1}{4}\right)\right]$$

$$= \frac{16}{3}\pi a^3\left[\frac{15}{4} - \frac{3}{8}\right] = \frac{16}{3}\pi a^3 \times \frac{27}{8} = 18\pi a^3.$$

TEST YOUR KNOWLEDGE

1. Find the volume of hemisphere.

   [**Hint.** Hemisphere is generated by revolving a quadrant of the circle $x^2 + y^2 = a^2$ about x-axis, thus $0 \le x \le a$.]

2. The part of the curve $y^2 = x^2(1 - x^2)$ between $x = 0$ and $x = 1$ rotates about the x-axis. Obtain the volume of the solid thus generated.

3. Find the volume of the solid obtained by rotating the ellipse $x^2 + 9y^2 = 9$ about the x-axis.

4. The area of the parabola $y^2 = 4ax$ lying between the ordinates $x = 0$ and $x = a$ is revolved about the x-axis. Find the volume of the solid thus generated.

5. Find the volume of the solid formed by the revolution of the curve $yx^2 + y = 1$ about its asymptote.

6. The curve $y^2(a + x) = x^2(a - x)$ revolves abut x-axis. Find the volume generated by the loop.

7. Find the volume of the solid formed by the revolution of the curve $xy^2 = 4(2 - x^2)$ about the y-axis.

8. The region enclosed by the curves $y = \sin x$, $y = \cos x$ and the x-axis, between $x = 0$ to $x = \frac{\pi}{2}$, is revolved about the x-axis. Find the volume of the solid thus obtained.

$$\left[\textbf{Hint. } V = \int_0^{\pi/4} \pi \sin^2 x \, dx + \int_{\pi/4}^{\pi/2} \pi \cos^2 x \, dx \right]$$

9. Find the volume of the solid generated by the revolution of the cardioid $r = a(1 - \cos \theta)$ about the initial line.

Answers

1. $\frac{2}{3} \pi a^3$ 2. $\frac{2\pi}{15}$ 3. 4π

4. $2\pi a^3$ 5. $\frac{\pi^2}{2}$ 6. $2\pi a^3 \left(\log 2 - \frac{2}{3} \right)$

7. $4\pi^2$ 8. $\frac{\pi(\pi - 2)}{4}$ 9. $\frac{8}{3} \pi a^3$

12

Surface of Solids of Revolution

12.1. INTRODUCTION

Surface area is the area of the given surface. It gives the sum of all the areas that cover the surface of the object. For example, the surface area of a cube is $6a^2$ if a is the length of the side of cube.

The area of surface or surface area can be obtained by revolving an arc of the curve about X-axis or Y-axis.

Prove that the curved surface of the solid generated by the evolution, about the x-axis, of the area bounded by the curve y = f(x), the x-axis and the ordinates x = a, x = b is

$$\int_{x=a}^{x=b} 2\pi\, y\, ds$$

where s is the length of the arc of the curve measured from a fixed point on it to any point (x, y).

Or

Show that the area of the surface of the solid obtained by revolving about x-axis the arc of the curve intercepted between the points whose abscissae are a and b is

$$\int_a^b 2\pi\, y \frac{ds}{dx}\, dx.$$

Let AB be the curve $y = f(x)$ and A, B the points corresponding to $x = a$, $x = b$ respectively. Draw AC, BD ⊥ s to x-axis.

Let P(x, y) be any point on the curve and let the arc AP be s. Draw PM ⊥ OX.

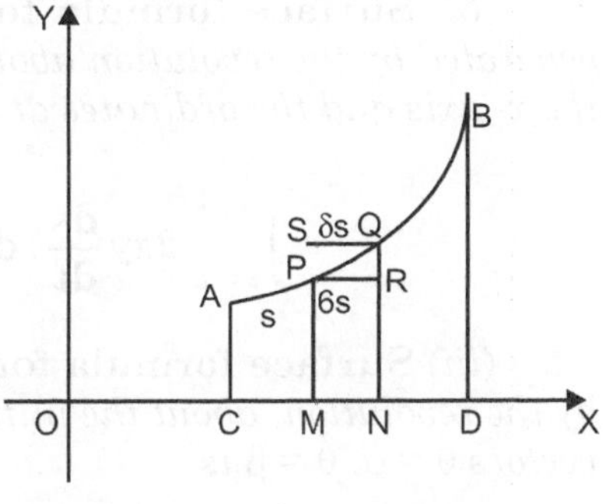

If S denotes the curved surface of the solid generated by the revolution of the area ACMP about the x-axis, then clearly S is a function of s.

Let Q ($x + \delta x$, $y + \delta y$) be a point on the curve in the neighbourhood of P. Draw QN ⊥ OX, and let the arc AQ be $s + \delta s$, so that arc PQ = δs.

The curved surface of the solid of revolution of the area PMNQ about the x-axis is δS. Draw PR and QS parallel to x-axis and each equal in length to the arc PQ = δs.

Now the lines PR and QS generated cylinder, when the arc PQ revolves about x-axis, whose base radii are PM and NQ. The area of the curved surface generated by the arc PQ lies between the areas of the curved surfaces of the cylinders whose base radii are PM and NQ.

i.e., δS lies between $2\pi y\ \delta s$ and $2\pi(y + \delta y)\ \delta s$.

$\therefore$ $\dfrac{\delta S}{\delta s}$ lies between $2\pi y$ and $2\pi\ (y + \delta y)$.

Proceeding to limit as $Q \to P$, *i.e.*, as $\delta x \to 0$ $(\because\ \delta y \to 0, \delta s \to 0)$

$\dfrac{\delta S}{\delta s}$ lies between $2\pi y$ and a quantity which approaches $2\pi y$.

$$\therefore \quad \frac{dS}{ds} = 2\pi y \qquad \text{...(1)}$$

$$\therefore \quad \int_{x=a}^{x=b} 2\pi y\ ds = \int_{x=a}^{x=b} \frac{dS}{ds}.ds = [S]_{x=a}^{x=b}$$

= (value of S when $x = b$) – value of S when $x = a$)

= area of the surface generated by the revolution of the area ACDB – 0

$\therefore$ Surface area of the solid generated by the revolution of the area

$$\text{ACDB} = \int_{x=a}^{x=b} 2\pi y\ ds.$$

Three practical forms of the surface formula :

(*i*) **Surface formula for Cartesian equations.** *The curved surface of the solid generated by the revolution about the x-axis, of the bounded by the curve y = f(x), the x-axis and the ordinates x = a, x = b is*

$$\int_{x=a}^{x=b} 2\pi\ y\frac{ds}{dx}\ dx, \text{where } \frac{ds}{dx} = \sqrt{1+\left(\frac{dy}{dx}\right)^2} \qquad | \textbf{ Remember}$$

(*ii*) **Surface formula for parametric equations.** *The curved surface of the solid generated by the revolution about the x-axis, of the area bounded by the curve x = f(t), y = φ(t), the x- axis and the ordinates at the points, where t = a, t = b is*

$$\int_{t=a}^{t=b} 2\pi y\frac{ds}{dt}.\ dt, \text{where } \frac{ds}{dt} = \sqrt{\left(\frac{dx}{dt}\right)^2\left(\frac{dy}{dt}\right)^2}. \qquad | \textbf{ Remember}$$

(*iii*) **Surface formula for polar equations.** *The curved surface of the solid generated by the revolution, about the initial line, of the area bounded by the curve r = f(θ) and the radii vectors θ = α, θ = β is*

$$\int_{\theta=\alpha}^{\theta=\beta} 2\pi\ y\frac{ds}{d\theta}\ d\theta, \text{where } \frac{ds}{d\theta} = \sqrt{r^2+\left(\frac{dr}{d\theta}\right)^2} \qquad | \textbf{ Remember}$$

and $\qquad y = r \sin \theta.$

ILLUSTRATIVE EXAMPLES

Example 1. (*a*) *Find the surface of a sphere of radius* a.

(*b*) *Show that the surface of the spherical zone contained between two parallel planes is* $2\pi ah$, *where* a *is the radius of the sphere and* h *the distance between the planes.*

Sol. (*a*) The sphere is generated by the revolution of a semi-circle ABA′ about its diameter AA′.

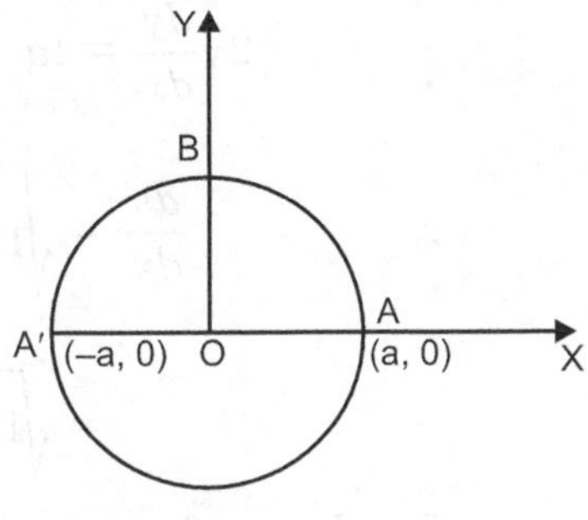

Let the equation of the generating curve (semi-circle) be

$$x^2 + y^2 = a^2$$

or $$y = \sqrt{a^2 - x^2}$$

Now $$\frac{dy}{dx} = \frac{1}{2}(a^2 - x^2)^{-1/2}(-2x) = \frac{-x}{\sqrt{a^2 - x^2}}$$

$$\therefore \quad \frac{ds}{dx} = \sqrt{1 + \left(\frac{dy}{dx}\right)^2} = \sqrt{1 + \frac{x^2}{a^2 - x^2}} = \sqrt{\frac{a^2 - x^2 + x^2}{a^2 - x^2}} = \frac{a}{\sqrt{a^2 - x^2}}$$

For the semi-circle ABA′, x varies from $-a$ to a.

$$\therefore \quad \text{Required volume} = \int_{x=-a}^{a} 2\pi y \frac{dx}{dx}\, dx = \int_{-a}^{a} 2\pi\sqrt{a^2 - x^2} \cdot \frac{a}{\sqrt{a^2 - x^2}} dx = 2\pi a \int_{-a}^{a} dx$$

$$= 2\pi a \left[x\right]_{-a}^{a} = 2\pi a(a + a) = 4\pi a^2.$$

(*b*) Let the sphere be generated by the revolution about the initial line, of the circle

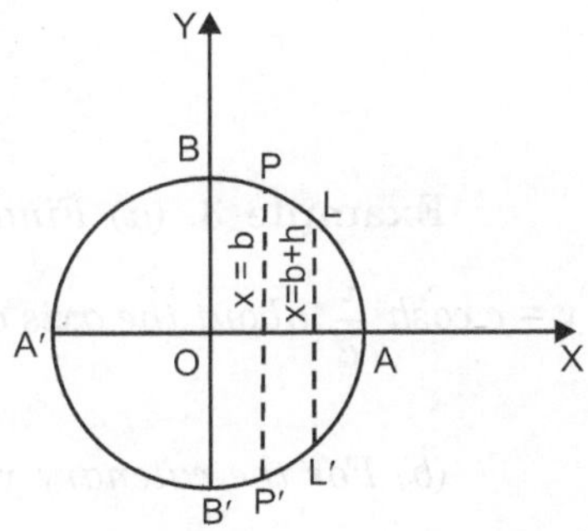

$$x^2 + y^2 = a^2 \quad \text{...(1)}$$

Let the two given planes forming the zone be at distances b and $b + h$ from the centre of the sphere.

From (1), differentiating w.r.t. x, we get

$$2x + 2y\frac{dy}{dx} = 0 \quad \text{or} \quad \frac{dy}{dx} = -\frac{x}{y}$$

$$\therefore \quad \frac{ds}{dx} = \sqrt{1 + \left(\frac{dy}{dx}\right)^2} = \sqrt{1 + \frac{x^2}{y^2}} = \sqrt{\frac{x^2 + y^2}{y^2}}$$

$$= \sqrt{\frac{a^2}{y^2}} = \frac{a}{y} \qquad | \text{ Using (1)}$$

$$\therefore \quad \text{Required surface} = \int_{x=b}^{b+h} 2\pi y \frac{ds}{dx}\, dx = 2\pi \int_{b}^{b+h} y\frac{a}{y}\, dx = 2\pi a \int_{b}^{b+h} dx$$

$$= 2\pi a \left[x\right]_{b}^{b+h} = 2\pi a(b + h - b) = 2\pi ah.$$

Example 2. *Find the area of the surface formed by the revolution of the parabola $y^2 = 4ax$ about the x-axis by the arc from the vertex to one end of the latus rectum.*

Sol. The equation of the parabola is

$$y^2 = 4ax \qquad \ldots(1)$$

Differentiating *w.r.t.* x, we get

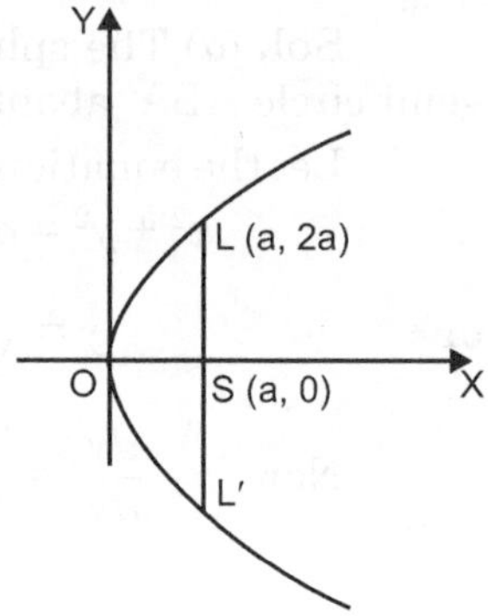

$$2y\frac{dy}{dx} = 4a \quad \text{or} \quad \frac{dy}{dx} = \frac{2a}{y}$$

$$\therefore \qquad \frac{ds}{dx} = \sqrt{1+\left(\frac{dy}{dx}\right)^2} = \sqrt{1+\frac{4a^2}{y^2}} = \sqrt{1+\frac{4a^2}{4ax}}$$

$$= \sqrt{1+\frac{a}{x}} = \sqrt{\frac{x+a}{x}}$$

For the arc from the vertex O to L, the end of the latus rectum, x varies from 0 to a.

$$\therefore \qquad \text{Required suface} = \int_{x=0}^{a} 2\pi y \frac{ds}{dx}\, dx$$

$$= \int_0^a 2\pi . \sqrt{4ax} . \sqrt{\frac{x+a}{x}}\, dx \qquad \left| \because \; y = \sqrt{4ax} \quad \text{From (1)} \right.$$

$$= 4\pi\sqrt{a}\int_0^a (x+a)^{1/2} dx = 4\pi\sqrt{a} . \frac{2}{3}\left[(x+a)^{3/2}\right]_0^a$$

$$= \frac{8\pi\sqrt{a}}{3} . [(2a)^{3/2} - a^{3/2}] = \frac{8\pi a^2}{3}(2\sqrt{2}-1).$$

Example 3. (*a*) *Find the surface generated by the revolution of arc of the catenary $y = c \cosh \frac{x}{c}$ about the axis of x.*

(*b*) *For the catenary $y = a \cosh \left(\frac{x}{a}\right)$, prove that aS = 2V = πa (ax + sy), where s is the length of the arc from the vertex, S and V are respectively the area of the curved surface and volume of the generated by the revolution of the arc about x-axis.*

Sol. (*a*) The equation of the curve is $y = c \cosh \frac{x}{c}$

$$\therefore \qquad \frac{dy}{dx} = c . \sinh\left(\frac{x}{c}\right) . \frac{1}{c} = \sinh\frac{x}{c}$$

$$\therefore \qquad \frac{ds}{dx} = \sqrt{1+\left(\frac{dy}{dx}\right)^2} = \sqrt{1+\sinh^2\frac{x}{c}} = \cosh\frac{x}{c}$$

If the arc be measured from the vertex (where $x = 0$) to point (x, y) then
The required surface

$$= \int_{x=0}^{x} 2\pi y \frac{ds}{dx}\, dx = \int_0^x 2\pi \,.\, c \cosh\frac{x}{c} \,.\, \cosh\frac{x}{x}\, dx = \int_0^x 2\pi c \cosh^2 \frac{x}{c}\, dx$$

$$= \pi c \int_0^x \left[1 + \cosh\left(\frac{2x}{c}\right)\right] dx \qquad \left|\ \because \cosh^2\theta = \frac{1+\cosh 2\theta}{2}\right.$$

$$= \pi c\left[x + \frac{c}{2}\sinh\left(\frac{2x}{c}\right)\right]_0^x = \pi c\left[x + \frac{c}{2}\sinh\left(\frac{2x}{c}\right)\right].$$

(*b*) The equation of the catenary is $y = a \cosh \dfrac{x}{a}$

$$\therefore \qquad \frac{dy}{dx} = \sinh\frac{x}{a}$$

$$\therefore \qquad \frac{ds}{dx} = \sqrt{1 + \left(\frac{dy}{dx}\right)^2} = \sqrt{1 + \sinh^2\frac{x}{a}} = \cosh\frac{x}{a}$$

At the vertex, $\quad x = a.$

$\therefore \qquad s$ = length of the arc from vertex to any point (x, y)

$$= \int_0^x \frac{ds}{dx}\,.\,dx = \int_0^x \cosh\frac{x}{a}\,dx = \left[a \sinh\frac{x}{a}\right]_0^x = a\left(\sinh\frac{x}{a} - 0\right) = a\sinh\frac{x}{a} \qquad \text{...(1)}$$

S = Surface area generated by the arc

$$= \int_0^x 2\pi\, y \frac{ds}{dx}\,.\,dx = \int_0^x 2\pi \,.\, a\cosh\frac{x}{a}\,.\,\cosh\frac{x}{a}\,.\,dx$$

$$= 2\pi a \int_0^x \cosh^2\frac{x}{a}dx = \pi a \int_0^x \left[1 + \cosh\left(\frac{2x}{a}\right)\right]dx = \pi a\left[x + \frac{a}{2}\sinh\left(\frac{2x}{a}\right)\right]_0^x$$

$$= \pi\, a\left[x + \frac{a}{2}\sinh\left(\frac{2x}{a}\right)\right] \qquad \text{...(2)}$$

and $\qquad$ V = the volume generated by the arc by the revolution about x-axis.

$$= \int_0^x \pi\, y^2 dx = \pi \int_0^x a^2 \cosh^2\frac{x}{a}dx$$

$$= \pi\, a^2 \int_0^x \frac{1}{2}\left[1 + \cosh\left(\frac{2x}{a}\right)\right]dx \qquad \left|\ \because \cosh^2\theta = \frac{1+\cosh 2\theta}{2}\right.$$

$$= \frac{\pi a^2}{2}\left[x + \frac{a}{2}\sinh\left(\frac{2x}{a}\right)\right]_0^x = \frac{\pi a^2}{2}\left[x + \frac{a}{2}\sinh\left(\frac{2x}{a}\right)\right] \qquad \text{...(3)}$$

From (2), $$aS = \pi a^2\left[x + \frac{a}{2}\sinh\left(\frac{2x}{a}\right)\right] \quad ...(4)$$

From (3), $$2V = \pi a^2\left[x + \frac{a}{2}\sinh\left(\frac{2x}{a}\right)\right] \quad ...(5)$$

From (1), $$\pi a(ax + sy) = \pi a\left[ax + a^2\sinh\frac{x}{a}\cosh\frac{x}{a}\right] = \pi a\left[ax + \frac{a^2}{2}.\,2\sinh\frac{x}{a}\cosh\frac{x}{a}\right]$$

$$= \pi a^2\left[x + \frac{a}{2}\sinh\left(\frac{2x}{a}\right)\right] \quad ...(6)$$

From (4), (5) and (6), it follows that $aS = 2V = a\pi\,(ax + sy)$

Example 4. (*a*) *Prove that the surface of the prolate spheroid formed by the revolution of the ellipse of eccentricity e about its major axis is equal to*

$$2 \times \textit{area of ellipse}.\left[\sqrt{1-e^2} + \frac{1}{e}\sin^{-1} e\right]$$

Another Statement. *Find the surface of the solid obtained by revolving the ellipse $b^2x^2 + a^2y^2 = a^2b^2$ about x-axis.*

(b) Find the surface of the solid generated by the revolution of the ellipse $x^2 + 4y^2 = 16$ about its major axis.

Sol. (*a*) [**Remember.** *Prolate spheroid is generated by the revolution of the ellipse about the major axis.]*

Let the equation of the ellipse be $\dfrac{x^2}{a^2} + \dfrac{y^2}{b^2} = 1$ ...(1)

Differentiating, $\dfrac{2x}{a^2} + \dfrac{2y}{b^2}.\dfrac{dy}{dx} = 0. \quad \therefore \quad \dfrac{dy}{dx} = -\dfrac{b^2x}{a^2y}.$

$$\therefore \qquad \frac{ds}{dx} = \sqrt{1 + \left(\frac{dy}{dx}\right)^2} = \sqrt{1 + \frac{b^4x^2}{a^4y^2}}$$

$$= \sqrt{1 + \frac{a^4(1-e^2)^2x^2}{a^4y^2}} \qquad \text{since } b^2 = a^2(1-e^2) \text{ for the ellipse}$$

$$= \sqrt{1 + \frac{(1-e^2)^2x^2}{\frac{b^2}{a^2}(a^2-x^2)}} \qquad \Big|\because \text{ From (1), } y^2 = \frac{b^2}{a^2}(a^2-x^2)$$

$$= \sqrt{1 + \frac{(1-e^2)x^2}{a^2-x^2}} \qquad \Big| \text{ Using } b^2 = a^2(1-e^2)$$

$$= \sqrt{\frac{a^2 - e^2x^2}{a^2 - x^2}}$$

Now for the upper half of the ellipse x varies from $-a$ to a.

$\therefore$ Required surface $= \int_{x=-a}^{a} 2\pi y \frac{ds}{dx} dx$

$$= 2\pi \int_{-a}^{a} \frac{b}{a}\sqrt{a^2 - x^2} \cdot \sqrt{\frac{a^2 - e^2x^2}{a^2 - x^2}}\, dx \qquad \Big| \because \text{From(1)},\ y = \frac{b}{a}\sqrt{a^2 - x^2}$$

$$= \frac{2\pi b}{a}\int_{-a}^{a} \sqrt{a^2 - e^2x^2}\, dx = \frac{2\pi be}{a}\int_{-a}^{a} \sqrt{\frac{a^2}{e^2} - x^2}\, dx$$

$$= \frac{2\pi be}{a}\left[\frac{x}{2}\sqrt{\frac{a^2}{e^2} - x^2} + \frac{a^2}{2e^2}\sin^{-1}\left(\frac{x}{a/e}\right)\right]_{-a}^{a}$$

$$= \frac{2\pi be}{a}\left[\frac{a}{2}\sqrt{\frac{a^2}{e^2} - a^2} + \frac{a^2}{2e^2}\sin^{-1} e - \left\{\frac{-a}{2}\sqrt{\frac{a^2}{e^2} - a^2} + \frac{a^2}{2e^2}\sin^{-1}(-e)\right\}\right]$$

$$= \frac{2\pi be}{a}\left[\frac{a^2}{2e}\sqrt{1 - e^2} + \frac{a^2}{2e^2}\sin^{-1} e + \frac{a^2}{2e}\sqrt{1 - e^2} + \frac{a^2}{2e^2}\sin^{-1} e\right]$$

$$\Big| \because \sin^{-1}(-x) = -\sin^{-1} x$$

$$= \frac{2\pi be}{a}\left[\frac{a^2}{e}\sqrt{1 - e^2} + \frac{a^2}{e^2}\sin^{-1} e\right] = 2\pi ab\left[\sqrt{1 - e^2} + \frac{1}{e}\sin^{-1} e\right]$$

$$= 2 \times \text{area of ellipse}\left[\sqrt{1 - e^2} + \frac{1}{e}\sin^{-1} e\right]$$

$\Big| \because$ area of ellipse $= \pi ab$ **[Remember]**

Hence the result.

(*b*) The equation of the ellipse is $x^2 + 4y^2 = 16$

or $\qquad 4y^2 = 16 - x^2, \quad y = \frac{\sqrt{16 - x^2}}{2}$ ...(1)

$$\therefore \quad \frac{dy}{dx} = \frac{1}{2}\cdot\frac{1}{2}(16 - x^2)^{-1/2}(-2x) = \frac{-x}{2\sqrt{16 - x^2}}$$

$$\therefore \quad \frac{ds}{dx} = \sqrt{1 + \left(\frac{dy}{dx}\right)^2} = \sqrt{1 + \frac{x^2}{4(16 - x^2)}} = \sqrt{\frac{64 - 3x^2}{4(16 - x^2)}}$$

The ellipse $x^2 + 4y^2 = 16$ meets x-axis, where putting $y = 0$, we get

$x^2 = 16$ or $x = \pm 4$.

Thus for the upper half of the ellipse in first quadrant x varies from 0 to 4.
Since the ellipse is symmetrical about y-axis.

$\therefore$ Required surface = 2 × surface generated by the arc in the first quadrant.

$$= 2\int_{x=0}^{4} 2\pi y \frac{ds}{dx}\,.\,dx = 4\pi\int_0^4 \frac{\sqrt{16-x^2}}{2}\,.\,\sqrt{\frac{64-3x^2}{4(16-x^2)}}\,dx$$

$$= \pi\int_0^4 \sqrt{64-3x^2}\,dx = \sqrt{3}\pi\int_0^4 \sqrt{\frac{64}{3}-x^2}\,dx$$

$$= \sqrt{3}\pi\left[\frac{x}{2}.\sqrt{\frac{64}{3}-x^2}+\frac{64}{3}.\frac{1}{2}\sin^{-1}\left(\frac{x}{8/\sqrt{3}}\right)\right]_0^4$$

$$= \sqrt{3}\,\pi\left[2\sqrt{\frac{64}{3}-16}+\frac{32}{3}\sin^{-1}\left(\frac{\sqrt{3}}{2}\right)\right]$$

$$= \sqrt{3}\,\pi\left[2.\frac{4}{\sqrt{3}}+\frac{32}{3}.\frac{\pi}{3}\right] \qquad \Big|\ \because\ \sin^{-1}\left(\frac{\sqrt{3}}{2}\right)=\frac{\pi}{3}$$

$$= 8\pi\left[1+\frac{4}{3\sqrt{3}}\right].$$

Example 5. *Find the volume and surface of the right circular cone formed by the revolution of right-angled triangle about a side which contains the right angle.*

Sol. Let the rt. $\angle$ d Δ OAB be formed by the line $y = x\tan\theta$, the x-axis and the ordinate $x = h$. Then OA = h is the height of the cone.

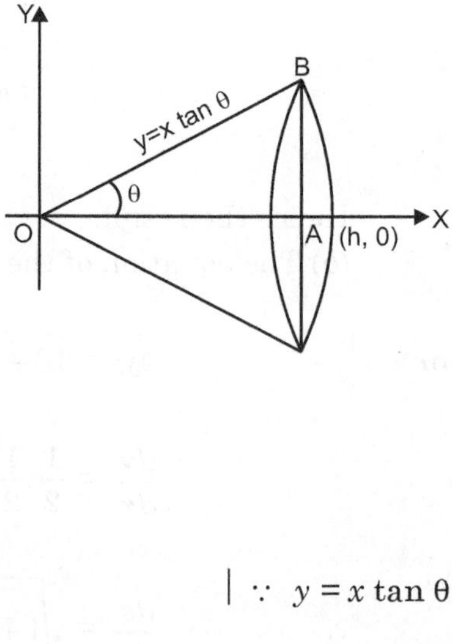

If r = AB is the radius of the base, then

$$\frac{r}{h} = \frac{AB}{OA} = \tan\theta \qquad ...(1)$$

and if l = OB is the slant side, then

$$\sec\theta = \frac{l}{h} \qquad ...(2)$$

$\therefore$ Required volume of the cone $= \int_0^h \pi y^2 dx$

$$= \pi\int_0^h x^2\tan^2\theta\,dx \qquad \Big|\ \because\ y = x\tan\theta$$

$$= \pi\tan^2\theta.\left[\frac{x^3}{3}\right]_0^h = \frac{\pi h^3}{3}\tan^2\theta = \frac{1}{3}\pi h^3.\frac{r^2}{h^2} = \frac{1}{3} = \pi r^2 h \qquad \Big|\ \text{Using (1)}$$

Now since $\quad y = x \tan\theta \quad \therefore \quad \dfrac{dy}{dx} = \tan\theta$

$$\therefore \quad \frac{ds}{dx} = \sqrt{1+\left(\frac{dy}{dx}\right)^2} = \sqrt{1+\tan^2\theta} = \sec\theta$$

$$\therefore \quad \text{Required surface area of cone} = \int_0^h 2\pi y \frac{ds}{dx} . dx$$

$$= 2\pi \int_0^h x \tan\theta . \sec\theta \, dx = 2\pi \tan\theta \sec\theta . \left[\frac{x^2}{2}\right]_0^h$$

$$= \pi h^2 \tan\theta \sec\theta = \pi h^2 . \frac{r}{h} . \frac{1}{h} \qquad | \text{ Using (1) and (2)}$$

$$= \pi r l.$$

Example 6. *Find the surface of the solid generated by the revolution of the astroid* $x^{2/3} + y^{2/3} = a^{2/3}$ *or* $x = a \cos^3 t,\ y = a \sin^3 t$ *about the x-axis.*

Sol. The parametric equations of the astroid are.

$$x = a \cos^3 t,\ y = a \sin^3 t$$

The curve is symmetrical about both the axes and it lies in the square bounded by the lines $x = \pm a$, $y = \pm a$.

For the portion of the curve in the first quadrant, t varies from 0 to $\dfrac{\pi}{2}$.

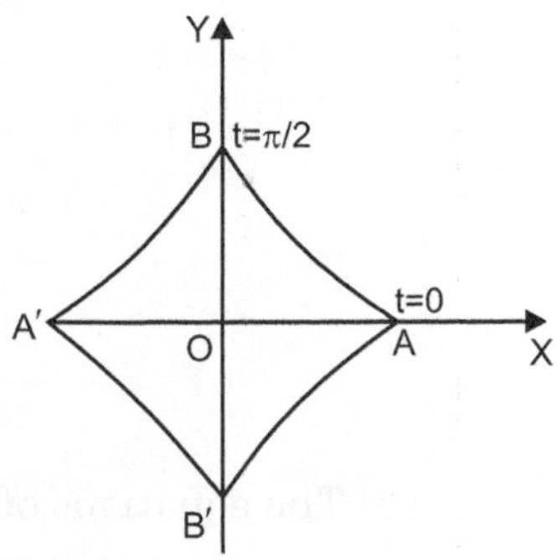

Now $\dfrac{dx}{dt} = -\,3a \cos^2 t \sin t$ and $\dfrac{dy}{dt} = 3a \sin^2 t \cos t$

$$\frac{ds}{dt} = \sqrt{9a^2 \cos^4 t \sin^2 t + 9a^2 \sin^4 t \cos^2 t}$$

$$= \sqrt{9a^2 \sin^2 t \cos^2 t (\cos^2 t + \sin^4 t)}$$

$$= 3a \sin t \cos t \qquad | \because \cos^2 t + \sin^2 t = 1$$

$$\therefore \quad \text{Required surface} = 2\int_{t=0}^{\pi/2} 2\pi y \frac{ds}{dt} dt$$

$$= 4\pi \int_0^{\pi/2} a \sin^3 t . 3a \sin t \cos t \, dt = 12\pi a^2 \int_0^{\pi/2} \sin^4 t \cos t \, dt$$

$$= 12\pi a^2 \left[\frac{\sin^5 t}{5}\right]_0^{\pi/2} = \frac{12}{5}\pi a^2 (1-0) = \frac{12}{5}\pi a^2.$$

Example 7. *Find the surface area of the solid generated by revolving the cycloid*
(i) $x = a(\theta - \sin\theta)$, $y = a(1 - \cos\theta)$ *about x-axis.*
(ii) $x = a(\theta + \sin\theta)$, $y = a(1 - \cos\theta)$ *about the tangent at the vertex.*

Sol. *(i)* The equations of the cycloid are

$$x = a(\theta - \sin\theta), \quad y = a(1 - \cos\theta),$$

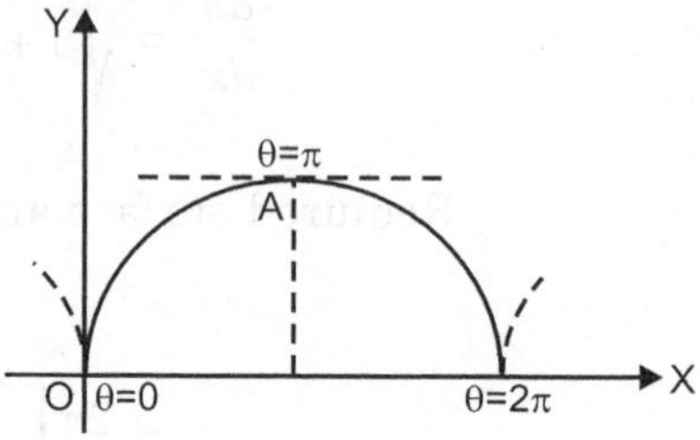

The cycloid is symmetrical about the line through the point $\theta = \pi$, $\perp$ to the x-axis.

For the first half of the cycloid, θ varies from 0 to π.

Now $\dfrac{dx}{d\theta} = a(1 - \cos\theta)$, $\dfrac{dy}{d\theta} = a\sin\theta$

$$\therefore \quad \frac{ds}{d\theta} = \sqrt{\left(\frac{dx}{d\theta}\right)^2 + \left(\frac{dy}{d\theta}\right)^2} = \sqrt{a^2(1-\cos\theta)^2 + a^2\sin^2\theta}$$

$$= a\sqrt{1 + \cos^2\theta - 2\cos\theta + \sin^2\theta} = a\sqrt{2(1-\cos\theta)} = a\sqrt{2.2\sin^2\theta/2} = 2a\sin\frac{\theta}{2}$$

$$\therefore \quad \text{Required surface} = 2\int_0^{\pi} 2\pi y\frac{ds}{d\theta}d\theta = 4\pi\int_0^{\pi} a(1-\cos\theta)\,.\,2a\sin\frac{\theta}{2}\,d\theta$$

$$= 4\pi\int_0^{\pi} a\left(2\sin^2\frac{\theta}{2}\right).\,2a\sin\frac{\theta}{2}\,d\theta$$

$$= 16\pi a^2\int_0^{\pi}\sin^3\frac{\theta}{2}\,d\theta \qquad \Big|\ \text{Put } \theta/2 = t \ \therefore\ d\theta = 2dt$$

$$= 16\pi a^2\int_0^{\pi/2}\sin^3 t\,.\,2\,dt = 32\pi a^2\frac{2}{3.1} = \frac{64}{3}\pi a^2.$$

(ii) The equations of the cycloid are

$$x = a(\theta + \sin\theta),\ y = a(1 - \cos\theta)$$

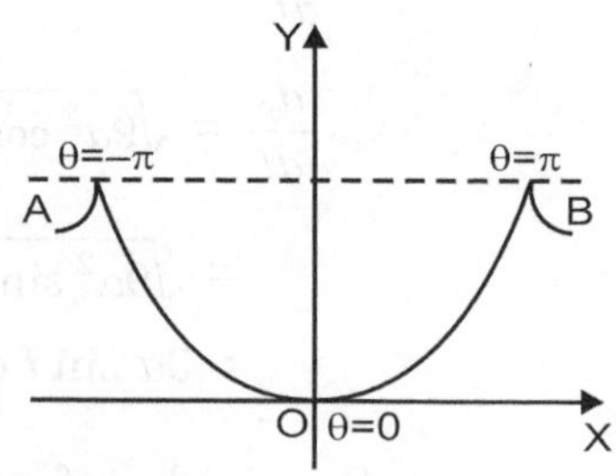

The cycloid is symmetrical about the y-axis and the tangent at the vertex is x-axis. For half of the curve, θ varies from 0 to π.

Now $\dfrac{dx}{d\theta} = a(1 + \cos\theta)$, $\dfrac{dy}{d\theta} = a\sin\theta$

$$\therefore \quad \frac{ds}{d\theta} = \sqrt{\left(\frac{dx}{d\theta}\right)^2 + \left(\frac{dy}{d\theta}\right)^2} = \sqrt{a^2(1+\cos\theta)^2 + a^2\sin^2\theta}$$

$$= a\sqrt{1 + \cos^2\theta + 2\cos\theta + \sin^2\theta} = a.\sqrt{2(1+\cos\theta)}$$

$$= a\sqrt{2.2\cos^2\frac{\theta}{2}} = 2a\cos\frac{\theta}{2}$$

$\therefore$ Required surface $= 2\int_0^{\pi} 2\pi y \frac{ds}{d\theta} d\theta = 4\pi \int_0^{\pi} a(1 - \cos\theta) \,.\, 2\cos\frac{\theta}{2}\, d\theta$

$$= 4\pi \int_0^{\pi} a\left(2\sin^2\frac{\theta}{2}\right) .2a\cos\frac{\theta}{2}\, d\theta = 16\pi a^2 \int_0^{\pi} \sin^2\frac{\theta}{2}\cos\frac{\theta}{2} d\theta$$

$$= 16\pi a^2 \left| \frac{\sin^3 \theta/2}{3 \,.\, \frac{1}{2}} \right|_0^{\pi} = \frac{32\pi a^2}{3}[1 - 0] = \frac{32}{3}\pi a^2.$$

Example 8. *The portion between two consecutive cusps of the cycloid $x = a(\theta + \sin\theta)$, $y = a(1 + \cos\theta)$ is evolved about the x-axis. Prove that the area of the surface so formed is to the area of the cycloid as 64 : 9.*

Sol. The equations of the cycloid are

$$x = a(\theta + \sin\theta), \quad y = a(1 + \cos\theta),$$

The cycloid is symmetrical about the y-axis and for half of the cycloid, θ varies from 0 to π.

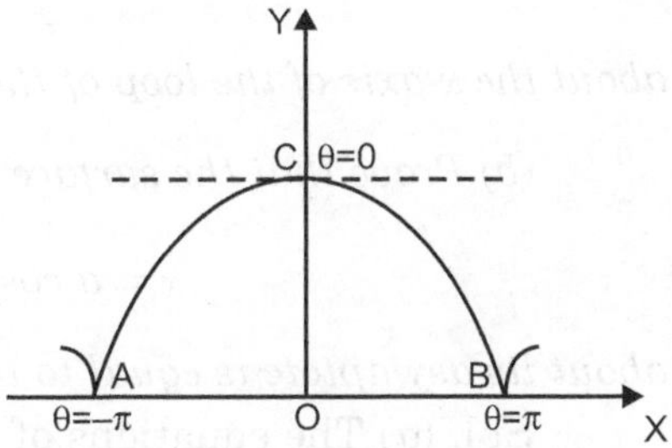

If Δ denotes the area of the cycloid, then

$$\Delta = 2\int_0^{\pi} y\frac{dx}{d\theta} d\theta = 2\int_0^{\pi} a(1 + \cos\theta)\, a(1 + \cos\theta)\, d\theta$$

$$= 2a^2 \int_0^{\pi} \left(2\cos^2\frac{\theta}{2}\right)^2 d\theta \qquad \because\ 1 + \cos\theta = 2\cos^2\frac{\theta}{2}$$

$$= 8a^2 \int_0^{\pi} \cos^4\frac{\theta}{2} d\theta$$

$$= 8a^2 \int_0^{\pi/2} \cos^4 t \,.\, 2dt, \text{ where } t = \frac{\theta}{2}$$

$$= 16a^2 \frac{3 \,.\, 1\, \pi}{4 \,.\, 2\, 2} = 3\pi a^2$$

Now $\quad \frac{dx}{d\theta} = a(1 + \cos\theta), \quad \frac{dy}{d\theta} = -a\sin\theta$

$$\therefore \quad \frac{ds}{d\theta} = \sqrt{\left(\frac{dx}{d\theta}\right)^2 + \left(\frac{dy}{d\theta}\right)^2} = \sqrt{a^2(1 + \cos\theta)^2 + a^2 \,.\, \sin^2\theta}$$

$$= a\sqrt{2 + 2\cos\theta} = a\sqrt{2(1 + \cos\theta)} = a\sqrt{2 \,.\, 2\cos^2\frac{\theta}{2}} = 2a\cos\frac{\theta}{2}$$

If S denotes the surface generated by the revolution of the cycloid about the x-axis, then

$$S = 2\int_0^{\pi} 2\pi y \frac{ds}{d\theta} d\theta = 4\pi \int_0^{\pi} a(1 + \cos\theta) \,.\, 2a\cos\frac{\theta}{2}\, d\theta$$

$$= 4\pi\int_0^{\pi} a\left(2\cos^2\frac{\theta}{2}\right).\,2a\cos\frac{\theta}{2}d\theta = 16\pi a^2\int_0^{\pi}\cos^3\frac{\theta}{2}d\theta$$

$$= 16\pi a^2\int_0^{\pi/2}\cos^3 t\,.\,2dt, \quad \text{where } t = \frac{\theta}{2}$$

$$= 32\pi a^2\frac{2}{3} + \frac{64}{3}\pi a^2.$$

$$\therefore \quad \text{Required ratio} = \frac{S}{\Delta} = \frac{\frac{64}{3}\pi a^2}{3\pi a^2} = \frac{64}{9}.$$

Hence the result.

Example 9. (*a*) *Prove that the surface area of the solid generated by the revolution, about the x-axis of the loop of the curve* $x = t^2$, $y = t - \frac{1}{3}t^3$ *is* 3π.

(*b*) *Prove that the surface of the solid generated by the revolution of the tractrix*

$$x = a\cos t + \frac{1}{2}a\log\tan^2\frac{t}{2},\ y = a\sin t,$$

about its asymptote is equal to the surface of a sphere of radius a.

Sol. (*a*) The equations of the curve are

$$x = t^2,\ y = t - \frac{1}{3}t^3$$

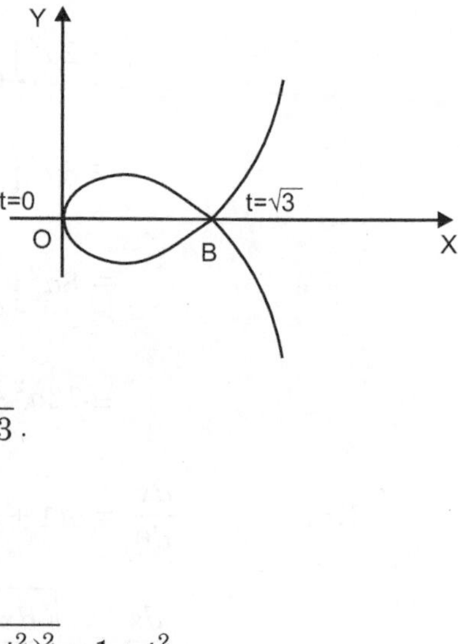

The curve is symmetrical about the x-axis, and for the loop, putting $y = 0$, we get

$$t - \frac{1}{3}t^3 = 0 \quad \text{or} \quad t(3 - t^2) = 0$$

$$t = 0 \quad \text{or} \quad \pm\sqrt{3}.$$

Thus for the upper half of the loop, t varies from 0 to $\sqrt{3}$.

Now $$\frac{dx}{dt} = 2t, \quad \frac{dy}{dt} = 1 - t^2$$

$$\therefore \quad \frac{ds}{dt} = \sqrt{\left(\frac{dx}{dt}\right)^2 + \left(\frac{dy}{dt}\right)^2} = \sqrt{4t^2 + (1 - t^2)^2} = 1 + t^2$$

$$\therefore \quad \text{Required surface} = \int_0^{\sqrt{3}} 2\pi y\frac{ds}{dt}\,.\,dt$$

$$= 2\pi\int_0^{\sqrt{3}}\left(t - \frac{1}{3}t^3\right).(1 + t^2)\,dt = 2\pi\int_0^{\sqrt{3}}\left(t + t^3 - \frac{1}{3}t^3 - \frac{1}{3}t^5\right)dt$$

$$= 2\pi\int_0^{\sqrt{3}}\left(t+\frac{2}{3}t^3-\frac{1}{3}t^5\right)dt = 2\pi\left[\frac{t^2}{2}+\frac{2}{3}\cdot\frac{t^4}{4}-\frac{1}{3}\cdot\frac{t^6}{6}\right]_0^{\sqrt{3}}$$

$$= 2\pi\left[\frac{3}{2}+\frac{1}{6}\cdot 9-\frac{1}{18}\cdot 27-0\right] = 2\pi\left[\frac{3}{2}+\frac{3}{2}-\frac{3}{2}\right] = 3\pi.$$

Hence the result.

(*b*) The equations of the tractrix are

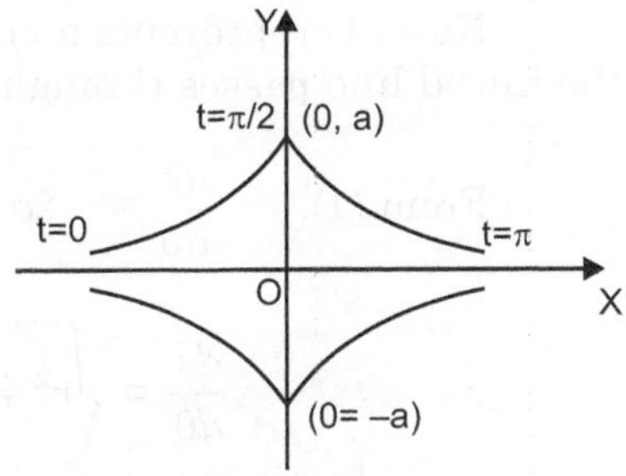

$$x = a\cos t + \frac{a}{2}\log\tan^2\frac{t}{2}$$

$$y = a\sin t \qquad \ldots(1)$$

The curve is symmetrical about both the axes and the asymptote is $y = 0$, *i.e.*, x-axis

From (1), $\dfrac{dx}{dt} = -a\sin t + \dfrac{a}{2}\cdot\dfrac{1}{\tan^2 t/2}\cdot 2\tan\dfrac{t}{2}\sec^2\dfrac{t}{2}\cdot\dfrac{1}{2}$

$$= -a\sin t + \frac{a}{2\sin t/2\cot t/2}$$

$$= -a\sin t + \frac{a}{\sin t}\cdot a\frac{(1-\sin^2 t)}{\sin t} = a\frac{\cos^2 t}{\sin t}$$

and $\dfrac{dy}{dt} = a\cos t$

$$\therefore \quad \frac{ds}{dt} = \sqrt{\left(\frac{dx}{dt}\right)^2+\left(\frac{dy}{dt}\right)^2} = \sqrt{a^2\frac{\cos^4 t}{\sin^2 t}+a^2\cos^2 t}$$

$$= \sqrt{\frac{a^2\cos^2 t(\cos^2 t+\sin^2 t)}{\sin^2 t}} = a\frac{\cos t}{\sin t} \qquad \Big|\because\ \cos^2 t+\sin^2 t = 1$$

For the curve in the second quadrant, t varies from 0 to $\dfrac{\pi}{2}$.

$$\therefore \quad \text{Required surface} = 2\int_0^{\pi/2} 2\pi y\frac{ds}{dt}\cdot dt = 4\pi\int_0^{\pi/2} a\sin t\cdot\frac{a\cos t}{\sin t}\,dt$$

$$= 4\pi a^2\int_0^{\pi/2}\cos t\,dt = 4\pi a^2\left[\sin t\right]_0^{\pi/2}$$

$= 4\pi a^2(1-0) = 4\pi a^2 =$ surface of sphere of radius a.

[$\because$ surface of a sphere $= 4\pi$ (radius)^2 **Remember**]

Example 10. (*a*) *Find the area of the surface of revolution formed by revolving the curve $r = 2a \cos \theta$ about the initial line.*

(*b*) *The curve $r = a(1 + \cos \theta)$ revolves about the initial line. Find the surface of the figure so formed.*

(*c*) *Find the surface of the solid generated by the revolution of the lemniscate $r^2 = a^2 \cos 2\theta$ about the initial line.*

Sol. (*a*) The equations of the curve is

$$x = 2a \cos \theta \qquad \text{...(1)}$$

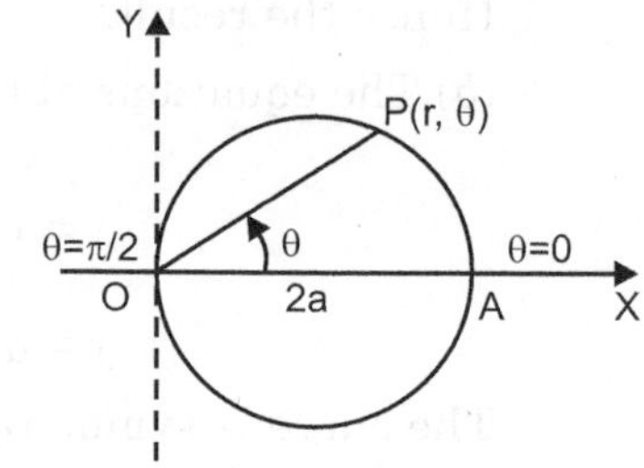

Now (1) represents a circle passing through the pole and the initial line passes through the centre of the circle.

From (1), $$\frac{dr}{d\theta} = -2a \sin \theta$$

$$\therefore \quad \frac{ds}{d\theta} = \sqrt{r^2 + \left(\frac{dr}{d\theta}\right)^2} = \sqrt{4a^2 \cos^2 \theta + 4a^2 \sin^2 \theta}$$

$$= 2a\sqrt{\cos^2 \theta + \sin^2 \theta} = 2a$$

$$\therefore \quad \text{Required surface} = \int_0^{\pi/2} 2\pi y \frac{ds}{d\theta} d\theta, \quad \text{where } y = r \sin \theta$$

$$= 2\pi \int_0^{\pi/2} r \sin \theta . 2a \, d\theta = 4a\pi \int_0^{\pi/2} 2a \cos \theta \sin \theta \, d\theta$$

$$= 4\pi a^2 \int_0^{\pi/2} 2 \sin \theta \cos \theta \, d\theta = 4\pi a^2 . \left[\sin^2 \theta\right]_0^{\pi/2} = 4\pi a^2 (1 - 0) = 4\pi a^2.$$

(*b*) The equation of the cardioid is

$$r = a(1 + \cos \theta) \qquad \text{...(1)}$$

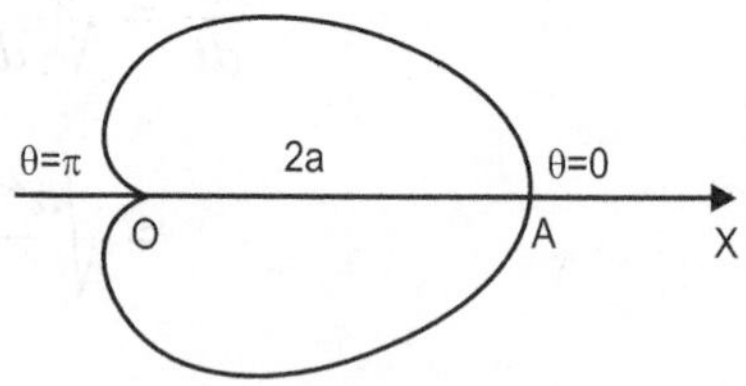

The cardioid is symmetrical about the initial line and for the upper half of the curve, θ varies from 0 to π.

Now from (1), $$\frac{dr}{d\theta} = -a \sin \theta$$

$$\therefore \quad \frac{ds}{d\theta} = \sqrt{r^2 + \left(\frac{dr}{d\theta}\right)^2} = \sqrt{a^2(1 + \cos \theta)^2 + a^2 \sin^2 \theta}$$

$$= a\sqrt{1 + \cos^2 \theta + 2 \cos \theta + \sin^2 \theta} = a\sqrt{2(1 + \cos \theta)}$$

$$= a\sqrt{2 . 2 \cos^2 \frac{\theta}{2}} = 2a \cos \frac{\theta}{2}$$

$\therefore$ Required volume $= \int 2\pi y \dfrac{ds}{d\theta}\, d\theta,$ where $y = r\sin\theta$

$$= 2\pi \int_0^{\pi} a\sin\theta(1+\cos\theta).\, 2a\cos\frac{\theta}{2} d\theta \qquad | \because r = a(1+\cos\theta)$$

$$= 2\pi \int_0^{\pi} a\,.\, 2\sin\frac{\theta}{2}\cos\frac{\theta}{2}\,.\, 2\cos^2\frac{\theta}{2}\,.\, 2a\cos\frac{\theta}{2}\, d\theta$$

$$= 16\pi a^2 \int_0^{\pi} \cos^4\frac{\theta}{2}\sin\frac{\theta}{2}\, d\theta$$

$$= 16\pi a^2 . \left[\frac{-\cos^5\theta/2}{5\,.\,\frac{1}{2}}\right]_0^{\pi} = -\frac{32}{5}\pi a^2(0-1) = \frac{32}{5}\pi a^2.$$

(c) The equation of the lemniscate is $r = a^2\cos 2\theta$ ...(1)

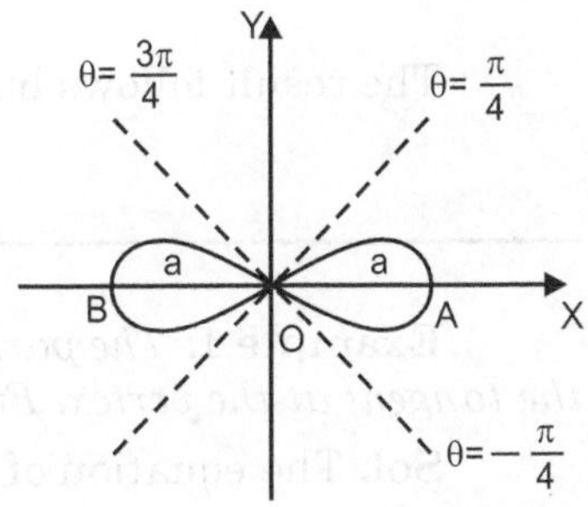

The curve is symmetrical about the initial line and the line

$$\theta = \frac{\pi}{2}$$

For a loop, putting $r = 0$, we get

$$\cos 2\theta = 0$$

$$2\theta = \pm\frac{\pi}{2} \quad \text{or} \quad \theta = -\frac{\pi}{4}, \frac{\pi}{4} \qquad \text{[two consecutive values]}$$

The curve consists of two equal loops and $\theta = \dfrac{-\pi}{4}, \theta = \dfrac{\pi}{4}$ are the tangents at the pole. In the first quadrant, for half of the loop θ varies from 0 to $\dfrac{\pi}{4}$.

From (1), $2r\dfrac{dr}{d\theta} = -2a^2\sin 2\theta$ or $\dfrac{dr}{d\theta} = \dfrac{-a^2\sin 2\theta}{r}$

$$\therefore \quad \frac{ds}{d\theta} = \sqrt{r^2 + \left(\frac{dr}{d\theta}\right)^2} = \sqrt{a^2\cos 2\theta + \frac{a^4\sin^2 2\theta}{r^2}}$$

$$= \sqrt{a^2\cos 2\theta + \frac{a^4\sin^2 2\theta}{a^2\cos 2\theta}} + \sqrt{\frac{a^2(\cos^2 2\theta + \sin^2 2\theta)}{\cos 2\theta}} = \frac{a}{\sqrt{\cos 2\theta}}$$

$$\therefore \text{ Required surface} = 2\int_0^{\pi/4} 2\pi y \frac{ds}{d\theta} \,.\, d\theta = 4\pi \int_0^{\pi/4} r \sin\theta \,.\, \frac{a}{\sqrt{\cos 2\theta}} \,.\, d\theta$$

$$= 4\pi \int_0^{\pi/4} a\sqrt{\cos 2\theta} \,.\, \sin\theta \,.\, \frac{a}{\sqrt{\cos 2\theta}} d\theta$$

$$= 4\pi a^2 \int_0^{\pi/4} \sin\theta \, d\theta = 4\pi a^2 [-\cos\theta]_0^{\pi/4} = 4\pi a^2 \left[-\frac{1}{\sqrt{2}}+1\right] = 4\pi a^2 \left(1-\frac{1}{\sqrt{2}}\right).$$

12.2. REVOLUTION ABOUT Y-AXIS

The curved surface of the solid generated by the revolution about the y-axis of the area bounded by the curve x = f(y), the y-axis and the abscissae y = a, y = b is

$$\int_{y=a}^{y=b} 2\pi \mathbf{x} \, \mathbf{ds}.$$

The result follows immediately on interchanging x and y in Article 7.

ILLUSTRATIVE EXAMPLES

Example 1. *The part of the parabola $y^2 = 4ax$ cut-off by the latus rectum revolves about the tangent at the vertex. Find the curved surface of the reel thus generated.*

Sol. The equation of the parabola is

$$y^2 = 4ax \qquad \ldots(1)$$

The required curved surface is generated by the revolution of the arc LAL′, cut-off by the latus rectum, about the tangent at the vertex *i.e.*, y-axis.

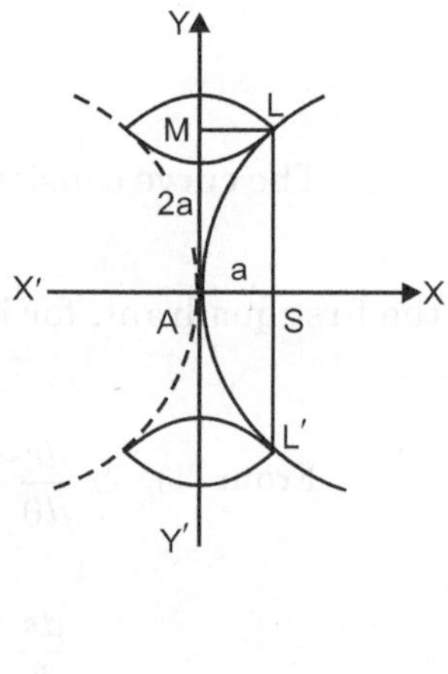

For half of the arc AL, x varies from 0 to a.

From (1), $\quad y = 2\sqrt{a}\sqrt{x} \quad \therefore \quad \frac{dy}{dx} = \frac{\sqrt{a}}{\sqrt{x}}$

$$\therefore \quad \frac{ds}{dx} = \sqrt{1+\left(\frac{dy}{dx}\right)^2} = \sqrt{1+\frac{a}{x}} = \sqrt{\frac{x+a}{x}}$$

$\therefore$ Required surface

$$= 2\int_0^a 2\pi x \frac{ds}{dx} dx = 4\pi \int_0^a x \,.\, \sqrt{\frac{x+a}{x}} \, dx$$

$$= 4\pi \int_0^a \sqrt{x^2+ax} \, dx = 4\pi \int_0^a \sqrt{\left(x+\frac{a}{2}\right)^2 - \left(\frac{a}{2}\right)^2} \, dx$$

$$= 4\pi\left[\frac{1}{2}\left(x+\frac{a}{2}\right)\sqrt{\left(x+\frac{a}{2}\right)^2-\left(\frac{a}{2}\right)^2}-\frac{1}{2}\cdot\frac{a^2}{4}\cosh^{-1}\left[\left(\frac{x+a/2}{a/2}\right)\right]\right]_0^a$$

$$\left|\because \int\sqrt{x^2-a^2}\,dx = \frac{x}{2}\sqrt{x^2-a^2}-\frac{a^2}{2}\cosh^{-1}\frac{x}{a}\right.$$

$$= 4\pi\left[\frac{1}{2}\cdot\frac{3a}{2}\sqrt{\frac{9a^2}{4}-\frac{a^2}{4}}-\frac{a^2}{8}(\cosh^{-1}3-\cosh^{-1}1)\right]$$

$$= 4\pi\left[\frac{3a}{4}\cdot\sqrt{2}a-\frac{a^2}{8}\left\{\log(3+\sqrt{9-1})-\log 1+\sqrt{(1-1)}\right\}\right]$$

$$\left|\because \cosh^{-1}x = \log(x+\sqrt{x^2-1})\right.$$

$$= 3\sqrt{2}\pi a^2-\frac{\pi a^2}{2}\log(3+2\sqrt{2}) = \pi a^2\left[3\sqrt{2}-\frac{1}{2}\log(\sqrt{2}+1)^2\right]$$

$$= \pi a^2\,[3\sqrt{2}-\log(\sqrt{2}+1)].$$

Example 2. *Prove that the surface of the oblate spheroid formed by the revolution of the ellipse of semi-major axis a and eccentricity e is* $2\pi a^2\left[1+\frac{1-e^2}{1+e^2}\log\left(\frac{1+e}{1-e}\right)\right]$.

Sol. [**Remember** : *Oblate spheroid is generated by the revolution of the ellipse about minor axis.*]

Let the parametric equations of the ellipse be

$$x = a\cos t,\ y = b\sin t, \text{ where } b^2 = a^2(1-e^2) \qquad ...(1)$$

$$\therefore \quad \frac{dx}{dt} = -a\sin t, \quad \frac{dy}{dt} = b\cos t$$

$$\therefore \quad \frac{ds}{dt} = \sqrt{\left(\frac{dx}{dt}\right)^2+\left(\frac{dy}{dt}\right)^2} = \sqrt{a^2\sin^2 t+b^2\cos^2 t}$$

For the upper half of the ellipse in first quadrant, t varies from 0 to $\frac{\pi}{2}$.

$$\therefore \quad \text{Required surface} = 2\int_0^{\pi/2} 2\pi x\frac{ds}{dt}\,dt = 4\pi\int_0^{\pi/2} a\cos t\sqrt{a^2\sin^2 t+b^2\cos^2 t}\,dt$$

$$= 4\pi a\int_0^{\pi/2}\cos t\sqrt{a^2\sin^2 t+a^2(1-e^2)\cos^2 t}\,dt$$

$$= 4\pi a^2 \int_0^{\pi/2} \cos t \sqrt{(\sin^2 t + \cos^2 t) - e^2 \cos^2 t}\, dt$$

$$= 4\pi a^2 \int_0^{\pi/2} \cos t \sqrt{1 - e^2 (1 - \sin^2 t)}\, dt$$

$$= 4\pi a^2 \int_0^{\pi/2} \sqrt{(1 - e)^2 + e^2 \sin^2 t}\, .\, \cos t\, dt$$

$$= \frac{4\pi a^2}{e} \int_0^{e} \sqrt{(1 - e^2) + z^2}\, dz, \quad \text{where } e \sin t = z$$

$$= \frac{4\pi a^2}{e} \left[\frac{z}{2} \sqrt{(1 - e^2) + z^2} + \frac{1}{2} (1 - e^2) \sinh^{-1} \frac{z}{\sqrt{1 - e^2}} \right]_0^e$$

$$= \frac{2\pi a^2}{e} \left[e + (1 - e^2) \sinh^{-1} \frac{e}{\sqrt{1 - e^2}} - 0 \right]$$

$$= \frac{2\pi a^2}{e} \left[e + (1 - e^2) \log \left(\frac{e}{\sqrt{1 - e^2}} + \sqrt{\frac{e^2}{1 - e^2} + 1} \right) \right]$$

$$\Big| \because \quad \sinh^{-1} x = \log (x + \sqrt{x^2 + 1})$$

$$= 2\pi a^2 \left[1 + \frac{1 - e^2}{e} \log \frac{e + 1}{\sqrt{1 - e^2}} \right] = 2\pi a^2 \left[1 + \frac{1 - e^2}{e} \log \left(\frac{1 + e}{1 - e} \right)^{1/2} \right]$$

$$= 2\pi a^2 \left[1 + \frac{1 - e^2}{2e} \log \frac{1 + e}{1 - e} \right].$$

Example 3. *The arc of the cardioid $r = a\ (1 + \cos \theta)$ included between $-\frac{\pi}{2} \le \theta \le \frac{\pi}{2}$ is rotated about the line $\theta = \frac{\pi}{2}$. Find the area of surface generated.*

Sol. The cardioid is $\qquad r = a\ (1 + \cos \theta) \qquad$ ...(1)

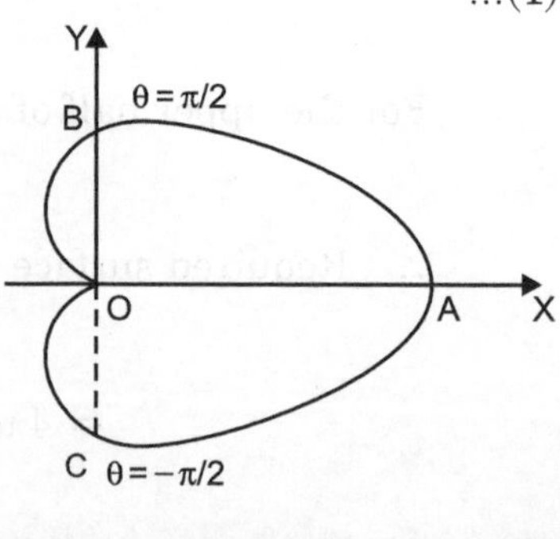

The arc CAB $\left(\text{from } \theta = -\frac{\pi}{2} \text{ to } \theta = \frac{\pi}{2}\right)$ revolves about the line $\theta = \frac{\pi}{2}$, *i.e.*, the y-axis.

Also the curve is symmetrical about the initial line or x-axis.

From (1), $\dfrac{dr}{d\theta} = -a\sin\theta$

$$\therefore \quad \frac{ds}{d\theta} = \sqrt{r^2 + \left(\frac{dr}{d\theta}\right)^2} = \sqrt{a^2(1+\cos\theta)^2 + a^2\sin^2\theta}$$

$$= a\sqrt{2(1+\cos\theta)} = a\sqrt{2 \,.\, 2\cos^2\frac{\theta}{2}} = 2a\cos\frac{\theta}{2}$$

$\therefore$ Required surface area

= 2 × surface generated by the revolution of arc AB

$$= 2\int_0^{\pi/2} 2\pi x \frac{ds}{d\theta}\, d\theta \qquad \Big|\because \text{ for the arc AB, } \theta \text{ varies from 0 to } \pi/2$$

$$= 4\pi\int_0^{\pi/2} r\cos\theta \,.\, 2a\cos\frac{\theta}{2}\, d\theta \qquad \Big|\because x = r\cos\theta$$

$$= 8\pi a\int_0^{\pi/2} a(1+\cos\theta)\,.\,\cos\theta\,.\,\cos\frac{\theta}{2}\, d\theta$$

$$= 8\pi a^2\int_0^{\pi/2}\left(2 - 2\sin^2\frac{\theta}{2}\right)\left(1 - 2\sin^2\frac{\theta}{2}\right)\cos\frac{\theta}{2}\, d\theta$$

$$= 16\pi a^2\int_0^{\pi/2}\left(1 - \sin^2\frac{\theta}{2}\right)\left(1 - 2\sin^2\frac{\theta}{2}\right)\cos\frac{\theta}{2}\, d\theta$$

$$= 16\pi a^2\int_0^{\pi/2}\left[1 - 3\sin^2\frac{\theta}{2} + 2\sin^4\frac{\theta}{2}\right]\cos\frac{\theta}{2}\, d\theta$$

| Put $\sin\theta/2 = t \quad \therefore \quad \dfrac{1}{2}\cos\theta/2\, d\theta = dt$

| when $\theta = 0, t = 0$; when $\theta = \pi/2, t = 1/\sqrt{2}$.

$$= 16\pi a^2\int_0^{1/\sqrt{2}} (1 - 3t^2 + 2t^4)\,.\,2dt$$

$$= 32\pi a^2\left[t - t^3 + \frac{2t^5}{5}\right]_0^{1/\sqrt{2}} = 32\pi a^2\left[\frac{1}{\sqrt{2}} - \frac{1}{2\sqrt{2}} + \frac{2}{5}\,.\,\frac{1}{4\sqrt{2}}\right]$$

$$= 32\pi a^2\left[\frac{20 - 10 + 2}{20\sqrt{2}}\right] = 32\pi a^2 \times \frac{12}{20\sqrt{2}} = \frac{96\pi a^2}{5\sqrt{2}}.$$

12.3. REVOLUTION ABOUT ANY AXIS

The curved surface of the solid generated by the revolution, about an axis CD of the area bounded by a curve AB, the axis CD and the perpendiculars AC, BD on the axis is

$$\int 2\pi\,(\mathrm{PM})\,ds \qquad \textbf{| Remember}$$

where PM is the ⊥ from any point P of the curve on the axis, and arc AP = s, the limits of integration being the values of the independent variable at the ends of the revolving arc.

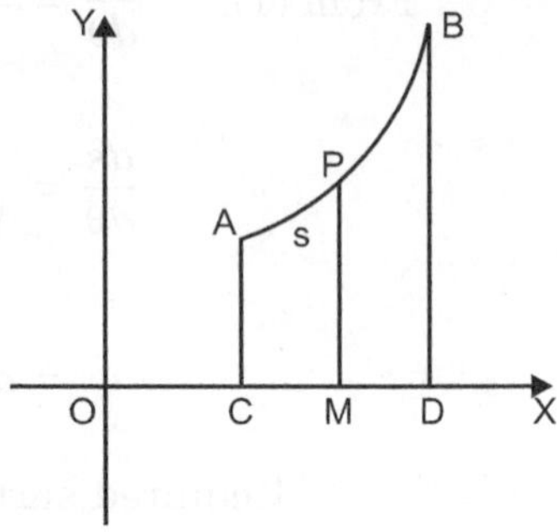

ILLUSTRATIVE EXAMPLES

Example 1. *A circular arc revolves about its chord. Find the area of the surface generated, when 2α is the angle subtended by the arc at the center.*

Sol. Let the radius of the circle be a. Then

$$OA = a \quad \text{and} \quad OD = a\cos\alpha,$$

where AC is the arc under consideration, subtending an angle 2α at the centre. B is the mid-poind of the arc AC.

Take O as the origin and OB as the x-axis, so that OY ⊥ to it is y-axis.

Let P ($a\cos\theta$, $a\sin\theta$) be any point on the curve. From P, draw PM ⊥ to the chord AC and PN ⊥ OX.

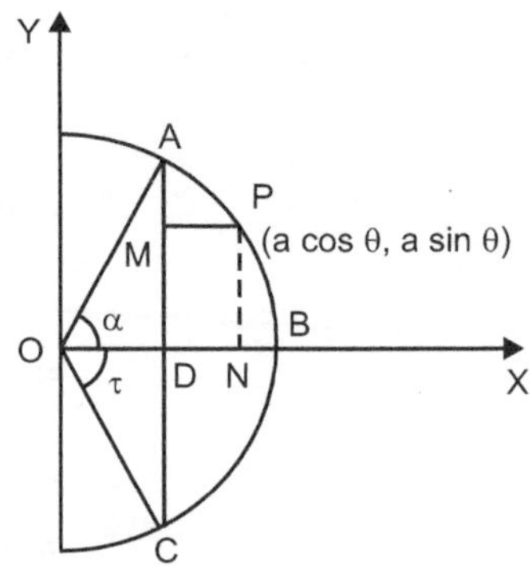

Then $\qquad PM = DN = ON - OD$

$$= a\cos\theta - a\cos\alpha = a(\cos\theta - \cos\alpha)$$

Also the parametric equations of the circle are

$$x = a\cos\theta,\ y = a\sin\theta$$

$$\therefore \qquad \frac{dx}{d\theta} = -a\sin\theta, \quad \frac{dy}{d\theta} = a\cos\theta$$

so that $\qquad \dfrac{ds}{d\theta} = \sqrt{\left(\dfrac{dx}{d\theta}\right)^2 + \left(\dfrac{dy}{d\theta}\right)^2} = \sqrt{a^2\sin^2\theta + a^2\cos^2\theta} = a\cdot$

Also for the arc BA, θ varies from 0 to α.

Hence the required surface

= 2 × surface generated by the revolution of the arc AB about the chord AC

$$= 2\int_{\theta=0}^{\alpha} 2\pi\,(\mathrm{PM})\,\frac{ds}{d\theta}\,d\theta = 4\pi\int_0^{\alpha} a(\cos\theta - \cos\alpha)\,.\,a\,.\,d\theta$$

$$= 4\pi a\left[\sin\theta - \theta\cos\alpha\right]_0^{\alpha} = 4\pi a^2(\sin\alpha - \alpha\cos\alpha)\,.$$

Example 2. *A quadrant of a circle of radius a revolves about its chord. Show that the surface of the spindle generated is* $2\sqrt{2}\,\pi a^2\left(1-\frac{\pi}{4}\right)$.

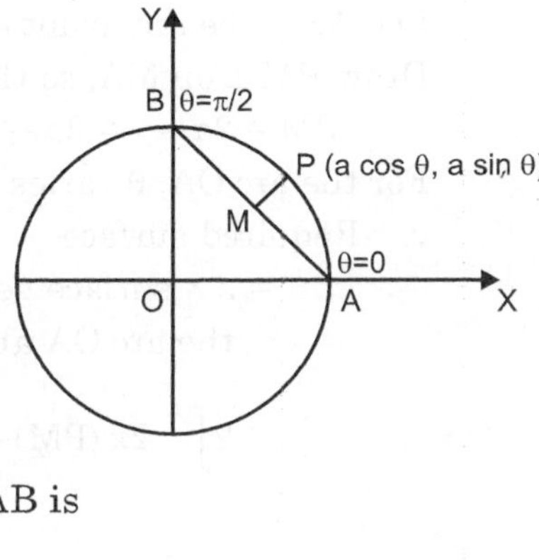

Sol. The parametric equations of the circle are

$$x = a\cos\theta,\quad y = a\sin\theta.$$

$$\therefore\quad \frac{ds}{d\theta} = \sqrt{\left(\frac{dx}{d\theta}\right)^2+\left(\frac{dy}{d\theta}\right)^2}$$

$$= \sqrt{a^2\sin^2\theta + a^2\cos^2\theta} = a.$$

Since $OA = OB = a$, $\therefore$ equation of the chord AB is

$$\frac{x}{a}+\frac{y}{a}=1,\quad\text{or}\quad x+y=a \qquad \text{...(1)} \mid \text{Intercept form}$$

Let P ($a\cos\theta$, $a\sin\theta$) be any point on the arc AB. From P draw PM ⊥ to chord AB, *i.e.*, (1).

$$\therefore\quad PM = \frac{a\cos\theta + a\sin\theta - a}{\sqrt{2}} = \frac{a}{\sqrt{2}}(\cos\theta+\sin\theta-1)$$

Also for the arc AB, θ varies from θ to $\frac{\pi}{2}$.

$$\therefore\quad \text{Required surface} = \int_0^{\pi/2} 2\pi\,(PM)\,\frac{ds}{d\theta}\,d\theta$$

$$= 2\pi\int_0^{\pi/2}\frac{a}{\sqrt{2}}(\cos\theta+\sin\theta-1).\,a\,d\theta = \sqrt{2}\pi a^2\left[\sin\theta-\cos\theta-\theta\right]_0^{\pi/2}$$

$$= \sqrt{2}\pi a^2\left[1-\frac{\pi}{2}+1\right] = \sqrt{2}\pi a^2\left(2-\frac{\pi}{2}\right) = 2\sqrt{2}\pi a^2\left(1-\frac{\pi}{4}\right).$$

Aliter. We can proceed as in Example 1 above, where $\alpha = \pi/4$.

Example 3. *Find the area of the surface generated if an arch of the cycloid*

$$x = a(\theta-\sin\theta),\ y = a(1-\cos\theta) \text{ revolves about the line } y = 2a.$$

Sol. The equations of the cycloid are $x = a(\theta-\sin\theta)$, $y = a(1-\cos\theta)$

The cycloid is symmetrical about the line ⊥ to x-axis, where $\theta = \pi$.

Now $\quad \frac{dx}{d\theta} = a(1-\cos\theta)$ and $\frac{dy}{d\theta} = a\sin\theta$

$$\therefore\quad \frac{ds}{d\theta} = \sqrt{\left(\frac{dx}{d\theta}\right)^2+\left(\frac{dy}{d\theta}\right)^2} = \sqrt{a^2(1-\cos\theta)^2+a^2\sin^2\theta}$$

$$= a\sqrt{2(1-\cos\theta)} = a\sqrt{4\sin^2\frac{\theta}{2}} = 2a\sin\frac{\theta}{2}$$

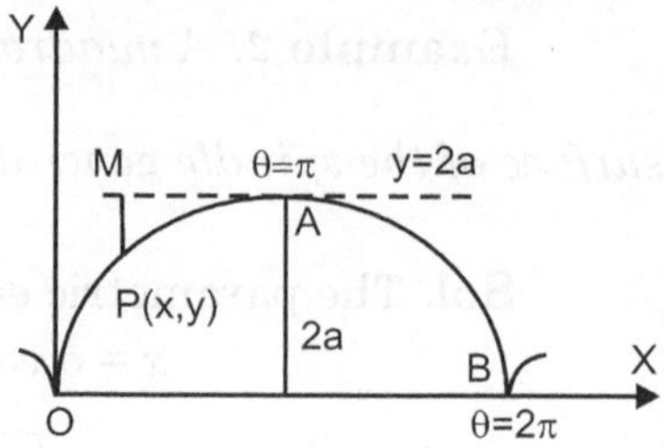

Now one arch OAB if the cycloid revolves about the line $y = 2a$, which is tangent to the cycloid at A, the vertex.

Let P(x, y) be any point on the arc OA.

Draw PM ⊥ on MA, so that

$$PM = 2a - y = 2a - a(1 - \cos\theta) = a(1 + \cos\theta)$$

For the arc OA, θ varies from 0 to π.

∴ Required surface

= 2 × surface generated by the revolution of the arc OA about the line $y = 2a$

$$= 2\int_0^{\pi} 2\pi\,(PM)\frac{ds}{d\theta}d\theta = 4\pi\int_0^{\pi} a(1+\cos\theta).\ 2a\sin\frac{\theta}{2}\,d\theta$$

$$= 4\pi\int_0^{\pi} a\left(2\cos^2\frac{\theta}{2}\right).\ 2a\sin\frac{\theta}{2}\,d\theta$$

$$= 16\pi a^2\int_0^{\pi}\cos^2\frac{\theta}{2}\sin\frac{\theta}{2}\,d\theta = -\,32\pi a^2\int_0^{\pi}\cos^2\frac{\theta}{2}\left(-\frac{1}{2}\sin\frac{\theta}{2}\right)d\theta$$

$$= -\,32\pi a^2\left[\frac{\cos^3\theta/2}{2}\right]_0^{\pi} = -\frac{32}{3}\pi a^2(0-1) = \frac{32}{3}\pi a^2.$$

TEST YOUR KNOWLEDGE

1. Find the curved surface of a hemisphere of radius a.
2. The circle $x^2 + y^2 = a^2$ is revolved about x-axis. Find the area of the sphere generated.
3. Find the area of the surface generated by rotating about x-axis the arc of the curve $y = x^3$ between $x = 0$ and $x = 1$.
4. Find the surface area of the solid obtained by revolving the arc of the curve $y = \sin x$ from $x = 0$ to $x = \pi$ about x-axis.
5. Find the curved surface of the solid generated by the revolution about the x-axis of the area bounded by the parabola $y^2 = 4ax$, the x-axis and the ordinate
   (*i*) $x = h$ (*ii*) $x = 3a$.
6. Find the surface generated by the revolution of the curve $y = c\cosh\left(\frac{x}{c}\right)$ about the x-axis, between $x = a$ and $x = b$.
7. Find the surface of the solid generated by revolving the curve
   $x = e^t\sin t,\ y = e^t\cos t,\ 0 \le t \le \frac{\pi}{2}$, about the axis of x.
   $$\left[\textbf{Hint.}\int e^{ax}\cos bx\,dx = \frac{e^{ax}}{a^2+b^2}(a\cos bx + b\sin bx)\right]$$
8. Find the surface of the solid formed by the revolution of the cardioid $r = a(1 - \cos\theta)$ about the initial line.

9. Find the area of the surface formed by the revolution of $y^2 = 4ax$ about y-axis by the arc from vertex to $x = \dfrac{a}{4}$.

10. Find the surface of the solid generated by the revolution of the ellipse $4x^2 + 5y^2 = 20$ about the minor axis.

[**Hint.** The equation of the ellipse can be written as $\dfrac{x^2}{5} + \dfrac{y^2}{4} = 1$. Here $a = \sqrt{5}, b = 2$

The parametric equations of the ellipse are $x = \sqrt{5}\cos t,\ y = 2\sin t$]

Answers

1. $2\pi a^2$

2. $4\pi a^2$

3. $\dfrac{\pi}{27}(10\sqrt{10} - 1)$

4. $2\pi\left[\sqrt{2} + \log(\sqrt{2} + 1)\right]$

5. (*i*) $\dfrac{8}{3}\pi\sqrt{a}\,[(h + a)^{3/2} - a^{3/2}]$ (*ii*) $\dfrac{56}{3}\pi a^2$

6. $\pi c\left[(b - a) + \dfrac{c}{2}\left(\sinh\dfrac{2b}{c} - \sinh\dfrac{2a}{c}\right)\right]$

7. $\dfrac{2\sqrt{2}}{5}\pi(e^{\pi} - 2)$

8. $\dfrac{32}{5}\pi a^2$

9. $\dfrac{\pi a^2}{16}\left[3\sqrt{5} - 8\log\dfrac{\sqrt{5} + 1}{2}\right]$

10. $10\pi\left[1 + \dfrac{2}{\sqrt{5}}\log\left(\dfrac{\sqrt{5} + 1}{\sqrt{5} - 1}\right)\right]$.

13

Multiple Integrals

13.1. DOUBLE INTEGRALS

The definite integral $\int_a^b f(x)\,dx$ is defined as the limits of the sum $f(x_1)\delta x_1 + f(x_2)\delta x_2 + \cdots + f(x_n)\delta x_n$ when $n \to \infty$ and each of the lengths $\delta x_1, \delta x_2, \ldots, \delta x_n$ tends to zero. Here $\delta x_1, \delta x_2, \ldots, \delta x_n$ are n sub-intervals into which the range $b - a$ has been divided and $x_1, x_2, \ldots, x_n$ are values of x lying respectively in the first, second, ..., nth sub-interval.

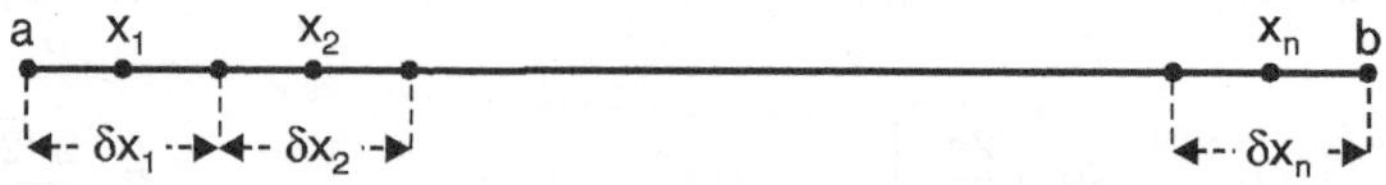

A double integral is its counterpart in two dimensions. Let a single-valued and bounded function $f(x, y)$ of two independent variables x, y be defined in a closed region R of the xy-plane. Divide the region R into sub-regions by drawing lines parallel to co-ordinate axes. Number the rectangles which lie entirely inside the region R, from 1 to n. Let (x_r, y_r) be any point inside the rth rectangle whose area is δA_r.

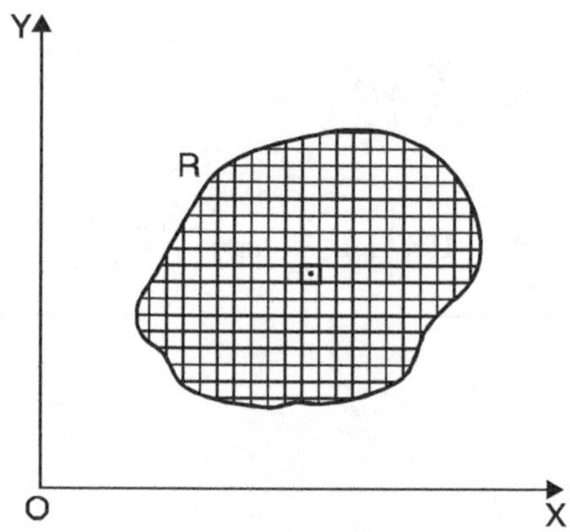

Consider the sum

$$f(x_1, y_1)\delta A_1 + f(x_2, y_2)\delta A_2 + \cdots + f(x_n, y_n)\delta A_n$$

$$= \sum_{r=1}^{n} f(x_r, y_r)\delta A_r \qquad \ldots(1)$$

Let the number of these sub-regions increase indefinitely, such that the largest linear dimension (*i.e.*, diagonal) of δA_r approaches zero. The limit of the sum (1), if it exists, irrespective of the mode of sub-division, is called the *double integral* of $f(x, y)$ over the region R and is denoted by $\iint_R f(x, y)\,dA$

In other words,

$$\lim_{\substack{n \to \infty \\ \delta A_r \to 0}} \sum_{r=1}^{n} f(x_r, y_r)\delta A_r = \iint_R f(x, y)\,dA$$

which is also expressed as

$$\iint_R f(x, y)\,dx\,dy \quad \text{or} \quad \iint_R f(x, y)\,dy\,dx$$

13.2. EVALUATION OF DOUBLE INTEGRALS

The methods of evaluating the double integrals depend upon the nature of the curves bounding the region R. Let the region R be bounded by the curves $x = x_1$, $x = x_2$ and $y = y_1$, $y = y_2$.

(*i*) **When x_1, x_2 are functions of y and y_1, y_2 are constants.** Let AB and CD be the curves $x_1 = \phi_1(y)$ and $x_2 = \phi_2(y)$.

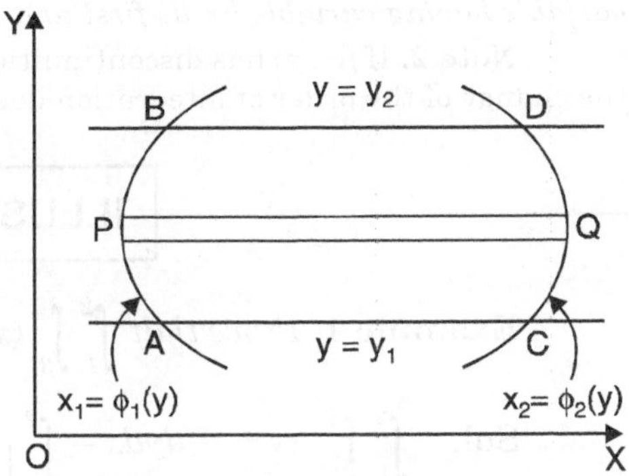

Take a horizontal strip PQ of width δy. Here the double integral is evaluated first w.r.t. x (treating y as a constant). The resulting expression which is a function of y is integrated w.r.t. y between the limits $y = y_1$ and $y = y_2$. Thus,

$$\iint_R f(x, y)\, dx\, dy = \int_{y_1}^{y_2} \left[\int_{x_1 = \phi_1(y)}^{x_2 = \phi_2(y)} f(x, y)\, dx \right] dy$$

the integration being carried from the inner to the outer rectangle. Geometrically, the integral in the inner rectangle indicates that the integration is performed along the horizontal strip PQ (keeping y constant) while the outer rectangle corresponds to the sliding of the strip PQ from AC to BD thus covering the entire region ABDC of integration.

(*ii*) **When y_1, y_2 are functions of x and x_1, x_2 are constants.** Let AB and CD be the curves $y_1 = \phi_1(x)$ and $y_2 = \phi_2(x)$. Take a vertical strip PQ of width δx. Here the double integral is evaluated first w.r.t. y (treating x as constant). The resulting expression which is a function of x is integrated w.r.t. x between the limits $x = x_1$ and $x = x_2$. Thus,

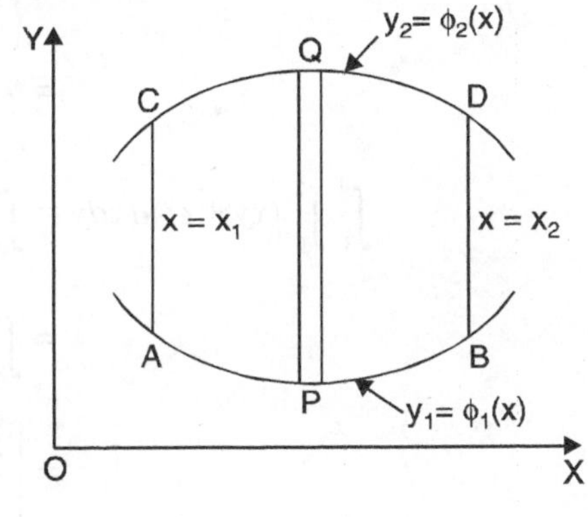

$$\iint_R f(x, y)\, dx\, dy = \int_{x_1}^{x_2} \left[\int_{y_1 = \phi_1(x)}^{y_2 = \phi_2(x)} f(x, y)\, dy \right] dx$$

the integration being carried from the inner to the outer rectangle. Geometrically, the integral in the inner rectangle indicates that the integration is performed along the vertical strip PQ (keeping x constant) while the outer rectangle corresponds to the sliding of the strip PQ from AC to BD thus covering the entire region ABDC of integration.

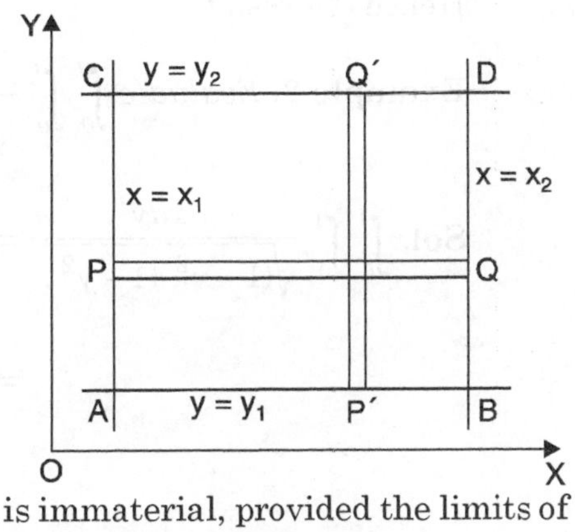

(*iii*) **When x_1, x_2, y_1, y_2 are constants.** Here the region of integration R is the rectangle ABDC. It is immaterial whether we integrate first along the horizontal strip PQ and then slide it from AB to CD ; or we integrate first along the vertical strip P′Q′ and then slide it from AC to BD. Thus the order of integration is immaterial, provided the limits of integration are changed accordingly.

$$\iint_R f(x, y)\,dxdy = \int_{y_1}^{y_2}\left[\int_{x_1}^{x_2} f(x, y)\,dx\right]dy = \int_{x_1}^{x_2}\left[\int_{y_1}^{y_2} f(x, y)\,dy\right]dx$$

Note 1. From cases (*i*) and (*ii*) above, we observe that *integration is to be performed w.r.t. that variable having variable limits first and then w.r.t. the variable with constant limits.*

Note 2. If $f(x, y)$ has discontinuities within or on the boundary of the region of integration, then the change of the order of integration does not result into the same integrals.

ILLUSTRATIVE EXAMPLES

Example 1. *Prove that* $\int_1^2\int_3^4 (xy + e^y)dydx = \int_3^4\int_1^2 (xy + e^y)dxdy.$

Sol. $\int_1^2\int_3^4 (xy + e^y)dydx = \int_1^2\left[\int_3^4 (xy + e^y)\,dy\right]dx$

$$= \int_1^2\left[\frac{xy^2}{2} + e^y\right]_3^4 dx = \int_1^2\left(8x + e^4 - \frac{9}{2}x - e^3\right)dx$$

$$= \int_1^2\left(\frac{7}{2}x + e^4 - e^3\right)dx = \left[\frac{7x^2}{4} + (e^4 - e^3)x\right]_1^2$$

$$= 7 + 2(e^4 - e^3) - \frac{7}{4} - (e^4 - e^3) = \frac{21}{4} + e^4 - e^3$$

$$\int_3^4\int_1^2 (xy + e^y)dxdy = \int_3^4\left[\int_1^2 (xy + e^y)dx\right]dy = \int_3^4\left[\frac{yx^2}{2} + xe^y\right]_1^2 dy$$

$$= \int_3^4\left(2y + 2e^y - \frac{y}{2} - e^y\right)dy = \int_3^4\left(\frac{3y}{2} + e^y\right)dy$$

$$= \left[\frac{3y^2}{4} + e^y\right]_3^4 = 12 + e^4 - \frac{27}{4} - e^3 = \frac{21}{4} + e^4 - e^3$$

Hence the result.

Example 2. *Evaluate* $\int_0^1\int_0^1 \frac{dxdy}{\sqrt{(1-x^2)(1-y^2)}}$

Sol. $\int_0^1\int_0^1 \frac{dxdy}{\sqrt{(1-x^2)(1-y^2)}} = \int_0^1\left[\int_0^1 \frac{dx}{\sqrt{(1-x^2)(1-y^2)}}\right]dy = \int_0^1 \frac{1}{\sqrt{1-y^2}}\left[\sin^{-1} x\right]_0^1 dy$

$$= \int_0^1 \frac{1}{\sqrt{1-y^2}} \cdot \frac{\pi}{2}\,dy = \frac{\pi}{2}\left[\sin^{-1} y\right]_0^1 = \frac{\pi}{2} \cdot \frac{\pi}{2} = \frac{\pi^2}{4}$$

Example 3. *Show that* $\int_0^1 dx \int_0^1 \frac{x-y}{(x+y)^3}\, dy \neq \int_0^1 dy \int_0^1 \frac{x-y}{(x+y)^3}\, dx.$

Sol. $$\text{LHS} = \int_0^1 dx \int_0^1 \frac{2x-(x+y)}{(x+y)^3}\, dy = \int_0^1 dx \int_0^1 \left[\frac{2x}{(x+y)^3} - \frac{1}{(x+y)^2}\right] dy$$

$$= \int_0^1 \left[2x \cdot \frac{(x+y)^{-2}}{-2} - \frac{(x+y)^{-1}}{-1}\right]_0^1 dx = \int_0^1 \left[\frac{-x}{(x+y)^2} + \frac{1}{x+y}\right]_0^1 dx$$

$$= \int_0^1 \left[\frac{-x}{(x+1)^2} + \frac{1}{x+1} + \frac{1}{x} - \frac{1}{x}\right] dx = \int_0^1 \frac{-x+x+1}{(x+1)^2}\, dx = \int_0^1 \frac{1}{(x+1)^2}\, dx$$

$$= \left[-\frac{1}{x+1}\right]_0^1 = -\frac{1}{2} + 1 = \frac{1}{2}$$

$$\text{RHS} = \int_0^1 dy \int_0^1 \frac{(x+y)-2y}{(x+y)^3}\, dx = \int_0^1 dy \int_0^1 \left[\frac{1}{(x+y)^2} - \frac{2y}{(x+y)^3}\right] dx$$

$$= \int_0^1 \left[\frac{(x+y)^{-1}}{-1} - 2y \cdot \frac{(x+y)^{-2}}{-2}\right]_0^1 dy = \int_0^1 \left[-\frac{1}{x+y} + \frac{y}{(x+y)^2}\right]_0^1 dy$$

$$= \int_0^1 \left[-\frac{1}{1+y} + \frac{y}{(1+y)^2} + \frac{1}{y} - \frac{1}{y}\right] dy = \int_0^1 \frac{-1-y+y}{(1+y)^2}\, dy = -\int_0^1 \frac{1}{(1+y)^2}\, dy$$

$$= \left[\frac{1}{1+y}\right]_0^1 = \frac{1}{2} - 1 = -\frac{1}{2}$$

$\therefore$ The two integrals are not equal.

Example 4. *Evaluate* $\int_0^1 \int_0^{\sqrt{1+x^2}} \frac{dy\, dx}{1+x^2+y^2}.$

Sol. $$I = \int_0^1 \left[\int_0^{\sqrt{1+x^2}} \frac{1}{(1+x^2)+y^2}\, dy\right] dx$$

$$= \int_0^1 \frac{1}{\sqrt{1+x^2}} \left[\tan^{-1} \frac{y}{\sqrt{1+x^2}}\right]_0^{\sqrt{1+x^2}} dx$$

$$= \int_0^1 \frac{1}{\sqrt{1+x^2}}\, [\tan^{-1} 1 - \tan^{-1} 0]\, dx = \frac{\pi}{4} \int_0^1 \frac{dx}{\sqrt{1+x^2}}$$

$$= \frac{\pi}{4} \left[\log (x + \sqrt{1+x^2})\right]_0^1 = \frac{\pi}{4}\, [\log (1+\sqrt{2}) - \log 1] = \frac{\pi}{4} \log (\sqrt{2}+1).$$

Example 5. *Evaluate* $\int_0^a \int_0^{\sqrt{a^2-y^2}} \sqrt{a^2-x^2-y^2}\, dxdy.$

Sol. $\mathrm{I} = \int_0^a \left[\int_0^{\sqrt{a^2 - y^2}} \sqrt{(a^2 - y^2) - x^2}\, dx \right] dy.$

$$= \int_0^a \left[\frac{x\sqrt{a^2 - y^2 - x^2}}{2} + \frac{a^2 - y^2}{2} \sin^{-1} \frac{x}{\sqrt{a^2 - y^2}} \right]_0^{\sqrt{a^2 - y^2}} dy$$

$$\left[\because \int \sqrt{a^2 - x^2}\, dx = \frac{x\sqrt{a^2 - x^2}}{2} + \frac{a^2}{2} \sin^{-1} \frac{x}{a} \right]$$

$$= \int_0^a \frac{a^2 - y^2}{2} \sin^{-1} 1\, dy = \frac{\pi}{4} \int_0^a (a^2 - y^2)\, dy$$

$$= \frac{\pi}{4} \left[a^2 y - \frac{y^3}{3} \right]_0^a = \frac{\pi}{4} \left[a^3 - \frac{a^3}{3} \right] = \frac{\pi a^3}{6}.$$

Example 6. *Evaluate* $\iint_R e^{2x+3y}\, dx\, dy$ *over the triangle bounded by* $x = 0,\ y = 0$ *and* $x + y = 1.$

Sol. The region R of integration is the triangle OAB. Here x varies from 0 to 1 and y varies from x-axis upto the line $x + y = 1$ *i.e.*, from 0 to $1 - x$.

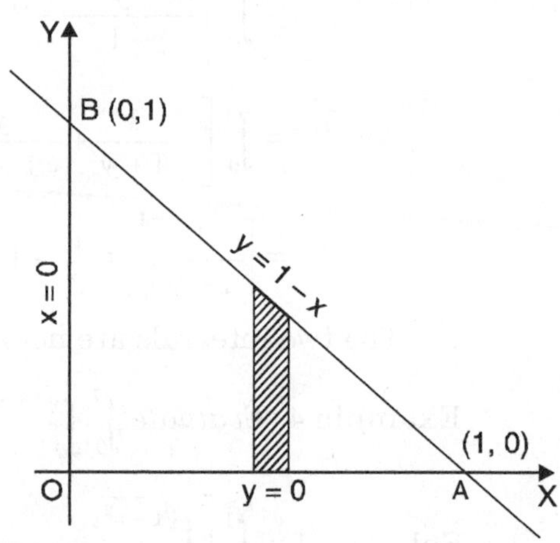

$\therefore$ The region R can be expressed as

$$0 \le x \le 1,\ 0 \le y \le 1 - x$$

$$\therefore \iint_R e^{2x+3y}\, dx\, dy$$

$$= \int_0^1 \int_0^{1-x} e^{2x+3y}\, dy\, dx$$

$$= \int_0^1 \left[\frac{1}{3} e^{2x+3y} \right]_0^{1-x} dx$$

$$= \frac{1}{3} \int_0^1 (e^{3-x} - e^{2x})\, dx$$

$$= \frac{1}{3} \left[-e^{3-x} - \frac{1}{2} e^{2x} \right]_0^1 = -\frac{1}{3} \left[\left(e^2 + \frac{1}{2} e^2 \right) - \left(e^3 + \frac{1}{2} \right) \right]$$

$$= -\frac{1}{3} \left[-e^2(e - 1) + \frac{1}{2}(e^2 - 1) \right]$$

$$= \frac{1}{6} (e - 1)\, [2e^2 - (e + 1)] = \frac{1}{6}\, (e - 1)(2e^2 - e - 1)$$

$$= \frac{1}{6} (e - 1)(e - 1)(2e + 1) = \frac{1}{6}\, (e - 1)^2(2e + 1).$$

Example 7. *Evaluate* $\iint_R y\,dx\,dy$, *where R is the region bounded by the parabolas* $y^2 = 4x$ *and* $x^2 = 4y$.

Sol. Solving $y^2 = 4x$ and $x^2 = 4y$, we have

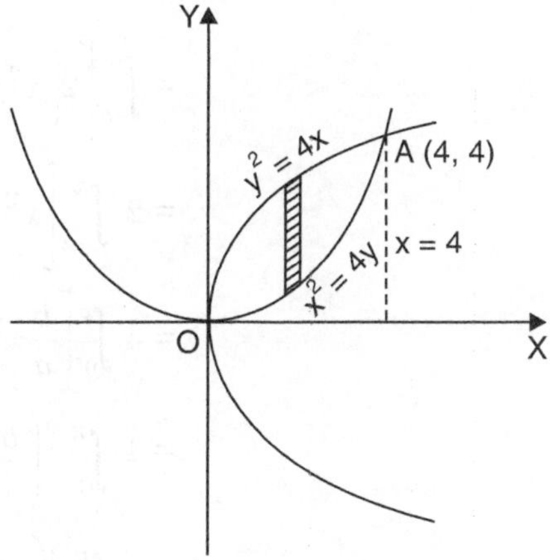

$$\left(\frac{x^2}{4}\right)^2 = 4x \quad \text{or} \quad x(x^3 - 64) = 0$$

$\therefore \qquad x = 0, 4$

When $\qquad x = 4,\ y = 4$

$\therefore$ Co-ordinates of A are (4, 4)

The region R can be expressed as

$$0 \le x \le 4,\ \frac{x^2}{4} \le y \le 2\sqrt{x}$$

$$\therefore \quad \iint_R y\,dx\,dy = \int_0^4 \int_{x^2/4}^{2\sqrt{x}} y\,dy\,dx$$

$$= \int_0^4 \frac{1}{2}\Big[y^2\Big]_{x^2/4}^{2\sqrt{x}} dx = \frac{1}{2}\int_0^4 \left(4x - \frac{x^4}{16}\right) dx$$

$$= \frac{1}{2}\left[2x^2 - \frac{x^5}{80}\right]_0^4 = \frac{1}{2}\left[32 - \frac{1024}{80}\right] = \frac{48}{5}.$$

Example 8. *Evaluate* $\iint (x + y)^2 dx\,dy$ *over the area bounded by the ellipse*

$$\frac{x^2}{a^2} + \frac{y^2}{b^2} = 1.$$

Sol. For the ellipse

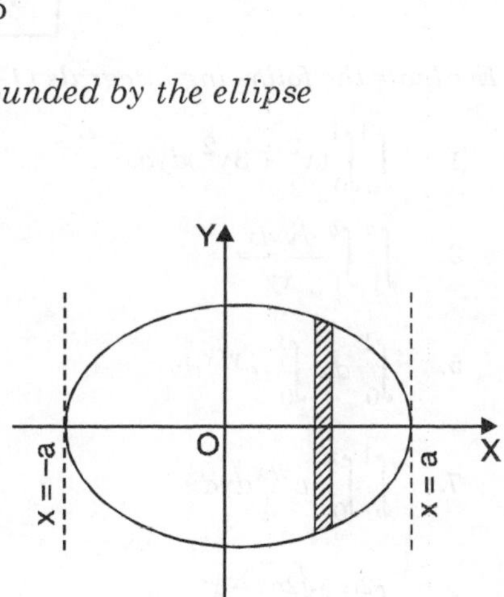

$$\frac{y}{b} = \pm\sqrt{1 - \frac{x^2}{a^2}} \quad \text{or} \quad y = \pm\frac{b}{a}\sqrt{a^2 - x^2}$$

$\therefore$ The region of integration R can be expressed as

$$-a \le x \le a,\ -\frac{b}{a}\sqrt{a^2 - x^2} \le y \le \frac{b}{a}\sqrt{a^2 - x^2}$$

$$\therefore \quad \iint_R (x + y)^2 dxdy = \iint_R (x^2 + y^2 + 2xy)dxdy$$

$$= \int_{-a}^{a} \int_{-\left(\frac{b}{a}\right)\sqrt{a^2 - x^2}}^{\left(\frac{b}{a}\right)\sqrt{a^2 - x^2}} (x^2 + y^2 + 2xy)dydx$$

$$= \int_{-a}^{a} \int_{-\left(\frac{b}{a}\right)\sqrt{a^2 - x^2}}^{\left(\frac{b}{a}\right)\sqrt{a^2 - x^2}} (x^2 + y^2)\,dy\,dx + \int_{-a}^{a} \int_{-\left(\frac{b}{a}\right)\sqrt{a^2 - x^2}}^{\left(\frac{b}{a}\right)\sqrt{a^2 - x^2}} 2xy\,dy\,dx$$

$$= \int_{-a}^{a} \int_{0}^{\left(\frac{b}{a}\right)\sqrt{a^2-x^2}} 2(x^2+y^2)dydx + 0$$

[since $(x^2 + y^2)$ is an even function of y and $2xy$ is an odd function of y]

$$= \int_{-a}^{a} \left[2\left(x^2 y + \frac{y^3}{3}\right)\right]_0^{\left(\frac{b}{a}\right)\sqrt{a^2-x^2}} dx$$

$$= 2\int_{-a}^{a} \left[x^2 \cdot \frac{b}{a}\sqrt{a^2-x^2} + \frac{1}{3}\cdot\frac{b^3}{a^3}(a^2-x^2)^{3/2}\right] dx$$

$$= 4\int_{0}^{a} \left[\frac{b}{a}x^2\sqrt{a^2-x^2} + \frac{b^3}{3a^3}(a^2-x^2)^{3/2}\right] dx$$

Put $x = a \sin \theta$
$\therefore \quad dx = a \cos \theta \, d\theta$

$$= 4\int_{0}^{\pi/2} \left(\frac{b}{a}\cdot a^2 \sin^2\theta \cdot a\cos\theta + \frac{b^3}{3a^3}\cdot a^3\cos^3\theta\right) a\cos\theta \, d\theta$$

$$= 4\int_{0}^{\pi/2} \left(a^3 b \sin^2\theta\cos^2\theta + \frac{ab^3}{3}\cos^4\theta\right) d\theta$$

$$= 4\left[a^3 b\cdot\frac{1\cdot 1}{4\cdot 2}\cdot\frac{\pi}{2} + \frac{ab^3}{3}\cdot\frac{3\cdot 1}{4\cdot 2}\cdot\frac{\pi}{2}\right] = \frac{\pi}{4}(a^3 b + ab^3) = \frac{\pi}{4}ab(a^2+b^2).$$

TEST YOUR KNOWLEDGE

Evaluate the following integrals (1–10):

1. $\int_0^3\int_0^1 (x^2+3y^2)dydx$

2. $\int_0^3\int_1^2 xy\,(1+x+y)dydx$

3. $\int_1^a\int_1^b \frac{dy\,dx}{xy}$

4. $\int_1^2\int_0^x \frac{dy\,dx}{x^2+y^2}$

5. $\int_0^1 dx \int_0^x e^{y/x}dy$

6. $\int_0^1\int_{x^2}^x (x^2+3y+2)dydx$

7. $\int_0^1\int_0^{x^2} e^{y/x}dydx$

8. $\int_0^1\int_y^{y^2+1} x^2 y\,dxdy$

9. $\int_0^{2a}\int_0^{\sqrt{2ax-x^2}} (x^2+y^2)dydx$

10. $\int_0^1\int_x^{\sqrt{x}} (x^2+y^2)dydx$

11. Evaluate $\iint (x^2+y^2)\,dx\,dy$ over the region in the positive quadrant for which $x + y \le 1$.

12. Evaluate $\iint x^2y^2\,dx\,dy$ over the circle $x^2 + y^2 = 1$.

13. Evaluate $\iint xy\,dx\,dy$ over the positive quadrant of the circle $x^2 + y^2 = a^2$.

14. Compute the value of $\iint_R y\,dx\,dy$, where R is the region in the first quadrant bounded by the ellipse $\frac{x^2}{a^2}+\frac{y^2}{b^2}=1$.

15. Evaluate $\iint_S \sqrt{xy-y^2}\,dx\,dy$, where S is a triangle with vertices (0, 0), (10, 1) and (1, 1).

16. Evaluate $\iint_R xy\,(x+y)\,dx\,dy$ over the area between $y = x^2$ and $y = x$.

17. Evaluate $\iint_A xy\,dx\,dy$, where A is the domain bounded by x-axis, ordinate $x = 2a$ and the curve $x^2 = 4ay$.

Answers

1. 12	**2.** $30\frac{3}{4}$	**3.** $\log a \log b$	**4.** $\frac{\pi}{4}\log 2$
5. $\frac{1}{2}(e-1)$	**6.** $\frac{7}{12}$	**7.** $\frac{1}{2}$	**8.** $\frac{67}{120}$
9. $\frac{3\pi a^4}{4}$	**10.** $\frac{3}{35}$	**11.** $\frac{1}{6}$	**12.** $\frac{\pi}{24}$
13. $\frac{a^4}{8}$	**14.** $\frac{ab^2}{3}$	**15.** 6	**16.** $\frac{3}{56}$
17. $\frac{a^4}{3}$			

13.3. EVALUATION OF DOUBLE INTEGRALS IN POLAR CO-ORDINATES

To evaluate $\int_{\theta_1}^{\theta_2}\int_{r_1}^{r_2} f(r,\theta)\,dr\,d\theta$ over the region bounded by the straight lines $\theta=\theta_1$, $\theta=\theta_2$ and the curves $r=r_1$, $r=r_2$, we first integrate w.r.t. r between the limits $r=r_1$ and $r=r_2$ (treating θ as a constant). The resulting expression is then integrated w.r.t. θ between the limits $\theta=\theta_1$ and $\theta=\theta_2$.

Geometrically, AB and CD are the curves $r=f_1(\theta)$ and $r=f_2(\theta)$ bounded by the lines $\theta=\theta_1$ and $\theta=\theta_2$ so that ACDB is the region of integration.

PQ is a wedge of angular thickness $\delta\theta$. Then $\int_{r=r_1}^{r=r_2} f(r,\theta)\,dr$ indicates that the integration is performed along PQ (*i.e.*, r varies, θ is constant) and the integration w.r.t. θ means rotation of this strip PQ from AC to BD.

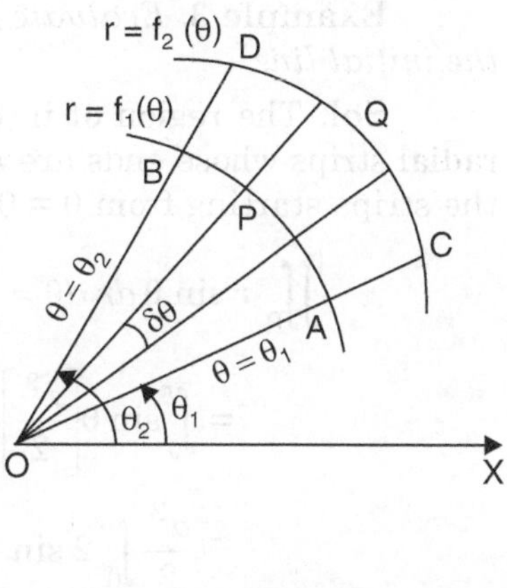

ILLUSTRATIVE EXAMPLES

Example 1. *Evaluate* $\int_0^{\pi/2}\left[\int_0^{a\cos\theta} r\sqrt{a^2-r^2}\,dr\right]d\theta$

Sol.
$$I=\int_0^{\pi/2}\left[\int_0^{a\cos\theta}-\frac{1}{2}(a^2-r^2)^{1/2}(-2r)dr\right]d\theta$$
$$=\int_0^{\pi/2}\left[-\frac{1}{2}\cdot\frac{(a^2-r^2)^{3/2}}{3/2}\right]_0^{a\cos\theta}d\theta$$
$$=-\frac{1}{3}\int_0^{\pi/2}(a^3\sin^3\theta-a^3)d\theta=-\frac{a^3}{3}\left[\frac{2}{3}-\frac{\pi}{2}\right]=\frac{a^3}{18}(3\pi-4).$$

Example 2. *Evaluate* $\iint r^3\,dr\,d\theta$, *over the area bounded between the circles* $r = 2\cos\theta$ *and* $r = 4\cos\theta$. [CUSAT 2009]

Sol. The region of integration R is shown shaded. Here r varies from $2\cos\theta$ to $4\cos\theta$ while θ varies from $-\frac{\pi}{2}$ to $\frac{\pi}{2}$.

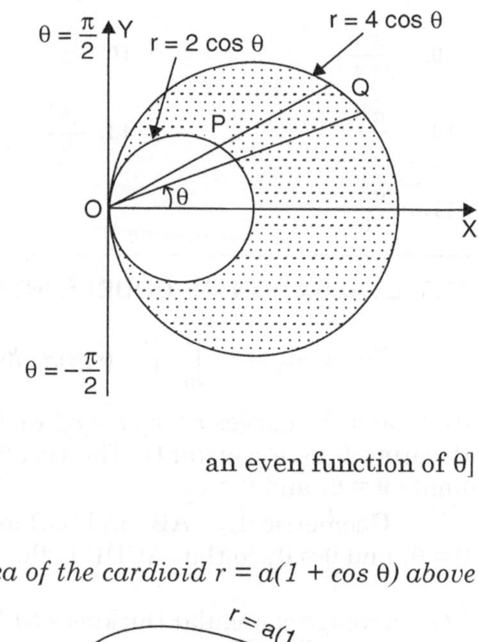

$$\therefore \quad \iint_R r^3drd\theta=\int_{-\pi/2}^{\pi/2}\int_{2\cos\theta}^{4\cos\theta}r^3drd\theta$$
$$=\int_{-\pi/2}^{\pi/2}\left[\frac{r^4}{4}\right]_{2\cos\theta}^{4\cos\theta}d\theta$$
$$=\int_{-\pi/2}^{\pi/2}\frac{1}{4}(256\cos^4\theta-16\cos^4\theta)d\theta$$
$$=60\int_{-\pi/2}^{\pi/2}\cos^4\theta\,d\theta$$
$$=120\int_0^{\pi/2}\cos^4\theta\,d\theta \quad [\text{since } \cos^4\theta \text{ is an even function of } \theta]$$
$$=120\times\frac{3\times1}{4\times2}\cdot\frac{\pi}{2}=\frac{45}{2}\pi.$$

Example 3. *Evaluate* $\iint r\sin\theta\,dr\,d\theta$ *over the area of the cardioid* $r = a(1+\cos\theta)$ *above the initial line.*

Sol. The region of integration R is covered by radial strips whose ends are $r = 0$ and $r = a(1+\cos\theta)$, the strips starting from $\theta = 0$ and ending at $\theta = \pi$.

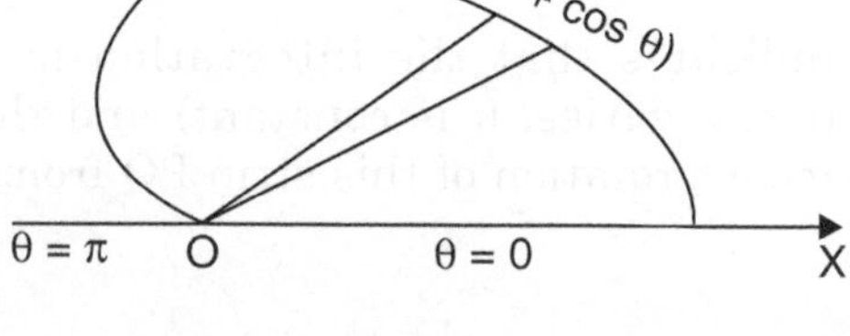

$$\therefore \quad \iint_R r\sin\theta\,dr\,d\theta=\int_0^{\pi}\int_0^{a(1+\cos\theta)}r\sin\theta\,drd\theta.$$
$$=\int_0^{\pi}\sin\theta\left[\frac{r^2}{2}\right]_0^{a(1+\cos\theta)}d\theta=\frac{1}{2}\int_0^{\pi}\sin\theta\,.\,a^2(1+\cos\theta)^2\,d\theta$$
$$=\frac{a^2}{2}\int_0^{\pi}2\sin\frac{\theta}{2}\cos\frac{\theta}{2}\cdot\left(2\cos^2\frac{\theta}{2}\right)^2d\theta=4a^2\int_0^{\pi}\sin\frac{\theta}{2}\cos^5\frac{\theta}{2}\,d\theta$$

$$= 4a^2 \int_0^{\pi/2} 2 \sin \phi \cos^5 \phi \, d\phi \qquad \left[\text{Putting } \frac{\theta}{2} = \phi \text{ and } d\theta = 2d\phi\right]$$

$$= -8a^2 \int_0^{\pi/2} \cos^5 \phi \, (-\sin \phi) \, d\phi = -8a^2 \cdot \left[\frac{\cos^6 \phi}{6}\right]_0^{\pi/2} = -\frac{4a^2}{3}(0-1) = \frac{4a^2}{3}.$$

Example 4. *Evaluate* $\iint \frac{r \, dr \, d\theta}{\sqrt{a^2 + r^2}}$ *over one loop of the lemniscate* $r^2 = a^2 \cos 2\theta$.

Sol. The region of integration R is covered by radial strips whose ends are $r = 0$ and $r = a\sqrt{\cos 2\theta}$, the strips starting from $\theta = -\frac{\pi}{4}$ and ending at $\theta = \frac{\pi}{4}$.

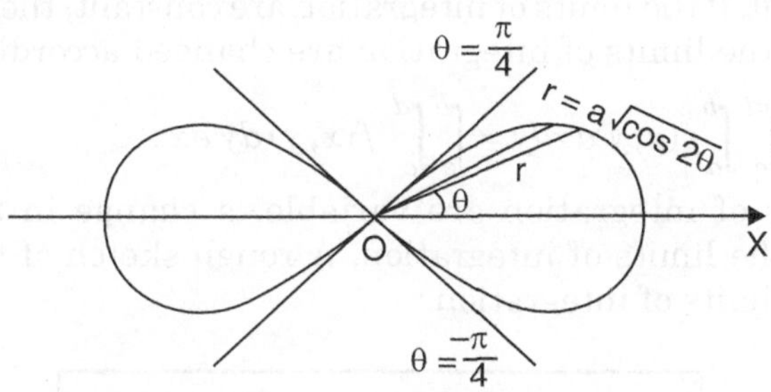

$$\therefore \quad \iint_R \frac{r \, dr \, d\theta}{\sqrt{a^2 + r^2}} = \int_{-\pi/4}^{\pi/4} \int_0^{a\sqrt{\cos 2\theta}} \tfrac{1}{2} (a^2 + r^2)^{-1/2} \cdot 2r \, dr \, d\theta$$

$$= \int_{-\pi/4}^{\pi/4} \left[\frac{1}{2} \cdot \frac{(a^2 + r^2)^{1/2}}{1/2}\right]_0^{a\sqrt{\cos 2\theta}} d\theta = \int_{-\pi/4}^{\pi/4} [(a^2 + a^2 \cos 2\theta)^{1/2} - a] \, d\theta$$

$$= a \int_{-\pi/4}^{\pi/4} [(1 + \cos 2\theta)^{1/2} - 1] \, d\theta = a \int_{-\pi/4}^{\pi/4} [(2 \cos^2 \theta)^{1/2} - 1] \, d\theta$$

$$= a \int_{-\pi/4}^{\pi/4} (\sqrt{2} \cos \theta - 1) \, d\theta = 2a \int_0^{\pi/4} (\sqrt{2} \cos \theta - 1) \, d\theta = 2a \left[\sqrt{2} \sin \theta - \theta\right]_0^{\pi/4}$$

$$= 2a \left[\sqrt{2} \cdot \frac{1}{\sqrt{2}} - \frac{\pi}{4}\right] = 2a\left(1 - \frac{\pi}{4}\right).$$

TEST YOUR KNOWLEDGE

Evaluate the following integrals (1–4):

1. $\int_0^{\pi} \int_0^{a \sin \theta} r \, dr \, d\theta$
2. $\int_0^{\pi/2} \int_0^{a \cos \theta} r \sin \theta \, dr \, d\theta$
3. $\int_0^{\pi/2} \int_{a(1+\cos \theta)}^{a} r \, dr \, d\theta$
4. $\int_0^{\pi} \int_0^{a(1+\cos \theta)} r^2 \cos \theta \, dr \, d\theta$

5. Show that $\iint_R r^2 \sin\theta \, dr \, d\theta = \frac{2a^3}{3}$, where R is the region bounded by the semi-circle $r = 2a\cos\theta$, above the initial line.

6. Evaluate $\iint r^3 \, dr \, d\theta$ over the area included between the circles $r = 2\sin\theta$ and $r = 4\sin\theta$.

[*CUSAT 2010*]

Answers

1. $\frac{\pi a^2}{4}$ 2. $\frac{a^2}{6}$ 3. $-a^2\left(1+\frac{\pi}{8}\right)$

4. $\frac{5}{8}\pi a^3$ 6. $\frac{45\pi}{2}$.

13.4. CHANGE OF ORDER OF INTEGRATION

In a double integral, if the limits of integration are constant, then the order of integration is immaterial, provided the limits of integration are changed accordingly. Thus

$$\int_c^d \int_a^b f(x, y) dx dy = \int_a^b \int_c^d f(x, y)\, dy\, dx.$$

But if the limits of integration are variable, a change in the order of integration necessitates change in the limits of integration. A rough sketch of the region of integration helps in fixing the new limits of integration.

ILLUSTRATIVE EXAMPLES

Example 1. *Change the order of integration in* $\int_0^a \int_y^a \frac{x dx dy}{x^2 + y^2}$ *and hence evaluate the same.* (CUSAT 2008)

Sol. From the limits of integration, it is clear that the region of integration is bounded by $x = y$, $x = a$, $y = 0$ and $y = a$. Thus the region of integration is the ΔOAB and is divided into horizontal strips. For changing the order of integration, we divide the region of integration into vertical strips. The new limits of integration become : y varies from 0 to x and x varies from 0 to a.

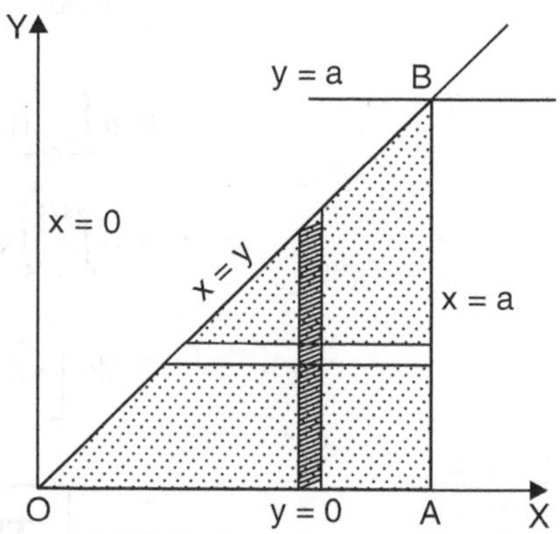

$$\int_0^a \int_y^a \frac{x\, dx\, dy}{x^2 + y^2} = \int_0^a \int_0^x \frac{x\, dy\, dx}{x^2 + y^2}$$

$$= \int_0^a x \cdot \left[\frac{1}{x}\tan^{-1}\frac{y}{x}\right]_0^x dx$$

$$= \int_0^a \frac{\pi}{4} dx = \frac{\pi}{4}\cdot\left[x\right]_0^a = \frac{\pi a}{4}.$$

Example 2. *Change the order of integration in* $I = \int_0^1 \int_{x^2}^{2-x} xy dy dx$ *and hence evaluate the same.* [CUSAT 2007]

Sol. From the limits of integration, it is clear that we have to integrate first with respect to y which varies from $y = x^2$ to $y = 2 - x$ and then with respect to x which varies from $x = 0$ to $x = 1$. The region of integration (shown shaded) is divided into vertical strips. For changing the order of integration, we divide the region of integration into horizontal strips.

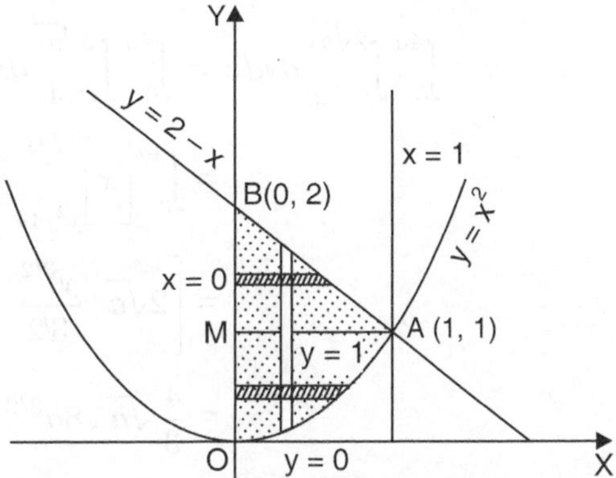

Solving $y = x^2$ and $y = 2 - x$, the co-ordinates of A are (1, 1). Draw AM ⊥ OY. The region of integration is divided into two parts, OAM and MAB.

For the region OAM, x varies from 0 to $\sqrt{y}$ and y varies from 0 to 1. For the region MAB, x varies from 0 to $2 - y$ and y varies from 1 to 2.

$$\therefore \quad \int_0^1 \int_{x^2}^{2-x} xy\,dx\,dy = \int_0^1 \int_0^{\sqrt{y}} xy\,dx\,dy + \int_1^2 \int_0^{2-y} xy\,dx\,dy$$

$$= \int_0^1 y \cdot \left[\frac{x^2}{2}\right]_0^{\sqrt{y}} dy + \int_1^2 y \cdot \left[\frac{x^2}{2}\right]_0^{2-y} dy = \frac{1}{2}\int_0^1 y^2 dy + \frac{1}{2}\int_1^2 y(2-y)^2 dy$$

$$= \frac{1}{2} \cdot \left[\frac{y^3}{3}\right]_0^1 + \frac{1}{2}\int_1^2 (4y - 4y^2 + y^3) dy = \frac{1}{6} + \frac{1}{2}\left[2y^2 - \frac{4y^3}{3} + \frac{y^4}{4}\right]_1^2$$

$$= \frac{1}{6} + \frac{1}{2}\left[\left(8 - \frac{32}{3} + 4\right) - \left(2 - \frac{4}{3} + \frac{1}{4}\right)\right] = \frac{3}{8}.$$

Example 3. *Change the order of integration in the following integral and evaluate:*

$$\int_0^{4a} \int_{x^2/4a}^{2\sqrt{ax}} dy\,dx.$$

Sol. From the limits of integration, it is clear that we have to integrate first w.r.t. y which varies from $y = \frac{x^2}{4a}$ to $y = 2\sqrt{ax}$ and then w.r.t. x which varies from $x = 0$ to $x = 4a$. Thus integration is first performed along the vertical strip PQ which extends from a point P on the parabola $y = \frac{x^2}{4a}$ (*i.e.*, $x^2 = 4ay$) to the point Q on the parabola $y = 2\sqrt{ax}$ (*i.e.*, $y^2 = 4ax$). Then the strip slides from O to A $(4a, 4a)$, the point of intersection of the two parabolas.

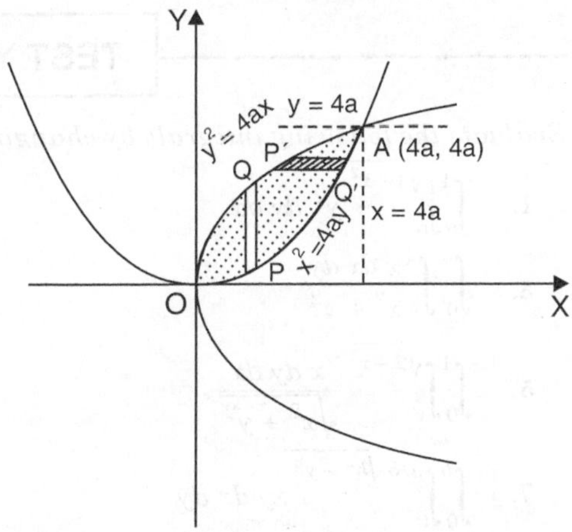

For changing the order of integration, we divide the region of integration OPAQO into horizontal strips P′Q′ which extend from P′ on the parabola $y^2 = 4ax$ *i.e.*, $x = \frac{y^2}{4a}$ to Q′ on the parabola $x^2 = 4ay$ *i.e.*, $x = 2\sqrt{ay}$. Then this strip slides from O to A$(4a, 4a)$, *i.e.*, varies from 0 to $4a$.

$$\therefore \quad \int_0^{4a}\int_{x^2/4a}^{2\sqrt{ax}} dydx = \int_0^{4a}\int_{y^2/4a}^{2\sqrt{ay}} dxdy$$

$$= \int_0^{4a}\left[x\right]_{y^2/4a}^{2\sqrt{ay}} dy = \int_0^{4a}\left(2\sqrt{ay} - \frac{y^2}{4a}\right)dy$$

$$= \left[2\sqrt{a}\,.\,\frac{y^{3/2}}{3/2} - \frac{y^3}{12a}\right]_0^{4a} = \frac{4}{3}\sqrt{a}\,.\,(4a)^{3/2} - \frac{64a^3}{12a}$$

$$= \frac{4}{3}\sqrt{a}\,.\,8a^{3/2} - \frac{16a^2}{3} = \frac{32a^2}{3} - \frac{16a^2}{3} = \frac{16a^2}{3}.$$

Example 4. *Change the order of integration in the integral* $\int_{-a}^{a}\int_0^{\sqrt{a^2-y^2}} f(x, y)\, dxdy.$

Sol. From the limits of integration, it is clear that we have to integrate first w.r.t. x which varies from $x = 0$ to $x = \sqrt{a^2 - y^2}$ and then w.r.t. y which varies from $y = -a$ to $y = a$. Thus integration is first performed along the horizontal strip PQ which extends from a point P on $x = 0$ (*i.e.*, y-axis) to the point Q on the circle $x = \sqrt{a^2 - y^2}$ (*i.e.*, $x^2 = a^2 - y^2$ or $x^2 + y^2 = a^2$). Then the strip slides from B′ to B.

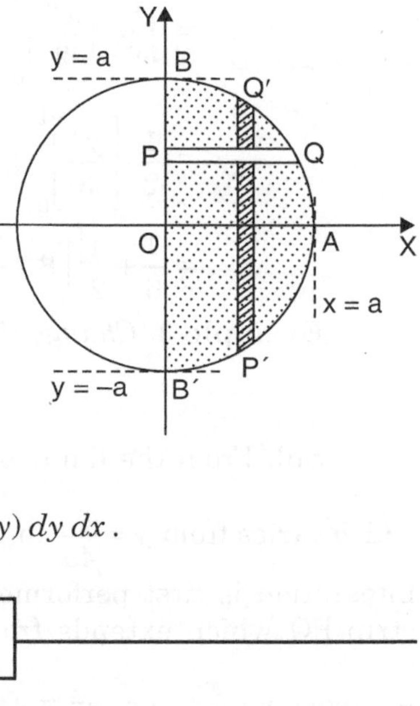

For changing the order of integration, we divide the region of integration B′AQBPB′ into vertical strips P′Q′ which extend from P′ on the circle $y = -\sqrt{a^2 - x^2}$ to Q′ on the circle $y = \sqrt{a^2 - x^2}$. [$x^2 + y^2 = a^2 \Rightarrow y = \pm\sqrt{a^2 - x^2}$; for points in the IV quadrant, $y = -\sqrt{a^2 - x^2}$ and for points in the first quadrant, $y = \sqrt{a^2 - x^2}$]. Then this strip slides from y-axis ($x = 0$) to A, where $x = a$.

$$\therefore \quad \int_{-a}^{a}\int_0^{\sqrt{a^2-x^2}} f(x, y)\,dx\,dy = \int_0^{a}\int_{-\sqrt{a^2-x^2}}^{\sqrt{a^2-x^2}} f(x, y)\,dy\,dx.$$

TEST YOUR KNOWLEDGE

Evaluate the following integrals by changing the order of integration:

1. $\int_0^1\int_0^{\sqrt{1-x^2}} y^2\,dy\,dx.$
2. $\int_0^1\int_{4y}^{4} e^{x^2}\,dx\,dy.$
3. $\int_0^4\int_y^4 \frac{x\,dx\,dy}{x^2 + y^2}.$
4. $\int_0^{a/\sqrt{2}}\int_y^{\sqrt{a^2-y^2}} x\,dx\,dy.$
5. $\int_0^1\int_x^{\sqrt{2-x^2}} \frac{x\,dy\,dx}{\sqrt{x^2 + y^2}}.$
6. $\int_0^a\int_{x^2/a}^{2a-x} xy\,dy\,dx.$
7. $\int_0^b\int_0^{a/b\,\sqrt{b^2-y^2}} xy\,dx\,dy.$
8. $\int_0^a\int_{\sqrt{ax}}^{a} \frac{y^2\,dy\,dx}{\sqrt{y^4 - a^2x^2}}.$

9. $\int_0^\infty \int_x^\infty \frac{e^{-y}}{y}\, dydx$. **10.** $\int_0^\infty \int_0^x xe^{-\frac{x^2}{y}}\, dydx$.

11. $\int_0^3 \int_1^{\sqrt{4-y}} (x+y)dxdy$.

12. $\int_0^a \int_{x/a}^{\sqrt{x/a}} (x^2+y^2)dydx$. **13.** $\int_0^{\frac{a}{\sqrt{2}}} \int_y^{\sqrt{a^2-y^2}} \log(x^2+y^2)dxdy, (a>0)$.

Answers

1. $\frac{\pi}{16}$	**2.** $\frac{1}{8}(e^{16}-1)$	**3.** π	**4.** $\frac{\sqrt{2}a^3}{6}$
5. $1-\frac{1}{\sqrt{2}}$	**6.** $\frac{3a^4}{8}$	**7.** $\frac{1}{8}a^2b^2$	**8.** $\frac{\pi a^2}{6}$
9. 1	**10.** $\frac{1}{2}$	**11.** $\frac{241}{60}$	**12.** $\frac{a^3}{28}+\frac{a}{20}$
13. $\frac{\pi a^2}{4}\left(\log a-\frac{1}{2}\right)$.			

13.5. TRIPLE INTEGRALS

Consider a function $f(x, y, z)$ which is continuous at every point of a finite region V of three dimensional space. Divide the region V into n sub-regions of respective volumes δV_1, δV_2, ..., δV_n. Let (x_r, y_r, z_r) be an arbitrary point in the rth sub-region. Consider the sum

$$\sum_{r=1}^{n} f(x_r, y_r, z_r)\, \delta V_r .$$

The limit of this sum as $n \to \infty$ and $\delta V_r \to 0$, if it exists, is called the *triple integral* of $f(x, y, z)$ over the region V and is denoted by $\iiint_V f(x, y, z)\, dV$.

For purposes of evaluation, it can be expressed as the repeated integral

$$\int_{z_1}^{z_2} \int_{y_1}^{y_2} \int_{x_1}^{x_2} f(x, y, z)\, dx\, dy\, dz, \qquad ...(1)$$

the order of integration depending upon the limits.

Let x_1, x_2 be function of y, z ; y_1, y_2 be function of z and z_1, z_2 be constants, *i.e.*,

Let $x_1 = f_1(y, z)$, $x_2 = f_2(y, z)$, $y_1 = \phi_1(z)$, $y_2 = \phi_2(z)$ and $z_1 = a$, $z_2 = b$.

Then the integral (1) is evaluated as follows:

$$\int_{z_1=a}^{z_2=b} \left[\int_{y_1=\phi_1(z)}^{y_2=\phi_2(z)} \left[\int_{x_1=f_1(y,z)}^{x_2=f_2(y,z)} f(x,y,z)\, dx \right] dy \right] dz$$

First $f(x, y, z)$ is integrated w.r.t. x (keeping y and z constant) between the limits x_1 and x_2. The resulting expression, which is a function of y and z is then integrated w.r.t. y (keeping z constant) between the limits y_1 and y_2. The resulting expression, which is a function of z only is then integrated w.r.t. z between the limits z_1 and z_2. The order of integration is from the innermost rectangle to the outermost rectangle.

Limits involving two variables are kept innermost, then the limits involving one variable and finally the constant limits.

If x_1, x_2 ; y_1, y_2 and z_1, z_2 are all constants, then the order of integration is immaterial, provided the limits are changed accordingly. Thus

$$\int_{z_1}^{z_2}\int_{y_1}^{y_2}\int_{x_1}^{x_2} f(x,y,z)dxdydz = \int_{x_1}^{x_2}\int_{z_1}^{z_2}\int_{y_1}^{y_2} f(x,y,z)dydzdx$$

$$= \int_{x_1}^{x_2}\int_{y_1}^{y_2}\int_{z_1}^{z_2} f(x,y,z)dzdydx \text{, etc.}$$

ILLUSTRATIVE EXAMPLES

Example 1. *Evaluate* $\int_0^1\int_0^{\sqrt{1-x^2}}\int_0^{\sqrt{1-x^2-y^2}} \frac{1}{\sqrt{1-x^2-y^2-z^2}}dzdydx$. [CUSAT 2009]

Sol.
$$I = \int_0^1\int_0^{\sqrt{1-x^2}}\int_0^{\sqrt{1-x^2-y^2}} \frac{1}{\sqrt{(1-x^2-y^2)-z^2}}dz\,dy\,dx$$

$$= \int_0^1\int_0^{\sqrt{1-x^2}}\left[\sin^{-1}\frac{z}{\sqrt{1-x^2-y^2}}\right]_0^{\sqrt{1-x^2-y^2}} dy\,dx$$

$$= \int_0^1\int_0^{\sqrt{1-x^2}} (\sin^{-1}1 - \sin^{-1}0)dydx$$

$$= \int_0^1\int_0^{\sqrt{1-x^2}} \frac{\pi}{2}dy\,dx = \int_0^1 \frac{\pi}{2}\Big[y\Big]_0^{\sqrt{1-x^2}} dx$$

$$= \frac{\pi}{2}\int_0^1 \sqrt{1-x^2}\,dx = \frac{\pi}{2}\left[\frac{x\sqrt{1-x^2}}{2}+\frac{1}{2}\sin^{-1}x\right]_0^1$$

$$= \frac{\pi}{4}[\sin^{-1}1] = \frac{\pi}{4}\cdot\frac{\pi}{2} = \frac{\pi^2}{8}.$$

Example 2. *Evaluate* $\int_1^e\int_1^{\log y}\int_1^{e^x} \log z\,dz\,dx\,dy$.

Sol.
$$I = \int_1^e\int_1^{\log y}\left[\int_1^{e^x}\log z\,dz\right]dx\,dy$$

$$= \int_1^e \int_1^{\log y} \left(\log z \cdot z - \int \frac{1}{z} \cdot z \, dz \right)_1^{e^x} dx\, dy$$

$$= \int_1^e \int_1^{\log y} \left\{ z (\log z - 1) \right\}_1^{e^x} dx\, dy$$

$$= \int_1^e \int_1^{\log y} \left\{ (x-1) e^x + 1) \right\} dx\, dy$$

Now $\int_1^{\log y} [(x-1)\, e^x + 1]\, dx = \int_1^{\log y} (x-1)e^x \, dx + \Big[x \Big]_1^{\log y}$

$$= \Big[(x-1)\, e^x \Big]_1^{\log y} - \int_1^{\log y} 1 \cdot e^x \, dx + \log y - 1$$

$$= (\log y - 1)\, e^{\log y} - \Big[e^x \Big]_1^{\log y} + \log y - 1$$

$$= y(\log y - 1) - (e^{\log y} - e) + \log y - 1 \qquad [\because \;\; e^{\log y} = y]$$

$$= y(\log y - 1) - y + e + \log y - 1 = (y+1) \log y - 2y + e - 1$$

$\therefore \qquad I = \int_1^e [\log y \cdot (y+1) - 2y + e - 1]\, dy$

$$= \left[\log y \cdot \left(\frac{y^2}{2} + y \right) \right]_1^e - \int_1^e \frac{1}{y} \left(\frac{y^2}{2} + y \right) dy - \Big[y^2 \Big]_1^e + (e-1) \Big[y \Big]_1^e$$

$$= \frac{e^2}{2} + e - \int_1^e \left(\frac{y}{2} + 1 \right) dy - (e^2 - 1) + (e-1)^2$$

$$= \frac{e^2}{2} + e - \left[\frac{y^2}{4} + y \right]_1^e - 2e + 2 = \frac{e^2}{2} + e - \left[\left(\frac{e^2}{4} + e \right) - \left(\frac{1}{4} + 1 \right) \right] - 2e + 2$$

$$= \frac{e^2}{4} - 2e + \frac{13}{4} = \frac{1}{4} (e^2 - 8e + 13).$$

TEST YOUR KNOWLEDGE

Evaluate the following integrals (1–11):

1. $\int_0^1 \int_0^1 \int_0^1 e^{x+y+z} \, dx\, dy\, dz.$

2. $\int_0^a \int_0^a \int_0^a (yz + zx + xy)\, dx\, dy\, dz.$

3. $\int_{-1}^1 \int_0^z \int_{x-z}^{x+z} (x + y + z)\, dy\, dx\, dz.$

4. $\int_1^3 \int_{1/x}^1 \int_0^{\sqrt{xy}} xyz \, dz\, dy\, dx.$

5. $\int_0^1 \int_0^{1-x} \int_0^{1-x-y} xyz \, dz\, dy\, dx.$

6. $\int_0^1 \int_{y^2}^1 \int_0^{1-x} x \, dz\, dx\, dy.$

7. $\int_0^1 \int_0^{\sqrt{1-x^2}} \int_0^{\sqrt{1-x^2-y^2}} xyz \, dz\, dy\, dx.$

8. $\int_0^{\log 2} \int_0^x \int_0^{x + \log y} e^{x+y+z} \, dz\, dy\, dx.$

9. $\int_0^{\pi/2} d\theta \int_0^{a \sin\theta} dr \int_0^{(a^2 - r^2)/a} r dz.$

10. $\int_0^{\pi/2} \int_0^{a\cos\theta} \int_0^{\sqrt{a^2 - r^2}} r dz\, dr\, d\theta.$

11. $\int_0^{a} \int_0^{x} \int_0^{x-y} e^{x+y+z}\, dz\, dy\, dx.$

Answers

1. $(e - 1)^3$
2. $\frac{3}{4}a^5$
3. 0
4. $\frac{13}{9} - \frac{1}{6}\log 3$
5. $\frac{1}{720}$
6. $\frac{4}{35}$
7. $\frac{1}{48}$
8. $\frac{8}{3}\log 2 - \frac{19}{9}$
9. $\frac{5\pi a^3}{64}$
10. $\frac{a^3}{3}\left(\frac{\pi}{2} - \frac{2}{3}\right)$
11. $\frac{1}{8}e^{4a} - \frac{3}{4}e^{2a} + e^a - \frac{3}{8}.$

13.6. CHANGE OF VARIABLES

Quite often, the evaluation of a double or triple integral is greatly simplified by a suitable change of variables.

Let the variables x, y in the double integral $\iint_R f(x, y)\, dx\, dy$ be changed to u, v by means of the relations $x = \phi(u, v)$, $y = \psi(u, v)$, then the double integral is transformed into

$$\iint_{R'} f\{\phi(u, v), \psi(u, v)\}\ |\ J\ |\ du\, dv, \text{ where } J = \frac{\partial(x, y)}{\partial(u, v)} = \begin{vmatrix} \frac{\partial x}{\partial u} & \frac{\partial x}{\partial v} \\ \frac{\partial y}{\partial u} & \frac{\partial y}{\partial v} \end{vmatrix}$$

is the Jacobian of transformation from (x, y) to (u, v) co-ordinates and R′ is the region in the uv-plane which corresponds to the region R in the xy-plane.

(i) To change cartesian co-ordinates (x, y) to polar co-ordinates (r, θ).

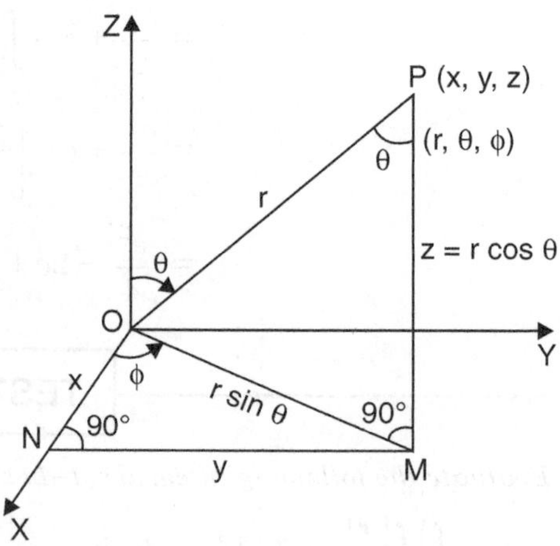

Here we have $x = r\cos\theta$, $y = r\sin\theta$ so that

$$x^2 + y^2 = r^2$$

$$J = \begin{vmatrix} \frac{\partial x}{\partial r} & \frac{\partial x}{\partial \theta} \\ \frac{\partial y}{\partial r} & \frac{\partial y}{\partial \theta} \end{vmatrix} = \begin{vmatrix} \cos\theta & -r\sin\theta \\ \sin\theta & r\cos\theta \end{vmatrix}$$

$$= r(\cos^2\theta + \sin^2\theta) = r$$

$$\therefore \iint_R f(x, y)\, dx\, dy = \iint_{R'} f(r\cos\theta, r\sin\theta)\, r\, dr\, d\theta$$

i.e., replace x by $r\cos\theta$, y by $r\sin\theta$ and $dx\, dy$ by $r\, dr\, d\theta$.

(ii) To change cartesian co-ordinates (x, y, z) to spherical polar co-ordinates (r, θ, ϕ).

Here, we have

$$x = r\sin\theta\cos\phi$$
$$y = r\sin\theta\sin\phi$$
$$z = r\cos\theta$$

so that $x^2 + y^2 + z^2 = r^2$

$$J = \begin{vmatrix} \frac{\partial x}{\partial r} & \frac{\partial x}{\partial \theta} & \frac{\partial x}{\partial \phi} \\ \frac{\partial y}{\partial r} & \frac{\partial y}{\partial \theta} & \frac{\partial y}{\partial \phi} \\ \frac{\partial z}{\partial r} & \frac{\partial z}{\partial \theta} & \frac{\partial z}{\partial \phi} \end{vmatrix}$$

$$= \begin{vmatrix} \sin\theta\cos\phi & r\cos\theta\cos\phi & -r\sin\theta\sin\phi \\ \sin\theta\sin\phi & r\cos\theta\sin\phi & r\sin\theta\cos\phi \\ \cos\theta & -r\sin\theta & 0 \end{vmatrix} = r^2\sin\theta$$

$$\therefore \quad \iiint_V f(x, y, z)\, dx\, dy\, dz = \iiint_{V'} f(r\sin\theta\cos\phi, r\sin\theta\sin\phi, r\cos\theta)\, r^2 \sin\theta\, dr\, d\theta\, d\phi.$$

(iii) To change cartesian co-ordinates (x, y, z) to cylindrical co-ordinates (r, φ, z).

Here, we have $x = r\cos\phi$

$y = r\sin\phi$

$z = z$

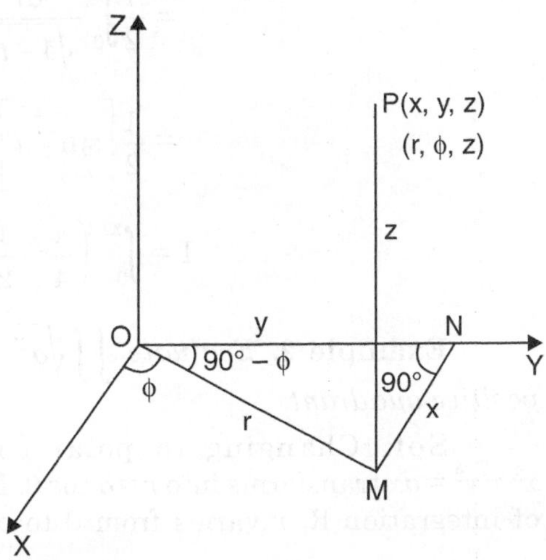

$$J = \begin{vmatrix} \frac{\partial x}{\partial r} & \frac{\partial x}{\partial \phi} & \frac{\partial x}{\partial z} \\ \frac{\partial y}{\partial r} & \frac{\partial y}{\partial \phi} & \frac{\partial y}{\partial z} \\ \frac{\partial z}{\partial r} & \frac{\partial z}{\partial \phi} & \frac{\partial z}{\partial z} \end{vmatrix}$$

$$= \begin{vmatrix} \cos\phi & -r\sin\phi & 0 \\ \sin\phi & r\cos\phi & 0 \\ 0 & 0 & 1 \end{vmatrix}$$

$$= r(\cos^2\phi + \sin^2\phi) = r$$

$$\therefore \quad \iiint_V f(x, y, z)\, dx\, dy\, dz$$

$$= \iiint_{V'} f(r\cos\phi, r\sin\phi, z)\, r\, dr\, d\phi\, dz.$$

ILLUSTRATIVE EXAMPLES

Example 1. *Evaluate* $\iint \sqrt{\frac{1 - x^2 - y^2}{1 + x^2 + y^2}}\, dx\, dy$ *over the positive quadrant of the circle* $x^2 + y^2 = 1$.

Sol. Changing to polar co-ordinates by putting $x = r\cos\theta$, $y = r\sin\theta$; $x^2 + y^2 = 1$ transforms into $r = 1$. For the region of integration R, r varies from 0 to 1 and θ varies from 0 to $\frac{\pi}{2}$.

$$I = \iint_R \sqrt{\frac{1 - x^2 - y^2}{1 + x^2 + y^2}}\, dx\, dy$$

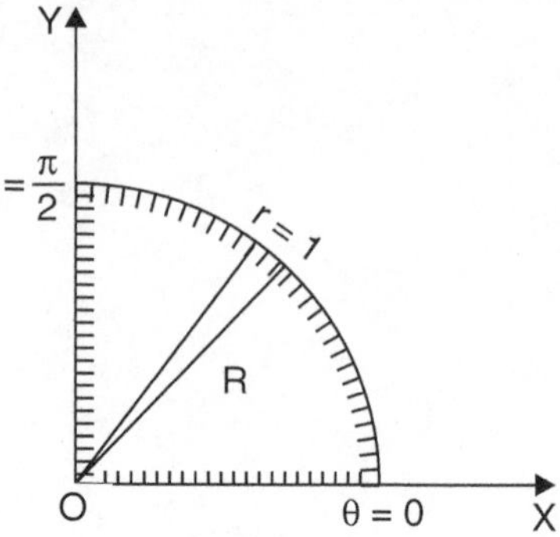

$$= \int_0^{\pi/2}\int_0^1 \sqrt{\frac{1-r^2}{1+r^2}}\, r dr d\theta$$

| $dx\, dy$ is replaced by $r dr d\theta$

$$= \int_0^{\pi/2}\int_0^1 \frac{r(1-r^2)}{\sqrt{1-r^4}} dr d\theta$$

Now $\int_0^1 \frac{r(1-r^2)}{\sqrt{1-r^4}} dr = \int_0^1 \left(\frac{r}{\sqrt{1-r^4}} - \frac{r^3}{\sqrt{1-r^4}}\right) dr$

$$= \frac{1}{2}\int_0^1 \frac{2r}{\sqrt{1-r^4}} dr + \frac{1}{4}\int_0^1 -4r^3 (1-r^4)^{-1/2} dr$$

$$= \frac{1}{2}\int_0^1 \frac{dt}{\sqrt{1-t^2}} + \frac{1}{4}\cdot\left[\frac{(1-r^4)^{1/2}}{1/2}\right]_0^1, \text{ where } t = r^2$$

$$= \frac{1}{2}\left[\sin^{-1} t\right]_0^1 + \frac{1}{2}(0-1) = \frac{1}{2}\left(\frac{\pi}{2}\right) - \frac{1}{2} = \frac{\pi}{4} - \frac{1}{2}$$

$$\therefore \quad I = \int_0^{\pi/2}\left(\frac{\pi}{4} - \frac{1}{2}\right) d\theta = \left(\frac{\pi}{4} - \frac{1}{2}\right)\left[\theta\right]_0^{\pi/2} = \left(\frac{\pi}{4} - \frac{1}{2}\right)\frac{\pi}{2} = \frac{\pi^2}{8} - \frac{\pi}{4}.$$

Example 2. *Evaluate* $\iint \sqrt{a^2 - x^2 - y^2}\, dxdy$ *over the semi-circle* $x^2 + y^2 = ax$ *in the positive quadrant.*

Sol. Changing to polar co-ordinates, $x^2 + y^2 = ax$ transforms into $r = a \cos \theta$. For the region of integration R, r varies from 0 to $a \cos \theta$ and θ varies from 0 to $\frac{\pi}{2}$.

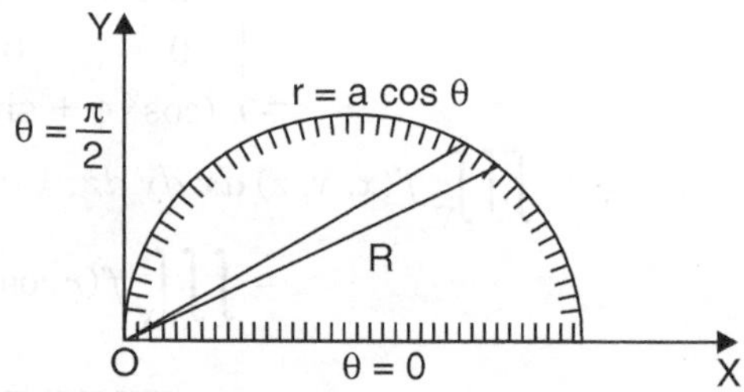

$$\therefore \quad \iint_R \sqrt{a^2 - x^2 - y^2}\, dxdy$$

$$= \int_0^{\pi/2}\int_0^{a \cos\theta} \sqrt{a^2 - r^2}\cdot r\, dr\, d\theta$$

$$= \int_0^{\pi/2}\int_0^{a \cos\theta} -\frac{1}{2}(a^2 - r^2)^{1/2}(-2r)\, dr\, d\theta$$

$$= \int_0^{\pi/2} -\frac{1}{2}\cdot\left[\frac{(a^2 - r^2)^{3/2}}{3/2}\right]_0^{a \cos\theta} d\theta$$

$$= -\frac{1}{3}\int_0^{\pi/2} (a^3 \sin^3\theta - a^3)\, d\theta = -\frac{a^3}{3}\int_0^{\pi/2} (\sin^3\theta - 1)\, d\theta$$

$$= -\frac{a^3}{3}\left[\frac{2}{3} - \frac{\pi}{2}\right] = \frac{a^3}{3}\left(\frac{\pi}{2} - \frac{2}{3}\right).$$

Example 3. *Change into polar co-ordinates and evaluate* $\int_0^\infty \int_0^\infty e^{-(x^2+y^2)}\, dy\, dx$.

Sol. For the region of integration in cartesian co-ordinates, y varies from 0 to ∞ and x also varies from 0 to ∞. Thus the region of integration is the plane XOY. Changing to polar co-ordinates by putting $x = r\cos\theta$, $y = r\sin\theta$ so that $x^2 + y^2 = r^2$; for the region of integration r varies from 0 to ∞ and θ varies from 0 to $\frac{\pi}{2}$.

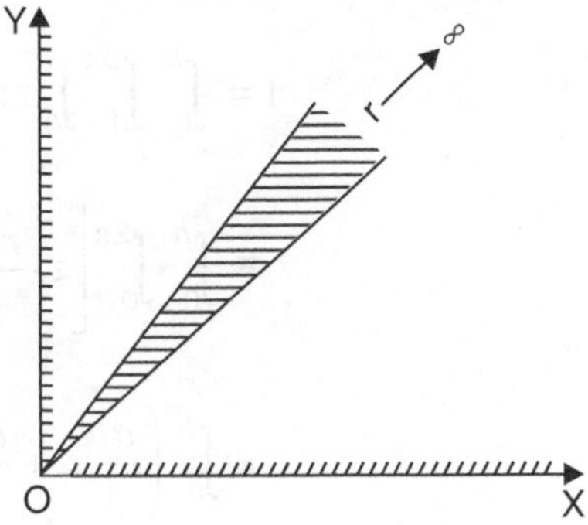

$$\therefore \quad \int_0^\infty \int_0^\infty e^{-(x^2+y^2)}\, dy\, dx = \int_0^{\pi/2} \int_0^\infty e^{-r^2} \cdot r\, dr\, d\theta$$

$$= \int_0^{\pi/2} \int_0^\infty \frac{1}{2} \cdot e^{-r^2} \cdot 2r\, dr\, d\theta$$

$$= \int_0^{\pi/2} \int_0^\infty \frac{1}{2} e^{-t} dt\, d\theta, \text{where } t = r^2$$

$$= \int_0^{\pi/2} \left[-\frac{1}{2} e^{-t} \right]_0^\infty d\theta = -\frac{1}{2} \int_0^{\pi/2} (0 - 1)\, d\theta = \frac{1}{2} \Big[\theta \Big]_0^{\pi/2} = \frac{\pi}{4}.$$

Example 4. *Evaluate* $\int_0^2 \int_0^{\sqrt{2x - x^2}} \frac{x\, dy\, dx}{\sqrt{x^2 + y^2}}$ *by changing to polar co-ordinates.*

Sol. In the given integral, y varies from 0 to $\sqrt{2x - x^2}$ and x varies from 0 to 2.

$$y = \sqrt{2x - x^2} \quad \Rightarrow \quad y^2 = 2x - x^2 \quad \Rightarrow \quad x^2 + y^2 = 2x.$$

In polar co-ordinates, we have $r^2 = 2r\cos\theta$ or $r = 2\cos\theta$.

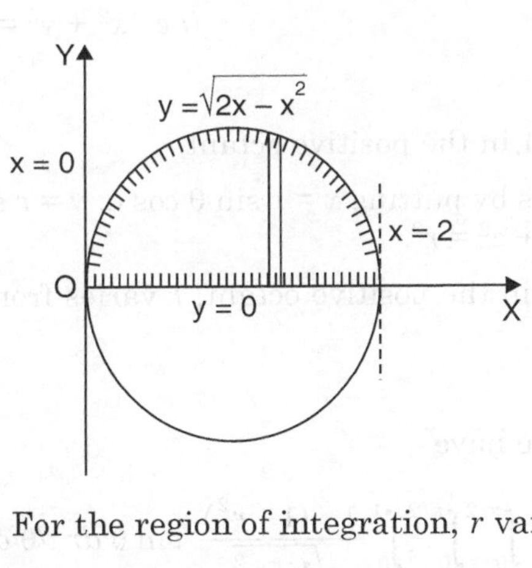

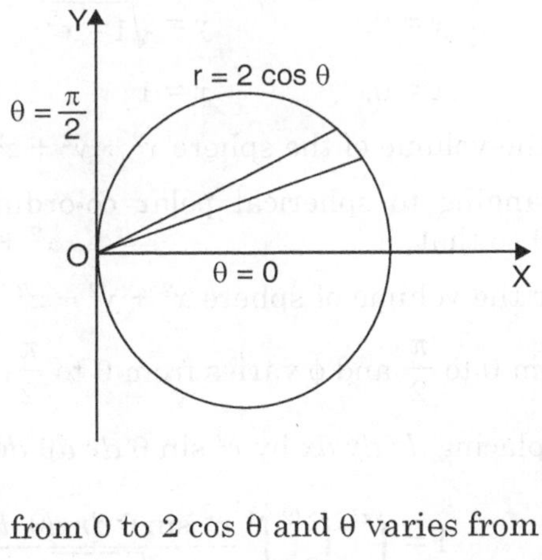

$\therefore$ For the region of integration, r varies from 0 to $2\cos\theta$ and θ varies from 0 to $\frac{\pi}{2}$.

In the given integral, replacing x by $r\cos\theta$, y by $r\sin\theta$, $dy\, dx$ by $r\, dr\, d\theta$, we have

$$\text{I} = \int_0^{\pi/2} \int_0^{2\cos\theta} \frac{r\cos\theta \,.\, r\, dr\, d\theta}{r} = \int_0^{\pi/2} \int_0^{2\cos\theta} r\cos\theta\, dr\, d\theta$$

$$= \int_0^{\pi/2} \cos\theta \left[\frac{r^2}{2} \right]_0^{2\cos\theta} d\theta = \int_0^{\pi/2} 2\cos^3\theta\, d\theta = 2 \cdot \frac{2}{3} = \frac{4}{3}.$$

Example 5. *Evaluate* $\iiint z(x^2 + y^2 + z^2)dxdydz$ *through the volume of the cylinder* $x^2 + y^2 = a^2$ *intercepted by the planes* $z = 0$ *and* $z = h$.

Sol. Changing to cylindrical co-ordinates by changing x to $r\cos\phi$, y to $r\sin\phi$ and replacing $dx\, dy\, dz$ by $r\, dr\, d\phi\, dz$

$$I = \int_0^h \int_0^{2\pi} \int_0^a z(r^2 + z^2) r dr d\phi dz = \int_0^h \int_0^{2\pi} \int_0^a (zr^3 + z^3 r)\, dr\, d\phi\, dz$$

$$= \int_0^h \int_0^{2\pi} \left[z \cdot \frac{r^4}{4} + z^3 \cdot \frac{r^2}{2} \right]_0^a d\phi\, dz = \int_0^h \int_0^{2\pi} \left(\frac{a^4}{4} z + \frac{a^2}{2} z^3 \right) d\phi\, dz$$

$$= \int_0^h \left(\frac{a^4}{4} z + \frac{a^2}{2} z^3 \right) \Big[\phi \Big]_0^{2\pi} dz = \int_0^h 2\pi \left(\frac{a^4}{4} z + \frac{a^2}{2} z^3 \right) dz$$

$$= 2\pi \left[\frac{a^4 z^2}{8} + \frac{a^2 z^4}{8} \right]_0^h = \frac{\pi}{4} (a^4 h^2 + a^2 h^4) = \frac{\pi}{4} a^2 h^2 (a^2 + h^2).$$

Example 6. *Evaluate* $\int_0^1 \int_0^{\sqrt{1-x^2}} \int_0^{\sqrt{1-x^2-y^2}} \frac{dz\, dy\, dx}{\sqrt{1 - x^2 - y^2 - z^2}}$, *by changing to spherical polar co-ordinates.* [CUSAT 2008]

Sol. Here the region of integration is bounded by

$z = 0, \quad z = \sqrt{1 - x^2 - y^2} \quad$ (*i.e.*, $x^2 + y^2 + z^2 = 1$)

$y = 0, \quad y = \sqrt{1 - x^2} \quad$ (*i.e.*, $x^2 + y^2 = 1$)

$x = 0, \quad x = 1$

which is the volume of the sphere $x^2 + y^2 + z^2 = 1$ in the positive octant.

Changing to spherical polar co-ordinates by putting $x = r\sin\theta\cos\phi$, $y = r\sin\theta\sin\phi$, $z = r\cos\theta$ so that $x^2 + y^2 + z^2 = r^2$.

For the volume of sphere $x^2 + y^2 + z^2 = 1$ in the positive octant, r varies from 0 to 1, θ varies from 0 to $\frac{\pi}{2}$ and ϕ varies from 0 to $\frac{\pi}{2}$.

Replacing $dz\, dy\, dx$ by $r^2 \sin\theta\, dr\, d\theta\, d\phi$, we have

$$I = \int_0^{\pi/2} \int_0^{\pi/2} \int_0^1 \frac{r^2 \sin\theta\, dr\, d\theta\, d\phi}{\sqrt{1 - r^2}} = \int_0^{\pi/2} \int_0^{\pi/2} \int_0^1 \frac{1 - (1 - r^2)}{\sqrt{1 - r^2}} \sin\theta\, dr\, d\theta\, d\phi$$

$$= \int_0^{\pi/2} \int_0^{\pi/2} \int_0^1 \left(\frac{1}{\sqrt{1 - r^2}} - \sqrt{1 - r^2} \right) \sin\theta\, dr\, d\theta\, d\phi$$

$$= \int_0^{\pi/2}\int_0^{\pi/2} \sin\theta \left[\sin^{-1} r - \left(\frac{r\sqrt{1-r^2}}{2} + \frac{1}{2}\sin^{-1} r \right) \right]_0^1 d\theta\, d\phi$$

$$= \int_0^{\pi/2}\int_0^{\pi/2} \sin\theta \left[\frac{\pi}{2} - \frac{1}{2}\cdot\frac{\pi}{2} \right] d\theta\, d\phi = \int_0^{\pi/2} \frac{\pi}{4}\Big[-\cos\theta\Big]_0^{\pi/2} d\phi = \int_0^{\pi/2} \frac{\pi}{4}\, d\phi$$

$$= \frac{\pi}{4}\cdot\frac{\pi}{2} = \frac{\pi^2}{8}.$$

Note. For the whole volume of the sphere $x^2 + y^2 + z^2 = a^2$.

$$0 \le r \le a,\ 0 \le \theta \le \pi,\ 0 \le \phi \le 2\pi.$$

Example 7. *Evaluate* $\iint_R (x+y)^2\, dxdy$, *where R is the parallelogram in the xy-plane with vertices (1, 0), (3, 1), (2, 2), (0, 1), using the transformation* $u = x + y$ *and* $v = x - 2y$.

Sol. The vertices A(1, 0), B(3, 1), C(2, 2), D(0, 1) of the parallelogram ABCD in the xy-plane become A′(1, 1), B′(4, 1), C′(4, –2), D′(1, –2) in the uv-plane under the given transformation.

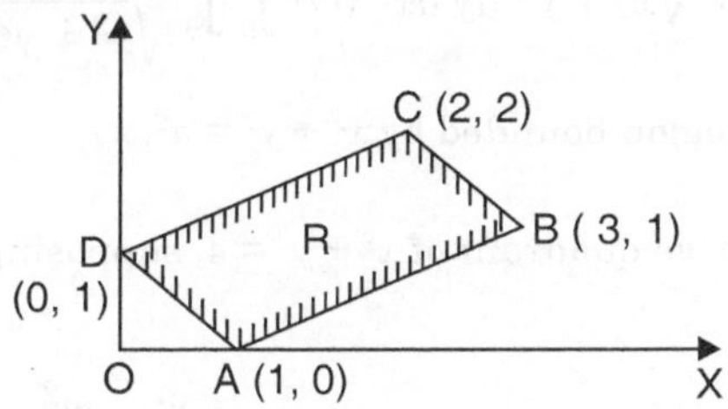

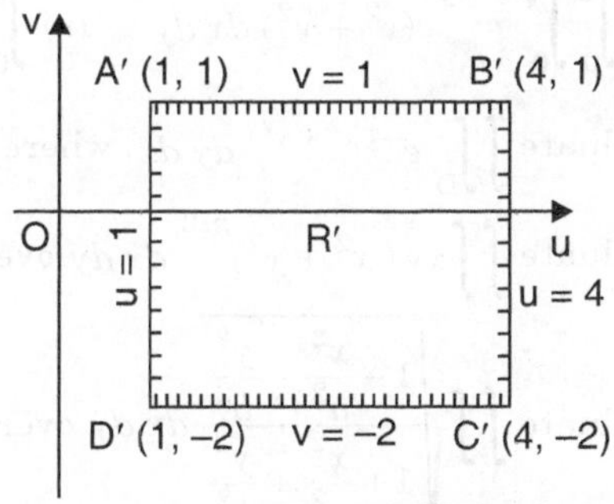

The region R in the xy-plane becomes the region R′ in the uv-plane which is a rectangle bounded by the line $u = 1$, $u = 4$ and $v = -2$, $v = 1$. Solving the given equations for x and y, we have $x = \frac{1}{3}(2u + v)$, $\quad y = \frac{1}{3}(u - v)$

$$\therefore \qquad J = \frac{\partial(x, y)}{\partial(u, v)} = \begin{vmatrix} \frac{\partial x}{\partial u} & \frac{\partial v}{\partial x} \\ \frac{\partial y}{\partial u} & \frac{\partial y}{\partial v} \end{vmatrix} = \begin{vmatrix} \frac{2}{3} & \frac{1}{3} \\ \frac{1}{3} & -\frac{1}{3} \end{vmatrix} = -\frac{1}{3}$$

$$\therefore \qquad \iint_R (x+y)^2\, dx\, dy = \iint_{R'} u^2\, |J|\, du\, dv = \int_{-2}^{1}\int_1^4 u^2 \cdot \frac{1}{3}\, du\, dv$$

$$= \int_{-2}^{1} \frac{1}{3}\left[\frac{u^3}{3}\right]_1^4 dv = \int_{-2}^{1} 7\, dv = 7\Big[v\Big]_{-2}^{1} = 7 \times 3 = 21.$$

TEST YOUR KNOWLEDGE

1. Evaluate $\iint \sin \pi (x^2 + y^2)\, dx\, dy$ over the region bounded by the circle $x^2 + y^2 = 1$ by changing to polar co-ordinates.
2. Evaluate $\iint (a^2 - x^2 - y^2)\, dx\, dy$ over the semi-circle $x^2 + y^2 = ax$ in the positive quadrant by changing to polar co-ordinates.
3. Evaluate $\iint (x^2 + y^2)^{7/2}\, dx\, dy$ over the circle $x^2 + y^2 = 1$.
4. Evaluate $\iint xy\, (x^2 + y^2)^{3/2}\, dx\, dy$ over the positive quadrant of the circle $x^2 + y^2 = 1$.
5. Evaluate the following by changing into polar co-ordinates:

   (i) $\int_0^a \int_y^a \frac{x\, dx\, dy}{x^2 + y^2}$ (ii) $\int_{-a}^{a} \int_{-\sqrt{a^2 - x^2}}^{\sqrt{a^2 - x^2}} dy\, dx$ (iii) $\int_0^{4a} \int_{y^2/4a}^{y} \frac{x^2 - y^2}{x^2 + y^2}\, dx\, dy$

   (iv) $\int_0^a \int_0^{\sqrt{a^2 - y^2}} (x^2 + y^2)\, dx\, dy$ (v) $\int_0^a \int_0^{\sqrt{a^2 - y^2}} y^2 \sqrt{x^2 + y^2}\, dy\, dx$ (vi) $\int_0^a \int_y^a \frac{x^2\, dx\, dy}{\sqrt{x^2 + y^2}}$.
6. Evaluate $\iint_D e^{-(x^2 + y^2)}\, dy\, dx$, where D is the region bounded by $x^2 + y^2 = a^2$.
7. Evaluate $\iint xy\, (x^2 + y^2)^{n/2}\, dx\, dy$ over the positive quadrant of $x^2 + y^2 = 4$, supposing $n + 3 > 0$.
8. Evaluate $\iint \sqrt{\frac{1 - \frac{x^2}{a^2} - \frac{y^2}{b^2}}{1 + \frac{x^2}{a^2} + \frac{y^2}{b^2}}}\, dx\, dy$ over the positive quadrant of the ellipse $\frac{x^2}{a^2} + \frac{y^2}{b^2} = 1$.
9. Transform the following to cartesian form and hence evaluate $\int_0^{\pi} \int_0^a r^3 \sin\theta \cos\theta\, dr\, d\theta$.
10. Evaluate $\iiint (x + y + z)\, dx\, dy\, dz$ over the tetrahedron bounded by the planes $x = 0, y = 0, z = 0$ and $x + y + z = 1$.
11. Evaluate $\iiint \frac{dx\, dy\, dz}{\sqrt{1 - x^2 - y^2 - z^2}}$, the integral being extended to the positive octant of the sphere $x^2 + y^2 + z^2 = 1$.
12. Evaluate $\iiint \frac{dx\, dy\, dz}{(x + y + z + 1)^3}$ over the tetrahedron bounded by the co-ordinate planes and the plane $x + y + z = 1$.
13. Evaluate $\iiint z(x^2 + y^2)\, dx\, dy\, dz$ over the volume of the cylinder $x^2 + y^2 = 1$ intercepted by the planes $z = 2$ and $z = 3$.
14. Evaluate the following integrals through the volume of the sphere $x^2 + y^2 + z^2 = 1$, by changing into spherical polar co-ordinates:

   (i) $\iiint z^2\, dx\, dy\, dz$ (ii) $\iiint (x^2 + y^2 + z^2)^m\, dx\, dy\, dz$. $(m > 0)$

Answers

1. 2	**2.** $\frac{5\pi a^4}{64}$	**3.** $\frac{2\pi}{9}$	**4.** $\frac{1}{14}$
5. (*i*) $\frac{\pi a}{4}$	(*ii*) πa^2	(*iii*) $8\left(\frac{\pi}{2}-\frac{5}{3}\right)a^2$	(*iv*) $\frac{\pi a^4}{8}$
(*v*) $\frac{\pi a^5}{20}$	(*vi*) $\frac{a^3}{3}\log(\sqrt{2}+1)$	**6.** $\pi(1-e^{-a^2})$	**7.** $\frac{2^{n+3}}{n+4}$
8. $\frac{\pi ab}{8}(\pi-2)$	**9.** 0	**10.** $\frac{1}{8}$	**11.** $\frac{\pi^2}{8}$
12. $\frac{1}{2}\log 2-\frac{5}{16}$	**13.** $\frac{5\pi}{4}$	**14.** (*i*) $\frac{4\pi}{15}$	(*ii*) $\frac{4\pi}{2m+3}$.

13.7. AREA BY DOUBLE INTEGRATION

(*a*) *Cartesian Co-ordinates.* The area A of the region bounded by the curves $y = f_1(x)$, $y = f_2(x)$ and the lines $x = a$, $x = b$ is given by $A = \int_a^b \int_{f_1(x)}^{f_2(x)} dy\, dx$

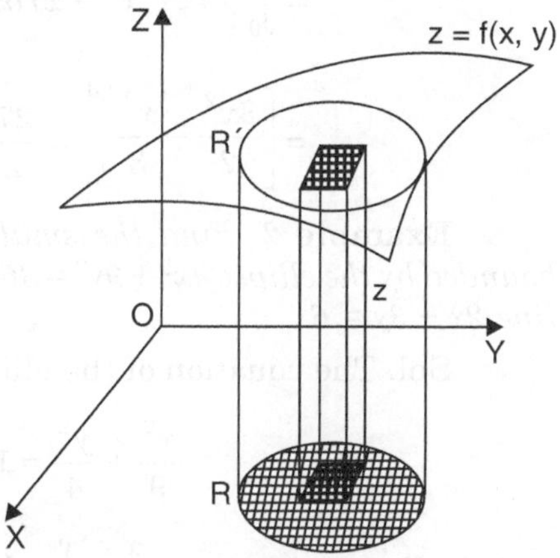

The area A of the region bounded by the curves $x = f_1(y)$, $x = f_2(y)$ and the lines $y = c$, $y = d$ is given by

$$A = \int_c^d \int_{f_1(y)}^{f_2(y)} dx\, dy\ .$$

(*b*) *Polar co-ordinates.* The area A of the region bounded by the curves $r = f_1(\theta)$, $r = f_2(\theta)$ and the lines $\theta = \alpha$, $\theta = \beta$ is given by $A = \int_\alpha^\beta \int_{f_1(\theta)}^{f_2(\theta)} r\, dr\, d\theta.$

13.8. VOLUME AS A DOUBLE INTEGRAL

(*a*) *Cartesian Co-ordinates.* Consider a surface $z = f(x, y)$ ...(1)

Let the orthogonal projection of its region R′ on the *xy*-plane be the region R given by $\phi(x, y) = 0$. ...(2)

Now (2) represents a cylinder with generators parallel to *z*-axis and the guiding curve given by (2). Let V be the volume of this cylinder between R and R′.

Divide R into elementary rectangles of area $\delta x\, \delta y$ by drawing lines parallel to the axis of *x* and *y*. On each of these rectangles, erect prisms of lengths parallel to *z*-axis. Volume of this prism between R and R′ is $z\, \delta x\, \delta y$. The volume V is composed of such prisms.

$$\therefore \quad V = \lim_{\substack{\delta x \to 0 \\ \delta y \to 0}} \sum z\, \delta x\, \delta y = \iint_R z\, dx\, dy\, .$$

(*b*) *Cylindrical Co-ordinates.* Let the equation of the surface be $z = f(r, \phi)$. Replacing $dx\, dy$ by $r\, dr\, d\phi$, we get $V = \iint_R zr\, dr\, d\phi\, .$

ILLUSTRATIVE EXAMPLES

Example 1. *Find the area lying between the parabola $y = 4x - x^2$ and the line $y = x$.*

Sol. The two curves intersect at points whose abscissae are given by $4x - x^2 = x$

or $\quad x^2 - 3x = 0 \quad$ *i.e.*, $\quad x = 0, 3.$

Using vertical strips, the required area lies between $x = 0$, $x = 3$ and $y = x$, $y = 4x - x^2$.

$$\therefore \quad \text{Required area} = \int_0^3 \int_x^{4x - x^2} dy\, dx$$

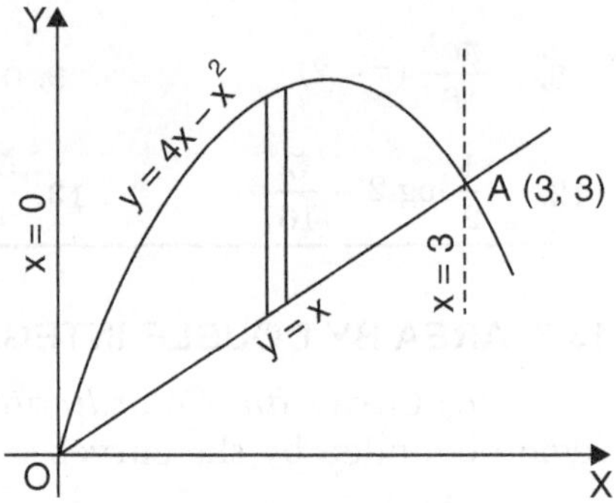

$$= \int_0^3 \left[y \right]_x^{4x - x^2} dx$$

$$= \int_0^3 (4x - x^2 - x)\, dx = \int_0^3 (3x - x^2)\, dx$$

$$= \left[\frac{3x^2}{2} - \frac{x^3}{3} \right]_0^3 = \frac{27}{2} - 9 = 4.5.$$

Example 2. *Find the smaller of the areas bounded by the ellipse $4x^2 + 9y^2 = 36$ and the straight line $2x + 3y = 6$.*

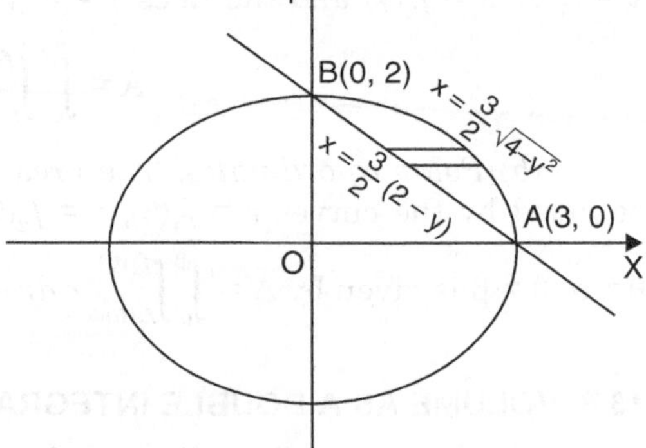

Sol. The equation of the ellipse is

$$\frac{x^2}{9} + \frac{y^2}{4} = 1 \qquad \ldots(1)$$

and the line is

$$\frac{x}{3} + \frac{y}{2} = 1 \qquad \ldots(2)$$

Both meet x-axis at A(3, 0) and y-axis at B(0, 2).

Using horizontal strips, the required area lies between

$$x = \frac{3}{2}(2 - y),\ x = \frac{3}{2}\sqrt{4 - y^2} \ \text{ and } \ y = 0, y = 2.$$

$$\therefore \ \text{Required area} = \int_0^2 \int_{(3/2)(2-y)}^{(3/2)\sqrt{4-y^2}} dx\, dy = \int_0^2 \left[x \right]_{(3/2)(2-y)}^{(3/2)\sqrt{4-y^2}} dy$$

$$= \int_0^2 \frac{3}{2} \left[\sqrt{4 - y^2} - (2 - y) \right] dy = \frac{3}{2} \left[\frac{y\sqrt{4 - y^2}}{2} + \frac{4}{2} \sin^{-1} \frac{y}{2} - 2y + \frac{y^2}{2} \right]_0^2$$

$$= \frac{3}{2} [2 \sin^{-1} 1 - 4 + 2] = \frac{3}{2} \left(2 \cdot \frac{\pi}{2} - 2 \right) = \frac{3}{2} (\pi - 2).$$

Example 3. *Find, by double integration, the area lying inside the circle r = a sin θ and outside the cardioid r = a(1 − cos θ).*

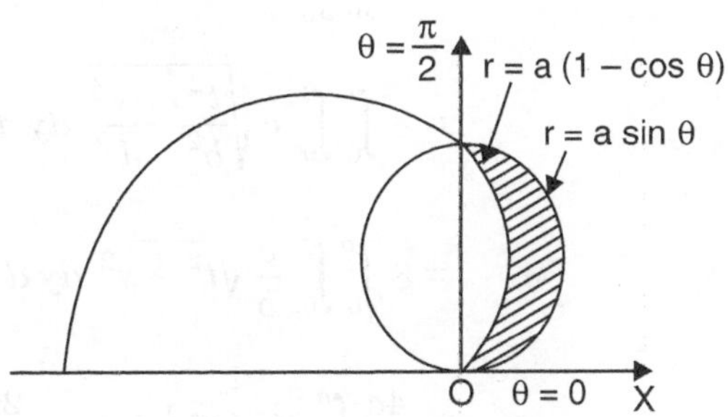

Sol. Eliminating r between the equations of two curves, $\sin\theta = 1 - \cos\theta$ or $\sin\theta + \cos\theta = 1$

Squaring $1 + \sin 2\theta = 1$ or $\sin 2\theta = 0$

$\therefore \quad 2\theta = 0$ or π

i.e., $\quad \theta = 0$ or $\dfrac{\pi}{2}$

For the required area, r varies from $a(1 - \cos\theta)$ to $a \sin\theta$ and θ varies from 0 to $\dfrac{\pi}{2}$.

$$\therefore \quad \text{Required area} = \int_0^{\pi/2} \int_{a(1-\cos\theta)}^{a\sin\theta} r\,dr\,d\theta$$

$$= \int_0^{\pi/2} \left[\frac{r^2}{2}\right]_{a(1-\cos\theta)}^{a\sin\theta} d\theta = \frac{1}{2}\int_0^{\pi/2} a^2\,[\sin^2\theta - (1-\cos\theta)^2]\,d\theta$$

$$= \frac{a^2}{2}\int_0^{\pi/2} (\sin^2\theta - 1 - \cos^2\theta + 2\cos\theta)\,d\theta$$

$$= \frac{a^2}{2}\int_0^{\pi/2} (-2\cos^2\theta + 2\cos\theta)\,d\theta = a^2\left[-\frac{1}{2}\cdot\frac{\pi}{2} + 1\right] = a^2\left(1 - \frac{\pi}{4}\right).$$

Example 4. *Find the volume of the ellipsoid* $\dfrac{x^2}{a^2} + \dfrac{y^2}{b^2} + \dfrac{z^2}{c^2} = 1.$ (CUSAT 2008)

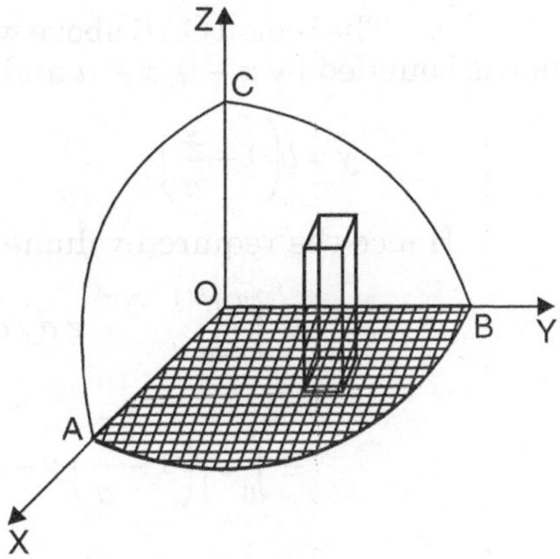

Sol. On account of symmetry, the required volume is 8 times the volume of the ellipsoid in the positive octant. The volume OABC in the positive octant lies between the ellipsoid $z = c\sqrt{1 - \dfrac{x^2}{a^2} - \dfrac{y^2}{b^2}}$ and the plane XOY, and is bounded on the sides by the planes $x = 0$, $y = 0$.

The ellipsoid cuts the plane XOY in the ellipse

$$\frac{x^2}{a^2} + \frac{y^2}{b^2} = 1,\ z = 0$$

$\therefore$ The region OAB above which the volume OABC lies is bounded by

$$x = 0,\ x = a \text{ and } y = 0,\ y = b\sqrt{1 - \frac{x^2}{a^2}}$$

Hence the required volume of the ellipsoid

$$= 8\int_0^a\int_0^{b\sqrt{1-x^2/a^2}} z\,dy\,dx = 8\int_0^a\int_0^{b\sqrt{1-x^2/a^2}} c\sqrt{1-\frac{x^2}{a^2}-\frac{y^2}{b^2}}\,dy\,dx$$

$$= 8\int_0^a\int_0^t c\sqrt{\frac{t^2}{b^2}-\frac{y^2}{b^2}}\,dy\,dx,\quad \text{where } b\sqrt{1-\frac{x^2}{a^2}} = t$$

$$= 8\int_0^a\int_0^t \frac{c}{b}\sqrt{t^2-y^2}\,dy\,dx = 8\int_0^a \frac{c}{b}\left[\frac{y\sqrt{t^2-y^2}}{2}+\frac{t^2}{2}\sin^{-1}\frac{y}{t}\right]_0^t dx$$

$$= \frac{4c}{b}\int_0^a t^2\sin^{-1}1\,dx = \frac{2\pi c}{b}\int_0^a t^2\,dx$$

$$= \frac{2\pi c}{b}\int_0^a b^2\left(1-\frac{x^2}{a^2}\right)dx = 2\pi bc\left[x-\frac{x^3}{3a^2}\right]_0^a = 2\pi bc\left(a-\frac{a}{3}\right) = \frac{4}{3}\pi abc.$$

Example 5. *Find the volume of the tetrahedron bounded by the co-ordinate planes and the plane* $\frac{x}{a}+\frac{y}{b}+\frac{z}{c} = 1$.

Sol. The required volume OABC lies between the plane $z = c\left(1-\frac{x}{a}-\frac{y}{b}\right)$ and the plane XOY, and is bounded on the sides by the planes $x = 0$, $y = 0$.

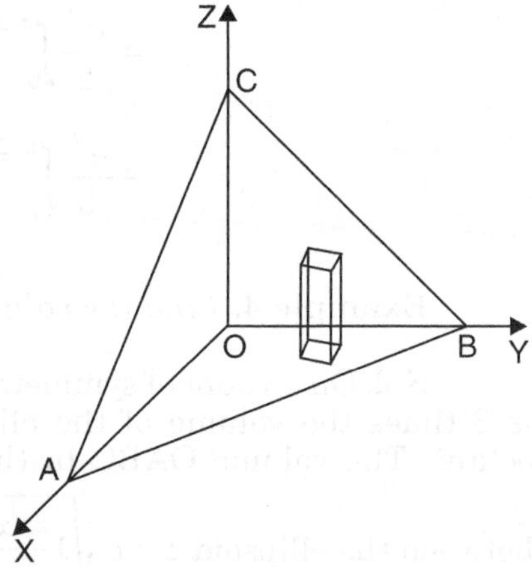

The plane $\frac{x}{a}+\frac{y}{b}+\frac{z}{c} = 1$ cuts the plane XOY in the line $\frac{x}{a}+\frac{y}{b} = 1, z = 0$

$\therefore$ The region OAB above which the volume OABC lies is bounded by $x = 0$, $x = a$ and $y = 0$,

$$y = b\left(1-\frac{x}{a}\right)$$

Hence the required volume of the tetrahedron

$$= \int_0^a\int_0^{b(1-x/a)} z\,dy\,dx = \int_0^a\int_0^{b(1-x/a)} c\left(1-\frac{x}{a}-\frac{y}{b}\right)dy\,dx$$

$$= \int_0^a c\left[\left(1-\frac{x}{a}\right)y-\frac{y^2}{2b}\right]_0^{b(1-x/a)} dx = c\int_0^a\left[b\left(1-\frac{x}{a}\right)^2-\frac{b}{2}\left(1-\frac{x}{a}\right)^2\right]dx$$

$$= \frac{bc}{2}\int_0^a\left(1-\frac{x}{a}\right)^2 dx = \frac{bc}{2}\cdot\left[\frac{\left(1-\frac{x}{a}\right)^3}{-\frac{3}{a}}\right]_0^a = -\frac{abc}{6}(0-1) = \frac{1}{6}abc.$$

Example 6. *Find the volume bounded by the cylinder $x^2 + y^2 = 4$ and the planes $y + z = 4$ and $z = 0$.* (CUSAT 2007)

Sol. From the figure, it is clear that $z = 4 - y$ is to be integrated over the circle $x^2 + y^2 = 4$ in the xy-plane. To cover the shaded half of this circle, x varies from 0 to $\sqrt{4 - y^2}$ while y varies from -2 to 2.

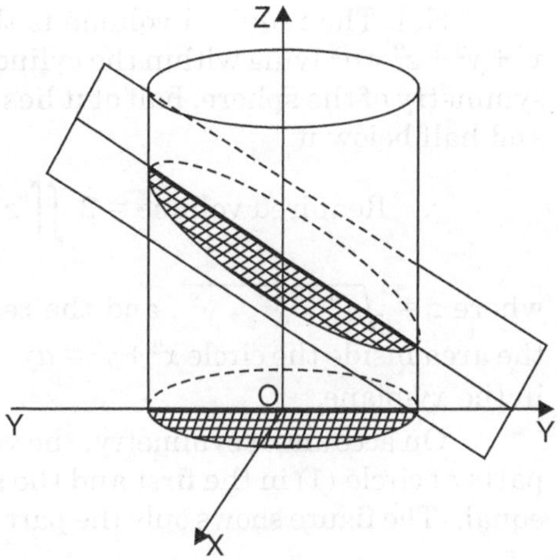

$\therefore$ Required volume

$$= 2 \int_{-2}^{2} \int_{0}^{\sqrt{4-y^2}} z \, dx \, dy$$

$$= 2 \int_{-2}^{2} \int_{0}^{\sqrt{4-y^2}} (4 - y) \, dx \, dy$$

$$= 2 \int_{-2}^{2} (4 - y) \Big[x \Big]_{0}^{\sqrt{4-y^2}} dy$$

$$= 2 \int_{-2}^{2} (4 - y)\sqrt{4 - y^2} \, dy = 2 \int_{-2}^{2} 4\sqrt{4 - y^2} \, dy - 2\int_{-2}^{2} y\sqrt{4 - y^2} \, dy$$

$$= 8 \int_{-2}^{2} \sqrt{4 - y^2} \, dy \qquad \left[\begin{array}{l}\text{The second integral is zero since} \\ y\sqrt{4-y^2} \text{ is an odd function of } y\end{array}\right]$$

$$= 16 \int_{0}^{2} \sqrt{4 - y^2} \, dy \qquad [\because \sqrt{4 - y^2} \text{ is an even function of } y]$$

$$= 16 \left[\frac{y\sqrt{4 - y^2}}{2} + \frac{4}{2} \sin^{-1} \frac{y}{2} \right]_{0}^{2} = 16\,[2 \sin^{-1} 1] = 32 \times \frac{\pi}{2} = 16\pi.$$

Example 7. *Find the volume common to the cylinders $x^2 + y^2 = a^2$ and $x^2 + z^2 = a^2$.*

Sol. The section of the cylinder $x^2 + y^2 = a^2$ is the circle $x^2 + y^2 = a^2$ in the xy-plane. In the figure, only one-eighth (in positive octant) of the required volume is shown.

From the figure, it is evident that $z = \sqrt{a^2 - x^2}$ is to be evaluated over the quadrant of the circle $x^2 + y^2 = a^2$ in the first quadrant for which x varies from 0 to a and y varies from 0 to $\sqrt{a^2 - x^2}$.

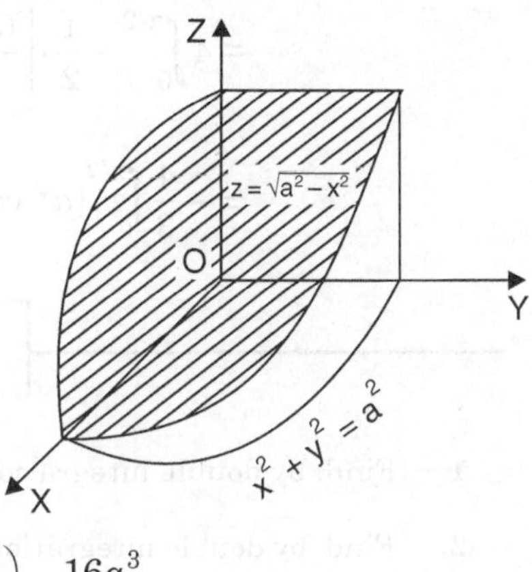

$$\therefore \quad \text{Required volume} = 8 \int_{0}^{a} \int_{0}^{\sqrt{a^2 - x^2}} z \, dy \, dx$$

$$= 8 \int_{0}^{a} \int_{0}^{\sqrt{a^2 - x^2}} \sqrt{a^2 - x^2} \, dy \, dx$$

$$= 8 \int_{0}^{a} \sqrt{a^2 - x^2} \Big[y \Big]_{0}^{\sqrt{a^2 - x^2}} dx$$

$$= 8 \int_{0}^{a} (a^2 - x^2) \, dx = 8 \left[a^2 x - \frac{x^3}{3} \right]_{0}^{a} = 8\left(a^3 - \frac{a^3}{3} \right) = \frac{16a^3}{3}.$$

Example 8. *Find the volume common to the sphere $x^2 + y^2 + z^2 = a^2$ and the cylinder $x^2 + y^2 = ay$.*

Sol. The required volume is the part of the sphere $x^2 + y^2 + z^2 = a^2$ lying within the cylinder. On account of the symmetry of the sphere, half of it lies above the plane XOY and half below it.

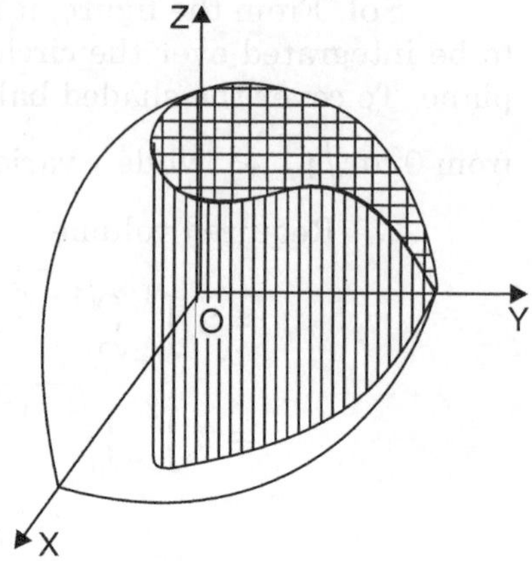

$$\therefore \quad \text{Required volume} = 2 \iint z \, dy \, dx$$

where $z = \sqrt{a^2 - x^2 - y^2}$, and the region of integration is the area inside the circle $x^2 + y^2 = ay$ ...(1)

in the xy-plane.

On account of symmetry, the volumes above the two parts of circle (1) in the first and the second quadrants are equal. (The figure shows only the part in the first quadrant).

$$\therefore \quad \text{Required volume} = 2 \times 2 \iint_R \sqrt{a^2 - x^2 - y^2} \, dydx$$

where R is the half of the circle (1) lying in the first quadrant.

Changing to polar co-ordinates by putting $x = r \cos \theta$, $y = r \sin \theta$ so that $x^2 + y^2 = r^2$, equation (1) becomes

$$r^2 = ar \sin \theta \quad \text{or} \quad r = a \sin \theta.$$

The region of integration is bounded by

$$r = 0, r = a \sin \theta \text{ and } \theta = 0, \theta = \frac{\pi}{2}.$$

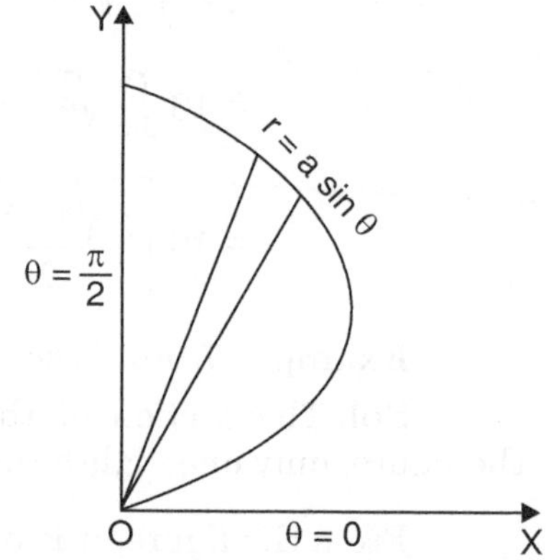

$\therefore$ Required volume

$$= 4 \int_0^{\pi/2} \int_0^{a \sin \theta} \sqrt{a^2 - r^2} \cdot r \, dr \, d\theta$$

$$= 4 \int_0^{\pi/2} \int_0^{a \sin \theta} -\frac{1}{2} (a^2 - r^2)^{1/2} (-2r) \, dr \, d\theta$$

$$= 4 \int_0^{\pi/2} -\frac{1}{2} \cdot \left[\frac{(a^2 - r^2)^{3/2}}{3/2} \right]_0^{a \sin \theta} d\theta$$

$$= -\frac{4}{3} \int_0^{\pi/2} (a^2 \cos^3 \theta - a^3) = -\frac{4a^3}{3} \left[\frac{2}{3} - \frac{\pi}{2} \right] = \frac{2}{9} a^3 (3\pi - 4).$$

TEST YOUR KNOWLEDGE

1. Find, by double integration, the area enclosed by the ellipse $\frac{x^2}{a^2} + \frac{y^2}{b^2} = 1$.

2. Find, by double integration, the smaller of the areas bounded by the circle $x^2 + y^2 = 9$ and the line $x + y = 3$.

3. Show, by double integration, that the area between the parabolas $y^2 = 4ax$ and $x^2 = 4ay$ is $\frac{16}{3}a^2$.

4. Find the area bounded by the parabola $y = x^2$ and the line $y = 2x + 3$.

5. Find the area bounded by the parabolas $y^2 = 4 - x$ and $y^2 = 4 - 4x$.

6. Find the area enclosed by the curves $y = \frac{3x}{x^2 + 2}$ and $4y = x^2$.

7. Find the area bounded by the circles $r = 2 \sin \theta$ and $r = 4 \sin \theta$.

8. Find, by double integration, the area of one loop of the lemniscate $r^2 = a^2 \cos 2\theta$.

9. Find the area outside the circle $r = a$ and inside the cardioid $r = a(1 + \cos \theta)$.

10. Find, by double integration, the volume of the sphere $x^2 + y^2 + z^2 = 9$.

11. Find the volume bounded by the xy-plane, the cylinder $x^2 + y^2 = 1$ and the plane $x + y + z = 3$.

12. A triangular prism is formed by the planes whose equations are $ay = bx$, $y = 0$ and $x = a$. Prove that the volume of this prism between the plane $z = 0$ and the surface $z = c + xy$ is $\frac{ab}{8}(4c + ab)$.

13. Find the volume bounded by the xy-plane, the paraboloid $2z = x^2 + y^2$ and the cylinder $x^2 + y^2 = 4$.

14. Find the volume of the region bounded by $z = x^2 + y^2$, $z = 0$, $x = -a$, $x = a$ and $y = -a$, $y = a$.

15. Prove that the volume enclosed between the cylinders $x^2 + y^2 = 2ax$ and $z^2 = 2ax$ is $\frac{128a^3}{15}$.

16. Find the volume bounded by the plane $z = 0$, surface $z = x^2 + y^2 + 2$ and the cylinder $x^2 + y^2 = 4$.

17. Find the volume under the plane $z = x + y$ and above the area cut from the first quadrant by the ellipse $4x^2 + 9y^2 = 36$.

18. Find the volume of the cylinder $x^2 + y^2 - 2ax = 0$ intercepted between the paraboloid $x^2 + y^2 = 2az$ and the xy-plane.

19. Find the volume bounded by the cylinder $x^2 + y^2 = 4$ and the hyperboloid $x^2 + y^2 - z^2 = 1$.

20. Find the volume bounded by the paraboloid $x^2 + y^2 = az$, the cylinder $x^2 + y^2 = 2ay$ and the plane $z = 0$.

Answers

1. πab	2. $\frac{9}{4}(\pi - 2)$	4. $10\frac{2}{3}$	5. 8
6. $\frac{3}{2}\log 3 - \frac{2}{3}$	7. 3π	8. $\frac{a^2}{2}$	9. $\frac{a^2}{4}(8 + \pi)$
10. 36π	11. $3\pi - \frac{4}{3}$	13. 4π	14. $\frac{8a^4}{3}$
16. 16π	17. 10	18. $\frac{3}{4}\pi a^3$	19. $4\sqrt{3}\pi$
20. $\frac{3\pi a^3}{2}$.			

13.9. VOLUME AS A TRIPLE INTEGRAL

The volume V of a three dimensional region is given by $V = \iiint_V dx\,dy\,dz$.

If the region is bounded by $x = f_1(y, z)$, $x = f_2(y, z)$; $y = \phi_1(z)$, $y = \phi_2(z)$ and $z = a$, $z = b$, then

$$V = \int_a^b \int_{\phi_1(z)}^{\phi_2(z)} \int_{f_1(y,z)}^{f_2(y,z)} dx\,dy\,dz$$

The order of integration may be changed with a suitable change in the limits of integration.

In cylindrical co-ordinates, we have $V = \iiint_V r\,dr\,d\phi\,dz$

In spherical polar co-ordinates, we have $V = \iiint_V r^2 \sin\theta\,dr\,d\theta\,d\phi$.

ILLUSTRATIVE EXAMPLES

Example 1. *Find by triple integration, the volume of the paraboloid of revolution $x^2 + y^2 = 4z$ cut off by the plane $z = 4$.*

Sol. By symmetry, the required volume is 4 times the volume in the positive octant. The volume in the positive octant is bounded on the sides by the zx and yz-planes ; from above by the plane $z = 4$ and below by the curved surface $x^2 + y^2 = 4z$.

The section of the paraboloid by the plane $z = 4$ is the circle $x^2 + y^2 = 16$, $z = 4$ and its projection on the xy-plane is the circle $x^2 + y^2 = 16$, $z = 0$.

The volume in the positive octant is bounded by $z = \dfrac{x^2 + y^2}{4}$, $z = 4$, $y = 0$, $y = \sqrt{16 - x^2}$ and $x = 0$, $x = 4$.

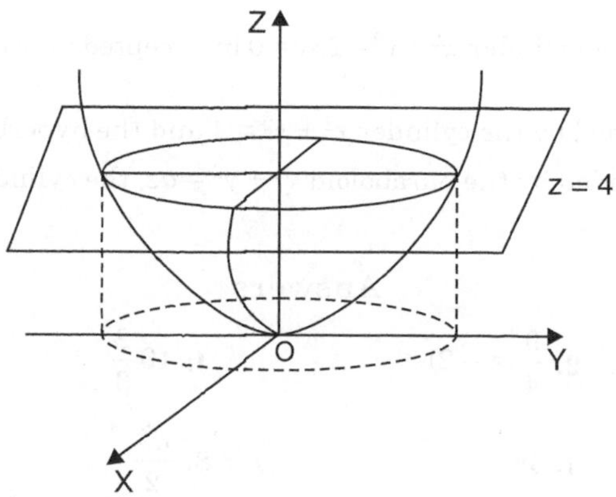

∴ Required volume

$$= 4\int_0^4 \int_0^{\sqrt{16-x^2}} \int_{(x^2+y^2)/4}^{4} dz\,dy\,dx = 4\int_0^4 \int_0^{\sqrt{16-x^2}} \Big[z \Big]_{(x^2+y^2)/4}^{4} dy\,dx$$

$$= 4\int_0^4 \int_0^{\sqrt{16-x^2}} \left(4 - \frac{x^2+y^2}{4}\right) dy\,dx = 4\int_0^4 \left[\left(4 - \frac{x^2}{4}\right) y - \frac{1}{4}\cdot\frac{y^3}{3}\right]_0^{\sqrt{16-x^2}} dx$$

$$= 4\int_0^4 \left[\left(4 - \frac{x^2}{4}\right)\sqrt{16 - x^2} - \frac{1}{12}(16 - x^2)^{3/2}\right] dx$$

$$= 4\int_0^4 \left[\frac{1}{4}(16-x^2)\sqrt{16-x^2} - \frac{1}{12}(16-x^2)^{3/2}\right] dx$$

$$= 4\int_0^4 \frac{1}{6}(16-x^2)^{3/2}\, dx = \frac{2}{3}\int_0^4 (16-x^2)^{3/2}\, dx$$

$$= \frac{2}{3}\int_0^{\pi/2} (16)^{3/2} \cdot \cos^3\theta \cdot 4\cos\theta\, d\theta, \text{ where } x = 4\sin\theta$$

$$= \frac{512}{3}\int_0^{\pi/2} \cos^4\theta\, d\theta = \frac{512}{3}\cdot\frac{3\cdot 1}{4\cdot 2}\cdot\frac{\pi}{2} = 32\pi.$$

Example 2. *Find, by triple integration, the volume of the region bounded by the paraboloid $az = x^2 + y^2$ and the cylinder $x^2 + y^2 = R^2$.*

Sol. Changing to cylindrical co-ordinates, by putting $x = r\cos\theta$, $y = r\sin\theta$ the equation of the paraboloid becomes $az = r^2$ or $z = \frac{r^2}{a}$ and the equation of the cylinder becomes $r^2 = R^2$ or $r = R$. On account of symmetry, the required volume is four times the volume in the positive octant. Thus, in the common region, z varies from 0 to $\frac{r^2}{a}$, r varies from 0 to R and θ varies from 0 to $\frac{\pi}{2}$.

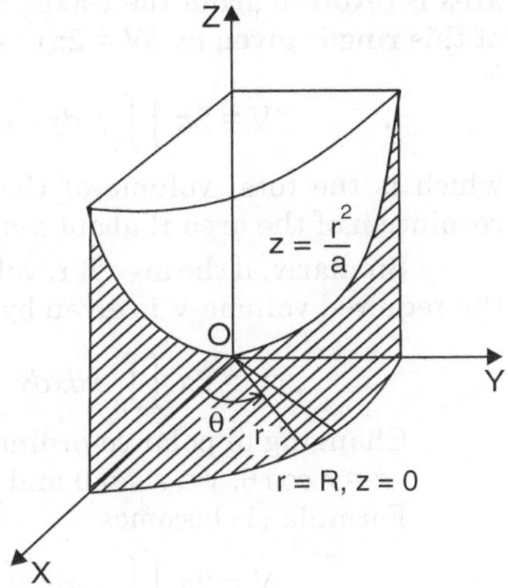

$$\therefore \quad \text{Required volume} = 4\int_0^{\pi/2}\int_0^R\int_0^{r^2/a} r\, dz\, dr\, d\theta$$

$$= 4\int_0^{\pi/2}\int_0^R r\left[z\right]_0^{r^2/a} dr\, d\theta$$

$$= 4\int_0^{\pi/2}\int_0^R r\cdot\frac{r^2}{a}\, dr\, d\theta = 4\int_0^{\pi/2}\left[\frac{r^4}{4a}\right]_0^R d\theta$$

$$= \frac{1}{a}\int_0^{\pi/2} R^4\, d\theta = \frac{R^4}{a}\cdot\frac{\pi}{2} = \frac{\pi R^4}{2a}.$$

Example 3. *Find, by triple integration, the volume of a sphere of radius a.*

Sol. Changing to spherical polar co-ordinates by putting $x = r\sin\theta\cos\phi$, $y = r\sin\theta\sin\phi$, $z = r\cos\theta$ so that $x^2 + y^2 + z^2 = r^2$.

The equation of a sphere of radius a in cartesian co-ordinates is $x^2 + y^2 + z^2 = a^2$

The same equation in spherical polar co-ordinates is $r^2 = a^2$ or $r = a$.

On account of symmetry, the required volume is 8 times the volume of the sphere in the positive octant for which r varies from 0 to a, θ varies from 0 to $\frac{\pi}{2}$ and ϕ varies from 0 to $\frac{\pi}{2}$.

$$\therefore \quad \text{Required volume} = 8\int_0^{\pi/2}\int_0^{\pi/2}\int_0^a r^2\sin\theta\, dr\, d\theta\, d\phi$$

$$= 8\int_0^{\pi/2}\int_0^{\pi/2} \sin\theta . \left[\frac{r^3}{3}\right]_0^a d\theta\, d\phi = 8\int_0^{\pi/2}\int_0^{\pi/2} \frac{a^3}{3}\sin\theta\, d\theta\, d\phi$$

$$= 8\int_0^{\pi/2} -\frac{a^3}{3}\left[\cos\theta\right]_0^{\pi/2} d\phi = \frac{8}{3}a^3\int_0^{\pi/2} d\phi = \frac{8}{3}a^3 . \frac{\pi}{2} = \frac{4}{3}\pi a^3.$$

13.10. VOLUMES OF SOLIDS OF REVOLUTION

Consider an elementary area $\delta x\delta y$ at the point P(x, y) of a plane area R. If this elementary area is revolved about the x-axis, it will generate a ring of radius y. The elementary volume δV of this ring is given by $\delta V = 2\pi y \cdot \delta y\, \delta x$

$$\therefore \qquad V = 2\pi \iint_R y\, dy\, dx \qquad ...(1)$$

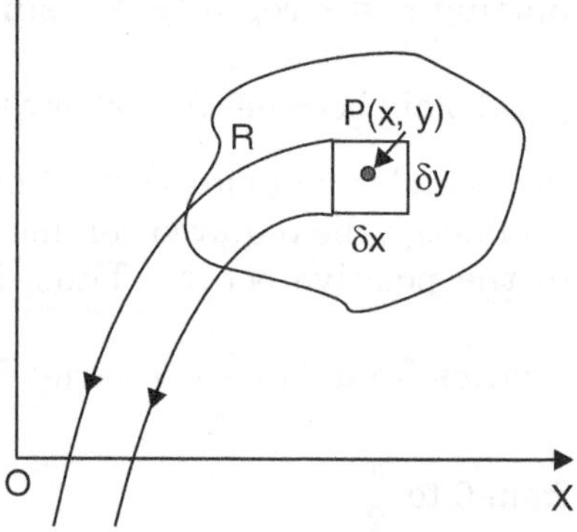

which is the total volume of the solid formed by the revolution of the area R about x-axis.

Similarly, if the area R revolves about y-axis, then the required volume V is given by

$$V = 2\pi \iint_R x dx dy \qquad ...(2)$$

Changing to polar co-ordinates by putting $x = r\cos\theta$, $y = r\sin\theta$ and $dy\, dx = r\, dr\, d\theta$

Formula (1) becomes

$$V = 2\pi \iint_R r\sin\theta \cdot r\, dr\, d\theta = 2\pi \iint_R r^2 \sin\theta\, dr\, d\theta$$

Formula (2) becomes $V = 2\pi \iint_R r^2 \cos\theta\, dr\, d\theta.$

ILLUSTRATIVE EXAMPLES

Example 1. *Find, by double integration, the volume of the solid generated by revolving the ellipse* $\frac{x^2}{a^2} + \frac{y^2}{b^2} = 1$ *about the y-axis.*

Sol. Since the ellipse $\frac{x^2}{a^2} + \frac{y^2}{b^2} = 1$ is symmetrical about the y-axis, the volumes generated by the left and the right halves overlap. As such we consider the revolution of the right half ABB′ only for which x varies from 0 to

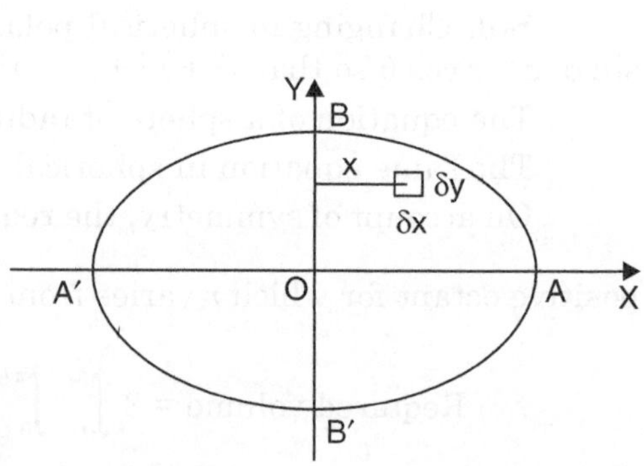

$$a\sqrt{1 - \frac{y^2}{b^2}} = \frac{a}{b}\sqrt{b^2 - y^2}$$

and y varies from $-b$ to b.

$\therefore$ Required volume of revolution

$$= 2\pi \int_{-b}^{b} \int_{0}^{(a/b)\sqrt{b^2 - y^2}} x \, dx \, dy$$

$$= 2\pi \int_{-b}^{b} \left[\frac{x^2}{2} \right]_0^{(a/b)\sqrt{b^2 - y^2}} dy = \pi \int_{-b}^{b} \frac{a^2}{b^2} (b^2 - y^2) \, dy = \frac{2\pi a^2}{b^2} \int_0^b (b^2 - y^2) \, dy$$

$$= \frac{2\pi a^2}{b^2} \left[b^2 y - \frac{y^3}{3} \right] = \frac{2\pi a^2}{b^2} \left[b^3 - \frac{b^3}{3} \right] = \frac{4}{3} \pi a^2 b.$$

Example 2. *Calculate, by double integration, the volume generated by the revolution of the cardioid r = a (1 – cos θ) about its axis.*

Sol. Since the upper and the lower halves of the cardioid generate the same volume, we consider the revolution of the upper half only for which r varies from 0 to $a(1 - \cos \theta)$ and θ varies from 0 to π.

$\therefore$ Required volume of revolution

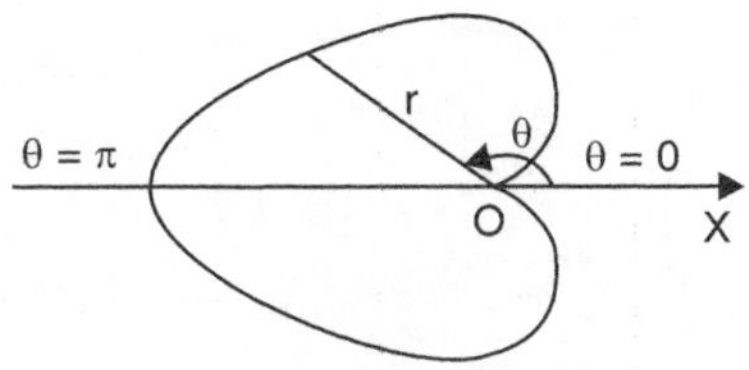

$$= 2\pi \int_0^{\pi} \int_0^{a(1 - \cos \theta)} r^2 \sin \theta \, dr \, d\theta$$

$$= 2\pi \int_0^{\pi} \sin \theta \left[\frac{r^3}{3} \right]_0^{a(1 - \cos \theta)} d\theta$$

$$= \frac{2\pi}{3} \int_0^{\pi} \sin \theta \cdot a^3 (1 - \cos \theta)^3 \, d\theta$$

$$= \frac{2\pi a^3}{3} \int_0^{\pi} (1 - \cos \theta)^3 \sin \theta \, d\theta = \frac{2\pi a^3}{3} \left[\frac{(1 - \cos \theta)^4}{4} \right]_0^{\pi} = \frac{\pi a^3}{6} [(1 - \cos \pi)^4]$$

$$= \frac{\pi a^3}{6} \cdot 2^4 = \frac{8}{3} \pi a^3.$$

TEST YOUR KNOWLEDGE

1. Find, by triple integration, the volume in the positive octant bounded by the co-ordinate planes and the plane $x + 2y + 3z = 4$.
2. Find the volume of the portion of the sphere $x^2 + y^2 + z^2 = a^2$ lying inside the cylinder $x^2 + y^2 = ax$.
3. Find the volume bounded above by the sphere $x^2 + y^2 + z^2 = 2a^2$ and below by the paraboloid $az = x^2 + y^2$.
4. Find, by double integration, the volume of the solid generated by revolving the ellipse $\frac{x^2}{a^2} + \frac{y^2}{b^2} = 1$ about the x-axis.
5. Find, by double integration, the volume generated by revolving the cardioid $r = a(1 + \cos \theta)$ about the initial line.

Answers

1. $\frac{16}{9}$ 2. $\frac{2a^3}{9} (3\pi - 4)$ 3. $\left(\frac{4}{3} \sqrt{2} - \frac{7}{6} \right) \pi a^3$

4. $\frac{4}{3} \pi ab^2$ 5. $\frac{8}{3} \pi a^3$.

APPENDIX—I

LIST OF IMPORTANT FORMULAE

1. $\int x^n \, dx = \dfrac{x^{n+1}}{n+1}$ $[n \neq -1]$.

2. $\int \dfrac{1}{x} \, dx = \log x$.

3. $\int e^x \, dx = e^x$.

4. $\int a^x \, dx = \dfrac{a^x}{\log a}$.

5. $\int \sin x \, dx = -\cos x$.

6. $\int \cos x \, dx = \sin x$.

7. $\int \sec^2 x \, dx = \tan x$.

8. $\int \text{cosec}^2 x \, dx = -\cot x$.

9. $\int \sec x \tan x \, dx = \sec x$.

10. $\int \text{cosec}\, x \cot x \, dx = -\text{cosec}\, x$

11. $\int 1 \, dx = x$.

12. If x be replaced by $ax + b$ [a and b being constants] on both sides of any standard result, the standard form remains true, provided the result on R.H.S. is divided by a, the co-efficient of x.

e.g., $\int \cos (ax + b) \, dx = \dfrac{\sin (ax + b)}{a}$.

13. $\int \sinh x \, dx = \cosh x$.

14. $\int \cosh x \, dx = \sinh x$.

15. $\int \text{sech}^2 x \, dx = \tanh x$.

16. $\int \text{cosech}^2 x \, dx = -\coth x$.

17. $\int \text{sech}\, x \tanh x \, dx = -\text{sech}\, x$.

18. $\int \text{cosech}\, x \coth x \, dx = -\text{cosech}\, x$.

19. $\int_a^b f(x) \, dx = [\phi(x)]_a^b = \phi(b) - \phi(a)$ where $\int f(x) \, dx = \phi(x)$.

20. While evaluating the definite integral $\int_a^b f(x) \, dx$ by the substitution $x = \phi(z)$, the limits of integration must also be changed.

Do not evaluate the integral in z between the old limits a and b which are the values of x and not of z.

21. If the integrand contains $\sqrt{a^2 - x^2}$, put $x = a \sin \theta$.

22. $\int \frac{dx}{\sqrt{a^2 - x^2}} = \sin^{-1} \frac{x}{a}$.

23. If the integrand contains $a^2 + x^2$, put $x = a \tan \theta$ (or $a \sinh \theta$).

24. $\int \frac{dx}{a^2 + x^2} = \frac{1}{a} \tan^{-1} \frac{x}{a}$; $\int \frac{dx}{\sqrt{a^2 + x^2}} = \sinh^{-1} \frac{x}{a}$.

25. If the integrand contains $\sqrt{x^2 - a^2}$, put $x = a \sec \theta$ (or $a \cosh \theta$).

26. $\int \frac{dx}{\sqrt{x^2 - a^2}} = \cosh^{-1} \frac{x}{a}$; $\int \frac{dx}{x\sqrt{x^2 - a^2}} = \frac{1}{a} \sec^{-1} \frac{x}{a}$.

27. If $\sqrt{a + x}$ occurs in the integrand, put $a + x = z^2$.

28. If $\sqrt{\frac{a - x}{a + x}}$ occurs in the integrand, put $x = a \cos 2\theta$.

29. If the integrand contains some linear expression of the variable, put that linear expression = z.

30. If the integrand is a rational function of e^x, put $e^x = z$.

31. If the integrand contains a t-ratio of $f(x)$ or logarithm of $f(x)$ or an exponential function in which the index is $f(x)$, put $f(x) = z$.

32. $\int \frac{f'(x)}{f(x)} dx = \log[f(x)]$.

33. $\int [f(x)]^n f'(x)\, dx = \frac{[f(x)]^{n+1}}{n+1}$ $[n \neq -1]$.

34. $\int \tan x\, dx = \log \sec x = -\log \cos x$.

35. $\int \cot x\, dx = \log \sin x$.

36. $\int \text{cosec}\, x\, dx = \log \tan \frac{x}{2} = \log (\text{cosec}\, x - \cot x)$.

37. $\int \sec x\, dx = \log \tan \left(\frac{\pi}{4} + \frac{x}{2}\right) = \log (\sec x + \tan x)$.

38. $\tan^{-1} a + \tan^{-1} b = \tan^{-1} \frac{a + b}{1 - ab}$, $\tan^{-1} a - \tan^{-1} b = \tan^{-1} \frac{a - b}{1 + ab}$.

39. **Rule of integration by parts** : If u and v be two functions of x, then

$$\int uv\, dx = u \int v\, dx - \int \frac{du}{dx} \left[\int v\, dx\right] dx$$

In words

Integral of the product of two functions

= 1st function × integral of 2nd function–integral of

[Derivative of 1st × integral of 2nd].

40. Care should be taken to choose the first function in the rule of integration by parts. If one factor of the product is a power of x, take it as the first function, provided the integral of the second function is handy. If, however, the integral of the second function is not readily available [in case of inverse circular function or inverse hyperbolic function or a logarithmic function] in that case take that factor as the first function.

If the integral is a single inverse circular [or hyperbolic] function or a single logarithm, take that function as the first function and unity (1) as the second function.

41. $\int \sqrt{a^2-x^2}\, dx = \frac{x\sqrt{a^2-x^2}}{2} + \frac{a^2}{2}\sin^{-1}\frac{x}{a}$. **42.** $\int \sqrt{a^2+x^2}\, dx = \frac{x\sqrt{a^2+x^2}}{2} + \frac{a^2}{2}\sinh^{-1}\frac{x}{a}$.

43. $\int \sqrt{x^2-a^2}\, dx = \frac{x\sqrt{x^2-a^2}}{2} - \frac{a^2}{2}\cosh^{-1}\frac{x}{a}$.

Rule to Remember Formulae 41–43.

Integral $= \frac{x \times \text{integrand}}{2}, \frac{+a^2 \text{ or } -a^2 \text{ as in integrand}}{2} \times$ integral of [reciprocal of integrand]

44. $\int e^{ax} \sin bx\, dx = \frac{e^{ax}}{a^2+b^2}(a \sin bx - b \cos bx)$

$\int e^{ax} \cos bx\, dx = \frac{e^{ax}}{a^2+b^2}(a \cos bx + b \sin bx)$ | First Form

45. $\int e^{ax} \sin bx\, dx = \frac{e^{ax}}{\sqrt{a^2+b^2}} \sin\left(bx - \tan^{-1}\frac{b}{a}\right)$

$\int e^{ax} \cos bx\, dx = \frac{e^{ax}}{\sqrt{a^2+b^2}} \cos\left(bx - \tan^{-1}\frac{b}{a}\right)$ | Second Form

46. $\int \frac{dx}{x^2-a^2} = \frac{1}{2a}\log\frac{x-a}{x+a}$. **47.** $\int \frac{dx}{a^2-x^2} = \frac{1}{2a}\log\frac{a+x}{a-x}$.

48. Rule to evaluate

$$\int \sqrt{\text{Quadratic}}\, dx \text{ or } \int \frac{dx}{\text{Quadratic}} \text{ or } \int \frac{dx}{\sqrt{\text{Quadratic}}}$$

* Make the co-efficient of x^2 unity by taking its numerical co-efficient outside.
* Complete the square in terms of x by adding and subtracting the square of half the co-efficient of x.
* Use proper standard form.

49. Rule to evaluate :

$$\int \text{Linear}\sqrt{\text{Quadratic}}\,dx \quad \text{or} \quad \int \frac{\text{Linear}}{\text{Quadratic}}\,dx \quad \text{or} \quad \int \frac{\text{Linear}}{\sqrt{\text{Quadratic}}}\,dx.$$

* Put linear $= \lambda \dfrac{d}{dx}$ (Quadratic) $+ \mu$.
* Equate co-efficients of x and constant terms on both sides to find λ and μ.
* Use proper formula.

50. When only even powers of x occur in both the numerator and denominator of a rational function, put $x^2 = y$ and resolve the resulting fraction in y into partial fractions. In the partial fractions, put $y = x^2$ and integrate.

51. If either the numerator or the denominator of a rational function consists of only odd powers of x, while the other consists of only even powers, put $x^2 = z$.

52. Rule of evaluate : $\displaystyle\int \frac{x^2 + a^2}{x^4 + kx^2 + a^4}\,dx$ or $\displaystyle\int \frac{x^2 - a^2}{x^4 + kx^2 + a^4}\,dx$

Divide the numerator and denominator by x^2.

Put $x - \dfrac{1}{x} = z$ or $x + \dfrac{1}{x} = z$, whichever substitution gives on differentiation the numerator of the resulting integrand.

53.
$$\frac{x^2}{x^4 + kx^2 + a^4} = \frac{1}{2}\cdot\frac{(x^2 + a^2) + (x^2 - a^2)}{x^4 + kx^2 + a^4}$$

$$\frac{1}{x^4 + kx^2 + a^4} = \frac{1}{2a^2}\cdot\frac{(x^2 + a^2) - (x^2 - a^2)}{x^4 + kx^2 + a^4}.$$

54. If the integrand contains $\sqrt{\dfrac{x - \alpha}{\beta - x}}$ or $\sqrt{(x - \alpha)(\beta - x)}$

Put $x = \alpha \cos^2 \theta + \beta \sin^2 \theta$.

If the integrand contains $\sqrt{\dfrac{\alpha - x}{x - \beta}}$ or $\sqrt{(\alpha - x)(x - \beta)}$

Put $x = \alpha \sin^2 \theta + \beta \cos^2 \theta$.

55. Rule to evaluate: $\displaystyle\int \frac{dx}{\text{Linear}\sqrt{\text{Linear}}}$

Put $\sqrt{\text{Linear}} = z$.

56. Rule to evaluate: $\displaystyle\int \frac{dx}{\text{Quadratic}\sqrt{\text{Linear}}}$

Put $\sqrt{\text{Linear}} = z$.

57. Rule to evaluate:

$$\int \frac{dx}{\text{Linear}\sqrt{\text{Quadratic}}} \quad \text{or} \quad \int \frac{dx}{(\text{Linear})^r \sqrt{\text{Quadratic}}} \quad \text{where } r \text{ is a +ve integer}$$

Put Linear $= \frac{1}{z}$.

58. Rule to evaluate :

$$\int \frac{xdx}{(\text{Pure Quadratic})\sqrt{\text{Pure Quadratic}}}$$

Put $\sqrt{\text{Pure Quadratic}} = z$.

59. Rule to evaluate :

$$\int \frac{dx}{(\text{Pure Quadratic})\sqrt{\text{Pure Quadratic}}}$$

Put $x = \frac{1}{y}$ and in the resulting integrand in y

Put $\sqrt{\text{Pure Quadratic}} = z$.

60. Rule to evaluate $\int \sin^n x\, dx$ or $\int \cos^n x\, dx$, where n is a +ve odd integer

When the index of sin x is a +ve odd integer, detach sin x and

Put $\cos x = z$

When the index of cos x is a +ve odd integer, detach cos x and

Put $\sin x = z$.

61. Rule to evaluate $\int \sin^n x\, dx$ or $\int \cos^n x\, dx$, where n is a +ve even integer.

Express $\sin^n x$ or $\cos^n x$ as a series of cosines of multiple angles of x, by using the formulae

$$2\cos^2 x = 1 + \cos 2x \ ; \quad 2\sin^2 x = 1 - \cos 2x$$

and then integrate term by term.

62. Rule to evaluate $\int \sin^p x \cos^q x\, dx$ when either p or q is a +ve odd integer, whatever the other may be :

When the index of sin x is a +ve odd integer, detach sin x and

Put $\cos x = z$

When the index of cos x is a +ve odd integer, detach cos x and

Put $\sin x = z$.

63. Rule to evaluate $\int \sin^p x \cos^q x\, dx$, where $p + q$ is a –ve even integer [p and q may or may not be integers].

Express the integrand as the product of the powers of tan x and sec x.

Put $\tan x = z$.

64. Rule to integrate $\int \sec^n x\, dx$

Put $\tan x = z$.

Rule to integrate $\int \operatorname{cosec}^n x\, dx$

Put $\cot x = z$.

65. If $a + b$ is + ve, then

$$\int \frac{dx}{a + b\cos x} = \frac{1}{\sqrt{a^2 - b^2}} \cos^{-1} \frac{b + a\cos x}{a + b\cos x} \quad \text{if } a^2 > b^2$$

$$= \frac{1}{\sqrt{b^2 - a^2}} \cosh^{-1} \frac{b + a\cos x}{a + b\cos x} \quad \text{if } a^2 < b^2.$$

66. $\int_0^{\pi/2} \sin^n x\, dx$ or $\int_0^{\pi/2} \cos^n x\, dx$

$$= \begin{cases} \dfrac{(n-1) \times \text{go on decreasing by 2}}{n \times \text{go on decreasing by 2}} & \text{if } n \text{ is a +ve odd integer} \\ \dfrac{(n-1) \times \text{go on decreasing by 2}}{n \times \text{go on decreasing by 2}} \times \dfrac{\pi}{2} & \text{if } n \text{ is a +ve even integer.} \end{cases}$$

67. $\int_0^{\pi/2} \sin^p x \cos^q x\, dx$

$$= \begin{cases} \dfrac{(p-1) \times \text{go on decreasing by 2} \times (q-1) \times \text{go on decreasing by 2}}{(p+q) \times \text{go on decreasing by 2}} \\ \text{if } p \text{ or } q \text{ or both are + ve odd integers} \\ \dfrac{(p-1) \times \text{go on decreasing by 2} \times (q-1) \times \text{go on decreasing by 2}}{(p+q) \times \text{go on decreasing by 2}} \times \dfrac{\pi}{2} \end{cases}$$

if p and q both are +ve even integers.

68. $\int_a^b f(x)\, dx = \int_a^b f(z)\, dz.$

69. $\int_a^b f(x)\, dx = -\int_b^a f(x)\, dx.$

70. $\int_a^b f(x)\, dx = \int_a^c f(x)\, dx + \int_c^b f(x)\, dx,$ where $a < c < b$

71. $\int_0^a f(x)\, dx = \int_0^a f(a - x)\, dx.$

72. $$\int_0^{2a} f(x)\, dx = \begin{cases} 2\int_0^a f(x)\, dx & \text{if } f(2a - x) = f(x) \\ 0 & \text{if } f(2a - x) = -f(x) \end{cases}$$

73. $$\int_{-a}^{a} f(x)\,dx = \begin{cases} 2\int_0^a f(x)\,dx & \text{if } f(x) \text{ is an even function of } x \\ 0 & \text{if } f(x) \text{ is an odd function of } x \end{cases}$$

74. (*i*) $1 - \frac{1}{2} + \frac{1}{3} - \frac{1}{4} + ... = \log 2$ (*ii*) $\frac{1}{1^2} + \frac{1}{2^2} + \frac{1}{3^2} + \frac{1}{4^2} + ... = \frac{\pi^2}{6}$

(*iii*) $\frac{1}{1^2} + \frac{1}{3^2} + \frac{1}{5^2} + \frac{1}{7^2} + ... = \frac{\pi^2}{8}$ (*iv*) $\frac{1}{1^2} - \frac{1}{2^2} + \frac{1}{3^2} - \frac{1}{4^2} + ... = \frac{\pi^2}{12}$

75. $\int_0^{\pi/2} \log \sin x\,dx = \int_0^{\pi/2} \log \cos x\,dx = -\frac{\pi}{2} \log 2.$

76. If $f(x)$ is continuous and single valued in the closed interval $[a, b]$, then

$$\int_a^b f(x)\,dx = \underset{h \to 0}{\text{Lt}}\ h[f(a) + f(a+h) + f(a+2h) + ... + f(a\,\overline{n-1}\,h)] \quad \text{where } nh = b - a$$

or $$\int_a^b f(x)\,dx = \underset{r \to 1}{\text{Lt}}\ (r-1)\,[a\,f(a) + ar\,f(ar) + ar^2\,f(ar^2) + ... + ar^{n-1}\,f(ar^{n-1})],$$

where $r^n = \frac{b}{a}$.

77. If a and b are independent of α, then $\frac{d}{d\alpha}\left[\int_a^b f(x,\alpha)\,dx\right] = \int_a^b \frac{\partial}{\partial\alpha}[f(x,\alpha)]\,dx.$

78. $$\frac{d}{d\alpha}\left[\int_{g(\alpha)}^{h(\alpha)} f(x,\alpha)\,dx\right] = \int_{g(\alpha)}^{h(\alpha)} \frac{\partial}{\partial\alpha}[f(x,\alpha)]\,dx + h'(\alpha)\,f[h(\alpha),\alpha] - g'(\alpha)\,f[g(\alpha),\alpha].$$

79. (*i*) $\int_a^{\infty} f(x)\,dx = \underset{X \to \infty}{\text{Lt}} \int_a^X f(x)\,dx$ (*ii*) $\int_{-\infty}^{b} f(x)\,dx = \underset{X \to -\infty}{\text{Lt}} \int_X^b f(x)\,dx$

(*iii*) $\int_{-\infty}^{b} f(x)\,dx = \underset{X_1 \to -\infty}{\text{Lt}} \int_{X_1}^{c} f(x)\,dx + \underset{X_2 \to -\infty}{\text{Lt}} \int_c^{X_2} f(x)\,dx$ where c is any real number.

(*iv*) If $f(x)$ becomes infinite at $x = b$ only, then

$$\int_a^b f(x)\,dx = \underset{\varepsilon \to 0}{\text{Lt}} \int_a^{b-\varepsilon} f(x)\,dx.$$

(*v*) If $f(x)$ becomes infinite at $x = a$ only, then

$$\int_a^b f(x)\,dx = \underset{\varepsilon \to 0}{\text{Lt}} \int_{a+\varepsilon}^{b} f(x)\,dx$$

(*vi*) If $f(x)$ becomes infinite at $x = c$, where $a < c < b$, then

$$\int_a^b f(x)\,dx = \underset{\varepsilon_1 \to 0}{\text{Lt}} \int_a^{c-\varepsilon_1} f(x)\,dx + \underset{\varepsilon_2 \to 0}{\text{Lt}} \int_{c+\varepsilon_2}^b f(x)\,dx$$

An Improper Integral is said to be convergent if the limit (when improper integral is expressed as the limit of a definite integral) exists finitely.

80. $B(m, n) = \int_0^1 x^{m-1}(1-x)^{n-1}\,dx$, where $m > 0, n > 0$.

81. $B(m, n) = B(n, m)$.

82. $B(m, n) = \int_0^\infty \frac{x^{m-1}}{(1+x)^{m+n}}\,dx = \int_0^\infty \frac{x^{n-1}}{(1-x)^{m+n}}\,dx.$

83. If m, n are both positive integers, then $B(m, n) = \frac{(m-1)!\,(n-1)!}{(m+n-1)!}$.

84. $\Gamma(n) = \int_0^\infty x^{n-1} e^{-x}\,dx$, where $n > 0$. **85.** $\Gamma(n) = (n-1)\,\Gamma(n-1)$, where $n > 1$.

86. $\Gamma(n) = (n-1)\,!$, where n is a +ve integer.

87. $B(m, n) = \frac{\Gamma(m)\,\Gamma(n)}{\Gamma(m+n)}$.

88. $\Gamma\left(\frac{1}{2}\right) = \sqrt{\pi}\,, \Gamma\left(\frac{3}{4}\right)\Gamma\left(\frac{1}{2}\right) = \sqrt{2}\,\pi, \Gamma\,(1) = 1.$

89. $\int_{-\infty}^\infty e^{-x^2}\,dx = \sqrt{\pi}.$

90. $\int_0^{\pi/2} \sin^p x \cos^q x\,dx = \frac{1}{2}\cdot\frac{\Gamma\left(\frac{p+1}{2}\right)\Gamma\left(\frac{q+1}{2}\right)}{\Gamma\left(\frac{p+1}{2}+\frac{q+1}{2}\right)}$ where $p > -1, q > -1$.

91. Duplication formula $\Gamma\,(m)\,\Gamma\left(m+\frac{1}{2}\right) = \frac{\sqrt{\pi}}{2^{2m-1}}\,\Gamma\,(2m)$.

92. Length of the arc of the curve $y = f(x)$, between the two points whose abscissae are a and b is given by

$$\int_a^b \sqrt{1+\left(\frac{dy}{dx}\right)^2}\,dx.$$

93. Length of the arc of the curve $x = f(t)$, $y = \phi(t)$ between the points, where $t = t_1$ to $t = t_2$ is given by

$$\int_{t_1}^{t_2} \sqrt{\left(\frac{dx}{dt}\right)^2 + \left(\frac{dy}{dt}\right)^2}\,dt.$$

94. Length of the arc of the curve $r = f(\theta)$ between the points whose vectorical angles are α

and β is

$$\int_{\alpha}^{\beta} \sqrt{r^2 + \left(\frac{dr}{d\theta}\right)^2}\, d\theta.$$

95. Area bounded by the curve $y = f(x)$, the x-axis and the ordinates $x = a$, $x = b$ is $\int_a^b y dx$.

96. Area bounded by the curves $y = f(x)$, $y = \phi(x)$ and the ordinates $x = a$, $x = b$ is $\int_a^b (y_1 - y_2)\, dx$

where y_1 is the 'y' of the upper curve and y_2 that of the lower curve.

97. Area bounded by the curve $x = f(t)$, $y = \phi(t)$, the x-axis and the ordinates at the points, where $t = a$, $t = b$ is given by $\int_a^b y \frac{dx}{dt}\, dx.$

98. Area bounded by the curve $r = f(\theta)$ and the radii vectors $\theta = \alpha$, $\theta = \beta$ is $\int_{\alpha}^{\beta} \frac{1}{2} r^2\, d\theta$

99. Area bounded by the curves $r = f(\theta)$, $r = F(\theta)$ and the radii vectors $\theta = \alpha$, $\theta = \beta$ is

$$\int_{\alpha}^{\beta} \frac{1}{2} (r_1^2 - r_2^2)\, d\theta$$

where r_1 is the 'r' of the outer curve and r_2 that of the inner curve.

100. Area of a closed curve $x = f(t)$, $y = \phi(t)$ is $\int_{t_1}^{t_2} \frac{1}{2}\left(x \frac{dy}{dt} - y \frac{dx}{dt}\right) dt$, t_1, t_2 being the values of t such that the point (x, y) travels once round the curve.

101. Simpson's Rule

If y is a function of x, then $\int_a^b y\, dx$ is approximately $= \frac{h}{3} [X + 2O + 4E]$ where $h = \frac{b-a}{2n}$

X = sum of extreme ordinates (*i.e.*, first and last)

O = sum of other odd ordinates (*i.e.*, leaving the first and last ordinates)

E = sum of even ordinates.

102. Volume of the solid generated by revolution about x-axis, of the area bounded by the curve $y = f(x)$, the x-axis and the bounding ordinates $x = a$, $x = b$ is $\int_a^b \pi y^2\, dx$.

103. Volume of the solid generated by revolution about x-axis, of the area bounded by the curve $x = f(t)$, $y = \phi(t)$ and the ordinates at the points where $t = a$, $t = b$ is given by

$$\int_a^b \pi y^2 \frac{dx}{dt}\, dt.$$

104. Volume of the solid generated by the revolution of the area bounded by the curve $r = f(\phi)$ and the radii vectors $\theta = \alpha$, $\theta = \beta$

(*i*) about the initial line OX ($\theta = 0$) = $\int_\alpha^\beta \frac{2}{3} \pi r^3 \sin\theta \, d\theta$

(*ii*) about the line $OY\left(\theta = \frac{\pi}{2}\right) = \int_\alpha^\beta \frac{2}{3} \pi r^3 \cos\theta \, d\theta.$

105. Surface Formulae

(*i*) **Cartesian Form** [for the curve $y = f(x)$]

$$s = \int 2\pi y \frac{ds}{dx} dx, \text{ where } \frac{ds}{dx} = \sqrt{1 + \left(\frac{dy}{dx}\right)^2}.$$

(*ii*) **Parametric Form** [for the curve $x = f(t), y = \phi(t)$]

$$s = \int 2\pi y \frac{ds}{dt} dt, \text{ where } \frac{ds}{dt} = \sqrt{\left(\frac{dx}{dt}\right)^2 + \left(\frac{dy}{dt}\right)^2}.$$

(*iii*) **Polar Form** [for the curve $r = f(\theta)$]

$$s = \int 2\pi y \frac{ds}{d\theta} d\theta, \text{ where } y = r\sin\theta \text{ and } \frac{ds}{d\theta} = \sqrt{r^2 + \left(\frac{dr}{d\theta}\right)^2}.$$

106. Pappu's Theorem. If a closed curve revolves about an axis in its plane, not intersecting it, then

(*i*) Volume of the solid generated = (Plane area which revolves) × length of the path of its C.G.

(*ii*) Curved surface of the solid generated = (Length of the arc of the curve) × length of the path of its C.G.

107. Change of variable in a double integral

Change of Polar Co-ordinates. Replace x by $r\cos\theta$, y by $r\sin\theta$ and $dx\,dy$ by $r\,dr\,d\theta$.

108. Change of variable in a triple integral

(*i*) **Change of Cylindrical Polar Co-ordinates.** Replace x by $r\cos\theta$, y by $r\sin\theta$, z by z and $dx\,dy\,dz$ by $r\,dr\,d\theta\,dz$.

(*ii*) **Change to spherical polar Co-ordinates.** Replace x by $r\sin\theta\cos\phi$, y by $r\sin\theta\sin\phi$, z by $r\cos\theta$ and $dx\,dy\,dz$ by $r^2\sin\theta\,dr\,d\theta\,d\phi$.

APPENDIX—II

SOME IMPORTANT CURVES

Cissoid (Imp.)

$y^2(2a - x) = x^3$

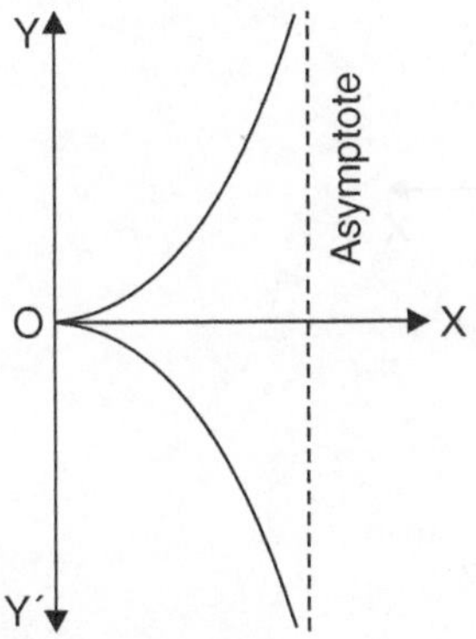

Catenary

$y = a \cosh \frac{x}{a}$

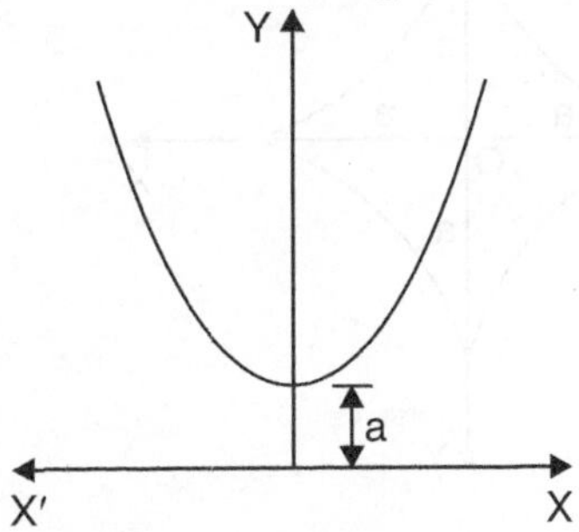

Cubical Parabola

$a^2y = x^3$

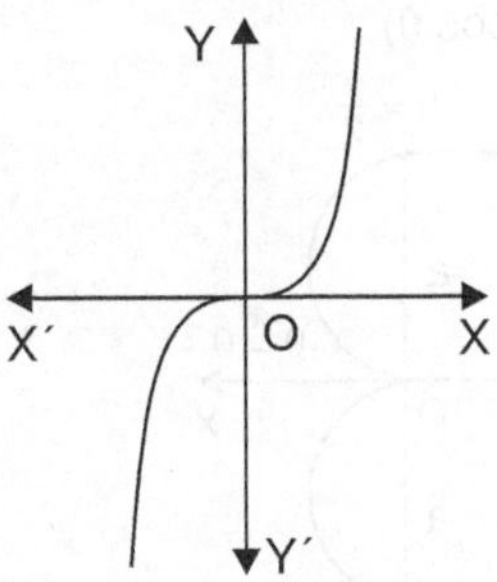

Semicubical Parabola

$ay^2 = x^3$

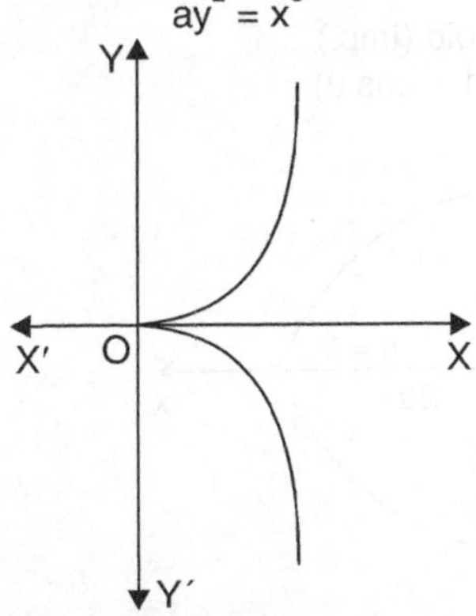

Strophoid

$y^2(a - x) = x^2(a + x)$

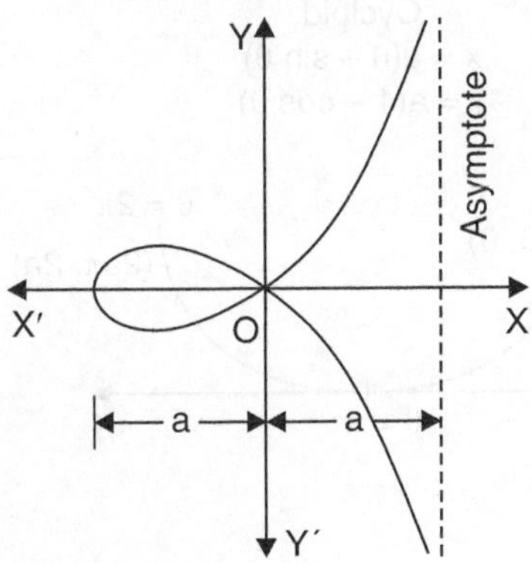

Folium of Diacartes

$x^3 + y^3 = 3axy$

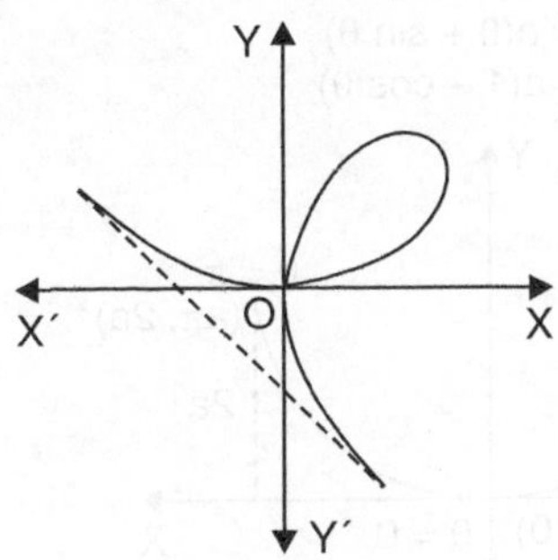

Hypocycloid (Imp.)

$x^{2/3} + y^{2/3} = a^{2/3}$ or

$x = a\cos^3\theta \quad y = a\sin^3\theta$

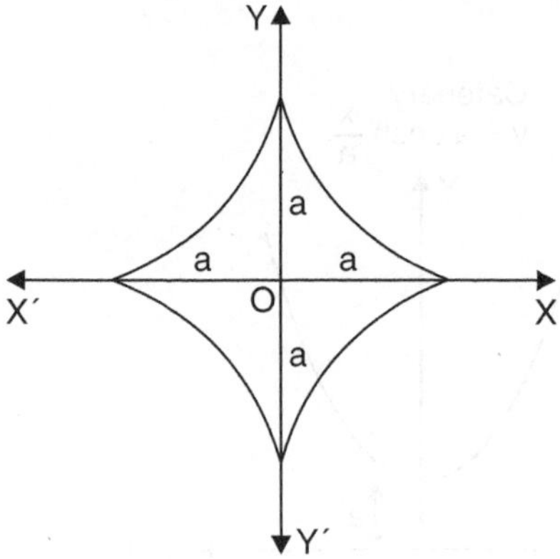

Witon of Agnesi

$xy^2 = 4a^2(2a - x)$

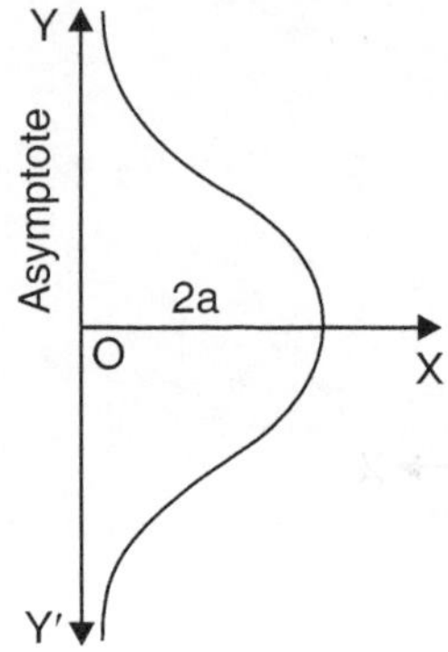

Cardioid (Imp.)

$r = a(1 + \cos\theta)$

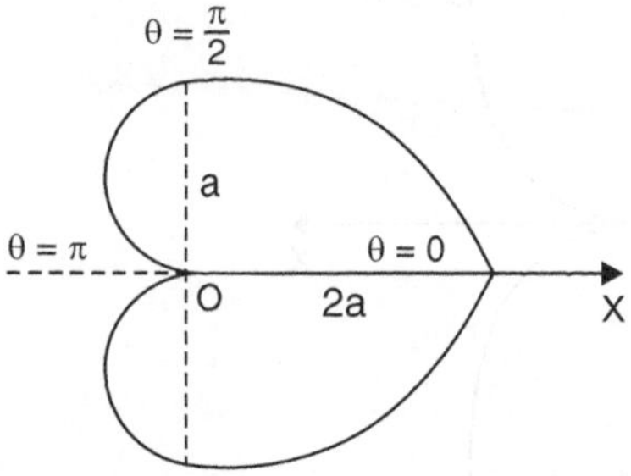

Cardioid

$r = a(1 + \cos\theta)$

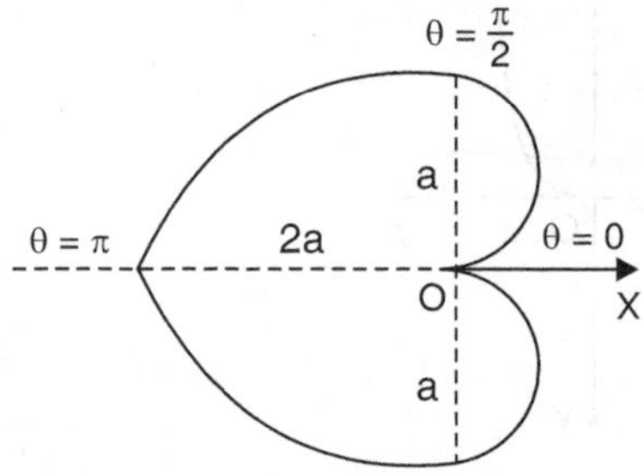

Cycloid

$x = a(\theta + \sin\theta)$

$y = a(1 - \cos\theta)$

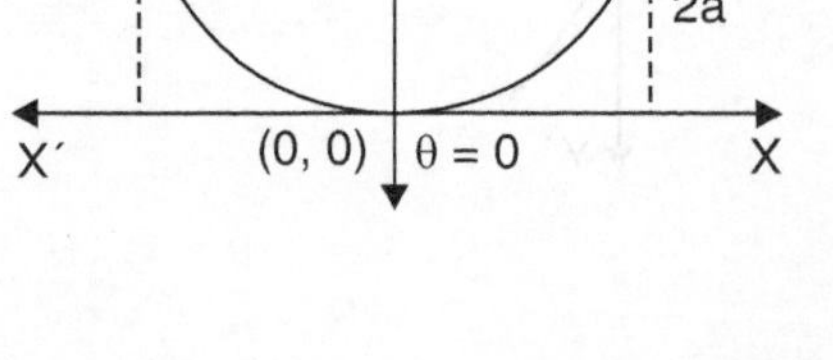

Cycloid

$x = a(\theta + \sin\theta)$

$y = a(1 - \cos\theta)$

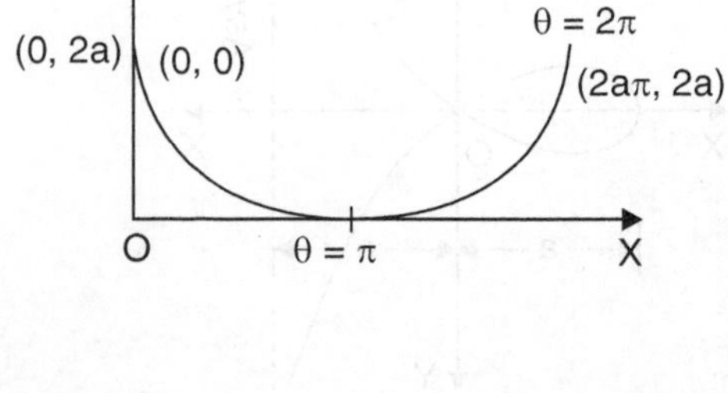

Cycloid

$x = a(\theta - \sin\theta)$

$y = a(1 - \cos\theta)$

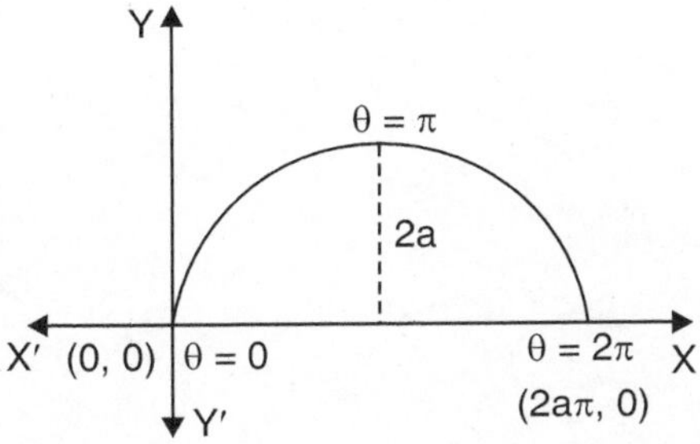

Cycloid

$x = a(\theta + \sin\theta)$

$y = a(1 + \cos\theta)$

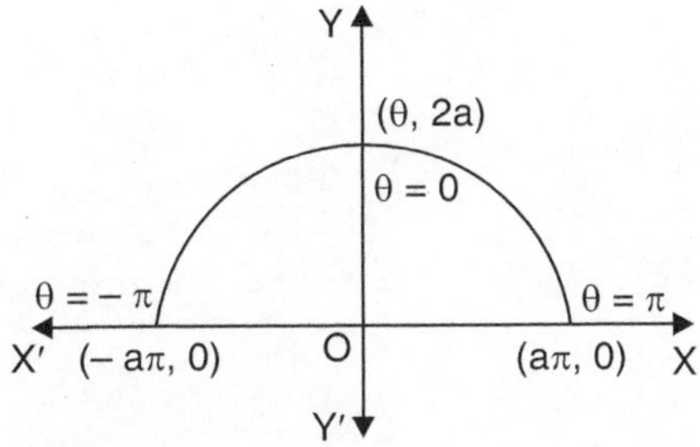

Lemniscate

$r^2 = a^2 \cos 2\theta$

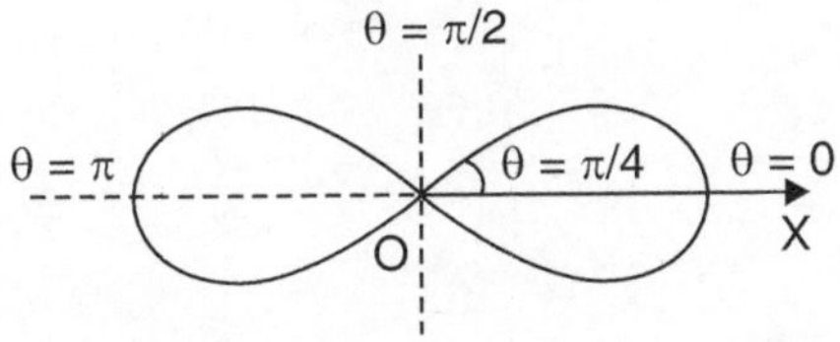

The three leaved rose

$r = a \sin 3\theta$

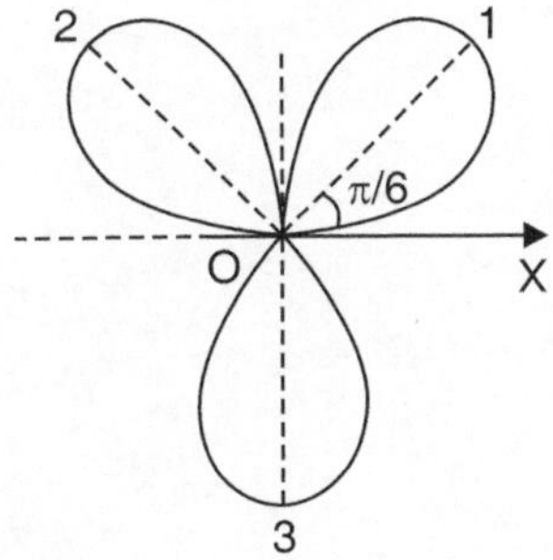